GEOGRAPHY

The World and Its People

VOLUME 2

NATIONAL GEOGRAPHIC

SENIOR AUTHOR
Richard G. Boehm, Ph.D.

David G. Armstrong, Ph.D.

Francis P. Hunkins, Ph.D.

Glencoe
McGraw-Hill

New York, New York
Columbus, Ohio
Woodland Hills, California
Peoria, Illinois

ABOUT THE AUTHORS

NATIONAL GEOGRAPHIC

The National Geographic Society, founded in 1888 for the increase and diffusion of geographic knowledge, is the world's largest nonprofit scientific and educational organization. Since its earliest days, the Society has used sophisticated communication technologies, from color photography to holography, to convey geographic knowledge to a worldwide membership. The School Publishing Division supports the Society's mission by developing innovative educational programs—ranging from traditional print materials to multimedia programs including CD-ROMS, videos, and software.

SENIOR AUTHOR

Richard G. Boehm

Richard G. Boehm, Ph.D., was one of seven authors of *Geography for Life,* national standards in geography, prepared under Goals 2000: Educate America Act. He was also one of the authors of the *Guidelines for Geographic Education,* in which the five themes of geography were first articulated. In 1990 Dr. Boehm was designated "Distinguished Geography Educator" by the National Geographic Society. In 1991 he received the George J. Miller award from the National Council for Geographic Education (NCGE) for distinguished service to geographic education. He was President of the NCGE and has twice won the *Journal of Geography* award for best article. He has received the NCGE's "Distinguished Teaching Achievement" award and presently holds the Jesse H. Jones Distinguished Chair in Geographic Education at Southwest Texas State University in San Marcos, Texas.

David G. Armstrong

David G. Armstrong, Ph.D., is Dean of the School of Education at the University of North Carolina at Greensboro. A social studies education specialist with additional advanced training in geography, Dr. Armstrong was educated at Stanford University, University of Montana, and University of Washington. He taught at the secondary level in the state of Washington before beginning a career in higher education. Dr. Armstrong has written books for students at the secondary and university levels, as well as for teachers and university professors. He maintains an active interest in travel, teaching, and social studies education.

Francis P. Hunkins

Francis P. Hunkins, Ph.D., is Professor of Education at the University of Washington. He began his professional career as a teacher in Massachusetts. He received his masters degree in education from Boston University and his doctorate from Kent State University with a major in general curriculum and a minor in geography. Dr. Hunkins has written numerous books and articles dealing with general curriculum, social studies, and questioning and thinking for students and educators at elementary, middle school, high school, and university levels.

Contributing Writer

Bob Haddad, ethnomusicologist, developed the Music features. He also chose the music selections for Glencoe's *World Music: A Cultural Legacy* program and authored the accompanying teacher guide. Mr. Haddad is founder of Music of the World and president of Owl's Head Music.

Glencoe/McGraw-Hill

A Division of The ***McGraw·Hill*** *Companies*

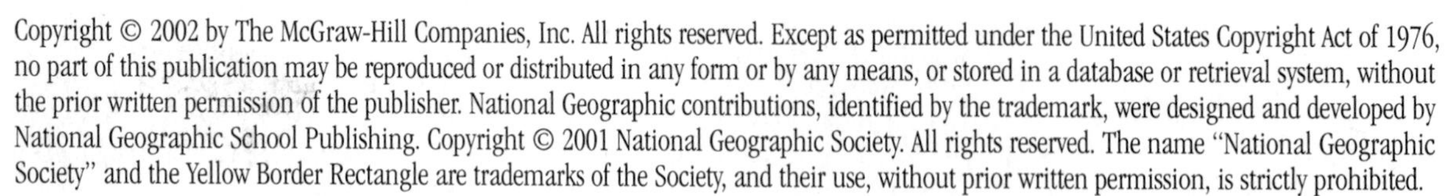

Printed in the United States of America.
Send all inquiries to:
Glencoe/McGraw-Hill
8787 Orion Place
Columbus, Ohio 43240-4027

ISBN 0-07-824941-4 (Student Edition) ISBN 0-07-824694-6 (Teacher Wraparound Edition)

6 7 8 9 027/043 11 10 09 08

CONSULTANTS

General Content Consultants

Dr. Sari Bennett
Director, Center for Geographic Education
University of Maryland Baltimore County
Baltimore, Maryland

Shabbir Mansuri
Founding Director

Susan L. Douglass
Affiliated Scholar
Council on Islamic Education
Fountain Valley, California

United States and Canada

Dr. David A. Lanegran
John S. Holl Professor of Geography
Macalester College
St. Paul, Minnesota

Latin America

Dr. Jose F. Betancourt
Professor of Geography
State University of New York—Oneonta
Oneonta, New York

Frank de Varona
Region Superintendent
Dade County Public Schools
Miami, Florida

Europe

Dr. Charles F. Kovacik
Director, Center of Excellence for Geographic Education
University of South Carolina
Columbia, South Carolina

Russia

Dr. Charles R. Gildersleeve
Professor of Geography
University of Nebraska at Omaha
Omaha, Nebraska

North Africa, Southwest Asia, and Central Asia

Dr. Jeffrey A. Gritzner
Department of Geography
The University of Montana
Missoula, Montana

Africa South of the Sahara

Dr. Richard S. Palm
Professor of Geography
University of Wisconsin—Eau Claire
Eau Claire, Wisconsin

Asia

Dr. Kenji Oshiro
Professor of Geography
Wright State University
Dayton, Ohio

Australia, Oceania, and Antarctica

Dr. Brock Brown
Associate Professor of Geography and Planning
Southwest Texas State University
San Marcos, Texas

TEACHER REVIEWERS

Brent Adcox
Cornersville High School
Cornersville, Tennessee

Peter John Arroyo
Space Coast Middle School
Port St. John, Florida

Nora Austin
Chestnut Hill Community School
Belchertown, Massachusetts

William Ball
West Central Junior/Senior High School
Francesville, Indiana

Beverly A. Blamer
Fremont Middle School
Fremont, Michigan

Sue Brinkley
Hampshire Unit School
Hampshire, Tennessee

Lee Ann Burrow
Bob Courtway Middle School
Conway, Arkansas

Kim Cavanaugh
Congress Middle School
Boynton Beach, Florida

Janet S. D'Meo-Townley
Park Middle School
Scotch Plains, New Jersey

Connie Dunn
Mulberry High School
Mulberry, Arkansas

Kristine Louise Edwards
Dodge Intermediate School
Twinsburg, Ohio

Gale Olp Ekiss
Powell Junior High School
Mesa, Arizona

Dawn Forman
Hawthorne Junior/Senior High
Gainesville, Florida

Paula Gordon
Nettleton Intermediate Center
Jonesboro, Arkansas

Destin L. Haas
Benton Central Junior/Senior High
Oxford, Indiana

Daniel Hanczar
West Allegheny Middle School
Imperial, Pennsylvania

Phillip G. Hays
Conrad Weiser Middle School
Robesonia, Pennsylvania

Julie Hill
Bob Courtway Middle School
Conway, Arkansas

Donna Jett
Powell Middle School
Powell, Tennessee

Jacquelyn J. Jones
International Teacher and Administrator of American Schools in Costa Rica
San José, Costa Rica

Camille King-Thompson
Williston Middle School
Williston, Florida

Rick Lyndsey
Roosevelt Full Service
West Palm Beach, Florida

Sara Monschein
Oakridge Middle School
Naples, Florida

Wanda J. Petersen
Landrum Middle School
Ponte Vedra, Florida

Mary Ann Polve
Mesa High School
Mesa, Arizona

Nancy Pund
Deltona Middle School
Deltona, Florida

Jean Serafino
Chestnut Hill Community School
Belchertown, Massachusetts

William Sim
L.L. Wright Middle School
Ironwood, Michigan

Ida Haskew Smith
South Pittsburg Elementary
South Pittsburg, Tennessee

Janice Suddith
Paul Laurence Dunbar Middle School for Innovation
Lynchburg, Virginia

Joseph Turso
Wayne Hills High School
Wayne, New Jersey

Volume One

Contents

Be an Active Reader! xvi

NATIONAL GEOGRAPHIC Reference Atlas

The World: Political RA2
The World: Physical RA4
North America: Political RA6
North America: Physical RA7
United States: Political RA8
United States: Physical RA10
Canada: Physical/Political RA12
Middle America: Physical/Political RA14
South America: Political RA16
South America: Physical RA17
Europe: Political RA18
Europe: Physical RA20
Africa: Political RA22
Africa: Physical RA23
Asia: Political RA24
Asia: Physical RA26
Oceania: Physical/Political RA28
Pacific Rim: Physical RA30
Ocean Floor RA32
Polar Regions RA34

NATIONAL GEOGRAPHIC Geography Handbook

How Do I Study Geography? 2
1 Using Globes and Maps 4
2 Learning Map Basics 10
3 Using Graphs, Charts, and Diagrams 14
Geographic Dictionary 18

Unit 1 Geography of the World 20

Chapter 1 Looking at the Earth 22
1 Thinking Like a Geographer 23
2 The Earth in Space 29
3 The Earth's Structure 34
4 Landforms 39

Chapter 2 Water, Climate, and Vegetation 48
1 The Water Planet 49
2 Climate 53
3 Climate Zones and Vegetation 63

Chapter 3 The World's People 76
1 Culture 77
2 Population 84
3 Resources and World Trade 90
4 People and the Environment 95

Unit 2 The United States and Canada 102

NATIONAL GEOGRAPHIC Atlas 104

Chapter 4 The United States 114
1 A Vast, Scenic Land 115
2 An Economic Leader 123
3 The Americans 130

Chapter 5 Canada 142
1 Landforms of the North 143
2 A Resource-Rich Country 150
3 The Canadians 155

Unit 3 Latin America 166

NATIONAL GEOGRAPHIC Atlas 168

Chapter 6 Mexico 180
1 Mexico's Land and Economy 181
2 Mexico's History and Government 188
3 Mexico Today 193

Chapter 7 Central America and the West Indies **204**
1 Central America 205
2 The West Indies 213

Chapter 8 Brazil and Its Neighbors **226**
1 Brazil 227
2 Argentina 235
3 Caribbean South America 241
4 Uruguay and Paraguay 245

Chapter 9 The Andean Countries **254**
1 Colombia 255
2 Peru and Ecuador 260
3 Bolivia and Chile 266

Unit 4 Europe **274**

NATIONAL GEOGRAPHIC **Atlas** **276**

Chapter 10 Western Europe **288**
1 The United Kingdom 289
2 The Republic of Ireland 296
3 France 300
4 Germany, Switzerland, and Austria 306
5 The Benelux Countries 313

Chapter 11 Southern Europe **322**
1 Spain and Portugal 323
2 Italy 330
3 Greece 336

Chapter 12 Northern Europe **344**
1 Norway, Sweden, and Finland 345
2 Denmark and Iceland 353

Chapter 13 Eastern Europe **364**
1 Poland 365
2 The Baltic Republics 371
3 Hungary, the Czech Republic, and Slovakia 375
4 The Balkan Countries 381
5 Ukraine, Belarus, and Moldova 386

Unit 5 Russia **394**

NATIONAL GEOGRAPHIC **Atlas** **396**

Chapter 14 Russia—A Eurasian Country **404**
1 The Russian Land 405
2 Russia's Economic Regions 412

Chapter 15 Russia—Past and Present **422**
1 A Troubled History 423
2 A New Russia 430

Exploring Other Regions ..... **REGIONS 1**
North Africa, Southwest Asia, and Central Asia .... REGIONS 2
Africa South of the Sahara .... REGIONS 14
Asia .... REGIONS 28
Australia, Oceania, and Antarctica .... REGIONS 40

APPENDIX **782**
What Is an Appendix and How Do I Use One? 782A
Gazetteer 783
Glossary 791
Spanish Glossary 799
Index 808
Photo Credits 827

Volume Two Contents

Be an Active Reader! xvi

NATIONAL GEOGRAPHIC Reference Atlas

The World: Political RA2
The World: Physical RA4
North America: Political RA6
North America: Physical RA7
United States: Political RA8
United States: Physical RA10
Canada: Physical/Political RA12
Middle America: Physical/Political RA14
South America: Political RA16
South America: Physical RA17
Europe: Political RA18
Europe: Physical RA20
Africa: Political RA22
Africa: Physical RA23
Asia: Political RA24
Asia: Physical RA26
Oceania: Physical/Political RA28
Pacific Rim: Physical RA30
Ocean Floor RA32
Polar Regions RA34

NATIONAL GEOGRAPHIC Geography Handbook

How Do I Study Geography? 2
1 **Using Globes and Maps** 4
2 **Learning Map Basics** 10
3 **Using Graphs, Charts, and Diagrams** 14
Geographic Dictionary 18

Unit 1 Geography of the World 20

Chapter 1 Looking at the Earth 22
1 Thinking Like a Geographer 23
2 The Earth in Space 29
3 The Earth's Structure 34
4 Landforms 39

Chapter 2 Water, Climate, and Vegetation 48
1 The Water Planet 49
2 Climate 53
3 Climate Zones and Vegetation 63

Chapter 3 The World's People 76
1 Culture 77
2 Population 84
3 Resources and World Trade 90
4 People and the Environment 95

Regional Review Handbook ..... Regions 1

The United States and Canada ..... Regions 2
Latin America ..... Regions 14
Europe ..... Regions 28
Russia ..... Regions 42

Unit 6 North Africa, Southwest Asia, and Central Asia 442

NATIONAL GEOGRAPHIC Atlas 444

Chapter 16 North Africa 456
1 Egypt 457
2 Libya and the Maghreb 464

Chapter 17 Southwest Asia 476
1 Turkey 477
2 Israel 482
3 Syria, Lebanon, and Jordan 488
4 The Arabian Peninsula 492
5 Iraq, Iran, and Afghanistan 497

Chapter 18 The Caucasus and Central Asia 508
1 Republics of the Caucasus 509
2 Central Asian Republics 515

Unit 7 Africa South of the Sahara 524
NATIONAL GEOGRAPHIC **Atlas 526**

Chapter 19 West Africa 540
1 Nigeria 541
2 The Sahel Countries 547
3 Coastal Countries 552

Chapter 20 Central Africa 562
1 Democratic Republic of the Congo 563
2 Other Countries of Central Africa 570

Chapter 21 East Africa 580
1 Kenya 581
2 Tanzania 587
3 Inland East Africa 592
4 The Horn of Africa 596

Chapter 22 South Africa and Its Neighbors 606
1 Republic of South Africa 607
2 Atlantic Countries 614
3 Inland Southern Africa 618
4 Indian Ocean Countries 623

Unit 8 Asia 630
NATIONAL GEOGRAPHIC **Atlas 632**

Chapter 23 South Asia 644
1 India 645
2 Pakistan and Bangladesh 652
3 Other Countries of South Asia 657

Chapter 24 China 668
1 China's Land and Climate 669
2 China's New Economy 675
3 China's People and Culture 680
4 China's Neighbors 685

Chapter 25 Japan and the Koreas 694
1 Japan 695
2 The Two Koreas 703

Chapter 26 Southeast Asia 714
1 Mainland Southeast Asia 715
2 Island Southeast Asia 721
3 Indonesia 726

Unit 9 Australia, Oceania, and Antarctica 734
NATIONAL GEOGRAPHIC **Atlas 736**

Chapter 27 Australia and New Zealand 746
1 Australia 747
2 New Zealand 754

Chapter 28 Oceania and Antarctica 764
1 Oceania 765
2 Antarctica 772

APPENDIX 782

Gazetteer 783
Glossary 791
Spanish Glossary 799
Index 808
Photo Credits 827

Features

Volume 1

The Oceans: Deep Trouble 74
Too Much Trash 140
Vanishing Rain Forests 252
Rain, Rain Go Away 362
Russia's Lethal Legacy 420

Volume 2

A Water Crisis 506
Endangered Spaces 604
The Himalaya at Risk 666
Earth's Natural Sunscreen 762

Volume 1

The St. Lawrence: A River of Trade 164
The Columbian Exchange 224
Germany: Together Again 320
Russia's Strategy: Freeze Your Foes 440

Volume 2

Birthplace of Three Religions 504
Please Pass the Salt: Africa's Salt Trade 560
The Silk Road 692
"Byrd's"-Eye View: Exploring Antarctica 780

Geography Skills

Volume 1

Using a Map Key 33
Using Latitude and Longitude 62
Reading a Special Purpose Map 83
Using Scale 129
Reading a Physical Map 149
Interpreting an Elevation Profile 220
Reading a Vegetation Map 305
Reading a Population Map 329
Reading a Transportation Map 416

Volume 2

Reading a Time Zones Map 487
Mental Mapping 569
Reading a Circle Graph 656
Reading a Contour Map 725

Technology Skills

Volume 1

Using the Internet 187
Using a Database 270

Volume 2

Using a Spreadsheet 470
Developing Multimedia Presentations 622

Critical Thinking Skills

Volume 1

Sequencing and Categorizing Information 234
Understanding Cause and Effect 436

Volume 2

Analyzing Information 520
Drawing Inferences and Conclusions 546
Making Predictions 586
Distinguishing Fact From Opinion 679
Making Comparisons 708

Study and Writing Skills

Volume 1

Using Library Resources 358
Taking Notes 370

Volume 2

Outlining 758
Writing a Report 771

Making Connections

Technology

Volume 1

Geographic Information Systems 28
Counting Heads 89
The Panama Canal Locks 212
Stonehenge 295

Volume 2

Egyptian Pyramids 463
The Three Gorges Dam 674
Antarctica's Environmental Stations 776

Science

Volume 1

Exploring Earth's Water 52
The Aztec Calendar Stone 192
The Galápagos Islands 265
Midnight Sun, Shine On! 352
Living in Space 429

Volume 2

Battling Sleeping Sickness 574
Mining and Cutting Diamonds 613
Australia's Amazing Animals 753

Literature

Volume 1

"Survival This Way" by Simon J. Ortiz 136
"I, Too" by Langston Hughes 136
The Gaucho Martín Fierro by José Hernández 240
"Is It Not Time for the Birds to Sing" by Boris Pasternak 411

Volume 2

"The World, My Friends, My Enemies, You, and the Earth" by Nazim Hikmet 481
"Where are those Songs?" by Micere Githae Mugo 591
Haiku 702

Art

Volume 1

Snow and Ice Sculpting 160
Leonardo da Vinci 335
Ukrainian Easter Eggs 390

Volume 2

Carpet Weaving 514
Great Mosque of Djenné 551
The Taj Mahal 651
Shadow Puppets 730

Volume 1

Surf's Up! 125
Time to Play 157
Digging for Treasure 190
What a Catch! 210
I Protest! 239
Time to Perform 299
Traveling Water Boulevards 334
Cozy Ballet? 357
Living in a Changed City 432

Volume 2

Bazaar! 466
Festival Time 480
I am a Samburu 585
What's for Dinner? 619
School's Out! 654
The Race Is On! 688
Hard Hats to School? 700
Life as a Monk 718
Dreamtime 752

Teen from Senegal ▶

Features

Who, What, Where in the World?

Volume 1

Solar Eclipse 30
Mt. Pinatubo 56
Saffron—A Valuable Resource....................91
San Xavier Del Bac 131
Bee Hummingbird 215
Roping a Capybara 243
Andean Dig....................264
Reims Cathedral 303
The Alhambra....................327
The Sami....................349
Transylvania....................382
Skiing the Strait 410
Peter the Great 425

Volume 2

Ramses II 461
Petra....................490
The Aral Sea....................519
Ken Saro-Wiwa 545
The Okapi....................568
Casting Votes....................598
Nelson Mandela....................612
Ship Breakers....................655
Clay Warriors....................681
Deep Sea Diver....................707
Komodo Dragon 729
The *Endeavour* 775

Volume 1

Architecture: Quake-Proof Structures......37
Music of North America....................134
Clothing: The Inuit....................156
Art: Diego Rivera and His Murals 194
Music of Latin America....................263
Architecture: Leaning Tower of Pisa......333
Music of Europe 379
Art: Fabergé Eggs....................426

Volume 2

Music of Southwest Asia 489
Clothing: The Tuareg....................549
Music of Africa South of the Sahara 610
Music of South Asia....................649
Architecture: Angkor Wat 719
Music of Australia 751

GeoLAB ACTIVITY

Volume 1

Tornadoes: Swirling Fury....................46
Built on Solid Ground? 202

Volume 2

Oil on the Ocean....................474
Water Finds a Way 578
Erosion: Saving the Soil 712

Komodo dragon ▶

Maps

Volumes 1 and 2

NATIONAL GEOGRAPHIC Reference Atlas

The World: Political RA2
The World: Physical RA4
North America: Political RA6
North America: Physical RA7
United States: Political RA8
United States: Physical RA10
Canada: Physical/Political RA12
Middle America: Physical/Political RA14
South America: Political RA16
South America: Physical RA17
Europe: Political RA18
Europe: Physical RA20
Africa: Political RA22
Africa: Physical RA23
Asia: Political RA24
Asia: Physical RA26
Oceania: Physical/Political RA28
Pacific Rim: Physical RA30
Ocean Floor RA32
Polar Regions RA34

NATIONAL GEOGRAPHIC Geography Handbook

Great Circle Route 6
Mercator Projection 7
Goode's Interrupted Projection 8
Robinson Projection 8
Winkel Tripel Projection 9
Spain: Political 10
Austin, Texas 11
Sri Lanka: Physical and Contour 12
Egypt: Population Density 13

Unit 1 Geography of the World

Washington, D.C. 33
World Continents and Oceans 41
Continental Drift 45
Prevailing Wind Patterns 55
El Niño 57
World Ocean Currents 58
Map of the World 62
World Climate Regions 65
World Natural Vegetation Regions 66
World Religions 78
World Culture Regions 81
Early Civilizations 83
World Population Density 86
World Economic Activity 92

Volume 1

Unit 2 The United States and Canada

The United States and Canada: Physical 108
The United States and Canada: Political 109
The United States and Canada: Food Production 110
United States and Canada: Land Comparison 110
Geo Extremes 111
The United States: Political 116
The United States: Physical 117
The United States: Climate 120
The United States: Economic Activity 124
Yellowstone National Park 129
The United States: Population Density 133
Canada: Political 144
Canada: Physical 145
Canada: Climate 147
British Columbia: Physical 149
Canada: Economic Activity 151
Canada: Population Density 158
The Great Lakes and St. Lawrence Seaway 165

Unit 3 Latin America

Latin America: Physical 172
Latin America: Political 173
Latin America: Urban Population Growth 174
United States and Latin America: Land Comparison 174
Geo Extremes 175
Mexico: Political 182
Mexico: Physical 183

Maps

Mexico: Climate 184
Mexico's Native American Civilizations 189
Mexico: Economic Activity 196
Mexico: Population Density 197
Central America and the West Indies: Political 206
Central America and the West Indies: Physical 207
Central America and the West Indies: Economic Activity....209
Central America and the West Indies: Climate....214
Central America and the West Indies: Population Density 216
The Spread of Plants and Animals 225
Brazil and Its Neighbors: Political....228
Brazil and Its Neighbors: Physical....229
Brazil and Its Neighbors: Climate....236
Brazil and Its Neighbors: Economic Activity 238
Brazil and Its Neighbors: Population Density 246
The Andean Countries: Political 256
The Andean Countries: Physical 257
The Andean Countries: Climate....261
The Andean Countries: Population Density....262
The Andean Countries: Economic Activity....267

Unit 4 Europe

Europe: Physical....280
Europe: Political 281
Europe: Languages 282
United States and Europe: Land Comparison 282
Geo Extremes 283
Western Europe: Political....290
Western Europe: Physical....291
Western Europe: Climate 297
Western Europe: Economic Activity 302
France: Vegetation....305
Western Europe: Population Density 314
Occupation of Germany 1945 319
The Two Germanies: 1945....321
Southern Europe: Political 324
Southern Europe: Physical 325
Spain and Portugal: Population Density....329
Southern Europe: Climate....331
Southern Europe: Population Density....337
Southern Europe: Economic Activity....338
Northern Europe: Political....346
Northern Europe: Physical....347
Northern Europe: Economic Activity 350
Northern Europe: Climate....354
Northern Europe: Population Density 356
Eastern Europe: Political 366
Eastern Europe: Physical....367
Eastern Europe: Climate....372
Eastern Europe: Economic Activity....376
Eastern Europe: Population Density 387
European Union 2000....393

Unit 5 Russia

Russia: Physical 400
Russia: Political....401
Russia: The Russian Winter....402
United States and Russia: Land Comparison....402
Geo Extremes 403
Russia: Political....406
Russia: Physical 407
Russia: Climate 409
Russia: Economic Activity....413
Russia: Transportation Routes....416
Expansion of Russia....424
Russia: Population Density....433
Chechnya 439
Average Winter Temperatures 441

Volume 2

Unit 6 North Africa, Southwest Asia, and Central Asia

North Africa, Southwest Asia, and Central Asia: Physical....448
North Africa, Southwest Asia, and Central Asia: Political....449
North Africa, Southwest Asia, and Central Asia: Oil and Gas Production and Distribution 450
United States and North Africa, Southwest Asia, and Central Asia: Land Comparison 450

Maps

Geo Extremes 451
North Africa: Political....458
North Africa: Physical 459
North Africa: Economic Activity....460
North Africa: Climate 465
North Africa: Population Density....467
Southwest Asia: Political 478
Southwest Asia: Physical 479
Southwest Asia: Climate....483
Israel and Its Neighbors....485
World Time Zones 487
Southwest Asia: Economic Activity....493
Southwest Asia: Population Density 498
Present-Day Lands Where Religions Began 505
The Caucasus and Central Asia: Political 510
The Caucasus and Central Asia: Physical 511
The Caucasus and Central Asia: Climate....512
The Caucasus and Central Asia: Economic Activity 516
The Caucasus and Central Asia: Population Density....518

Unit 7 Africa South of the Sahara

Africa South of the Sahara: Physical....530
Africa South of the Sahara: Political....531
Africa South of the Sahara: Gems and Minerals....532
United States and Africa South of the Sahara: Land Comparison....532
Geo Extremes 533
West Africa: Political 542
West Africa: Physical 543
West Africa: Climate....548
West Africa: Economic Activity....553
West Africa: Population Density....554
Salt Trade Routes 561
Central Africa: Political 564
Central Africa: Physical 565
Central Africa: Economic Activity....566
Mental Map of Chicago....569
Central Africa: Population Density....571
Central Africa: Climate....572
East Africa: Political....582
East Africa: Physical 583
East Africa: Climate 588
East Africa: Economic Activity 593
East Africa: Population Density....597
Major African Ethnic Groups....599
Kenya 603
South Africa and Its Neighbors: Political 608
South Africa and Its Neighbors: Physical....609
South Africa and Its Neighbors: Climate....615
African Independence Dates 616
South Africa and Its Neighbors: Economic Activity 620
South Africa and Its Neighbors: Population Density....625

Unit 8 Asia

Asia: Physical....636
Asia: Political 637
Asia: Monsoons 638
United States and Asia: Land Comparison 638
Geo Extremes 639
South Asia: Political....646
South Asia: Physical....647
South Asia: Climate 653
South Asia: Economic Activity 659
South Asia: Population Density....661
China: Political 670
China: Physical 671
China: Climate....672
China: Economic Activity....676
China: Population Density 682
Silk Road Routes 693
Japan and the Koreas: Political....696
Japan and the Koreas: Physical 697
Japan and the Koreas: Population Density....699
Japan and the Koreas: Economic Activity 705
Japan and the Koreas: Climate 706
Asia's Pacific Rim 708
Southeast Asia: Political....716
Southeast Asia: Physical 717
Southeast Asia: Climate 722
Borneo: Contour Map 725
Southeast Asia: Economic Activity....727
Southeast Asia: Population Density....728

Maps

Unit 9 Australia, Oceania, and Antarctica

Australia, Oceania, and Antarctica: Physical 740
Australia, Oceania, and Antarctica: Political.........741
Australia, Oceania, and Antarctica: Endangered Environments 742
United States and Australia, Oceania, and Antarctica: Land Comparison 742
Geo Extremes 743
Australia and New Zealand: Political.................748
Australia and New Zealand: Physical 749
Australia and New Zealand: Climate 750
Australia and New Zealand: Economic Activity 755
Australia and New Zealand: Population Density.......................................756
Oceania and Antarctica: Political 766
Oceania and Antarctica: Physical.......................767
Oceania and Antarctica: Population Density.......................................768
Oceania and Antarctica: Climate........................773
Oceania and Antarctica: Economic Activity.........774
Byrd's First Flight to the South Pole 781

Charts and Graphs

Volumes 1 and 2

NATIONAL GEOGRAPHIC Geography Handbook

Hemispheres..4
Latitude and Longitude..5
Comparing World Languages...............................14
U.S. Farms, 1940–1998..15
World Population, 1999 15
Major Automobile-Producing Countries.............16
Climograph: Moscow, Russia...............................16
Africa: Elevation Profile.......................................17
Landforms and Water Bodies..............................18

Unit 1 Geography of the World

The Solar System...30
Seasons...31
Earth's Layers 35
Tectonic Plate Boundaries..................................36
The Water Cycle 50
Rain Shadow 59
Number of Hurricanes in a Year.........................73
Population Growth 85
Most Populous Countries 85
Exports by Culture Region 101

Volume 1

Unit 2 The United States and Canada

Elevation Profile..108
Comparing Population: United States and Canada.............................111
Ethnic Groups: United States and Canada..111
Country Profiles...112
U.S. State Names: Meaning and Origin...............112
Canadian Province and Territory Names: Meaning and Origin...........113
Branches of the United States Government 113
Top 6 Tourist Destinations 139
St. Lawrence Seaway 152
Highest Peaks of the Americas 163

Unit 3 Latin America

Elevation Profile..172
Comparing Population: United States and Selected Countries of Latin America..175
Ethnic Groups: Selected Countries of Latin America...175

Charts and Graphs

Country Profiles....176
Mexico's Altitude Zones....185
The Panama Canal Locks....212
Jamaica: Elevation Profile....220
Percentage of Workers in Agriculture and Manufacturing....223
Leading Coffee-Producing Countries....230

Unit 4 Europe

Elevation Profile....280
Comparing Population: United States and Selected Countries of Europe....283
Religions: Selected Countries of Europe....283
Country Profiles....284
Number of Personal Computers per 1,000 People....361
Language Families of Europe....388

Unit 5 Russia

Elevation Profile....400
Country Profile....401
Comparing Population: United States and Russia....403
Comparing Area and Population: Russia East and West of the Ural Mountains....403
Mir Space Station Core Module....429

Volume 2

Unit 6 North Africa, Southwest Asia, and Central Asia

Elevation Profile....448
Comparing Population: United States and Selected Countries of North Africa, Southwest Asia, and Central Asia....451
Urban Populations: Selected Cities of North Africa, Southwest Asia, and Central Asia....451
Country Profiles....452
Percentage of North Africa's People Living in Each Country....473
World Oil Production....494
Countries With the Largest Oil Reserves....503

Unit 7 Africa South of the Sahara

Elevation Profile....530
Comparing Population: United States and Selected Countries of Africa South of the Sahara....533
Selected Rural and Urban Populations: Africa South of the Sahara....533
Country Profiles....534
Leading Cacao-Producing Countries....559
Leading Diamond-Producing Countries....567
Tourism in Kenya....586
Diamond Cutting....613
Percentage of Workers in Agriculture....629

Unit 8 Asia

Elevation Profile....636
Comparing Population: United States and Selected Countries of Asia....639
World Population: Asia's Share of the World's People....639
Country Profiles....640
Religions of South Asia....656
Highest Mountain on Each Continent....660
Comparing Population....665
Leading Rice-Producing Countries....677
China's Rank in World Agricultural Production....691
Asia's Pacific Rim: Exports/Imports....708
Top Tin Producers....733

Unit 9 Australia, Oceania, and Antarctica

Elevation Profile....740
Comparing Population: United States and Selected Countries of Australia, Oceania, and Antarctica....743
Population Growth: Australia, 1958–2008....743
Country Profiles....744
Leading Wool-Producing Countries....761

Be an Active Reader!

How Should I Read My Textbook? Reading your social studies book is different than other reading you might do. Your textbook has a great amount of information in it. It is an example of nonfiction writing—it describes real-life events, people, ideas, and places.

Here are some reading strategies that will help you become an active textbook reader. Choose the strategies that work best for you. If you have trouble as you read your textbook, look back at these strategies for help.

Before You Read

Set a Purpose

- Why are you reading the textbook?
- How might you be able to use what you learn in your own life?

Preview

- Read the chapter title to find out what the topic will be.
- Read the subtitles to see what you will learn about the topic.
- Skim the photos, charts, graphs, or maps.
- Look for vocabulary words that are boldfaced. How are they defined?

Draw From Your Own Background

- What do you already know about the topic?
- How is the new information different from what you already know?

If You Don't Know What A Word Means...

- think about the setting, or *context,* in which the word is used.
- check if prefixes such as *un, non,* or *pre* can help you break down the word.
- look up the word's definition in a dictionary or glossary.

As You Read

Question

- What is the main idea?
- How well do the details support the main idea?
- How do the photos, charts, graphs, and maps support the main idea?

Connect

- Think about people, places, and events in your own life. Are there any similarities with those in your textbook?

Predict

- Predict events or outcomes by using clues and information that you already know.
- Change your predictions as you read and gather new information.

Visualize

- Use your imagination to picture the settings, actions, and people that are described.
- Create graphic organizers to help you see relationships found in the information.

Reading Do's

Do . . .

- ✔ establish a purpose for reading.
- ✔ think about how your own experiences relate to the topic.
- ✔ try different reading strategies.

Reading Don'ts

Don't . . .

- ⊘ ignore how the textbook is organized.
- ⊘ allow yourself to be easily distracted.
- ⊘ hurry to finish the material.

After You Read

Summarize

- Describe the main idea and how the details support it.
- Use your own words to explain what you have read.

Assess

- What was the main idea?
- Did the text clearly support the main idea?
- Did you learn anything new from the material?
- Can you use this new information in other school subjects or at home?

Reference Atlas

The World: Political RA2
The World: Physical RA4
North America: Political RA6
North America: Physical RA7
United States: Political RA8
United States: Physical RA10
Canada: Physical/Political RA12
Middle America: Physical/Political . . .RA14
South America: Political RA16
South America: Physical RA17
Europe: Political RA18
Europe: Physical RA20
Africa: Political RA22
Africa: Physical RA23
Asia: Political RA24
Asia: Physical .RA26
Oceania: Physical/Political RA28
Pacific Rim: Physical RA30
Ocean Floor RA32
Polar Regions .RA34

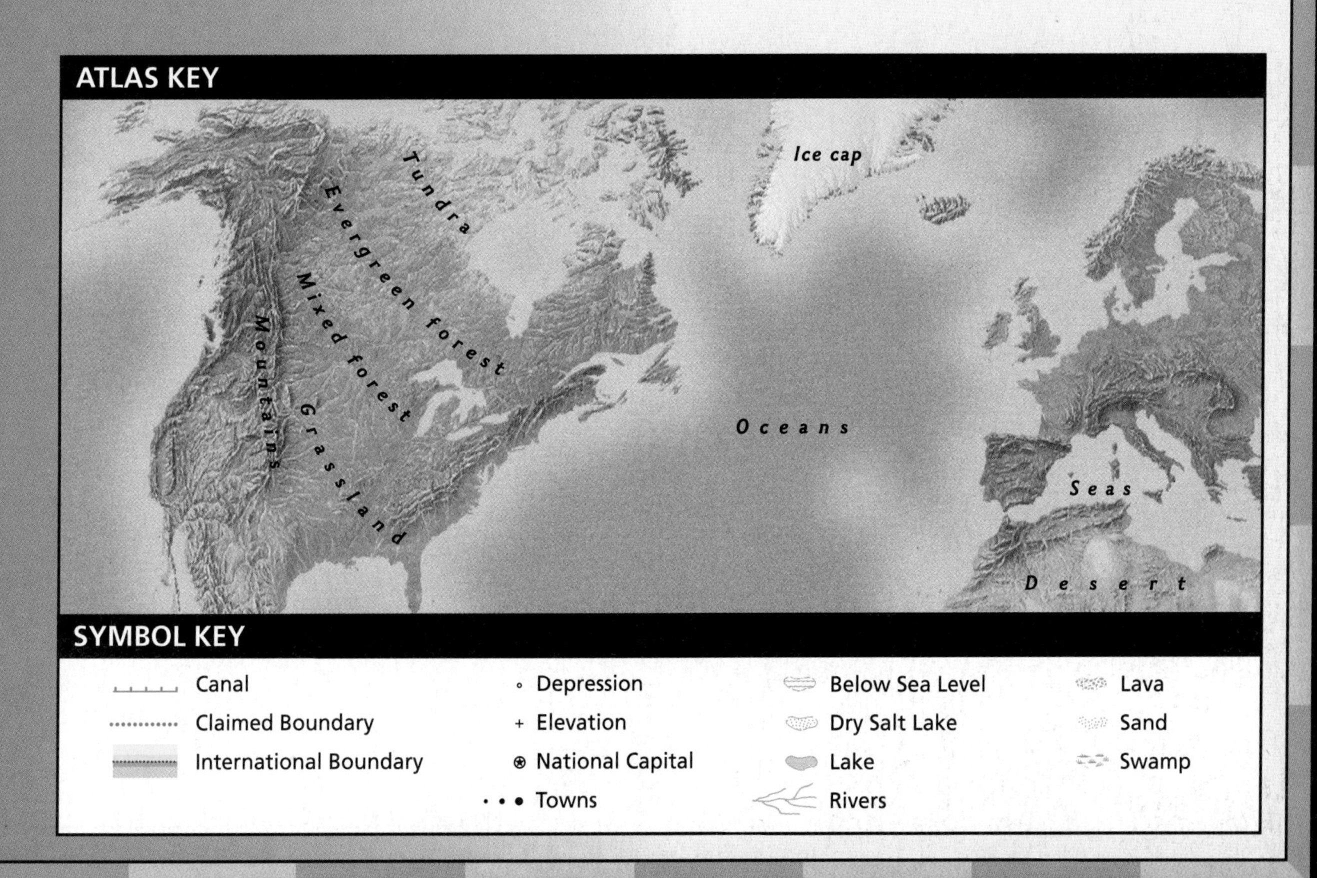

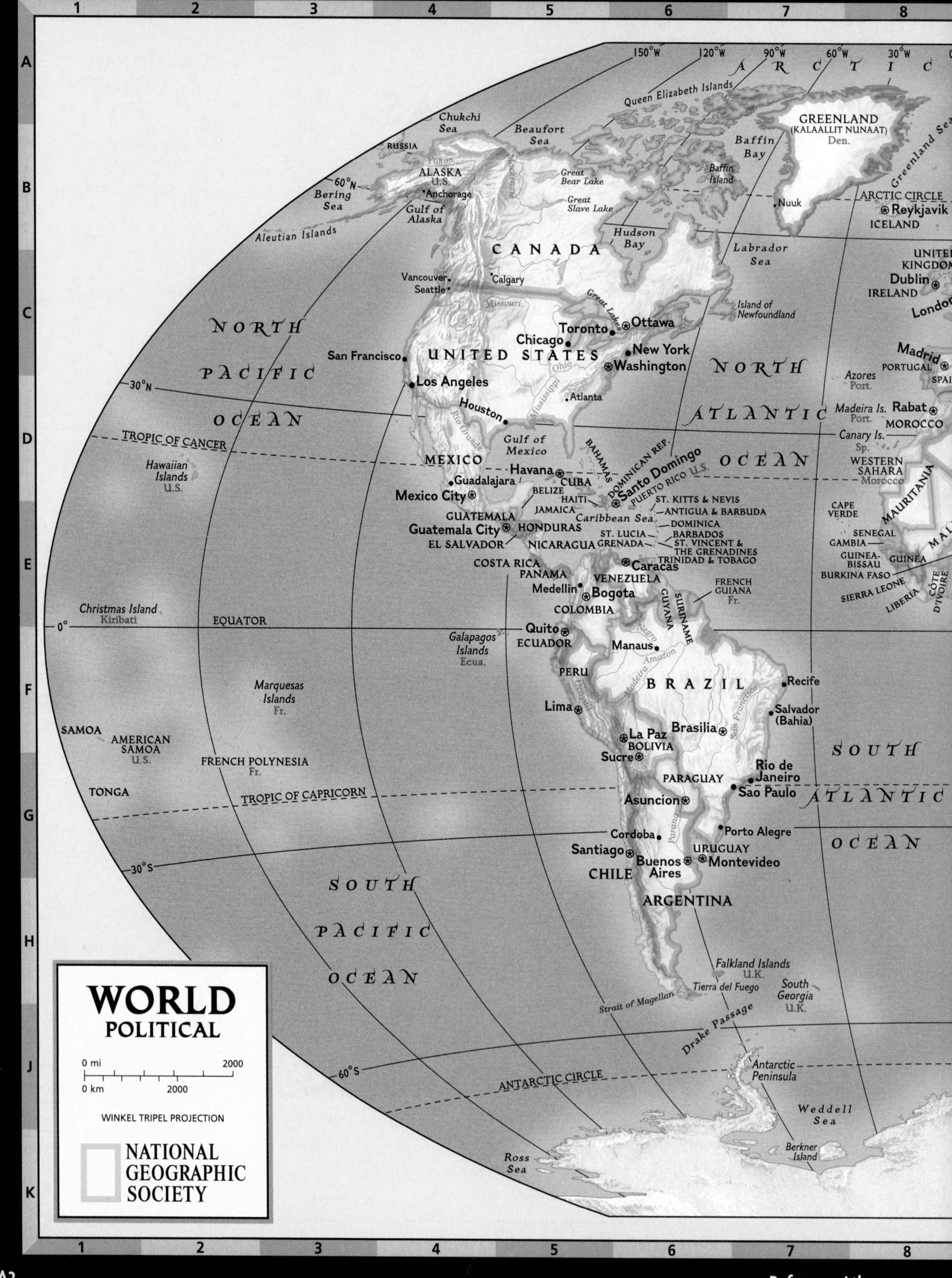

WORLD
POLITICAL
0 mi 2000
0 km 2000
WINKEL TRIPEL PROJECTION
NATIONAL GEOGRAPHIC SOCIETY
NORTH PACIFIC OCEAN
SOUTH PACIFIC OCEAN
NORTH ATLANTIC OCEAN
SOUTH ATLANTIC OCEAN
ARCTIC
CANADA
UNITED STATES
MEXICO
BRAZIL
ARGENTINA
CHILE
GREENLAND (KALAALLIT NUNAAT) Den.
ICELAND
Reykjavik
Nuuk
ALASKA U.S.
Anchorage
RUSSIA
Chukchi Sea
Beaufort Sea
Bering Sea
Gulf of Alaska
Aleutian Islands
Queen Elizabeth Islands
Baffin Bay
Baffin Island
Hudson Bay
Labrador Sea
Greenland Sea
ARCTIC CIRCLE
Great Bear Lake
Great Slave Lake
Vancouver
Seattle
Calgary
San Francisco
Los Angeles
Chicago
Toronto
Ottawa
New York
Washington
Atlanta
Houston
Island of Newfoundland
Great Lakes
Gulf of Mexico
TROPIC OF CANCER
Hawaiian Islands U.S.
Havana
CUBA
BAHAMAS
DOMINICAN REP.
Santo Domingo
PUERTO RICO U.S.
HAITI
JAMAICA
BELIZE
Guadalajara
Mexico City
GUATEMALA
Guatemala City
HONDURAS
EL SALVADOR
NICARAGUA
COSTA RICA
PANAMA
Caribbean Sea
ST. KITTS & NEVIS
ANTIGUA & BARBUDA
DOMINICA
ST. LUCIA
BARBADOS
GRENADA
ST. VINCENT & THE GRENADINES
TRINIDAD & TOBAGO
Caracas
VENEZUELA
Medellin
Bogota
COLOMBIA
GUYANA
SURINAME
FRENCH GUIANA Fr.
Quito
ECUADOR
Galapagos Islands Ecua.
EQUATOR
Christmas Island Kiribati
Manaus
PERU
Lima
Recife
Salvador (Bahia)
Brasilia
La Paz
BOLIVIA
Sucre
PARAGUAY
Asuncion
Rio de Janeiro
Sao Paulo
Porto Alegre
Cordoba
Santiago
Buenos Aires
URUGUAY
Montevideo
Marquesas Islands Fr.
SAMOA
AMERICAN SAMOA U.S.
FRENCH POLYNESIA Fr.
TONGA
TROPIC OF CAPRICORN
Falkland Islands U.K.
Tierra del Fuego
Strait of Magellan
South Georgia U.K.
Drake Passage
Antarctic Peninsula
ANTARCTIC CIRCLE
Weddell Sea
Berkner Island
Ross Sea
UNITED KINGDOM
Dublin
IRELAND
London
Madrid
PORTUGAL
SPAIN
Azores Port.
Madeira Is. Port.
Rabat
MOROCCO
Canary Is. Sp.
WESTERN SAHARA Morocco
MAURITANIA
CAPE VERDE
SENEGAL
GAMBIA
GUINEA-BISSAU
GUINEA
BURKINA FASO
SIERRA LEONE
LIBERIA
CÔTE D'IVOIRE
MALI
150°W 120°W 90°W 60°W 30°W
60°N 30°N 0° 30°S 60°S
1 2 3 4 5 6 7 8
A B C D E F G H J K

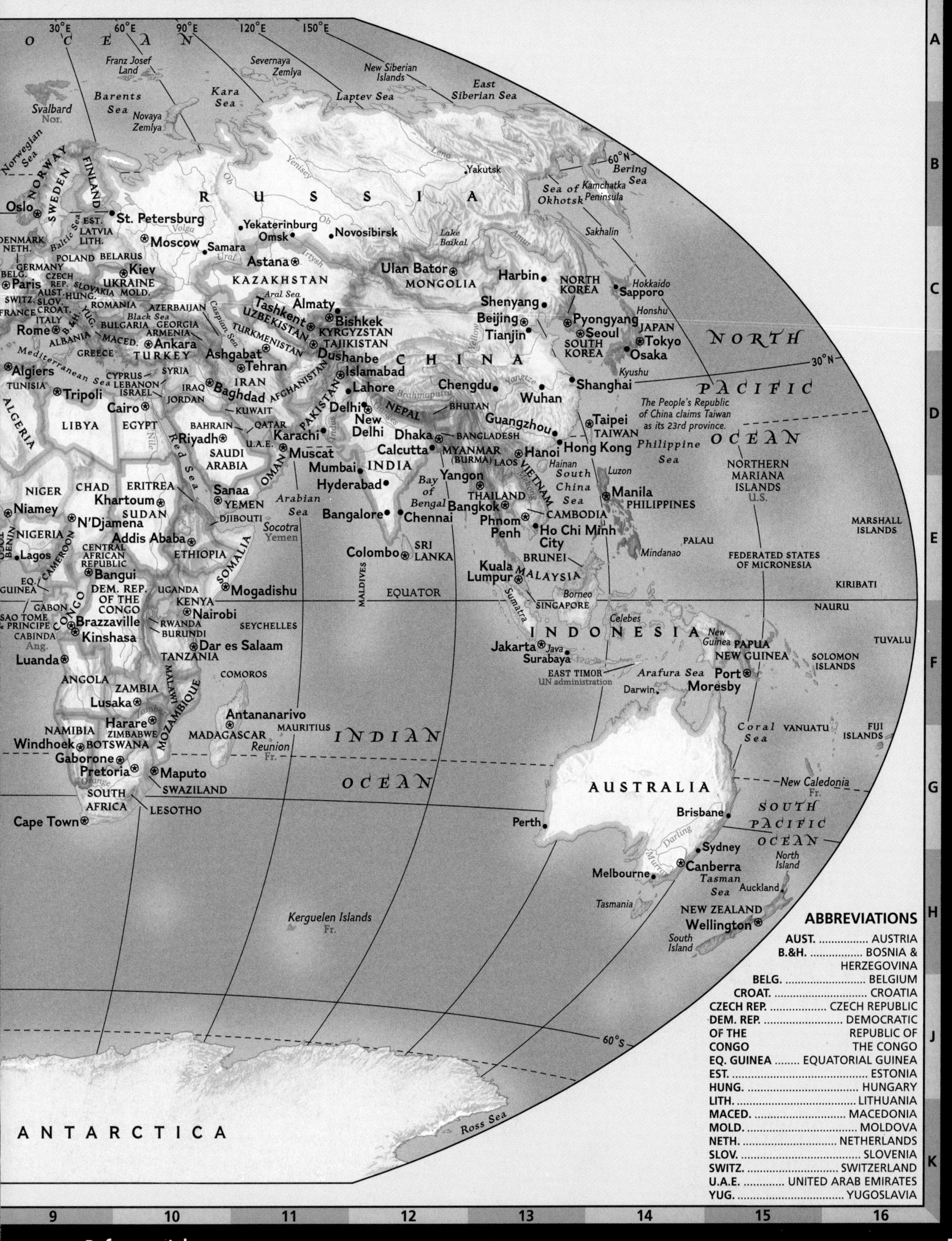

ABBREVIATIONS

AUST.	AUSTRIA
B.&H.	BOSNIA & HERZEGOVINA
BELG.	BELGIUM
CROAT.	CROATIA
CZECH REP.	CZECH REPUBLIC
DEM. REP. OF THE CONGO	DEMOCRATIC REPUBLIC OF THE CONGO
EQ. GUINEA	EQUATORIAL GUINEA
EST.	ESTONIA
HUNG.	HUNGARY
LITH.	LITHUANIA
MACED.	MACEDONIA
MOLD.	MOLDOVA
NETH.	NETHERLANDS
SLOV.	SLOVENIA
SWITZ.	SWITZERLAND
U.A.E.	UNITED ARAB EMIRATES
YUG.	YUGOSLAVIA

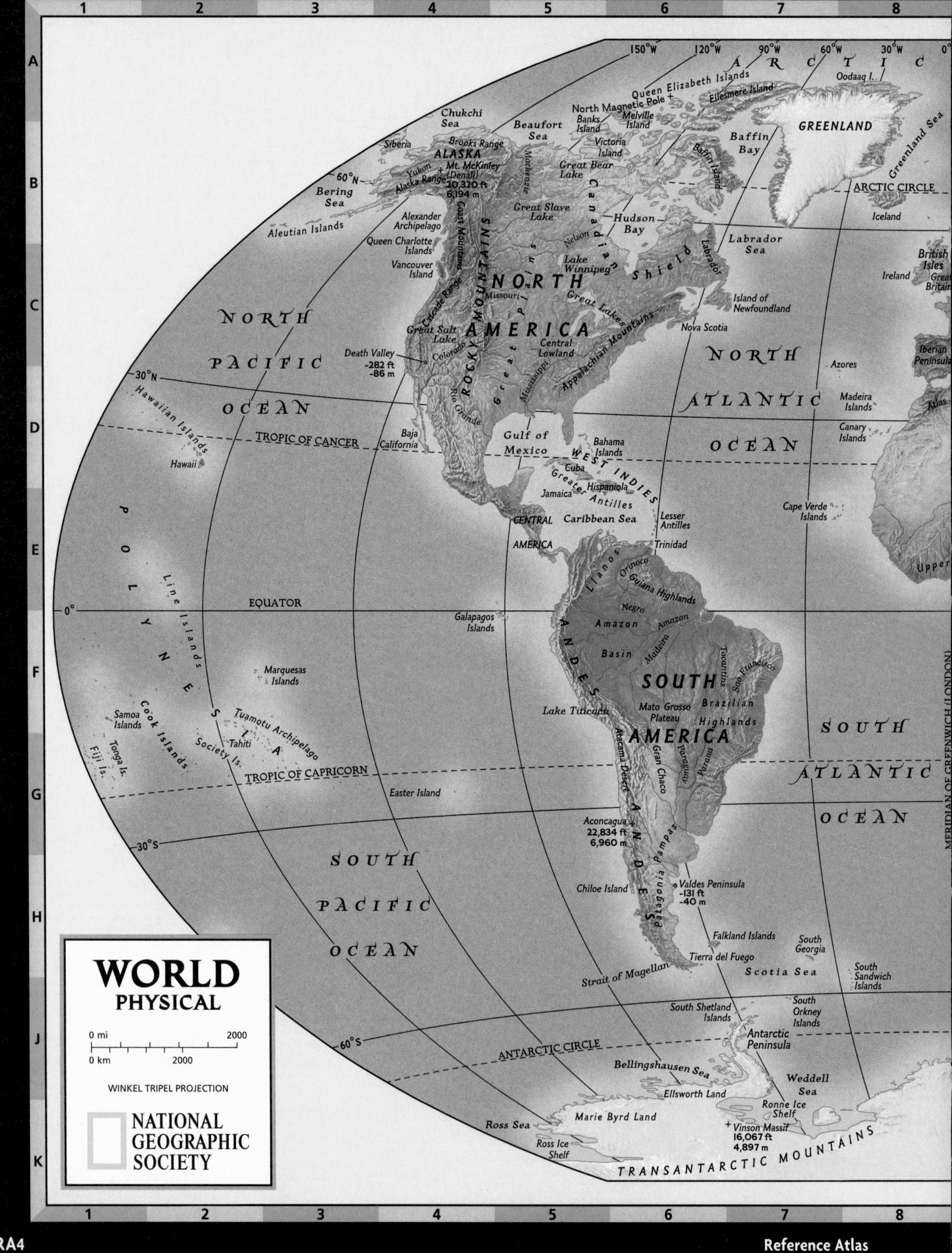
WORLD
PHYSICAL
0 mi
2000
0 km
2000
WINKEL TRIPEL PROJECTION
NATIONAL GEOGRAPHIC SOCIETY
NORTH AMERICA
SOUTH AMERICA
NORTH PACIFIC OCEAN
SOUTH PACIFIC OCEAN
NORTH ATLANTIC OCEAN
SOUTH ATLANTIC OCEAN
POLYNESIA
GREENLAND
ALASKA
ROCKY MOUNTAINS
ANDES
TRANSANTARCTIC MOUNTAINS
Canadian Shield
Great Plains
Mt. McKinley (Denali) 20,320 ft 6,194 m
Death Valley -282 ft -86 m
Aconcagua 22,834 ft 6,960 m
Valdes Peninsula -131 ft -40 m
Vinson Massif 16,067 ft 4,897 m
Gulf of Mexico
Caribbean Sea
Hudson Bay
Baffin Bay
Labrador Sea
Bering Sea
Beaufort Sea
Chukchi Sea
Scotia Sea
Weddell Sea
Ross Sea
Bellingshausen Sea
Amazon Basin
Brazilian Highlands
Mato Grosso Plateau
Gran Chaco
Pampas
Patagonia
TROPIC OF CANCER
TROPIC OF CAPRICORN
EQUATOR
ARCTIC CIRCLE
ANTARCTIC CIRCLE
MERIDIAN OF GREENWICH (LONDON)

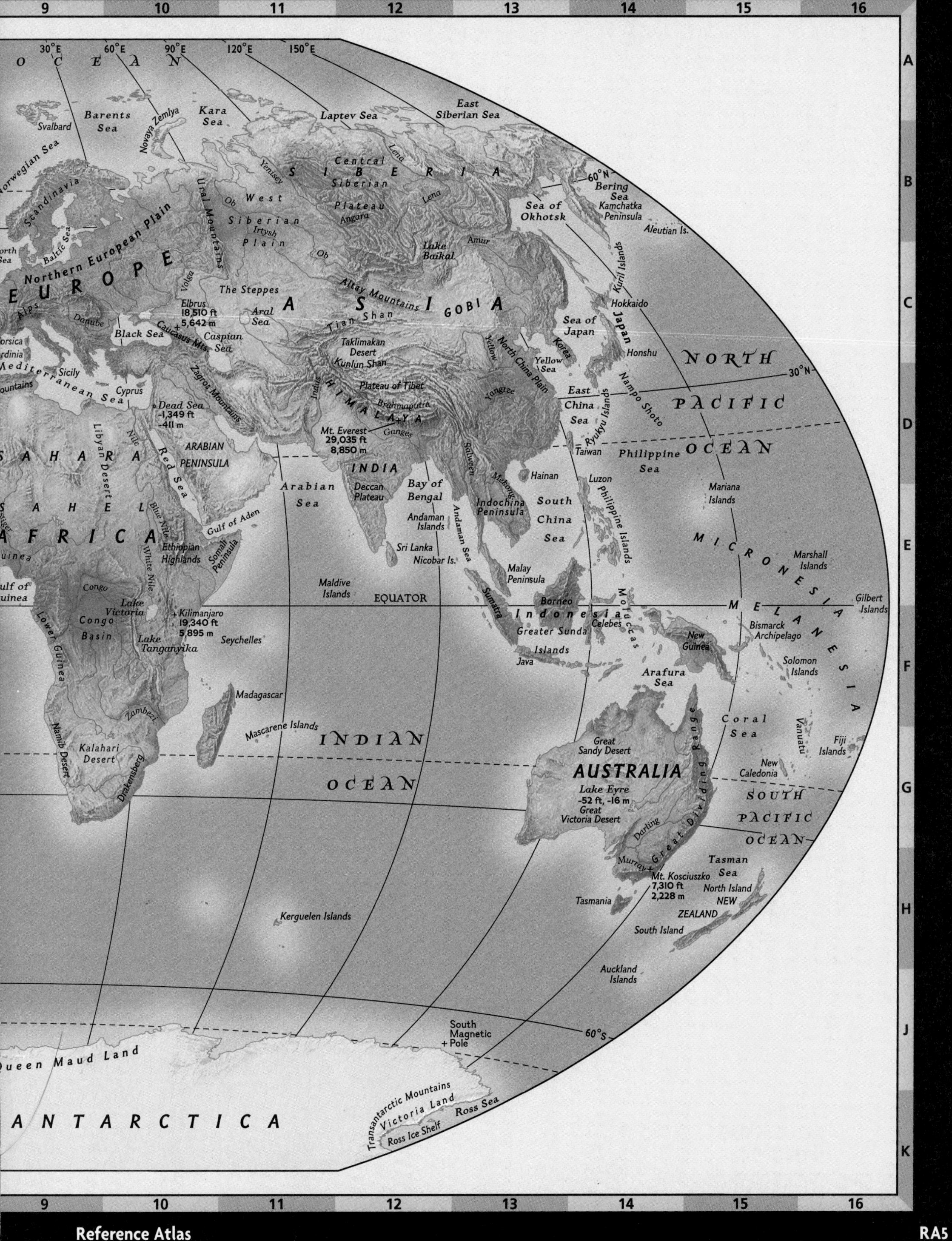

EUROPE
ASIA
AFRICA
AUSTRALIA
ANTARCTICA
INDIAN OCEAN
NORTH PACIFIC OCEAN
SOUTH PACIFIC OCEAN
SIBERIA
SAHARA
SAHEL
MICRONESIA
MELANESIA
INDIA
Barents Sea
Kara Sea
Laptev Sea
East Siberian Sea
Bering Sea
Sea of Okhotsk
Sea of Japan
Yellow Sea
East China Sea
South China Sea
Philippine Sea
Arabian Sea
Bay of Bengal
Andaman Sea
Arafura Sea
Coral Sea
Tasman Sea
Ross Sea
Caspian Sea
Aral Sea
Black Sea
Mediterranean Sea
Red Sea
Gulf of Aden
Baltic Sea
Norwegian Sea
Svalbard
Novaya Zemlya
Scandinavia
Northern European Plain
West Siberian Plain
Central Siberian Plateau
Ural Mountains
Caucasus Mts.
Elbrus 18,510 ft 5,642 m
Zagros Mountains
Altay Mountains
Tian Shan
Kunlun Shan
Taklimakan Desert
GOBI
Plateau of Tibet
HIMALAYA
Mt. Everest 29,035 ft 8,850 m
Deccan Plateau
ARABIAN PENINSULA
Dead Sea -1,349 ft -411 m
Ethiopian Highlands
Somali Peninsula
Kilimanjaro 19,340 ft 5,895 m
Congo Basin
Kalahari Desert
Namib Desert
Drakensberg
Libyan Desert
Lake Baikal
Lake Victoria
Lake Tanganyika
Lake Eyre -52 ft, -16 m
Great Sandy Desert
Great Victoria Desert
Great Dividing Range
Mt. Kosciuszko 7,310 ft 2,228 m
Kamchatka Peninsula
Aleutian Is.
Kuril Islands
Hokkaido
Honshu
Japan
Korea
North China Plain
Nampo Shoto
Ryukyu Islands
Taiwan
Hainan
Luzon
Philippine Islands
Indochina Peninsula
Malay Peninsula
Sumatra
Borneo
Java
Celebes
Moluccas
Indonesia
Greater Sunda Islands
New Guinea
Bismarck Archipelago
Solomon Islands
Mariana Islands
Marshall Islands
Gilbert Islands
Vanuatu
Fiji Islands
New Caledonia
Tasmania
North Island
NEW ZEALAND
South Island
Auckland Islands
Sri Lanka
Andaman Islands
Nicobar Is.
Maldive Islands
Seychelles
Madagascar
Mascarene Islands
Kerguelen Islands
Cyprus
Sicily
Queen Maud Land
Transantarctic Mountains
Victoria Land
Ross Ice Shelf
South Magnetic Pole
The Steppes
Alps
Danube
Volga
Ob
Irtysh
Yenisey
Lena
Angara
Amur
Yellow
Yangtze
Indus
Ganges
Brahmaputra
Salween
Mekong
Nile
Blue Nile
White Nile
Congo
Lower Guinea
Zambezi
Darling
Murray
EQUATOR
30°E 60°E 90°E 120°E 150°E
60°N 30°N 60°S

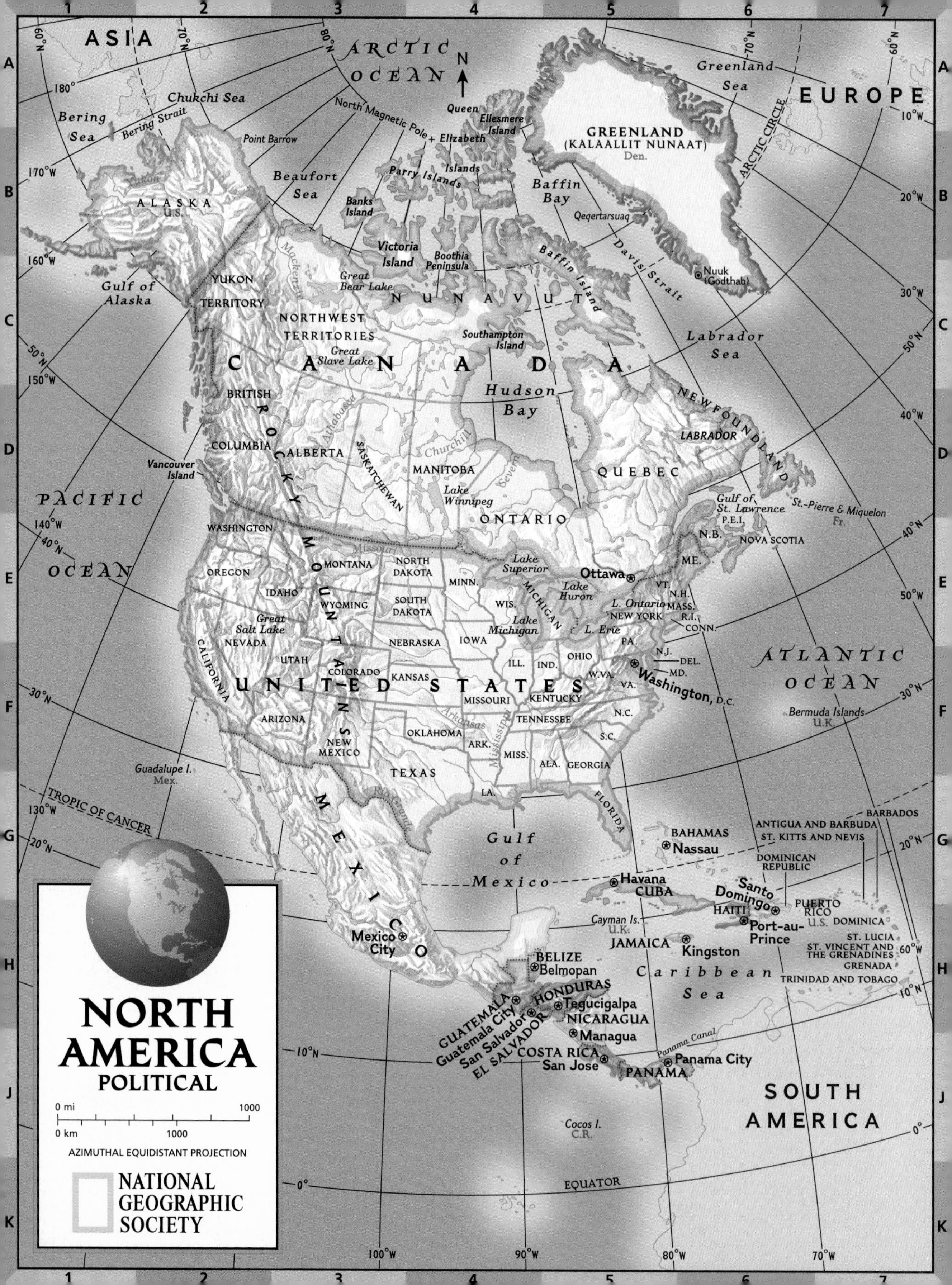
NORTH AMERICA
POLITICAL
0 mi 1000
0 km 1000
AZIMUTHAL EQUIDISTANT PROJECTION
NATIONAL GEOGRAPHIC SOCIETY
ASIA
ARCTIC OCEAN
EUROPE
PACIFIC OCEAN
ATLANTIC OCEAN
SOUTH AMERICA
Greenland Sea
GREENLAND (KALAALLIT NUNAAT) Den.
Chukchi Sea
Bering Sea
Bering Strait
Point Barrow
North Magnetic Pole
Queen Elizabeth Islands
Ellesmere Island
Beaufort Sea
Parry Islands
Banks Island
Victoria Island
Boothia Peninsula
Baffin Bay
Qeqertarsuaq
Baffin Island
Davis Strait
Nuuk (Godthab)
ARCTIC CIRCLE
ALASKA U.S.
Yukon
Gulf of Alaska
YUKON TERRITORY
Mackenzie
Great Bear Lake
NUNAVUT
NORTHWEST TERRITORIES
Great Slave Lake
Southampton Island
Labrador Sea
CANADA
Hudson Bay
BRITISH COLUMBIA
ROCKY MOUNTAINS
Athabasca
ALBERTA
SASKATCHEWAN
Churchill
MANITOBA
Severn
Lake Winnipeg
ONTARIO
QUEBEC
LABRADOR
NEWFOUNDLAND
Vancouver Island
Gulf of St. Lawrence
St.-Pierre & Miquelon Fr.
P.E.I.
N.B.
NOVA SCOTIA
WASHINGTON
Missouri
MONTANA
NORTH DAKOTA
OREGON
IDAHO
WYOMING
SOUTH DAKOTA
MINN.
Lake Superior
Ottawa
ME.
VT.
N.H.
MASS.
R.I.
CONN.
WIS.
MICHIGAN
Lake Huron
L. Ontario
NEW YORK
Great Salt Lake
NEVADA
Lake Michigan
L. Erie
PA.
NEBRASKA
IOWA
N.J.
DEL.
MD.
CALIFORNIA
UTAH
COLORADO
KANSAS
ILL.
IND.
OHIO
W.VA.
Washington, D.C.
VA.
UNITED STATES
MISSOURI
KENTUCKY
ARIZONA
Arkansas
TENNESSEE
N.C.
Bermuda Islands U.K.
OKLAHOMA
NEW MEXICO
ARK.
Mississippi
S.C.
TEXAS
MISS.
ALA.
GEORGIA
Guadalupe I. Mex.
TROPIC OF CANCER
Rio Grande
LA.
FLORIDA
MEXICO
Gulf of Mexico
BAHAMAS
Nassau
ANTIGUA AND BARBUDA
ST. KITTS AND NEVIS
BARBADOS
DOMINICAN REPUBLIC
Havana
CUBA
Santo Domingo
HAITI
PUERTO RICO U.S.
DOMINICA
Cayman Is. U.K.
Port-au-Prince
JAMAICA
Kingston
ST. LUCIA
ST. VINCENT AND THE GRENADINES
GRENADA
TRINIDAD AND TOBAGO
Mexico City
BELIZE
Belmopan
Caribbean Sea
HONDURAS
Tegucigalpa
GUATEMALA
Guatemala City
San Salvador
EL SALVADOR
NICARAGUA
Managua
COSTA RICA
San Jose
Panama Canal
Panama City
PANAMA
Cocos I. C.R.
EQUATOR
180°
170°W
160°W
150°W
140°W
130°W
100°W
90°W
80°W
70°W
60°W
10°W
20°W
30°W
40°W
50°W
60°N
70°N
80°N
50°N
40°N
30°N
20°N
10°N
0°
1 2 3 4 5 6 7
A B C D E F G H J K

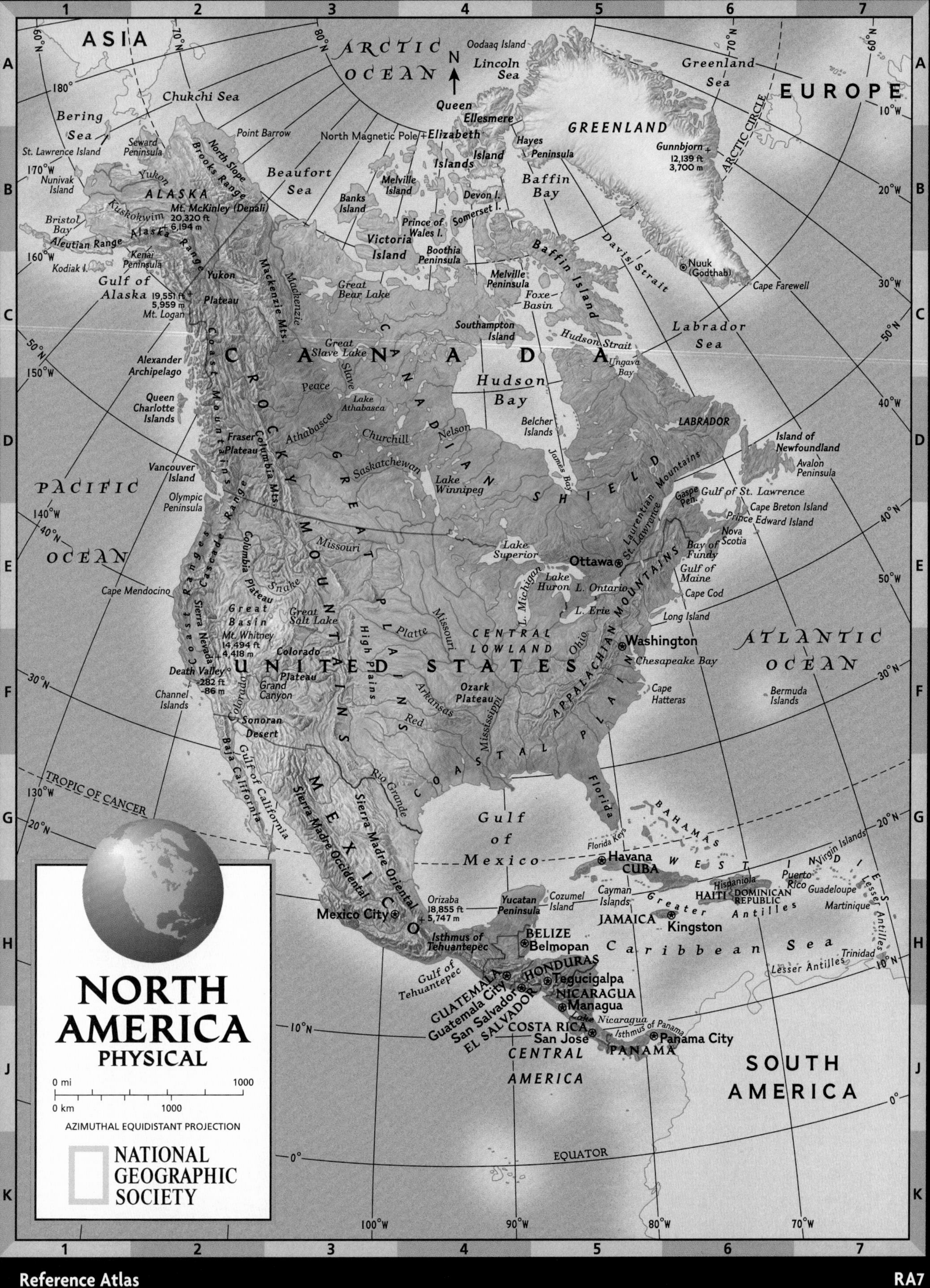
NORTH AMERICA
PHYSICAL
0 mi
1000
0 km
1000
AZIMUTHAL EQUIDISTANT PROJECTION
NATIONAL GEOGRAPHIC SOCIETY
ASIA
EUROPE
SOUTH AMERICA
CENTRAL AMERICA
ARCTIC OCEAN
PACIFIC OCEAN
ATLANTIC OCEAN
CANADA
UNITED STATES
MEXICO
GREENLAND
ALASKA
LABRADOR
CUBA
HAITI
DOMINICAN REPUBLIC
JAMAICA
BELIZE
HONDURAS
GUATEMALA
EL SALVADOR
NICARAGUA
COSTA RICA
PANAMA
BAHAMAS
WEST INDIES
Ottawa
Washington
Mexico City
Havana
Kingston
Belmopan
Tegucigalpa
Guatemala City
San Salvador
Managua
San Jose
Panama City
Nuuk (Godthab)
Chukchi Sea
Bering Sea
Beaufort Sea
Lincoln Sea
Greenland Sea
Baffin Bay
Labrador Sea
Hudson Bay
Gulf of Alaska
Gulf of Mexico
Caribbean Sea
Gulf of St. Lawrence
Gulf of California
Gulf of Tehuantepec
Gulf of Maine
Bay of Fundy
Chesapeake Bay
Davis Strait
Hudson Strait
Ungava Bay
James Bay
Bristol Bay
Foxe Basin
Oodaaq Island
North Magnetic Pole
Point Barrow
Seward Peninsula
St. Lawrence Island
Nunivak Island
Kenai Peninsula
Kodiak I.
Aleutian Range
Alaska Range
Brooks Range
North Slope
Yukon
Kuskokwim
Mt. McKinley (Denali)
20,320 ft
6,194 m
Mt. Logan
19,551 ft
5,959 m
Yukon Plateau
Mackenzie Mts.
Mackenzie
Queen Elizabeth Islands
Ellesmere Island
Hayes Peninsula
Devon I.
Somerset I.
Melville Island
Banks Island
Victoria Island
Prince of Wales I.
Boothia Peninsula
Melville Peninsula
Baffin Island
Southampton Island
Belcher Islands
Gunnbjorn
12,139 ft
3,700 m
Arctic Circle
Cape Farewell
Great Bear Lake
Great Slave Lake
Slave
Peace
Lake Athabasca
Athabasca
Churchill
Nelson
Saskatchewan
Lake Winnipeg
Canadian Shield
Laurentian Mountains
St. Lawrence
Gaspe Pen.
Island of Newfoundland
Avalon Peninsula
Cape Breton Island
Prince Edward Island
Nova Scotia
Cape Cod
Long Island
Alexander Archipelago
Queen Charlotte Islands
Coast Mountains
Rocky Mountains
Fraser Plateau
Columbia Mts.
Vancouver Island
Olympic Peninsula
Cascade Range
Coast Ranges
Columbia Plateau
Columbia
Snake
Cape Mendocino
Great Basin
Sierra Nevada
Mt. Whitney
14,494 ft
4,418 m
Great Salt Lake
Death Valley
-282 ft
-86 m
Colorado Plateau
Colorado
Grand Canyon
Channel Islands
Sonoran Desert
Baja California
Great Plains
High Plains
Missouri
Platte
Arkansas
Red
Rio Grande
Mississippi
Ozark Plateau
Central Lowland
Lake Superior
L. Michigan
Lake Huron
L. Ontario
L. Erie
Ohio
Appalachian Mountains
Coastal Plain
Cape Hatteras
Florida
Florida Keys
Bermuda Islands
Sierra Madre Occidental
Sierra Madre Oriental
Orizaba
18,855 ft
5,747 m
Yucatan Peninsula
Isthmus of Tehuantepec
Cozumel Island
Cayman Islands
Greater Antilles
Hispaniola
Puerto Rico
Virgin Islands
Guadeloupe
Martinique
Lesser Antilles
Trinidad
Lake Nicaragua
Isthmus of Panama
Tropic of Cancer
Equator
180°
170°W
160°W
150°W
140°W
130°W
100°W
90°W
80°W
70°W
10°W
20°W
30°W
40°W
60°N
70°N
80°N
50°N
40°N
30°N
20°N
10°N
0°

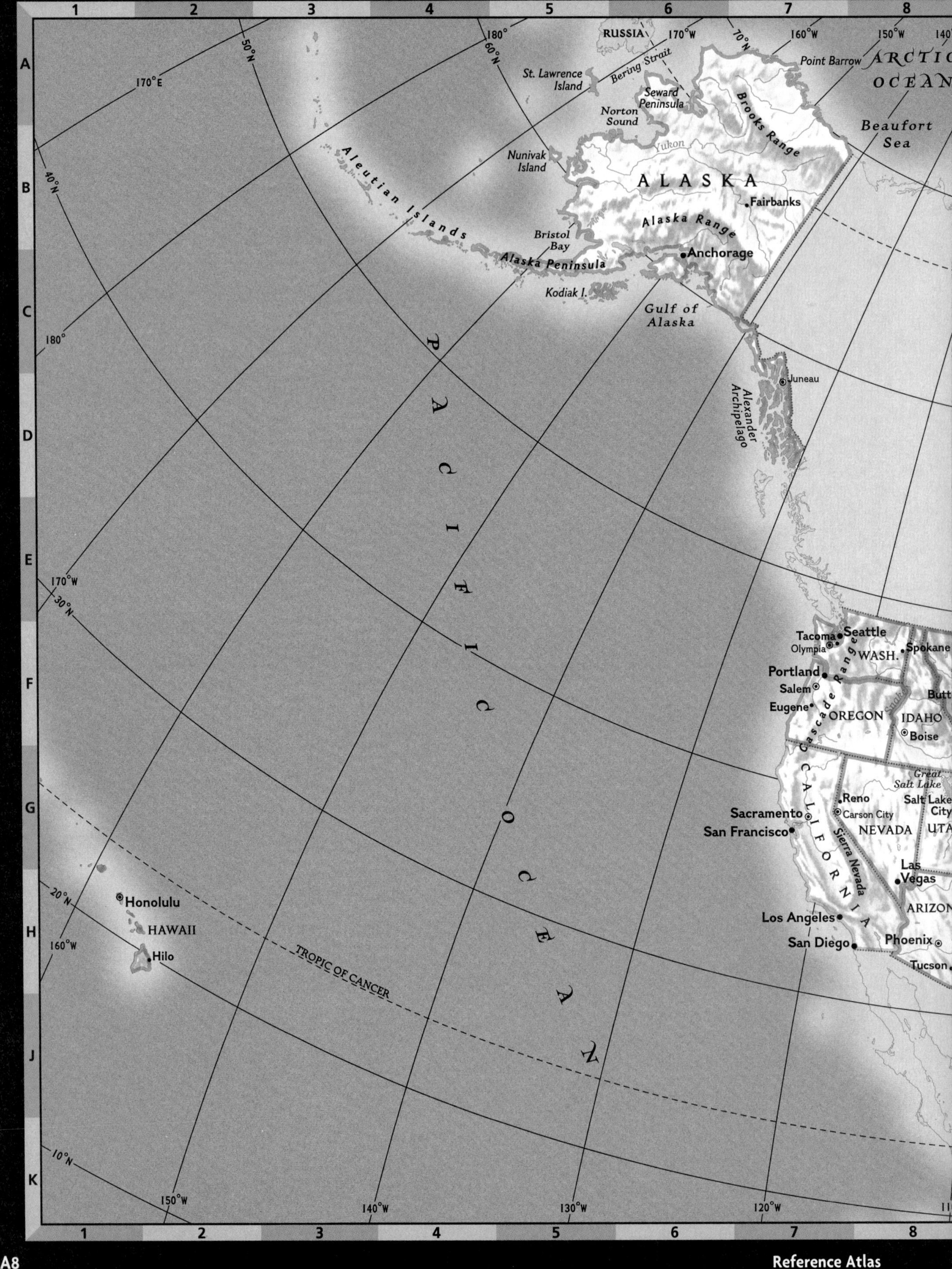

RUSSIA
Bering Strait
St. Lawrence Island
Seward Peninsula
Norton Sound
Point Barrow
ARCTIC OCEAN
Beaufort Sea
Brooks Range
Yukon
Nunivak Island
ALASKA
Fairbanks
Aleutian Islands
Alaska Range
Bristol Bay
Anchorage
Alaska Peninsula
Kodiak I.
Gulf of Alaska
Juneau
Alexander Archipelago
PACIFIC OCEAN
Tacoma
Seattle
Olympia
WASH.
Spokane
Portland
Salem
Eugene
Cascade Range
OREGON
IDAHO
Boise
Great Salt Lake
Salt Lake City
Reno
Carson City
Sacramento
San Francisco
NEVADA
CALIFORNIA
Sierra Nevada
Las Vegas
ARIZON
Los Angeles
San Diego
Phoenix
Tucson
Honolulu
HAWAII
Hilo
TROPIC OF CANCER

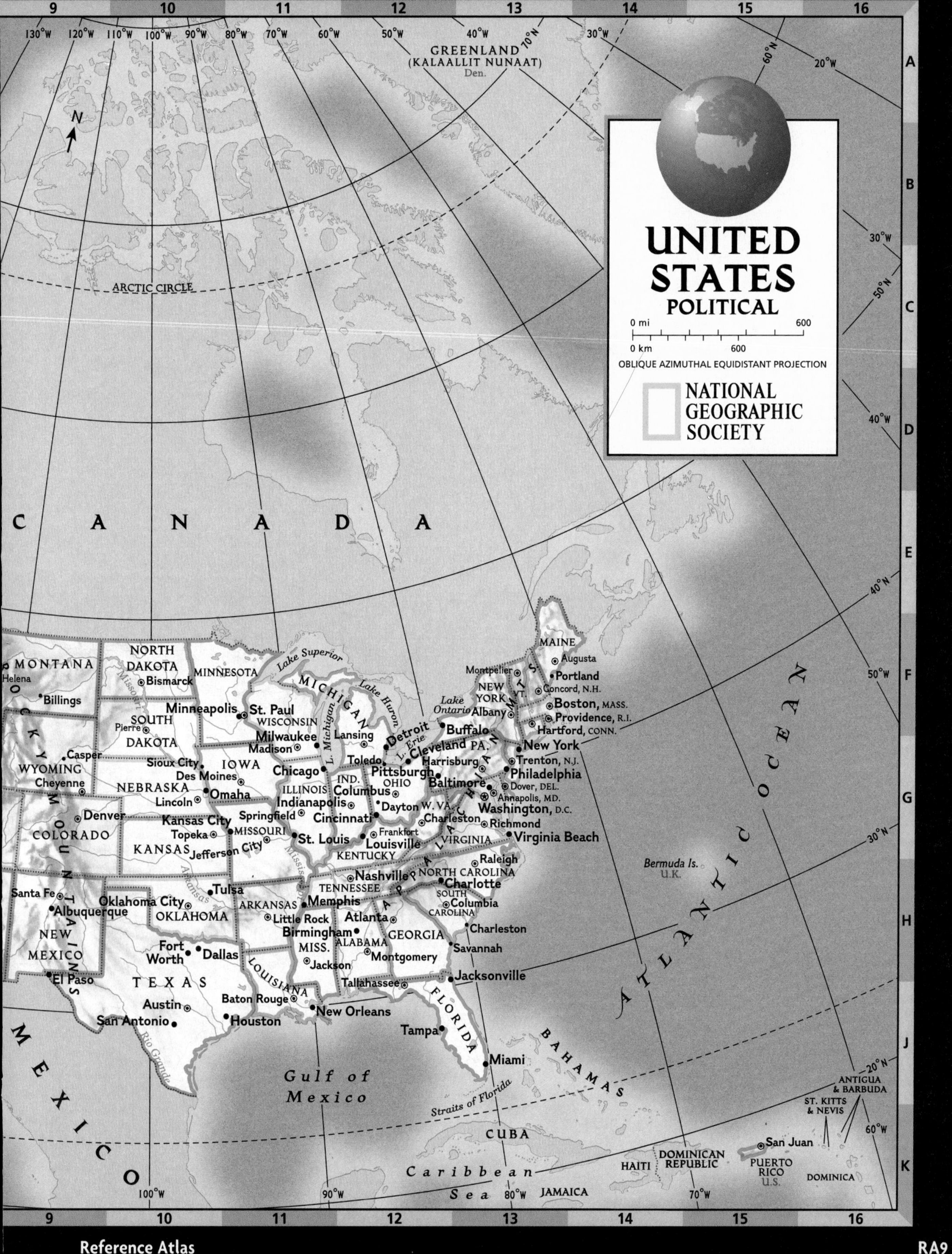
UNITED STATES
POLITICAL
0 mi 600
0 km 600
OBLIQUE AZIMUTHAL EQUIDISTANT PROJECTION
NATIONAL GEOGRAPHIC SOCIETY
GREENLAND (KALAALLIT NUNAAT) Den.
ARCTIC CIRCLE
CANADA
MONTANA
Helena
Billings
NORTH DAKOTA
Bismarck
MINNESOTA
SOUTH DAKOTA
Pierre
Minneapolis
St. Paul
WISCONSIN
Milwaukee
Madison
MICHIGAN
Lansing
Detroit
Lake Superior
Lake Huron
L. Michigan
L. Erie
Lake Ontario
WYOMING
Casper
Cheyenne
Sioux City
Des Moines
IOWA
Chicago
NEBRASKA
Omaha
Lincoln
ILLINOIS
IND.
OHIO
Toledo
Cleveland
Pittsburgh
Columbus
Indianapolis
Dayton
Cincinnati
Springfield
Denver
COLORADO
Kansas City
Topeka
MISSOURI
KANSAS
Jefferson City
St. Louis
Louisville
Frankfort
KENTUCKY
W. VA.
Charleston
VIRGINIA
Richmond
Virginia Beach
Washington, D.C.
Baltimore
Annapolis, MD.
Dover, DEL.
Philadelphia
Harrisburg
PA.
Trenton, N.J.
New York
Buffalo
NEW YORK
Albany
Montpelier
VT.
MAINE
Augusta
Portland
Concord, N.H.
Boston, MASS.
Providence, R.I.
Hartford, CONN.
ROCKY MOUNTAINS
APPALACHIAN MTS.
Santa Fe
Albuquerque
NEW MEXICO
El Paso
Oklahoma City
OKLAHOMA
Tulsa
Arkansas
ARKANSAS
Little Rock
Memphis
TENNESSEE
Nashville
NORTH CAROLINA
Raleigh
Charlotte
SOUTH CAROLINA
Columbia
Charleston
Atlanta
GEORGIA
Savannah
Birmingham
ALABAMA
Montgomery
MISS.
Jackson
Fort Worth
Dallas
TEXAS
Austin
San Antonio
Houston
Rio Grande
LOUISIANA
Baton Rouge
New Orleans
Tallahassee
Jacksonville
FLORIDA
Tampa
Miami
Missouri
Mississippi
MEXICO
Gulf of Mexico
Straits of Florida
BAHAMAS
CUBA
Caribbean Sea
JAMAICA
HAITI
DOMINICAN REPUBLIC
PUERTO RICO U.S.
San Juan
ANTIGUA & BARBUDA
ST. KITTS & NEVIS
DOMINICA
Bermuda Is. U.K.
ATLANTIC OCEAN
130°W 120°W 110°W 100°W 90°W 80°W 70°W 60°W 50°W 40°W 30°W 20°W
70°N 60°N 50°N 40°N 30°N 20°N
9 10 11 12 13 14 15 16
A B C D E F G H J K

PACIFIC OCEAN
CASCADE RANGE
COAST RANGES
SIERRA NEVADA
COLUMBIA PLATEAU
GREAT BASIN
ROCKY MOUNTAINS
GREAT PLAINS
High Plains
CANADA
MEXICO
Cape Flattery
Mt. Olympus
7,966 ft
2,428 m
Seattle
Columbia
Blue Mts.
Clearwater Mts.
Bitterroot Range
Salmon River Mts.
Great Sandy Desert
Snake
Snake River Plain
Shoshone Falls
Absaroka Range
Bighorn Mts.
Wind River Range
Black Hills
Missouri
Laramie Mts.
N. Platte
Sand Hills
Platte
Cape Mendocino
Great Salt Lake
Wasatch Range
Uinta Mts.
Lake Tahoe
Central Valley
San Francisco
Denver
14,433 ft
4,399 m
Mt. Elbert
Mt. Whitney
14,494 ft
4,418 m
Death Valley
-282 ft, -86 m
Lake Powell
Colorado
Lake Mead
San Juan Mts.
Sangre de Cristo Mts.
Arkansas
Point Conception
Mojave Desert
Grand Canyon
Colorado Plateau
Los Angeles
Channel Islands
Salton Sea
San Diego
Phoenix
Sonoran Desert
Rio Grande
Sacramento Mts.
Llano Estacado
Red
Dalla
Brazo
Edwards Plateau
TROPIC OF CANCER
ARCTIC OCEAN
Point Barrow
Beaufort Sea
Chukchi Sea
RUSSIA
North Slope
Brooks Range
Bering Strait
ARCTIC CIRCLE
Seward Pen.
ALASKA
CANADA
St. Lawrence Island
Yukon
Tanana
Kuskokwim
Alaska Range
Mt. McKinley (Denali)
20,320 ft, 6,194 m
Anchorage
Nunivak Island
Bering Sea
Bristol Bay
Gulf of Alaska
Alexander Archipelago
Kodiak I.
Alaska Peninsula
PACIFIC OCEAN
0 mi 300
0 km 300

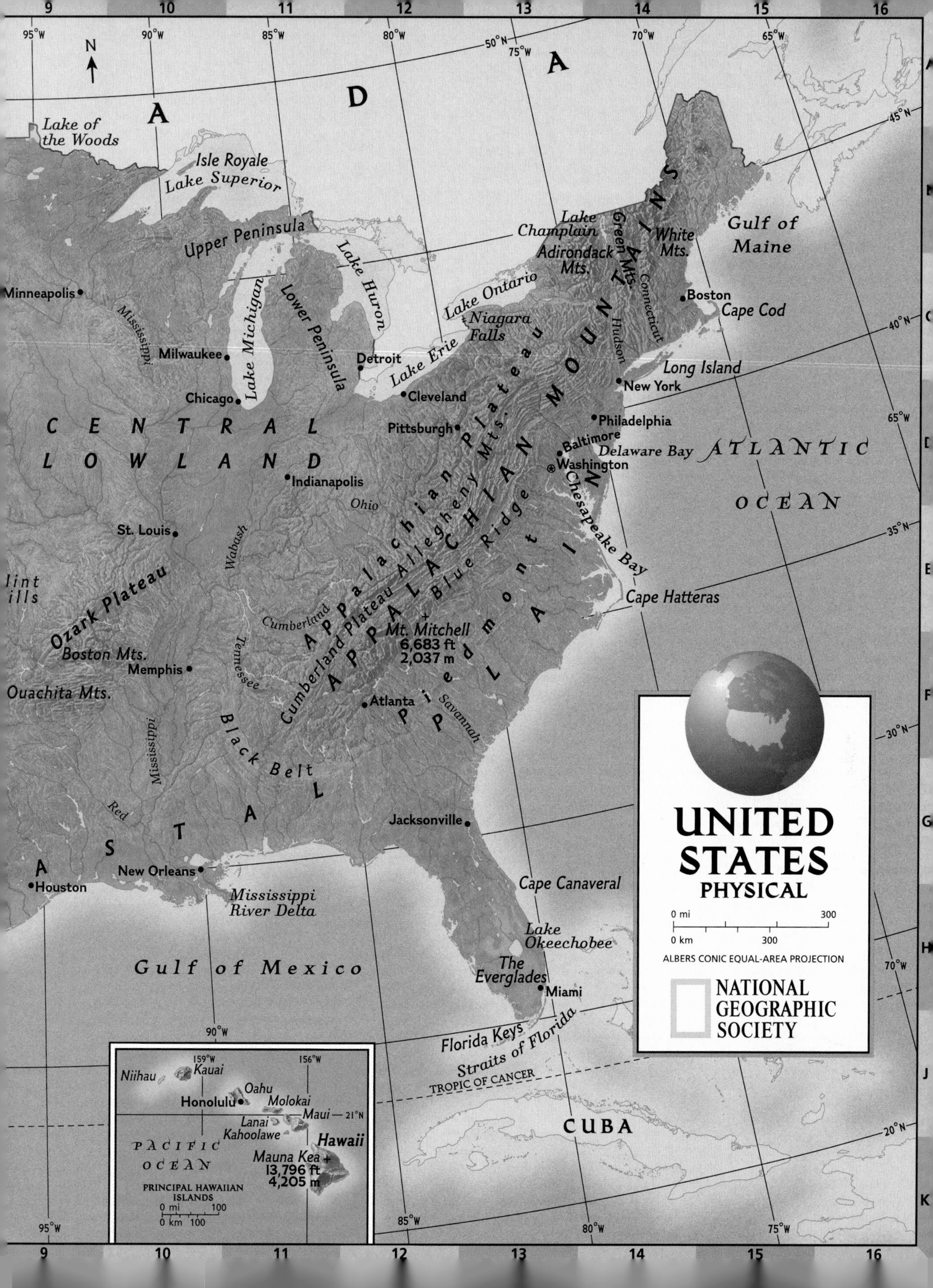
UNITED STATES
PHYSICAL
0 mi 300
0 km 300
ALBERS CONIC EQUAL-AREA PROJECTION
NATIONAL GEOGRAPHIC SOCIETY
CANADA
Lake of the Woods
Isle Royale
Lake Superior
Upper Peninsula
Lake Michigan
Lower Peninsula
Lake Huron
Lake Ontario
Lake Erie
Niagara Falls
Lake Champlain
Adirondack Mts.
Green Mts.
White Mts.
Gulf of Maine
Boston
Cape Cod
Connecticut
Hudson
Long Island
New York
Philadelphia
Baltimore
Washington
Delaware Bay
Chesapeake Bay
ATLANTIC OCEAN
Cape Hatteras
APPALACHIAN MOUNTAINS
Appalachian Plateau
Allegheny Mts.
Blue Ridge
Cumberland Plateau
Piedmont
Mt. Mitchell 6,683 ft 2,037 m
Minneapolis
Mississippi
Milwaukee
Chicago
Detroit
Cleveland
Pittsburgh
CENTRAL LOWLAND
Indianapolis
Ohio
St. Louis
Wabash
Ozark Plateau
Boston Mts.
Ouachita Mts.
Memphis
Tennessee
Cumberland
Atlanta
Savannah
Black Belt
COASTAL PLAIN
Red
Jacksonville
New Orleans
Houston
Mississippi River Delta
Gulf of Mexico
Cape Canaveral
Lake Okeechobee
The Everglades
Miami
Florida Keys
Straits of Florida
TROPIC OF CANCER
CUBA
Niihau
Kauai
Oahu
Honolulu
Molokai
Maui
Lanai
Kahoolawe
Hawaii
Mauna Kea 13,796 ft 4,205 m
PACIFIC OCEAN
PRINCIPAL HAWAIIAN ISLANDS
0 mi 100
0 km 100

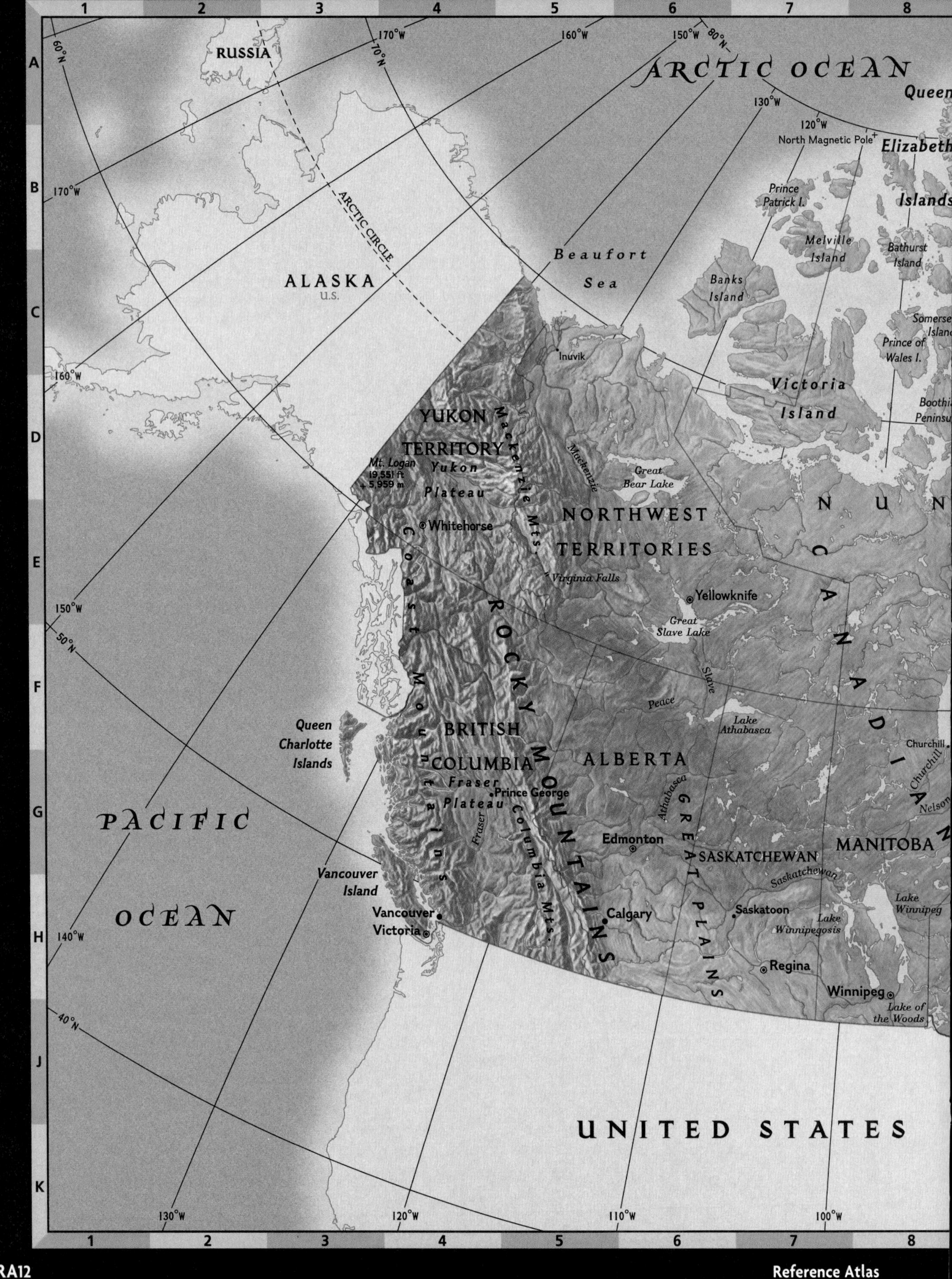

ARCTIC OCEAN
RUSSIA
ALASKA
U.S.
ARCTIC CIRCLE
Beaufort Sea
Queen
Elizabeth Islands
North Magnetic Pole
Prince Patrick I.
Melville Island
Bathurst Island
Banks Island
Somerset Island
Prince of Wales I.
Victoria Island
Boothia Peninsula
Inuvik
YUKON TERRITORY
Mackenzie Mts.
Mackenzie
Mt. Logan 19,551 ft 5,959 m
Yukon Plateau
Whitehorse
Great Bear Lake
NORTHWEST TERRITORIES
NUNAVUT
Virginia Falls
Yellowknife
Great Slave Lake
Slave
Peace
Coast Mountains
ROCKY MOUNTAINS
CANADIAN
Lake Athabasca
Churchill
Queen Charlotte Islands
BRITISH COLUMBIA
ALBERTA
Fraser Plateau
Prince George
Fraser
Columbia Mts.
Athabasca
GREAT PLAINS
Nelson
PACIFIC OCEAN
Edmonton
SASKATCHEWAN
MANITOBA
Saskatchewan
Vancouver Island
Vancouver
Victoria
Calgary
Saskatoon
Lake Winnipegosis
Lake Winnipeg
Regina
Winnipeg
Lake of the Woods
UNITED STATES
80°N
70°N
60°N
50°N
40°N
170°W
160°W
150°W
140°W
130°W
120°W
110°W
100°W

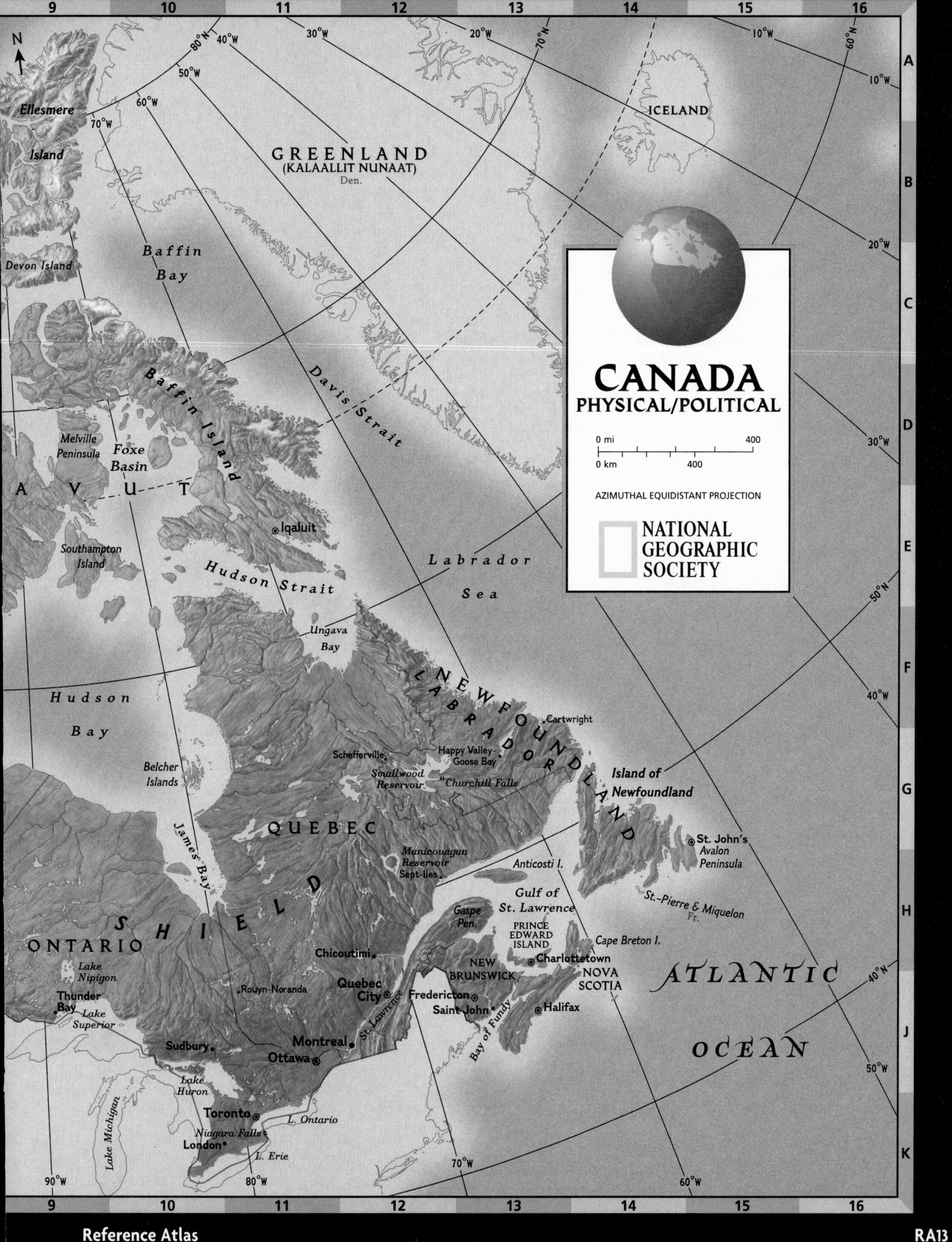
CANADA
PHYSICAL/POLITICAL
0 mi
400
0 km
400
AZIMUTHAL EQUIDISTANT PROJECTION
NATIONAL
GEOGRAPHIC
SOCIETY
GREENLAND
(KALAALLIT NUNAAT)
Den.
ICELAND
Ellesmere
Island
Devon Island
Baffin
Bay
Baffin Island
Davis Strait
Melville
Peninsula
Foxe
Basin
Iqaluit
Southampton
Island
Hudson Strait
Labrador
Sea
Ungava
Bay
Hudson
Bay
NEWFOUNDLAND
LABRADOR
Cartwright
Schefferville
Happy Valley-
Goose Bay
Smallwood
Reservoir
Churchill Falls
Belcher
Islands
Island of
Newfoundland
QUEBEC
James Bay
Manicouagan
Reservoir
Sept-Iles
Anticosti I.
St. John's
Avalon
Peninsula
St.-Pierre & Miquelon
Fr.
Gulf of
St. Lawrence
Gaspe
Pen.
PRINCE
EDWARD
ISLAND
Cape Breton I.
Charlottetown
SHIELD
ONTARIO
Lake
Nipigon
Chicoutimi
NEW
BRUNSWICK
NOVA
SCOTIA
ATLANTIC
OCEAN
Thunder
Bay
Lake
Superior
Rouyn-Noranda
Quebec
City
St. Lawrence
Fredericton
Saint John
Halifax
Bay of Fundy
Sudbury
Montreal
Ottawa
Lake
Huron
Toronto
L. Ontario
Niagara Falls
London
L. Erie
Lake Michigan
90°W
80°W
70°W
60°W
50°W
40°W
30°W
20°W
10°W
80°N
70°N
60°N
50°N
40°N
A
B
C
D
E
F
G
H
J
K
9
10
11
12
13
14
15
16

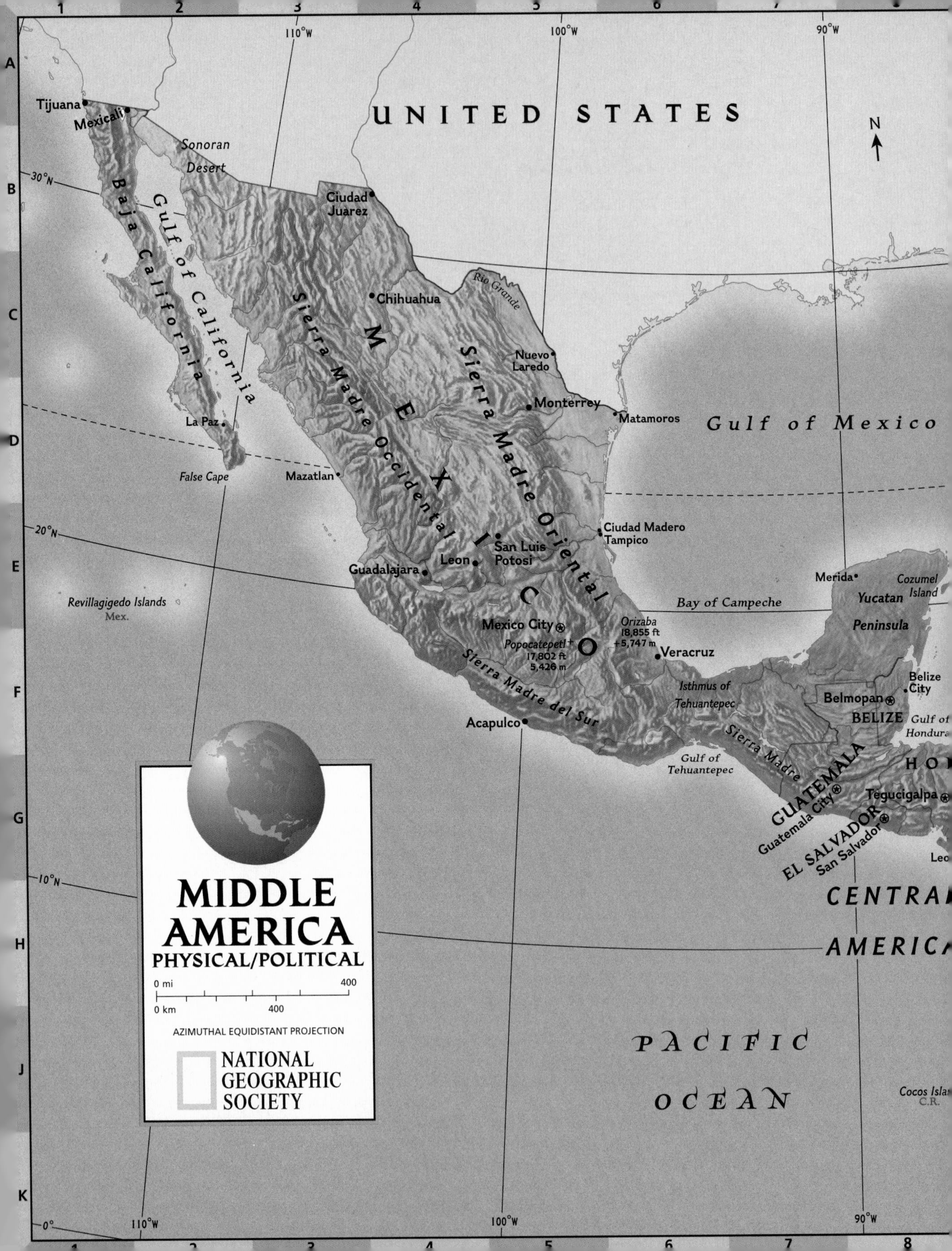

MIDDLE AMERICA
PHYSICAL/POLITICAL
0 mi 400
0 km 400
AZIMUTHAL EQUIDISTANT PROJECTION
NATIONAL GEOGRAPHIC SOCIETY
UNITED STATES
MEXICO
Tijuana
Mexicali
Sonoran Desert
Baja California
Gulf of California
Ciudad Juarez
Chihuahua
Rio Grande
Sierra Madre Occidental
Sierra Madre Oriental
Nuevo Laredo
Monterrey
Matamoros
Gulf of Mexico
La Paz
False Cape
Mazatlan
Ciudad Madero
Tampico
San Luis Potosi
Leon
Guadalajara
Revillagigedo Islands
Mex.
Bay of Campeche
Merida
Cozumel Island
Yucatan Peninsula
Mexico City
Orizaba
18,855 ft
5,747 m
Popocatepetl
17,802 ft
5,426 m
Veracruz
Sierra Madre del Sur
Isthmus of Tehuantepec
Acapulco
Gulf of Tehuantepec
Sierra Madre
Belmopan
Belize City
BELIZE
GUATEMALA
Guatemala City
Tegucigalpa
EL SALVADOR
San Salvador
CENTRA
AMERIC
PACIFIC OCEAN
Cocos Isla
C.R.
N
30°N
20°N
10°N
0°
110°W
100°W
90°W

ATLANTIC OCEAN
BAHAMAS
Freeport
Nassau
Andros Island
Straits of Florida
Turks & Caicos Islands U.K.
TROPIC OF CANCER
WEST INDIES
Havana
CUBA
Camaguey
Holguin
Santiago de Cuba
Isle of Youth
Cayman Islands U.K.
Hispaniola
Santiago
HAITI
Port-au-Prince
Santo Domingo
DOMINICAN REPUBLIC
San Juan
Puerto Rico U.S.
Virgin Islands U.S. U.K.
ST. KITTS & NEVIS
ANTIGUA & BARBUDA
Guadeloupe Fr.
DOMINICA
Martinique Fr.
ST. LUCIA
BARBADOS
ST. VINCENT & THE GRENADINES
GRENADA
TRINIDAD & TOBAGO
Port of Spain
Tobago
Trinidad
Lesser Antilles
Bird I. Venez.
Greater Antilles
Montego Bay
JAMAICA
Kingston
Caribbean Sea
Aruba Neth.
Curacao
Bonaire
Neth.
Lesser Antilles
HONDURAS
Coco
Mosquito Coast
NICARAGUA
Managua
Lake Nicaragua
COSTA RICA
San Jose
Puerto Limon
Gulf of Mosquitos
Isthmus of Panama
Panama City
PANAMA
David
Gulf of Panama
SOUTH AMERICA
EQUATOR
80°W
70°W
60°W
30°N
20°N
10°N
0°

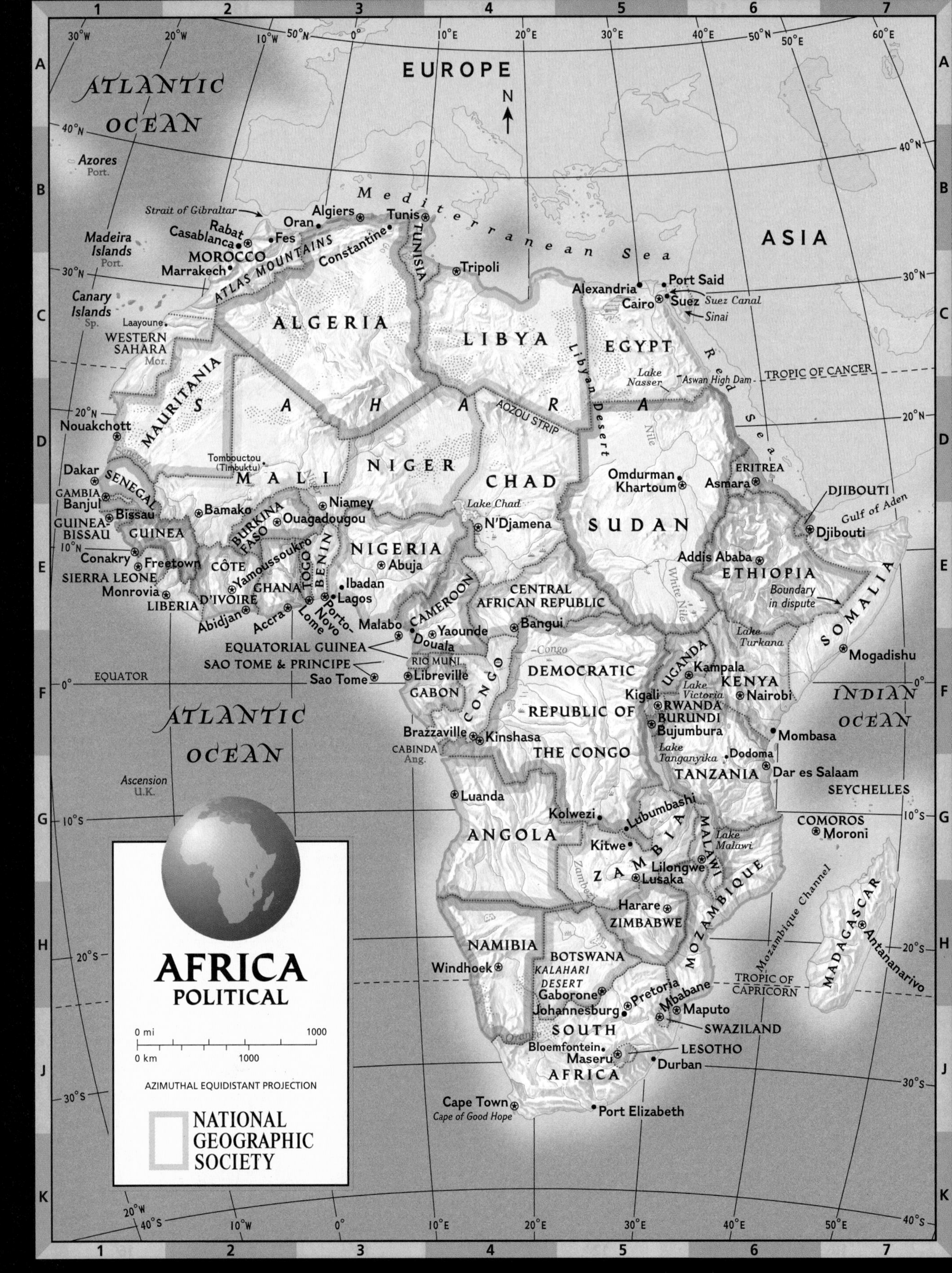

AFRICA
POLITICAL
0 mi 1000
0 km 1000
AZIMUTHAL EQUIDISTANT PROJECTION
NATIONAL GEOGRAPHIC SOCIETY
EUROPE
ASIA
ATLANTIC OCEAN
INDIAN OCEAN
Mediterranean Sea
Red Sea
Gulf of Aden
Mozambique Channel
Strait of Gibraltar
Azores Port.
Madeira Islands Port.
Canary Islands Sp.
Ascension U.K.
ATLAS MOUNTAINS
SAHARA
Libyan Desert
KALAHARI DESERT
Cape of Good Hope
Lake Chad
Lake Nasser
Aswan High Dam
Lake Turkana
Lake Victoria
Lake Tanganyika
Lake Malawi
Nile
White Nile
Niger
Senegal
Congo
Zambezi
Orange
Suez Canal
Sinai
TROPIC OF CANCER
EQUATOR
TROPIC OF CAPRICORN
AOZOU STRIP
Boundary in dispute
MOROCCO
WESTERN SAHARA Mor.
ALGERIA
TUNISIA
LIBYA
EGYPT
MAURITANIA
MALI
NIGER
CHAD
SUDAN
ERITREA
DJIBOUTI
ETHIOPIA
SOMALIA
SENEGAL
GAMBIA
GUINEA-BISSAU
GUINEA
SIERRA LEONE
LIBERIA
CÔTE D'IVOIRE
BURKINA FASO
GHANA
TOGO
BENIN
NIGERIA
CAMEROON
CENTRAL AFRICAN REPUBLIC
EQUATORIAL GUINEA
RIO MUNI
SAO TOME & PRINCIPE
GABON
CONGO
DEMOCRATIC REPUBLIC OF THE CONGO
CABINDA Ang.
UGANDA
KENYA
RWANDA
BURUNDI
TANZANIA
SEYCHELLES
COMOROS
ANGOLA
ZAMBIA
MALAWI
MOZAMBIQUE
ZIMBABWE
NAMIBIA
BOTSWANA
SOUTH AFRICA
SWAZILAND
LESOTHO
MADAGASCAR
Rabat
Casablanca
Fes
Marrakech
Oran
Algiers
Constantine
Tunis
Tripoli
Alexandria
Cairo
Port Said
Suez
Laayoune
Nouakchott
Tombouctou (Timbuktu)
Dakar
Banjul
Bissau
Conakry
Freetown
Monrovia
Bamako
Niamey
Ouagadougou
Yamoussoukro
Abidjan
Accra
Lome
Porto-Novo
Lagos
Ibadan
Abuja
N'Djamena
Omdurman
Khartoum
Asmara
Djibouti
Addis Ababa
Malabo
Douala
Yaounde
Bangui
Libreville
Sao Tome
Brazzaville
Kinshasa
Kampala
Nairobi
Kigali
Bujumbura
Mogadishu
Mombasa
Dodoma
Dar es Salaam
Moroni
Luanda
Kolwezi
Lubumbashi
Kitwe
Lilongwe
Lusaka
Harare
Antananarivo
Windhoek
Gaborone
Pretoria
Johannesburg
Mbabane
Maputo
Bloemfontein
Maseru
Durban
Cape Town
Port Elizabeth

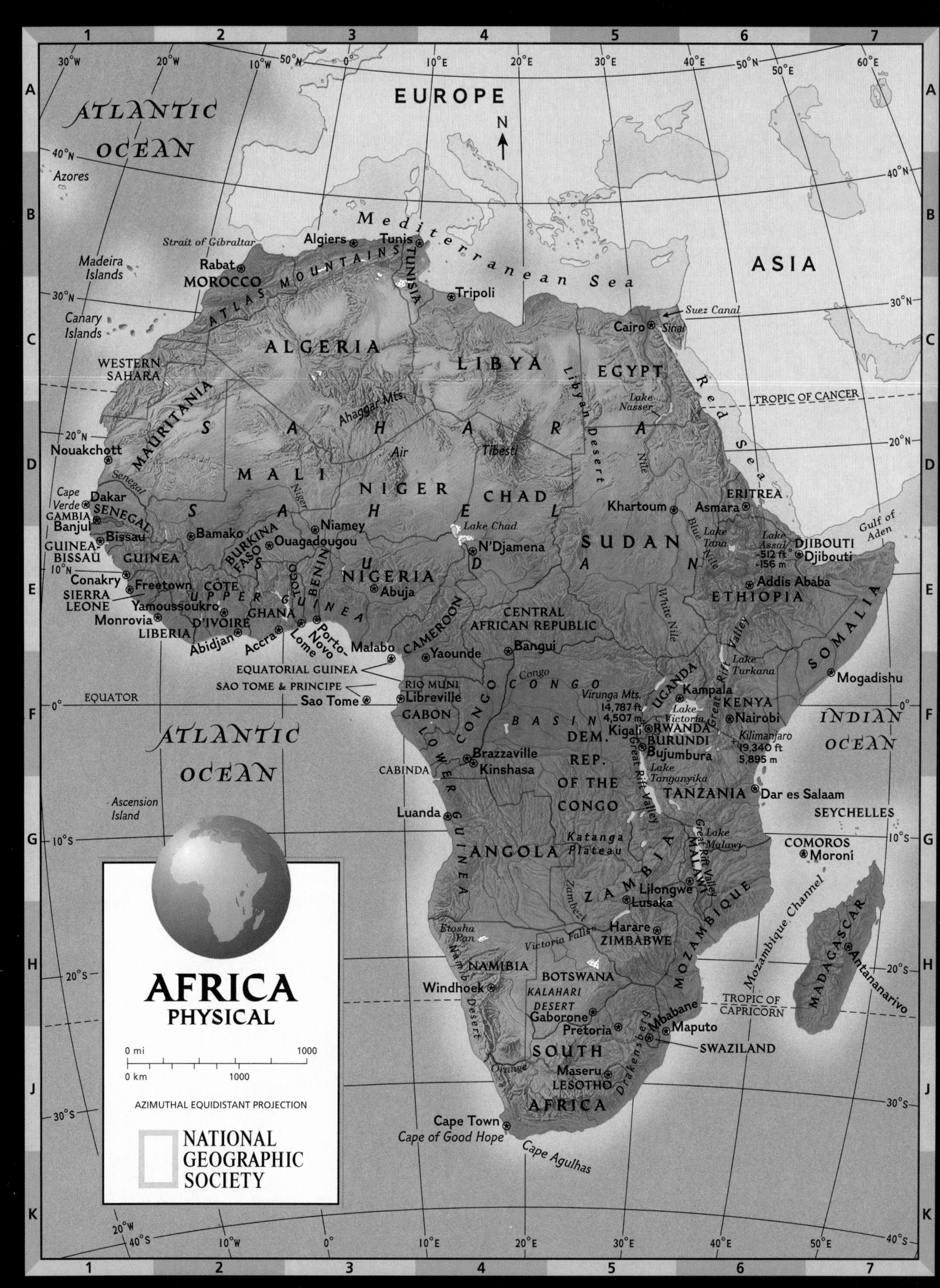
AFRICA
PHYSICAL
EUROPE
ASIA
ATLANTIC OCEAN
ATLANTIC OCEAN
INDIAN OCEAN
Mediterranean Sea
Red Sea
Gulf of Aden
Mozambique Channel
Strait of Gibraltar
Suez Canal
Sinai
Azores
Madeira Islands
Canary Islands
Cape Verde
Ascension Island
ATLAS MOUNTAINS
SAHARA
SAHEL
Ahaggar Mts.
Air
Tibesti
Libyan Desert
Lake Nasser
Nile
Blue Nile
White Nile
Lake Tana
Lake Assal -512 ft -156 m
Lake Chad
Senegal
Niger
UPPER GUINEA
LOWER GUINEA
CONGO BASIN
Congo
Virunga Mts. 14,787 ft 4,507 m
Lake Victoria
Lake Turkana
Lake Tanganyika
Lake Malawi
Great Rift Valley
Kilimanjaro 19,340 ft 5,895 m
Katanga Plateau
Zambezi
Victoria Falls
Etosha Pan
Namib Desert
KALAHARI DESERT
Orange
Drakensberg
Cape of Good Hope
Cape Agulhas
TROPIC OF CANCER
EQUATOR
TROPIC OF CAPRICORN
MOROCCO
Rabat
Algiers
Tunis
TUNISIA
Tripoli
ALGERIA
LIBYA
EGYPT
Cairo
WESTERN SAHARA
MAURITANIA
Nouakchott
MALI
NIGER
CHAD
SUDAN
Khartoum
ERITREA
Asmara
DJIBOUTI
Djibouti
ETHIOPIA
Addis Ababa
SOMALIA
Mogadishu
Dakar
SENEGAL
GAMBIA
Banjul
GUINEA-BISSAU
Bissau
GUINEA
Conakry
SIERRA LEONE
Freetown
LIBERIA
Monrovia
CÔTE D'IVOIRE
Yamoussoukro
Abidjan
Bamako
BURKINA FASO
Ouagadougou
Niamey
GHANA
Accra
TOGO
Lome
BENIN
Porto-Novo
NIGERIA
Abuja
N'Djamena
CAMEROON
Yaounde
CENTRAL AFRICAN REPUBLIC
Bangui
EQUATORIAL GUINEA
Malabo
SAO TOME & PRINCIPE
Sao Tome
RIO MUNI
Libreville
GABON
CONGO
Brazzaville
Kinshasa
CABINDA
DEM. REP. OF THE CONGO
UGANDA
Kampala
KENYA
Nairobi
RWANDA
Kigali
BURUNDI
Bujumbura
TANZANIA
Dar es Salaam
SEYCHELLES
COMOROS
Moroni
Luanda
ANGOLA
ZAMBIA
Lusaka
MALAWI
Lilongwe
MOZAMBIQUE
Harare
ZIMBABWE
NAMIBIA
Windhoek
BOTSWANA
Gaborone
Pretoria
Mbabane
Maputo
SWAZILAND
SOUTH AFRICA
Maseru
LESOTHO
Cape Town
MADAGASCAR
Antananarivo
0 mi 1000
0 km 1000
AZIMUTHAL EQUIDISTANT PROJECTION
NATIONAL GEOGRAPHIC SOCIETY

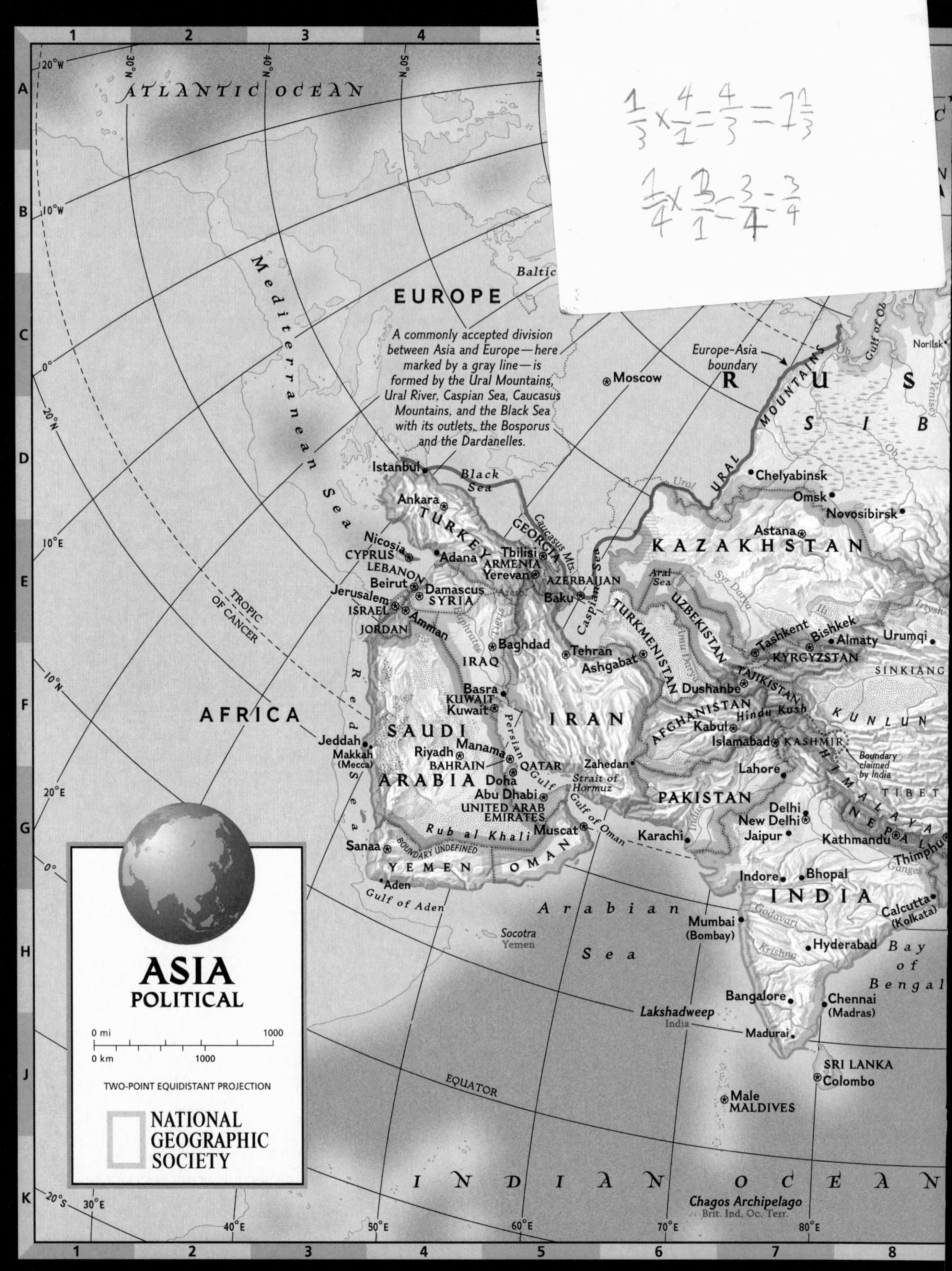
ASIA
POLITICAL
0 mi
1000
0 km
1000
TWO-POINT EQUIDISTANT PROJECTION
NATIONAL
GEOGRAPHIC
SOCIETY
ATLANTIC OCEAN
EUROPE
A commonly accepted division between Asia and Europe—here marked by a gray line—is formed by the Ural Mountains, Ural River, Caspian Sea, Caucasus Mountains, and the Black Sea with its outlets, the Bosporus and the Dardanelles.
Europe-Asia boundary
Mediterranean Sea
Baltic
Moscow
URAL MOUNTAINS
Ural
Gulf of Ob
Ob
Norilsk
Yenisey
Chelyabinsk
Omsk
Novosibirsk
Astana
KAZAKHSTAN
Istanbul
Black Sea
Ankara
TURKEY
GEORGIA
Caucasus Mts.
Tbilisi
ARMENIA
Yerevan
AZERBAIJAN
Baku
Caspian Sea
Nicosia
CYPRUS
Adana
LEBANON
Beirut
Damascus
SYRIA
Jerusalem
ISRAEL
Amman
JORDAN
Euphrates
Tigris
Baghdad
IRAQ
Tehran
Ashgabat
TURKMENISTAN
UZBEKISTAN
Aral Sea
Syr Darya
Amu Darya
Tashkent
Bishkek
Almaty
KYRGYZSTAN
Urumqi
SINKIANG
Irtysh
Ili
TAJIKISTAN
Dushanbe
AFGHANISTAN
Kabul
Hindu Kush
Islamabad
KASHMIR
KUNLUN
Boundary claimed by India
TIBET
HIMALAYA
Lahore
PAKISTAN
Indus
Karachi
Delhi
New Delhi
Jaipur
Kathmandu
NEPAL
Thimphu
Ganges
Basra
KUWAIT
Kuwait
IRAN
SAUDI ARABIA
Persian Gulf
Riyadh
Manama
BAHRAIN
QATAR
Doha
Abu Dhabi
UNITED ARAB EMIRATES
Zahedan
Strait of Hormuz
Gulf of Oman
Muscat
OMAN
Rub al Khali
BOUNDARY UNDEFINED
YEMEN
Sanaa
Aden
Gulf of Aden
Jeddah
Makkah (Mecca)
Red Sea
AFRICA
TROPIC OF CANCER
Arabian Sea
Socotra
Yemen
Indore
Bhopal
INDIA
Calcutta (Kolkata)
Mumbai (Bombay)
Godavari
Krishna
Hyderabad
Bay of Bengal
Bangalore
Chennai (Madras)
Lakshadweep
India
Madurai
SRI LANKA
Colombo
Male
MALDIVES
EQUATOR
INDIAN OCEAN
Chagos Archipelago
Brit. Ind. Oc. Terr.
20°W
10°W
0°
10°E
20°E
30°E
40°E
50°E
60°E
70°E
80°E
30°N
40°N
50°N
20°N
10°N
0°
20°S

NORTH AMERICA
Bering Strait
North Pole
Chukchi Sea
Wrangel I.
Gulf of Anadyr
Anadyr
Bering Sea
East Siberian Sea
North Land
New Siberian Islands
Laptev Sea
Kolyma Range
Cherskiy Range
Verkhoyansk Range
Commander Is.
Kamchatka Peninsula
Magadan
Sea of Okhotsk
Yakutsk
Sakhalin
Kuril Islands
Lake Baikal
Irkutsk
MANCHURIA
Hokkaido
Sapporo
Sea of Japan
Vladivostok
JAPAN
Tokyo
Honshu
Kyoto
Osaka
Hiroshima
Kyushu
Ulan Bator
Changchun
MONGOLIA
ALTAY MTS.
GOBI
Shenyang
NORTH KOREA
Pyongyang
Seoul
SOUTH KOREA
Beijing
Qingdao
Shijiazhuang
Yellow Sea
Xuzhou
East China Sea
Lanzhou
Xian
Nanjing
Shanghai
CHINA
Ryukyu Islands
Okinawa
Nanchang
Chengdu
Fuzhou
Changsha
Taipei
TAIWAN
Boundary claimed by China
Guiyang
Kunming
Guangzhou
Hong Kong
Macau
The People's Republic of China claims Taiwan as its 23rd province.
MYANMAR (BURMA)
Hanoi
Haiphong
Hainan
South China Sea
Luzon
Quezon City
Manila
Philippine Sea
Vientiane
LAOS
Yangon (Rangoon)
Da Nang
THAILAND
Mindoro
Samar
PHILIPPINES
Leyte
Panay
Negros
Bangkok
CAMBODIA
VIETNAM
Palawan
Andaman Islands
India
Phnom Penh
Ho Chi Minh City
Mindanao
Gulf of Thailand
Andaman Sea
Bandar Seri Begawan
BRUNEI
SABAH
SARAWAK
MALAYSIA
Morotai
Halmahera
Moluccas
Nicobar Islands
Kuala Lumpur
Medan
SINGAPORE
Borneo
Celebes
Buru
Ceram
INDONESIA
Sumatra
Jambi
GREATER SUNDA ISLANDS
Mentawai Islands
Java Sea
Jakarta
Java
Dili
EAST TIMOR
UN admin.
Timor
Kupang
Timor Sea
Tanimbar Is.
Aru Is.
Dolak
Merauke
Kepi
New Guinea
Biak
Jayapura
AUSTRALIA
Following East Timor's vote for independence from Indonesia, the United Nations, in October 1999, established a UN transitional administration to help East Timor's 870,000 people toward independence.
PACIFIC OCEAN
Marcus I.
Jap.
TROPIC OF CANCER
Bonin Is.
Volcano Is.
Parece Vela
EQUATOR

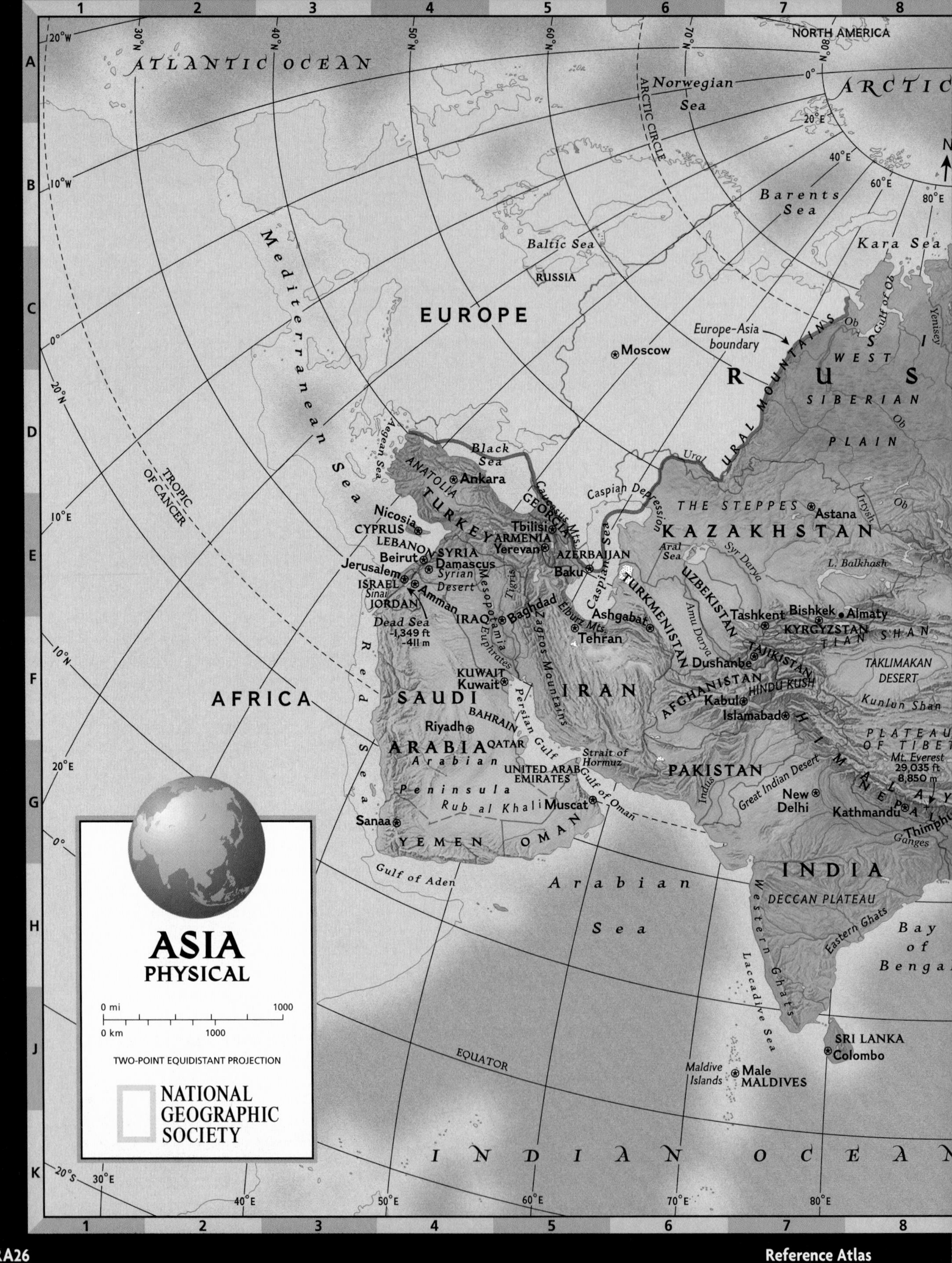
ASIA
PHYSICAL
0 mi
1000
0 km
1000
TWO-POINT EQUIDISTANT PROJECTION
NATIONAL GEOGRAPHIC SOCIETY
ATLANTIC OCEAN
ARCTIC
NORTH AMERICA
Norwegian Sea
ARCTIC CIRCLE
Barents Sea
Kara Sea
Baltic Sea
RUSSIA
EUROPE
Moscow
Mediterranean Sea
Europe-Asia boundary
URAL MOUNTAINS
WEST SIBERIAN PLAIN
Black Sea
ANATOLIA
Ankara
TURKEY
Aegean Sea
TROPIC OF CANCER
Nicosia
CYPRUS
LEBANON
SYRIA
Beirut
Damascus
Syrian Desert
Jerusalem
ISRAEL
Amman
JORDAN
Sinai
Dead Sea -1,349 ft -411 m
GEORGIA
Caucasus Mts.
Tbilisi
ARMENIA
Yerevan
AZERBAIJAN
Baku
Caspian Depression
Caspian Sea
THE STEPPES
Astana
KAZAKHSTAN
Aral Sea
Syr Darya
L. Balkhash
UZBEKISTAN
TURKMENISTAN
Amu Darya
Tashkent
Bishkek
Almaty
KYRGYZSTAN
TIAN SHAN
TAJIKISTAN
Dushanbe
Ashgabat
Elburz Mts.
Tehran
IRAQ
Baghdad
Mesopotamia
Tigris
Euphrates
Zagros Mountains
KUWAIT
Kuwait
IRAN
AFRICA
Red Sea
SAUDI ARABIA
BAHRAIN
Riyadh
QATAR
Persian Gulf
Strait of Hormuz
UNITED ARAB EMIRATES
Arabian Peninsula
Rub al Khali
Muscat
Gulf of Oman
OMAN
Sanaa
YEMEN
Gulf of Aden
AFGHANISTAN
Kabul
HINDU KUSH
Islamabad
PAKISTAN
Indus
Great Indian Desert
New Delhi
Kathmandu
Thimphu
Ganges
HIMALAYA
TAKLIMAKAN DESERT
Kunlun Shan
PLATEAU OF TIBET
Mt. Everest 29,035 ft 8,850 m
INDIA
DECCAN PLATEAU
Western Ghats
Eastern Ghats
Bay of Bengal
Arabian Sea
Laccadive Sea
SRI LANKA
Colombo
Maldive Islands
Male
MALDIVES
EQUATOR
INDIAN OCEAN
Ob
Gulf of Ob
Yenisey
Irtysh
Ural

NORTH AMERICA
Bering Strait
North Pole
OCEAN
Chukchi Sea
Chukchi Peninsula
Wrangel Island
Gulf of Anadyr
East Siberian Sea
Bering Sea
Aleutian Islands
North Land
New Siberian Islands
Laptev Sea
Taymyr Peninsula
Kolyma
Kolyma Range
Commander Islands
Cherskiy Range
Kamchatka Peninsula
Verkhoyansk Range
Lena
Aldan
CENTRAL SIBERIAN PLATEAU
SIBERIA
ASIA
Sea of Okhotsk
Sakhalin
Kuril Islands
Angara
Lake Baikal
Yablonovyy Range
Amur
Sikhote Alin Range
Hokkaido
Yenisey
Greater Khingan Range
Manchurian Plain
Sea of Japan
Ulan Bator
MONGOLIA
ALTAY MOUNTAINS
GOBI
JAPAN
Tokyo
Honshu
NORTH KOREA
Pyongyang
Seoul
SOUTH KOREA
Beijing
North China Plain
Yellow
Shikoku
Kyushu
Nampo Shoto
TROPIC OF CANCER
Yellow Sea
East China Sea
Ryukyu Islands
Qaidam Basin
CHINA
Yangtze
Mekong
Salween
Sichuan Basin
Gongga Shan 24,790 ft 7,556 m
Mariana Islands
Taipei
TAIWAN
Philippine Sea
PACIFIC OCEAN
BHUTAN
Brahmaputra
BANGLADESH
Dhaka
MYANMAR (BURMA)
Hanoi
LAOS
Luzon
PHILIPPINE ISLANDS
Caroline Islands
South China Sea
Hainan
Vientiane
Yangon (Rangoon)
Manila
PHILIPPINES
THAILAND
Bangkok
CAMBODIA
VIETNAM
Phnom Penh
EQUATOR
Andaman Islands
Sulu Sea
Mindanao
Gulf of Thailand
Andaman Sea
Malay Peninsula
Bandar Seri Begawan
BRUNEI
MALAYSIA
Celebes Sea
Moluccas
New Guinea
Nicobar Islands
Kuala Lumpur
SINGAPORE
Borneo
Celebes
INDONESIA
Sumatra
Arafura Sea
GREATER SUNDA ISLANDS
Mentawai Islands
Java Sea
Timor
Timor Sea
AUSTRALIA
Jakarta
Java
80°N
70°N
60°N
50°N
40°N
30°N
20°N
10°N
0°
10°S
20°S
180°
160°E
120°E
100°E
110°E
130°E
140°E
150°E
170°E
160°W
170°W
9
10
11
12
13
14
15
16
A
B
C
D
E
F
G
H
J
K

NORTH PACIFIC OCEAN
TROPIC OF CANCER
NORTHERN MARIANA ISLANDS
U.S.
Saipan
GUAM
U.S.
Hagatna
MICRONESIA
ASIA
Wake Island
U.S.
Bikini Atoll
MARSHALL ISLANDS
Ratak Chain
Ralik Chain
Majuro
PALAU
Koror
Yap Islands
Truk Islands
Caroline Islands
Palikir
Pohnpei (Ponape)
FEDERATED STATES OF MICRONESIA
EQUATOR
Tarawa
Gilbert Islands
Yaren
NAURU
MELANESIA
New Guinea
Mt. Wilhelm
14,793 ft
4,509 m
New Britain
PAPUA NEW GUINEA
Port Moresby
Torres Strait
Solomon Is.
SOLOMON ISLANDS
Honiara
TUVALU
Funafuti
Santa Cruz Is.
Darwin
Gulf of Carpentaria
Coral Sea
CORAL SEA ISLANDS TERRITORY
Austral.
VANUATU
Port-Vila
Suva
FIJI ISLANDS
NEW CALEDONIA
Fr.
Noumea
TROPIC OF CAPRICORN
Kimberley Plateau
GREAT DIVIDING RANGE
Macdonnell Ranges
AUSTRALIA
Brisbane
GREAT VICTORIA DESERT
Lake Eyre
-52 ft
-16 m
Norfolk Island
Austral.
Perth
Darling
Lord Howe Island
Austral.
Great Australian Bight
Sydney
Adelaide
Murray
Canberra
Mt. Kosciuszko
7,310 ft
2,228 m
Melbourne
North Island
Auckland
NEW ZEALAND
Tasman Sea
INDIAN OCEAN
Wellington
Tasmania
Hobart
Mt. Cook
12,316 ft
3,754 m
Christchurch
South Island
Stewart Island
120°E
130°E
140°E
150°E
160°E
170°E
20°N
10°N
0°
10°S
20°S
30°S
40°S
50°S
1
2
3
4
5
6
7
8
A
B
C
D
E
F
G
H
J
K

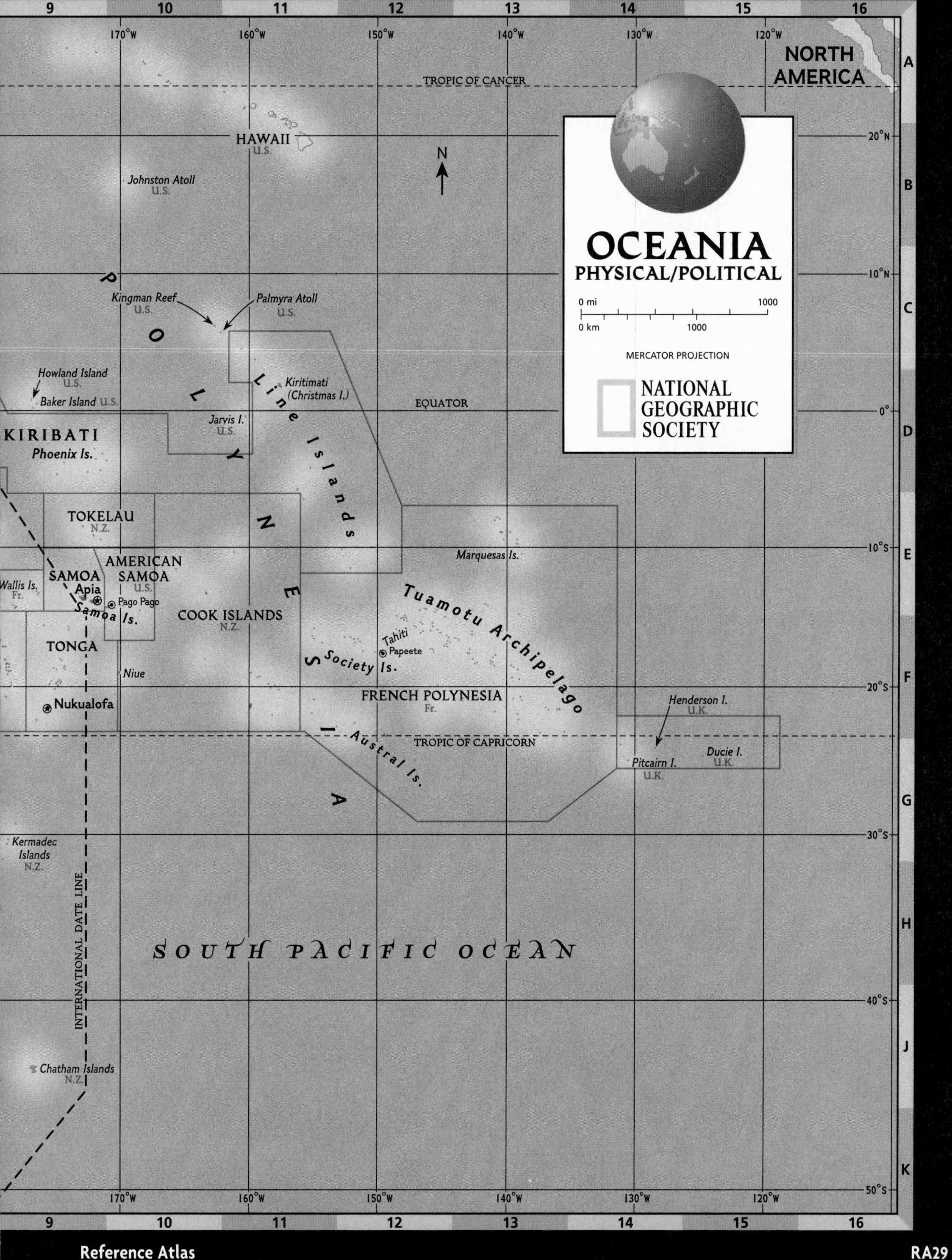
OCEANIA
PHYSICAL/POLITICAL
0 mi
1000
0 km
1000
MERCATOR PROJECTION
NATIONAL GEOGRAPHIC SOCIETY
NORTH AMERICA
TROPIC OF CANCER
HAWAII
U.S.
Johnston Atoll
U.S.
N
Kingman Reef
U.S.
Palmyra Atoll
U.S.
Howland Island
U.S.
Baker Island U.S.
Kiritimati
(Christmas I.)
EQUATOR
Jarvis I.
U.S.
KIRIBATI
Phoenix Is.
Line Islands
POLYNESIA
TOKELAU
N.Z.
Marquesas Is.
AMERICAN SAMOA
U.S.
SAMOA
Apia
Pago Pago
Wallis Is.
Fr.
Samoa Is.
COOK ISLANDS
N.Z.
Tuamotu Archipelago
Tahiti
Papeete
Society Is.
TONGA
Niue
Nukualofa
FRENCH POLYNESIA
Fr.
Henderson I.
U.K.
TROPIC OF CAPRICORN
Austral Is.
Pitcairn I.
U.K.
Ducie I.
U.K.
Kermadec Islands
N.Z.
INTERNATIONAL DATE LINE
SOUTH PACIFIC OCEAN
Chatham Islands
N.Z.
170°W
160°W
150°W
140°W
130°W
120°W
20°N
10°N
0°
10°S
20°S
30°S
40°S
50°S
9
10
11
12
13
14
15
16
A
B
C
D
E
F
G
H
J
K

RUSSIA
Lake Baikal
Yablonovyy Range
Ulan Bator
MONGOLIA
ALTAY MOUNTAINS
GOBI
Greater Khingan Range
Manchurian Plain
Amur
Sikhote Alin Range
Sakhalin
KAMCHATKA PENINSULA
Sea of Okhotsk
Bering Sea
ALEUTIAN ISLANDS
KURIL ISLANDS
Hokkaido
Beijing
Pyongyang
NORTH KOREA
Seoul
SOUTH KOREA
Sea of Japan
Honshu
JAPAN
Tokyo
Yellow
Yellow Sea
CHINA
Yangtze
Mekong
Kyushu
Shikoku
NAMPO SHOTO
NORTH PACIF
East China Sea
RYUKYU ISLANDS
Taipei
TAIWAN
Hawaiian
INDIA
MYANMAR (BURMA)
Hanoi
Hainan
Vientiane
LAOS
VIETNAM
THAILAND
Yangon (Rangoon)
Bangkok
CAMBODIA
Phnom Penh
Andaman Sea
South China Sea
Luzon
Manila
PHILIPPINES
PHILIPPINE ISLANDS
Philippine Sea
MARIANA ISLANDS
NORTHERN MARIANA ISLANDS
U.S.
GUAM
U.S.
MICRONESIA
MARSHALL ISLANDS
Ratak Chain
Ralik Chain
Majuro
PALAU
Koror
CAROLINE ISLANDS
Palikir
FEDERATED STATES OF MICRONESIA
Sulu Sea
Mindanao
Bandar Seri Begawan
BRUNEI
Kuala Lumpur
MALAYSIA
SINGAPORE
Borneo
Celebes Sea
Sumatra
INDONESIA
GREATER SUNDA ISLANDS
Celebes
MOLUCCAS
Java Sea
Jakarta
Java
LESSER SUNDA ISLANDS
NEW GUINEA
PAPUA NEW GUINEA
Port Moresby
Arafura Sea
Tarawa
Gilbert Islands
Yaren
NAURU
KIRIBATI
Phoenix Is.
MELANESIA
SOLOMON ISLANDS
Solomon Is.
Honiara
Santa Cruz Islands
TUVALU
Funafuti
Tokelau
N.Z.
WALLIS AND FUTUNA IS.
Fr.
SAMOA
Apia
AMERICAN SAMOA
U.S.
CORAL SEA ISLANDS TERRITORY
Austral.
VANUATU
Port-Vila
NEW CALEDONIA
Fr.
Suva
FIJI ISLANDS
TONGA
Nuku'alofa
Niue
N.Z.
TROPIC OF CAPRICORN
Coral Sea
AUSTRALIA
Norfolk Island
Austral.
Lord Howe Island
Austral.
Kermadec Islands
N.Z.
Darling
Great Australian Bight
Canberra
Tasman Sea
NEW ZEALAND
Wellington
INDIAN OCEAN
Tasmania
Chatham Island
N.Z.
105°E
120°E
135°E
150°E
165°E
180°

PACIFIC RIM
PHYSICAL/POLITICAL
0 mi 1500
0 km 1500
MILLER CYLINDRICAL PROJECTION
NATIONAL GEOGRAPHIC SOCIETY
Gulf of Alaska
Kodiak I.
Alexander Archipelago
Queen Charlotte Islands
Vancouver Island
Coast Mountains
ROCKY MOUNTAINS
GREAT PLAINS
CANADIAN SHIELD
CANADA
Hudson Bay
Cascade Range
Missouri
Ottawa
Great Lakes
CENTRAL LOWLAND
APPALACHIAN MTS.
Washington
UNITED STATES
Mississippi
COASTAL PLAIN
ATLANTIC OCEAN
PACIFIC OCEAN
Baja California
Gulf of California
Sierra Madre Occidental
Sierra Madre Oriental
MEXICO
Gulf of Mexico
TROPIC OF CANCER
Nassau
BAHAMAS
Havana
CUBA
DOMINICAN REPUBLIC
JAMAICA
HAITI
Santo Domingo
Mexico City
BELIZE
HONDURAS
Tegucigalpa
GUATEMALA
Guatemala City
San Salvador
EL SALVADOR
NICARAGUA
Managua
Caribbean Sea
Caracas
San Jose
COSTA RICA
Panama City
PANAMA
LLANOS
VENEZ.
Bogota
COLOMBIA
Hawaiian Islands
HAWAII
U.S.
Kiritimati
Line Islands
EQUATOR
POLYNESIA
Galapagos Islands
Ecua.
Quito
ECUADOR
AMAZON BASIN
BRAZIL
PERU
Lima
ANDES
Marquesas Is.
Tuamotu Archipelago
Society Is.
COOK ISLANDS
N.Z.
FRENCH POLYNESIA
Fr.
Austral Is.
Pitcairn Island U.K.
La Paz
BOLIVIA
Santiago
CHILE
ARGENTINA
PATAGONIA
Chiloe Island
SOUTH PACIFIC OCEAN
45°N
30°N
15°N
0°
15°S
30°S
45°S
165°W
150°W
135°W
120°W
105°W
90°W
75°W
9
10
11
12
13
14
15
16
A
B
C
D
E
F
G
H
J
K

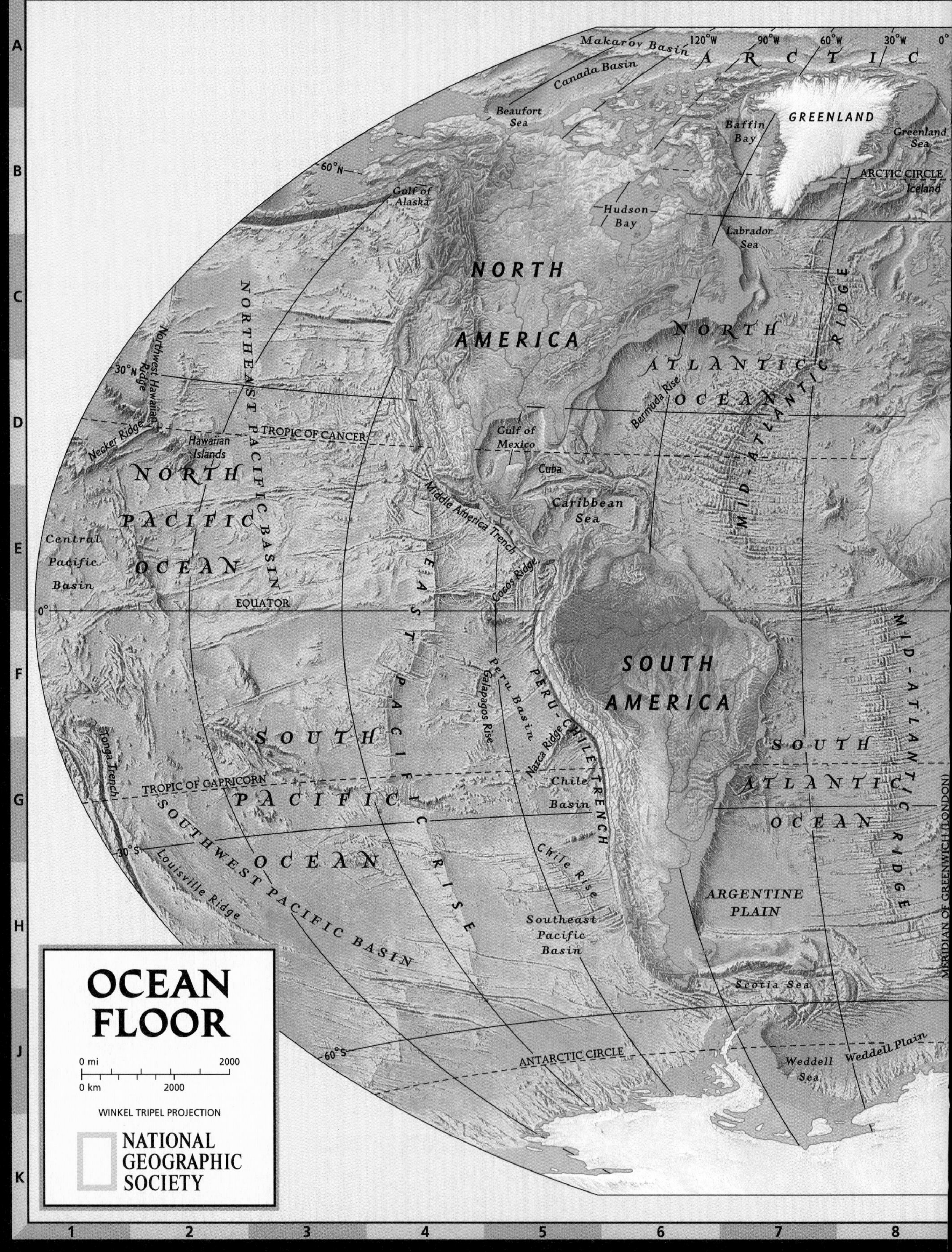
OCEAN
FLOOR
0 mi
2000
0 km
2000
WINKEL TRIPEL PROJECTION
NATIONAL
GEOGRAPHIC
SOCIETY
NORTH AMERICA
SOUTH AMERICA
GREENLAND
ARCTIC
NORTH PACIFIC OCEAN
SOUTH PACIFIC OCEAN
NORTH ATLANTIC OCEAN
SOUTH ATLANTIC OCEAN
Makarov Basin
Canada Basin
Beaufort Sea
Baffin Bay
Greenland Sea
ARCTIC CIRCLE
Iceland
Gulf of Alaska
Hudson Bay
Labrador Sea
NORTHEAST PACIFIC BASIN
Northwest Hawaiian Ridge
Necker Ridge
Hawaiian Islands
TROPIC OF CANCER
Gulf of Mexico
Cuba
Caribbean Sea
Bermuda Rise
MID-ATLANTIC RIDGE
Middle America Trench
Cocos Ridge
Central Pacific Basin
EQUATOR
EAST PACIFIC RISE
Peru Basin
Galapagos Rise
PERU-CHILE TRENCH
Nazca Ridge
Chile Basin
Chile Rise
Tonga Trench
TROPIC OF CAPRICORN
SOUTHWEST PACIFIC BASIN
Louisville Ridge
Southeast Pacific Basin
ARGENTINE PLAIN
Scotia Sea
Weddell Sea
Weddell Plain
ANTARCTIC CIRCLE
MERIDIAN OF GREENWICH (LONDON)
120°W
90°W
60°W
30°W
0°
60°N
30°N
0°
30°S
60°S
1 2 3 4 5 6 7 8
A B C D E F G H J K

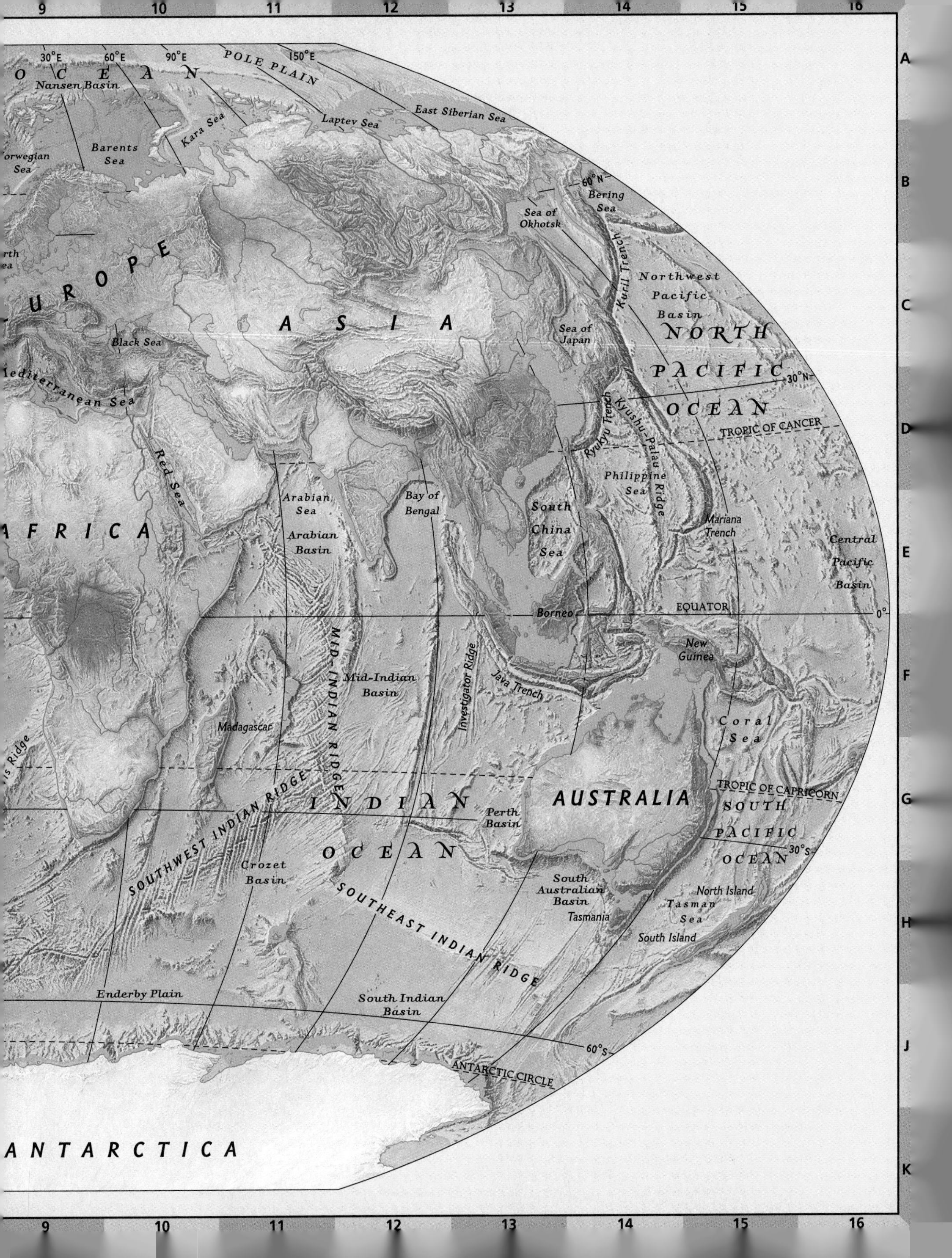

9
10
11
12
13
14
15
16
A
B
C
D
E
F
G
H
J
K
30°E
60°E
90°E
150°E
POLE PLAIN
O C E A N
Nansen Basin
Kara Sea
Laptev Sea
East Siberian Sea
Barents Sea
60°N
Bering Sea
Sea of Okhotsk
Kuril Trench
EUROPE
Northwest Pacific Basin
A S I A
Sea of Japan
NORTH PACIFIC OCEAN
Black Sea
Mediterranean Sea
30°N
Ryukyu Trench
Kyushu-Palau Ridge
TROPIC OF CANCER
Red Sea
Philippine Sea
Arabian Sea
Bay of Bengal
South China Sea
Mariana Trench
AFRICA
Arabian Basin
Central Pacific Basin
Borneo
EQUATOR
0°
MID-INDIAN RIDGE
New Guinea
Investigator Ridge
Mid-Indian Basin
Java Trench
Madagascar
Coral Sea
INDIAN OCEAN
AUSTRALIA
TROPIC OF CAPRICORN
SOUTH PACIFIC OCEAN
Perth Basin
30°S
SOUTHWEST INDIAN RIDGE
Crozet Basin
South Australian Basin
North Island
Tasman Sea
Tasmania
SOUTHEAST INDIAN RIDGE
South Island
Enderby Plain
South Indian Basin
60°S
ANTARCTIC CIRCLE
ANTARCTICA

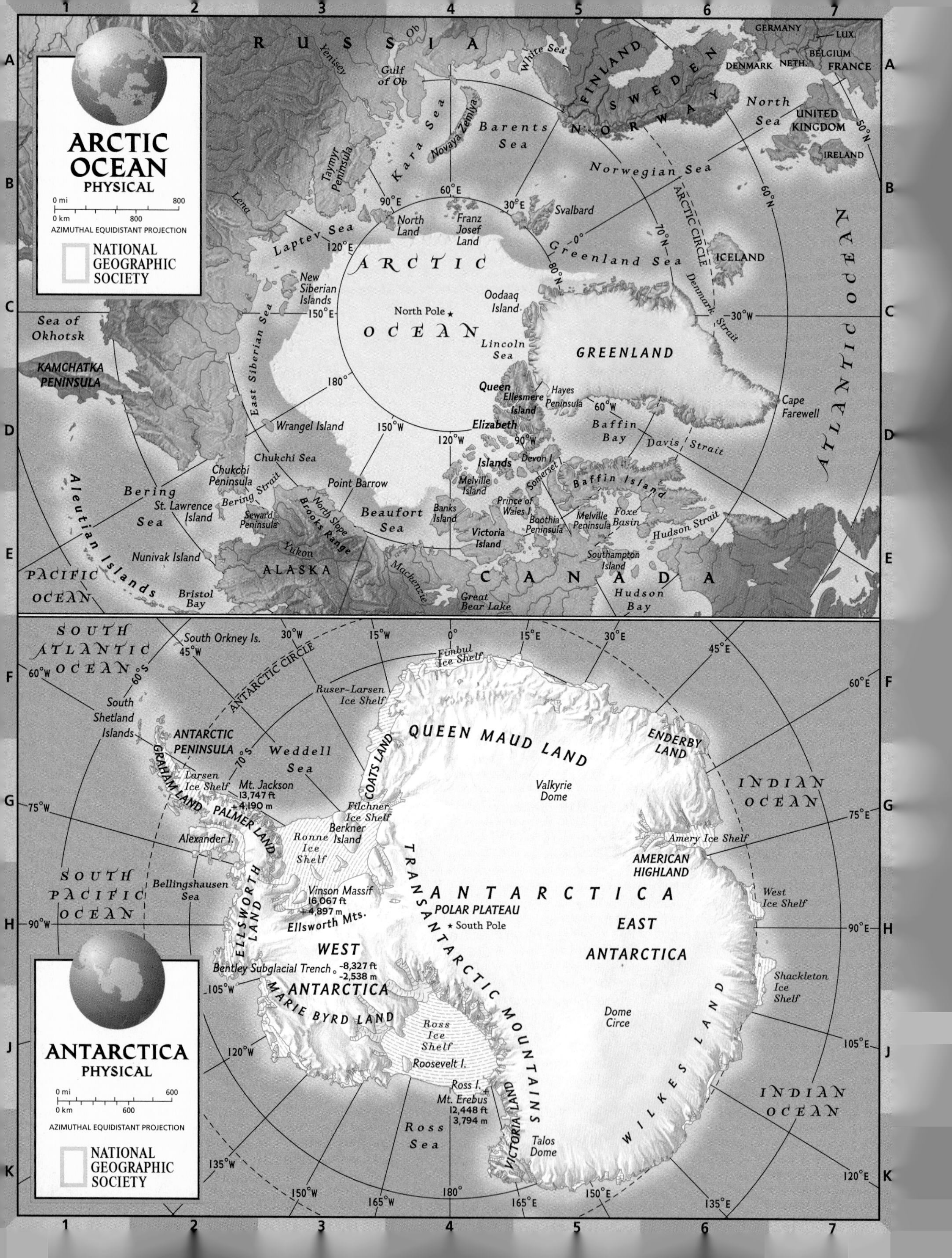

ARCTIC OCEAN
PHYSICAL
AZIMUTHAL EQUIDISTANT PROJECTION
NATIONAL GEOGRAPHIC SOCIETY
RUSSIA
FINLAND
SWEDEN
NORWAY
GERMANY
LUX.
BELGIUM
DENMARK
NETH.
FRANCE
UNITED KINGDOM
IRELAND
ICELAND
GREENLAND
CANADA
ALASKA
ARCTIC OCEAN
ATLANTIC OCEAN
PACIFIC OCEAN
North Pole
Barents Sea
Kara Sea
Laptev Sea
North Sea
Norwegian Sea
Greenland Sea
White Sea
Beaufort Sea
Chukchi Sea
Bering Sea
East Siberian Sea
Lincoln Sea
Baffin Bay
Hudson Bay
Sea of Okhotsk
Svalbard
Franz Josef Land
North Land
Novaya Zemlya
Taymyr Peninsula
New Siberian Islands
Wrangel Island
Oodaaq Island
Queen Elizabeth Islands
Ellesmere Island
Hayes Peninsula
Cape Farewell
Davis Strait
Denmark Strait
Hudson Strait
Bering Strait
Baffin Island
Devon I.
Somerset I.
Melville Island
Banks Island
Victoria Island
Prince of Wales I.
Boothia Peninsula
Melville Peninsula
Foxe Basin
Southampton Island
Point Barrow
Chukchi Peninsula
St. Lawrence Island
Seward Peninsula
Nunivak Island
Bristol Bay
Aleutian Islands
KAMCHATKA PENINSULA
North Slope
Brooks Range
Yukon
Mackenzie
Great Bear Lake
Lena
Yenisey
Ob
Gulf of Ob
ARCTIC CIRCLE
ANTARCTICA
PHYSICAL
AZIMUTHAL EQUIDISTANT PROJECTION
NATIONAL GEOGRAPHIC SOCIETY
SOUTH ATLANTIC OCEAN
SOUTH PACIFIC OCEAN
INDIAN OCEAN
ANTARCTIC CIRCLE
South Orkney Is.
South Shetland Islands
ANTARCTIC PENINSULA
GRAHAM LAND
PALMER LAND
Larsen Ice Shelf
Mt. Jackson 13,747 ft 4,190 m
Weddell Sea
COATS LAND
Filchner Ice Shelf
Berkner Island
Ronne Ice Shelf
Alexander I.
Bellingshausen Sea
ELLSWORTH LAND
Vinson Massif 16,067 ft 4,897 m
Ellsworth Mts.
WEST ANTARCTICA
Bentley Subglacial Trench -8,327 ft -2,538 m
MARIE BYRD LAND
Ross Ice Shelf
Roosevelt I.
Ross I.
Mt. Erebus 12,448 ft 3,794 m
Ross Sea
VICTORIA LAND
Talos Dome
TRANSANTARCTIC MOUNTAINS
ANTARCTICA
POLAR PLATEAU
South Pole
EAST ANTARCTICA
Dome Circe
WILKES LAND
Shackleton Ice Shelf
West Ice Shelf
AMERICAN HIGHLAND
Amery Ice Shelf
Valkyrie Dome
QUEEN MAUD LAND
ENDERBY LAND
Fimbul Ice Shelf
Ruser-Larsen Ice Shelf

GEOGRAPHY HANDBOOK

A geographer is a person who studies the earth and its people. Have you ever wondered if you could be a geographer? One way to learn how is to use this **Geography Handbook.** It will show you that geography is more than studying facts and figures. It also means doing some of your own exploring of the earth. By learning how to use geographic tools, such as globes, maps, and graphs, you will get to know and appreciate the wonders of our planet.

Geologists studying a volcano ▲

Table of Contents

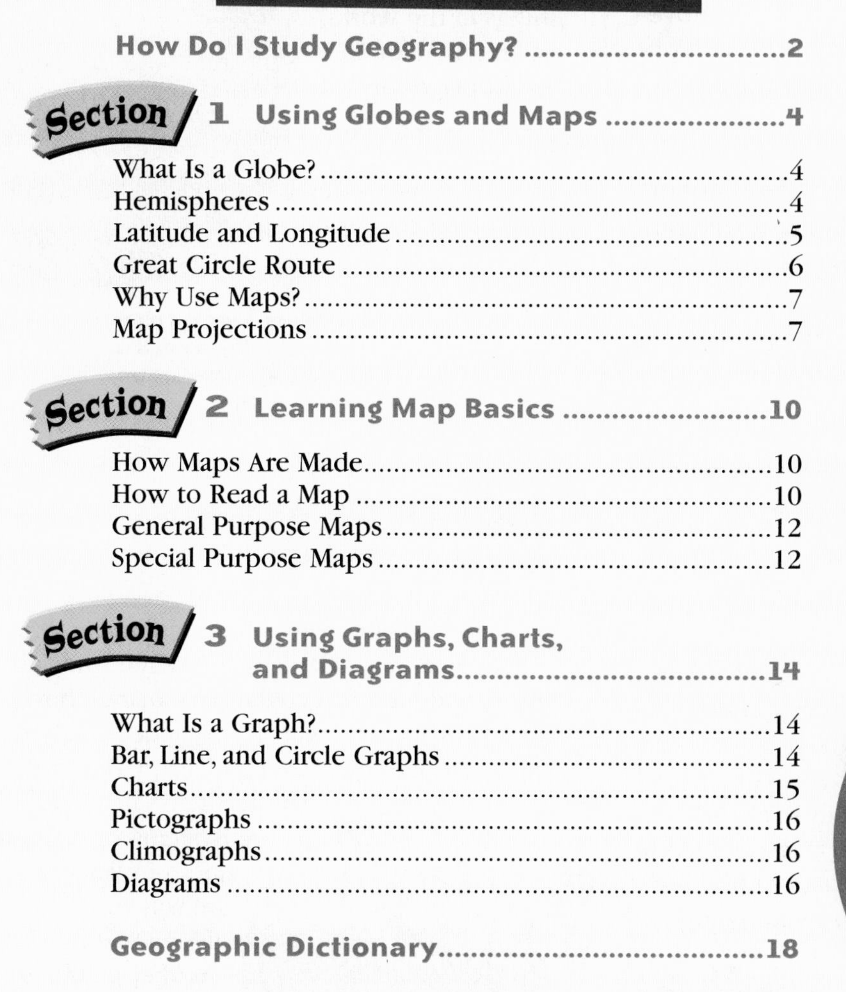

How Do I Study Geography? 2

Section 1 Using Globes and Maps 4

What Is a Globe? 4
Hemispheres 4
Latitude and Longitude 5
Great Circle Route 6
Why Use Maps? 7
Map Projections 7

Section 2 Learning Map Basics 10

How Maps Are Made 10
How to Read a Map 10
General Purpose Maps 12
Special Purpose Maps 12

Section 3 Using Graphs, Charts, and Diagrams 14

What Is a Graph? 14
Bar, Line, and Circle Graphs 14
Charts 15
Pictographs 16
Climographs 16
Diagrams 16

Geographic Dictionary 18

Polar ice cap ▼

◀ Lava flow

How Do I Study

Everything you see, touch, use, and even hear is related to geography—the study of the world's people, places, and environments. How can you possibly study such a huge amount of information in your geography class? Where do you start?

Geographers—people who study geography—ask themselves this question, too. To understand how our world is connected, some geographers have broken down the study of geography into five themes. The **Five Themes of Geography** are (1) location, (2) place, (3) human/environment interaction, (4) movement, and (5) regions. These themes are highlighted in blue throughout this textbook.

Most recently, as suggested in the ***Geography Standards for Life,*** geographers have begun to look at geography a different way. They break down the study of geography into **Six Essential Elements,** which are explained for you below. Being aware of these elements will help you sort out what you are learning about geography.

ELEMENT 1 The World in Spatial Terms

Geographers first take a look at where a place is located. **Location** serves as a starting point by asking "Where is it?" Knowing the location of places helps you to orient yourself in space and to develop an awareness of the world around you.

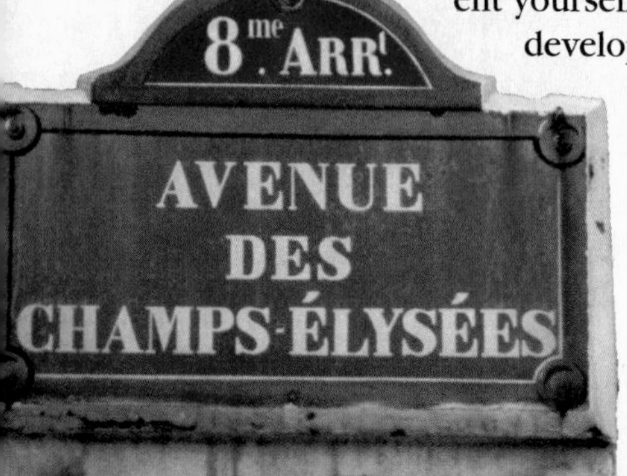

This street sign is located in Paris, France.

ELEMENT 2 Places and Regions

Geographers also look at places and regions. **Place** includes those features and characteristics that give an area its own identity or personality. These can be physical characteristics—such as landforms, climate, plants, and animals—or human characteristics—such as language, religion, architecture, music, politics, and way of life.

To make sense of all the complex things in the world, geographers often group places or areas into regions. **Regions** are united by one or more common characteristics.

Des Moines, Iowa, is a place characterized by farms. It is also part of a region known as the Corn Belt.

ELEMENT 3 Physical Systems

Why do some places have mountains and other places have flat deserts? When studying places and regions, geographers analyze how physical systems—such as volcanoes, glaciers, and hurricanes—interact and shape the earth's surface. They also look at ecosystems, or communities of plants and animals that are dependent upon one another and their particular surroundings for survival.

A glacier carved this deep valley in New Zealand.

Geography?

ELEMENT 4 Human Systems

Geographers also examine human systems, or how people have shaped our world. They look at how boundary lines are determined and analyze why people settle in certain places and not in others. An ongoing theme in geography is the continual **movement** of people, ideas, and goods.

People, vehicles, and goods move quickly through Asmara, the capital of Eritrea.

ELEMENT 5 Environment and Society

The study of geography includes looking at **human/environment interaction,** or how and why people change their surroundings. Throughout history, people have cut forests and dammed rivers to build farms and cities. Some activities have led to air and water pollution. The physical environment affects human activities as well. The type of soil and amount of water in a place determines if crops can be grown. Earthquakes and floods also affect human life.

Romanian farmers work in a field near a nuclear power plant.

ELEMENT 6 The Uses of Geography

Understanding geography, and knowing how to use the tools and technology available to study it, prepares you for life in our modern society. Individuals, businesses, and governments use geography and maps of all kinds on a daily basis. For example, geographic information systems (GIS) are special software programs that help geographers gather and use information about a place.

A cartographer uses GIS to make a map.

Using Globes and Maps

Guide To Reading

Main Idea

Globes and maps provide different ways of showing features of the earth.

Terms to Know

- globe
- hemisphere
- latitude
- longitude
- grid system
- absolute location
- great circle route
- projection

What Is a Globe?

A **globe** is a model of the earth that shows the earth's shape, lands, distances, and directions as they truly relate to one another. A world globe can help you find your way around the earth. By using one, you can locate places and determine distances.

Hemispheres

To locate places on the earth, geographers use a system of imaginary lines that crisscross the globe. One of these lines, the Equator, circles the middle of the earth like a belt. It

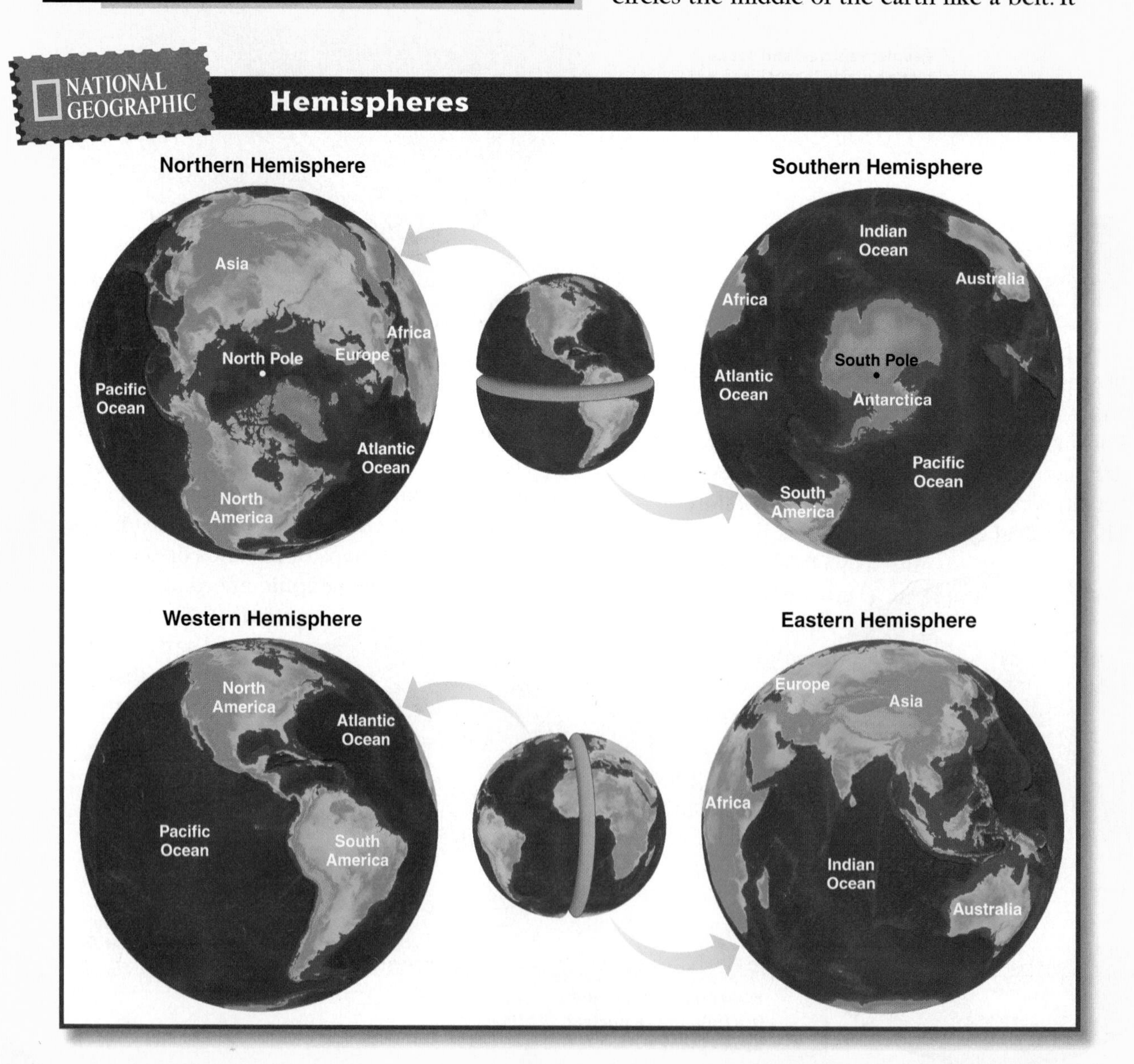

divides the earth into "half spheres," or **hemispheres.** Everything north of the Equator is in the Northern Hemisphere. Everything south of the Equator is in the Southern Hemisphere.

Another imaginary line runs from north to south and helps divide the earth into half spheres in the other direction. Find this line—called the Prime Meridian or the Meridian of Greenwich—on a globe. Everything east of the Prime Meridian for 180 degrees is in the Eastern Hemisphere. Everything west of the Prime Meridian for 180 degrees is in the Western Hemisphere. In which hemispheres is North America located? It is found in both the Northern Hemisphere and the Western Hemisphere.

Latitude and Longitude

The Equator and the Prime Meridian are the starting points for two sets of lines used to find any location. *Parallels* circle the earth like stacked rings and show **latitude,** or distance measured in degrees north and south of the Equator. The letter *N* or *S* following the degree symbol tells you if the location is north or south of the Equator. The North Pole, for example, is at 90°N (North) latitude, and the South Pole is at 90°S (South) latitude.

Two important parallels in between the poles are the Tropic of Cancer at 23½°N latitude and the Tropic of Capricorn at 23½°S latitude. You can also find the Arctic Circle at

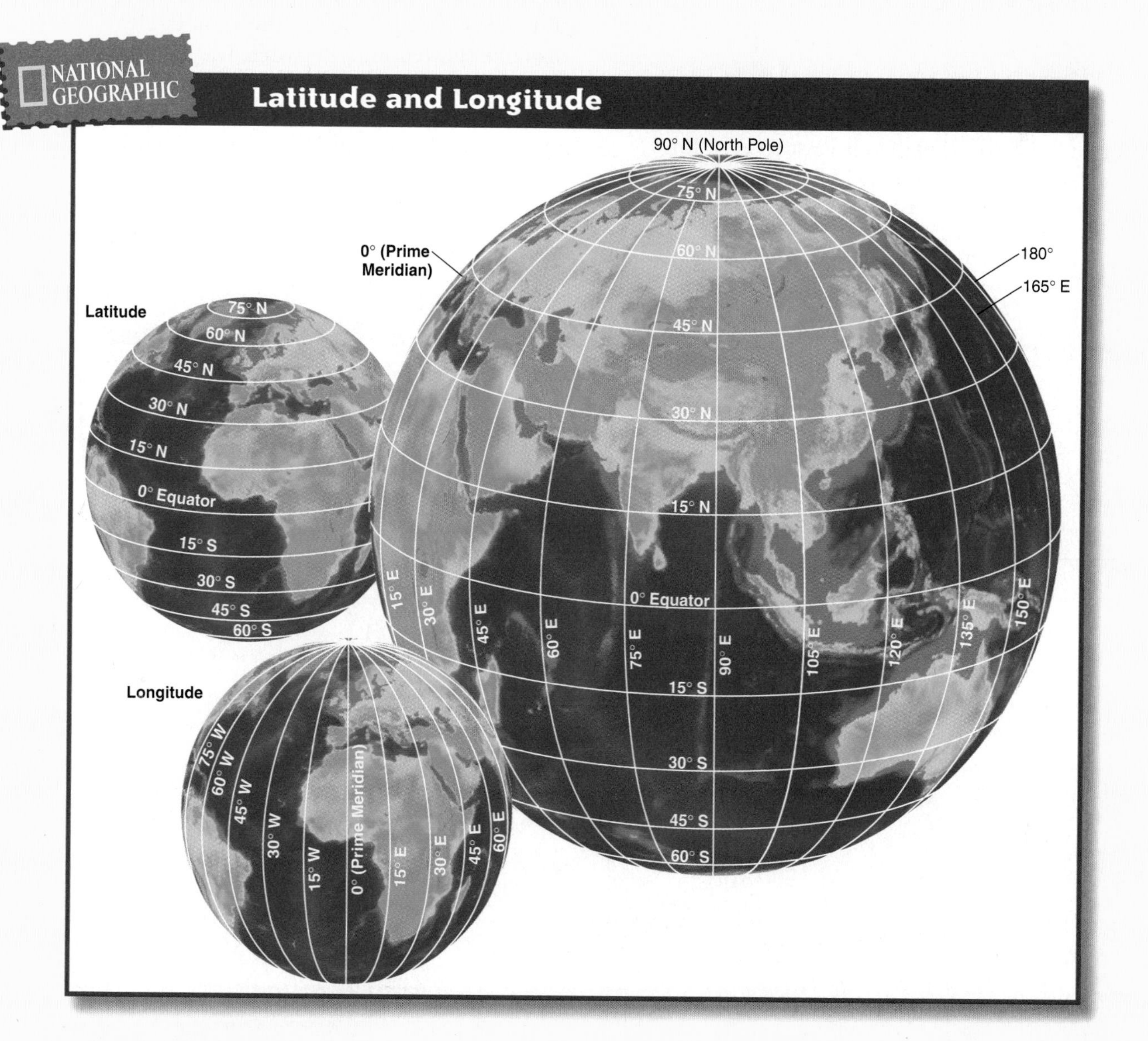

66½°N latitude and the Antarctic Circle at 66½°S latitude.

Meridians run from pole to pole and crisscross parallels. Meridians signify **longitude,** or distance measured in degrees east *(E)* or west *(W)* of the Prime Meridian. The Prime Meridian, or 0° longitude, runs through Greenwich, England. On the opposite side of the earth is the 180° meridian, also called the International Date Line.

Lines of latitude and longitude cross each other in the form of a **grid system.** You can find a place's **absolute location** by naming the latitude and longitude lines that cross exactly at that place. For example, the city of Tokyo, Japan, is located at 36°N latitude and 140°E longitude.

Great Circle Route

A straight line of true direction—one that moves directly from west to east, for example—is not always the shortest distance between two points on the earth. To find the shortest distance between any two places, take a piece of string and stretch it around a globe from one point to another. The string will form part of a *great circle,* or an imaginary line that follows the curve of the earth. A line drawn along the Equator is an example of a great circle. Traveling along a great circle is called following a **great circle route.** Airplane pilots use great circle routes to reduce travel time and to save fuel.

The idea of a great circle shows one important difference between using a globe and using a map. Because a globe is round, it accurately shows great circles. However, on a flat map the great circle route between two points may not appear to be the shortest distance. On map A below, the great circle distance (dotted line) between Tokyo and Los Angeles appears to be far longer than the true direction distance (solid line). In fact, the great circle distance is 345 miles (555 km) shorter, which is evident on map B.

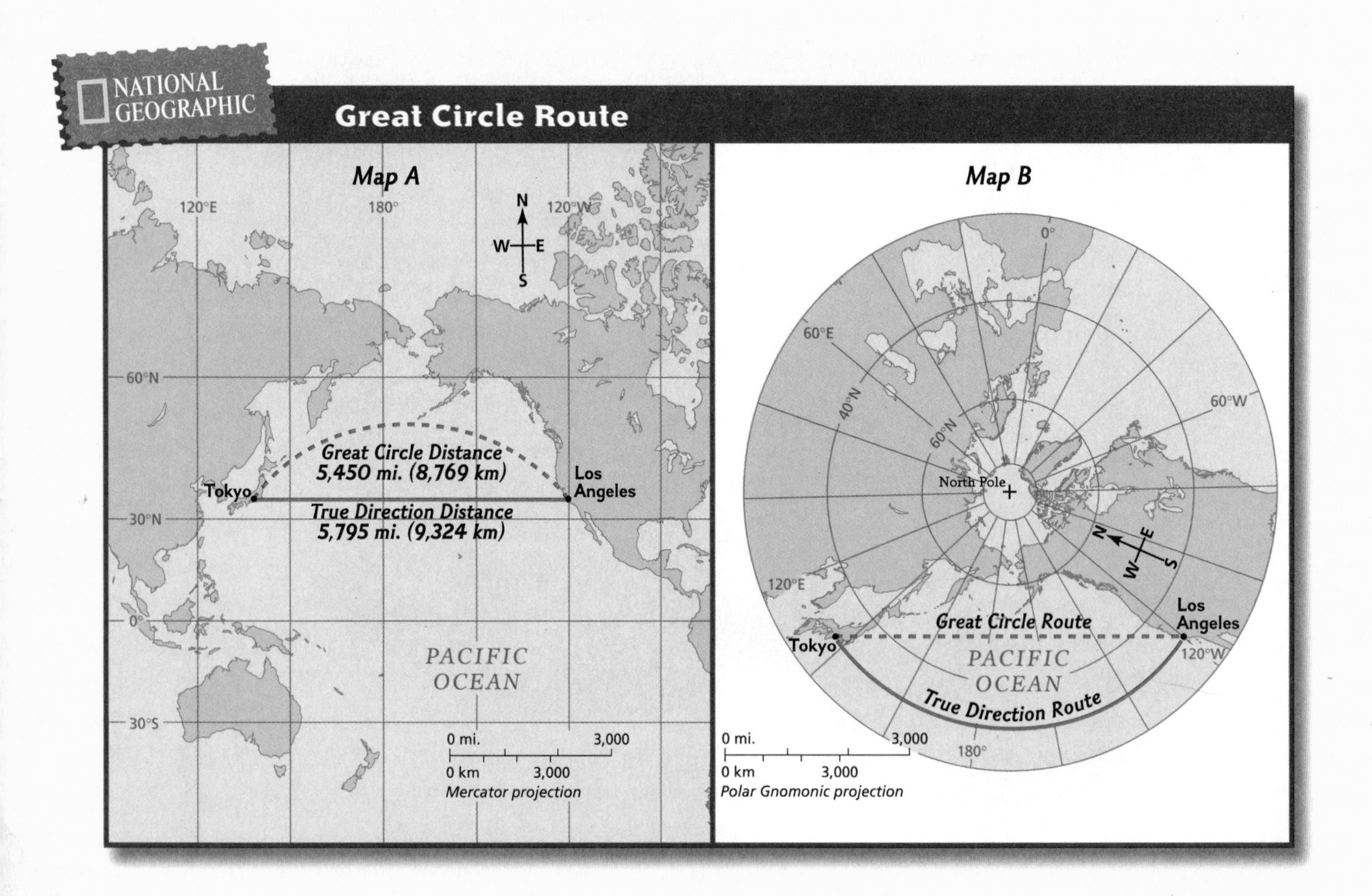

Why Use Maps?

Globes are the best, most accurate way to show the round earth. Using a globe has its difficulties, though. First, a globe is too big and awkward to carry around. Also, a globe cannot show you the whole world at one time. For these reasons, geographers use maps instead. A map is made by taking data from a round globe and placing it on a flat surface. It is important to remember that the earth's features, which are shown accurately on a globe, become distorted when the curves of a globe become straight lines on a flat map.

Map Projections

Imagine taking the whole peel from an orange and trying to flatten it on a table. You would either have to cut it or stretch parts of it. Mapmakers face a similar problem in showing the surface of the round earth on a flat map. When the earth's surface is flattened, big gaps open up. To fill in the gaps, mapmakers stretch parts of the earth. They choose to show either the correct shapes of places or their correct sizes. It is impossible to show both. As a result, mapmakers have developed different **projections,** or ways of showing the earth on a flat piece of paper. Each projection has its strengths and weaknesses. None is a completely accurate representation of the earth, but all prove useful in one way or another.

Mercator Projection The *Mercator projection* shows land shapes fairly accurately, but not size or distance. Areas that are located far from the Equator are quite distorted on this projection. Alaska, for example, appears much larger on a Mercator map than it does on a globe. The Mercator projection does show true directions, however. This makes it very useful for sea travel.

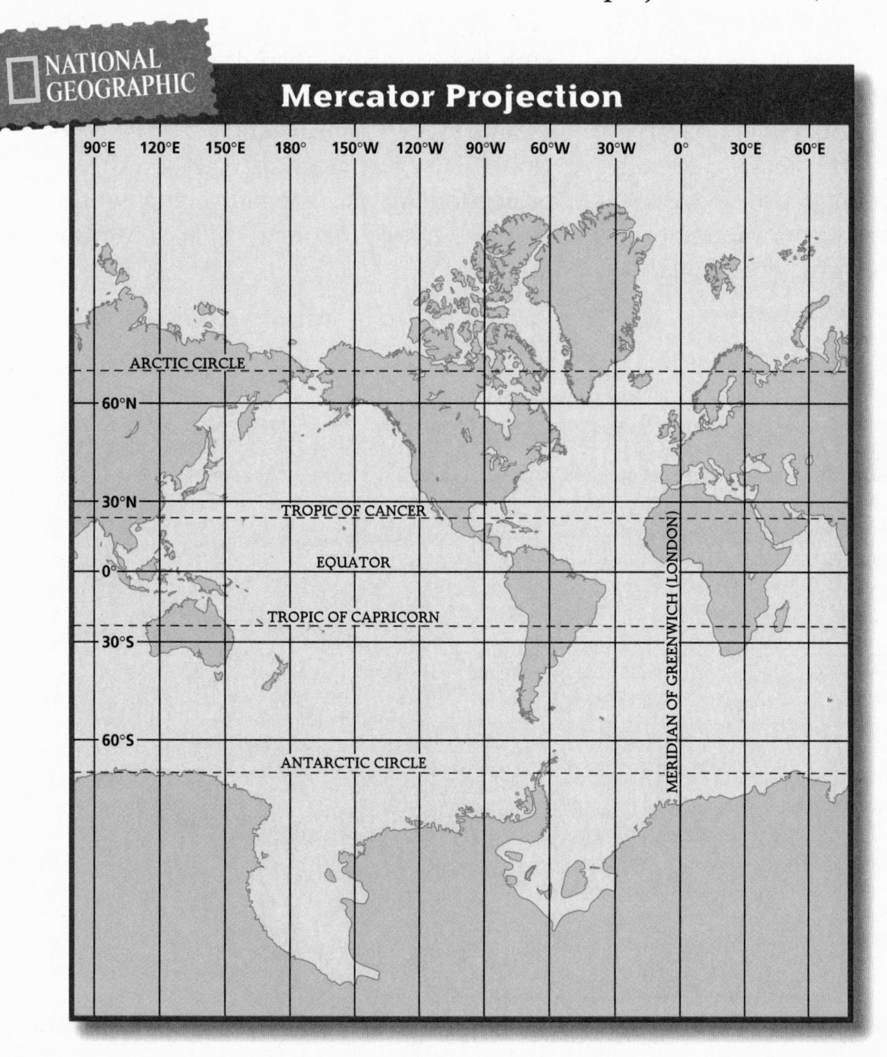

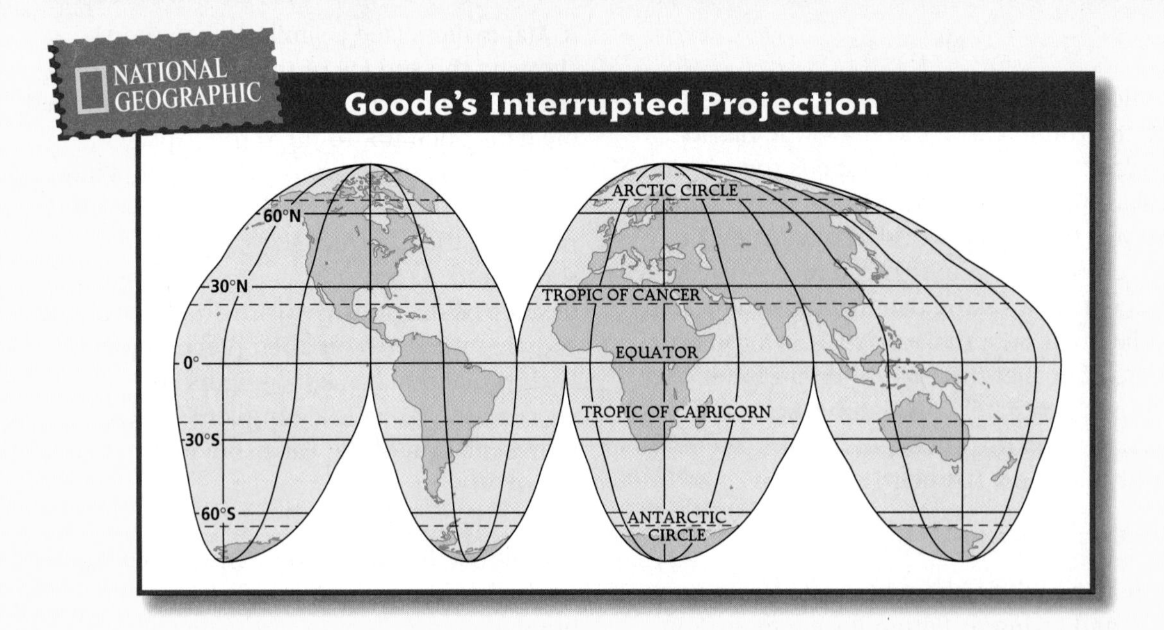

Goode's Interrupted Projection Take a second look at your peeled, flattened orange. You might have something that looks like a map based on *Goode's interrupted projection.* A map with this projection shows continents close to their true shapes and sizes. Distances—especially in the oceans—are less accurate. The Goode's projection would be helpful if you wanted to compare land area data about the continents.

Robinson Projection A map using the *Robinson projection* shows size and shape with less distortion than does a Mercator map. Land on the western and eastern sides of the Robinson map appear much as they do on a globe. The areas most distorted on this projection are near the North and South poles. You may notice that many atlases use Robinson projections.

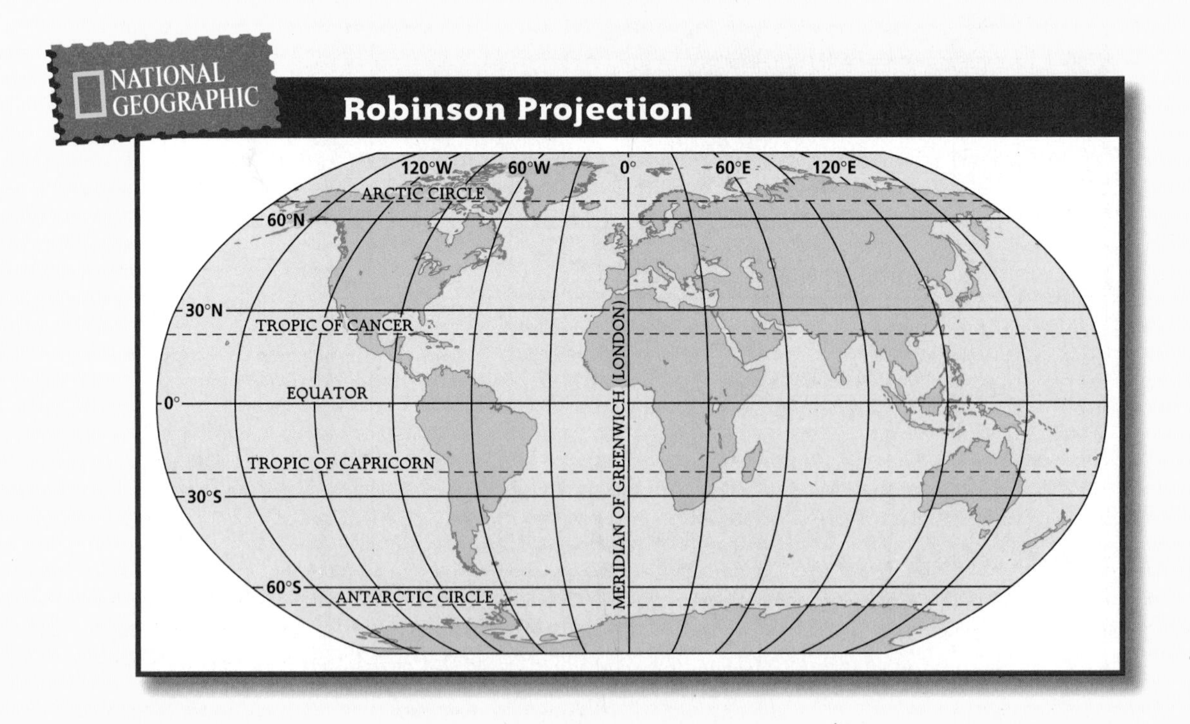

Winkel Tripel Projection

Mapmakers are always looking for a more accurate way to show the round earth on flat paper. The *Winkel Tripel projection* gives a good overall view of the continents' shapes and sizes. You may notice a close similarity between the Winkel Tripel and Robinson projections. Land areas in a Winkel Tripel projection are not as distorted near the poles as they are in the Robinson projection. In 1998 the National Geographic Society began using the Winkel Tripel projection for its reference maps of the world.

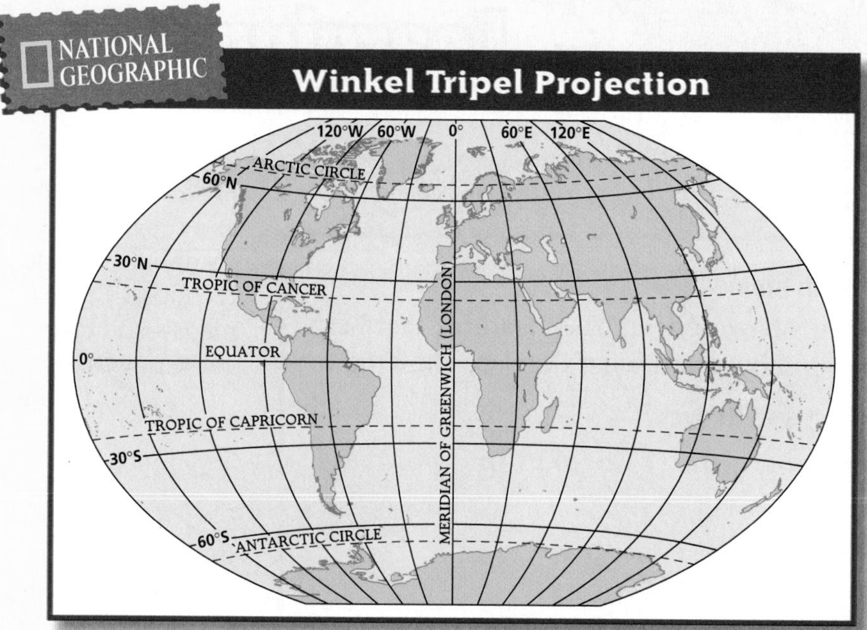

GEOGRAPHY HANDBOOK

Assessment

Defining Terms

1. **Define** globe, hemisphere, latitude, longitude, grid system, absolute location, great circle route, projection.

Recalling Facts

2. What imaginary line divides the earth into the Northern and the Southern Hemispheres?
3. What imaginary line divides the earth into the Eastern and the Western Hemispheres?
4. What is the best way to find the shortest distance between two places?

Critical Thinking

5. **Drawing Conclusions** Why do map projections distort some parts of the earth?
6. **Making Comparisons** What map projection has fairly accurate shapes in the center but increasing distortion toward the edges?

Graphic Organizer

7. **Organizing Information** Create a chart like this one. On the left, list three words or phrases that describe globes. On the right, do the same for maps.

Globes	Maps

Applying Geography Skills

8. **Analyzing Globes** Look at the globe showing latitude on page 5. Where on the globe do parallels, or lines of latitude, become shorter?

Learning Map Basics

Guide To Reading

Main Idea

Maps have several basic features that help you understand what they are showing.

Terms to Know

- geographic information systems (GIS)
- map key
- cardinal directions
- compass rose
- intermediate directions
- scale bar
- scale
- relief
- elevation
- contour line

How Maps Are Made

For more than 4,000 years, people have made maps to organize their knowledge of the world. The reason for producing maps has not changed over the centuries, but the tools of mapmaking have. Today satellites located thousands of miles in space gather data about the earth below. The data are then sent back to the earth, where computers change the data into images of the earth's surface. Mapmakers analyze and use these images to produce maps.

For modern mapmakers, computers have replaced pen and paper. Most mapmakers use computers with software programs called geographic information systems (GIS). With GIS, each kind of information on a map is kept as a separate electronic "layer" in the map's computer files. Because of this modern technology, mapmakers are able to make maps—and change them—more quickly and easily than before.

How to Read a Map

Maps can direct you down the street, across the country, or around the world. An ordinary map holds all kinds of information. Learn the map's code, and you can read it like a book.

Map Key The map key explains the lines, symbols, and colors used on a map. Look at the map of Spain below. Its key shows that dots mark major cities. A circled star indicates the national capital—in Spain's case, the city of Madrid. Some keys tell which lines stand for national boundaries, roads, or railroads. Other

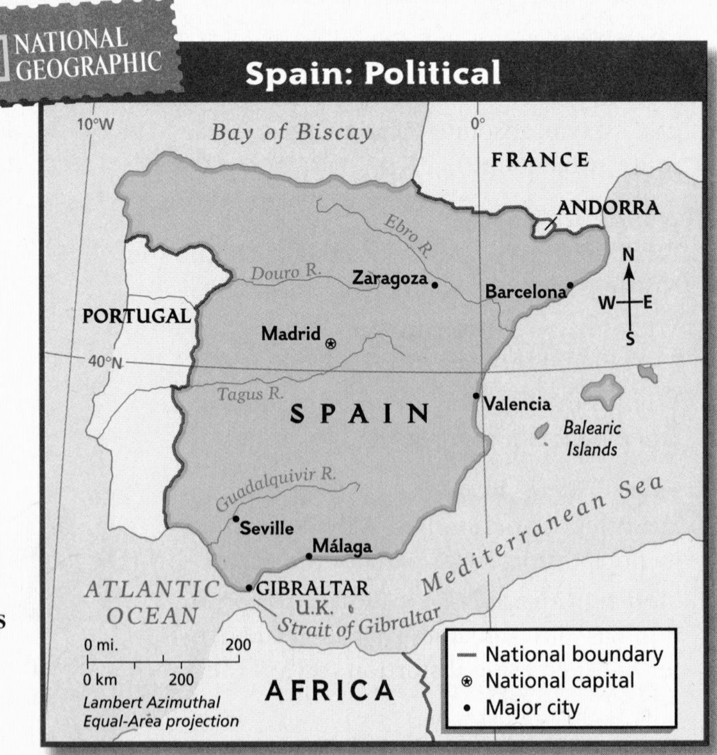

map symbols may represent human-made or natural features, such as canals, forests, or natural gas deposits.

Compass Rose An important step in reading any map is to find the direction marker. A map has a symbol that tells you where the **cardinal directions**—north, south, east, and west—are positioned. Sometimes all of these directions are shown with a **compass rose.** An **intermediate direction,** such as southeast, may also be on the compass rose. Intermediate directions fall between the cardinal directions.

Latitude and Longitude Lines Like globes, maps have lines of latitude and longitude that form a grid. Every place on the earth has a unique position or "address" on this grid. Knowing this address makes it easier for you to locate cities and other places on a map. For example, what is the grid address of Madrid, Spain? The map on page 10 shows you that the address is 41°N latitude, 4°W longitude.

Scale A measuring line, often called a **scale bar,** helps you determine distance on a map. The map's **scale** tells you what distance on the earth is represented by the measurement on the scale bar. For example, 200 miles on the earth may be represented by 1 inch on the map. Knowing the scale allows you to see how large an area is. Map scale is usually given in both miles and kilometers.

Each map has its own scale. What scale a mapmaker uses depends on the size of the area shown on the map. If you were drawing a map of your backyard, you might use a scale of 1 inch equals 5 feet. In contrast, the scale bar on the map above of Austin, Texas, shows that about ⅝ inch represents 8 miles. Scale is important when you are trying to compare the size of one area to another.

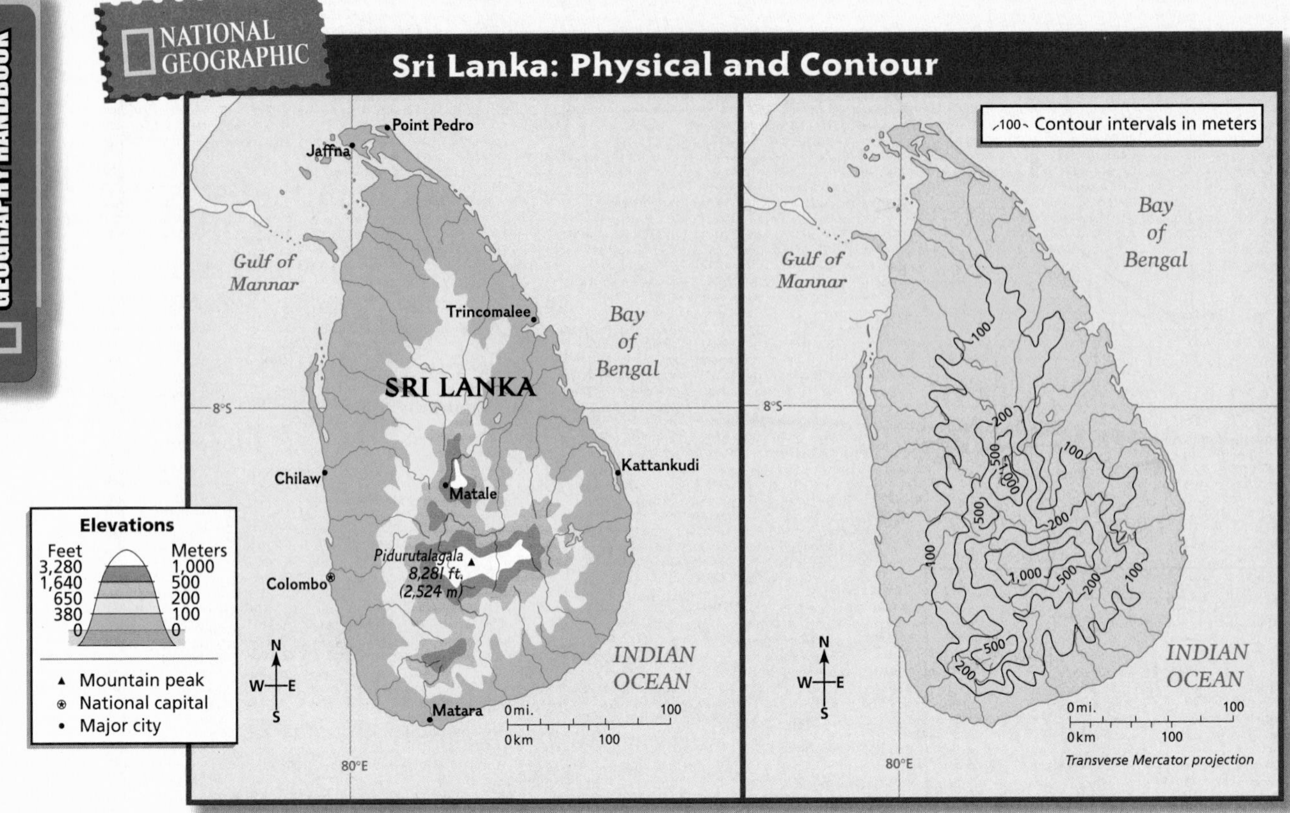

General Purpose Maps

Maps are amazingly useful tools. You can use them to preserve information, to display data, and to make connections between seemingly unrelated things. Geographers use many different types of maps. Maps that show a wide range of general information about an area are called *general purpose maps.* Two of the most common general purpose maps are political and physical maps.

Political Maps *Political maps* show the names and boundaries of countries and often identify major physical features. The political map of Spain on page 10, for example, shows the boundaries between Spain and other countries. It also shows cities and rivers within Spain and bodies of water surrounding Spain.

Physical Maps *Physical maps* call out landforms and water features. The physical map of Sri Lanka above shows rivers and mountains. The colors used on physical maps include brown or green for land, and blue for water. These colors and shadings may show **relief**—or how flat or rugged the land surface is. In addition, physical maps may use colors to show **elevation**—the height of an area above sea level. A key explains what each color and symbol stands for.

Contour Maps One kind of physical map, called a *contour map,* also shows elevation. A contour map has **contour lines**—one line for each major level of elevation. All the land at the same elevation is connected by a line. These lines usually form circles or ovals—one inside the other. If contour lines come very close together, the surface is steep. If the lines are spread apart, the land is flat or rises very gradually. Compare the contour map of Sri Lanka above to its physical map.

Special Purpose Maps

Some maps are made to present specific kinds of information. These are called *special purpose maps.* They usually show patterns, often emphasizing one subject or theme. Special purpose maps may present climate, natural resources, or population density. They may also display historical information, such as battles or territorial changes. The map's title tells what kind of special information it shows. Colors and symbols in the map key are especially important on these types of maps.

One type of special purpose map uses colors to show population density, or the average number of people living in a square mile or square kilometer. As with other maps, it is important to first read the title and the key. The population density map of Egypt to the right gives a striking picture of differences in population density. The Nile River valley and delta are very densely populated. In contrast, the desert areas east and west of the river are home to few people.

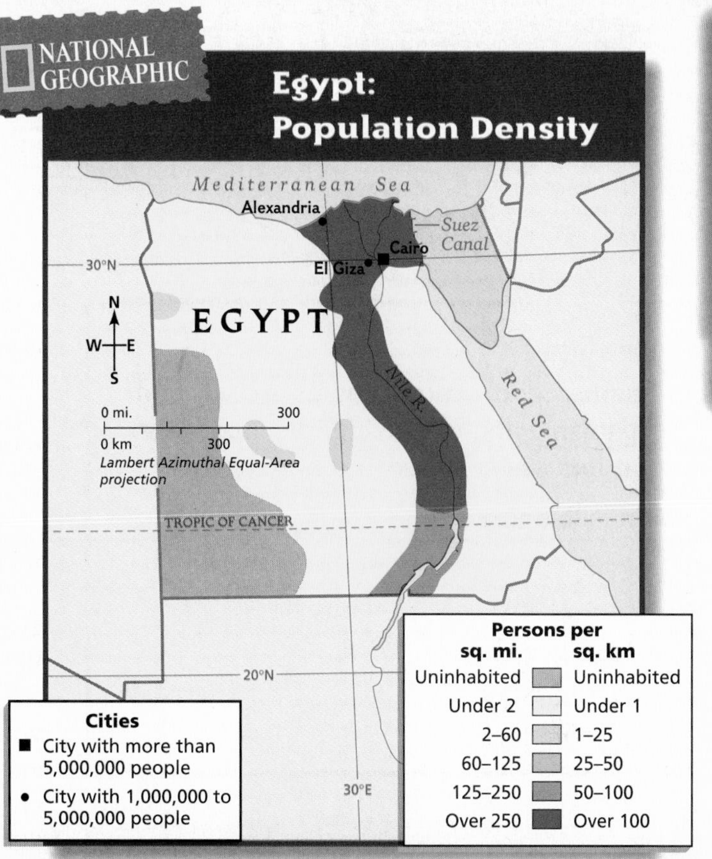

Defining Terms

1. Define geographic information systems (GIS), map key, cardinal directions, compass rose, intermediate directions, scale bar, scale, relief, elevation, contour line.

Recalling Facts

2. Why do people make maps?

3. What are the four cardinal directions?

4. What are two of the most common types of general purpose maps?

Critical Thinking

5. Comparing and Contrasting Describe the similarities and differences between physical maps and contour maps.

6. Synthesizing Information Draw a compass rose that shows the cardinal and intermediate directions.

Graphic Organizer

7. Organizing Information Create a diagram like the one below. In each of the outer ovals, write an example of a feature that you would find on a typical physical map.

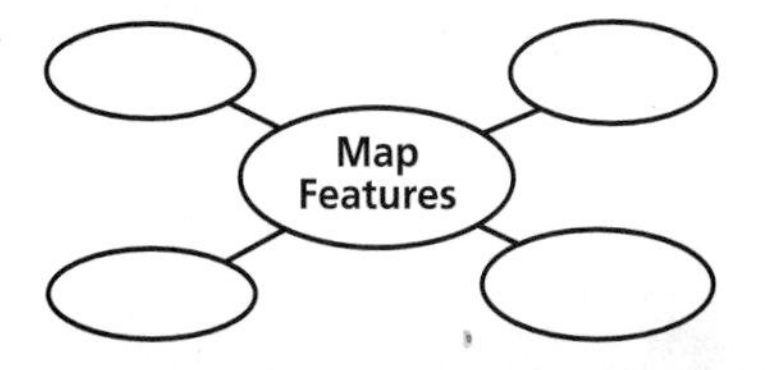

Applying Geography Skills

8. Analyzing Maps Look at the map of Egypt above. At what latitude and longitude is Alexandria located? Use the key to describe the population density of Alexandria and its surrounding area.

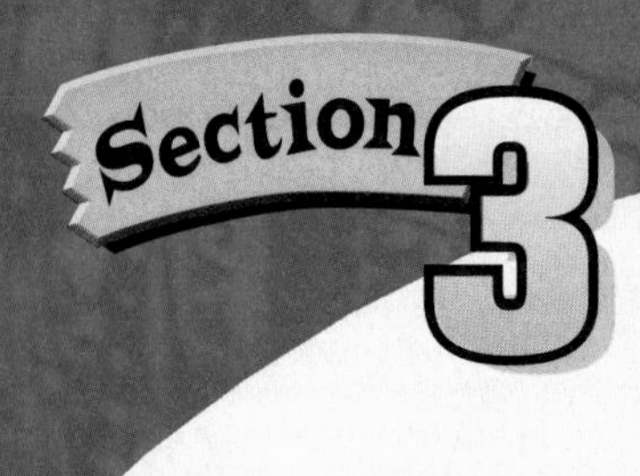

Section 3 Using Graphs, Charts, and Diagrams

Guide To Reading

Main Idea

Graphs, charts, and diagrams are ways of organizing and displaying information so that it is easier to see and understand.

Terms to Know

- axis
- bar graph
- line graph
- circle graph
- pictograph
- climograph
- chart
- diagram
- elevation profile

What Is a Graph?

A graph is a way of summarizing and presenting information visually. Each part of a graph gives useful information. First read the graph's title to find out its subject. Then read the labels along the graph's **axes**—the vertical line along the left side of the graph and the horizontal line along the bottom of the graph. One axis will tell you what is being measured. The other axis tells what units of measurement are being used.

Bar, Line, and Circle Graphs

Bar Graphs Graphs that use bars or wide lines to compare data visually are called **bar graphs.** Look carefully at the bar graph below, which compares world languages. The vertical axis lists the languages. The horizontal axis gives speakers of the language, in millions. By comparing the lengths of the bars, you can quickly tell which language is spoken by the most people. Bar graphs are especially useful for comparing quantities, and they may show the bars rising up from the bottom of the graph or extending out from the vertical axis.

NATIONAL GEOGRAPHIC

Comparing World Languages

Languages	Numbers of Speakers (in millions)
Chinese (Mandarin)	885
English	322
Spanish	266
Bengali	189
Hindi	182
Portuguese	170
Russian	170
Japanese	125
German	98
Chinese (Wu)	77

Source: *National Geographic Atlas of the World,* 1999.

Line Graphs A **line graph** is a useful tool for showing changes over a period of time. The amounts being measured are plotted on the grid above each year, and then are connected by a line. Line graphs sometimes have two or more lines plotted on them. The line graph on page 15 shows that the number of farms in the United States has decreased since 1940. The vertical axis lists the number of farms in millions. The horizontal axis shows the passage of time in ten-year periods from 1940 to 1998.

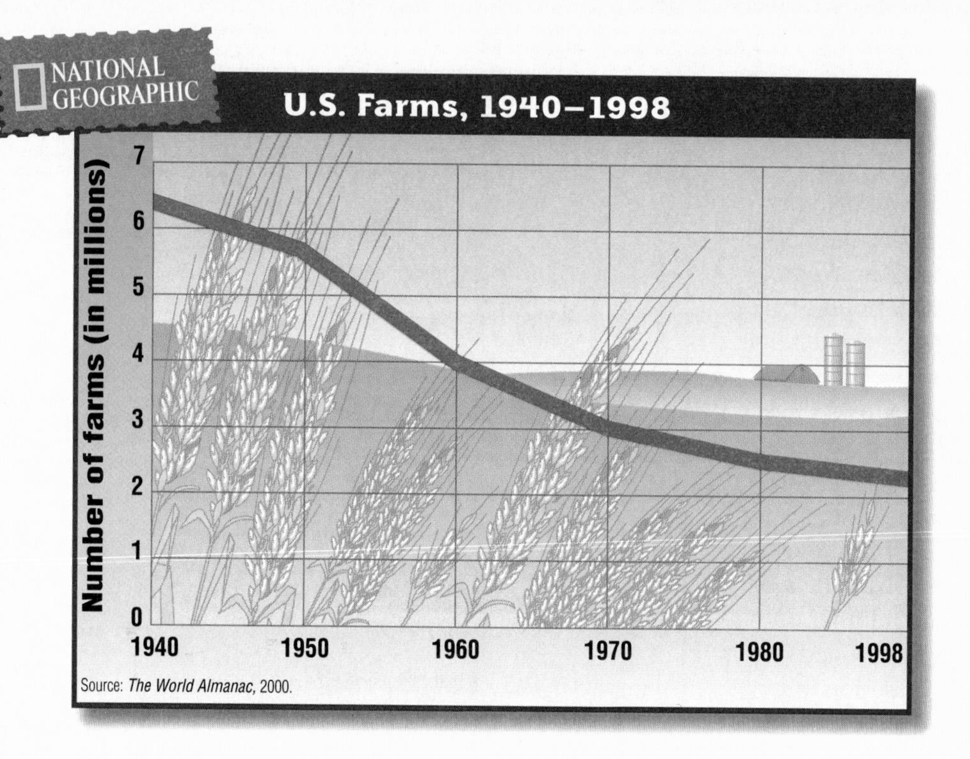

Circle Graphs You can use **circle graphs** when you want to show how the *whole* of something is divided into its *parts.* Because of their shape, circle graphs are often called pie graphs. Each "slice" represents a part or percentage of the whole "pie." On the circle graph below, the whole circle represents the world's population in 1999. The slices show how this population is divided among the world's five largest continents.

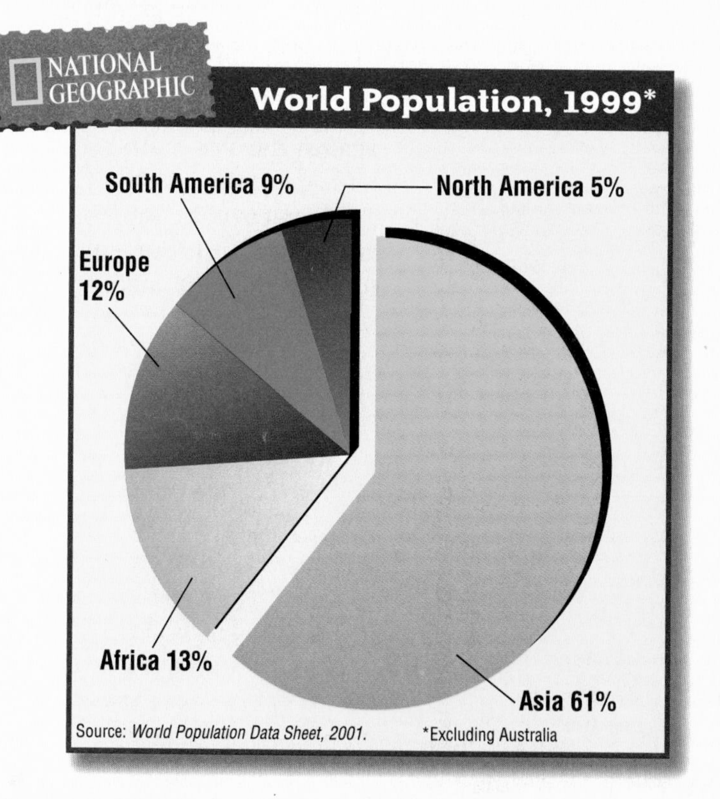

Charts

Charts present related facts and numbers in an organized way. They arrange data, especially numbers, in rows and columns for easy reference. Look at the chart on page 85. To interpret the chart, first read the title. It tells you what information the chart contains. Next, read the labels at the top of each column and on the left side of the chart. They explain what the numbers or data on the chart are measuring. One kind of chart, a *flowchart,* joins certain elements of a chart and a diagram. It can show the order of how things happen or how they are related to each other. The flowchart on page 132 presents the branches of the United States government. Notice how the chart shows the relationship among the tasks and the offices or bodies of each branch.

Pictographs

Like bar and circle graphs, pictographs are good for making comparisons. **Pictographs** use rows of small pictures or symbols, with each picture or symbol representing an amount. The pictograph on the right shows the number of automobiles produced in the world's five major automobile-producing countries. The key tells you that one car symbol stands for 1 million automobiles. Pictographs are read like a bar graph. The total number of car symbols in a row adds up to the auto production in each selected country.

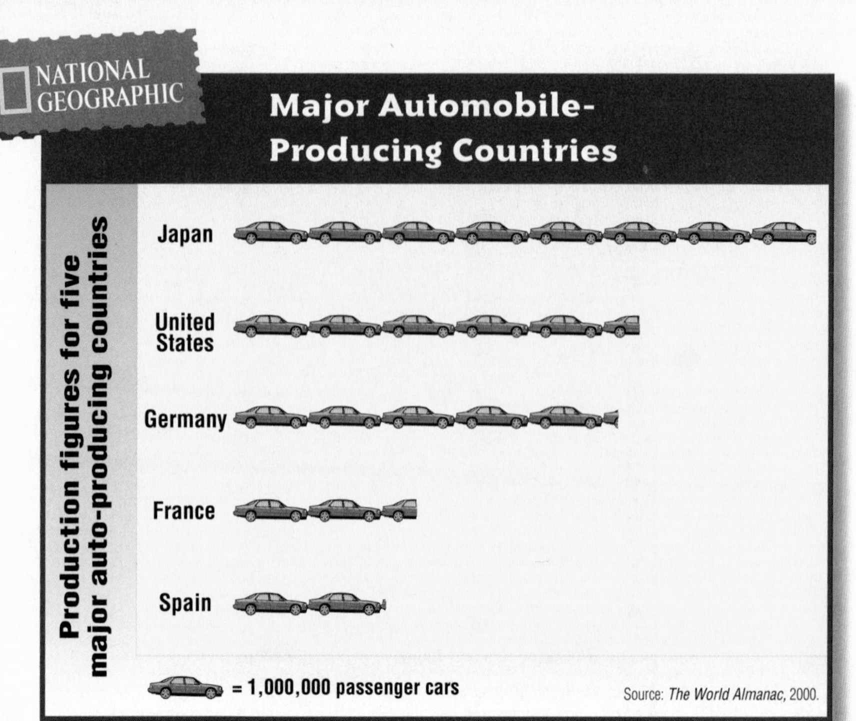

Climographs

A **climograph,** or climate graph, combines a line graph and a bar graph. It gives an overall picture of the climate—the long-term weather patterns—in a specific place. Because climographs include several kinds of information, you need to read them carefully.

Note that the vertical bars on the climograph below represent average amounts of precipitation (rain, snow, or sleet) in each month of the year. These bars are measured against the axis on the right side of the graph. The line plotted above the bars represents changes in the average monthly temperature. You measure this line against the axis on the left side of the graph. The names of the months are shown in shortened form on the bottom axis of the graph.

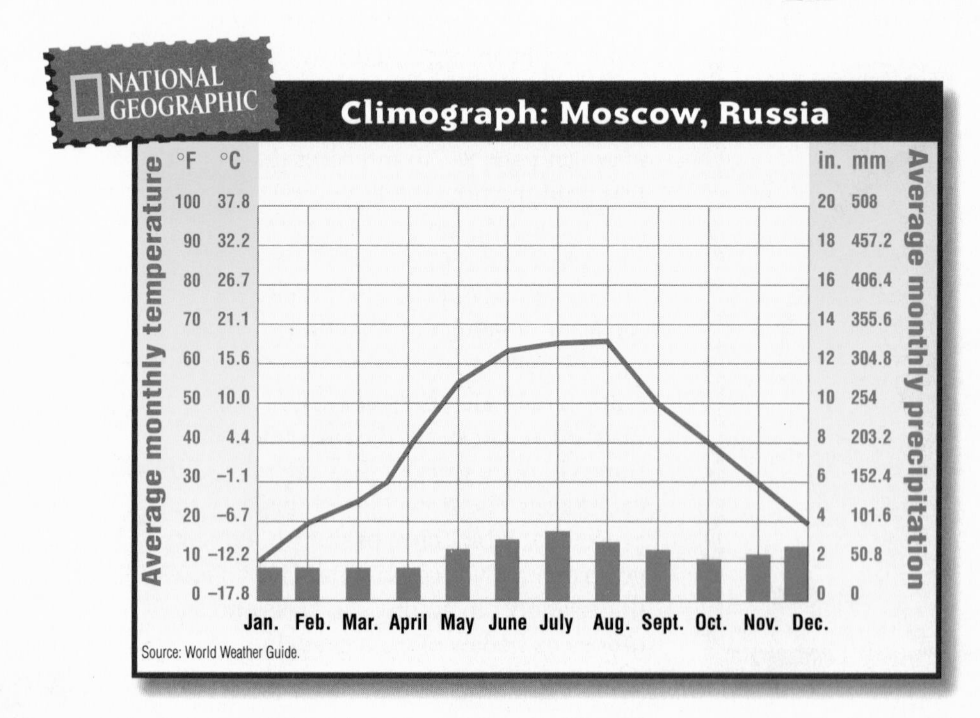

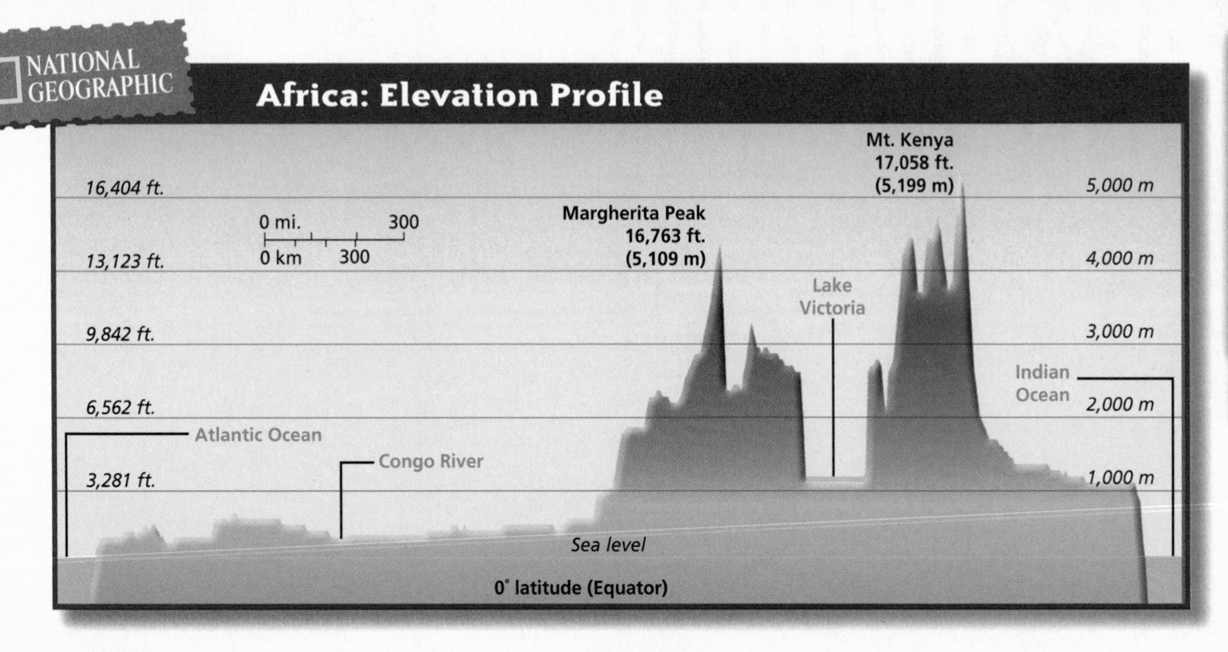

Diagrams

Diagrams are drawings that show steps in a process, point out the parts of an object, or explain how something works. You can use a diagram to assemble a stereo. The diagram on page 212 shows how locks enable ships to move through a canal. An **elevation profile** is a type of diagram that can be helpful when comparing the elevations of an area. It shows a profile, or side view, of the land as if it were sliced and you were viewing it from the side. (For reasons of scale, the profile is exaggerated.) The elevation profile of Africa above clearly shows low areas and mountains. The line of latitude at the bottom tells you where this profile was "sliced."

Defining Terms

1. **Define** axis, bar graph, line graph, circle graph, pictograph, climograph, chart, diagram, elevation profile.

Recalling Facts

2. How does a bar graph differ from a line graph?

3. What percentage does the whole circle in a circle graph always represent?

4. What two features does a climograph show?

Critical Thinking

5. **Synthesizing Information** Draw and label a flowchart showing the steps in some simple process—for example, making a sandwich or doing laundry.

Graphic Organizer

6. **Organizing Information** Create a chart like the one below. In the left column, list the types of graphs that are discussed in this section. In the right column, list what each type of graph is most useful for showing.

Types of Graphs	Useful for showing . . .

Applying Geography Skills

7. **Analyzing Graphs** Look at the bar graph on page 14. Which language is the most widely spoken? About how many people speak it?

GEOGRAPHIC DICTIONARY

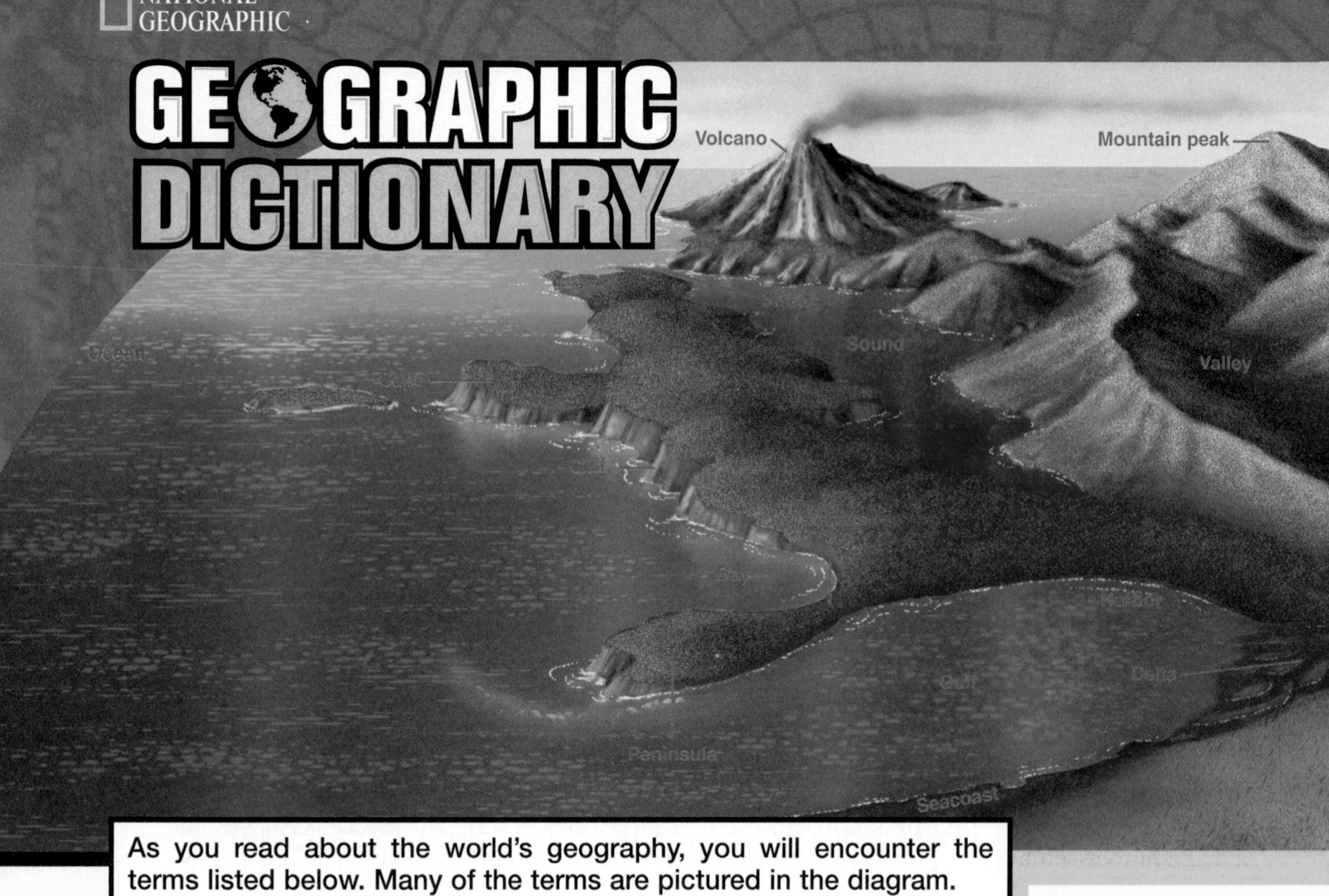

As you read about the world's geography, you will encounter the terms listed below. Many of the terms are pictured in the diagram.

absolute location exact location of a place on the earth described by global coordinates

basin area of land drained by a given river and its branches; area of land surrounded by lands of higher elevations

bay part of a large body of water that extends into a shoreline, generally smaller than a gulf

canyon deep and narrow valley with steep walls

cape point of land that extends into a river, lake, or ocean

channel wide strait or waterway between two landmasses that lie close to each other; deep part of a river or other waterway

cliff steep, high wall of rock, earth, or ice

continent one of the seven large landmasses on the earth

delta flat, low-lying land built up from soil carried downstream by a river and deposited at its mouth

divide stretch of high land that separates river systems

downstream direction in which a river or stream flows from its source to its mouth

elevation height of land above sea level

Equator imaginary line that runs around the earth halfway between the North and South Poles; used as the starting point to measure degrees of north and south latitude

glacier large, thick body of slowly moving ice

gulf part of a large body of water that extends into a shoreline, generally larger and more deeply indented than a bay

harbor a sheltered place along a shoreline where ships can anchor safely

highland elevated land area such as a hill, mountain, or plateau

hill elevated land with sloping sides and rounded summit; generally smaller than a mountain

island land area, smaller than a continent, completely surrounded by water

isthmus narrow stretch of land connecting two larger land areas

lake a sizable inland body of water

latitude distance north or south of the Equator, measured in degrees

longitude distance east or west of the Prime Meridian, measured in degrees

lowland land, usually level, at a low elevation

map drawing of the earth shown on a flat surface

meridian one of many lines on the global grid running from the North Pole to the South Pole; used to measure degrees of longitude

mesa broad, flat-topped landform with steep sides; smaller than a plateau

mountain land with steep sides that rises sharply (1,000 feet or more) from surrounding land; generally larger and more rugged than a hill

mountain peak pointed top of a mountain

mountain range a series of connected mountains

mouth (of a river) place where a stream or river flows into a larger body of water

ocean one of the four major bodies of salt water that surround the continents

ocean current stream of either cold or warm water that moves in a definite direction through an ocean

parallel one of many lines on the global grid that circles the earth north or south of the Equator; used to measure degrees of latitude

peninsula body of land jutting into a lake or ocean, surrounded on three sides by water

physical feature characteristic of a place occurring naturally, such as a landform, body of water, climate pattern, or resource

plain area of level land, usually at low elevation and often covered with grasses

plateau area of flat or rolling land at a high elevation, about 300–3,000 feet high

Prime Meridian line of the global grid running from the North Pole to the South Pole through Greenwich, England; starting point for measuring degrees of east and west longitude

relief changes in elevation over a given area of land

river large natural stream of water that runs through the land

sea large body of water completely or partly surrounded by land

seacoast land lying next to a sea or an ocean

sound broad inland body of water, often between a coastline and one or more islands off the coast

source (of a river) place where a river or stream begins, often in highlands

strait narrow stretch of water joining two larger bodies of water

tributary small river or stream that flows into a large river or stream; a branch of the river

upstream direction opposite the flow of a river; toward the source of a river or stream

valley area of low land between hills or mountains

volcano mountain created as liquid rock and ash erupt from inside the earth

Unit 1

Athlete at Olympic Games, Sydney, Australia

City of Hong Kong, China

The World

You are about to journey to dense rain forests, bleak deserts, bustling cities and marketplaces, and remote villages. You are entering the world of geography—the study of the earth and all its people. Imagine that you could visit any place in the world. Where would you want to go? What would you want to see?

Hot air balloon floating over cultivated fields, Egypt

NGS ONLINE
www.nationalgeographic.com/education

Chapter 1

Looking at the Earth

The World and Its People NATIONAL GEOGRAPHIC

To learn more about Earth's structure and landforms, view ***The World and Its People*** **Chapter 1** video.

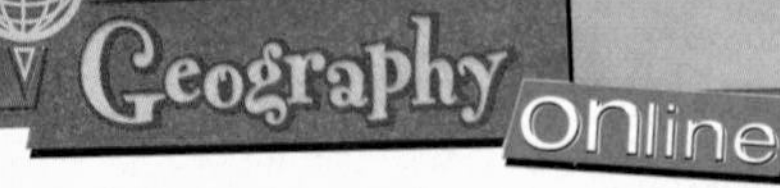

Chapter Overview Visit the ***Geography: The World and Its People*** Web site at gwip.glencoe.com and click on **Chapter 1—Chapter Overviews** to preview information about Earth.

Section 1

Thinking Like a Geographer

Guide to Reading

Main Idea

Geographers use various tools to understand the world.

Terms to Know

- geography
- landform
- environment
- region
- Global Positioning System (GPS)
- geographic information systems (GIS)

Places to Locate

- China
- Yangtze River
- United States
- Andes
- Antarctica
- California

Reading Strategy

Create a chart like this one and write three details or examples for each heading.

How Geographers View the World 1. 2. 3.
Tools of Geography 1. 2. 3.
Uses of Geography 1. 2. 3.

NATIONAL GEOGRAPHIC **Exploring Our World**

How would *you* go about making an accurate map of the world? Scientists decided the best way to map the earth was to see it from space. In February 2000, the space shuttle *Endeavour* used a radar camera to take pictures of the land below. By using radar, the camera was not hampered by clouds or darkness.

Why do geographers want to know exactly what the earth looks like? Think about the following: In **China,** the spring flooding of the **Yangtze** (YANG•SEE) **River** threatens people and crops every year. In 1998 the floods killed more than 4,000 people. After studying the land and the climate, officials in China's government sought solutions. They built dams to hold some of the floodwaters back. Recently, the number of people who died from the floods fell to 700.

This is just one example of how people around the world use geographic knowledge collected from various sources. **Geography** is the study of the earth in all its variety. When you study geography, you learn about the earth's land, water, plants, and animals. This is physical geography. You also study people—where they live, how they live, how they change and are influenced by their environment, and how different groups compare to one another. This is human geography.

◀ Skydiving over Key West, Florida

A Geographer's View of Place

Geographers look at major issues—like the flooding of the Yangtze, which affects millions of people. They also look at local issues—like where the best place is for a company to build a new store in town. Whether major or local, geographers try to understand both the physical and human characteristics, or features, of an issue.

Physical Characteristics Geographers study places. They look at *where* something is located on the earth. They also try to understand what the place is *like*. They ask: What features make a place similar to or different from other places?

To answer this question, geographers identify the landforms of a place. **Landforms** are individual features of the land, like mountains and valleys. Geographers also look at water. Is the place near the ocean or on a river? Does it have plentiful or very little freshwater? They consider whether the soil will produce crops. They see how much rain the place usually receives and how hot or cold the area is. They find out whether the place has minerals, metals, trees, or other resources.

Human Characteristics Geographers also look at the human characteristics of the people living in the place. Do many or only a few people live there? Do they live close together or far apart? Why? What kind of government do they have? What religion do they follow? What kinds of work do they do? What languages do they speak? From where did the people's ancestors come?

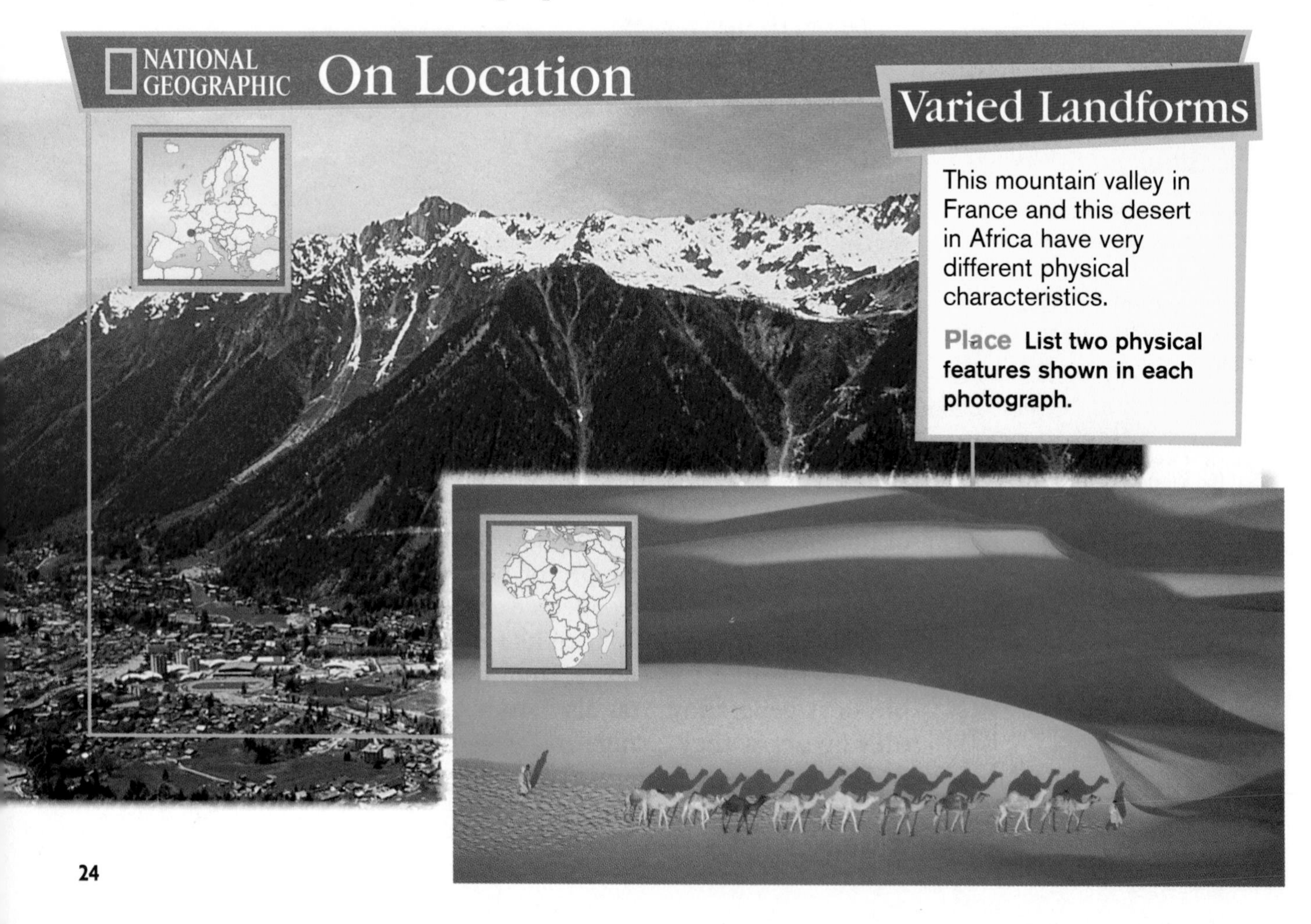

This mountain valley in France and this desert in Africa have very different physical characteristics.

Place **List two physical features shown in each photograph.**

People and Places Geographers are especially interested in how people interact with their **environment,** or natural surroundings. People can have a major impact on the environment. Remember how the Chinese built dams along the Yangtze? When they did so, they changed the way the river behaved in flood season.

Where people live often has a strong influence on *how* they live. People near the sea might catch fish and build ships for trade. Those living inland might farm or take up ranching. Of course, as the people of the world use more computers and other technology in their work, the surrounding environment may have less of an influence on what types of work they do.

Regions Geographers carefully study individual cities, rivers, and other landforms. They also look at the big picture, or how individual places relate to other places. In other words, geographers look at a **region,** or an area that shares common characteristics. Regions can be relatively small—like your state or town. They can also be huge—like all of the western **United States.** Some regions may even include several countries because these countries have similar environments, or because their people follow similar ways of life and speak the same language. The countries along the western side of South America are often discussed as a region—the Andean countries—because the **Andes,** a series of mountain ranges, run through all of them.

✓Reading Check **What do geographers study to determine the human characteristics of a place?**

The Tools of Geography

Geographers need tools to study people and places. Maps and globes are the main tools they use. As you read in the **Geography Handbook** on page 11, geographers use many different types of maps. Each type gives geographers a particular kind of information about a place.

Collecting Data for Mapping Earth How do geographers gather information so they can make accurate maps? One way is to take photographs from high above the earth. These are called LANDSAT photos and show details such as the shape of the land, what plants cover an area, and how land is being used. Special cameras can even reveal hidden information. Photos of **Antarctica** taken from radar cameras show rivers of ice 500 miles (805 km) long—all hidden by snow.

How do geographers accurately label the exact locations of places on a map? Believe it or not, the best way to find a location is from outer space. A group of satellites traveling around the earth make up the **Global Positioning System (GPS).** A GPS receiver is a special device that receives signals from these satellites. When the receiver is put at a location, the GPS satellite can tell the exact latitude and longitude of that place. As a result, a mapmaker can know where exactly on the earth the particular area is located. GPS devices are even installed in vehicles to help drivers find their way.

NATIONAL GEOGRAPHIC **On Location**

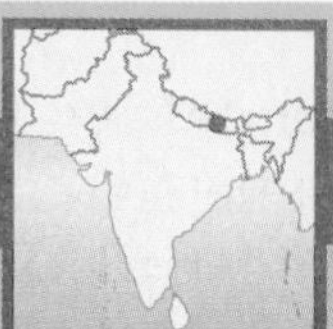

Mt. Everest

GPS satellites locate and *measure* places on the earth. A GPS receiver placed on top of Mt. Everest, the tallest mountain in the world, showed that it is 7 feet (2.1 m) higher than people had previously thought.

Location **Why is it important for geographers to know exactly where places are located on the earth?**

Web Activity Visit the ***Geography: The World and Its People*** Web site at gwip.glencoe.com and click on **Chapter 1—Student Web Activities** to learn more about geographic information systems.

Geographic Information Systems Today geographers use another powerful tool in their work—computers. Special computer software called geographic information systems (GIS) helps geographers gather many different kinds of information about the same place. After typing in all the data they collect, geographers use the software to combine and overlap the information on special maps.

As a result, different kinds of information about the same place can be compared. People then use that information to make decisions. For example, in the 1990s, GIS technology was used to help settle an argument over how to use resources in northern **California.** A logging company wanted to cut down parts of a forest. Environmental groups said that doing so would destroy the nesting areas of some rare birds. Geographers created maps that showed both the forest and the nesting sites—and then overlapped these maps. People could then see which areas had to be protected and which could be cut.

✓Reading Check **What is the difference between GPS and GIS?**

Uses of Geography

Have you ever gone on a long-distance trip in a car or taken a subway ride? If you used a road map or subway map to figure out where you were going, you were using geography. This is just one of the many uses of geographic information.

Geographic information is used in planning. Government leaders use geographic information to plan new services in their communities. They might plan how to handle disasters or how much new housing to allow in an area. Companies can see where people are moving in a region to make their plans for expanding.

In addition, the availability of geographic information helps people make sound decisions. Perhaps a question arises over whether a new building should be constructed. City leaders look at street use to see if the area can handle additional traffic. They make sure the area has the power, water, and sewage systems the building will need.

Businesses use geographic information for making decisions, too. Many businesses study population trends to see what areas need new products or services. Some businesses offer geographic information to their customers. Suppose you were looking for an apartment to rent. Some realtors have a computer program that can identify all the apartments of a certain size and price in an area. The program can create a map so you can see exactly where each apartment is located.

Finally, geographic information helps people manage resources. Many natural resources, such as oil or coal, are available only in limited supply. Geographic information can help people find more of those resources. Other resources, such as trees or water, can be replaced or renewed. People can use geographic information to show them how to manage these resources so they are not all used up.

Reading Check **Why do people have to manage resources carefully?**

Assessment

Defining Terms

1. **Define** geography, landform, environment, region, Global Positioning System (GPS), geographic information systems (GIS).

Recalling Facts

2. **Place** What two kinds of characteristics of a place do geographers study?
3. **Technology** What are the main tools of geography?
4. **Human/Environment Interaction** What are three uses for geography?

Critical Thinking

5. **Understanding Cause and Effect** How have the physical characteristics of your region affected the way people live there?
6. **Categorizing Information** Give five examples of regions. Begin with an area near you that shares common characteristics, then look for larger and larger regions.

Graphic Organizer

7. **Organizing Information** Draw a diagram like this one. In the center, write the name of a place you would like to visit. In the outer ovals, identify the types of geographic information you would like to learn about this place.

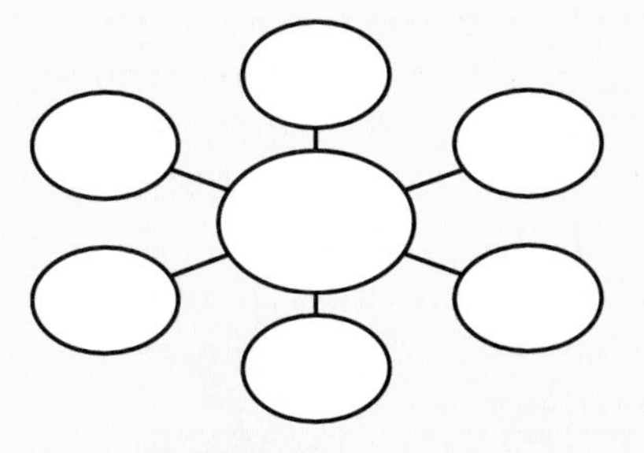

Applying Geography Skills

8. **Analyzing Maps** Find Egypt on the map on page RA23 of the **Reference Atlas.** Along what physical feature do you think most Egyptians live? Why? Turn to the population map of Egypt on page 13 of the **Geography Handbook** to see if you are correct.

Making Connections

ART | SCIENCE | LITERATURE | TECHNOLOGY

Geographic Information Systems

What if a farmer could save money by applying fertilizer only to the crops that needed it? Today, thanks to computer technology called geographic information systems (GIS), farmers can do just that.

The Technology

Geographic information systems (GIS) use computer software to combine and display a wide range of information about an area. GIS programs start with a map showing a specific location on the earth. This map is then linked with other information about that same place, such as satellite photos, amounts of rainfall, or where houses are located.

Think of geographic information systems as a stack of overhead transparencies. Each transparency shows the same general background but highlights different information. The first transparency may show a base map of an area. Only the borders may appear. The second transparency may show only rivers and highways. The third may highlight mountains and other physical features, buildings, or cities.

In a similar way, GIS technology places layers of information onto a base map. It can then switch each layer of information on or off, allowing data to be viewed in many different ways. In the case of the farmer mentioned above, GIS software combines information about soil type, plant needs, and last year's crop to pinpoint exact areas that need fertilizer.

How It Is Used

GIS technology allows users to quickly pull together data from many different sources and construct maps tailored to specific needs. This helps people analyze past events, predict future scenarios, and make sound decisions.

A person who is deciding where to build a new store can use GIS technology to help select the best location. The process might begin with a list of possible sites. The store owner gathers information about the areas surrounding each place. This could include shoppers' ages, incomes, and educations; where shoppers live; traffic patterns; and other stores in the area. The GIS software then builds a computerized map composed of these layers of information. The store owner can use the information to decide on a new store location.

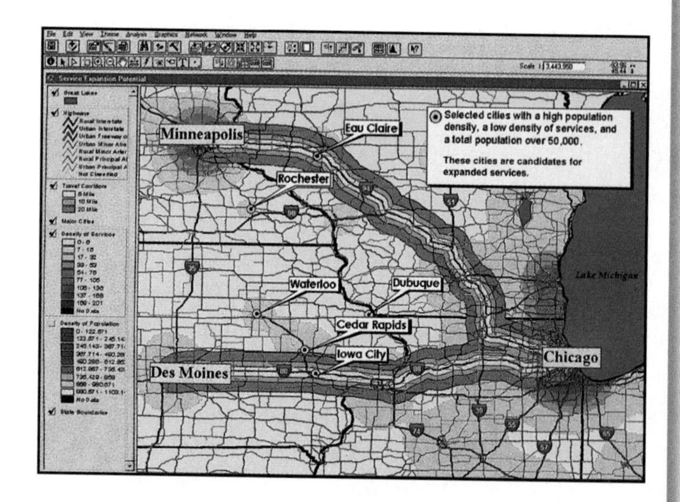

Graphic image created using ArcView® GIS software, and provided courtesy of Environmental Systems Research Institute, Inc.

Making the Connection

1. What is GIS technology?
2. How do GIS programs analyze data in a variety of ways?
3. **Drawing Conclusions** How could a school district use GIS technology to locate the best place to add a new school?

Section 2

The Earth in Space

Guide to Reading

Main Idea

The earth has life because of the sun. The earth has different seasons because of the way it tilts and revolves around the sun.

Terms to Know

- solar system
- orbit
- atmosphere
- axis
- revolution
- leap year
- summer solstice
- winter solstice
- equinox

Places to Locate

- Earth
- sun
- moon

Reading Strategy

Draw a diagram like this one and list three facts about the sun in the first column. In the second, write how these facts contribute to life on the earth.

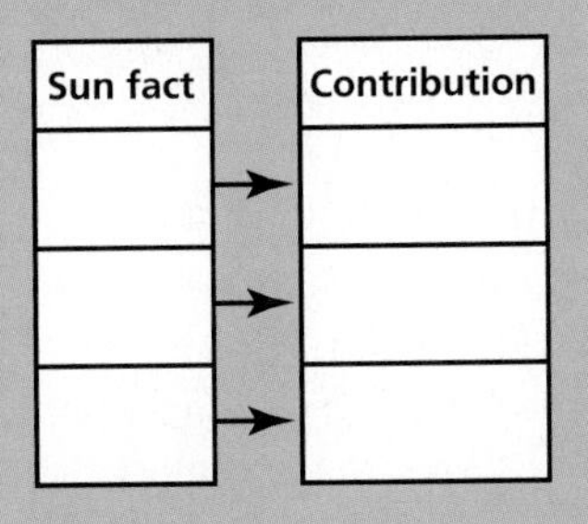

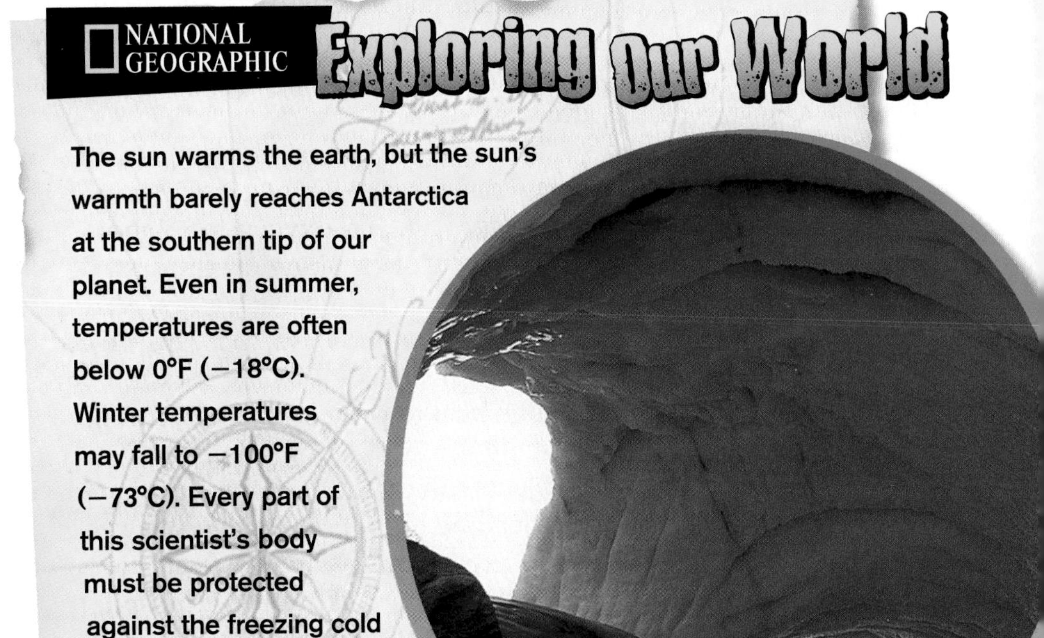

The sun warms the earth, but the sun's warmth barely reaches Antarctica at the southern tip of our planet. Even in summer, temperatures are often below 0°F (−18°C). Winter temperatures may fall to −100°F (−73°C). Every part of this scientist's body must be protected against the freezing cold as he moves through an enormous ice tunnel in Antarctica.

The sun's heat provides life on our planet. **Earth,** eight other planets, and thousands of smaller bodies all revolve around the **sun.** Together with the sun, these bodies form the solar system. Look at the diagram of the **solar system** on page 30. As you can see, Earth is the third planet from the sun.

The Solar System

Each planet travels along its own path, or **orbit,** around the sun. The paths they travel are ellipses, which are like stretched-out circles. Each planet takes a different amount of time to complete one full trip around the sun. Earth makes one trip in 365¼ days. Mercury orbits the sun in just 88 days. Far-off Pluto takes almost 250 years!

Planets can be classified into two types—those that are like Earth and those that are like Jupiter. Earthlike planets are Mercury, Venus, Mars, and Pluto. These planets are solid and small. They have few or no moons. They also rotate, or spin, fairly slowly.

The other four planets—Jupiter, Saturn, Neptune, and Uranus—are huge. Uranus, the smallest of the four, is 15 times the size of Earth.

Solar Eclipse

One of the most spectacular sights in the sky is a solar eclipse. This event takes place when the moon passes between Earth and the sun and covers some or all of the sun. The photograph here shows a total eclipse, when the moon completely blocks the sun. When the moon blocks the sun's light, a large shadow is cast on part of Earth.

These planets are more like balls of gas than rockier Earthlike planets. They spin rapidly and have many moons. Surrounding each one is a series of rings made of bits of rock and dust.

Sun, Earth, and Moon The sun—about 93 million miles (150 million km) from Earth—is made mostly of intensely hot gases. Reactions that occur inside the sun make it as hot as 27 million degrees Fahrenheit (about 15 million degrees Celsius). As a result, the sun gives off light and warmth. Life on Earth could not exist without the sun.

The layer of air surrounding Earth—the **atmosphere**—also supports life. This cushion of gases measures about 1,000 miles (1,609 km) thick. Nitrogen and oxygen form about 99 percent of the atmosphere, with other gases making up the rest.

Humans and animals need oxygen to breathe. The atmosphere is important in other ways, too. This protective layer holds in enough of the sun's heat to make life possible, just as a greenhouse keeps in enough heat to protect plants. Without this protection, Earth would be too cold for most living things. At the same time, the atmosphere also reflects some heat back into space with the result that Earth does not become too warm. Finally, the atmosphere shields living things. It screens out some rays from the sun that are dangerous. You will learn more about the atmosphere in Chapter 2.

Earth's nearest neighbor in the solar system is its **moon.** The moon orbits Earth, taking about 30 days to complete each trip. A cold, rocky sphere, the moon has no water and no atmosphere. The moon also gives off no light of its own. When you see the moon shining, it is actually reflecting light from the sun.

NATIONAL GEOGRAPHIC

The Solar System

Analyzing the Diagram

Earth and eight other planets in our solar system travel around the sun.

Movement **Between what two planets' orbits is Earth's orbit?**

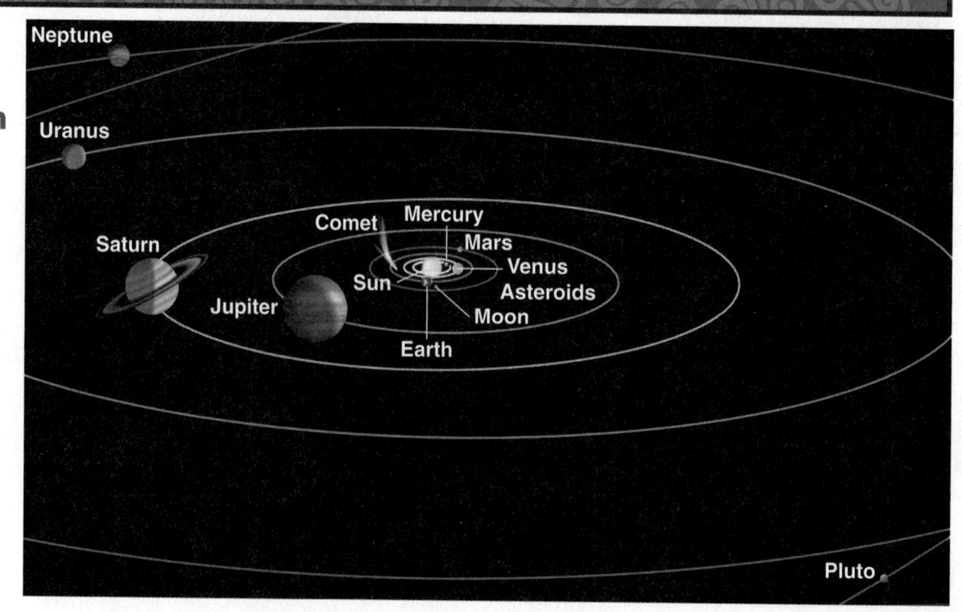

NATIONAL GEOGRAPHIC

Seasons

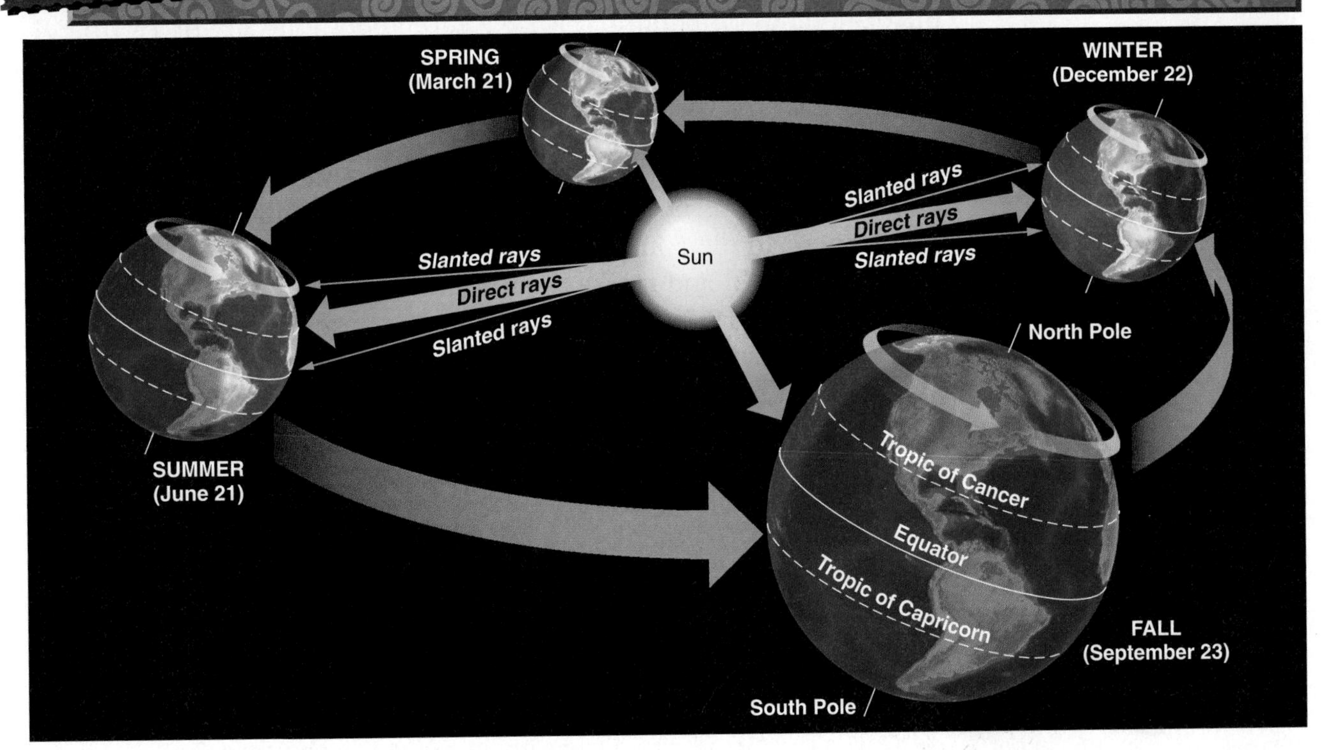

Analyzing the Diagram

Because Earth is tilted, different areas receive direct rays from the sun at different times of the year.

Movement **How does this fact cause changes in seasons?**

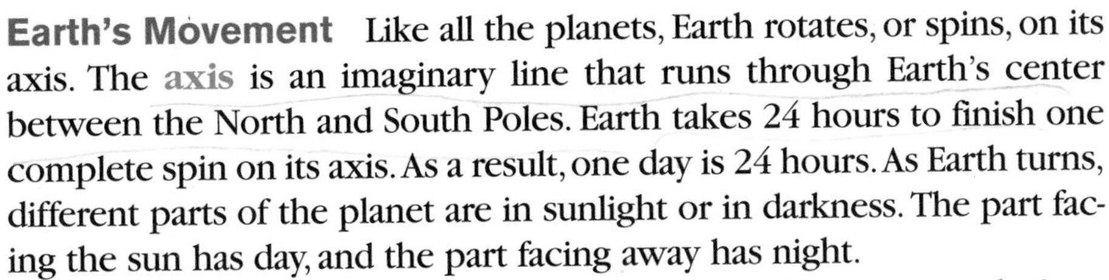

Earth's Movement Like all the planets, Earth rotates, or spins, on its axis. The **axis** is an imaginary line that runs through Earth's center between the North and South Poles. Earth takes 24 hours to finish one complete spin on its axis. As a result, one day is 24 hours. As Earth turns, different parts of the planet are in sunlight or in darkness. The part facing the sun has day, and the part facing away has night.

Earth has another motion, too. The planet makes one **revolution,** or complete orbit around the sun, in 365¼ days. This period is what we define as one year. Every four years, the extra one-fourths of a day are combined and added to the calendar as February 29. A year that contains one of these extra days is called a **leap year.**

✓Reading Check **How does Earth's orbit affect you?**

The Sun and the Seasons

Earth is tilted 23½ degrees on its axis. As a result, seasons change as Earth makes its year-long orbit around the sun. To see why this happens, look at the four globes in the diagram above. Notice how sunlight falls directly on the northern or southern halves of Earth at different times of the year. Direct rays from the sun bring more warmth than the slanted rays. When the people in a hemisphere receive those direct rays from the sun, they enjoy the warmth of summer. When they receive only indirect rays, they experience winter, which is colder.

Solstices and Equinoxes Four days in the year have special names because of the position of the sun in relation to Earth. These days mark the beginnings of the four seasons. On or about June 21, the North Pole is tilted toward the sun. On noon of this day, the sun appears directly overhead at the line of latitude called the Tropic of Cancer (23½°N latitude). This day is the **summer solstice,** the day in the Northern Hemisphere with the most hours of sunlight and the fewest hours of darkness. It is the beginning of summer—but only in the Northern Hemisphere. In the Southern Hemisphere, it is the day with the fewest hours of sunlight and marks the beginning of winter.

Six months later—on or about December 22—the North Pole is tilted away from the sun. At noon, the sun's direct rays strike the line of latitude known as the Tropic of Capricorn (23½°S latitude). In the Northern Hemisphere, this day is the **winter solstice**—the day with the fewest hours of sunlight. This same day, though, marks the beginning of summer in the Southern Hemisphere.

Spring and autumn begin midway between the two solstices. These are the **equinoxes,** when day and night are of equal length in both hemispheres. On or about March 21, the *vernal equinox* (spring) occurs. On or about September 23, the *autumnal equinox* occurs. On both of these days, the noon sun shines directly over the Equator.

Reading Check **Which seasons begin on the two equinoxes?**

Assessment

Defining Terms

1. Define solar system, orbit, atmosphere, axis, revolution, leap year, summer solstice, winter solstice, equinox.

Recalling Facts

2. Region What bodies make up the solar system?

3. Science List two gases that are found in the atmosphere.

4. Movement What two motions does Earth make in space?

Critical Thinking

5. Analyzing Information How does the position of Earth determine whether a day is one of the solstice or equinox days?

6. Summarizing Information In a paragraph, describe why you experience seasonal changes.

Graphic Organizer

7. Organizing Information Draw two diagrams like those below. In the first, list the effects of Earth's rotation on human, plant, and animal life. In the second, list effects if Earth were to stop rotating.

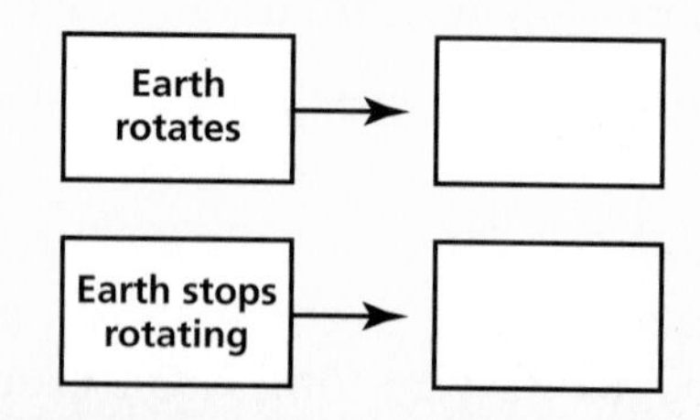

Applying Geography Skills

8. Analyzing Diagrams Look at the diagram on page 31. When the sun's direct rays hit the Tropic of Capricorn, what season is it in the Northern Hemisphere?

Geography Skill

Using a Map Key

To understand what a map is showing, you must read the **map key,** or legend. The map key explains the meaning of special colors, symbols, and lines on the map.

Learning the Skill

Colors in the map key may represent different elevations or heights of land, climate areas, or languages. Lines may stand for rivers, streets, or boundaries.

Maps also have a **compass rose** showing directions. The **cardinal directions** are north, south, east, and west. North and south are the directions of the North and South Poles. If you stand facing north, east is the direction to your right. West is the direction on your left. The compass rose might also show **intermediate directions,** or those that fall between the cardinal directions. For example, the intermediate direction *northeast* falls between north and east. To use a map key, follow these steps:

- Read the map title.
- Read the map key to find out what special information it gives.
- Find examples of each map key color, line, or symbol on the map.
- Use the compass rose to identify the four cardinal directions.

Practicing the Skill

Look at the map of Washington, D.C., below to answer the following questions.

1. What does the red square represent?
2. What does the blue square represent?
3. Does the Washington Monument lie east or west of the Lincoln Memorial?
4. From the White House, in what direction would you go to get to the Capitol?

Applying the Skill

Find a map in a newspaper or magazine. Use the map key to explain three things the map is showing.

GO TO Practice key skills with **Glencoe Skillbuilder Interactive Workbook, Level 1.**

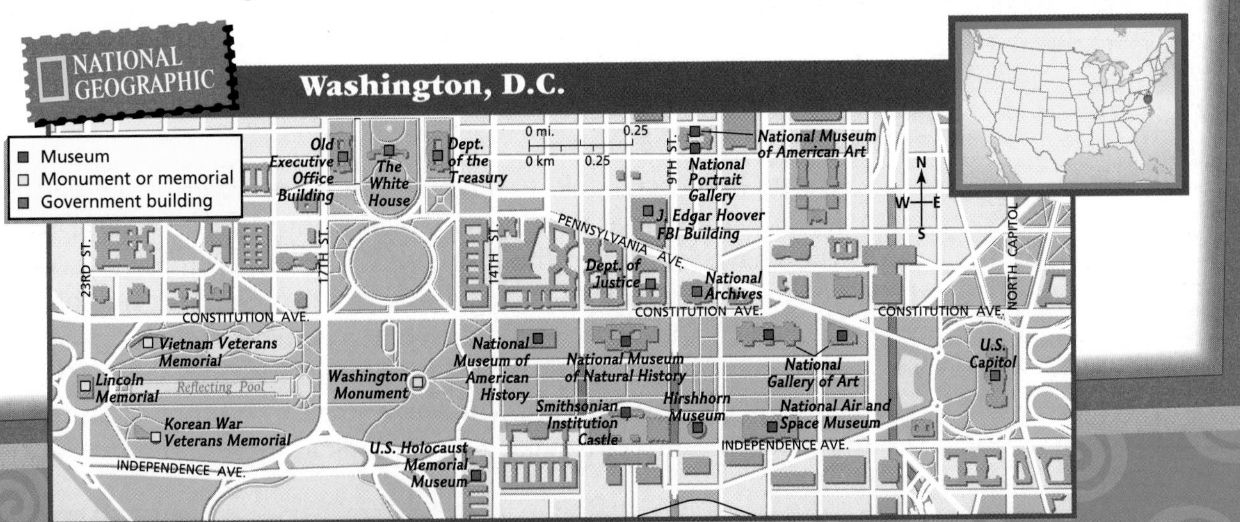

Section 3

The Earth's Structure

Guide to Reading

Main Idea

Forces both inside the earth and on its surface affect the shape of the land.

Terms to Know

- core
- mantle
- magma
- crust
- continent
- plate tectonics
- earthquake
- tsunami
- fault
- weathering
- erosion
- glacier

Places to Locate

- South America
- Africa
- Himalaya
- Asia

Reading Strategy

Create a chart like this one. Write five forces that change the shape of the land. Then write five effects these forces can have on the earth.

Forces	Effects
→	
→	
→	
→	
→	

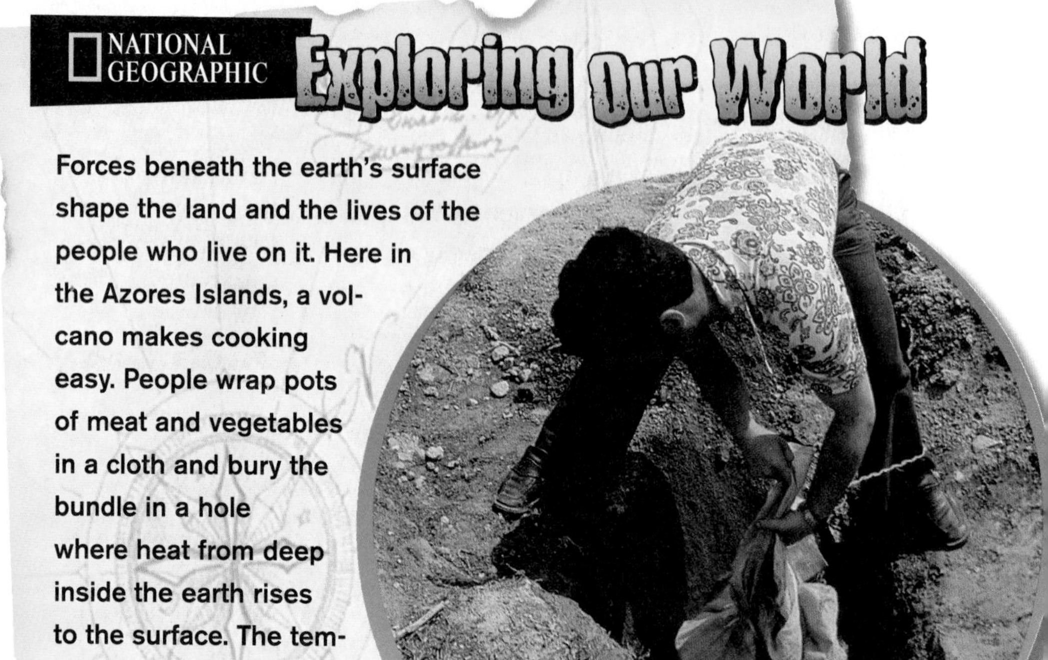

Forces beneath the earth's surface shape the land and the lives of the people who live on it. Here in the Azores Islands, a volcano makes cooking easy. People wrap pots of meat and vegetables in a cloth and bury the bundle in a hole where heat from deep inside the earth rises to the surface. The temperature reaches 200°F (93°C), which is hot enough to steam the food.

It is amazing to think that the earth thousands of miles beneath your feet is so hot that it has turned metal into liquid. Although you may not feel these forces, what lies inside the earth affects what lies on top. Mountains, deserts, and other landscapes were formed over millions of years by forces acting below the earth's surface—and they are still changing today. Some forces work slowly and show no results for thousands of years. Others appear suddenly and have dramatic, and sometimes very destructive, effects.

Inside the Earth

Scientists have been able to study only the top layer of the earth firsthand. Still, they have developed a picture of what lies inside the earth. They have found that the inside of the earth has three layers—the core, the mantle, and the crust. Have you ever seen a cantaloupe cut in half? The earth's core is like the center of the cantaloupe, where you find the seeds. The mantle is like the flesh of the fruit, sandwiched between the core and the skin. The earth's crust, or topmost layer, is like the melon's skin. Let us take a closer look at these three layers.

In the center of the earth is a dense core of hot iron mixed with other metals. The very center is solid, but the outer core is so hot that the metal has melted into liquid. Surrounding the core is the mantle, a layer of rock about 1,800 miles (2,897 km) thick. Like the core, the mantle also has two parts. The section nearest the core remains solid, but the rock in the outer mantle sometimes melts. If you have seen photographs of an active volcano, then you have seen this melted rock, called magma, when it flowed to the surface in a volcanic eruption.

The uppermost layer of the earth, the crust, is relatively thin. It reaches only 31 to 62 miles (50 to 100 km) deep. The crust includes the ocean floors. It also includes seven massive land areas known as continents. The crust is thinnest on the ocean floor. It is thicker below the continents. Turn to the map on page 41 to see where the earth's seven continents are located.

Reading Check **What layer of the earth is thinnest?**

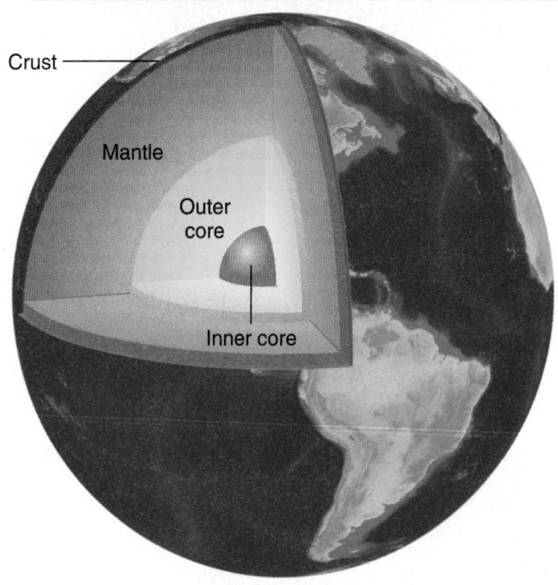

Analyzing the Diagram

Hot rock and metal—some of it liquid—fill the center of the earth.

Region **What is the innermost layer inside the earth called? In what layer do you find the continents?**

Forces Beneath the Earth's Crust

Most of you have probably watched science shows about earthquakes and volcanoes. You have probably also seen news on television discussing the destruction caused by earthquakes. These events result from forces at work inside the earth.

Plate Movements Scientists have developed a theory about the earth's structure called plate tectonics. This theory states that the crust is not an unbroken shell but consists of plates, or huge slabs of rock, that move. The plates float on top of the liquid rock in the upper part of the mantle. They move—but often in different directions. Oceans and continents sit on these gigantic plates, as the diagram on page 36 shows.

Have you ever noticed that the eastern part of **South America** seems to fit into the western side of **Africa?** That is because these two continents were once joined together in a landmass that scientists call Pangaea. Millions of years ago, however, the continents moved apart. Tectonic activity caused them to move. The plates are still moving today, but they move so slowly that you do not feel it. The plate under the Pacific Ocean moves to the west at the rate of about 4 inches (10 cm) a year. That is about the same rate that a man's beard grows. The plate along the western edge of South America moves east at the rate of about 1.8 inches (5 cm) a year. That is a little faster than your finger nails grow. Turn to page 45 to see what Pangaea looked like before and after it experienced this movement, known as *continental drift.*

Tectonic Plate Boundaries

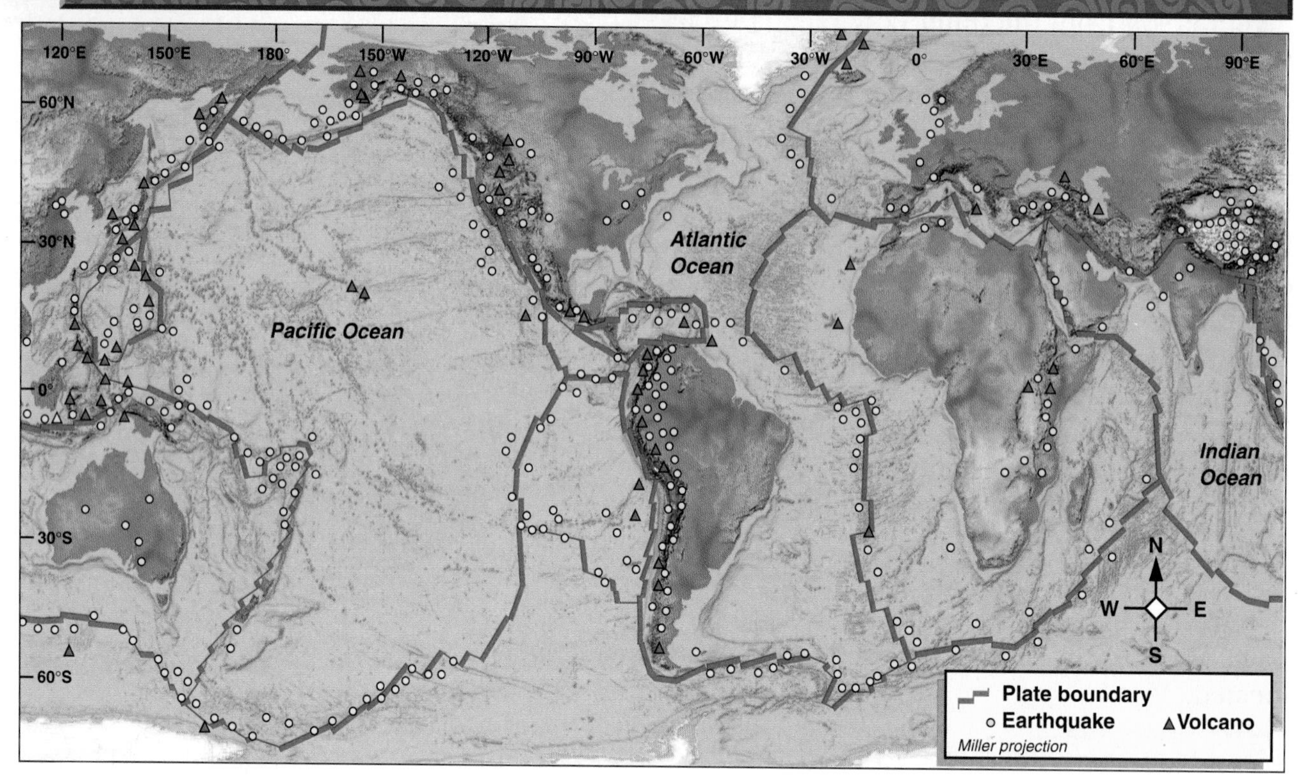

Analyzing the Diagram

Most of North America sits on one plate.

Region What pattern do you see among plate boundaries, earthquakes, and volcanoes?

most earthquakes and volcanoes occur along plate boundaries.

When Plates Meet The movements of the earth's plates have actually shaped the surface of the earth. Sometimes the plates spread, or pull away from each other. That type of tectonic action separated South America and Africa millions of years ago. Sometimes, though, the plates push against each other. When this happens, one of three events occurs, depending on what kinds of plates are involved.

If two continental plates smash into each other, the collision produces high mountain ranges. This kind of collision produced the **Himalaya** in South **Asia.**

If a continental plate and an ocean plate move against each other, the thicker continental plate slides over the thinner ocean plate. The edge of the lower plate melts into the mantle, causing molten rock to build up and perhaps erupt in a volcano. Another result may occur from the pressure that builds up between the two sliding plates. This pressure may cause one plate to move suddenly. The result is called an **earthquake,** or a violent and sudden movement of the earth's crust.

Earthquakes can be very damaging to both physical structures and human lives. They can collapse buildings, destroy bridges, and break apart underground water or gas pipes. Undersea earthquakes can cause huge waves called **tsunamis** (tsu•NAH•mees). These waves may reach as high as 98 feet (30 m). Such waves can cause severe flooding of coastal towns.

Sometimes two plates do not meet head-on but move alongside each other. To picture this, put your hands together and then move them in opposite directions. When this action occurs in the earth, the two plates slide against each other. This movement creates **faults**, or cracks in the earth's crust. Violent earthquakes can happen near these faults. In 1988, for example, an earthquake struck the country of Armenia. About 25,000 people were killed, and another 500,000 lost their homes. One of the most famous faults in the United States is the San Andreas Fault in California. The earth's movement along this fault caused a severe earthquake in San Francisco in 1906 and another less serious quake in 1989.

✓Reading Check **What happens when two continental plates collide?**

Forces Shaping Landforms

The forces under the earth's crust that move tectonic plates cause volcanoes and earthquakes to change the earth's landforms. Once formed, however, these landforms will continue to change because of forces that work on the earth's surface.

Weathering **Weathering** is the process of breaking surface rock into boulders, gravel, sand, and soil. Water and frost, chemicals, and even plants cause weathering. Water seeps into cracks of rocks and then freezes. As it freezes, the ice expands and splits the rock. Sometimes entire sides of cliffs fall off because frost has wedged the rock apart. Chemicals, too, cause weathering when acids in air pollution mix with rain and fall back to the earth. The chemicals eat away the surfaces of stone structures and natural rocks. Even tiny seeds that fall into cracks can spread out roots, causing huge boulders to eventually break apart.

Architecture

In earthquake-prone parts of the world, engineers design new buildings to stand up to tremors, or shaking of the earth. Flexible structures allow buildings to sway rather than break apart. Placing a building on pads or rollers cushions the structure from the motion of the ground. Some so-called intelligent buildings automatically respond to tremors, shifting their weight or tightening and loosening joints.

Looking Closer **How can studying earthquake-damaged buildings help designers improve future construction?**

San Francisco, California, 1989 ▶

Erosion Erosion is the process of wearing away or moving weathered material. Water, wind, and ice are the greatest factors that erode, or wear away, surface material. Moving water in oceans, rivers, streams, and rain can erode even the hardest stone over time. Rainwater working its way to streams and rivers picks up and moves soil and sand. These particles make the river water similar to a giant scrub brush that grinds away at riverbanks and any other surface in the water's path.

Wind is also a major cause of erosion as it lifts weathered soil and sand. The areas that lose soil often become unable to grow crops and support life. The areas that receive the windblown soil often benefit from the additional nutrients to the land. When wind carries sand, however, it acts as sandpaper. Rock and other structures are carved into smooth shapes.

The third cause of erosion is ice. Giant, slow-moving sheets of ice are called glaciers. Forming high in mountains, glaciers change the land as they inch over it. Similar to windstorms, glaciers act like sandpaper as they pick up and carry rocks down the mountainside, grinding smooth everything beneath them. Some glaciers are thousands of feet thick. The weight and pressure of thousands of feet of ice also cut deep valleys at the mountain's base.

✓Reading Check **List three things that can cause weathering.**

Assessment

Defining Terms

1. **Define** core, mantle, magma, crust, continent, plate tectonics, earthquake, tsunami, fault, weathering, erosion, glacier.

Recalling Facts

2. **Region** What are the three layers inside the earth?
3. **Movement** In what three ways can tectonic plates move?
4. **Science** What are the three greatest factors that cause erosion?

Critical Thinking

5. **Making Comparisons** How does water play a role in both processes of weathering and erosion?
6. **Understanding Cause and Effect** How does erosion hurt some areas yet benefit others?

Graphic Organizer

7. **Organizing Information** Draw a diagram like this one, then label the inner arrows with inside forces that shape landforms. Label the outer arrows with surface forces that change the earth's landforms.

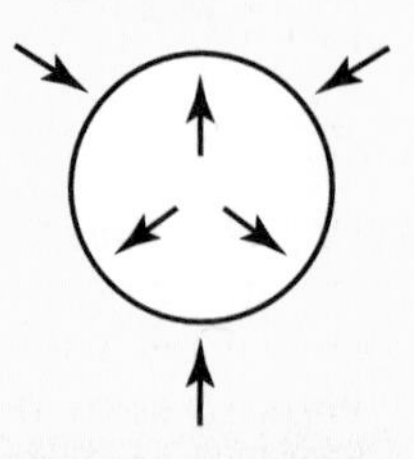

Applying Geography Skills

8. **Analyzing Diagrams** Look at the diagram of tectonic plate boundaries on page 36. Why might it be a problem that most of the world's population lives along the western edge of the Pacific Ocean?

Guide to Reading

Main Idea

Landforms in all their variety affect how people live.

Terms to Know

- elevation
- plain
- plateau
- isthmus
- peninsula
- island
- continental shelf
- trench
- strait
- channel
- delta

Places to Locate

- Mt. Everest
- North European Plain
- Plateau of Tibet
- Grand Canyon
- Mariana Trench
- Strait of Magellan

Reading Strategy

Draw a diagram like this one. In each of the surrounding circles, write the name of a landform and a fact about it.

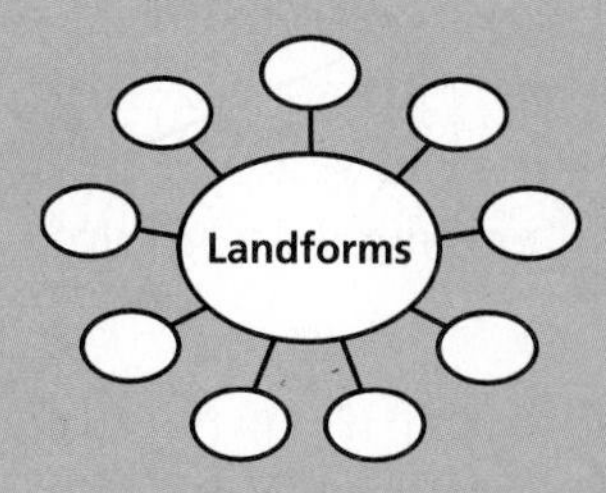

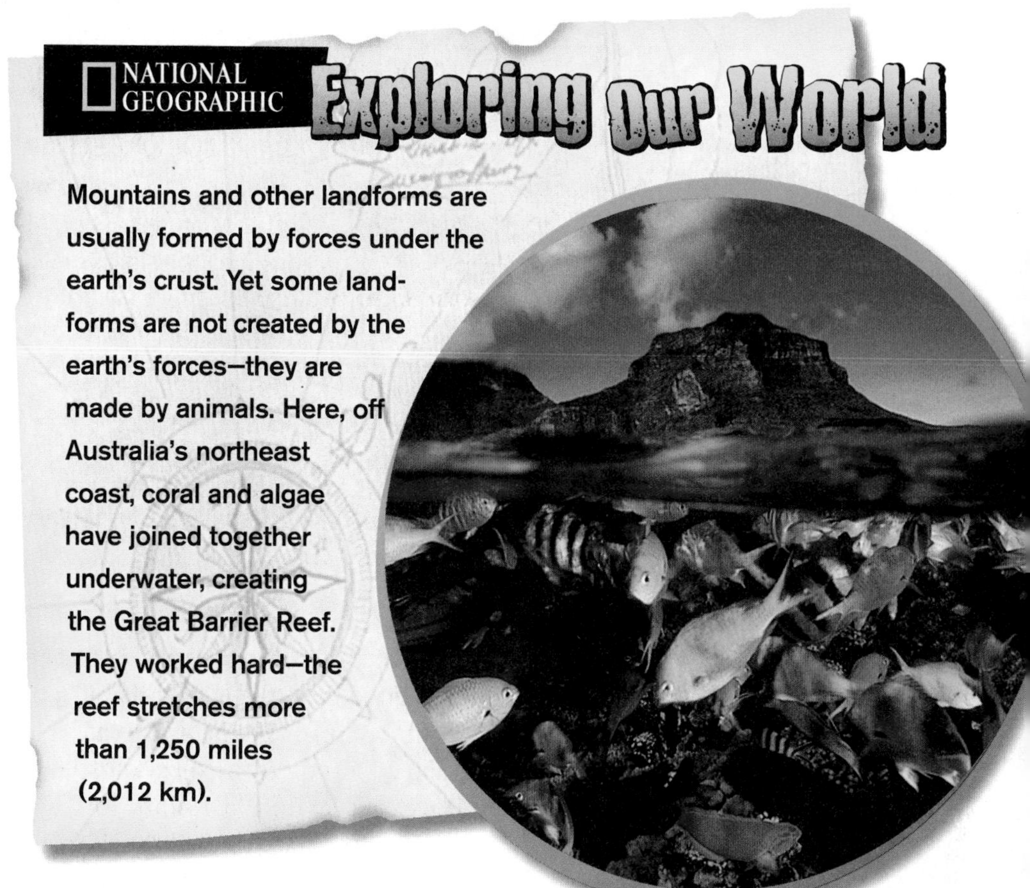

Mountains and other landforms are usually formed by forces under the earth's crust. Yet some landforms are not created by the earth's forces—they are made by animals. Here, off Australia's northeast coast, coral and algae have joined together underwater, creating the Great Barrier Reef. They worked hard—the reef stretches more than 1,250 miles (2,012 km).

The earth's land surface consists of seven continents. These are North America, South America, Europe, Africa, Asia, Australia, and Antarctica. All have a variety of landforms, or individual features of the land—even icy Antarctica.

Types of Landforms

Look at the illustration on page 18 of the **Geography Handbook.** Notice the many different forms that the land may take. Which ones are familiar to you? Which ones are new to you?

On Land Mountains are huge towers of rock formed by the collision of the earth's tectonic plates or by volcanoes. Some mountains may be a few thousand feet high. Others can soar higher than 20,000 feet (6,096 m). The world's tallest mountain is **Mt. Everest,** located in South Asia's Himalaya mountain ranges. It towers at 29,035 feet (8,850 m)—nearly 5.5 miles (8.9 km) high.

Mountains often have high peaks and steep, rugged slopes. Hills are lower and more rounded, though they are still higher than the land

NATIONAL GEOGRAPHIC **On Location**

Valleys vs. Canyons

The Great Rift Valley in Africa is surrounded by mountains (above). Canyons, like the Grand Canyon in Arizona (right), are carved from plateaus.

Place **How are valleys and canyons similar?**

Both are lowlands, but canyons are more steep-sided.

around them. Some hills form at the foot, or base, of mountains. As a result, these hills are called foothills.

In contrast, plains and plateaus are mostly flat. What makes them different from one another is their **elevation,** or height above sea level. **Plains** are low-lying stretches of flat or gently rolling land. Many plains reach from the middle of a continent to the coast. The **North European Plain** is an example. **Plateaus** are also flat but have higher elevation. With some plateaus, a steep cliff forms on one side where the plateau rises above nearby lowlands. With others, such as the **Plateau of Tibet** in Asia, the plateau is surrounded by mountains.

Between mountains and hills lie valleys. A valley is a long stretch of land lower than the land on either side. You often find rivers at the bottom of valleys. Canyons are steep-sided lowlands that rivers have cut through a plateau. One of the most famous canyons is the **Grand Canyon** in Arizona. For millions of years, the Colorado River flowed over a plateau and carved through rock, forming the Grand Canyon.

Geographers describe some landforms by their relationship to larger land areas or to bodies of water. An **isthmus** is a narrow piece of land that connects two larger pieces of land. A **peninsula** is a piece of land with water on three sides. A body of land smaller than a continent and completely surrounded by water is an **island.**

Under the Oceans If you were to explore the oceans, you would see landforms under the water that are similar to those on land. The map on page RA32 of the **Reference Atlas** shows you what the ocean floors look like. Off each coast of a continent lies a plateau called a continental shelf that stretches for several miles underwater. At the edge of the shelf, steep cliffs drop down to the ocean floor.

Tall mountains and very deep valleys line the ocean floor. Valleys here are called trenches, and they are the lowest spots in the earth's crust. The deepest one, in the western Pacific Ocean, is called the **Mariana Trench.** This trench plunges 35,840 feet (10,924 m) below sea level. How deep is that? If Mt. Everest were placed into this trench, the mountain would have to grow 1.3 miles (2 km) higher just to reach the ocean's surface.

Landforms and People Humans have settled on all types of landforms. Some people live at high elevations in the Andes mountain ranges of South America. The people of Bangladesh live on a low coastal plain. Farmers in Ethiopia work the land on a plateau called the Ethiopian Highlands.

Why do people decide to live in a particular area? Climate—the average temperature and rainfall of a region—is one reason. You will read more about climate in the next chapter. The availability of resources is another reason. People settle where they can get freshwater and where they can grow food, catch fish, or raise animals. They might settle in an area because it has good supplies of useful items such as trees for building, iron for manufacturing, or petroleum for making energy. You will read more about resources in Chapter 3.

✓Reading Check **How are plains and plateaus similar? How are they different?**

Applying Map Skills

1. What are the names of the seven large landmasses on the earth?
2. What are the earth's four major oceans?

Find NGS online map resources @ www.nationalgeographic.com/maps

NATIONAL GEOGRAPHIC

World Continents and Oceans

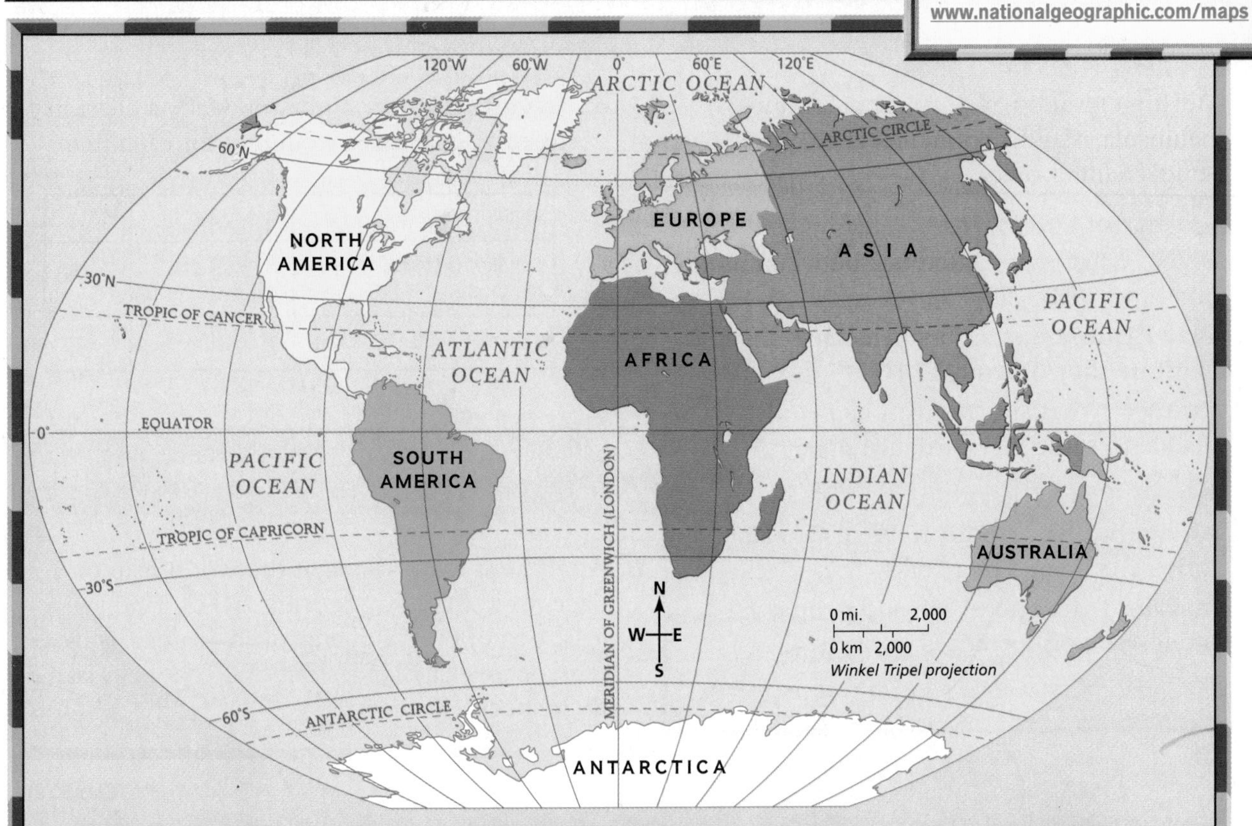

Bodies of Water

About 70 percent of the earth's surface is water. Most of that water is salt water, which people and most animals cannot drink. Only a small percentage is freshwater, which is drinkable.

Oceans, consisting of salt water, are the earth's largest bodies of water. Smaller bodies of salt water are connected to oceans but are at least partly enclosed by land. These bodies include seas, gulfs, and bays.

Two other kinds of water form passages that connect two larger bodies of water. A **strait** is a narrow body of water between two pieces of land. The **Strait of Magellan** flows between the southern tip of South America and an island called Tierra del Fuego (tee•EHR•uh DEHL fyu•AY•goh). This strait connects the Atlantic and the Pacific Oceans. A wider passage is called a **channel.** The Mozambique Channel separates southeastern Africa from the island of Madagascar.

Bodies of freshwater appear on the world's continents and islands. They include larger bodies like lakes and rivers as well as smaller ones such as ponds and streams. The point at which a river originates—usually high in the mountains—is called its source. The mouth of a river is where it empties into another body of water. As you learned in Section 3, rivers carry soil and sand. They eventually deposit this soil at the mouth, which builds up over time to form a **delta.**

✓Reading Check **What is the difference between the source and the mouth of a river?**

Assessment

Defining Terms

1. **Define** elevation, plain, plateau, isthmus, peninsula, island, continental shelf, trench, strait, channel, delta.

Recalling Facts

2. **Place** What is the difference between mountains and hills?
3. **Place** How are straits and channels similar? How are they different?
4. **Culture** What are two reasons people decide to settle in a particular area?

Critical Thinking

5. **Analyzing Information** What two landforms are created by rivers?
6. **Making Inferences** Why do people often settle on the edges of rivers?

Graphic Organizer

7. **Organizing Information** Make a chart like this and give three examples for each item.

Landforms			
Landforms Under the Ocean			
Types of Bodies of Water			

Applying Geography Skills

8. **Analyzing Maps** Look at the world map on page RA4 of the **Reference Atlas.** Find at least two of the following: plain, plateau, isthmus, peninsula, island, strait, and channel. List their specific names.

Reading Review

Section 1 Thinking Like a Geographer

Terms to Know
geography
landform
environment
region
Global Positioning System (GPS)
geographic information systems (GIS)

Main Idea
Geographers use various tools to understand the world.

✓ **Place** Geographers study the physical and human characteristics of places.

✓ **Culture** Geographers are especially interested in how people interact with their environment.

✓ **Technology** To study the earth, geographers use maps, globes, photographs, the Global Positioning System, and geographic information systems.

✓ **Economics** People can use information from geography to plan, make decisions, and manage resources.

Section 2 The Earth in Space

Terms to Know
solar system
orbit
atmosphere
axis
revolution
leap year
summer solstice
winter solstice
equinox

Main Idea
The earth has life because of the sun. The earth has different seasons because of the way it tilts and revolves around the sun.

✓ **Science** The sun's light and warmth allow life to exist on Earth.

✓ **Science** The atmosphere is a cushion of gases that protects Earth and provides air to breathe.

✓ **Movement** Earth spins on its axis to make day and night.

✓ **Movement** The tilt of Earth and its revolution around the sun cause changes in seasons.

Section 3 The Earth's Structure

Terms to Know
core
mantle
magma
crust
continent
earthquake
plate tectonics
tsunami
fault
weathering
erosion
glacier

Main Idea
Forces both inside the earth and on its surface affect the shape of the land.

✓ **Region** Earth has an inner and outer core, a mantle, and a crust.

✓ **Movement** The continents are on large plates of rock that move.

✓ **Movement** Earthquakes and volcanoes can reshape the land.

✓ **Movement** Wind, water, and ice can change the look of the land.

Section 4 Landforms

Terms to Know
elevation
plain
plateau
isthmus
peninsula
island
continental shelf
trench
strait
channel
delta

Main Idea
Landforms in all their variety affect how people live.

✓ **Location** Mountains, plateaus, valleys, and other landforms are found on land and under the oceans.

✓ **Science** About 70 percent of the earth's surface is water.

✓ **Culture** People have adapted in order to live on various landforms.

Chapter 1

Assessment and Activities

Using Key Terms

Match the terms in Part A with their definitions in Part B.

A.

1. elevation
2. landform
3. summer solstice
4. plate tectonics
5. geographic information systems
6. Global Positioning System
7. erosion
8. equinox
9. fault
10. weathering

B.

a. height above sea level
b. wearing away of the earth's surface
c. theory that the earth's crust consists of huge slabs of rock that move
d. a group of satellites around the earth
e. special software that helps geographers gather and use information
f. when day and night are of equal length
g. a process that breaks surface rocks into gravel, sand, or soil
h. a crack in the earth's crust
i. the day with the most hours of sunlight
j. particular features of the land

Reviewing the Main Ideas

Section 1 Thinking Like a Geographer

11. **Place** Give three examples of the physical characteristics of a place.
12. **Region** How is a region different from a place?
13. **Human/Environment Interaction** Give an example of how people use geographic knowledge.

Section 2 The Earth in Space

14. **Region** How many planets are in the solar system?
15. **Movement** What movement of Earth causes day and night?
16. **Movement** How does Earth's revolution around the sun relate to the seasons?

Section 3 The Earth's Structure

17. **Movement** How do the plates in the earth's crust move?
18. **Movement** Give an example of erosion.

Section 4 Landforms

19. **Place** Which has a higher elevation—plains or plateaus?
20. **Movement** What two reasons lead people to settle in a particular region?

NATIONAL GEOGRAPHIC

The World

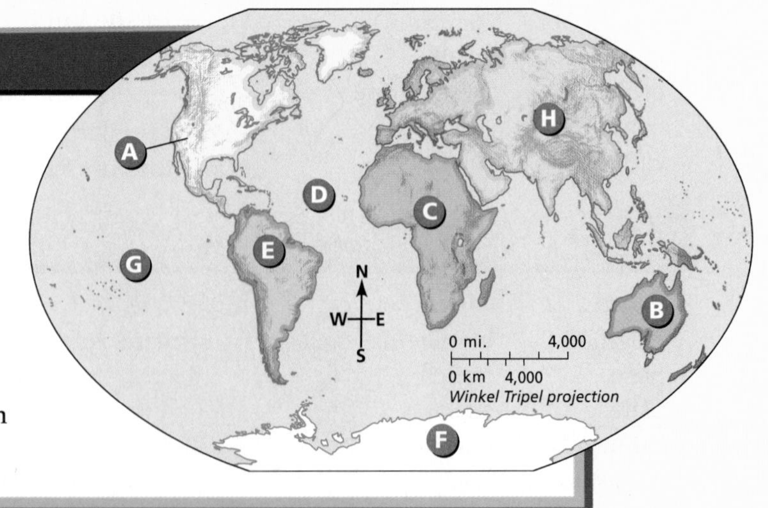

Place Location Activity

On a separate sheet of paper, match the letters on the map with the numbered places listed below.

1. North America
2. Pacific Ocean
3. Africa
4. South America
5. Antarctica
6. Australia
7. Atlantic Ocean
8. Asia

Self-Check Quiz Visit the ***Geography: The World and Its People*** Web site at gwip.glencoe.com and click on **Chapter 1—Self-Check Quizzes** to prepare for the Chapter Test.

Critical Thinking

21. **Making Comparisons** Why do people in Australia snow-ski during the Northern Hemisphere's summer months?
22. **Understanding Cause and Effect** Create a diagram like this one. In the left box, write "plate movements." In the right box, describe the effect that this force has on the earth. Draw five more pairs of boxes and do the same for the other forces that shape the earth: earthquakes, volcanoes, weathering, and erosion.

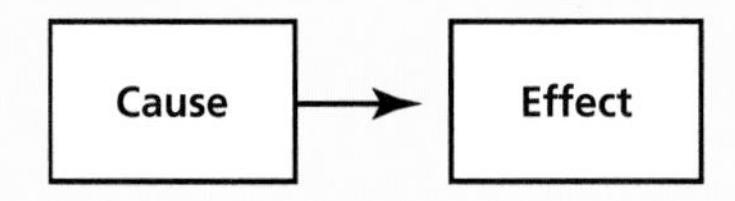

GeoJournal Activity

23. **Writing a Paragraph** Write a description of how the earth looks in your area. Then sketch a picture of your area showing all the landforms and bodies of water you described. Label each landform and body of water that you show.

Mental Mapping Activity

24. **Focusing on the Region** Draw a simple outline map of the earth, then label the following:
    - core
    - mantle
    - crust
    - atmosphere

Technology Skills Activity

25. **Building a Database** Use a word processing program to make a database like the following. In each row of the left column, list one of the following landforms: mountain, hill, plain, plateau, valley, canyon, isthmus, peninsula, and island. In the other columns, write a fact about each landform and give an example.

Landform	Facts	Example

Standardized Test Practice

Directions: Study the maps below, then answer the question that follows.

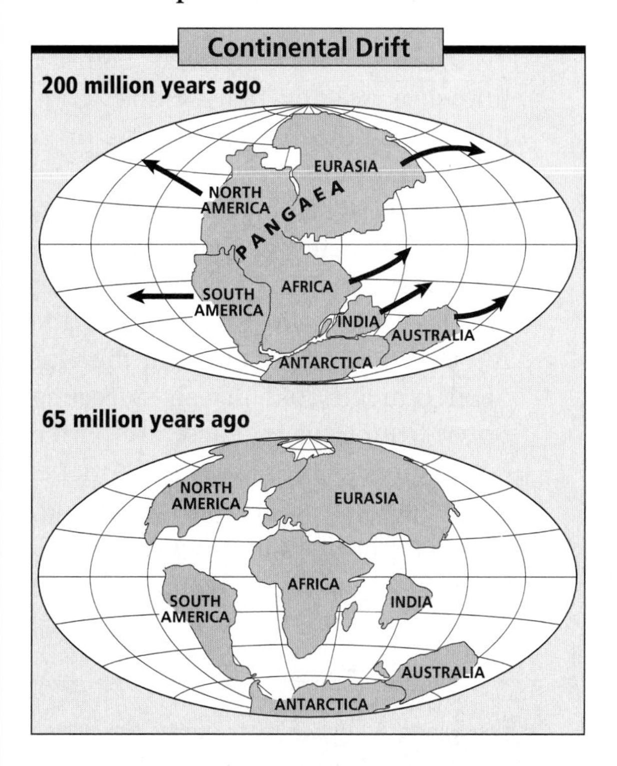

1. **What "supercontinent" do many scientists believe existed 200 million years ago?**
   A Eurasia
   B Pangaea
   C Gondwana
   D Antarctica

Test-Taking Tip: Use information *on the maps* to answer this question. Read the title above the maps and then the two subtitles. If you reread the question, you see it is asking about a certain time period. Make sure you use the correct map above to answer the question.

GeoLAB ACTIVITY

Tornadoes: Swirling Fury

1 Background

It is a stormy, humid day. You see a funnel of twisting, black clouds coming down from the sky and stirring up dust on the ground below. Tornado warning: Take cover! The United States has more tornadoes each year than anywhere else on the planet. Over the central part of our country—nicknamed Tornado Alley—warm, moist air from the Gulf of Mexico meets cold, dry air sweeping down from Canada. These are perfect conditions for tornadoes to form. Learn more about these storms by making and experimenting with a model of a tornado.

2 Materials

- **2 empty 2-liter clear soda bottles**
- **black marker**
- **1 rubber washer that is the same size as the bottle openings**
- **duct tape**
- **water**
- **paper towels**
- **glitter**

A Tornado Hits Oakfield, Wisconsin

Believe It or Not!

The United States has the most violent weather in the world. Our country experiences 1,000 tornadoes, 5,000 floods, and 10,000 strong thunderstorms in an average year.

What to Do

1. Use the marker to label one bottle "A" and the other "B."
2. Fill Bottle A three-fourths full of water. Dry the top and outside of the bottle completely. Add a small amount of glitter to the water.
3. Tape the rubber washer to the opening of Bottle A. Do not cover the washer hole with tape.
4. Put Bottle B upside down on Bottle A. Tape the mouths of both bottles together. Make sure the seal is tight and use several layers of tape.
5. Turn the bottles over so that Bottle A is on top. Then quickly swirl the bottles a couple of times, as you would stir something in a bowl.
6. Set the bottles on a table with Bottle A on top and observe the "tornado." The spinning funnel you see is called a *vortex.*
7. Repeat the experiment several times. Compare how water near the center and edges of the vortex moves each time. Observe what happens when you spin the bottles faster or slower.

4 LAB ACTIVITY REPORT

1. Describe what you saw when you swirled the bottles and set them down.
2. Does water near the center of the vortex move any differently than water at the edges of the vortex? Explain.
3. What did you observe when you swirled the bottles faster or slower?
4. **Drawing Conclusions** Did the vortex always spin in the same direction? What should you do to make it spin in the opposite direction?

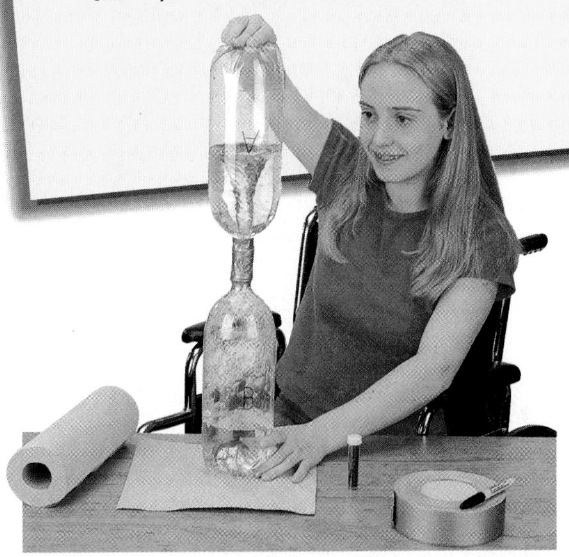

▲ **Glitter added to the water allows you to see the vortex more clearly.**

Extending the Lab

Activity

Find out what you should do if a tornado warning is announced in your area. Discover what to do if you are outside, in a car, or inside a building. Use the information you gather to create a tornado safety poster. Present your poster to the class.

Chapter 2

Water, Climate, and Vegetation

The World and Its People NATIONAL GEOGRAPHIC

To learn more about water, climate, and vegetation, view ***The World and Its People*** **Chapter 2** video.

Geography online

Chapter Overview Visit the ***Geography: The World and Its People*** Web site at gwip.glencoe.com and click on **Chapter 2—Chapter Overviews** to preview information about water, climate, and vegetation.

The Water Planet

Guide to Reading

Main Idea

Water is one of the earth's most precious resources.

Terms to Know

- water vapor
- water cycle
- evaporation
- condensation
- precipitation
- collection
- glacier
- groundwater
- aquifer

Places to Locate

- Pacific Ocean
- Atlantic Ocean
- Indian Ocean
- Arctic Ocean

Reading Strategy

Make a diagram like this one. Starting at the top, write the steps of the water cycle—each in a separate square—in the correct sequence.

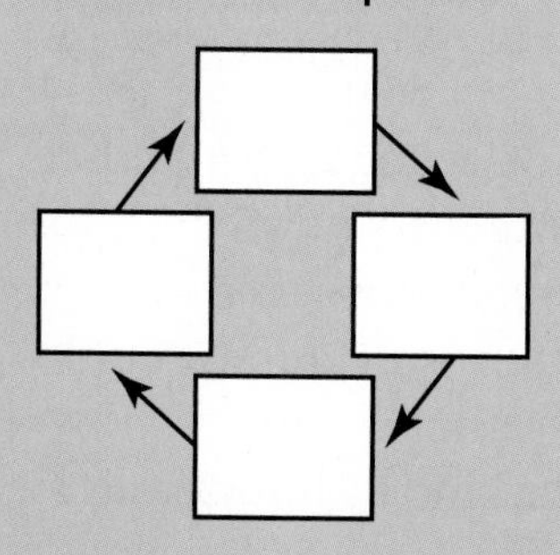

NATIONAL GEOGRAPHIC **Exploring Our World**

Humans, trees, other plants, and animals all need water. We cannot survive without it. Here, a swimmer enjoys a remarkable sight—a permanent pool of water in the middle of Mexico's Chihuahuan Desert. The water bubbles to the surface from an underground spring. The clear water does more than attract swimmers, though. It supports a variety of animal and plant life.

Some people call Earth "the water planet." Why? Water covers about 70 percent of the earth's surface. Water exists all around you in many different forms. Streams, rivers, lakes, seas, and oceans contain water in liquid form. The atmosphere holds **water vapor,** or water in the form of gas. Glaciers and ice sheets are masses of water that have been frozen solid. As a matter of fact, the human body itself is about 60 percent water.

The Water Cycle

The total amount of water on the earth does not change, but it does not stay in one place, either. Instead, the water moves constantly. In a process called the **water cycle,** the water goes from the oceans to the air to the ground and finally back to the oceans.

Look at the diagram on page 50 to see how the water cycle works. The sun drives the cycle by evaporating water mostly from the surface of oceans, but also from lakes and streams. In **evaporation,** the sun's heat turns liquid water into water vapor—also called *humidity.* The amount of water vapor that the air holds depends on the air

◀ The Yellowstone River in Paradise Valley, Montana

The Water Cycle

Analyzing the Diagram

The water cycle involves evaporation, condensation, precipitation, and the collection of water above and below the ground.

Movement How does water get from the ground to the oceans?

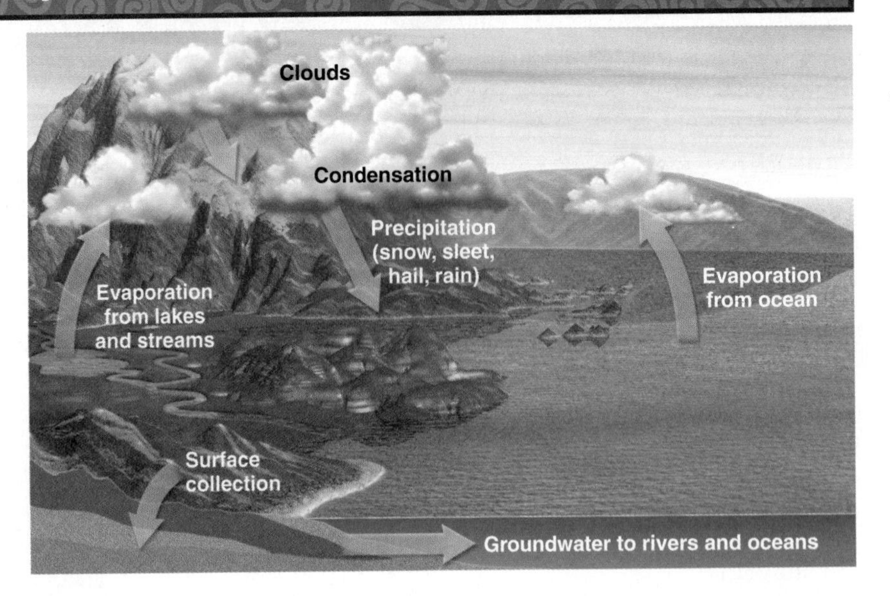

temperature. Warm air can hold more humidity than cool air, which you have probably felt on warm, muggy summer days.

In addition, warm air tends to rise. As warm air rises higher in the atmosphere, it cools, thus losing its ability to hold as much humidity. As a result, the water vapor changes back into a liquid in a process called **condensation.** Tiny droplets of water come together to form clouds. Eventually, the water falls back to the earth as some form of **precipitation**—rain, snow, sleet, or hail—depending on the temperature of the surrounding air.

When this precipitation reaches the earth's surface, it soaks into the ground and collects in streams and lakes. During **collection,** streams and rivers both above and below the ground carry the water back to the oceans, and the cycle begins again.

✓Reading Check **What kind of air—warm or cold—holds the most water vapor?**

Water Resources

It is a hot day, and you rush home for a glass of water. Like all other people—and all plants and animals—you need water to survive. Think about the many ways you use water in just a single day—to bathe, to brush your teeth, to cook your food, and to quench your thirst. People and most animals need freshwater to live. Many other creatures, however, make their homes in the earth's much larger kind of liquid water: salt water.

Freshwater Only about 2 percent of the water on the earth is freshwater. Eighty percent of that freshwater is frozen in **glaciers,** or giant sheets of ice. Only a tiny fraction of the world's freshwater—not even four-hundredths of a percent—is found in lakes and rivers.

When you think of freshwater, you probably think of mighty rivers and huge lakes. People can get freshwater from another source, though. Groundwater is water that fills tiny cracks and holes in the rock layers below the surface of the earth. This is a vital source of water because there is 10 times more groundwater than there is water in rivers and lakes. Groundwater can be tapped by wells. Some areas have aquifers, or underground rock layers that water flows through. In regions with little rainfall, both farmers and city dwellers sometimes have to depend on aquifers and other groundwater for most of their water supply.

Salt Water All the oceans on the earth are part of a huge, continuous body of salt water—almost 98 percent of the planet's water. Look at the map on page 58. You will see that the four major oceans are the **Pacific Ocean,** the **Atlantic Ocean,** the **Indian Ocean,** and the **Arctic Ocean.**

The Pacific Ocean is the largest and deepest of these four oceans. It covers almost 64 million square miles (166 million sq. km)—more than all the land areas of the earth combined. As you learned in Chapter 1, bodies of salt water smaller than the oceans are called seas, gulfs, bays, or straits. Look back at the diagram on page 18 of the Geography Handbook to see these features again.

✓Reading Check **What is the difference between groundwater and aquifers?**

Assessment

Defining Terms

1. **Define** water vapor, water cycle, evaporation, condensation, precipitation, collection, glacier, groundwater, aquifer.

Recalling Facts

2. **Region** What percentage of the earth is covered by water?
3. **Movement** In which part of the water cycle does water return to the earth?
4. **Region** What are the world's four oceans?

Critical Thinking

5. **Understanding Cause and Effect** How does the temperature of the air affect the amount of humidity that you feel? How does the air's temperature also influence the form of precipitation that falls?
6. **Drawing Conclusions** Why do you think it is important to keep groundwater free of dangerous chemicals?

Graphic Organizer

7. **Organizing Information** Draw a diagram like this one. List at least four sources of freshwater and salt water on the lines under each heading.

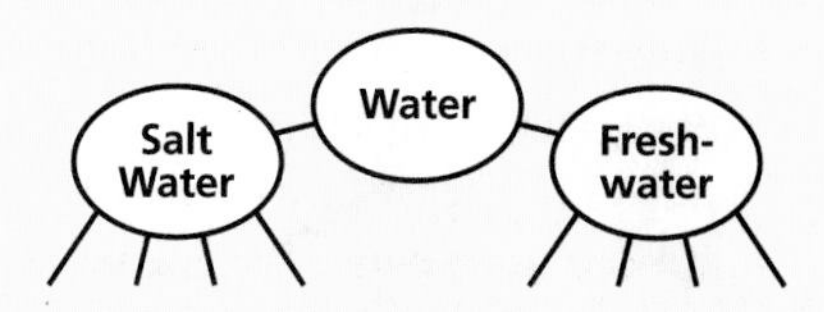

Applying Geography Skills

8. **Analyzing Diagrams** Look at the diagram of the water cycle on page 50. From where does water evaporate?

Making Connections

ART | SCIENCE | LITERATURE | TECHNOLOGY

Exploring Earth's Water

More than two-thirds of the earth's surface is covered with water, yet scientists know more about the surface of the moon than they do about the ocean floor. Using an AUV, or autonomous underwater vehicle, called *Autosub,* researchers hope to gain new understanding about the earth's watery surface.

What It Does

It looks like a giant torpedo, but *Autosub* is really a battery-powered robotic submarine that is 23 feet (7 m) long. Its mission is to explore parts of the ocean that are beyond the reach of other research vessels or are too dangerous for humans. Although it is still being tested, *Autosub* has already conducted hundreds of underwater missions.

Exploring Ice Shelves

One of the most promising areas of research for *Autosub* lies in seawater under the ice shelves near Greenland in the Arctic and near Antarctica at the southern extreme of the globe. Traditional submarines are unable to explore these places safely. Satellite photographs show that the area of the ice shelves is changing. Scientists want to use *Autosub*'s technology to measure changes in the thickness of sea ice. They believe that this information may give important clues about the possible rise in the earth's temperature.

Sea ice plays an important role in keeping the earth's climate stable. It acts as insulation—a kind of protection—between the ocean and the atmosphere. Sea ice reflects light, so it limits the amount of heat absorbed into the water and keeps the ocean from getting too warm. In winter, sea ice helps prevent heat from escaping the warmer oceans into the atmosphere.

What the Future Holds

So far, *Autosub*'s missions have been fairly short. Scientists hope to someday program *Autosub* to make long voyages, sampling seawater and collecting data from ocean floors. The information that *Autosub* provides will help scientists make better predictions about the earth's climate.

▲ ***Autosub* can be launched from shore, towed out to sea by a small boat, or lowered by a crane into the water.**

Making the Connection

1. What is *Autosub*?
2. Why do scientists want to use *Autosub* to explore under the ice shelves?
3. **Understanding Cause and Effect** How could a loss of sea ice affect the earth's climate?

Climate

Guide to Reading

Main Idea

Wind and water carry rainfall and the sun's warmth around the world to create different climates.

Terms to Know

- weather
- climate
- tropics
- monsoon
- tornado
- hurricane
- typhoon
- drought
- El Niño
- La Niña
- current
- local wind
- rain shadow
- greenhouse effect
- rain forest

Places to Locate

- Equator
- Tropic of Cancer
- Tropic of Capricorn

Reading Strategy

Make a chart like this one. Write at least two details that explain how each force contributes to climate.

Sun	Wind	Water

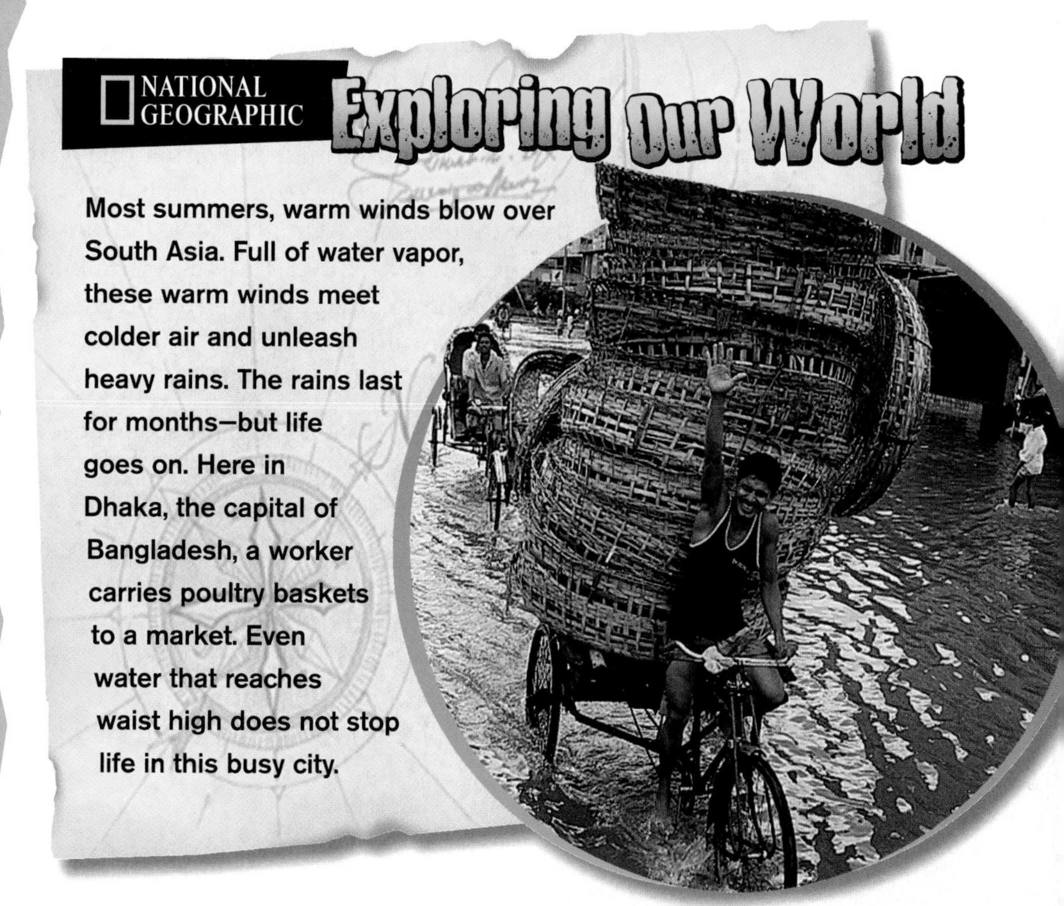

Most summers, warm winds blow over South Asia. Full of water vapor, these warm winds meet colder air and unleash heavy rains. The rains last for months—but life goes on. Here in Dhaka, the capital of Bangladesh, a worker carries poultry baskets to a market. Even water that reaches waist high does not stop life in this busy city.

Why are some areas of the world full of lush forests, while others are covered by bone-dry deserts? Why do some people struggle through chilling winters, while others enjoy a day at the beach? To understand these mysteries, you need to unlock the secrets of climate.

Weather and Climate

As you learned in Chapter 1, the earth is surrounded by the atmosphere, which holds a combination of gases we call air. The atmosphere's many layers protect life on the earth from harmful rays of the sun. The layer of atmosphere immediately surrounding the earth is also where you will find weather patterns. Suppose a friend calls you and asks what it is like outside. You might say, "It's a beautiful day—warm and sunny!" You are describing the weather. **Weather** refers to the unpredictable changes in air that take place over a short period of time.

Suppose that a cousin from another part of the country asks what summers and winters are like in your area. You might say, "Summers are usually hot and rainy, and winters are cool but dry." Your answer describes not the weather but your area's climate. **Climate** is the usual, predictable pattern of weather in an area over a long period of time. It is affected by the sun, the wind, the oceans and other bodies of water, landforms, and even people.

You might also say to your cousin, "Usually we get a little rain in the summer. Last year, though, we went two months without any rain!" You recognize that the weather does not always follow its *usual* climate pattern. Some years it rains more than others; some years the temperature is lower than others.

To understand climate, scientists have to look at the extremes of *temperature* and *precipitation* that an area can have. Consider a country that has an average precipitation of 38 inches (97 cm) a year. In any year, it may be as low as 13 inches (33 cm) or as high as 63 inches (160 cm). Imagine if you were a farmer trying to grow food there. How would your crops be affected if you received only one-third of the rain you expected?

✓Reading Check **What is the difference between weather and climate?**

The Sun and Climate

What causes climate? The original source of climate is the sun. It gives off energy and light that all plants and animals need to survive. The sun's rays warm the air, water, and land on our planet. Warm gases and liquids are lighter than cool gases and liquids. Because they are lighter, the warmer gases and liquids rise. Then wind and water carry this warmth around the globe, distributing the sun's heat.

NATIONAL GEOGRAPHIC **On Location**

Washington, D.C.

A cross-country skier braves a blizzard in our nation's capital.

Place When scientists study climate, what two factors do they analyze?

Prevailing Wind Patterns

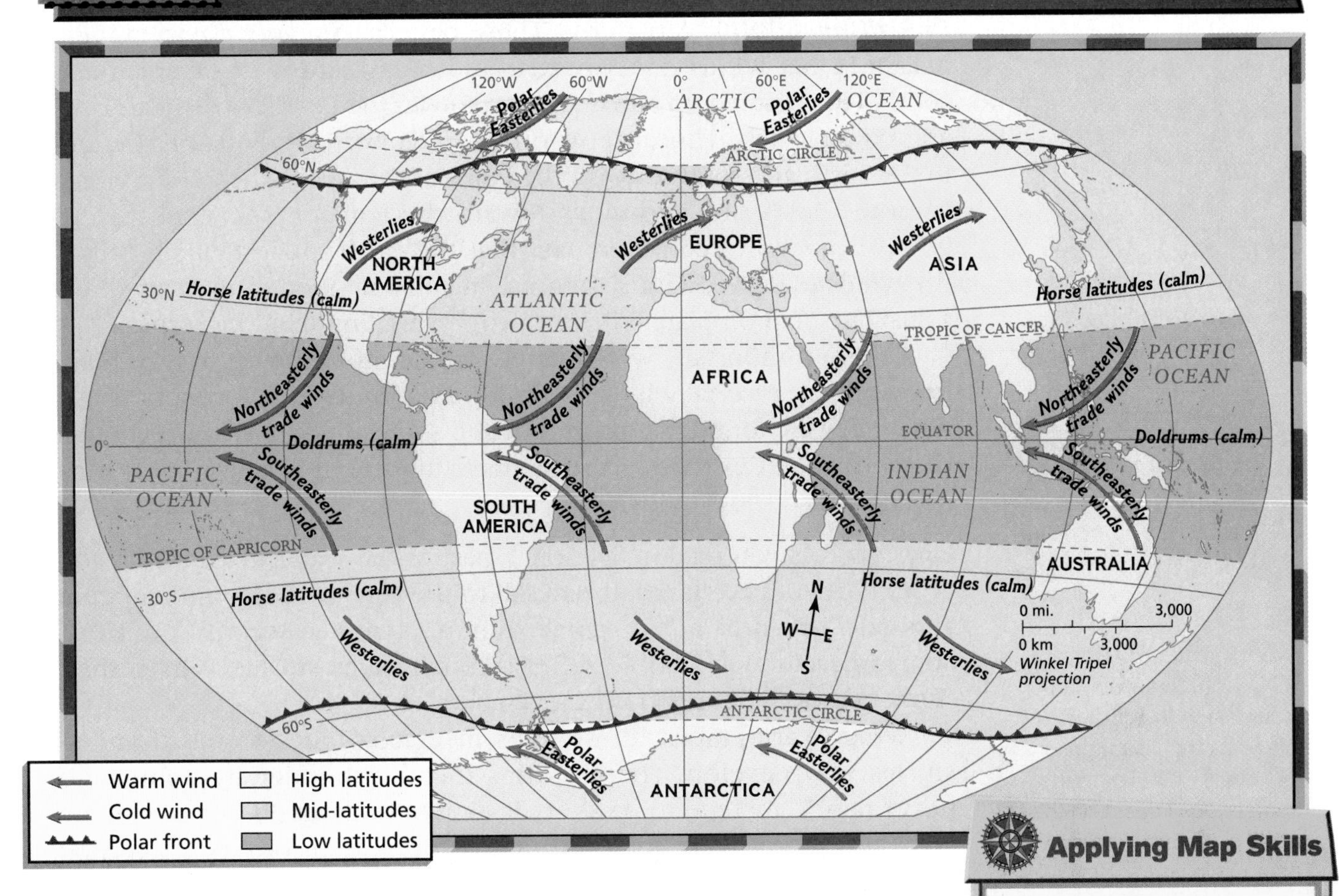

Applying Map Skills

1. In which general direction does the wind blow over North America?
2. What winds did European sailors use to get to South America and the islands north of it?

Find NGS online map resources @ www.nationalgeographic.com/maps

Latitude and Climate Climate is also affected by the angle at which the sun's rays hit the earth. As you learned in Chapter 1, because of the earth's tilt and revolution around the sun, the sun's rays hit various places at different angles at different times of the year. The sun's rays hit places in low latitudes—regions near the **Equator**—more directly than places at higher latitudes. The areas near the Equator, known as the **tropics,** lie between the **Tropic of Cancer** (23½°N latitude) and the **Tropic of Capricorn** (23½°S latitude). If you lived in the tropics, you would almost always experience a hot climate, unless you lived high in the mountains where temperatures are cooler. Find the low-latitude tropics on the map above. (To learn how to use latitude and longitude, turn to page 62.)

Outside the tropics, the sun is never directly overhead. The mid-latitudes extend from the tropics to about 60° both north and south of the Equator. When the North Pole is tilted toward the sun, the sun's rays fall more directly on the Northern Hemisphere. This affects our climate by giving us warm summer days. Six months later, the South Pole is tilted toward the sun, and the seasons are reversed. At the high latitudes near the North and South Poles, the sun's rays hit *very* indirectly. Climates in these regions are always cool or cold.

✓Reading Check **How does the tilt of the earth affect climate?**

The Wind's Effect on Climate

Movements of air are called winds. From year to year, winds follow *prevailing,* or typical, patterns. These patterns are very complex. One reason is that winds do more than move east and west or north and south. They also go up and down. As you learned earlier, warm air rises and cold air falls. Thus, the warmer winds near the Equator rise and move north and south toward the Poles of the earth. The colder winds from the Poles sink and move toward the Equator. This exchange of winds is complicated by the fact that the earth rotates, which causes the winds to curve. Winds, then, are in constant motion in many directions. The map on page 55 shows you prevailing wind patterns.

Another important wind pattern is the monsoon. **Monsoons** are tremendous seasonal winds that blow over continents for months at a time. They are found mainly in Asia and some areas in Africa. Although they often are destructive, the summer monsoons in South Asia bring much-needed heavy rains.

Storms As you read in Section 1, part of the water cycle is rain and other types of precipitation that fall to the earth. A little rain may ruin a picnic or spoil a ball game, but it is not a serious problem. Sometimes, though, people suffer through fierce storms. Why is that? What causes these destructive events?

When warm, moist air systems meet cold air systems, thunderstorms may develop. These storms include thunder, lightning, and heavy rain. They tend to be short, lasting only about 30 minutes. Some areas are more likely to see thunderstorms than others. In central Florida, as many as 90 days a year may experience thunderstorms.

A thunderstorm may produce another danger—a tornado. **Tornadoes** are funnel-shaped windstorms that sometimes form during severe thunderstorms. They occur all over the world, but the United States has more tornadoes than any other area. Winds in tornadoes often reach 250 miles (402 km) per hour.

Hurricanes, or violent tropical storm systems, form over the warm Atlantic Ocean in late summer and fall. Hurricanes bring high winds that can reach more than 150 miles (241 km) per hour. They also produce rough seas and carry drenching rain. Hurricanes strike North America and the islands in the Caribbean Sea. They also rip through Asia, although in that region they are called **typhoons.** These storms can do tremendous damage. Their strong winds destroy buildings and snap power lines. Heavy rains can flood low-lying areas.

El Niño and La Niña In 1998 the world experienced unusual weather. Heavy rains brought floods to Peru, washing away whole villages. Europe, eastern Africa, and most of the southern United States also had severe flooding. In the western Pacific, normally heavy rains never came. Indonesia suffered a **drought,** a long period of extreme dryness. The land there became so dry that forest fires burned thousands of acres of trees. Thick smoke from the fires forced drivers to put their headlights on at noon!

Mt. Pinatubo

Mt. Pinatubo (PEE•nah•TOO•boh) is a volcanic mountain in the Philippine Islands. Its eruption in the early 1990s had a tremendous impact on the world's climate. The powerful explosion shot massive amounts of ash and sulfur dioxide into the earth's atmosphere. The ash and chemicals blocked some of the sun's rays from reaching the earth. As a result, the world's climate was cooler for two years after the volcano's blast.

Why did these disasters take place? They resulted from a combination of temperature, wind, and water effects in the Pacific Ocean called **El Niño** (ehl NEE•nyoh). The name "El Niño" was coined by early Spanish explorers in the Pacific. They use the phrase—which refers to the Christ child—because the effect hits South America around Christmas.

El Niños form when cold winds from the east are weak. Without these cold winds, the central Pacific Ocean grows warmer than usual. More water evaporates, and more clouds form. The thick band of clouds changes wind and rain patterns. Some areas receive heavier than normal rains and others have less than normal rainfall.

Does El Niño come every year? Scientists have found that El Niño occurs about every three years. They also found that in some years, the opposite kind of unusual weather takes place. This event is called **La Niña** (lah NEE•nyah), Spanish for "girl," because the effects are the opposite of those in El Niño. Winds from the east become very strong, cooling more of the Pacific. When this happens, heavy clouds form in the western Pacific.

Reading Check Why do El Niños occur?

Ocean Currents

Winds carry large masses of warm and cool air around the earth. At the same time, moving streams of water called **currents** carry warm or cool water through the world's oceans. Look at the map on page 58. As you can see, these currents follow certain patterns. Notice how the warm currents tend to move along the Equator or from the Equator to the Poles. The cold currents carry cold polar water toward the Equator.

These currents affect the climate of land areas. Look at the warm current called the Gulf Stream. It flows from the Gulf of Mexico along the east coast of North America. Then it crosses the Atlantic Ocean toward Europe, where it is called the North Atlantic Current. Winds that blow over these warm waters bring warm air to western Europe. Because these winds blow from west to east, areas in Europe enjoy warmer weather than areas lying west of the Gulf Stream in Canada.

Reading Check What areas of the world would be affected by a change in the Gulf Stream?

NORMAL CONDITIONS

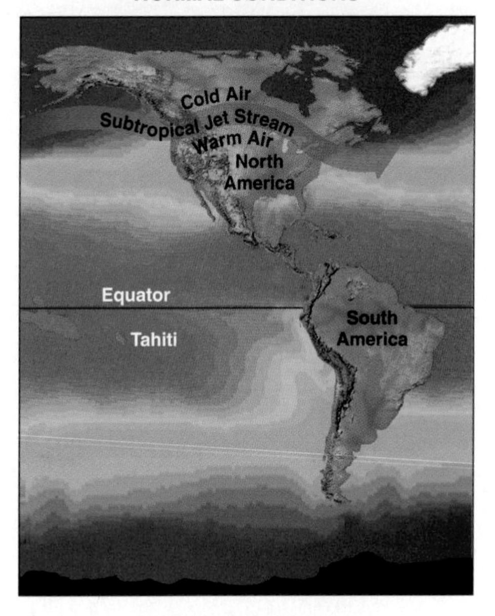

EL NIÑO CONDITIONS

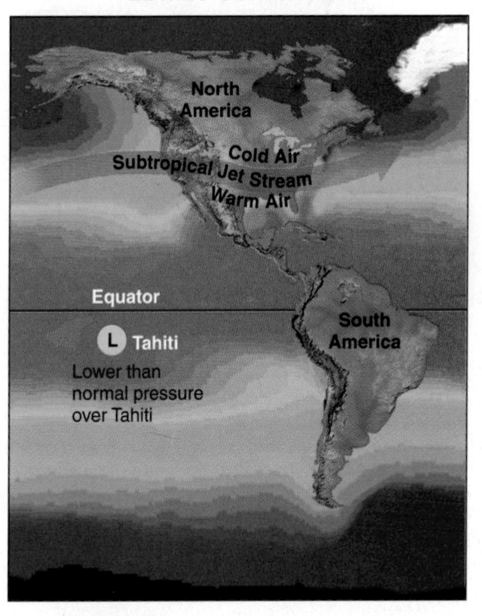

Analyzing the Diagram

The temperature of the oceans varies from warm (dark red) to very cold (dark purple).

Movement What happens to the Jet Stream during El Niño conditions?

NATIONAL GEOGRAPHIC

World Ocean Currents

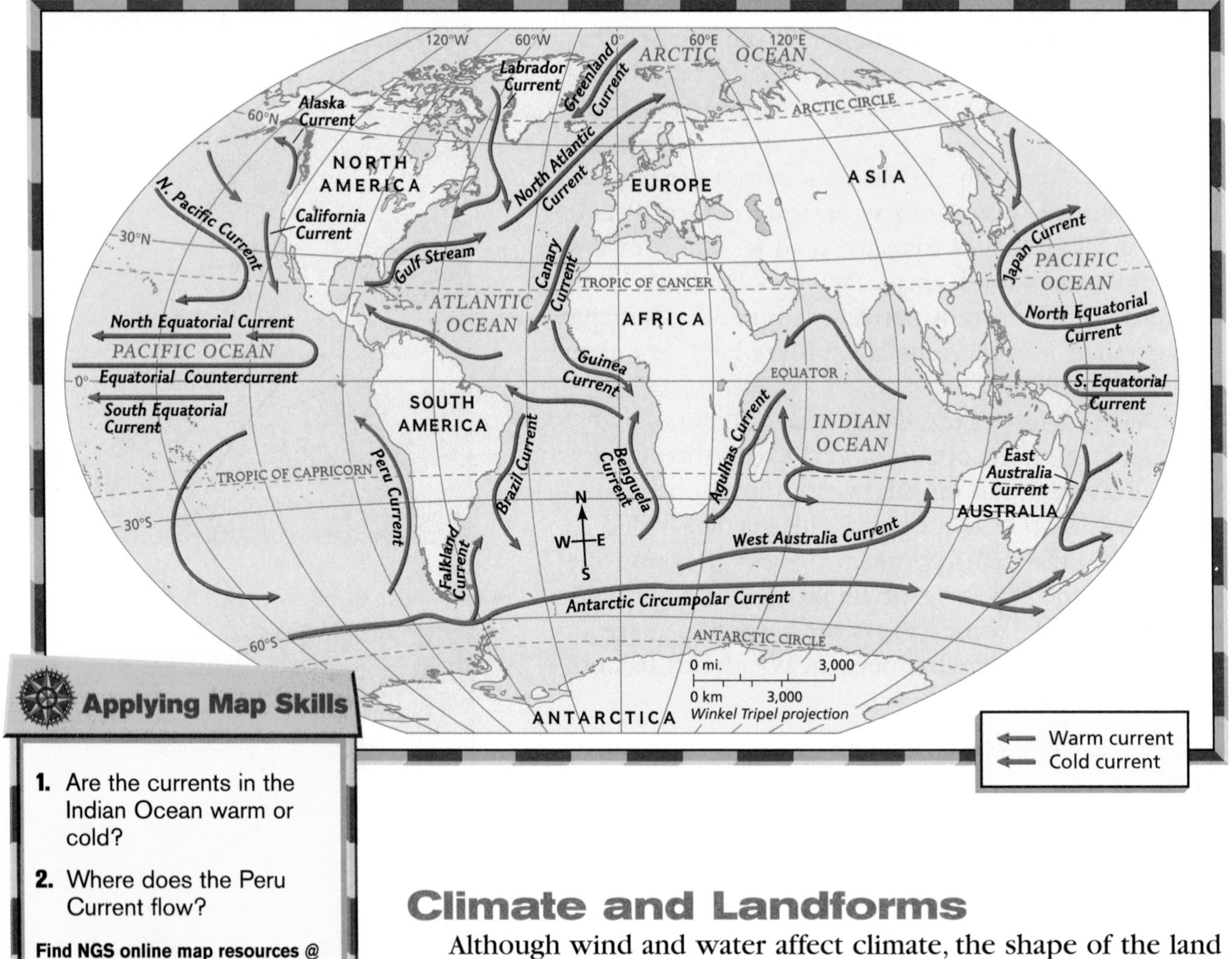

Applying Map Skills

1. Are the currents in the Indian Ocean warm or cold?
2. Where does the Peru Current flow?

Find NGS online map resources @ www.nationalgeographic.com/maps

Climate and Landforms

Although wind and water affect climate, the shape of the land has an effect on climate as well. *Where* landforms are in relation to one other and to water influences climate, too.

Landforms and Local Winds Geographers study major wind patterns that blow over the earth. When they study the climate of a given area, however, geographers also look at what they call local winds. **Local winds** are patterns of wind caused by landforms in a particular area.

Some local winds occur because land warms and cools more quickly than water does. As a result, cool sea breezes keep coastal areas cool during the day. After the sun sets, the opposite occurs. The air over the land cools more quickly than the air over water. At night, then, a land breeze blows from the land out to the sea.

A similar effect occurs near mountains. Air warmed by the sun rises up mountain slopes during the day. At night, cooler air moves down the mountain into the valley below. Have you ever seen fog lying on a valley floor on a cool morning? That fog was caused by the cool air that came down the mountain during the night.

Mountains, Temperature, and Rainfall The higher the elevation of a particular place, the lower the temperature that place will have. In high mountains, the air becomes thinner and cannot hold as much heat from the sun. The temperature drops. Even near the Equator, where the sun's rays strike the earth directly, snow covers the peaks of high mountains.

The cooling effect of elevation is not as strong on high plateaus, however. Why is this so? As you learned in Chapter 1, plateaus are flat surfaces, so there are no rising air currents as there are along mountainsides. Some plateaus—as in the southwestern United States—can reach as much as a mile above sea level. Even these high plateaus can hold the sun's warmth.

Mountains also have an effect on rainfall. When warm moist winds blow inland from the ocean toward a coastal mountain range, the winds are forced upward over the mountains. As these warm winds rise, the air cools and loses its moisture. Rain or snow falls on the mountains. The climate on this *windward*—or wind-facing—side of mountain ranges is moist and often foggy. Trees and vegetation are thick and green.

By the time the air moves over the mountain peaks, it is cool and dry. This creates a **rain shadow,** a dry area on the side of the mountains facing away from the wind. Geographers call this side the *leeward* side. The dry air of a rain shadow warms up again as it moves down the leeward side, giving the region a dry or desert climate.

A rain shadow occurs along the western coast of the United States. Winds moving east from the Pacific Ocean lose their moisture as they move upward on the windward slopes of the coastal mountains. Great deserts and dry basins are located on the leeward side of these ranges.

✓Reading Check **Why are areas of higher elevation often cooler?**

Rain Shadow

Analyzing the Diagram

Rain shadows usually occur on the inland sides of mountain ranges near oceans.

Location **What is the term for the side of a mountain where precipitation falls?**

Shanghai, China (right), like other modern cities, is generally warmer than the rural areas around it (above).

Human/Environment Interaction **Why are cities warmer than surrounding areas?**

The Impact of People on Climate

People's actions can affect climate. You may have noticed that temperatures in large cities are generally higher than those in nearby rural areas. Why is that? The city's streets and buildings absorb more of the sun's rays than do the plants and trees of rural areas.

Cities are warmer even in winter. People burn fuels to warm houses, power industry, and move cars and buses along the streets. This burning raises the temperature in the city. The burning also releases a cloud of chemicals into the air. These chemicals blanket the city and hold in more of the sun's heat, creating a so-called *heat island.*

The Greenhouse Effect The burning of fuels may be creating a worldwide problem. In the past two hundred years, people have burned coal, oil, and natural gas as sources of energy. Burning these fuels releases certain gases into the air.

Like the glass in a greenhouse, gases in the atmosphere trap the sun's warmth. Without this **greenhouse effect,** the earth would be too cold for most living things. Some scientists, however, say that the increased buildup of gases is strengthening the greenhouse effect. They claim that more of the sun's heat is being trapped near the Earth's surface. As a result, temperatures around the planet are rising. Some scientists warn that this global warming could bring heavier rains to some areas and drought to others. They also say that ice trapped in polar areas will melt. Then ocean levels will rise and flood low-lying coastal areas.

Not all scientists agree about the effects of atmospheric gases. Some argue that the world is not warming. Others say that even if it is, the predictions of disaster are extreme. Many scientists are studying world temperature trends closely. They hope to be able to discover whether greenhouse gases are a real threat.

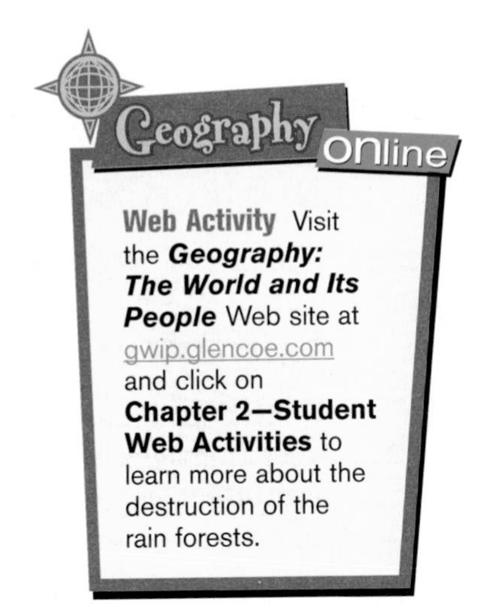

Geography
online
Web Activity Visit the Geography: The World and Its People Web site at gwip.glencoe.com and click on Chapter 2—Student Web Activities to learn more about the destruction of the rain forests.

Clearing the Rain Forests Along the Equator, you can see dense forests called **rain forests** that receive high amounts of rain each year. In some countries, people are clearing large areas of these forests. They want to sell the lumber from the trees. They also want to use the land to grow crops or as pasture for cattle. Clearing the rain forests, though, can hurt the world's climate.

One reason is related to the greenhouse effect. People often clear the forests by burning down the trees. This burning releases more gases into the air, just like burning oil or natural gas does. Another danger of clearing the rain forests is related to rainfall. Remember the water cycle discussed in Section 1? Water on the earth's surface evaporates into the air and then falls as rain. In the rain forests, much of this water evaporates from the leaves of trees. If the trees are cut, less water will evaporate. As a result, less rain will fall. Scientists worry that, over time, the area that now holds rain forests will actually become dry and unable to grow anything.

✓Reading Check **What is the greenhouse effect?**

Section 2 Assessment

Defining Terms

1. **Define** weather, climate, tropics, monsoon, tornado, hurricane, typhoon, drought, El Niño, La Niña, current, local wind, rain shadow, greenhouse effect, rain forest.

Recalling Facts

2. **Movement** What five elements affect climate?
3. **Location** Between what two lines of latitude are the tropics?
4. **Place** Provide an example of how landforms influence climate.

Critical Thinking

5. **Making Comparisons** How does the amount of rainfall on the windward side of a mountain differ from that on the leeward side?
6. **Summarizing Information** What general patterns do wind and currents follow?

Graphic Organizer

7. **Organizing Information** Draw a diagram like this one. In the first box, list three human actions that lead to the greenhouse effect. In the third box, list four possible results of the greenhouse effect.

Applying Geography Skills

8. **Analyzing Maps** Look at the world ocean currents map on page 58. Which two continents lie completely outside the tropics?

Geography Skill

Using Latitude and Longitude

Learning the Skill

To find an exact location, geographers use a set of imaginary lines. One set of lines—**latitude** lines—circles the earth's surface east and west. The starting point for numbering latitude lines is the Equator, which is 0° latitude. Every other line of latitude is numbered from 1° to 90° and is followed by an N or S to show whether it is north or south of the Equator. Latitude lines are also called parallels.

A second set of lines—**longitude** lines—runs vertically from the North Pole to the South Pole. Each of these lines is also called a meridian. The starting point—0° longitude—is called the Prime Meridian (or Meridian of Greenwich). Longitude lines are numbered from 1° to 180° followed by an E or W—to show whether they are east or west of the Prime Meridian.

To find latitude and longitude, choose a place on a map. Identify the nearest parallel, or line of latitude. Is it located north or south of the Equator? Now identify the nearest meridian, or line of longitude. Is it located east or west of the Prime Meridian?

Practicing the Skill

1. On the map below, what is the exact location of Washington, D.C.?
2. What cities on the map lie south of 0° latitude?
3. What city is located near 30°N, 30°E?

Applying the Skill

Turn to page RA2 of the **Reference Atlas.** List the latitude and longitude for one city. Ask a classmate to use the information to find and name the city.

NATIONAL GEOGRAPHIC

Map of the World

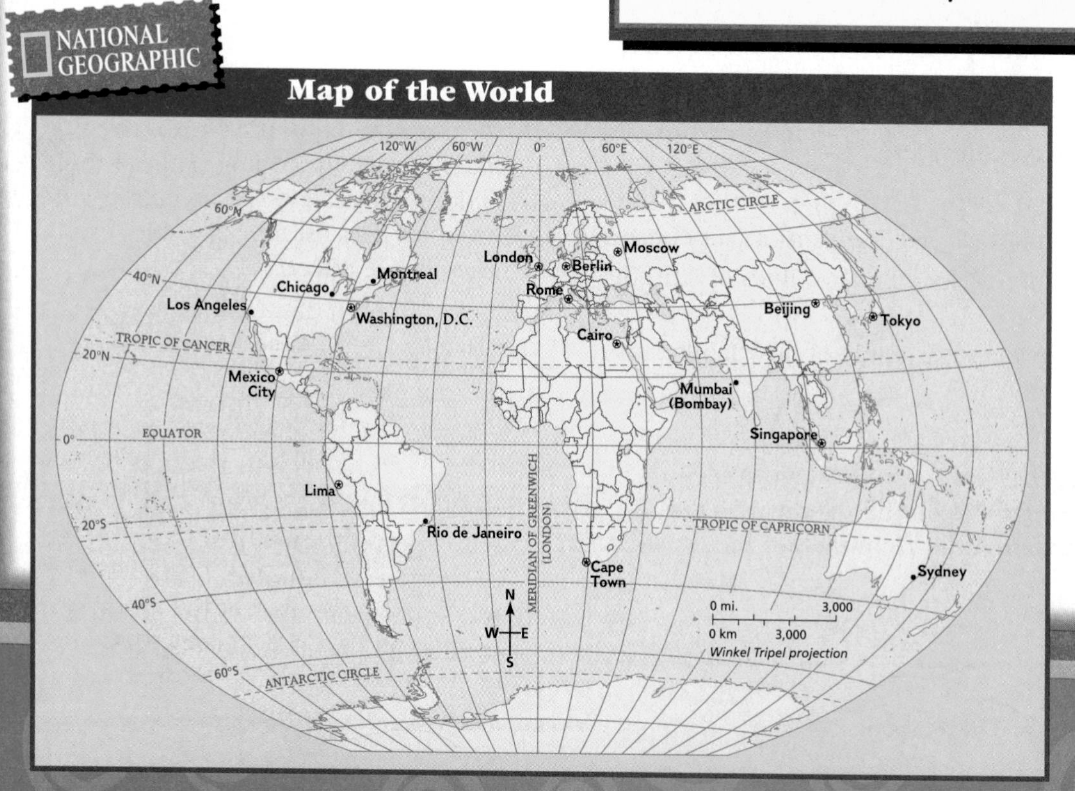

Section 3

Climate Zones and Vegetation

Guide to Reading

Main Idea

Geographers divide the world into different climate zones, each of which has special characteristics.

Terms to Know

- canopy
- savanna
- marine west coast climate
- Mediterranean climate
- humid continental climate
- humid subtropical climate
- subarctic
- tundra
- permafrost
- steppe
- timberline

Reading Strategy

Complete a chart like the one below by listing the categories of each type of climate next to the correct headings.

Climate Type	Categories
Tropical	
Mid-Latitude	
High Latitude	
Dry	
Highland	

NATIONAL GEOGRAPHIC **Exploring Our World**

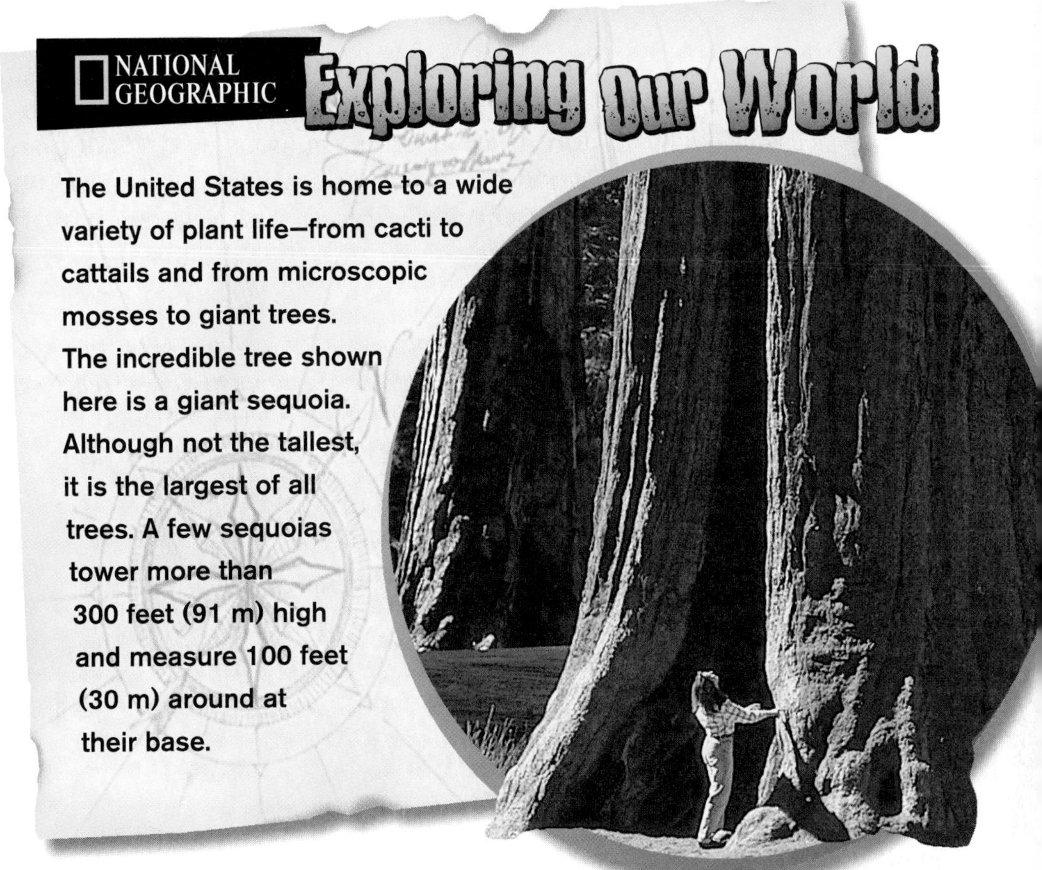

The United States is home to a wide variety of plant life—from cacti to cattails and from microscopic mosses to giant trees. The incredible tree shown here is a giant sequoia. Although not the tallest, it is the largest of all trees. A few sequoias tower more than 300 feet (91 m) high and measure 100 feet (30 m) around at their base.

Why do you think a photograph of a giant tree is in a chapter on climate? The reason is that climate and vegetation go together. Consider this: The state of Washington sits next to the state of Idaho. The plant life in western Washington, however, is much more similar to that of the United Kingdom, which is thousands of miles away, than it is to the plant life in eastern Washington and Idaho, which touch each other. Why? The patterns of temperature, wind, and precipitation in western Washington and the United Kingdom are similar.

Scientists use these patterns to group climates into many different types. They have put the world's climates into five major groups: tropical, mid-latitude, high latitude, dry, and highland. Three of these groups—tropical, mid-latitude, and high latitude—are based on an area's latitude, or distance from the Equator. Some of these major groups have subcategories of climate zones within them. In addition, each climate zone has particular kinds of plants that grow in it.

Tropical Climates

The tropical climate gets its name from the tropics—the areas along the Equator reaching from 23½°N to 23½°S. If you like warm weather all during the year, you would love a tropical climate. The tropical climate region can be separated into two types—tropical rain forest and tropical savanna. The tropical rain forest climate receives up to 100 inches (254 cm) of rain a year. As a result, the rain forest climate is wet in most months. The tropical savanna climate has two distinct seasons—one wet and one dry.

Tropical Rain Forest Climate Year-round rains in some parts of the tropics produce lush vegetation and thick rain forests. These forests are home to millions of kinds of plant and animal life. Tall hardwood trees such as mahogany, teak, and ebony form the **canopy,** or top layer of the forest. The vegetation at the canopy layer is so thick that little sunlight reaches the forest floor. The Amazon Basin in South America is the world's largest rain forest area.

Tropical Savanna Climate In other parts of the tropics, such as southern India and eastern Africa, most of the year's rain falls in just a few months of the year. This is called the wet season. The rest of the year is hot and dry. **Savannas,** or broad grasslands with few trees, occur in this climate region. Find the tropical savanna climate areas on the map on page 65.

Reading Check **Where are the tropical climate zones found?**

Tropic of Cancer to Tropic of Capricorn

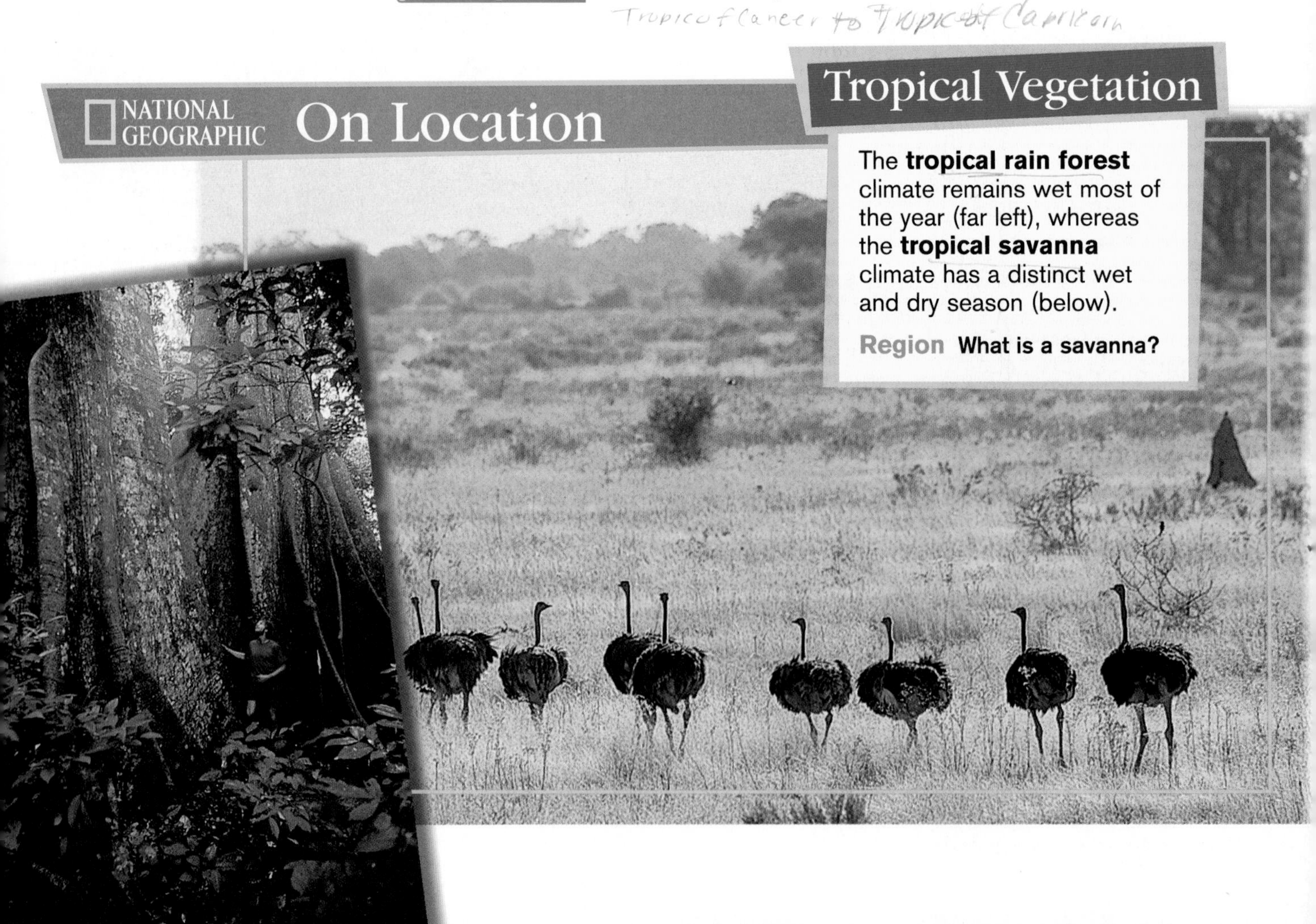

NATIONAL GEOGRAPHIC **On Location**

Tropical Vegetation

The **tropical rain forest** climate remains wet most of the year (far left), whereas the **tropical savanna** climate has a distinct wet and dry season (below).

Region **What is a savanna?**

World Climate Regions

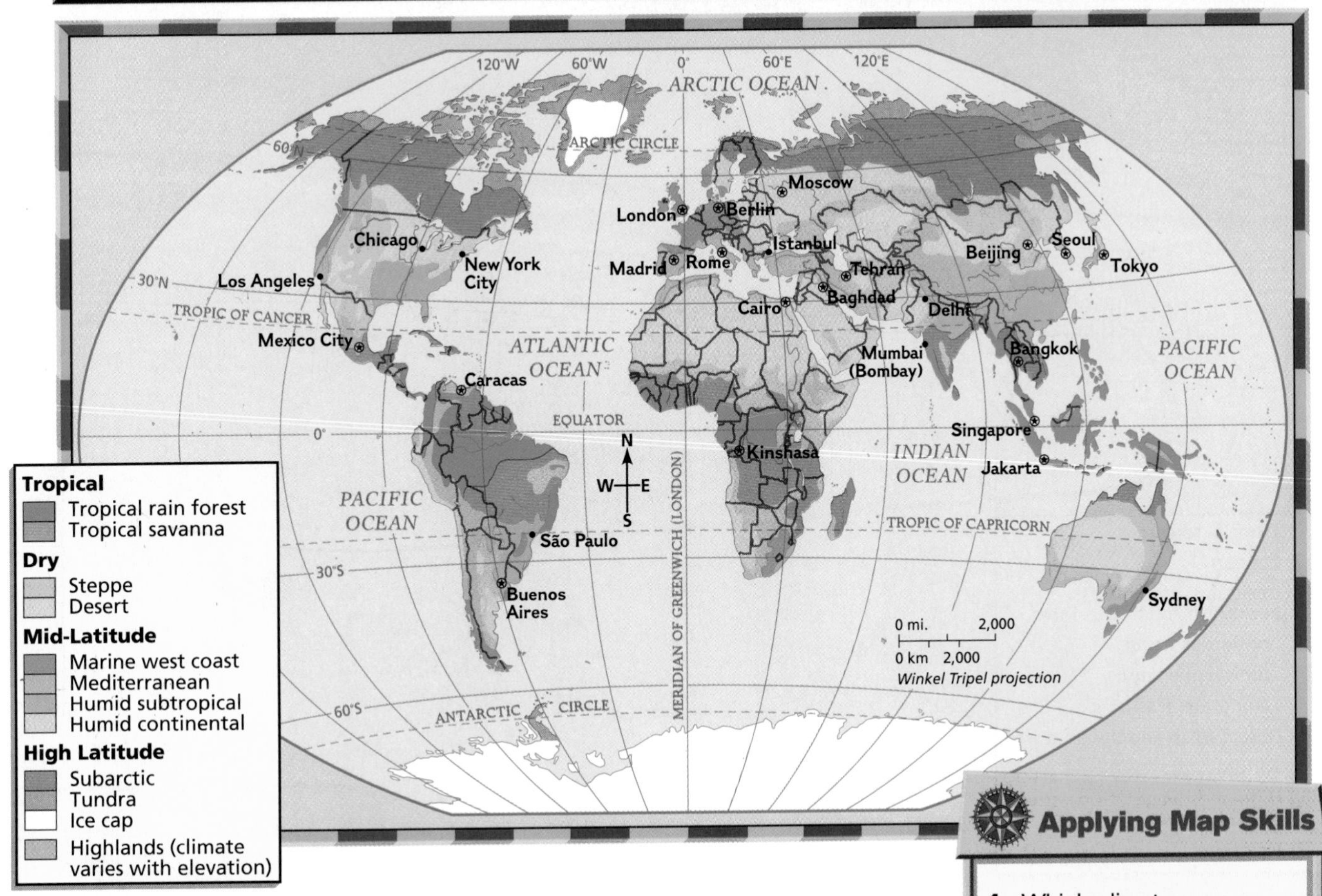

Applying Map Skills

1. Which climate covers most of the southeastern United States?
2. Which climate is most common in countries directly on the Equator?

Find NGS online map resources @ www.nationalgeographic.com/maps

Mid-Latitude Climates

Mid-latitude, or moderate, climates are found in the middle latitudes of the Northern and Southern Hemispheres. They extend from about 23½° to 60° both north and south of the Equator. Most of the world's people—probably including you—live within these two bands around the earth. They are called mid-latitude because they are in the middle of both the Northern Hemisphere and the Southern Hemisphere. The mid-latitude climates are neither as close to the Equator as the tropics nor as close to the Poles as the high latitude climates.

The mid-latitude region includes more and different climate zones than other regions. This variety results from a mix of air masses. As you remember from Section 2, warm air comes from the tropics, and cool air comes from the polar regions. In most mid-latitude climates, the temperature changes with the seasons. Sometimes the climate zones in this region are called *temperate climates.*

Marine West Coast Climate Coastal areas that receive winds from the ocean usually have a mild **marine west coast climate.** If you lived in one of these areas, your winters would be rainy and mild, and your summers would be cool. These areas usually have strong growth

World Natural Vegetation Regions

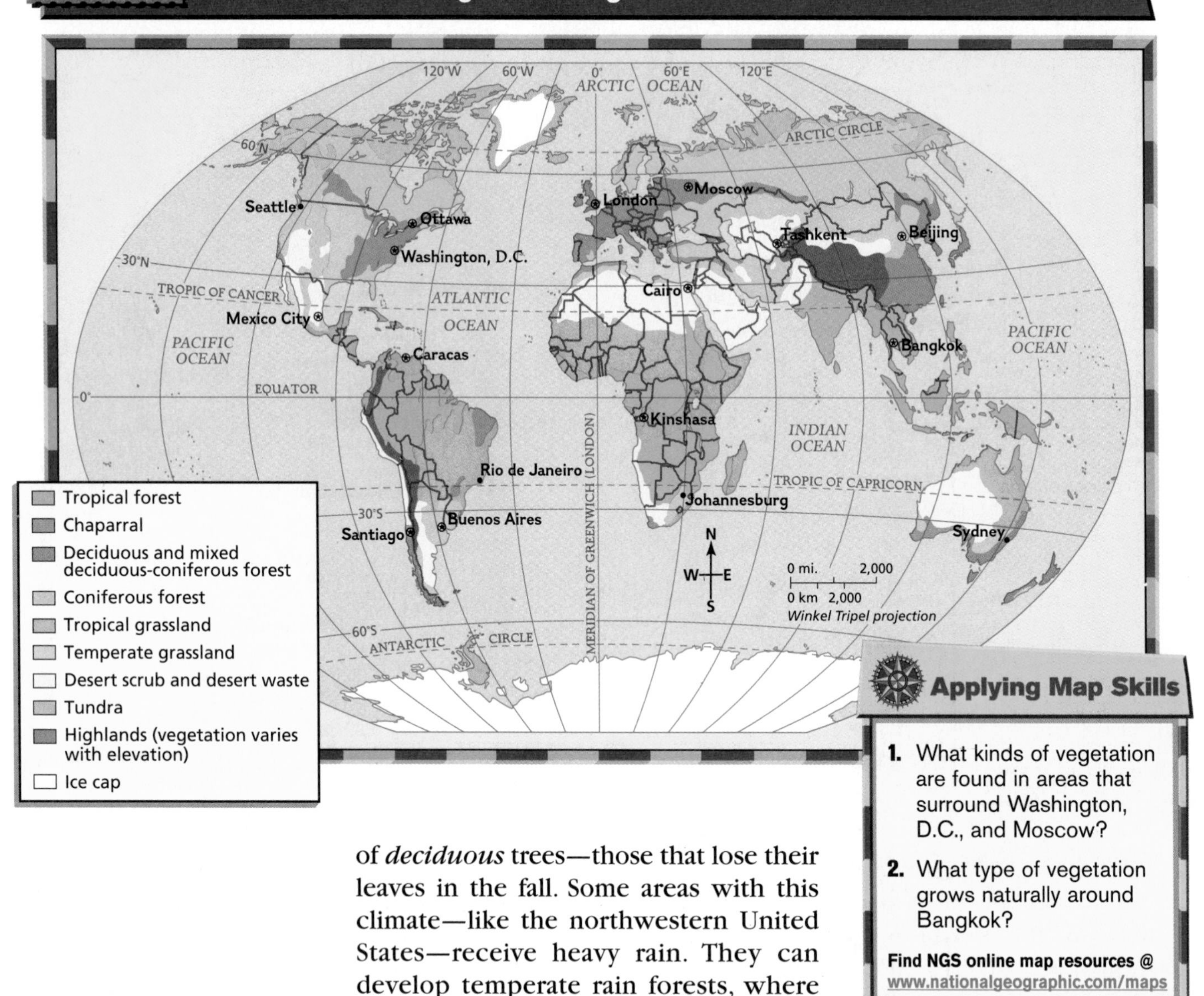

Applying Map Skills

1. What kinds of vegetation are found in areas that surround Washington, D.C., and Moscow?
2. What type of vegetation grows naturally around Bangkok?

Find NGS online map resources @ www.nationalgeographic.com/maps

of *deciduous* trees—those that lose their leaves in the fall. Some areas with this climate—like the northwestern United States—receive heavy rain. They can develop temperate rain forests, where evergreens like cedar, fir, and redwood grow.

Mediterranean Climate Another mid-latitude coastal climate is called a **Mediterranean climate** because it is similar to the climate found around the Mediterranean Sea. This climate has mild, rainy winters like the marine west coast climate. Instead of cool summers, however, people living in a Mediterranean climate experience hot, dry summers. The vegetation that grows in this climate includes shrubs and short trees. Some are evergreens, but others lose their leaves in the dry season.

Humid Continental Climate If you live in inland areas of North America, Europe, or Asia, you usually face a harsher **humid continental climate.** In these areas, winters can be long, cold, and snowy. Summers are short but may be very hot. Deciduous trees grow in forests, and vast grasslands flourish in some areas of this zone.

Humid Subtropical Climate Mid-latitude regions close to the tropics have a **humid subtropical climate.** Rain falls throughout the year but is heaviest during the hot and humid summer months. Humid subtropical winters are generally short and mild. Trees like oak, magnolia, and palms grow in this zone.

✓ Reading Check **What causes the mid-latitude region to have more and different climate zones than other regions?** Because a mix of warm air masses from the Equator and cold ones from the poles.

High Latitude Climates

High latitude climate regions lie mostly in the high latitudes of each hemisphere, from 60°N to the North Pole and 60°S to the South Pole. These climates are generally cold, but some are more severely cold than others.

Subarctic Climate In the high latitudes nearest the mid-latitude zones, you will find the **subarctic** climate. The few people living here face severely cold and bitter winters, but temperatures do rise above freezing during summer

Mid-Latitude Vegetation

Fir trees (bottom left) thrive in a **marine west coast** climate. Shrubs and olive trees (top right) grow in a **Mediterranean** climate. Deciduous trees (bottom right) flourish in a **humid continental** climate. Palm trees are common in **humid subtropical** zones.

Place **Which type of vegetation is most common in your area?**

months. Huge evergreen forests called *taiga* (TY•guh) grow in the subarctic region, especially in northern Russia.

Tundra Climate Closer to the Poles than the subarctic zone lie the **tundra** areas. The tundra is a vast rolling plain without trees. The climate in this zone is harsh and dry. In parts of the tundra and subarctic regions, the lower layers of soil are called **permafrost** because they stay permanently frozen. Only the top few inches of the ground thaw during summer months. Because the surface is flat and the soil below is frozen, water tends to stay on the land when it melts. This provides the moisture that plants need to grow. You will find sturdy grasses and low-growing berry bushes in the tundra.

Ice Cap Climate On the polar ice caps and the great ice sheets of Antarctica and Greenland, the climate is bitterly cold. Monthly temperatures average below freezing. Temperatures in Antarctica have been measured at −128°F (−89°C)! Although no other vegetation grows here, *lichens*—or funguslike plants and mosses—can live on rocks.

✓Reading Check **What are the three types of high latitude climates?**

NATIONAL GEOGRAPHIC On Location

High Latitude Vegetation

Vegetation in high latitude climates includes **tundra** grasses and bushes (top left), **subarctic** evergreen forests (bottom left), and **ice cap** lichens (background).

Region **How does permafrost affect vegetation in the tundra and subarctic regions?**

NATIONAL GEOGRAPHIC On Location

Dry Climate Vegetation

The vegetation that survives **desert** and **steppe** climates includes cacti (above) and short grasses (left).

Region **Are very dry climates always hot? Explain.** No, because dryness has to do with precipitation not temperature

Dry Climates

Dry climate refers to dry or partially dry areas that receive little or no rainfall. Temperatures can be extremely hot during the day and very cold at night. Dry climates can also have severely cold winters. You can find dry climate regions at any latitude.

Desert Climate The driest climates have less than 10 inches (25 cm) of rainfall a year. Regions with such climates are called deserts. Only scattered plants such as cacti can survive a desert climate. With roots close to the surface, cacti can collect any rain that falls. Most cacti are found only in North America. In other countries, however, small areas of thick plant life dot the deserts. These arise along rivers or where underground springs reach the surface.

Steppe Climate Many deserts are surrounded by partly dry grasslands and prairies known as **steppes.** The word *steppe* comes from a Russian word meaning "treeless plain." The steppes get more rain than deserts, averaging 10 to 20 inches (25 to 51 cm) a year. Bushes and short grasses cover the steppe landscape. The Great Plains of the United States has a steppe climate.

✓Reading Check **Where are steppe climate zones often located?**

On the outskirts of the desert

Highland Vegetation

The wildflowers and shrubs that grow in meadows above the timberline are often called *alpine* vegetation. This name refers to the Alps mountain ranges in Europe.

Location **How does elevation affect climate?**

Highland Climate

As you read in Section 2, the elevation of a place changes its climate dramatically. Mountains tend to have cool climates—and the highest mountains have very cold climates. This is true even for mountains that are on the Equator. A highland, or mountain, climate has cool or cold temperatures year-round.

If you climb a mountain, you will reach an area called the timberline. The **timberline** is the elevation above which no trees grow. Once you reach the timberline, you will find only small shrubs and wildflowers growing in meadows.

Reading Check **What is the timberline?**

the elevation above which no trees grow.

Assessment

Defining Terms

1. **Define** canopy, savanna, marine west coast climate, Mediterranean climate, humid continental climate, humid subtropical climate, subarctic, tundra, permafrost, steppe, timberline.

Recalling Facts

2. **Region** What are the five types of climate regions?
3. **Region** How do the climate zones in the mid-latitude region differ?
4. **Region** What kind of vegetation grows in the tundra climate zone?

Critical Thinking

5. **Making Comparisons** What do the tropical savanna and humid continental climates have in common?
6. **Drawing Conclusions** How can snow exist in the tropics along the Equator?

Graphic Organizer

7. **Organizing Information** Draw a globe like this one. Label the three climate regions that are based on latitude, then identify the lines of latitude that separate the climate regions.

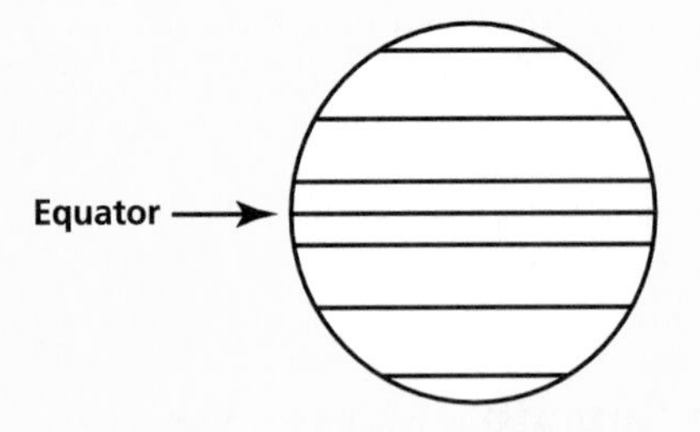

Applying Geography Skills

8. **Analyzing Maps** Look at the world natural vegetation regions map on page 66. What type of natural vegetation thrives around Cairo?

Chapter 2 Reading Review

Section 1 The Water Planet

Terms to Know

water vapor
water cycle
evaporation
condensation
precipitation
collection
glacier
groundwater
aquifer

Main Idea

Water is one of the earth's most precious resources.

✓ **Region** Water covers about 70 percent of the earth's surface.

✓ **Movement** Water follows a cycle of evaporation, condensation, precipitation, and collection on and beneath the ground.

✓ **Science** Humans and most animals need freshwater to live. Only a small fraction of the world's water is found in rivers and lakes.

Section 2 Climate

Terms to Know

weather
climate
tropics
monsoon
tornado
hurricane
typhoon
drought
El Niño
La Niña
current
local wind
rain shadow
greenhouse effect
rain forest

Main Idea

Wind and water carry rainfall and the sun's warmth around the world to create different climates.

✓ **Region** Climate is the usual pattern of weather over a long period of time. It includes extremes in temperature and rainfall.

✓ **Region** The tropics, near the Equator, receive more of the sun's warmth than other regions.

✓ **Location** Landforms and position near water affect climate in a local area.

✓ **Culture** Human actions like building cities, burning fuels, and clearing the rain forests can affect climate.

Section 3 Climate Zones and Vegetation

Terms to Know

canopy
savanna
marine west coast climate
Mediterranean climate
humid continental climate
humid subtropical climate
subarctic
tundra
permafrost
steppe
timberline

Main Idea

Geographers divide the world into different climate zones, each of which has special characteristics.

✓ **Region** The world has five main climate regions that are based on latitude, amount of moisture, and/or elevation. These regions are tropical, mid-latitude, high latitude, dry, and highland.

✓ **Region** Each climate zone has particular kinds of vegetation.

Fishing on the Pacific Ocean ▶

Chapter 2 Assessment and Activities

Using Key Terms

Match the terms in Part A with their definitions in Part B.

A.

1. evaporation
2. savanna
3. monsoon
4. tundra
5. condensation
6. greenhouse effect
7. rain forest
8. El Niño
9. precipitation
10. current

B.

a. moving streams of water in the oceans
b. treeless plain in which only the top few inches of ground thaw in summer
c. weather pattern in the Pacific Ocean
d. seasonal wind that blows over a continent
e. buildup of certain gases in the atmosphere that hold the sun's warmth
f. water that falls back to the earth
g. dense forest that receives much rain
h. water vapor changes back into a liquid
i. sun's heat turns water into water vapor
j. broad grassland in the tropics

Reviewing the Main Ideas

Section 1 The Water Planet

11. **Movement** What are the four steps in the water cycle?
12. **Region** What percentage of the world's water is freshwater?
13. **Region** What has more freshwater—lakes and rivers or groundwater?

Section 2 Climate

14. **Movement** How do wind and water affect climate?
15. **Location** How do mountains affect rainfall?
16. **Human/Environment Interaction** Why are cities warmer than nearby rural areas?

Section 3 Climate Zones and Vegetation

17. **Region** Which climate region has the most climate zones? Why?
18. **Place** What kind of vegetation grows in Mediterranean climates?
19. **Place** In what climate zone would you find large grasslands?

NATIONAL GEOGRAPHIC

World Oceans and Currents

Place Location Activity

On a separate sheet of paper, match the letters on the map with the numbered places listed below.

1. Arctic Ocean
2. Atlantic Ocean
3. California Current
4. Japan Current
5. Indian Ocean
6. Gulf Stream

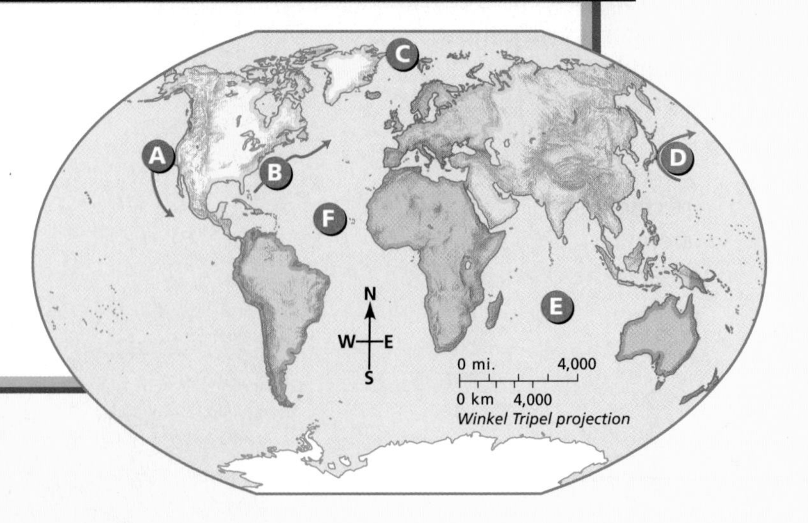

Self-Check Quiz Visit the ***Geography: The World and Its People*** Web site at gwip.glencoe.com and click on **Chapter 2–Self-Check Quizzes** to prepare for the Chapter Test.

Critical Thinking

20. **Analyzing Information** From where does the freshwater in your community come? How can you find out?
21. **Categorizing Information** Create five webs like the one shown here. In each large oval, write the name of a climate region. In the medium-sized ovals, write the name of each climate zone in that region. For each zone, fill in the three small ovals with the usual weather in summer, the usual weather in winter, and the kind of vegetation.

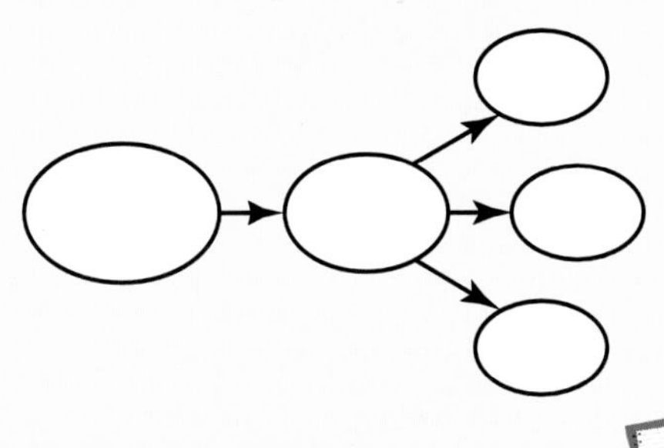

GeoJournal Activity

22. **Writing a Poem** Make a list of all the types of precipitation that you can think of. Write a poem about precipitation and climate using the words in your list.

Mental Mapping Activity

23. **Focusing on the Region** Draw a freehand map of the world's oceans and continents. Label the following items:
    - Equator
    - North America
    - Pacific Ocean
    - Africa
    - high latitude climate regions
    - tropical climate regions

Technology Activity

24. **Using the Internet** Research a recent hurricane or tornado. Find out when and where it occurred, how much force the storm had, and what damage it caused.

Standardized Test Practice

Directions: Study the graph, then answer the following question.

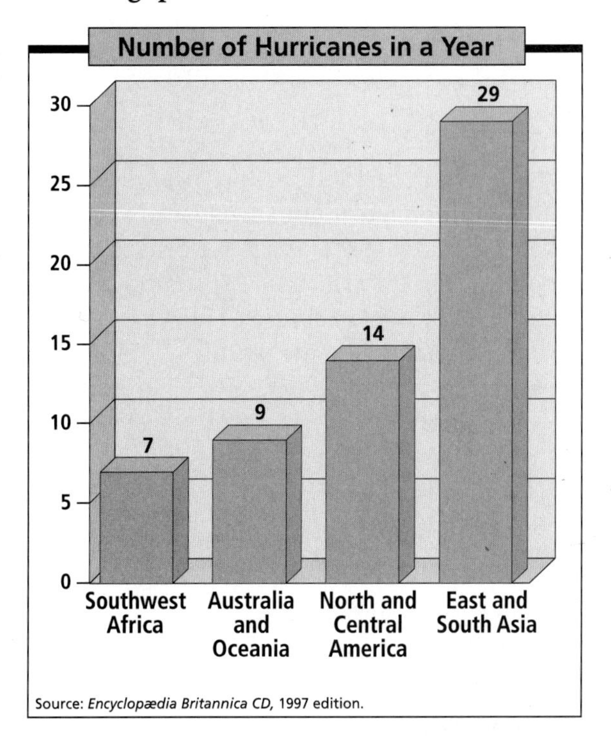

1. How many more hurricanes does East and South Asia experience in a year than North and Central America?

F 29
G 14
H 9
J 15

Test-Taking Tip: Make sure you read the question carefully. It is not asking for the total number of hurricanes in East and South Asia. Instead, the question asks how many *more* hurricanes one region has than another.

EYE on the Environment

THE OCEANS
Deep Trouble

Endangered Seas Next time you are in the grocery store, walk past the fresh seafood section. You will see fish, shrimp, and other seafood on ice. The abundance and variety of seafood available makes it seem as if the oceans can provide us with food forever. Unfortunately, that is not true. Oceans worldwide are in deep trouble because of human activities. There are several trouble spots around the world.

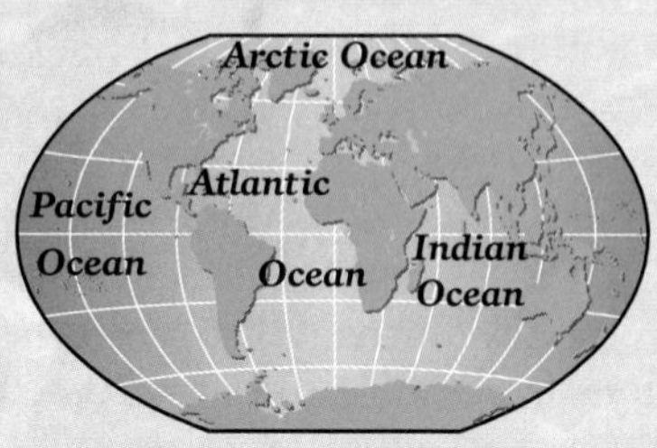

- Grand Banks — Overfishing in the northwest Atlantic Ocean has seriously harmed once-rich populations of cod, halibut, and other fish.
- Gulf of Mexico — Pollution from the United States's rivers causes a "dead zone" in the Gulf each summer. Fish die due to a lack of oxygen in the water.
- Yellow Sea — China's industries have dumped poisons into these Pacific Ocean waters. This sea may have the world's highest levels of heavy metals.
- Kara Sea — The former Soviet Union dumped huge quantities of radioactive wastes into these Arctic waters.

Saving Oceans It is not too late to preserve the oceans. Some steps have been taken.

- More than 1,200 protected zones exist along coasts worldwide.
- Countries are assigning individual catch limits to fishing boats.
- Science organizations are collecting data on oceans and urging government leaders to pass laws to protect marine environments.

Sea stars

Workers clean a beach after an oil spill.

Bottlenose dolphin

Making a Difference

Underwater Explorer Imagine exploring where no one has ever been before. Dr. Sylvia Earle has! Earle explores deep in the oceans, discovering new kinds of animals, plants, and systems of life. Earle is a marine biologist and former chief scientist of the National Oceanic and Atmospheric Administration (NOAA). Now she is leading a research project called the Sustainable Seas Expeditions. Using a special, one-person submersible called *DeepWorker*, Earle and her team are studying deepwater habitats in the United States's National Marine Sanctuary system. This system includes 12 underwater parks that lie in coastal waters around the continental United States and in Hawaii and American Samoa.

DeepWorker has a clear acrylic dome that covers the pilot's head and shoulders. Mechanical arms allow the pilot to work underwater. The expedition team takes photographs and collects data on marine organisms and their habitats in the sanctuaries. Sylvia Earle and her team hope that their research will help people worldwide realize how important the oceans are and why we need to protect them.

Marine biologist Sylvia Earle

Earle studying animals in Hawaiian waters

What Can You Do?

Get Involved

How can you help the oceans? Recycle and do not pollute! About 77 percent of all pollutants in the oceans originate on land. Plastic and other trash in ocean waters kill fish, sea birds, turtles, seals, and whales.

Find Out More

Learn more about the 12 national marine sanctuaries. Find out where the sanctuaries are located and the types of marine organisms found in each. Visit the Web site www.sanctuaries.nos.noaa.gov. Locate the sanctuaries on a world map.

Use the Internet

Follow the progress of the Sustainable Seas Expeditions. Check the project's "Sanctuary Log" at http://sustainableseas.noaa.gov/missions/missions.html. You can also see what it is like to pilot *DeepWorker*, by visiting www.nationalgeographic.com/monterey

Chapter 3

The World's People

The World and Its People NATIONAL GEOGRAPHIC

To learn more about the world's culture regions, view ***The World and Its People*** **Chapter 3** video.

Geography Online

Chapter Overview Visit the ***Geography: The World and Its People*** Web site at gwip.glencoe.com and click on **Chapter 3–Chapter Overviews** to preview information about the world's people.

Section 1 Culture

Guide to Reading

Main Idea

People usually live with others who follow similar beliefs learned from the past.

Terms to Know

- culture
- ethnic group
- dialect
- monarchy
- dictator
- democracy
- economic system
- cultural diffusion
- civilization
- culture region

Places to Locate

- Iraq
- Egypt
- India
- China

Reading Strategy

Draw a diagram like this one. In each section, write one of the eight elements of culture and give an example of it from the United States today.

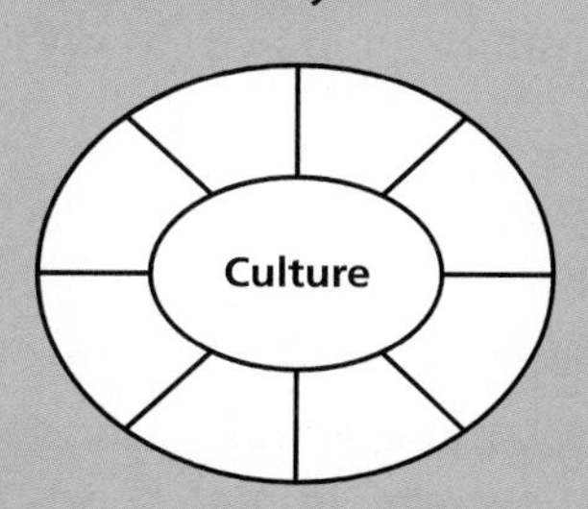

NATIONAL GEOGRAPHIC **Exploring Our World**

Three thousand years ago, the Olmec people lived in Mexico. They sometimes wore skins of jaguars, cats that were sacred to them. This young boy lives in an area where the jaguar is still honored. He is preparing for a jaguar dance. An object from modern culture—a soft drink bottle—is used to make the "jaguar" spots of ash on the boy's clay-covered skin.

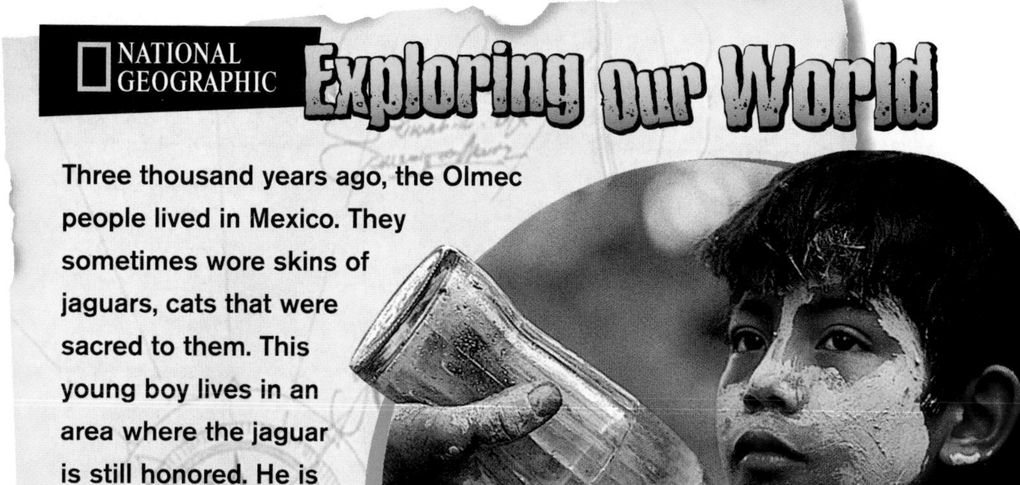

If you wake up to rock music, put on denim jeans, speak English, and celebrate the Fourth of July, those things are part of your culture. If you eat a tortilla, or flat bread, for breakfast, speak Spanish, and take part in celebrations honoring the jaguar, those things are part of your culture.

What Is Culture?

As used in geography, **culture** is the way of life of a group of people who share similar beliefs and customs. In studying a society's culture, geographers look at eight elements called *traits.* They study what groups the society is divided into, what language the people speak, and what religion they follow. They examine people's daily lives. They consider what history the people share and what artworks they have created. They also look at how the society is governed and how the people make a living.

Social Groups One way of studying cultures is by looking at the different groups of people in the society. For instance, geographers study how many people are rich, poor, and in the middle class. They

◀ Thousands of Indians meet for an annual Hindu festival.

NATIONAL GEOGRAPHIC

World Religions

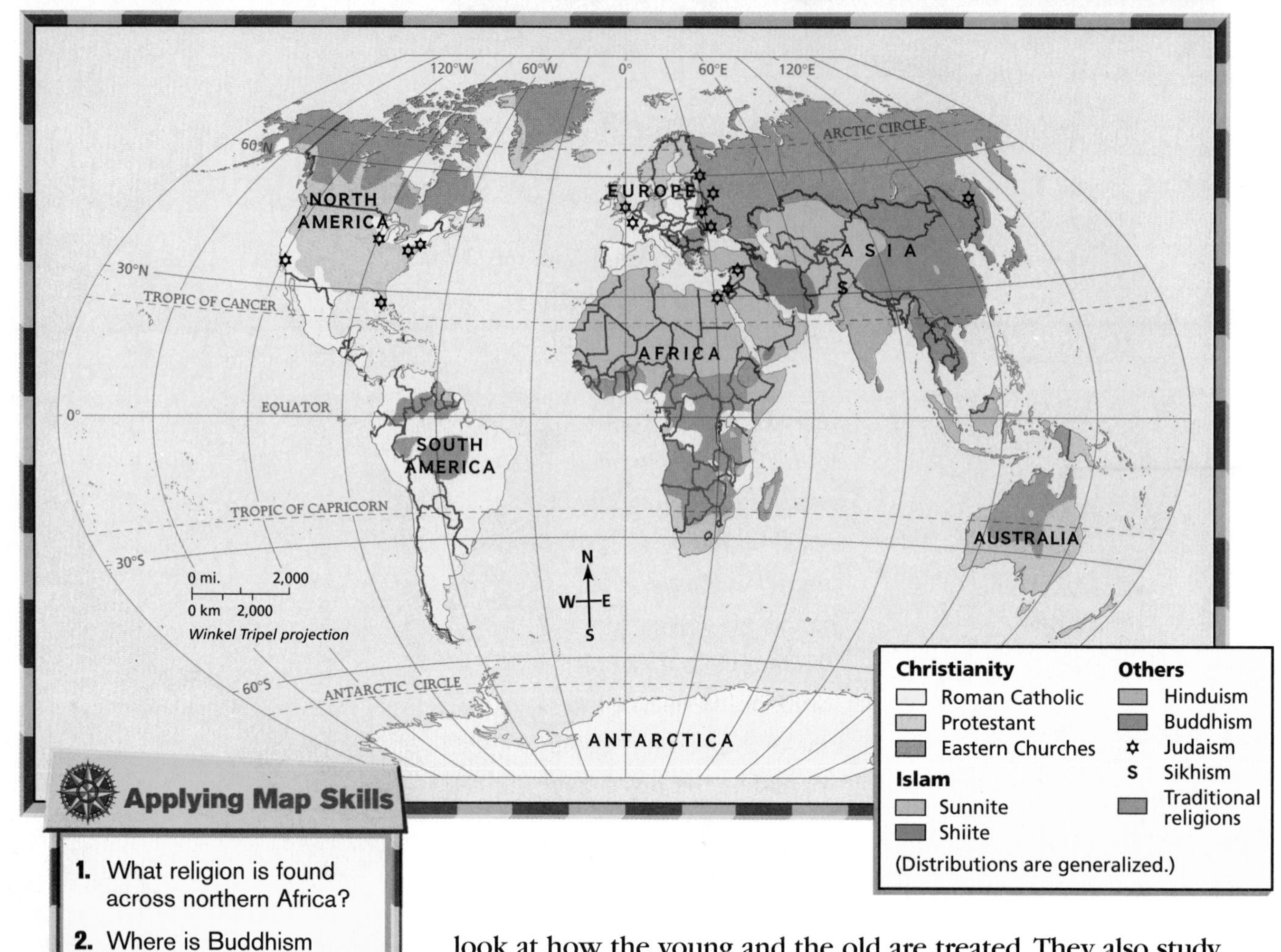

Applying Map Skills

1. What religion is found across northern Africa?
2. Where is Buddhism practiced?

Find NGS online map resources @ www.nationalgeographic.com/maps

look at how the young and the old are treated. They also study the roles of men and women. In many societies, males and females have different rights and responsibilities. Understanding those differences is a vital part of understanding a particular culture.

Geographers also examine how people in the culture treat others who are different from themselves. Most countries include people who belong to different ethnic groups. An **ethnic group** is a group of people who share a common culture, language, or history. Many societies have people who have moved there from another place. Many have people who practice different religions.

Language People use language to share information. Sharing a language is one of the strongest unifying forces for a culture. Even within a culture, though, geographers find language differences. Some people may speak a **dialect,** or a local form of a language that differs from the language in other areas. The differences may include pronunciation and the meaning of words. For example, people in the northeastern United States say "soda," whereas people in the Midwest say "pop." Both groups are referring to soft drinks, however.

Religion Another important part of culture is religion. In many cultures, religion helps people answer basic questions about life's meaning. Religious beliefs vary significantly around the world. Struggles over religious differences are a problem in many countries. Some of the major world religions are Buddhism, Christianity, Hinduism, Islam, and Judaism. The map on page 78 shows you the main areas where these religions are practiced.

Daily Life Do you enjoy eating pizza, tacos, yogurt, and eggrolls? All of these foods came from different cultures. What people eat and *how* they eat it—whether with their fingers, silverware, chopsticks, and so on—reflect their culture. What people wear also reflects cultural differences. The same is true of how people build traditional homes in their societies.

History A culture group has a shared history, and that history shapes how they view the world. People remember the successes of the past and often celebrate holidays to honor the heroes and heroines who brought about those successes. Stories about these heroes reveal the personal characteristics that the people think are important. A group also remembers the dark periods of history, when they met with disaster or defeat. These experiences, too, influence how a group of people sees itself.

Celebrations

In most cultures, people often bring out their most beautiful clothes for events like weddings. This wedding guest is from Morocco (upper left), and this bride is from Mauritius.

Place What aspects of daily life besides clothing reflect culture?

Arts People express their culture through the arts. Art is not just paintings and sculptures, but also architecture, dance, music, theater, and literature. By viewing the arts of a culture, you can gain insight into what the people of that culture think is beautiful and important.

Government The kind of government, or political system, a society has reflects its culture. Some countries are led by individuals. In a **monarchy,** kings and queens inherit the right to rule. In others, **dictators** take control of the government and rule the country as they wish.

In many countries today, power rests with the people of the nation. Citizens choose their leaders by voting for them. When the people of a country hold the power of government, we call that government a **democracy.** Some places have mixed forms of government. For example, the United Kingdom is both a monarchy and a democracy. The queen is the symbolic head of the country, but the power to rule is in the hands of elected leaders.

The Economy People must make a living, whether in farming or in industry or by providing services such as designing a Web page or preparing food. Geographers look at how people in a culture earn a living. They also look at the culture's economic system.

An **economic system** sets rules for how people decide what goods and services to produce and how they are exchanged. In a *traditional economy,* things are done "the way they have always been done." Economic decisions are based on customs and beliefs—often religious—handed down from generation to generation. For example, if your grandparents and parents fished for a living, you will fish for a living.

In a *market economy,* individuals make decisions about what to produce. People who own businesses make what they think customers want. Customers have the freedom to choose what products they will buy. A market economy is based on *free enterprise.* This is the idea that you have the right to own property or businesses and make a profit without the government interfering. People are free to choose what jobs they will do and for whom they will work.

In a *command economy,* however, the government owns businesses and controls decisions about what goods and services will be produced and who will receive them. A command economy is often called *socialism* or *communism,* depending on how much the government is involved. In some command economies, the government even decides which people receive training for particular jobs.

✓Reading Check **What is culture?**

Sri Lanka

Some people in Sri Lanka harvest tea (left). Others work in factories making dolls to be sold around the world (below).

Place **On what is a market economy based?**

NATIONAL GEOGRAPHIC

World Culture Regions

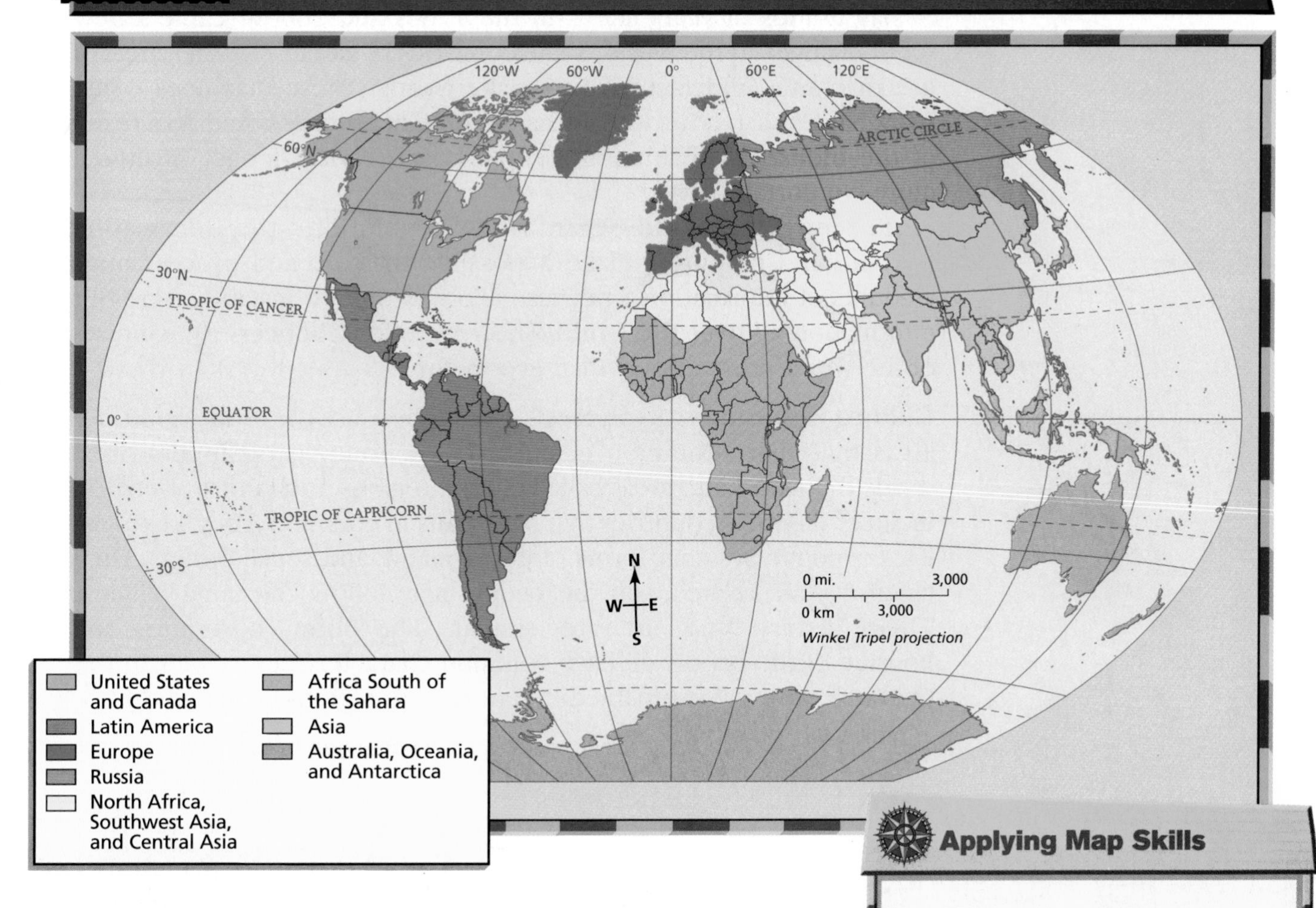

Applying Map Skills

1. Which culture region includes most nations of Africa?
2. What culture region is on the continents of both Africa and Asia?

Find NGS online map resources @ www.nationalgeographic.com/maps

Cultural Change

Cultures do not remain the same. Humans constantly invent new ideas and technologies and create new solutions to problems. Trade, the movement of people, and war can spread these changes to other cultures. The process of spreading new knowledge and skills to other cultures is called **cultural diffusion.** Today the Internet is making cultural diffusion take place more rapidly than ever before.

Culture Over Time Historians have traced the tremendous changes that humans have made in their cultures. In the first human societies, people lived by hunting animals and gathering fruits and vegetables. They were *nomadic,* or lived in small groups that moved from place to place to follow sources of food.

Starting about 10,000 years ago, people learned to grow food by planting seeds. This change brought about the Agricultural Revolution. Groups stayed in one place and built settlements. Their societies became more complex. As a result, four **civilizations,** or highly developed cultures, arose in river valleys in present-day **Iraq, Egypt, India,** and **China.** These civilizations included cities, complex governments and religions, and systems of writing. The map on page 83 shows you where these civilizations were located.

Thousands of years later—in the 1700s and 1800s—came a new set of changes in the world. Some countries began to *industrialize,* or use machines and factories to make goods. These machines could work harder, faster, and longer than people or animals could. As a result of the Industrial Revolution, people began to live longer, healthier, more comfortable lives.

Recently, the world began a new revolution—the Information Revolution. Computers make it possible to store and process huge amounts of information. They also allow people to instantly send this information all over the world. This revolution connects the cultures of the world more closely than ever before.

Culture Regions As you recall, geographers use the term *regions* for areas that share common characteristics. Today geographers often divide the world into areas called culture regions. Each **culture region** includes different countries that have traits in common. They share similar economic systems, forms of government, and social groups. Their languages are related, and the people may follow the same religion. Their history and art are similar. The food, costumes, and housing of the people all have common characteristics as well. In this textbook, you will study the different culture regions of the world.

Reading Check **What three revolutions have changed the world?**

Section 1 Assessment

Defining Terms

1. **Define** culture, ethnic group, dialect, monarchy, dictator, democracy, economic system, cultural diffusion, civilization, culture region.

Recalling Facts

2. **Culture** What kinds of social groups do geographers study?
3. **Government** What are the different forms of government a society may have?
4. **Culture** In what ways does cultural diffusion occur?

Critical Thinking

5. **Understanding Cause and Effect** How does history shape a people's culture?
6. **Making Comparisons** Describe two kinds of economic systems.

Graphic Organizer

7. **Organizing Information** Create a diagram like this one that describes features of your culture. On the lines write the types of food, clothing, language, music, and so on.

Applying Geography Skills

8. **Analyzing Maps** Look at the map on page 81. In which culture region do you live? In which culture region(s) did your ancestors live?

Geography Skill

Reading a Special Purpose Map

Special purpose maps concentrate on a single theme. This theme may be to show the battles of a particular war or endangered species, for example.

Learning the Skill

To read a special purpose map, follow these steps:

- Read the map title. It tells what kind of special information the map shows.
- Find the map's scale to determine the general size of the area.
- Read the key. Colors and symbols in the map key are especially important on this type of map.
- Analyze the areas on the map that are highlighted in the key. Look for patterns.

Practicing the Skill

Look at the map below to answer the following questions.

1. What is the title of the map?
2. Read the key. What four civilizations are shown on this map?
3. Which civilization was farthest west? East?
4. What do the locations of each of these civilizations have in common?

Applying the Skill

Find a special purpose map in a newspaper or magazine. Write three questions about the map's purpose, then have a classmate answer the questions.

GO TO Practice key skills with **Glencoe Skillbuilder Interactive Workbook, Level 1.**

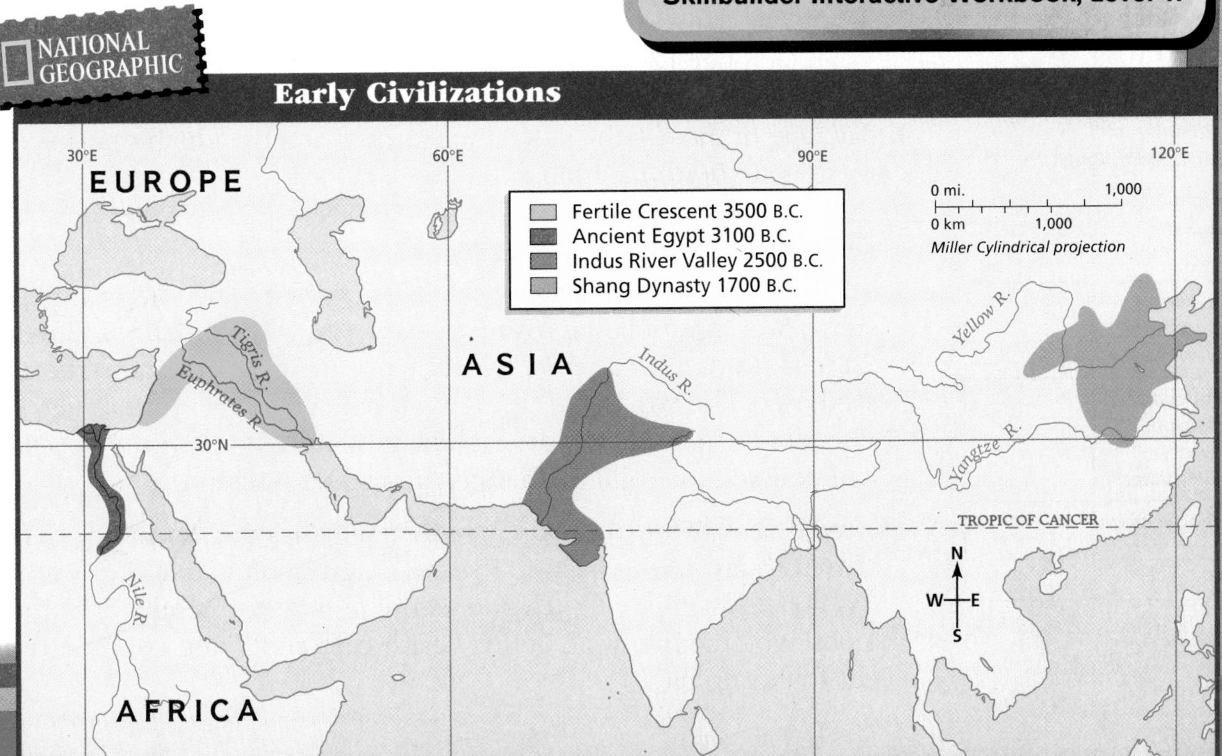

Section 2

Population

Guide to Reading

Main Idea

The world's population is growing rapidly, and how and where people live are changing, too.

Terms to Know

- death rate
- birthrate
- famine
- population density
- urbanization
- emigrate
- refugee

Places to Locate

- Afghanistan
- Nepal
- Mexico City
- Buenos Aires

Reading Strategy

Draw a chart like this one. In the right column, write a result of the fact listed in the left column.

Fact	Result
World population is increasing.	
Population is not evenly distributed.	
People move from place to place.	

NATIONAL GEOGRAPHIC **Exploring Our World**

Imagine that you and your friends are in Berlin, Germany. Can you hear the music? Every summer, hundreds of thousands of young people gather here for a music festival. Although most of these young people are here only to visit, many thousands of others come to find jobs and new lives. Germany faces challenges in finding room for its newcomers.

On October 12, 1999, the world reached a significant turning point in its history. About 370,000 babies were born around the world that day. One of those babies—no one knows exactly which one—was the world's six *billionth* human being.

Population Growth

How fast has the earth's population grown? The chart on page 85 shows world population over the years. You will see that for more than fifteen hundred years, the world's population remained about the same size. The world did not have 1 billion people until about 1800. The second billion was not reached until 1930. By 1974 the population had doubled to 4 billion. In 1999, it reached 6 billion.

Reasons for Population Growth Why has the world's population grown so fast in the past 200 years? One reason is that the death rate has gone down. The **death rate** is the number of people out of every 1,000 who die in a year. Better health care and living conditions have cut the death rate.

World Population

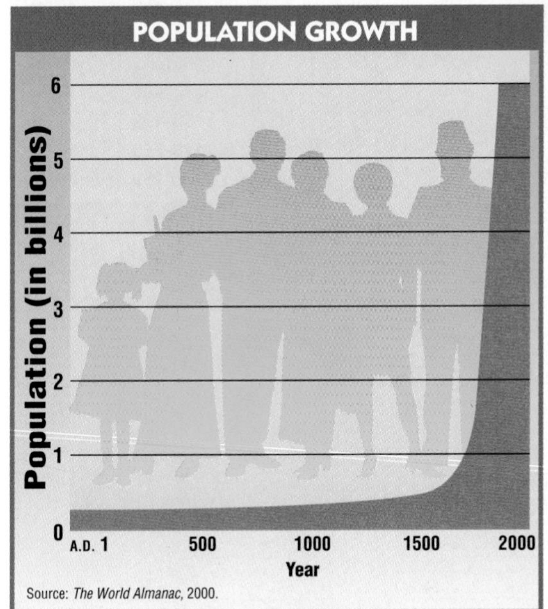

MOST POPULOUS COUNTRIES

Country	Millions of People
China	1,254.1
India	1,000.8
United States	281.0
Indonesia	211.8
Brazil	168.0
Pakistan	146.5
Russia	146.5

Sources: *National Geographic Atlas of the World; The World Almanac*, 2000.

Analyzing the Graph and Chart

The world's population is expected to reach about 9 billion by 2050.

Place **Which country has the second-largest number of people?**

Visit gwip.glencoe.com and click on **Chapter 3—Textbook Updates.**

Another reason for the fast growth in the world's population is that in some regions of the world the birthrate is high. The **birthrate** is the number of children born each year for every 1,000 people. In Asia, Africa, and Latin America, families traditionally are large because children help with farming. High numbers of births have combined with low death rates to increase population growth in these areas. As a result, population in these continents has doubled every 25 years or so.

Challenges From Population Growth Rapid population growth presents many challenges. An increase in the number of people means that more food is needed. Fortunately, since 1950 world food production has increased faster than population on all continents except Africa. Because so many people there need food, bad weather or war can ruin crops and bring disaster. Millions may suffer from **famine,** or lack of food.

Also, populations that grow rapidly may use resources more quickly than populations that do not grow as fast. Some countries face shortages of water and housing. Population growth also puts a strain on economies. More people means a country must create more jobs. Some experts claim that rapid population growth could badly affect the planet. Others are optimistic. They predict that as the number of humans rises, the levels of technology and creativity will also rise.

✓Reading Check **How do the definitions of death rate and birthrate differ?**

Where People Live

Where do all the people live? Actually, the world's people live on a surprisingly small part of the earth. As you learned in Chapter 2, land covers only about 30 percent of the earth's surface. Half of this land is

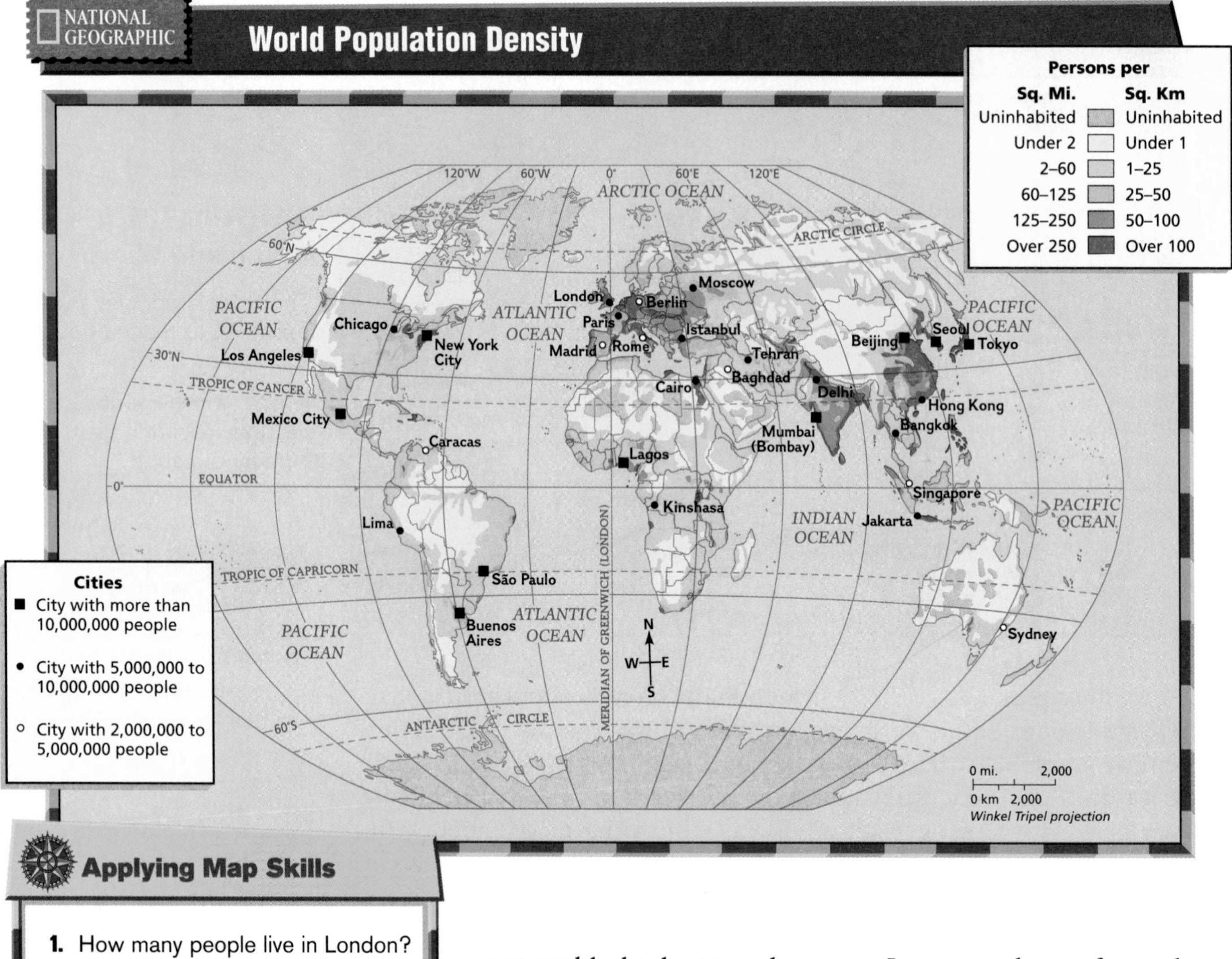

Applying Map Skills

1. How many people live in London?
2. What cities have more than 10 million people?

Find NGS online map resources @ www.nationalgeographic.com/maps

not usable by humans, however. Large numbers of people cannot survive on land covered with ice, deserts, or high mountains. The world's people, then, live on a small fraction of the earth's surface.

Population Distribution Even on the usable land, population is not distributed, or spread, evenly. People naturally prefer to live in places that have plentiful water, good land, and a favorable climate. During the industrial age, people moved to places that had important resources such as coal or iron ore to run or make machines. People gather in other areas because these places hold religious significance or because they are government and transportation centers. The table on page 85 shows you the most populous countries in the world. Four of these countries are located on the Asian continent.

Population Density Geographers have a way of determining how crowded a country or region is. They measure **population density**—the average number of people living in a square mile or square kilometer. To arrive at this figure, the total population is divided by the total land area. For example, the countries of **Afghanistan** and **Nepal**

have about the same number of people. They are very different in terms of population density, though. With a smaller land area, Nepal has 447 people per square mile (173 people per sq. km). Afghanistan has an average of only 103 people per square mile (40 people per sq. km). Nepal, then, is more crowded than Afghanistan.

Remember that population density is an *average.* It assumes that people are distributed evenly throughout a country. Of course, this seldom happens. A country may have several large cities where most of the people actually live. In Egypt, for example, overall population density is 173 people per square mile (67 people per sq. km). In reality, about 99 percent of Egypt's people live within 20 miles (32 km) of the Nile River. The rest of Egypt is desert. Thus, some geographers prefer to figure a country's population density in terms of farmable or usable land rather than total land area. When Egypt's population density is measured this way, it equals about *5,550 people* per square mile. The map on page 13 of the **Geography Handbook** shows how population density can vary within a country. The areas with high density in Egypt follow the path of the Nile River.

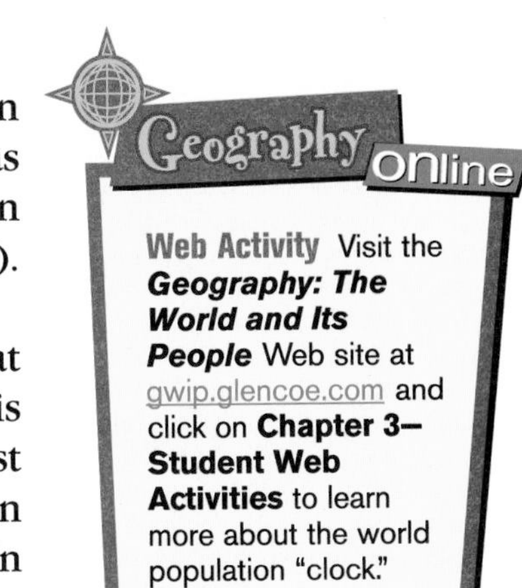

Web Activity Visit the ***Geography: The World and Its People*** Web site at gwip.glencoe.com and click on **Chapter 3—Student Web Activities** to learn more about the world population "clock."

Reading Check **What is population density?**

Population Movement

Throughout the world, people are moving in great numbers from place to place. Some people move from city to city, or suburb to suburb. More and more people are leaving villages and farms and moving to cities. This movement to cities is called **urbanization.**

NATIONAL GEOGRAPHIC **On Location**

Kosovo, Yugoslavia

In 1999 a civil war exploded in Kosovo, a province of Yugoslavia. Thousands of people were forced from their homes.

Movement **What causes people to become refugees?**

People move to cities for many reasons. The biggest one is to find jobs. Rural populations have grown, but the amount of land that can be farmed has not increased to meet the growing number of people who need to work and to eat. As a result, many people find city jobs in manufacturing or in services like tourism.

Nearly half the world's people live in cities—a far higher percentage than ever before. Between 1960 and 2000, the population of **Mexico City** more than tripled. Other cities in Latin America, as well as cities in Asia and Africa, have seen similar growth. Some of these cities hold a large part of a country's entire population. About one-third of Argentina's people, for instance, live in the city of **Buenos Aires.**

Some population movement is between countries. Some people **emigrate,** or leave the country where they were born and move to another. They are called *emigrants* in their homeland and referred to as *immigrants* in their new country. In the past 40 years, millions have left Africa, Asia, and Latin America to find jobs in the richer nations of Europe and North America. Some people were forced to flee their country because of wars, political unrest, food shortages, or other problems. They are **refugees,** or people who flee to another country to escape persecution or disaster.

Reading Check **Why do so many people move from rural areas to cities?**

Assessment

Defining Terms

1. **Define** death rate, birthrate, famine, population density, urbanization, emigrate, refugee.

Recalling Facts

2. **Culture** What are three problems caused by overpopulation?
3. **Human/Environment Interaction** Why do people live on only a small fraction of the earth?
4. **Economics** What is the main reason for growing urbanization?

Critical Thinking

5. **Making Comparisons** What is the difference between an emigrant and an immigrant?
6. **Understanding Cause and Effect** Why have populations in areas of Asia, Africa, and Latin America doubled every 25 years or so?

Graphic Organizer

7. **Organizing Information** Draw a diagram like this one, and list three causes of population growth.

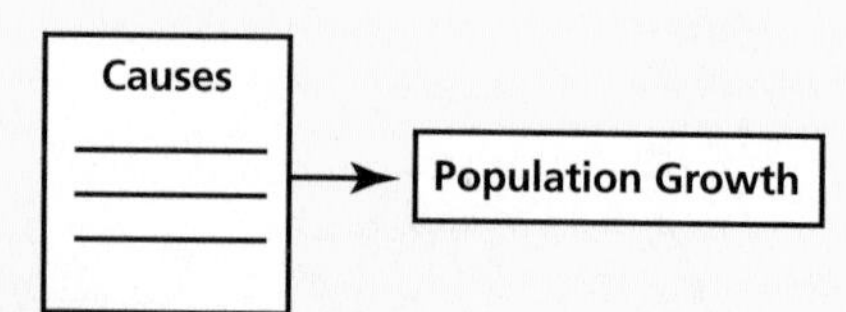

Applying Geography Skills

8. **Analyzing Maps** Look at the population density map on page 86. How would you describe the population density around Tokyo?

Making Connections

ART | SCIENCE | 

| TECHNOLOGY

Counting Heads

How do we know there are more than 280 million people in the United States? Who counts the people? Every 10 years since 1790, the United States Census Bureau has counted heads in this country. Why and how do they do this?

The First Census

After the American colonies fought the Revolutionary War and won their independence, the new government ordered a census. By knowing how many people were in each state, the government could divide the war expenses fairly. The census would also determine the number of people that each state could send to Congress.

This census began in August 1790, about a year after George Washington became president. The law defined who would be counted and required that every household be visited by census takers. These workers walked or rode on horseback to gather their data. By the time it was completed, the census counted 3.9 million people.

The first census asked for little more than one's name and address. Over time, the census added questions to gather more than just population data. By 1820, there were questions about a person's job. Soon after, questions about crime, education, and wages appeared.

Changing Technology

As the country's population grew and the quantity of data increased, new technology helped census workers. In 1890 clerks began to use a keypunch device, invented by a Census Bureau worker, to add the numbers. The Tabulating Machine, as it was called, used an electric current to sense holes in punched cards and to keep a running total of the data. In 1950 the census used its first computer to process data. Now census data are released over the Internet.

Remarkably, one technology slow to change has been the way the government takes the census. Not until 1960 did the U.S. Postal Service become the major means of conducting the census. Even today, census takers go door-to-door to gather information from those who do not return their census forms in the mail.

▲ **The Electric Tabulating Machine processed the 1890 census in 2½ years, a job that would have taken nearly 10 years to complete by hand.**

Making the Connection

1. In what two ways were population data from the first census used?
2. How has technology changed the way census data are collected and processed?
3. **Drawing Conclusions** Why do you think the national and state governments want information about people's education and jobs?

Section 3

Resources and World Trade

Guide to Reading

Main Idea

Because many resources are limited and distributed unevenly, countries must trade for goods.

Terms to Know

- natural resource
- renewable resource
- nonrenewable resource
- export
- import
- tariff
- quota
- free trade
- developed country
- developing country

Places to Locate

- Brazil
- China

Reading Strategy

Draw a chart like the one below, then write in the names of different resources and how they are used.

Resource	Use

NATIONAL GEOGRAPHIC **Exploring Our World**

About 7,000 windmills stand on an 80-square-mile patch of hilly land near San Francisco. They turn in the strong winds that blow through a nearby pass in California's mountains. Why were they put there? These wind vanes generate electricity. In fact, they churn out enough electricity every year to meet the needs of all the homes in San Francisco.

Natural Resources

As you learned in Section 2, people settle in some areas to gain access to resources. **Natural resources** are products of the earth that people use to meet their needs. Wind, water, and oil are resources that provide energy to power machines. Good soil and fish are resources that people use to produce food. Stones like granite and ores like iron ore are resources people can use for making products.

The value of resources changes as people discover new technology. Trees, for example, have been a valuable resource throughout history. People used wood to stay warm, cook food, and build homes. Oil, on the other hand, was a gooey nuisance until the Industrial Revolution. People soon began to use oil products to run cars and heat homes. Today many countries in Southwest Asia have great pools of underground oil. As a result, they are among the richest countries on the earth.

Renewable Resources People can use some natural resources as much as they want. These **renewable resources** cannot be used up or can be replaced naturally or grown again. Wind and sun cannot be used up—the wind will continue to blow even if we use a windmill to catch some of its power. Forests, grasslands, plants and animals, and soil can be replaced—*if* people manage them carefully. A lumber company concerned about future growth can plant as many new trees as it cuts. Fishing fleets can limit the number of fish they catch to make sure that enough fish remain to reproduce.

Today many countries are trying to find efficient ways of using renewable energy sources. Some produce *hydroelectric power,* the energy generated by falling water. Even if a river does not flow quickly, it is possible to turn it into hydroelectric power. Engineers can build a dam and then release the water in a powerful stream.

Do you have a solar-powered calculator? If so you know that the sun can provide energy to run people's machines. *Solar energy* is power produced by the heat of the sun. Making use of this energy on a large scale requires huge pieces of equipment. As a result, this energy source is not yet economical to use.

Nonrenewable Resources Humans use a great variety of minerals. They make steel from iron and aluminum from bauxite. Can you think of other examples? Metals and other minerals found in the earth's crust are also resources. They are **nonrenewable resources** because the earth provides limited supplies of them and they cannot be replaced. These resources were formed over millions of years by forces within the earth. Thus, it simply takes too long to generate new supplies.

One major nonrenewable source of energy is *fossil fuels*—coal, oil, and natural gas. People burn oil and gas to heat homes or run cars. They burn fossil fuels to generate electricity. Oil and coal are also used as raw materials to make plastics and medicines.

Another nonrenewable energy source is nuclear energy. *Nuclear energy* is power made by creating a controlled atomic reaction. Nuclear energy can be used to produce electricity, but some people fear its use. Nuclear reactions produce dangerous waste products that are difficult to dispose of. They need thousands of years to become safe. Some people worry that there is no safe way to transport and store this waste. Still, some countries rely on nuclear energy to generate power. France, Japan, South Korea, and Taiwan are examples.

✓Reading Check **What are three fossil fuels?** coal, oil, gas

World Trade

Resources, like people, are not distributed evenly around the world. Some areas have large amounts of one resource. Others have none of that resource but are rich in another one. These differences affect the economies of the world's countries.

Saffron—A Valuable Resource

A resource does not have to produce energy to be valued. The people in the Indian region of Kashmir are picking a resource that is precious to cooks—crocus flowers. Inside each crocus are three tiny orange stalks. When dried, the stalks become a spice called saffron. Cooks use it to add a delicate orange color and flavor to food. Saffron—the world's most expensive spice—is in short supply, though. Producers need nearly 4,700 flowers to produce just 1 ounce (28 g) of saffron.

World Economic Activity

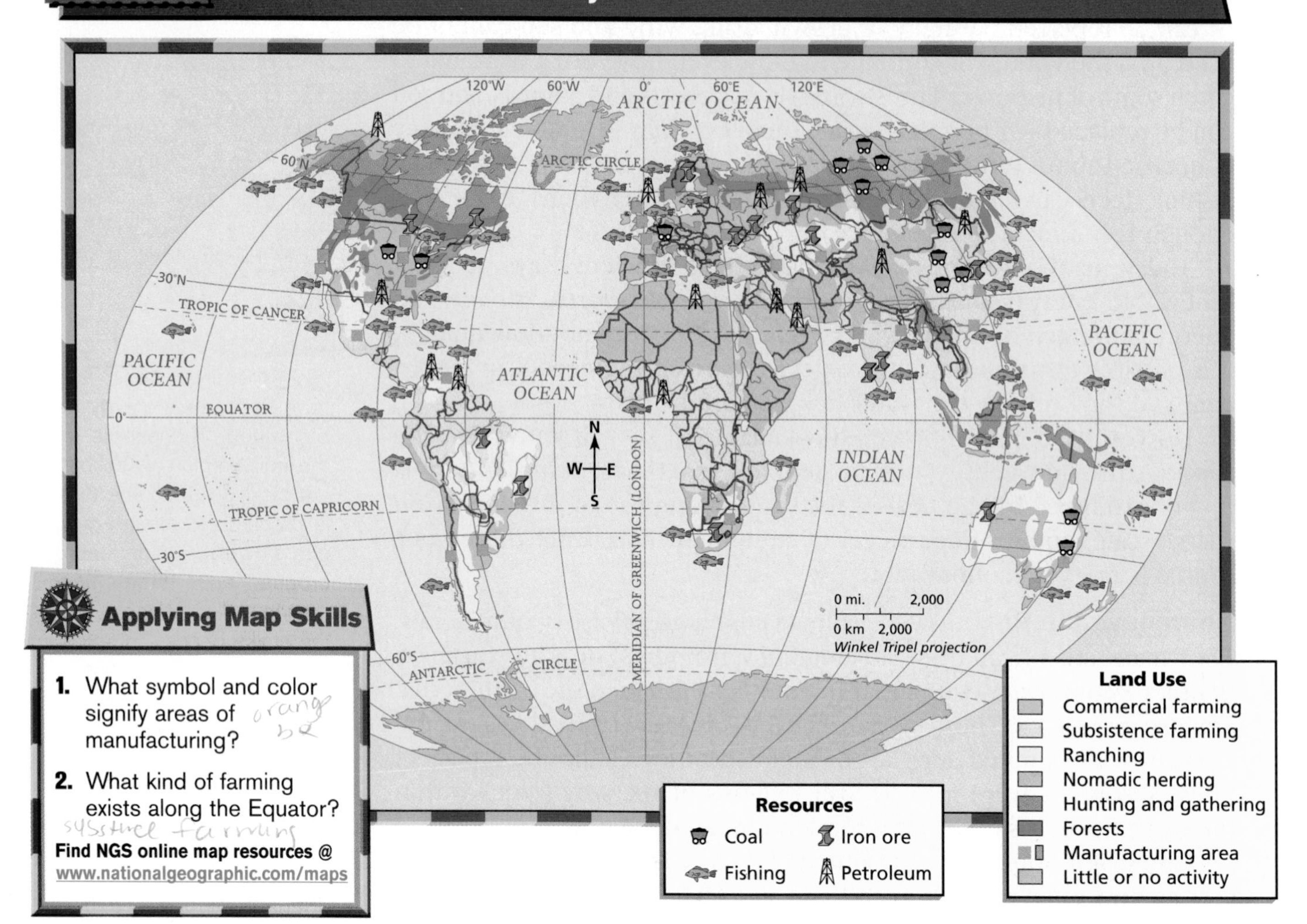

Applying Map Skills

1. What symbol and color signify areas of manufacturing?
2. What kind of farming exists along the Equator?

Find NGS online map resources @ www.nationalgeographic.com/maps

Look at the map above. Do you see the centers of manufacturing in the northern and eastern United States? There are large supplies of coal in the region and deposits of iron ore nearby. These areas became industrial centers because the people here took advantage of the resources they had.

In the western United States, you see another picture. People use much of the land for ranching. The soil and climate are well suited to raising livestock. *Commercial farming*—or growing food for sale in markets—occurs throughout much of the United States.

These examples show how the world's people respond to the unequal distribution of resources. They *specialize,* or focus on the economic activities best suited to their resources. Parts of **Brazil** have the perfect soil and climate for growing coffee. As a result, Brazil produces more coffee beans than any other country. Cotton grows well in parts of **China,** the world's top producer of that product.

Countries often cannot use all that they produce. What do they do with the extra? They **export** what they do not need, trading it to other countries. When they cannot produce as much as they need of a good, they **import** it, or buy it from another country. The world's countries, then, are connected with one another in a complex web of trade.

Barriers to Trade Governments try to manage their country's trade to benefit their country's economy. Some charge a **tariff,** or a tax added to the price of goods that are imported. If there is a tariff on cars, for instance, all people who buy an imported car pay more than just the purchase price. Governments often create tariffs hoping to persuade their own people to buy products made in their own country instead of buying imported goods.

Governments sometimes create other barriers to trade. They might put a strict **quota,** or number limit, on how many items of a particular product can be imported from a particular country. A government may even stop trading with another country altogether as a way to punish it.

Free Trade In recent years, governments around the world have moved toward free trade. **Free trade** means taking down trade barriers so that goods flow freely among countries. Several countries have joined together to create free trade agreements in certain parts of the world. For instance, the United States, Mexico, and Canada have agreed to eliminate all trade barriers to one another's goods. These three countries set up the North American Free Trade Agreement (NAFTA). The largest free trade agreement—the European Union (EU)—includes most of the countries of Western Europe.

✓Reading Check **What are three kinds of barriers to trade?**

Earning a Living

Factory workers in China sew clothes (left). Most people still farm in Kenya (above). In the United States, more jobs are in service or technology industries.

Region **What regions are considered developed?**

Differences in Development

Look again at the economic activity map on page 92. You see that countries differ in how much manufacturing they have. Countries that have a great deal of manufacturing are called **developed countries.** The countries of Europe and North America are developed countries. So are Australia and Japan. Other countries have only a few—or no—manufacturing centers. Many people in these countries practice *subsistence farming,* or grow only enough food for their own families. These countries, which are working toward industrialization, are called **developing countries.** Developing countries are found throughout Africa, Asia, and Latin America.

Why do countries want manufacturing? Workers in industry typically earn more than farmers. Industrial companies can generally charge more for their products than food companies can. As a result, industrial countries are wealthier than agricultural countries.

In recent decades, the economies of many countries have changed dramatically. Developing countries have invited large companies from developed countries to build factories in their lands. The companies find that developing countries have a valuable resource—people. The spread of industry has created booming economies in places like Hong Kong, Singapore, South Korea, Taiwan, and China.

✓Reading Check **Why do developing countries want more industry?**

Industry brings greater wealth.

Section 3 Assessment

Defining Terms

1. **Define** natural resource, renewable resource, nonrenewable resource, export, import, tariff, quota, free trade, developed country, developing country.

Recalling Facts

2. **History** How can the value of resources change over time?
3. **Economics** What is the difference between commercial farming and subsistence farming?
4. **Economics** Why do countries specialize in producing certain goods?
5. **Economics** What is the difference between a developing country and a developed country?

Critical Thinking

6. **Drawing Conclusions** Why are tariffs and quotas called "barriers" to trade?
7. **Analyzing Information** How has the world's economy changed in recent decades?

Graphic Organizer

8. **Organizing Information** Draw a chart like this one, listing three examples for each type of resource.

Renewable resources	Nonrenewable resources

Applying Geography Skills

9. **Analyzing Maps** Look at the economic activity map on page 92. What two types of farming are shown on the map?

Section 4

People and the Environment

Guide to Reading

Main Idea

The actions that people take have a huge effect on the environment.

Terms to Know

- desalinization
- conservation
- pesticide
- ecosystem
- crop rotation
- irrigation
- erosion
- deforestation
- acid rain

Places to Locate

- Saudi Arabia

Reading Strategy

Draw a diagram like this one. On the lines write at least two problems that arise with human use of water, land, and air.

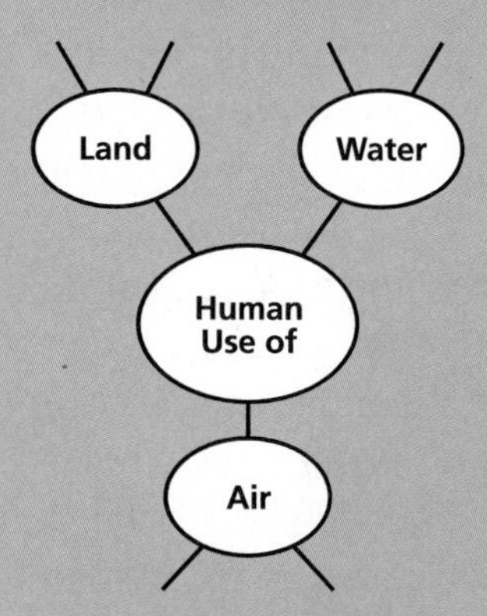

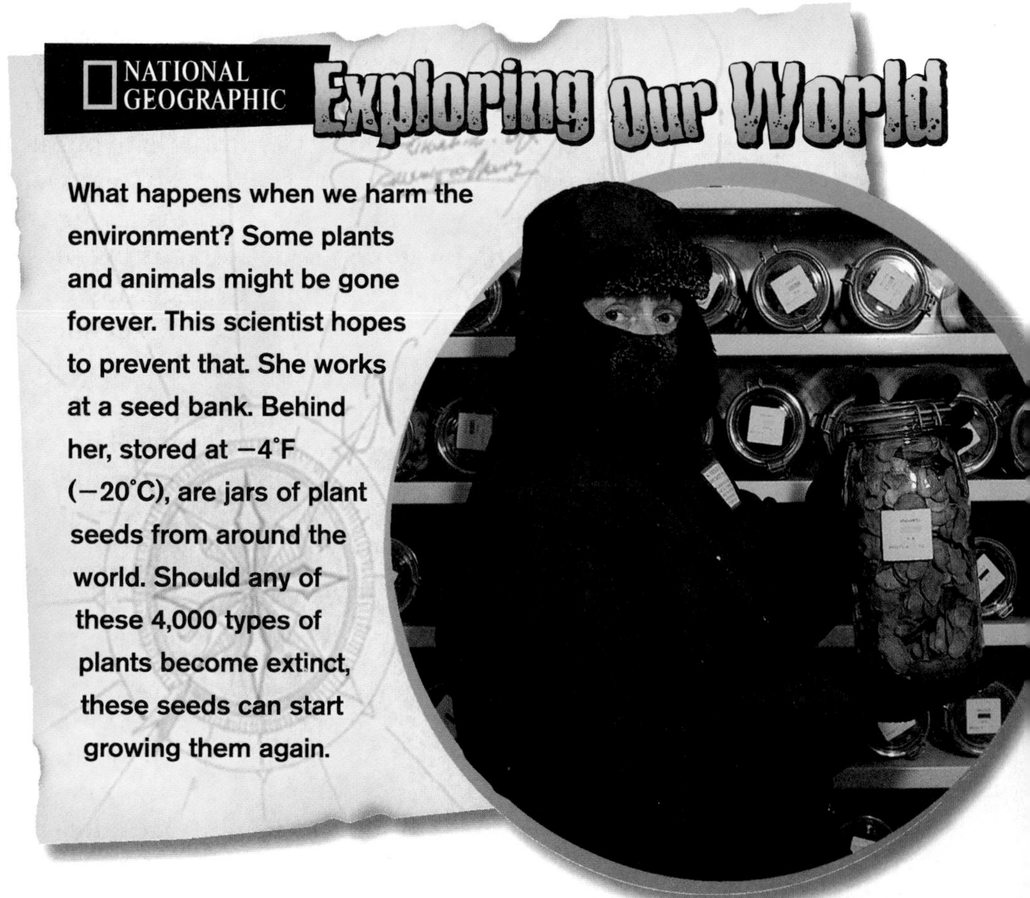

What happens when we harm the environment? Some plants and animals might be gone forever. This scientist hopes to prevent that. She works at a seed bank. Behind her, stored at −4°F (−20°C), are jars of plant seeds from around the world. Should any of these 4,000 types of plants become extinct, these seeds can start growing them again.

The rapidly growing number of people threatens the delicate balance of life in the world. More and more people use more water. They need more land to live on and to grow more food. Spreading industry fouls the air. Humans must act carefully to be sure not to destroy the earth that gives us life.

Water Use

As you recall from Chapter 2, people, plants, and animals need freshwater to live. People need clean water to drink. They also need water for their crops and their animals. In fact, as much as 70 percent of the water used is for farming. Only a small fraction of the world's water is freshwater, though. Some countries, such as **Saudi Arabia,** obtain drinkable water through **desalinization**—or removing salt from seawater. This is a process in which seawater is boiled in huge chambers, and the resulting steam is condensed as freshwater. For now, though, this process is expensive and cannot be widely used. Because the earth's supply of water is limited, people must manage freshwater carefully.

Water Management Some regions receive heavy rainfall in some months of the year and little, or none, in other months. They can manage their water supply by building storage areas to hold the heavy rains for later use. People have to build these storage areas carefully to prevent too much of the water from evaporating.

Managing water supplies involves two main steps. The first step is **conservation,** or the careful use of resources so they are not wasted. Did you know that 6 or 7 gallons (23 to 27 liters) of water go down the drain every *minute* that you shower? Taking shorter showers is an easy way to prevent wasting water.

The second approach to managing the water supply is to avoid polluting water. Some manufacturing processes use water. Sometimes those processes result in dangerous chemicals or dirt entering the water supply. Many farmers use fertilizers to help their crops grow. Many also use **pesticides,** or powerful chemicals that kill crop-destroying insects. These substances can seep into the water supply and cause harm.

✓Reading Check **How can manufacturing and farming harm the water supply?**

NATIONAL GEOGRAPHIC On Location

Disappearing Rain Forests

Workers harvest trees in a rain forest in Ecuador.

Human/Environment Interaction **Why is cutting down the rain forests a problem?**

Land Use

In addition to managing water resources, people must carefully manage the land they use. As humans expand their communities, they invade **ecosystems.** These are places where the plants and animals are dependent upon one another—and their particular surroundings—for survival. Ecosystems can be found in every climate and vegetation region of the world. For example, some people may want to drain a wet, marshy area to get rid of disease-carrying mosquitoes and make the soil useful for farming. When the area is drained, however, the ecosystem is destroyed. The delicate balance among the insects, reptiles, birds, and water plants is torn apart. In managing land, humans must recognize that some soils are not suited for growing certain crops—and some land is not useful for growing any crops at all.

Soil To grow food, soil needs to have certain minerals. Farmers add fertilizers to the soil to supply some of these minerals. Some also practice **crop rotation,** or changing what they plant in a field to avoid using up all the minerals in the soil. Some crops—like beans—actually restore valuable minerals to the soil. Many farmers now plant bean crops every three years to build the soil back up.

In many dry areas, farmers use **irrigation** to deliver water to their crops. Over time, the small amounts of salt in this freshwater can build up on the land. Once that happens, the land is no longer fertile.

If people do not carefully manage the soil, it can erode away. In **erosion,** wind or water carries soil away, leaving the land less fertile than before. Have you ever seen a group of trees alongside a farmer's field? The farmer may have planted those trees to block the wind and prevent erosion. In the tropics, erosion by water presents a problem—especially if farmers plant their crops on sloping land. When heavy rains come, the soil may simply wash down the hillside.

Forests Some areas of the world have thick forests of tall trees. Growing populations in these countries often turn to these lush forests as a source of land to grow food. Yet **deforestation,** or cutting down forests, is a problem—especially in the tropics. Rains in these areas are extremely heavy. When the tree roots are no longer there to hold the soil, the water can wash it away. Often, the cleared land is fertile for just a few years. Then farmers have to move to a new area and cut the trees in that section of forest. As a result, more and more of the forest is lost over time.

Taiwan

Burning fossil fuels adds harmful chemicals to the air.

Human/Environment Interaction What are some effects of air pollution?

✓Reading Check **What problem can result from irrigation?**

Air Pollution

Industries and vehicles that burn fossil fuels are the main sources of air pollution. Throughout the world, fumes from cars and other vehicles pollute the air. The chemicals in air pollution can seriously damage people's health.

These chemicals also combine with precipitation. They then fall as **acid rain,** or rain containing high amounts of chemical pollutants. Acid rain kills fish and eats away at the surfaces of buildings. It can even destroy entire forests.

Some scientists believe that increasing amounts of pollutants in the atmosphere will cause the earth to warm. You learned about this greenhouse effect in Chapter 2. While not all experts agree, some scientists say that the increase in temperature can have disastrous effects. Glaciers and ice caps may melt, raising the level of the world's seas. Higher seas could flood coastal cities. Warmer temperatures can also make some land no longer able to produce food.

✓Reading Check **What causes acid rain?**

Balancing People and Resources

Water, land, and air are among people's most precious resources. We need water and air to live. We need land to grow food. Only by caring for these resources can we be sure that we will still have them to use in the future.

Sometimes, though, protecting the environment for the future seems to clash with feeding people in the present. For example, farmers destroy the rain forests not because they want to but because they need to feed their families. They dislike being told by people in other countries that they should save the rain forests. Before they stop cutting down the rain forests, these farmers will need to find new ways to meet their needs.

✓Reading Check **How does saving the rain forests clash with current human needs?**

Assessment

Defining Terms

1. Define desalinization, conservation, pesticide, ecosystem, crop rotation, irrigation, erosion, deforestation, acid rain.

Recalling Facts

2. Human/Environment Interaction How do human activities affect ecosystems?

3. Human/Environment Interaction What are three ways of managing water?

4. Economics Why do farmers practice crop rotation?

Critical Thinking

5. Making Comparisons Which resource—water, soil, or air—do you think is most precious to people? Why?

6. Analyzing Information What ecosystems were affected by the growth of your community?

Graphic Organizer

7. Organizing Information Draw a diagram like this one and list three results of global warming.

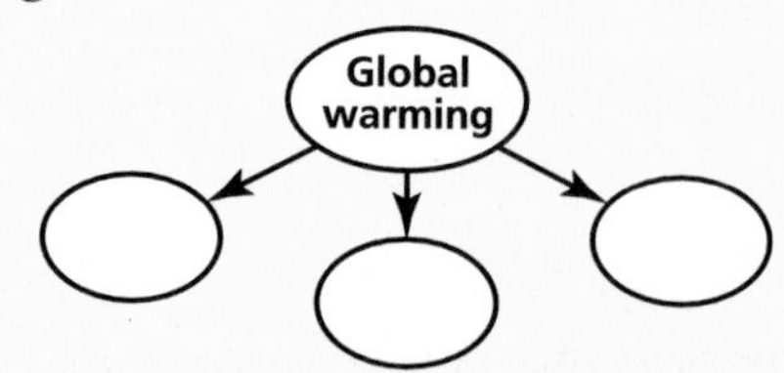

Applying Geography Skills

8. Analyzing Maps Look at the vegetation map in Chapter 2 on page 66. In what parts of the world do you see tropical rain forests?

Reading Review

Section 1 Culture

Terms to Know
culture
ethnic group
dialect
monarchy
dictator
democracy
economic system
cultural diffusion
civilization
culture region

Main Idea
People usually live with others who follow similar beliefs learned from the past.

✓ Culture *Culture* is the way of life of a group of people who share similar beliefs and customs.

✓ Culture Culture includes eight elements or traits: social groups, language, religion, daily life, history, arts, a government system, and an economic system.

✓ Culture Cultures change over time and influence other regions.

Section 2 Population

Terms to Know
death rate
birthrate
famine
population density
urbanization
emigrate
refugee

Main Idea
The world's population is growing rapidly, and how and where people live are changing, too.

✓ History In the past 200 years, the world's population has grown at a very rapid rate.

✓ Movement Some areas are more densely populated than others.

✓ Culture About 50 percent of the world's people live in cities.

Section 3 Resources and World Trade

Terms to Know
natural resource
renewable resource
nonrenewable resource
export
import
tariff
quota
free trade
developed country
developing country

Main Idea
Because many resources are limited and distributed unevenly, countries must trade for goods.

✓ Human/Environment Interaction Renewable resources cannot be used up or can be replaced fairly quickly.

✓ Human/Environment Interaction Some resources—such as fossil fuels and minerals—are nonrenewable.

✓ Economics Countries specialize by producing what they can produce best with the resources they have.

✓ Economics Countries export their specialized products and import what they need.

Section 4 People and the Environment

Terms to Know
desalinization
conservation
pesticide
ecosystem
crop rotation
irrigation
erosion
deforestation
acid rain

Main Idea
The actions that people take have a profound effect on the environment.

✓ Human/Environment Interaction People need to manage water resources because freshwater is not available everywhere.

✓ Human/Environment Interaction Air pollution has damaging effects on the land and on people's health.

Chapter 3

Assessment and Activities

Using Key Terms

Match the terms in Part A with their definitions in Part B.

A.

1. culture
2. developed country
3. irrigation
4. crop rotation
5. population density
6. culture region
7. tariff
8. quota
9. developing country
10. cultural diffusion

B.

a. collecting water and bringing it to crops
b. spreading knowledge to other cultures
c. countries working toward industrialization
d. many different countries with cultural traits in common
e. a number limit on imports from a country
f. the average number of people living in a square mile
g. country where much manufacturing is carried out
h. the way of life of a group of people who share similar beliefs and customs
i. a tax added to the price of imported goods
j. alternating what is planted in a field

Reviewing the Main Ideas

Section 1 Culture

11. **Culture** What are the major religions?
12. **Economics** What is the difference between a market economy and socialism?
13. **Movement** Give an example of cultural diffusion.

Section 2 Population

14. **Culture** What has created rapid population growth?
15. **Culture** How do you calculate population density?
16. **Movement** Why have many people moved to cities?

Section 3 Resources and World Trade

17. **Human/Environment Interaction** What are three kinds of renewable energy sources?
18. **Economics** How do countries respond to the unequal distribution of resources?

Section 4 People and the Environment

19. **Human/Environment Interaction** How can farmers restore the minerals in the soil?
20. **Human/Environment Interaction** What two problems can result from air pollution?

World Culture Regions

Place Location Activity

On a separate sheet of paper, match the letters on the map with the numbered places listed below.

1. Latin America
2. North Africa, Southwest Asia, and Central Asia
3. Europe
4. Russia
5. East Asia
6. United States and Canada
7. Australia, Oceania, and Antarctica
8. Africa South of the Sahara

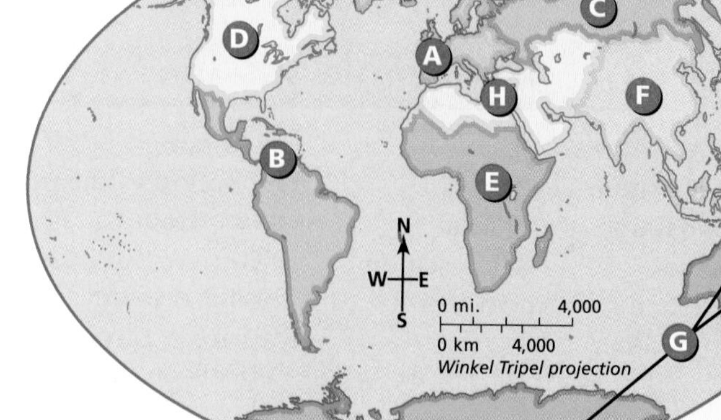

Self-Check Quiz Visit the ***Geography: The World and Its People*** Web site at gwip.glencoe.com and click on **Chapter 3—Self-Check Quizzes** to prepare for the Chapter Test.

Critical Thinking

21. **Making Predictions** In what ways do you think a company investing in a developing country could help the people there? How could that same company harm the culture?
22. **Sequencing Information** Make a chart like the one below, and list the ways you use electricity from the moment you wake up until you go to sleep. In the second column, write how you would perform the same activity if you had no electricity to rely on.

Activities With Electricity	Without Electricity

GeoJournal Activity

23. **Writing a Paragraph** Write a paragraph about the settlement of your community. Answer such questions as: Why did people originally settle in the area? How has the culture of your area changed?

Mental Mapping Activity

24. **Focusing on the Region** Draw a simple outline map of the United States. On your map, label the areas where the following activities take place:
    - Commercial farming
    - Manufacturing
    - Raising livestock
    - Fishing
    - Obtaining oil

Technology Skills Activity

25. **Developing Multimedia Presentations** Research how your state's climate influences its culture, including tourist attractions, types of clothing, and the economy. Use your research to develop a commercial promoting your state.

Standardized Test Practice

Directions: Study the graph, then answer the following question.

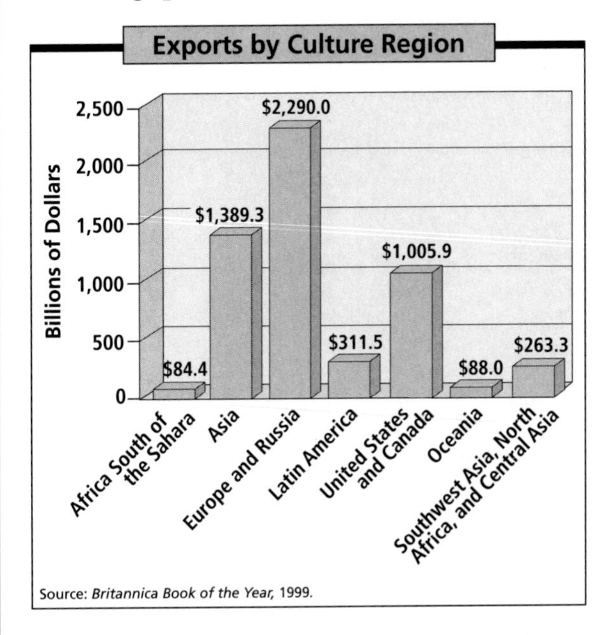

Source: *Britannica Book of the Year*, 1999.

1. **According to the graph, how much do the United States and Canada export?**
   A $1,005,900,000,000
   B $1,005,900,000
   C $1,005,900
   D $1,005

Test-Taking Tip: In order to understand any type of graph, look carefully around the graph for keys that show how it is organized. On this bar graph, the numbers along the left side represent billions of dollars. Therefore, you need to multiply the number on the graph by 1,000,000,000 to get your answer.

Unit 2

◀ Skier in Idaho's stretch of the Rocky Mountains

Grocer in Chinatown, New York City

Farm on the Manitoba plains

The United States and Canada

Which of the world's culture regions do you call home? It is probably the United States and Canada. If you look at a globe, you will see that the United States and Canada cover most of North America. These two nations share many of the same landforms, including rugged mountains in the west, rounded mountains in the east, and rolling plains in the center.

NGS ONLINE
www.nationalgeographic.com/education

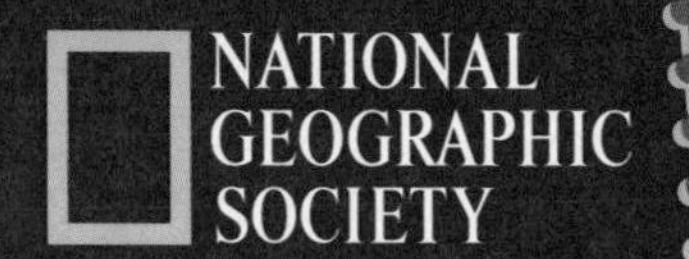

Focus on:

The United States and Canada

SPANNING MORE THAN 7 MILLION square miles (18 million sq. km), the United States and Canada cover much of North America. These huge countries share many of the same landscapes, climates, and natural resources.

The Land

The United States and Canada make up a region bordered by the very cold Arctic Ocean in the north and bathed by the Gulf of Mexico's warm currents in the south. The western coast faces the Pacific Ocean. Eastern shores are edged by the Atlantic.

Rugged mountains are found in the western part of each country. The Pacific ranges follow the coastline. Farther inland are the massive, jagged peaks of the Rocky Mountains. Relatively young as mountains go, the Rockies stretch more than 3,000 miles (4,828 km) from Alaska to the southwestern United States.

East of the Rockies are the wide and windswept Great Plains. This gently rolling landscape covers the central part of both the United States and Canada. In the United States, the Mississippi River—the largest river system in North America—flows through the heart of these plains.

The Appalachian range, much older than the Rockies, is the dominant landform in the eastern part of the region. East and south of the Appalachians' low, rounded peaks are coastal plains that end at the Atlantic shores.

The Climate

This region's vast size and varied landforms help give it great diversity in climate and vegetation. In the far northern parts of Alaska and Canada, amid the treeless tundra and dense evergreen forests, brief summers and bitterly cold winters prevail. The Pacific coast, from southern Alaska to northern California, has a mild, wet climate. Rain clouds blowing in from the ocean are blocked by the Pacific ranges. Robbed of moisture, the land immediately east of these mountains is dry.

Hot, humid summers and cold, snowy winters are the rule on the Great Plains. This humid continental climate extends from the plains across southeastern Canada and the northeastern United States. The southeastern states, however, enjoy much milder winters. The mildest of all are found on Florida's southern tip, the only part of the mainland that has a tropical climate.

UNIT 2

Parachutist plunging toward the Appalachian Mountains, West Virginia ▶

◀ Polar bear snoozing in the Canadian Arctic

Regional Atlas

The Economy

The United States and Canada are prosperous countries. Abundant natural resources and plenty of skilled workers have been key ingredients in creating two of the most successful economies in the world. Both countries operate under the free enterprise system, in which individuals and groups—not the government—control businesses and industries.

The region's strong economy was built on agriculture, which remains important today. Fertile soil, numerous waterways, a favorable climate, and high-tech equipment have made the United States and Canada two of the world's top food producers. Livestock, grains, vegetables, and fruits all are raised by the region's farmers.

Rich oil, coal, and natural gas deposits occur in this region. So do deposits of valuable minerals, including copper, iron ore, nickel, silver, and gold. These energy sources and raw materials have made it possible for the United States and Canada to develop large industrial economies. Today, however, people are more likely to work in offices than in factories. Service industries such as banking, communications, entertainment, insurance, and health care employ most people in the region.

The People

The United States and Canada have a rich mix of cultures. Native Americans were the nations' first inhabitants. Centuries later, settlers from Europe arrived. Immigrants from Africa, Asia, Latin America, and almost every other part of the world eventually followed. Some came looking for religious or political freedom. Some came as enslaved laborers. Some came for a fresh start in these immense lands of boundless opportunity.

Today more than 310 million people call this region home. Thirty-one million of them live in Canada, while the remaining 281 million live in the United States. On either side of the border, most people live in urban areas. Toronto, Vancouver, and Montreal are among Canada's largest cities. In the United States, New York City, Los Angeles, and Chicago are the most populous cities.

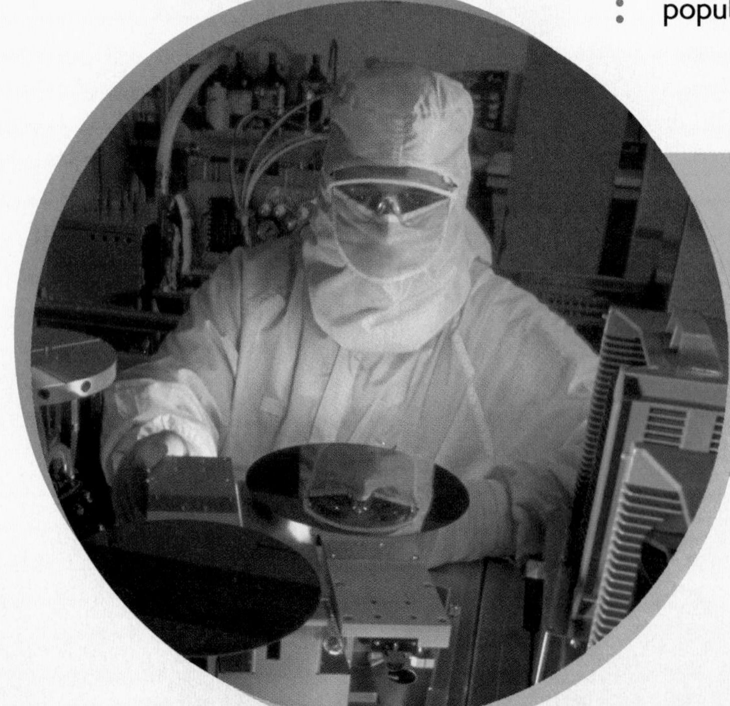

Exploring the Region

1. **Which oceans border the region?**
2. **Why is the climate dry just east of the Pacific ranges?**
3. **What factors have helped make the region prosperous?**
4. **In which country do most of the region's people live?**

◀ **Worker in sterile gown manufacturing computer chips in Texas**

UNIT 2

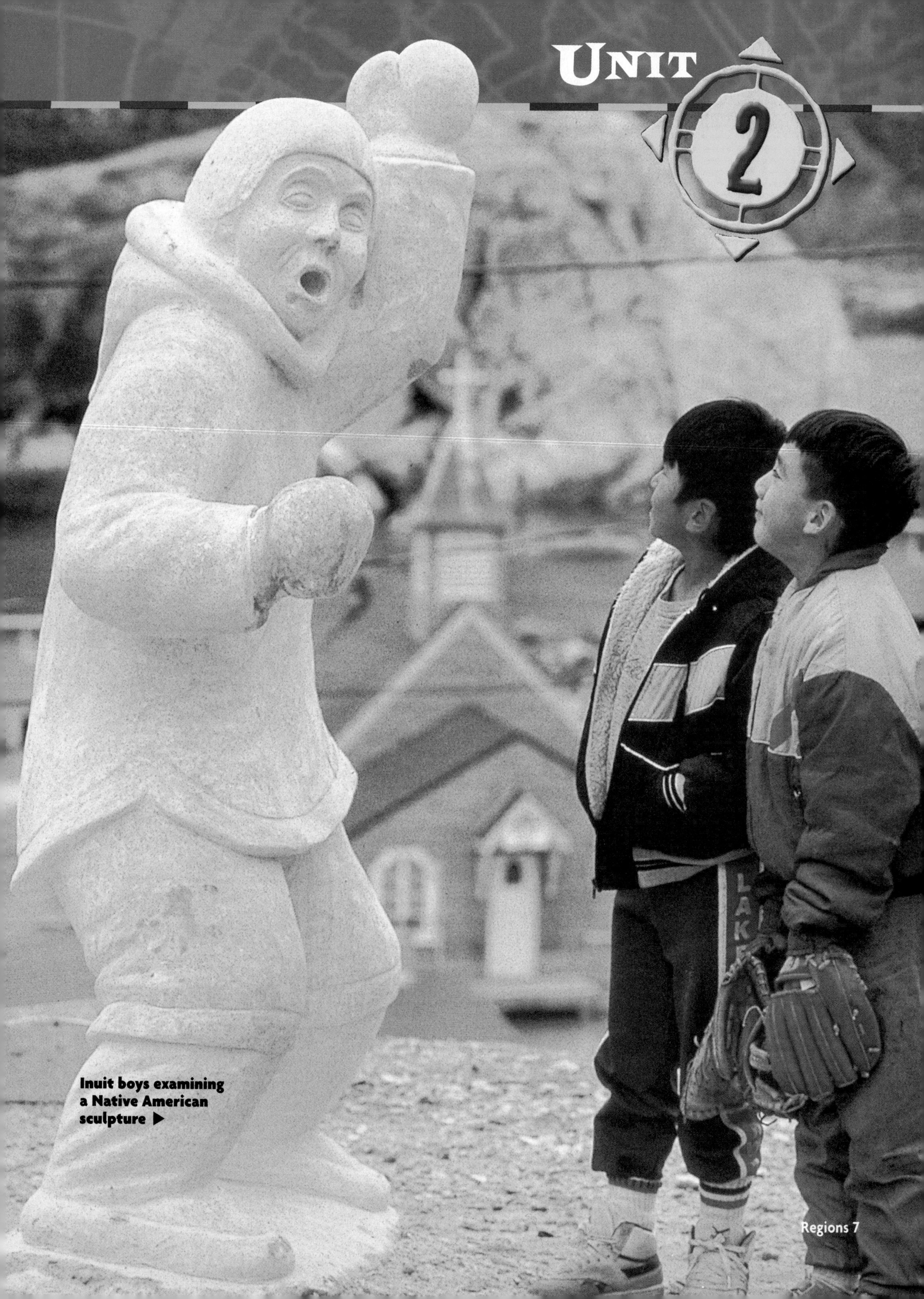

Inuit boys examining a Native American sculpture ▶

The United States and Canada

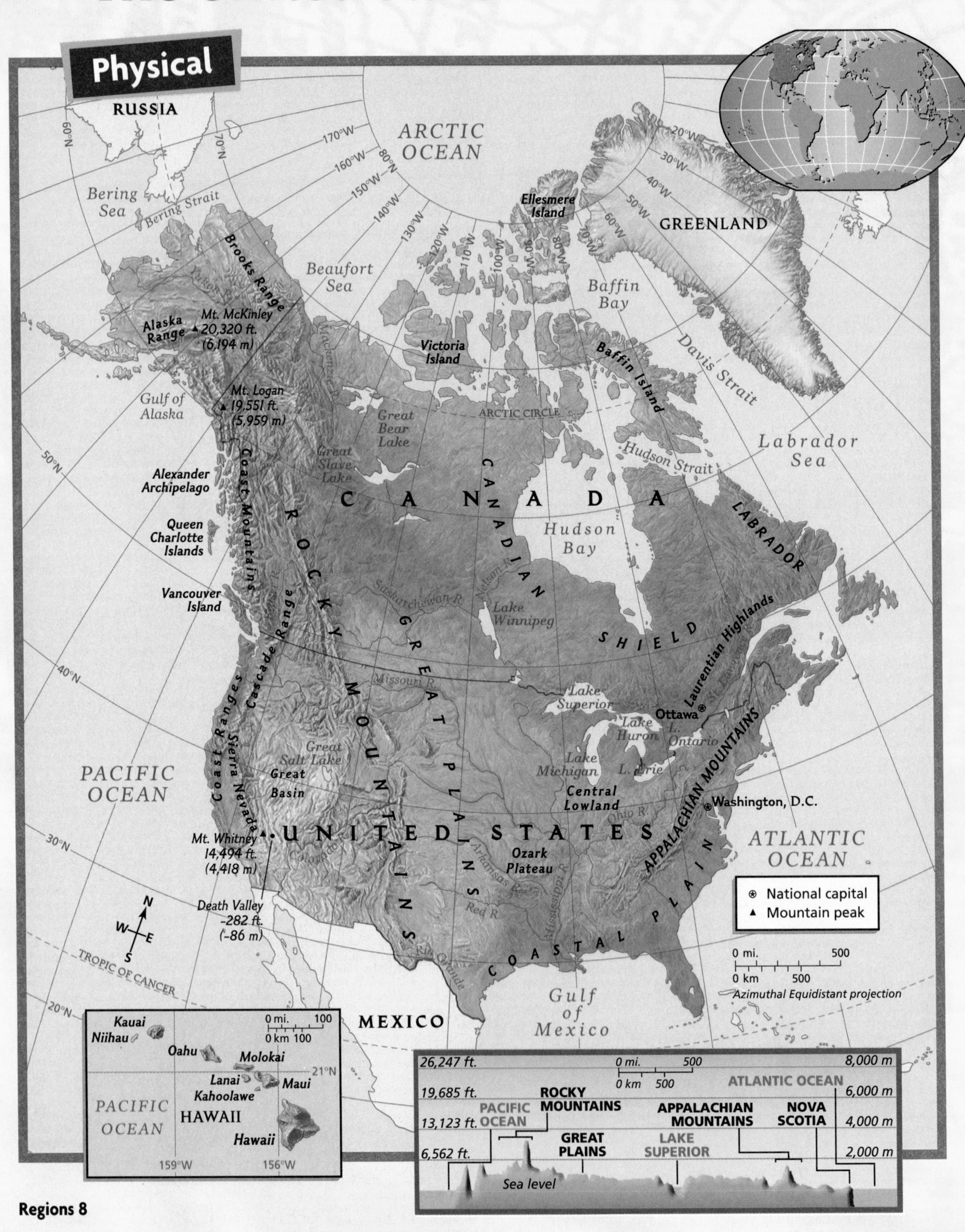

UNIT 2

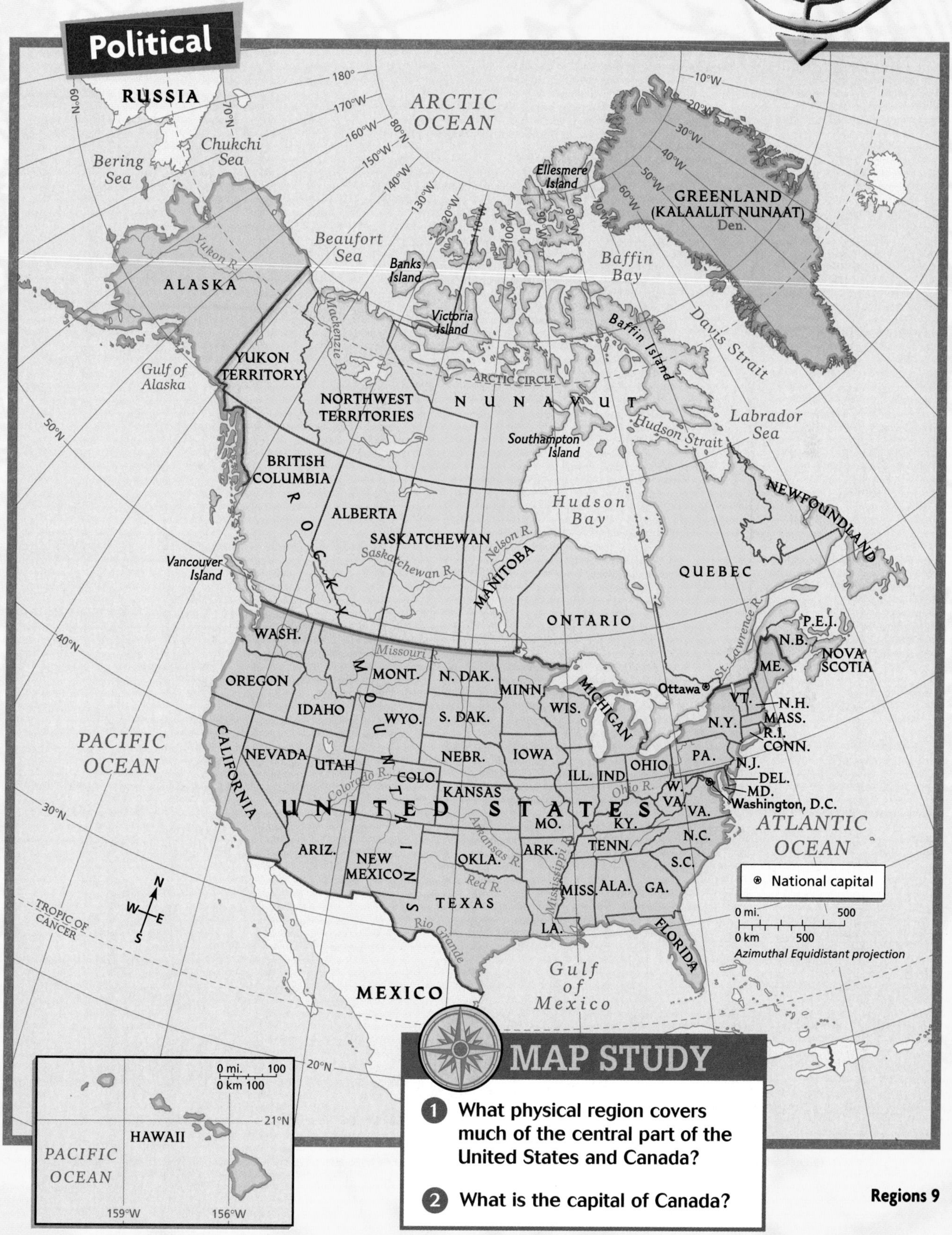

Political

National capital

0 mi. 500
0 km 500
Azimuthal Equidistant projection

MAP STUDY

1. What physical region covers much of the central part of the United States and Canada?

2. What is the capital of Canada?

REGIONAL ATLAS

The United States and Canada

Food Production

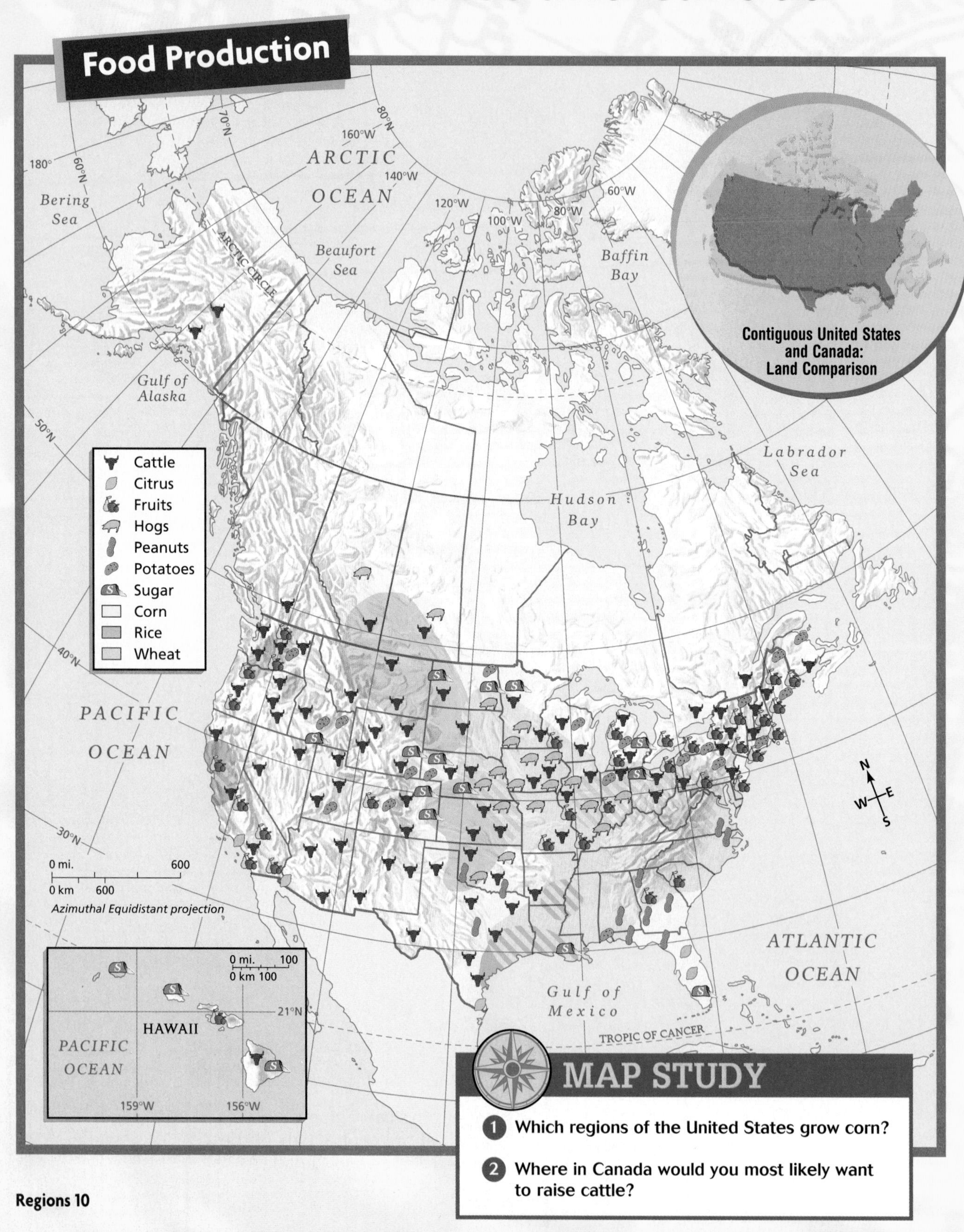

MAP STUDY

1. Which regions of the United States grow corn?
2. Where in Canada would you most likely want to raise cattle?

Geo Extremes

① HIGHEST POINT
Mount McKinley (Alaska)
20,320 ft. (6,194 m) high

② LOWEST POINT
Death Valley (California)
282 ft. (86 m)
below sea level

③ LONGEST RIVER
Mississippi-Missouri
(United States)
3,710 mi. (5,971 km) long

④ LARGEST LAKE
Lake Superior
31,700 sq. mi.
(82,103 sq. km)

⑤ LARGEST CANYON
Grand Canyon (Arizona)
277 mi. (446 km) long
1 mi. (1.6 km) deep

⑥ GREATEST TIDES
Bay of Fundy (Nova Scotia)
52 ft. (16 m)

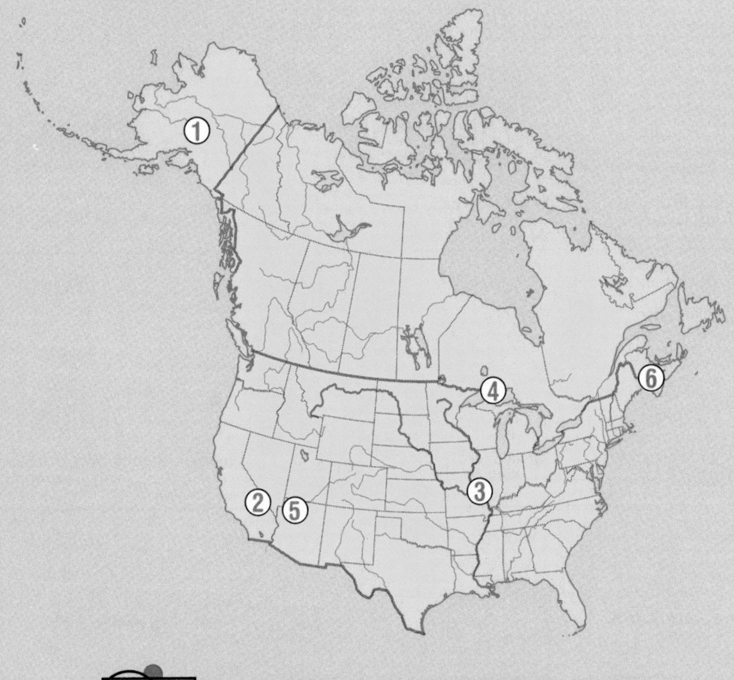

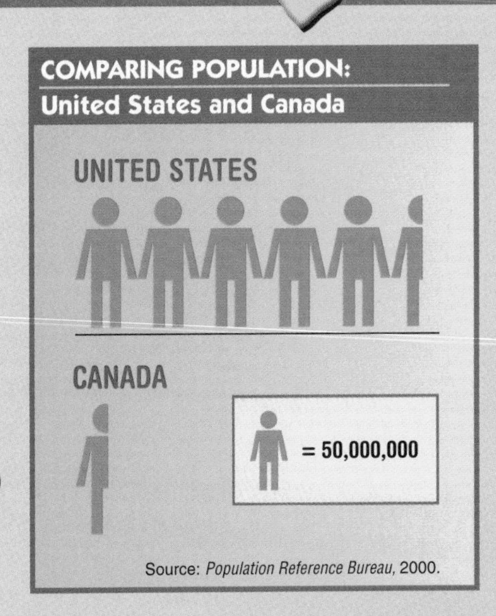

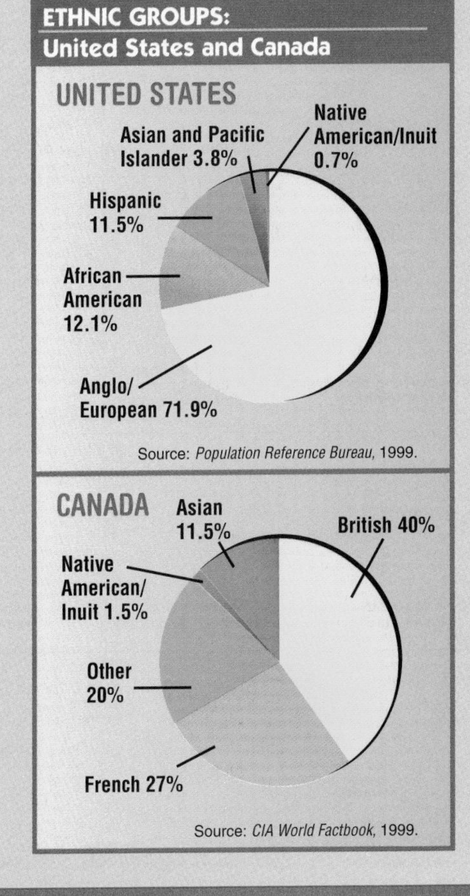

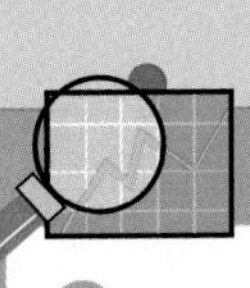

GRAPHIC STUDY

1. In what area of the United States do you find both the lowest point and largest canyon?
2. How does the percentage of Native American/Inuit population in the United States compare with their percentage of the population in Canada?

REGIONAL ATLAS

Country Profiles

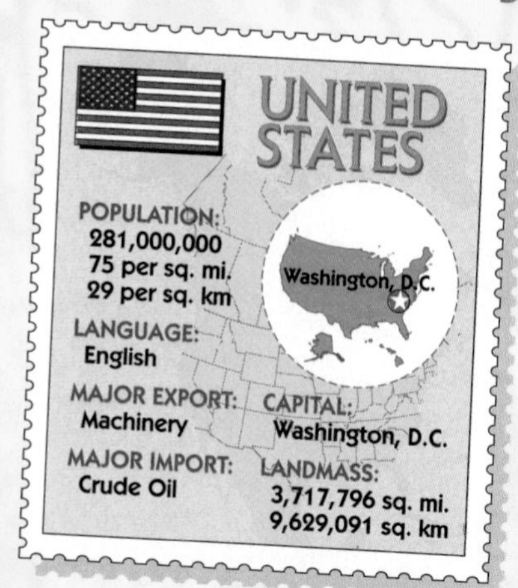

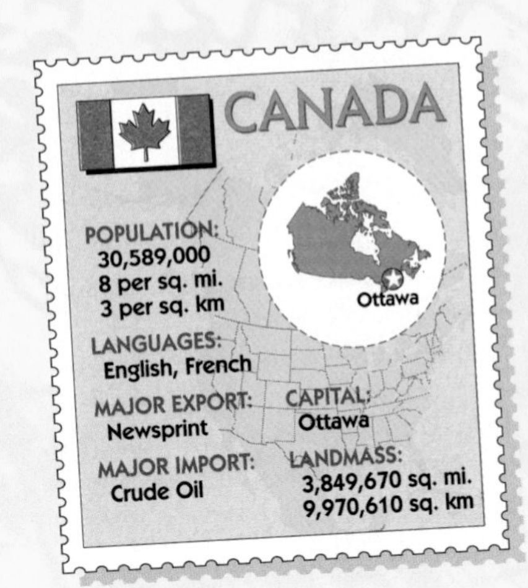

CANADA
POPULATION:
30,589,000
8 per sq. mi.
3 per sq. km
LANGUAGES:
English, French
MAJOR EXPORT:
Newsprint
MAJOR IMPORT:
Crude Oil
Ottawa
CAPITAL:
Ottawa
LANDMASS:
3,849,670 sq. mi.
9,970,610 sq. km

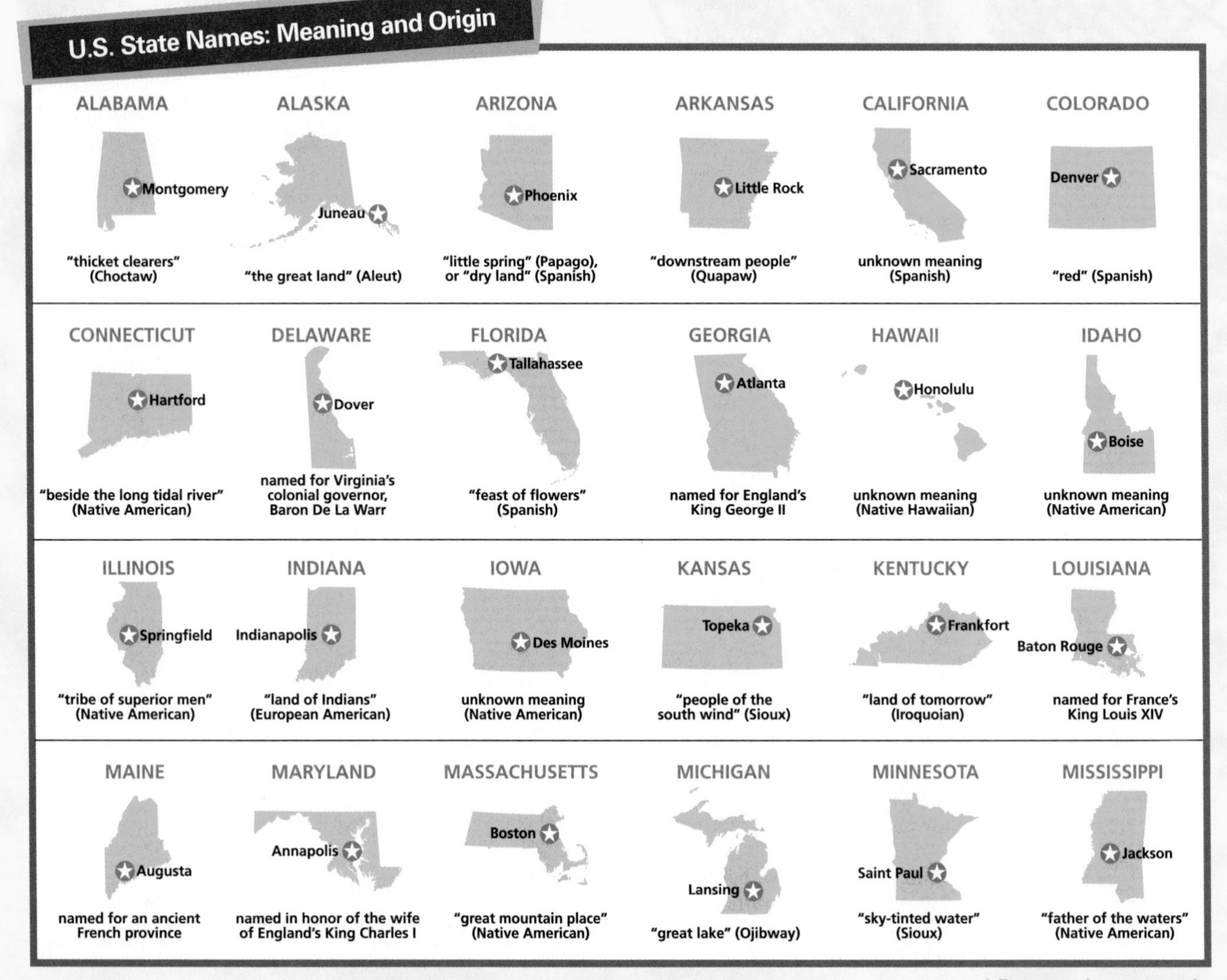

U.S. State Names: Meaning and Origin
ALABAMA
Montgomery
"thicket clearers" (Choctaw)
ALASKA
Juneau
"the great land" (Aleut)
ARIZONA
Phoenix
"little spring" (Papago), or "dry land" (Spanish)
ARKANSAS
Little Rock
"downstream people" (Quapaw)
CALIFORNIA
Sacramento
unknown meaning (Spanish)
COLORADO
Denver
"red" (Spanish)
CONNECTICUT
Hartford
"beside the long tidal river" (Native American)
DELAWARE
Dover
named for Virginia's colonial governor, Baron De La Warr
FLORIDA
Tallahassee
"feast of flowers" (Spanish)
GEORGIA
Atlanta
named for England's King George II
HAWAII
Honolulu
unknown meaning (Native Hawaiian)
IDAHO
Boise
unknown meaning (Native American)
ILLINOIS
Springfield
"tribe of superior men" (Native American)
INDIANA
Indianapolis
"land of Indians" (European American)
IOWA
Des Moines
unknown meaning (Native American)
KANSAS
Topeka
"people of the south wind" (Sioux)
KENTUCKY
Frankfort
"land of tomorrow" (Iroquoian)
LOUISIANA
Baton Rouge
named for France's King Louis XIV
MAINE
Augusta
named for an ancient French province
MARYLAND
Annapolis
named in honor of the wife of England's King Charles I
MASSACHUSETTS
Boston
"great mountain place" (Native American)
MICHIGAN
Lansing
"great lake" (Ojibway)
MINNESOTA
Saint Paul
"sky-tinted water" (Sioux)
MISSISSIPPI
Jackson
"father of the waters" (Native American)

Countries, states, provinces, and flags not drawn to scale

UNIT
2

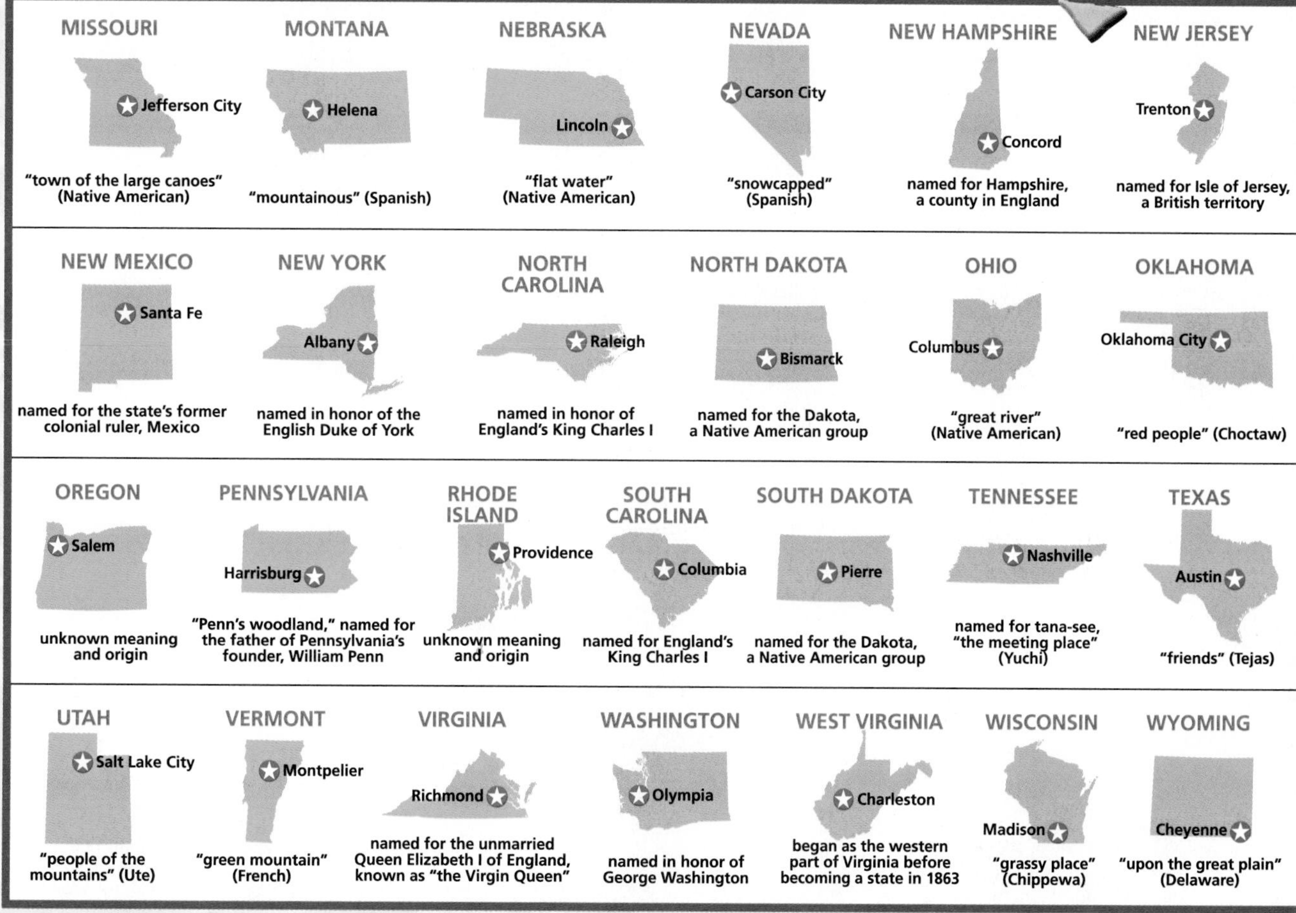
MISSOURI
Jefferson City
"town of the large canoes" (Native American)
MONTANA
Helena
"mountainous" (Spanish)
NEBRASKA
Lincoln
"flat water" (Native American)
NEVADA
Carson City
"snowcapped" (Spanish)
NEW HAMPSHIRE
Concord
named for Hampshire, a county in England
NEW JERSEY
Trenton
named for Isle of Jersey, a British territory
NEW MEXICO
Santa Fe
named for the state's former colonial ruler, Mexico
NEW YORK
Albany
named in honor of the English Duke of York
NORTH CAROLINA
Raleigh
named in honor of England's King Charles I
NORTH DAKOTA
Bismarck
named for the Dakota, a Native American group
OHIO
Columbus
"great river" (Native American)
OKLAHOMA
Oklahoma City
"red people" (Choctaw)
OREGON
Salem
unknown meaning and origin
PENNSYLVANIA
Harrisburg
"Penn's woodland," named for the father of Pennsylvania's founder, William Penn
RHODE ISLAND
Providence
unknown meaning and origin
SOUTH CAROLINA
Columbia
named for England's King Charles I
SOUTH DAKOTA
Pierre
named for the Dakota, a Native American group
TENNESSEE
Nashville
named for tana-see, "the meeting place" (Yuchi)
TEXAS
Austin
"friends" (Tejas)
UTAH
Salt Lake City
"people of the mountains" (Ute)
VERMONT
Montpelier
"green mountain" (French)
VIRGINIA
Richmond
named for the unmarried Queen Elizabeth I of England, known as "the Virgin Queen"
WASHINGTON
Olympia
named in honor of George Washington
WEST VIRGINIA
Charleston
began as the western part of Virginia before becoming a state in 1863
WISCONSIN
Madison
"grassy place" (Chippewa)
WYOMING
Cheyenne
"upon the great plain" (Delaware)

Canadian Province and Territory Names: Meaning and Origin

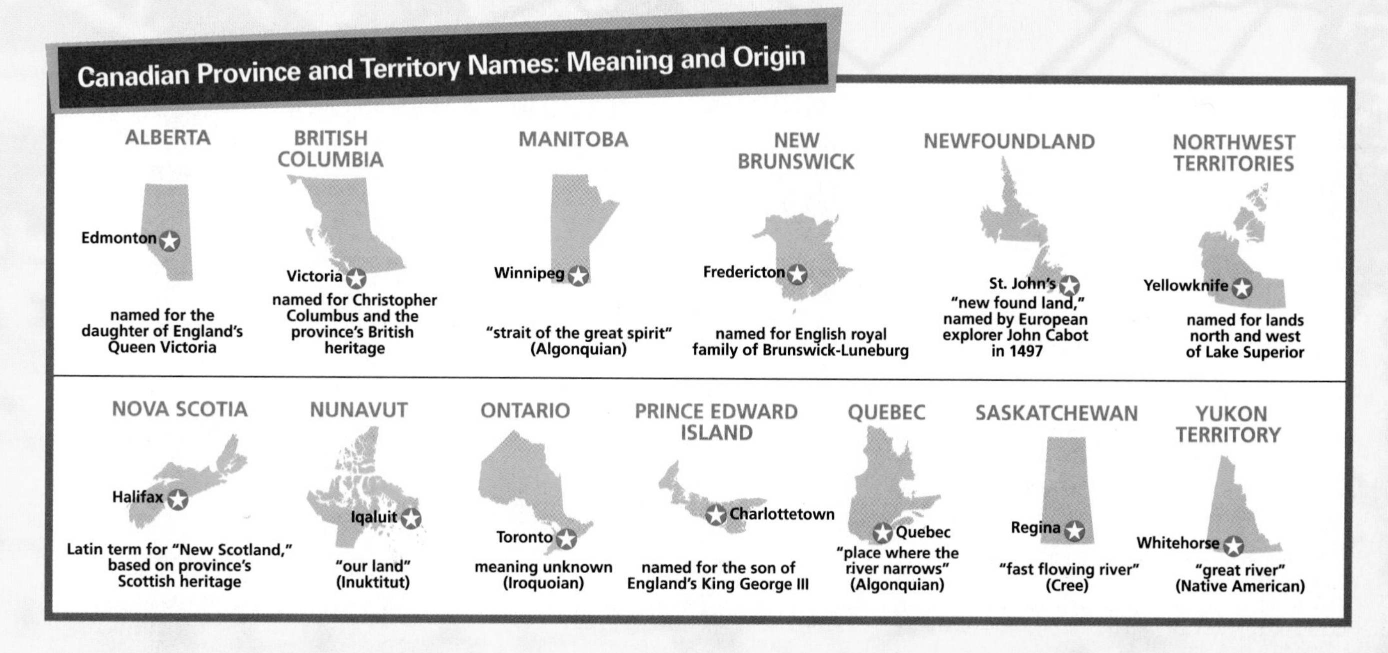
ALBERTA
Edmonton
named for the daughter of England's Queen Victoria
BRITISH COLUMBIA
Victoria
named for Christopher Columbus and the province's British heritage
MANITOBA
Winnipeg
"strait of the great spirit" (Algonquian)
NEW BRUNSWICK
Fredericton
named for English royal family of Brunswick-Luneburg
NEWFOUNDLAND
St. John's
"new found land," named by European explorer John Cabot in 1497
NORTHWEST TERRITORIES
Yellowknife
named for lands north and west of Lake Superior
NOVA SCOTIA
Halifax
Latin term for "New Scotland," based on province's Scottish heritage
NUNAVUT
Iqaluit
"our land" (Inuktitut)
ONTARIO
Toronto
meaning unknown (Iroquoian)
PRINCE EDWARD ISLAND
Charlottetown
named for the son of England's King George III
QUEBEC
Quebec
"place where the river narrows" (Algonquian)
SASKATCHEWAN
Regina
"fast flowing river" (Cree)
YUKON TERRITORY
Whitehorse
"great river" (Native American)

Unit 3

Peruvian Indian woman and child

Spanish colonial architecture in Guatemala

Latin America

Where can you find steamy tropical forests, frigid mountain peaks, thundering waterfalls, and peaceful island beaches? All of these contrasts can be found in Latin America—a huge part of the world made up of 33 nations on two continents. This region stretches from the Mexico–United States border, in North America, to the southernmost tip of South America.

NGS ONLINE
www.nationalgeographic.com/education

Spider monkey and Mayan ruins, Mexico ▲

Focus on: Latin America

COMMON THREADS OF LANGUAGE AND RELIGION unite this region. Once claimed as European colonies, most Latin American countries still use either Spanish or Portuguese as the official language. These two languages are based on Latin, which is how the region gets its name. Most Latin Americans are Roman Catholic, another influence from colonial times.

The Land

Latin America stretches from the Rio Grande south to Tierra del Fuego, just 600 miles (966 km) from Antarctica's frozen shores. Three times larger than the continental United States, the region includes Mexico, Central America, the Caribbean islands, and South America.

Mountains are prominent features in many parts of Latin America. Some Caribbean islands are actually the exposed tops of ancient, submerged volcanoes. In Mexico, the branches of the Sierra Madre spread like welcoming arms to hug a central highland known as the Mexican Plateau. Mist-covered peaks stretch through the interior of Central America. Most impressive, however, are the Andes. The longest series of mountain ranges in the world, the lofty Andes follow the western coast of South America for 4,500 miles (7,242 km).

Narrow coastal plains line the edges of Mexico and Central America. South America has vast inland plains. These include the pampas of Argentina and the llanos of Colombia and Venezuela. The largest lowland area on this continent is the basin of the Amazon River, the longest river in the Western Hemisphere.

The Climate

Most of this region has a tropical climate. Daily showers drench the rain forests, which thrive in the lowlands. In Brazil the Amazon River and its tributaries snake through the largest area of rain forest regions, which covers roughly one-third of South America. Rain forests contain more species of plants and animals than any other ecosystem on Earth.

The climate tends to be drier and cooler at higher elevations and farther away from the Equator. Under these conditions, tall grasses and scattered trees flourish.

Drier still are parts of northern Mexico and southern Argentina. Here, rainfall is sparse and so is vegetation. Yet even these places are lush compared to the Atacama Desert, along Chile's coast. The barren Atacama is among the world's driest places.

UNIT 3

Three-toed sloth in rain forest, Panama

Peaks of the Andes, Chile

Regional Atlas

The Economy

Latin America is rich in natural resources. Gold drew many of the first European conquerors. Copper, silver, iron ore, tin, and lead also are abundant in the region. Some Latin American countries are among the world's leading producers of oil and natural gas.

Agriculture plays an important role in the region's economy. Coffee, bananas, and sugarcane thrive in the moist, fertile lowlands. On higher ground, farmers raise grain and fruit, while cowhands known as gauchos drive huge herds of cattle across rolling grasslands.

Industrialization is increasing in Latin America. However, some countries are moving along this path more quickly than others. In recent years, Mexico, Brazil, and Chile have become major producers of manufactured goods. Some things have hindered industrial development in other parts of the region. These include shortages of money, skilled labor, and reliable transportation, along with geographic barriers such as rugged mountains and thick forests.

The People

Long before Europeans crossed the Atlantic Ocean, great Native American civilizations developed in Latin America. The Maya flourished in the Guatemalan and Honduran lowlands and across Mexico's Yucatán Peninsula. The central highlands of Mexico were the site of the Aztec Empire. In South America, the Inca established an empire that stretched from southern Colombia to central Chile.

Beginning in the 1500s, Spain and Portugal ruled most of Latin America. These European invaders destroyed the Native American civilizations. They also brought enslaved Africans to work alongside Native Americans on plantations.

Independence came for many Latin American countries in the early 1800s. These countries remain a cultural mixture—Native Americans, Europeans, Africans, and others all have left their mark.

Today most Latin Americans live in urban areas along the coasts of South America or in a band reaching from Mexico into Central America. Some of the largest cities in the world are in this region, including Mexico City, Rio de Janeiro, and Buenos Aires.

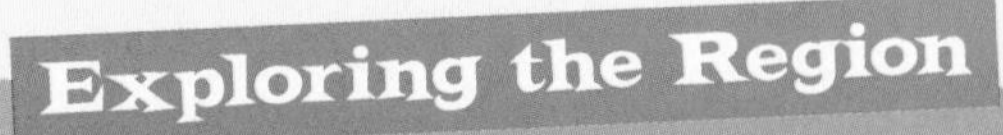

Exploring the Region

1. **What is Latin America's longest series of mountain ranges?**
2. **What type of climate is found across most of the region?**
3. **What European countries once ruled Latin America?**
4. **Which Latin American countries are industrializing most rapidly?**

◀ **Mexican boy carrying decorated cross for religious celebration**

Unit 3

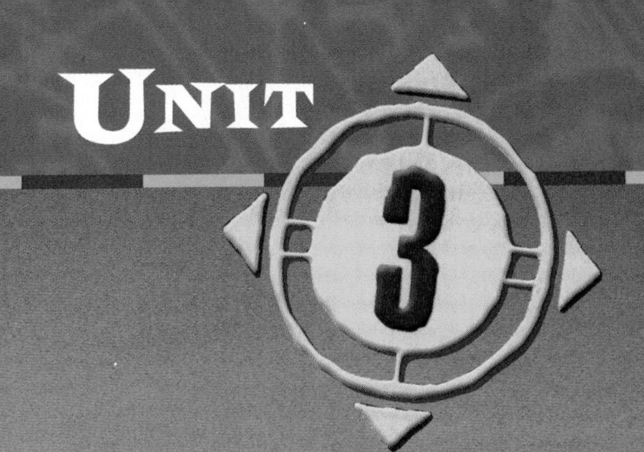

Rio de Janeiro, Brazil ▼

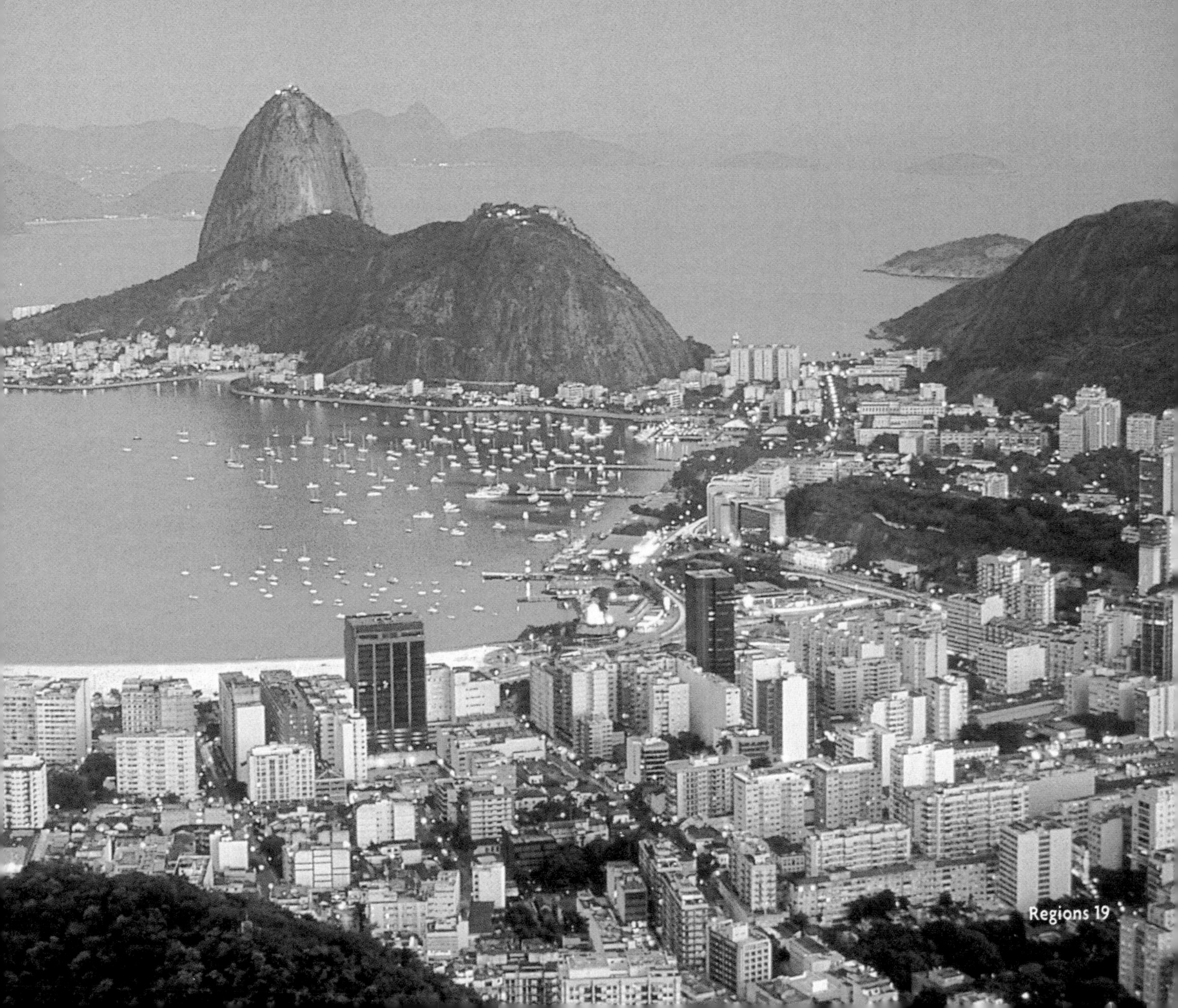

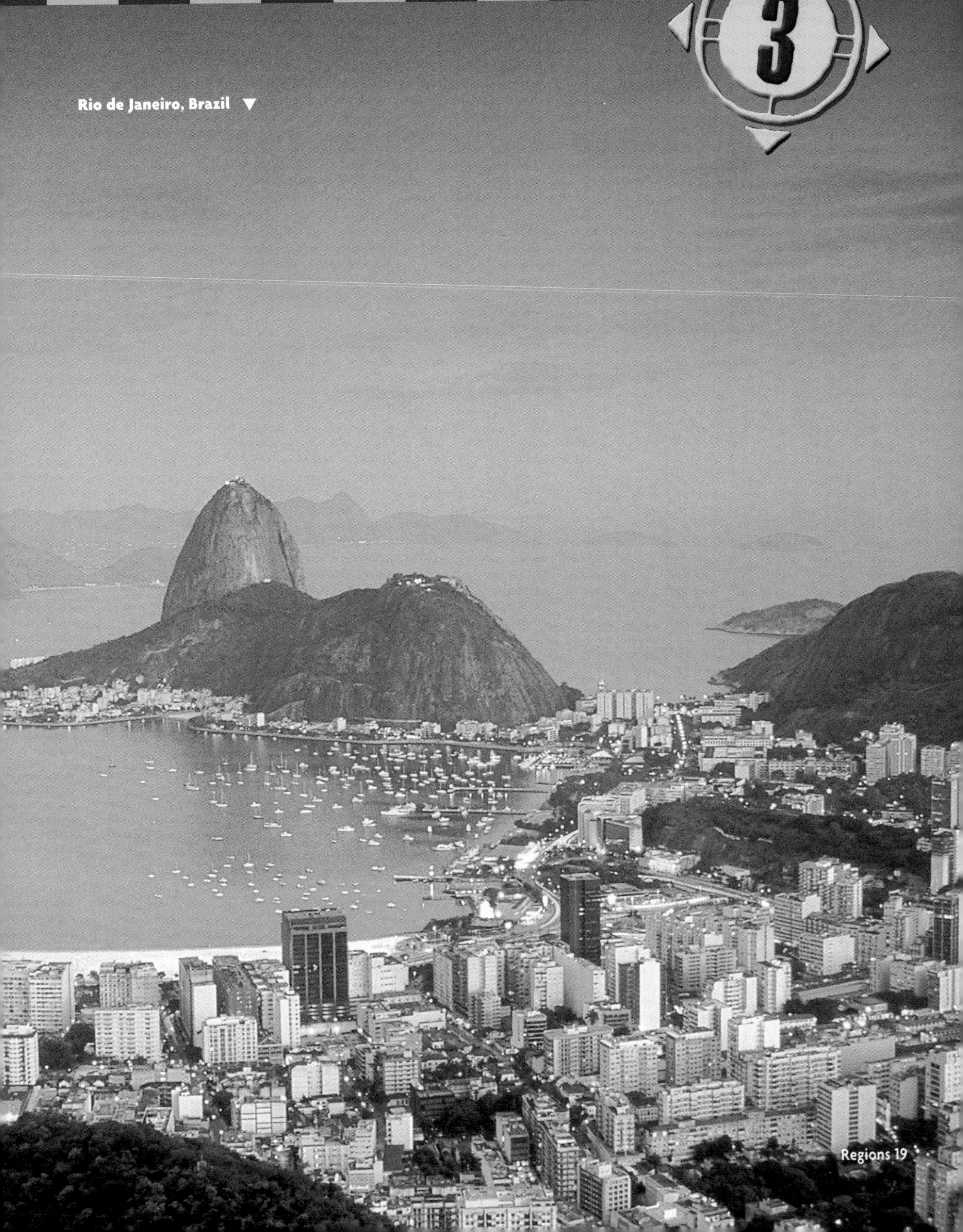

NATIONAL
GEOGRAPHIC
SOCIETY

REGIONAL ATLAS

Latin America

Physical

120°W
110°W
100°W
90°W
80°W
70°W
60°W
50°W
30°N
20°N
10°N
0°
10°S
20°S
30°S
40°S
UNITED STATES
Rio Grande
Bermuda Is.
Baja California
Plateau of Mexico
SIERRA MADRE OCCIDENTAL
SIERRA MADRE ORIENTAL
Gulf of Mexico
BAHAMAS
TROPIC OF CANCER
CUBA
WEST INDIES
Yucatán Peninsula
Greater Antilles
HAITI
DOM. REP.
MEXICO
Sierra Madre del Sur
BELIZE
JAMAICA
Puerto Rico
HONDURAS
ATLANTIC OCEAN
GUATEMALA
NICARAGUA
Caribbean Sea
Lesser Antilles
EL SALVADOR
Lake Maracaibo
COSTA RICA
Isthmus of Panama
VENEZUELA
PANAMA
Orinoco R.
Llanos
GUYANA
SURINAME
FRENCH GUIANA
COLOMBIA
Guiana Highlands
EQUATOR
AMAZON
Negro R.
Amazon R.
Galápagos Islands
ECUADOR
BASIN
PERU
BRAZIL
Madeira R.
ANDES
São Francisco R.
PACIFIC OCEAN
Lake Titicaca
Mato Grosso Plateau
BRAZILIAN HIGHLANDS
BOLIVIA
Atacama Desert
Altiplano
PARAGUAY
Paraná R.
TROPIC OF CAPRICORN
Gran Chaco
Paraguay R.
▲ Mountain peak
CHILE
0 mi. 1,000
0 km 1,000
Lambert Azimuthal Equal-Area projection
ARGENTINA
Aconcagua 22,834 ft. (6,960 m)
URUGUAY
Pampas
Río de la Plata
ATLANTIC OCEAN
PATAGONIA
Strait of Magellan
Tierra del Fuego
Falkland Islands
Cape Horn
South Georgia I.
26,247 ft.
19,685 ft.
13,123 ft.
6,562 ft.
0 mi. 500
0 km 500
8,000 m
6,000 m
4,000 m
2,000 m
ANDES
AMAZON BASIN
MATO GROSSO PLATEAU
BRAZILIAN HIGHLANDS
LIMA
Sea level
SALVADOR

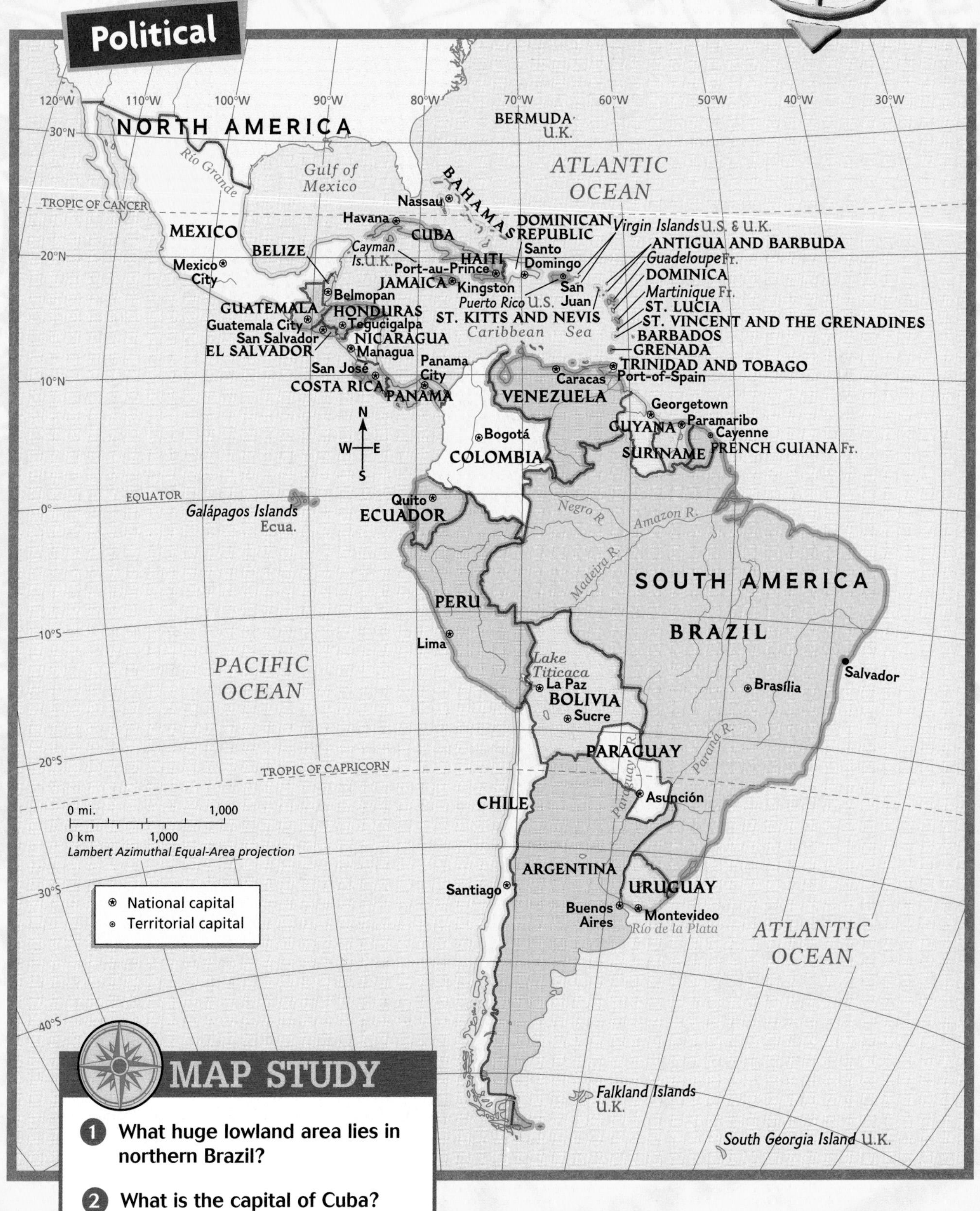

MAP STUDY

1. What huge lowland area lies in northern Brazil?
2. What is the capital of Cuba?

REGIONAL ATLAS

Latin America

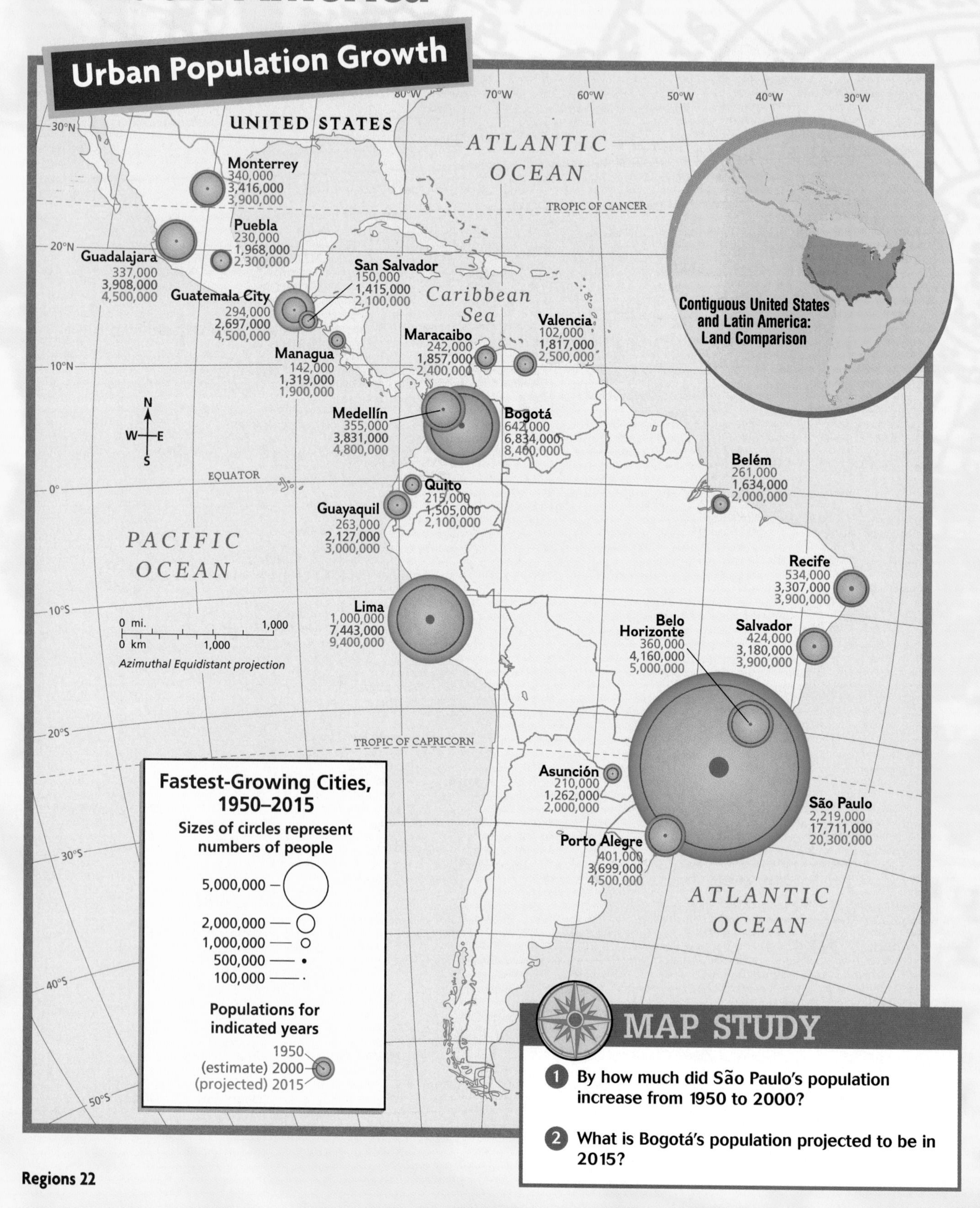

MAP STUDY

1. By how much did São Paulo's population increase from 1950 to 2000?
2. What is Bogotá's population projected to be in 2015?

Geo Extremes

① **HIGHEST POINT**
Aconcagua (Argentina)
22,834 ft. (6,960 m) high

② **LOWEST POINT**
Valdés Peninsula (Argentina)
131 ft. (40 m) below sea level

③ **LONGEST RIVER**
Amazon River
(Brazil and Peru)
4,000 mi. (6,437 km) long

④ **LARGEST LAKE**
Lake Maracaibo (Venezuela)
5,217 sq. mi. (13,512 sq. km)

⑤ **HIGHEST LARGE NAVIGABLE LAKE**
Lake Titicaca
(Peru and Bolivia)
12,500 ft. (3,810 m) high

⑥ **HIGHEST WATERFALL**
Angel Falls (Venezuela)
3,212 ft. (979 m) high

⑦ **DRIEST PLACE**
Atacama Desert (Chile)
rainfall barely measurable

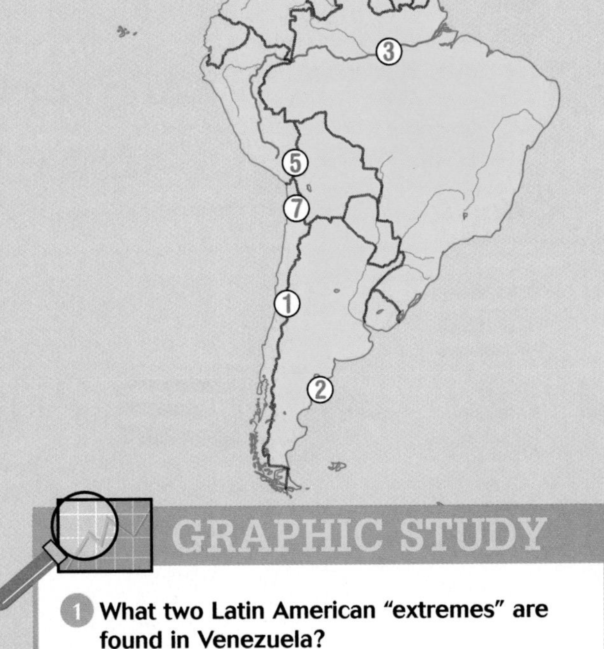

COMPARING POPULATION: United States and Selected Countries of Latin America

UNITED STATES

BRAZIL

MEXICO

ARGENTINA

GUATEMALA

BOLIVIA

(figure) = 25,000,000

Source: *Population Reference Bureau*, 2000.

ETHNIC GROUPS: Selected Countries of Latin America

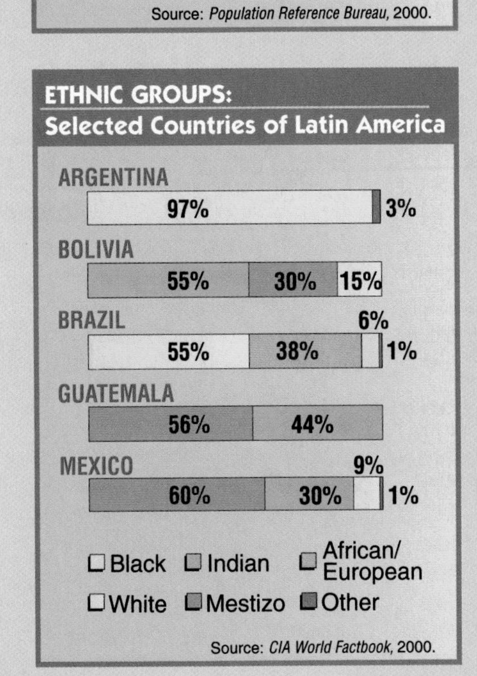

Source: *CIA World Factbook*, 2000.

GRAPHIC STUDY

1. What two Latin American "extremes" are found in Venezuela?
2. What countries have a majority of mestizos (people of mixed European and Native American ancestry)?

Country Profiles

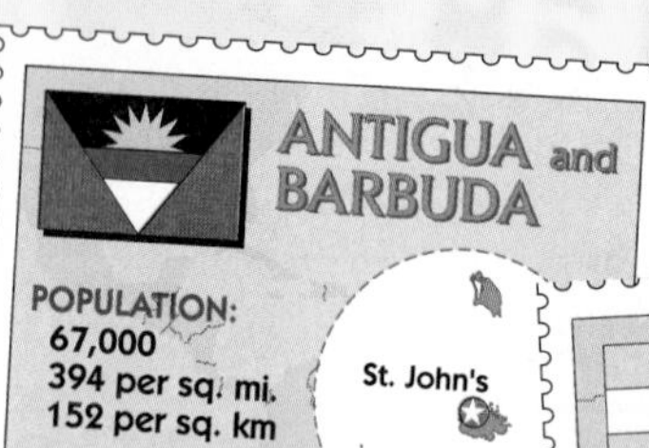

ANTIGUA and BARBUDA

POPULATION:
67,000
394 per sq. mi.
152 per sq. km

LANGUAGE:
English

MAJOR EXPORT:
Petroleum Products

MAJOR IMPORTS:
Foods and Livestock

CAPITAL:
St. John's

LANDMASS:
170 sq. mi.
440 sq. km

ARGENTINA

POPULATION:
36,568,000
34 per sq. mi.
13 per sq. km

LANGUAGE:
Spanish

MAJOR EXPORT:
Meat

MAJOR IMPORT:
Machinery

CAPITAL:
Buenos Aires

LANDMASS:
1,068,302 sq. mi.
2,766,889 sq. km

BAHAMAS

POPULATION:
301,000
56 per sq. mi.
22 per sq. km

LANGUAGES:
English, Creole

MAJOR EXPORT:
Pharmaceuticals

MAJOR IMPORT:
Foods

CAPITAL:
Nassau

LANDMASS:
5,382 sq. mi.
13,939 sq. km

BARBADOS

POPULATION:
269,000
1,620 per sq. mi.
626 per sq. km

LANGUAGE:
English

MAJOR EXPORT:
Sugar

MAJOR IMPORT:
Manufactured Goods

CAPITAL:
Bridgetown

LANDMASS:
166 sq. mi.
430 sq. km

BELIZE

POPULATION:
248,000
28 per sq. mi.
11 per sq. km

LANGUAGE:
English

MAJOR EXPORT:
Sugar

MAJOR IMPORT:
Machinery

CAPITAL:
Belmopan

LANDMASS:
8,867 sq. mi.
22,965 sq. km

BOLIVIA

POPULATION:
8,090,000
19 per sq. mi.
7 per sq. km

LANGUAGES:
Spanish, Quechua, Aymara

MAJOR EXPORT:
Metals

MAJOR IMPORT:
Machinery

CAPITALS:
La Paz, Sucre

LANDMASS:
424,164 sq. mi.
1,098,581 sq. km

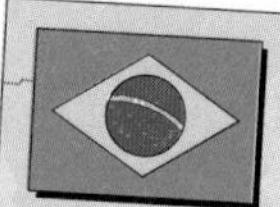

BRAZIL

POPULATION:
167,988,000
51 per sq. mi.
20 per sq. km

LANGUAGE:
Portuguese

MAJOR EXPORT:
Iron Ore

MAJOR IMPORT:
Crude Oil

CAPITAL:
Brasília

LANDMASS:
3,286,488 sq. mi.
8,511,965 sq. km

CHILE

POPULATION:
15,018,000
51 per sq. mi.
20 per sq. km

LANGUAGE:
Spanish

MAJOR EXPORT:
Copper

MAJOR IMPORT:
Machinery

CAPITAL:
Santiago

LANDMASS:
292,135 sq. mi.
756,626 sq. km

COLOMBIA

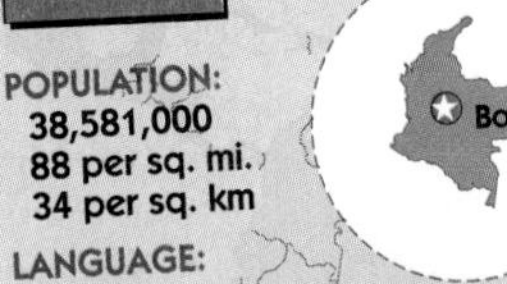

POPULATION:
38,581,000
88 per sq. mi.
34 per sq. km

LANGUAGE:
Spanish

MAJOR EXPORT:
Petroleum

MAJOR IMPORT:
Machinery

CAPITAL:
Bogotá

LANDMASS:
439,737 sq. mi.
1,138,914 sq. km

COSTA RICA

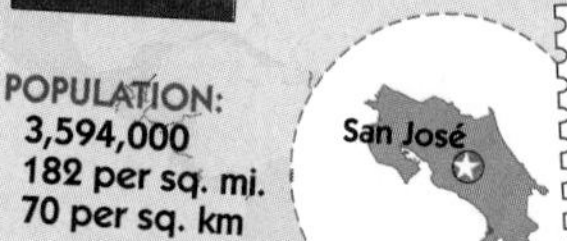

POPULATION:
3,594,000
182 per sq. mi.
70 per sq. km

LANGUAGE:
Spanish

MAJOR EXPORT:
Coffee

MAJOR IMPORT:
Raw Materials

CAPITAL:
San José

LANDMASS:
19,730 sq. mi.
51,100 sq. km

CUBA

POPULATION:
11,178,000
261 per sq. mi.
101 per sq. km

LANGUAGE:
Spanish

MAJOR EXPORT:
Sugar

MAJOR IMPORT:
Petroleum

CAPITAL:
Havana

LANDMASS:
42,804 sq. mi.
110,861 sq. km

Countries and flags not drawn to scale

DOMINICA

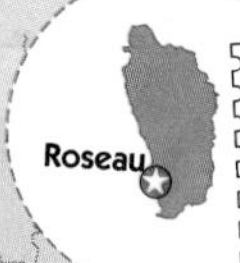

POPULATION: 71,000; 245 per sq. mi.; 95 per sq. km
LANGUAGES: English, French
MAJOR EXPORT: Bananas
MAJOR IMPORT: Manufactured Goods
CAPITAL: Roseau
LANDMASS: 290 sq. mi.; 751 sq. km

DOMINICAN REPUBLIC

POPULATION: 8,299,000; 441 per sq. mi.; 170 per sq. km
LANGUAGE: Spanish
MAJOR EXPORT: Ferronickel
MAJOR IMPORT: Foods
CAPITAL: Santo Domingo
LANDMASS: 18,816 sq. mi.; 48,734 sq. km

ECUADOR

POPULATION: 12,411,000; 113 per sq. mi.; 44 per sq. km
LANGUAGES: Spanish, Quechua
MAJOR EXPORT: Petroleum
MAJOR IMPORT: Transport Equipment
CAPITAL: Quito
LANDMASS: 109,484 sq. mi.; 283,561 sq. km

EL SALVADOR

POPULATION: 5,859,000; 721 per sq. mi.; 278 per sq. km
LANGUAGE: Spanish
MAJOR EXPORT: Coffee
MAJOR IMPORT: Raw Materials
CAPITAL: San Salvador
LANDMASS: 8,124 sq. mi.; 21,041 sq. km

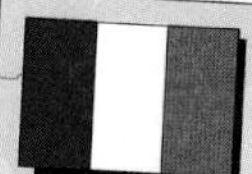

FRENCH GUIANA*

POPULATION: 185,000; 5 per sq. mi.; 2 per sq. km
LANGUAGE: French
MAJOR EXPORT: Shrimp
MAJOR IMPORT: Foods
CAPITAL: Cayenne
LANDMASS: 34,749 sq. mi.; 89,999 sq. km

* Territory of France

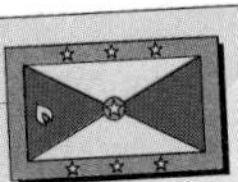

GRENADA

POPULATION: 97,000; 729 per sq. mi.; 282 per sq. km
LANGUAGES: English, French
MAJOR EXPORT: Bananas
MAJOR IMPORT: Foods
CAPITAL: St. George's
LANDMASS: 133 sq. mi.; 344 sq. km

GUATEMALA

POPULATION: 12,336,000; 293 per sq. mi.; 113 per sq. km
LANGUAGES: Spanish, Mayan Languages
MAJOR EXPORT: Coffee
MAJOR IMPORT: Petroleum
CAPITAL: Guatemala City
LANDMASS: 42,042 sq. mi.; 108,889 sq. km

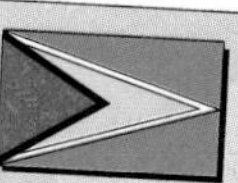

GUYANA

POPULATION: 705,000; 8 per sq. mi.; 3 per sq. km
LANGUAGE: English
MAJOR EXPORT: Sugar
MAJOR IMPORT: Manufactured Goods
CAPITAL: Georgetown
LANDMASS: 83,000 sq. mi.; 214,969 sq. km

HAITI

POPULATION: 7,751,000; 723 per sq. mi.; 279 per sq. km
LANGUAGES: French, Creole
MAJOR EXPORT: Manufactured Goods
MAJOR IMPORT: Machinery
CAPITAL: Port-au-Prince
LANDMASS: 10,714 sq. mi.; 27,750 sq. km

HONDURAS

POPULATION: 5,901,000; 136 per sq. mi.; 53 per sq. km
LANGUAGE: Spanish
MAJOR EXPORT: Bananas
MAJOR IMPORT: Machinery
CAPITAL: Tegucigalpa
LANDMASS: 43,277 sq. mi.; 112,088 sq. km

JAMAICA

POPULATION: 2,621,000; 618 per sq. mi.; 238 per sq. km
LANGUAGES: English, Creole
MAJOR EXPORT: Alumina
MAJOR IMPORT: Machinery
CAPITAL: Kingston
LANDMASS: 4,244 sq. mi.; 10,991 sq. km

Country Profiles

MEXICO

POPULATION:
99,734,000
132 per sq. mi.
51 per sq. km

LANGUAGES:
Spanish,
Native American Languages

MAJOR EXPORT:
Crude Oil

MAJOR IMPORT:
Machinery

CAPITAL:
Mexico City

LANDMASS:
756,066 sq. mi.
1,958,201 sq. km

NICARAGUA

POPULATION:
4,952,000
99 per sq. mi.
38 per sq. km

LANGUAGE:
Spanish

MAJOR EXPORT:
Coffee

MAJOR IMPORT:
Manufactured Goods

CAPITAL:
Managua

LANDMASS:
50,193 sq. mi.
129,999 sq. km

PANAMA

POPULATION:
2,809,000
94 per sq. mi.
36 per sq. km

LANGUAGE:
Spanish

MAJOR EXPORT:
Bananas

MAJOR IMPORT:
Machinery

CAPITAL:
Panama City

LANDMASS:
29,762 sq. mi.
77,082 sq. km

PARAGUAY

POPULATION:
5,219,000
33 per sq. mi.
13 per sq. km

LANGUAGES:
Spanish, Guaraní

MAJOR EXPORT:
Cotton

MAJOR IMPORT:
Machinery

CAPITAL:
Asunción

LANDMASS:
157,048 sq. mi.
406,752 sq. km

PERU

POPULATION:
26,624,000
54 per sq. mi.
21 per sq. km

LANGUAGES:
Spanish, Quechua, Aymara

MAJOR EXPORT:
Copper

MAJOR IMPORT:
Machinery

CAPITAL:
Lima

LANDMASS:
496,225 sq. mi.
1,285,217 sq. km

PUERTO RICO*

POPULATION:
3,887,652
1,132 per sq. mi.
437 per sq. km

LANGUAGES:
Spanish, English

MAJOR EXPORT:
Pharmaceuticals

MAJOR IMPORT:
Chemical Products

CAPITAL:
San Juan

LANDMASS:
3,435 sq. mi.
8,897 sq. km

* U.S. Commonwealth

ST. KITTS and NEVIS

POPULATION:
39,000
386 per sq. mi.
149 per sq. km

LANGUAGE:
English

MAJOR EXPORT:
Machinery

MAJOR IMPORT:
Electronic Goods

CAPITAL:
Basseterre

LANDMASS:
101 sq. mi.
261 sq. km

ST. LUCIA

POPULATION:
154,000
647 per sq. mi.
250 per sq. km

LANGUAGES:
English, French

MAJOR EXPORT:
Bananas

MAJOR IMPORT:
Foods

CAPITAL:
Castries

LANDMASS:
238 sq. mi.
617 sq. km

ST. VINCENT and the GRENADINES

POPULATION:
114,000
760 per sq. mi.
294 per sq. km

LANGUAGES:
English, French

MAJOR EXPORT:
Bananas

MAJOR IMPORT:
Foods

CAPITAL:
Kingstown

LANDMASS:
150 sq. mi.
388 sq. km

SURINAME

POPULATION:
431,000
7 per sq. mi.
3 per sq. km

LANGUAGE:
Dutch

MAJOR EXPORT:
Bauxite

MAJOR IMPORT:
Machinery

CAPITAL:
Paramaribo

LANDMASS:
63,037 sq. mi.
163,265 sq. km

TRINIDAD and TOBAGO

POPULATION:
1,285,000
649 per sq. mi.
250 per sq. km

LANGUAGE:
English

MAJOR EXPORT:
Petroleum

MAJOR IMPORT:
Machinery

CAPITAL:
Port-of-Spain

LANDMASS:
1,981 sq. mi.
5,131 sq. km

Countries and flags not drawn to scale

Questions From Buzz Bee!

The following questions are taken from National Geographic GeoBees. Use your textbook, the Internet, and other library resources to find the answers.

1. In 1911 Hiram Bingham found the site of a major Incan city in the Peruvian Andes. Name this site.
2. Marajó Island, which is just south of the Equator and about the same size as Denmark, is bordered on the northwest by what river?
3. Cowhands who tend cattle on the grasslands of Argentina are part of the country's folklore. What are these cowhands called?
4. Name the world's most populous Spanish-speaking country.
5. Bananas grown on coastal lowlands north of Tegucigalpa are a major export of which Central American country?

Unit 4

Woman in Hungary creating folk art

Ancient ruins in Delphi, Greece

Europe

You have learned about the Americas. Now let us spin the globe and travel to the Eastern Hemisphere. The first region you will learn about is Europe—relatively small as continents go, but rich in history and culture. Like the United States, most nations in Europe are industrialized and have high standards of living. Unlike the United States, however, the people of Europe do not share a common language or government.

▲ The Louvre museum, Paris, France

NGS ONLINE
www.nationalgeographic.com/education

Focus on:

Europe

BOTH A CONTINENT and a region, Europe has a wide range of cultures—and a history of conflict among its people. Recently, connections in trade, communication, and transportation have helped to create greater unity among European nations.

The Land

Jutting westward from Asia, Europe is a great peninsula that breaks into smaller peninsulas and is bordered by several large islands. Europe's long, jagged coastline is washed by many bodies of water, including the Arctic and Atlantic Oceans, and the North, Baltic, and Mediterranean Seas. Deep bays and well-protected inlets shelter fine harbors.

Mountains sweep across much of the continent. Those in the British Isles and large parts of northern Europe are low and rounded. Higher and more rugged are the Pyrenees, between France and Spain, and the Carpathians, in eastern Europe. The snow-capped Alps are Europe's highest mountains, towering over the central and southern parts of the continent.

Curving around these mountain ranges are broad, fertile plains. In the north, the North European Plain stretches from the British Isles to the Russian border. Cities, towns, and farms dot the gently rolling landscape.

For centuries, Europe's rivers have provided links between coastal ports and inland population centers. In western Europe, the Rhine flows northwest from the Alps until it empties into the North Sea. The Danube winds through eastern Europe, bound for the Black Sea.

The Climate

Despite its northern location, Europe enjoys a relatively mild climate. The secret lies in the region's closeness to the Atlantic. An ocean current known as the North Atlantic Current brings warm waters and winds to bathe Europe's western shores. As a result, northwestern Europe enjoys mild temperatures all year, along with plentiful rainfall. Farther south, countries along the Mediterranean Sea have hot, dry summers and mild winters. The region's northernmost countries have longer, colder winters than their southern neighbors. Winters are also cold in Europe's interior, which lies far from the influence of the Atlantic.

The vegetation varies from one climate zone to another. In Scandinavia's far north, you would find mostly mosses and small shrubs blanketing a tundra-like landscape. In northwestern and eastern Europe, grasslands and forests cover the rolling land. Farther south, drought-resistant shrubs and small trees cover rugged hills.

Village at the foot of the Alps, Switzerland

Fisherman mending nets in Malta

Regional Atlas

The Economy

An abundance of key natural resources, waterways, and ports has helped make Europe a global economic power. Agriculture, manufacturing, and service industries dominate the region's economies.

Some of the most productive farmland in the world can be found on the European continent. From the fertile black soil, farmers gather bountiful harvests of grains, fruits, and vegetables. Cattle and sheep graze through lush European pastures.

Vast reserves of oil and natural gas lie offshore. Rich deposits of iron ore, coal, and other minerals have provided the raw materials for heavy industry and manufacturing. Europe was the birthplace of the Industrial Revolution, which transformed the region from an agricultural society into an industrial one. Today, countries such as France, Germany, Italy, Poland, and the United Kingdom rank among the world's top manufacturing centers.

The People

After Asia, Europe is the most densely populated continent on the earth. In some European countries, such as Sweden, most people belong to the same ethnic group. The populations of other countries, however, are made up of several ethnic groups. Some ethnic groups live together peacefully. Other groups often face tension and conflict.

Europeans enjoy a rich cultural heritage that stretches back thousands of years. Walk through the heart of any large European city and you might see ancient Roman ruins, Gothic cathedrals built in the Middle Ages, and sculptures created by Renaissance masters.

Throughout their long history, Europeans have explored and settled other lands. They have spread their culture to every part of the globe. Competition among European nations in the past century led to two world wars and a bitter division into Communist and non-Communist areas. Setting aside their differences, many European nations have recently joined the European Union. They are moving into the new century as a united economic force.

Exploring the Region

1. **What bodies of water border Europe?**
2. **Why is Europe's climate relatively mild?**
3. **What has helped make Europe a global economic power?**
4. **How did European culture spread to other parts of the world?**

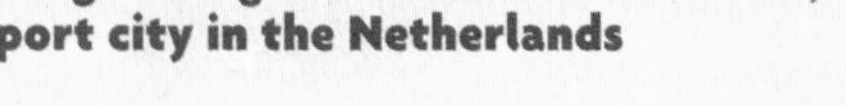

◀ **Cargo lining the docks of Rotterdam, a port city in the Netherlands**

UNIT 4

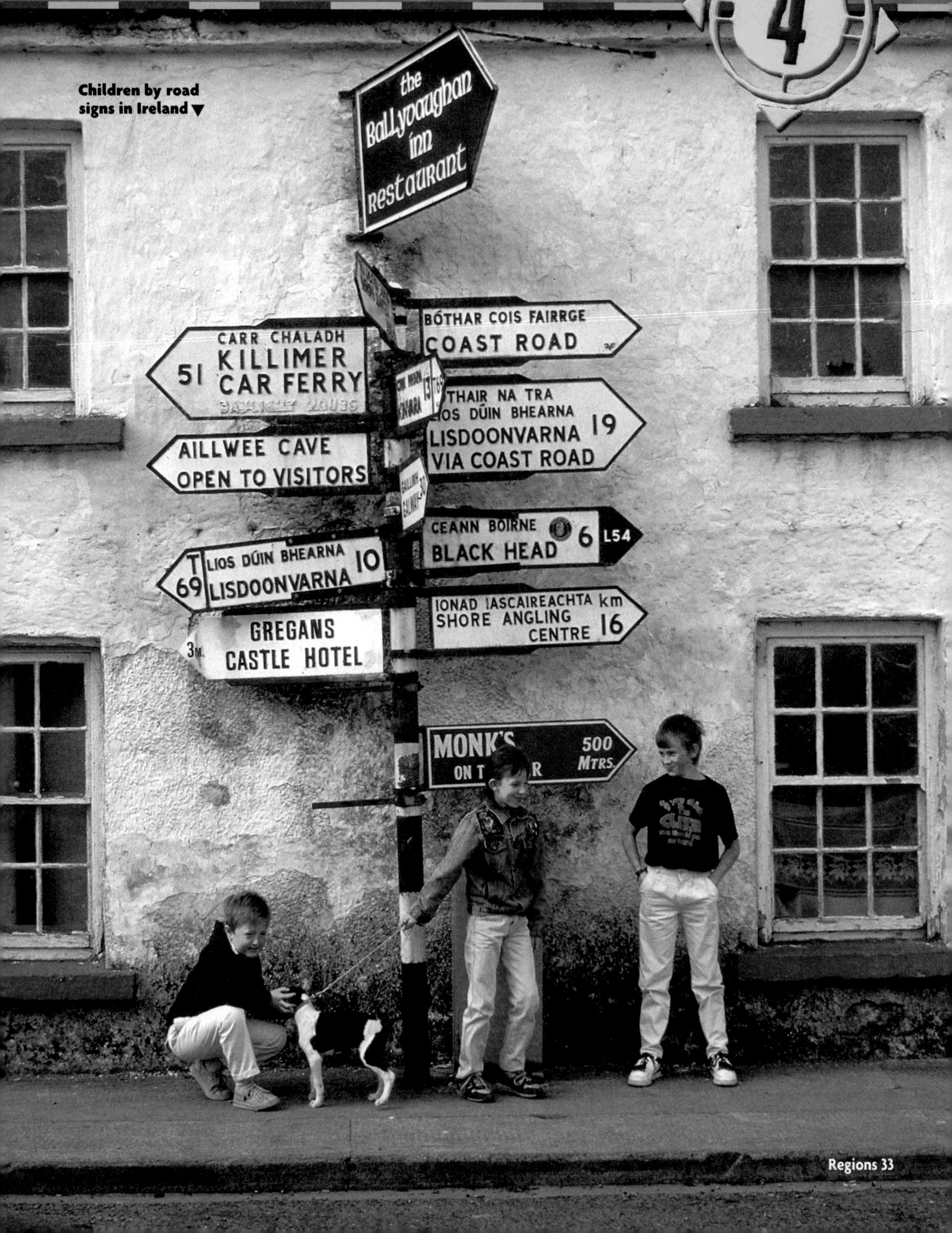

Children by road signs in Ireland ▼

NATIONAL
GEOGRAPHIC
SOCIETY

REGIONAL ATLAS

Europe

Physical

20°W
10°W
0°
10°E
20°E
30°E
ICELAND
ARCTIC CIRCLE
60°N
MERIDIAN OF GREENWICH (LONDON)
Norwegian Sea
▲ Mountain peak
0 mi. 500
0 km 500
Lambert Azimuthal Equal-Area projection
Faroe Is.
Shetland Is.
Orkney Is.
NORWAY
SCANDINAVIA
SWEDEN
FINLAND
ATLANTIC OCEAN
IRELAND
UNITED KINGDOM
British Isles
Thames R.
North Sea
Jutland
DENMARK
Baltic Sea
ESTONIA
LATVIA
LITHUANIA
RUSSIA
RUSSIA
BELARUS
50°N
NETH.
GERMANY
Elbe R.
NORTH EUROPEAN PLAIN
POLAND
Vistula R.
Oder R.
BELG.
LUX.
Rhine R.
Seine R.
Loire R.
CZECH REP.
Carpathian Mountains
SLOVAKIA
UKRAINE
Dnieper R.
FRANCE
LIECH.
SWITZ.
AUSTRIA
Hungarian Plain
HUNGARY
MOLDOVA
Bay of Biscay
ALPS
Mt. Blanc 15,771 ft. (4,807 m)
Matterhorn 14,690 ft. (4,478 m)
Po R.
SLOV.
CROATIA
ROMANIA
ANDORRA
Pyrenees
SAN MARINO
BOSN. & HERZG.
YUG.
Danube R.
Black Sea
40°N
PORTUGAL
Douro R.
Ebro R.
Tagus R.
SPAIN
IBERIAN PENINSULA
MONACO
Corsica
Adriatic Sea
Apennines
Balkan Peninsula
BULGARIA
ITALY
MACED.
Sardinia
ALBANIA
GREECE
Aegean Sea
Strait of Gibraltar
Mediterranean Sea
Sicily
MALTA
Crete
CYPRUS
26,247 ft.
0 mi. 500
0 km 500
8,000 m
19,685 ft.
6,000 m
ALPS
PYRENEES
13,123 ft.
4,000 m
6,562 ft.
2,000 m
LISBON
Sea level
WARSAW

UNIT 4

Political

MAP STUDY

1. What body of water lies between Scandinavia and Poland?
2. What is the capital of the United Kingdom?

REGIONAL ATLAS

Europe

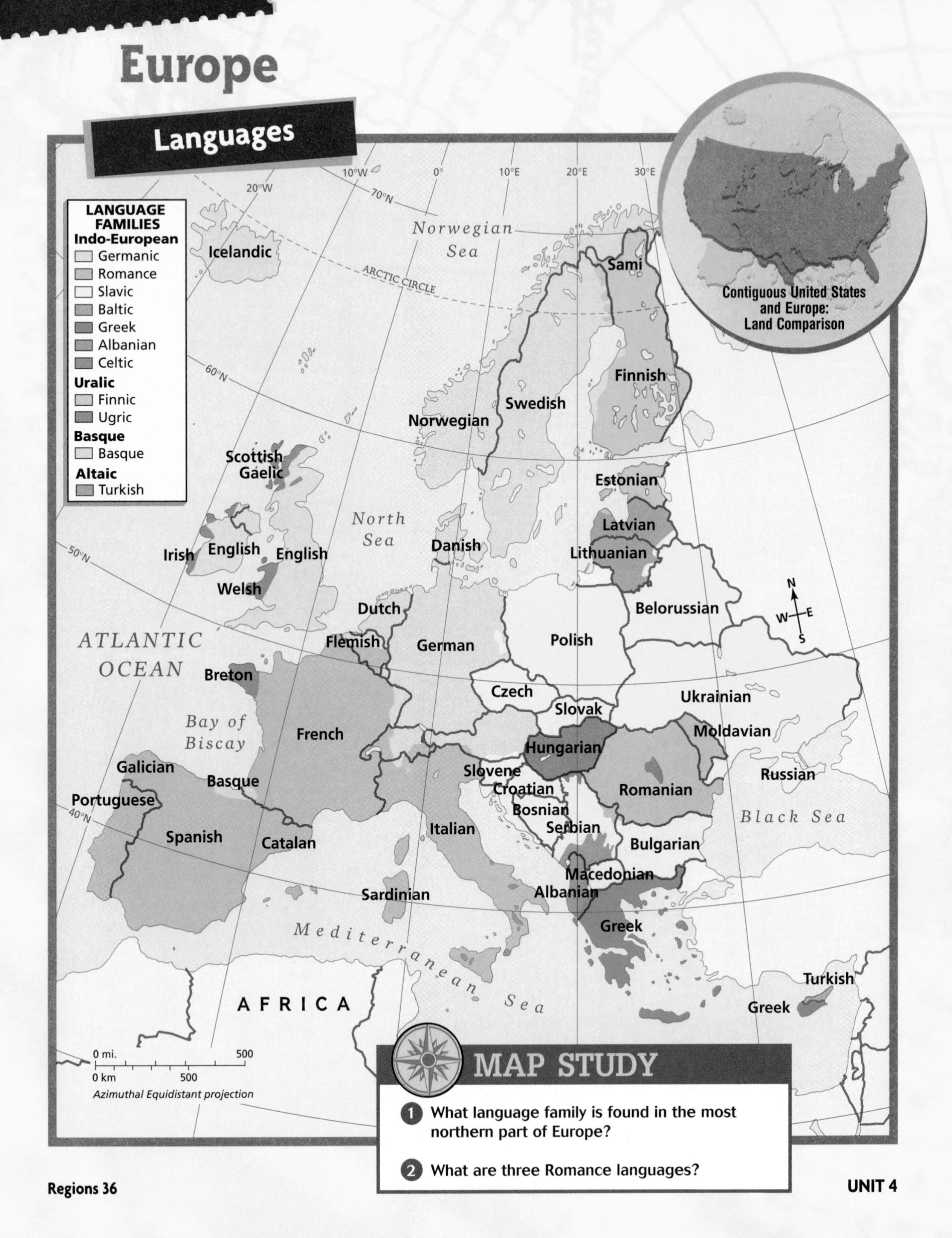

MAP STUDY

1. What language family is found in the most northern part of Europe?
2. What are three Romance languages?

Geo Extremes

① **HIGHEST POINT**
Mont Blanc (France and Italy)
15,771 ft. (4,807 m) high

② **LOWEST POINT**
Nieuwerkerk aan den IJssel (Netherlands)
22 ft. (7 m) below sea level

③ **LONGEST RIVER**
Danube (central Europe)
1,776 mi. (2,858 km) long

④ **LARGEST LAKE**
Lake Vänern (Sweden)
2,156 sq. mi. (5,584 sq. km)

⑤ **HIGHEST WATERFALL**
Mardalsfossen, Southern (Norway)
2,149 ft. (655 m) high

⑥ **LARGEST ISLAND**
Great Britain
84,210 sq. mi.
(218,103 sq. km)

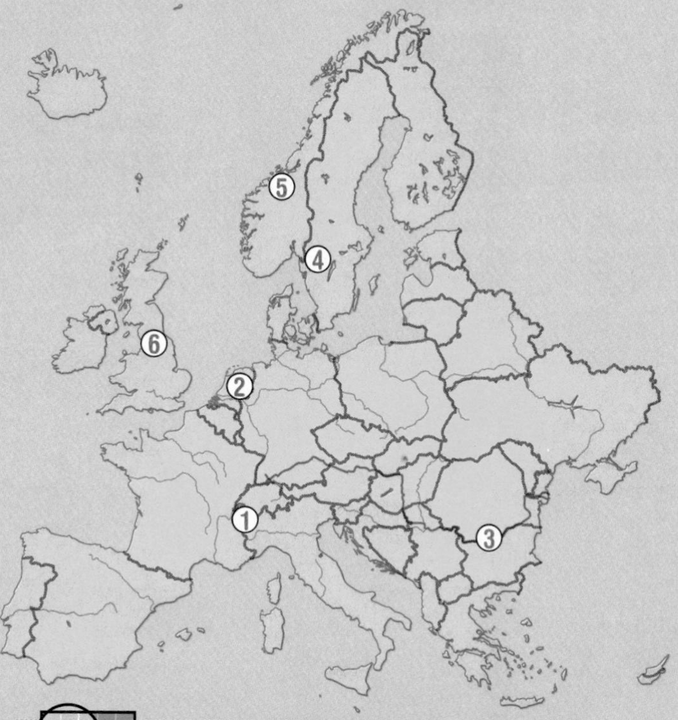

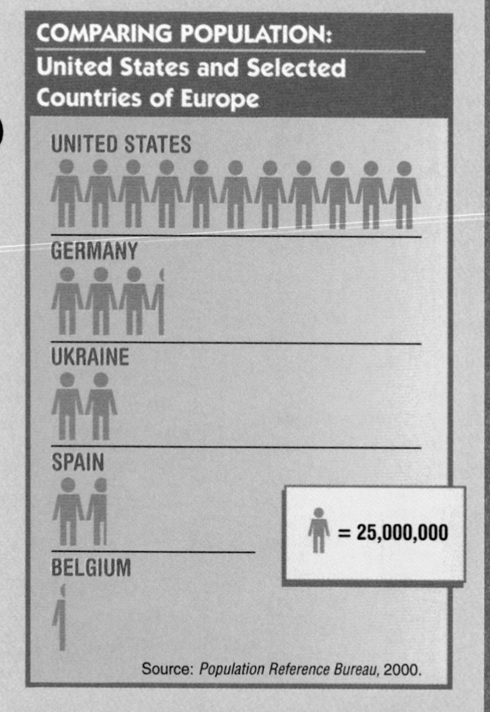

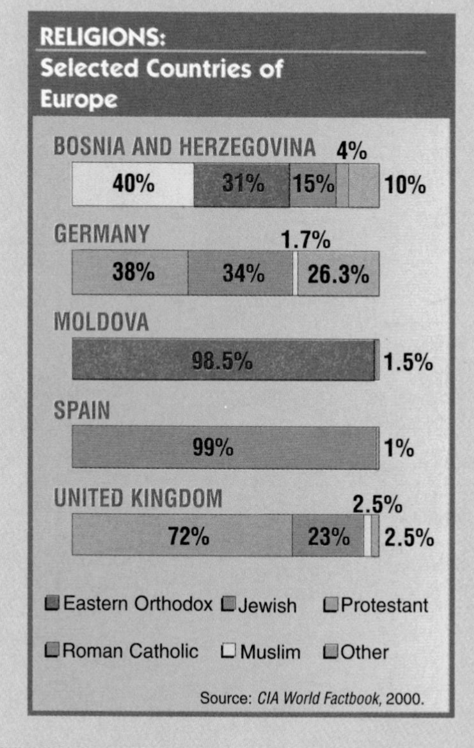

GRAPHIC STUDY

1. What two countries share the highest point in Europe?
2. Roughly what is the population of Germany? What percentage of the population is Protestant?

Country Profiles

ALBANIA

POPULATION:
3,460,000
312 per sq. mi.
120 per sq. km

LANGUAGE:
Albanian

MAJOR EXPORT:
Asphalt

MAJOR IMPORT:
Machinery

CAPITAL:
Tirana

LANDMASS:
11,100 sq. mi.
28,748 sq. km

ANDORRA

POPULATION:
66,000
377 per sq. mi.
146 per sq. km

LANGUAGES:
Catalan, French, Spanish

MAJOR EXPORT:
Electricity

MAJOR IMPORT:
Manufactured Goods

CAPITAL:
Andorra la Vella

LANDMASS:
175 sq. mi.
453 sq. km

AUSTRIA

POPULATION:
8,087,000
250 per sq. mi.
96 per sq. km

LANGUAGE:
German

MAJOR EXPORT:
Machinery

MAJOR IMPORT:
Petroleum

CAPITAL:
Vienna

LANDMASS:
32,377 sq. mi.
83,856 sq. km

BELARUS

POPULATION:
10,167,000
127 per sq. mi.
49 per sq. km

LANGUAGES:
Belarussian, Russian

MAJOR EXPORT:
Machinery

MAJOR IMPORT:
Fuels

CAPITAL:
Minsk

LANDMASS:
80,154 sq. mi.
207,598 sq. km

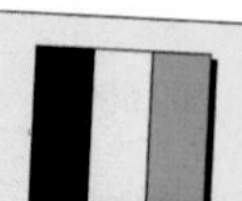

BELGIUM

POPULATION:
10,225,000
868 per sq. mi.
335 per sq. km

LANGUAGES:
Flemish, French

MAJOR EXPORTS:
Iron and Steel

MAJOR IMPORT:
Fuels

CAPITAL:
Brussels

LANDMASS:
11,783 sq. mi.
30,518 sq. km

BOSNIA and HERZEGOVINA

POPULATION:
3,839,000
194 per sq. mi.
75 per sq. km

LANGUAGE:
Serbo-Croatian

MAJOR EXPORT:
N/A

MAJOR IMPORT:
N/A

CAPITAL:
Sarajevo

LANDMASS:
19,741 sq. mi.
51,129 sq. km

BULGARIA

POPULATION:
8,188,000
191 per sq. mi.
74 per sq. km

LANGUAGE:
Bulgarian

MAJOR EXPORT:
Machinery

MAJOR IMPORT:
Fuels

CAPITAL:
Sofia

LANDMASS:
42,823 sq. mi.
110,912 sq. km

CROATIA

POPULATION:
4,600,000
211 per sq. mi.
81 per sq. km

LANGUAGE:
Serbo-Croatian

MAJOR EXPORT:
Transport Equipment

MAJOR IMPORT:
Machinery

CAPITAL:
Zagreb

LANDMASS:
21,829 sq. mi.
56,538 sq. km

CYPRUS

POPULATION:
875,000
384 per sq. mi.
148 per sq. km

LANGUAGES:
Greek, Turkish

MAJOR EXPORT:
Citrus Fruits

MAJOR IMPORT:
Manufactured Goods

CAPITAL:
Nicosia

LANDMASS:
2,277 sq. mi.
5,897 sq. km

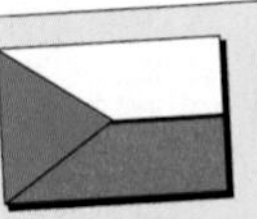

CZECH REPUBLIC

POPULATION:
10,284,000
338 per sq. mi.
130 per sq. km

LANGUAGES:
Czech, Slovak

MAJOR EXPORT:
Machinery

MAJOR IMPORT:
Crude Oil

CAPITAL:
Prague

LANDMASS:
30,450 sq. mi.
78,864 sq. km

DENMARK

POPULATION:
5,325,000
320 per sq. mi.
124 per sq. km

LANGUAGE:
Danish

MAJOR EXPORT:
Machinery

MAJOR IMPORT:
Machinery

CAPITAL:
Copenhagen

LANDMASS:
16,638 sq. mi.
43,092 sq. km

ESTONIA

POPULATION:
1,441,000
83 per sq. mi.
32 per sq. km

LANGUAGE:
Estonian

MAJOR EXPORT:
Textiles

MAJOR IMPORT:
Machinery

CAPITAL:
Tallinn

LANDMASS:
17,413 sq. mi.
45,099 sq. km

Countries and flags not drawn to scale

FINLAND

POPULATION:
5,170,000
40 per sq. mi.
15 per sq. km

LANGUAGES:
Finnish, Swedish

MAJOR EXPORT:
Paper

MAJOR IMPORT:
Foods

CAPITAL:
Helsinki

LANDMASS:
130,558 sq. mi.
338,145 sq. km

FRANCE

POPULATION:
59,067,000
281 per sq. mi.
109 per sq. km

LANGUAGE:
French

MAJOR EXPORT:
Machinery

MAJOR IMPORT:
Crude Oil

CAPITAL:
Paris

LANDMASS:
210,026 sq. mi.
543,965 sq. km

GERMANY

POPULATION:
81,950,000
594 per sq. mi.
230 per sq. km

LANGUAGE:
German

MAJOR EXPORT:
Machinery

MAJOR IMPORT:
Machinery

CAPITAL:
Berlin

LANDMASS:
137,857 sq. mi.
357,046 sq. km

GREECE

POPULATION:
10,539,000
207 per sq. mi.
80 per sq. km

LANGUAGE:
Greek

MAJOR EXPORT:
Foods

MAJOR IMPORT:
Machinery

CAPITAL:
Athens

LANDMASS:
50,962 sq. mi.
131,990 sq. km

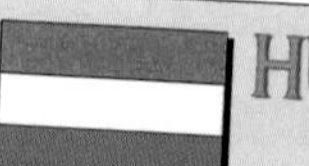

HUNGARY

POPULATION:
10,076,000
281 per sq. mi.
108 per sq. km

LANGUAGE:
Hungarian

MAJOR EXPORT:
Machinery

MAJOR IMPORT:
Crude Oil

CAPITAL:
Budapest

LANDMASS:
35,919 sq. mi.
93,030 sq. km

ICELAND

POPULATION:
277,000
7 per sq. mi.
3 per sq. km

LANGUAGE:
Icelandic

MAJOR EXPORT:
Fish

MAJOR IMPORT:
Machinery

CAPITAL:
Reykjavík

LANDMASS:
39,769 sq. mi.
103,001 sq. km

POPULATION:
3,734,000
138 per sq. mi.
53 per sq. km

LANGUAGES:
English, Irish Gaelic

MAJOR EXPORT:
Chemical Products

MAJOR IMPORT:
Foods

CAPITAL:
Dublin

LANDMASS:
27,137 sq. mi.
70,284 sq. km

ITALY

POPULATION:
57,717,000
496 per sq. mi.
192 per sq. km

LANGUAGE:
Italian

MAJOR EXPORT:
Metals

MAJOR IMPORT:
Machinery

CAPITAL:
Rome

LANDMASS:
116,324 sq. mi.
301,277 sq. km

LATVIA

POPULATION:
2,430,000
97 per sq. mi.
38 per sq. km

LANGUAGES:
Latvian, Russian

MAJOR EXPORT:
Wood

MAJOR IMPORT:
Fuels

CAPITAL:
Rīga

LANDMASS:
24,942 sq. mi.
64,599 sq. km

POPULATION:
32,000
516 per sq. mi.
200 per sq. km

LANGUAGE:
German

MAJOR EXPORT:
Machinery

MAJOR IMPORT:
Machinery

CAPITAL:
Vaduz

LANDMASS:
62 sq. mi.
160 sq. km

POPULATION:
3,700,000
147 per sq. mi.
57 per sq. km

LANGUAGES:
Lithuanian, Polish, Russian

MAJOR EXPORTS:
Foods and Livestock

MAJOR IMPORT:
Minerals

CAPITAL:
Vilnius

LANDMASS:
25,174 sq. mi.
65,200 sq. km

LUXEMBOURG

POPULATION:
432,000
433 per sq. mi.
167 per sq. km

LANGUAGES:
Luxembourgian, German, French

MAJOR EXPORT:
Steel Products

MAJOR IMPORT:
Minerals

CAPITAL:
Luxembourg

LANDMASS:
998 sq. mi.
2,586 sq. km

Russia

Physical

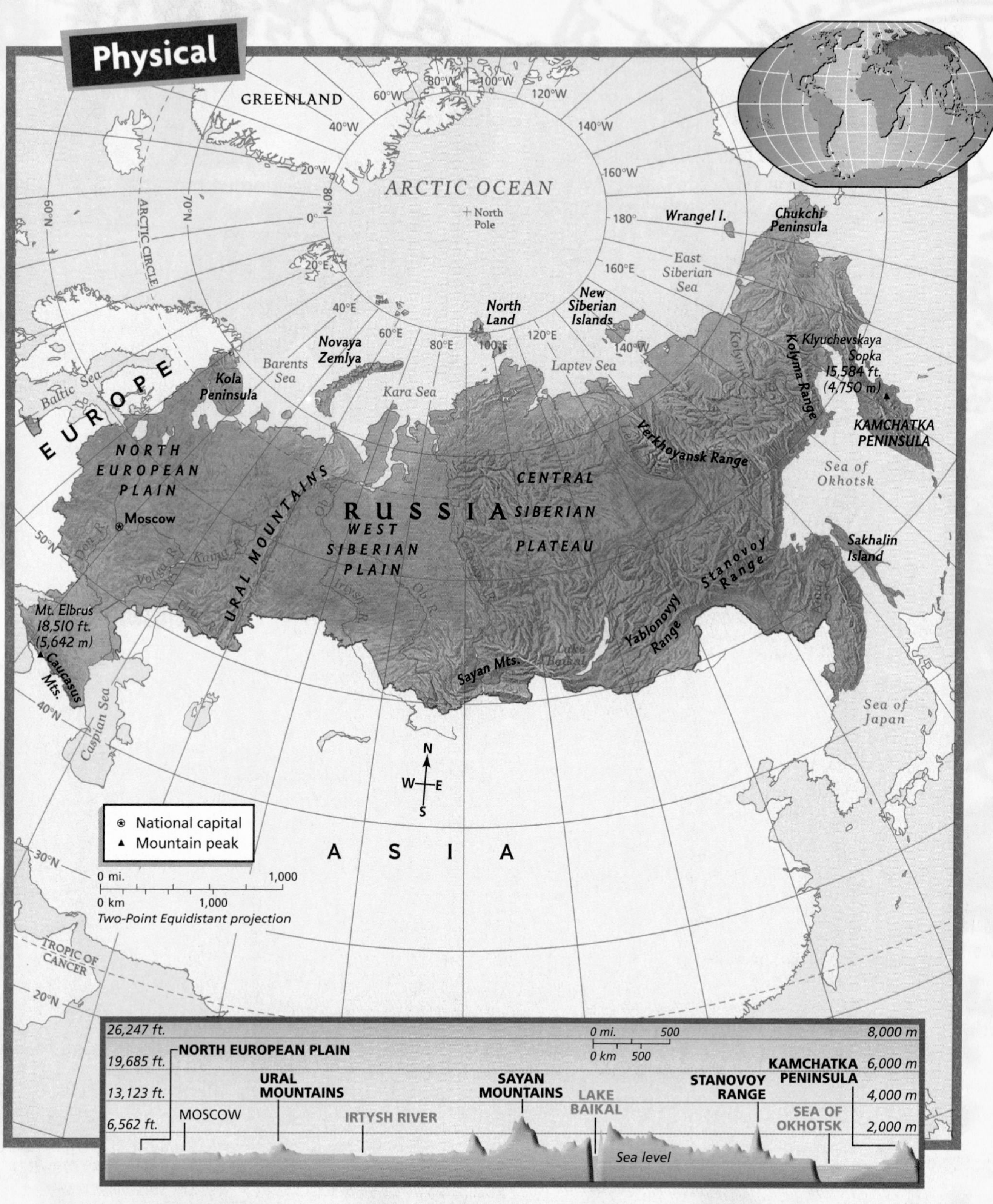

FINLAND

POPULATION:
5,170,000
40 per sq. mi.
15 per sq. km

LANGUAGES:
Finnish, Swedish

MAJOR EXPORT:
Paper

MAJOR IMPORT:
Foods

CAPITAL:
Helsinki

LANDMASS:
130,558 sq. mi.
338,145 sq. km

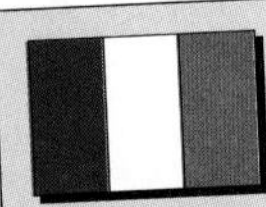

FRANCE

POPULATION:
59,067,000
281 per sq. mi.
109 per sq. km

LANGUAGE:
French

MAJOR EXPORT:
Machinery

MAJOR IMPORT:
Crude Oil

CAPITAL:
Paris

LANDMASS:
210,026 sq. mi.
543,965 sq. km

GERMANY

POPULATION:
81,950,000
594 per sq. mi.
230 per sq. km

LANGUAGE:
German

MAJOR EXPORT:
Machinery

MAJOR IMPORT:
Machinery

CAPITAL:
Berlin

LANDMASS:
137,857 sq. mi.
357,046 sq. km

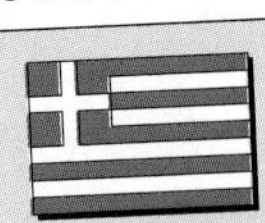

GREECE

POPULATION:
10,539,000
207 per sq. mi.
80 per sq. km

LANGUAGE:
Greek

MAJOR EXPORT:
Foods

MAJOR IMPORT:
Machinery

CAPITAL:
Athens

LANDMASS:
50,962 sq. mi.
131,990 sq. km

HUNGARY

POPULATION:
10,076,000
281 per sq. mi.
108 per sq. km

LANGUAGE:
Hungarian

MAJOR EXPORT:
Machinery

MAJOR IMPORT:
Crude Oil

CAPITAL:
Budapest

LANDMASS:
35,919 sq. mi.
93,030 sq. km

ICELAND

POPULATION:
277,000
7 per sq. mi.
3 per sq. km

LANGUAGE:
Icelandic

MAJOR EXPORT:
Fish

MAJOR IMPORT:
Machinery

CAPITAL:
Reykjavík

LANDMASS:
39,769 sq. mi.
103,001 sq. km

IRELAND

POPULATION:
3,734,000
138 per sq. mi.
53 per sq. km

LANGUAGES:
English, Irish Gaelic

MAJOR EXPORT:
Chemical Products

MAJOR IMPORT:
Foods

CAPITAL:
Dublin

LANDMASS:
27,137 sq. mi.
70,284 sq. km

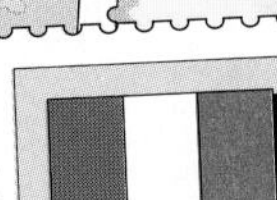

ITALY

POPULATION:
57,717,000
496 per sq. mi.
192 per sq. km

LANGUAGE:
Italian

MAJOR EXPORT:
Metals

MAJOR IMPORT:
Machinery

CAPITAL:
Rome

LANDMASS:
116,324 sq. mi.
301,277 sq. km

LATVIA

POPULATION:
2,430,000
97 per sq. mi.
38 per sq. km

LANGUAGES:
Latvian, Russian

MAJOR EXPORT:
Wood

MAJOR IMPORT:
Fuels

CAPITAL:
Rīga

LANDMASS:
24,942 sq. mi.
64,599 sq. km

LIECHTENSTEIN

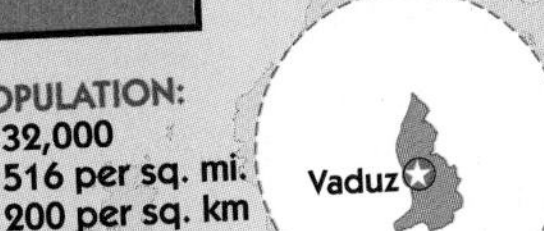

POPULATION:
32,000
516 per sq. mi.
200 per sq. km

LANGUAGE:
German

MAJOR EXPORT:
Machinery

MAJOR IMPORT:
Machinery

CAPITAL:
Vaduz

LANDMASS:
62 sq. mi.
160 sq. km

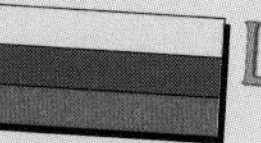

LITHUANIA

POPULATION:
3,700,000
147 per sq. mi.
57 per sq. km

LANGUAGES:
Lithuanian, Polish, Russian

MAJOR EXPORTS:
Foods and Livestock

MAJOR IMPORT:
Minerals

CAPITAL:
Vilnius

LANDMASS:
25,174 sq. mi.
65,200 sq. km

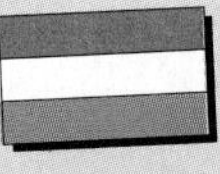

LUXEMBOURG

POPULATION:
432,000
433 per sq. mi.
167 per sq. km

LANGUAGES:
Luxembourgian, German, French

MAJOR EXPORT:
Steel Products

MAJOR IMPORT:
Minerals

CAPITAL:
Luxembourg

LANDMASS:
998 sq. mi.
2,586 sq. km

Country Profiles

MACEDONIA, Former Yugoslav Republic of

POPULATION:
2,019,000
203 per sq. mi.
79 per sq. km

LANGUAGES:
Macedonian, Albanian

MAJOR EXPORT:
Manufactured Goods

MAJOR IMPORT:
Fuels

CAPITAL:
Skopje

LANDMASS:
9,928 sq. mi.
25,713 sq. km

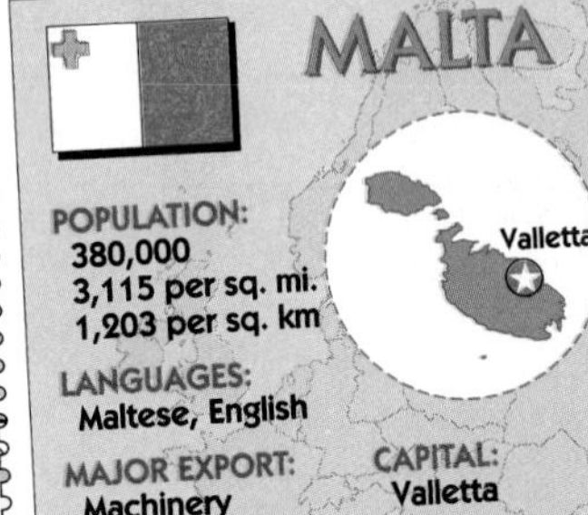

MALTA

POPULATION:
380,000
3,115 per sq. mi.
1,203 per sq. km

LANGUAGES:
Maltese, English

MAJOR EXPORT:
Machinery

MAJOR IMPORT:
Foods

CAPITAL:
Valletta

LANDMASS:
122 sq. mi.
316 sq. km

MOLDOVA

POPULATION:
4,284,000
324 per sq. mi.
126 per sq. km

LANGUAGES:
Moldovan, Russian

MAJOR EXPORT:
Foods

MAJOR IMPORT:
Petroleum

CAPITAL:
Chişinău

LANDMASS:
13,217 sq. mi.
33,999 sq. km

MONACO

POPULATION:
33,000
55,000 per sq. mi.
17,368 per sq. km

LANGUAGE:
French

MAJOR EXPORT:
N/A

MAJOR IMPORT:
N/A

CAPITAL:
Monaco

LANDMASS:
0.6 sq. mi.
1.9 sq. km

NETHERLANDS

POPULATION:
15,799,000
986 per sq. mi.
381 per sq. km

LANGUAGE:
Dutch

MAJOR EXPORT:
Manufactured Goods

MAJOR IMPORT:
Raw Materials

CAPITAL:
Amsterdam

LANDMASS:
16,023 sq. mi.
41,499 sq. km

NORWAY

POPULATION:
4,462,000
36 per sq. mi.
14 per sq. km

LANGUAGE:
Norwegian

MAJOR EXPORT:
Petroleum

MAJOR IMPORT:
Machinery

CAPITAL:
Oslo

LANDMASS:
125,182 sq. mi.
324,220 sq. km

POLAND

POPULATION:
38,674,000
320 per sq. mi.
124 per sq. km

LANGUAGE:
Polish

MAJOR EXPORT:
Manufactured Goods

MAJOR IMPORT:
Machinery

CAPITAL:
Warsaw

LANDMASS:
120,725 sq. mi.
312,677 sq. km

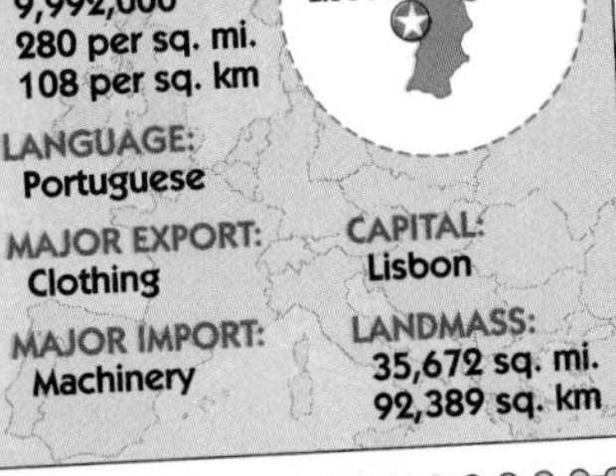

PORTUGAL

POPULATION:
9,992,000
280 per sq. mi.
108 per sq. km

LANGUAGE:
Portuguese

MAJOR EXPORT:
Clothing

MAJOR IMPORT:
Machinery

CAPITAL:
Lisbon

LANDMASS:
35,672 sq. mi.
92,389 sq. km

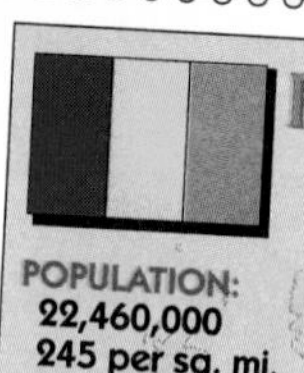

ROMANIA

POPULATION:
22,460,000
245 per sq. mi.
95 per sq. km

LANGUAGES:
Romanian, Hungarian, German

MAJOR EXPORT:
Textiles

MAJOR IMPORT:
Fuels

CAPITAL:
Bucharest

LANDMASS:
91,699 sq. mi.
237,499 sq. km

SAN MARINO

POPULATION:
26,000
1,083 per sq. mi.
426 per sq. km

LANGUAGE:
Italian

MAJOR EXPORT:
Building Stone

MAJOR IMPORT:
Manufactured Goods

CAPITAL:
San Marino

LANDMASS:
24 sq. mi.
61 sq. km

SLOVAKIA

POPULATION:
5,401,000
285 per sq. mi.
110 per sq. km

LANGUAGES:
Slovak, Hungarian

MAJOR EXPORT:
Transport Equipment

MAJOR IMPORT:
Machinery

CAPITAL:
Bratislava

LANDMASS:
18,921 sq. mi.
49,006 sq. km

SLOVENIA

POPULATION:
1,978,000
253 per sq. mi.
98 per sq. km

LANGUAGES:
Slovene, Serbo-Croatian

MAJOR EXPORT:
Transport Equipment

MAJOR IMPORT:
Machinery

CAPITAL:
Ljubljana

LANDMASS:
7,819 sq. mi.
20,251 sq. km

Countries and flags not drawn to scale

UNIT 4

SPAIN

POPULATION:
39,418,000
202 per sq. mi.
78 per sq. km

LANGUAGES:
Spanish, Catalan, Galician, Basque

MAJOR EXPORTS:
Cars and Trucks

MAJOR IMPORT:
Machinery

CAPITAL:
Madrid

LANDMASS:
194,897 sq. mi.
504,782 sq. km

SWEDEN

POPULATION:
8,856,000
51 per sq. mi.
20 per sq. km

LANGUAGE:
Swedish

MAJOR EXPORT:
Paper Products

MAJOR IMPORT:
Crude Oil

CAPITAL:
Stockholm

LANDMASS:
173,732 sq. mi.
449,964 sq. km

SWITZERLAND

POPULATION:
7,119,000
447 per sq. mi.
172 per sq. km

LANGUAGES:
German, French, Italian, Romansch

MAJOR EXPORT:
Precision Instruments

MAJOR IMPORT:
Machinery

CAPITAL:
Bern

LANDMASS:
15,941 sq. mi.
41,288 sq. km

UKRAINE

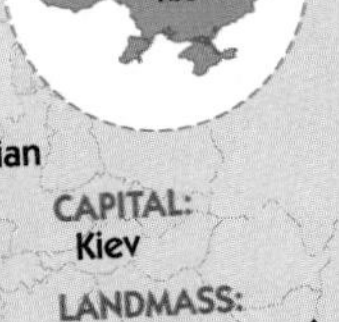

POPULATION:
49,910,000
214 per sq. mi.
83 per sq. km

LANGUAGES:
Ukrainian, Russian

MAJOR EXPORT:
Metals

MAJOR IMPORT:
Machinery

CAPITAL:
Kiev

LANDMASS:
233,206 sq. mi.
604,001 sq. km

UNITED KINGDOM

POPULATION:
59,364,000
630 per sq. mi.
243 per sq. km

LANGUAGES:
English, Welsh, Scottish Gaelic

MAJOR EXPORT:
Manufactured Goods

MAJOR IMPORT:
Foods

CAPITAL:
London

LANDMASS:
94,248 sq. mi.
244,101 sq. km

VATICAN CITY

POPULATION:
1,000

LANGUAGES:
Italian, Latin

MAJOR EXPORT:
N/A

MAJOR IMPORT:
N/A

CAPITAL:
N/A

LANDMASS:
0.2 sq. mi.
0.4 sq. km

YUGOSLAVIA
(Serbia and Montenegro)

POPULATION:
10,646,000
270 per sq. mi.
104 per sq. km

LANGUAGES:
Serbo-Croatian, Albanian

MAJOR EXPORT:
Manufactured Goods

MAJOR IMPORT:
Machinery

CAPITAL:
Belgrade

LANDMASS:
39,450 sq. mi.
102,173 sq. km

GEO BEE

Questions From Buzz Bee!

The following questions are taken from National Geographic GeoBees. Use your textbook, the Internet, and other library resources to find the answers.

1. Which European capital city located on the Seine River is known as the City of Light?
2. Which country leased Hong Kong from China for 99 years?
3. Slovakia and which other present-day central European country became independent in 1993?

Unit 5

Workers on the statue *Motherland Calls*, Volgograd

Russians in front of St. Basil's Cathedral, Moscow

Russia

If you had to describe Russia in one word, that word would be "BIG"! Russia is the largest country in the world in area. Its almost 6.6 million square miles (17 million sq. km) are spread across two continents—Europe and Asia. As you can imagine, such a large country faces equally large challenges. Since 1991, when Russia emerged as an independent country, it has been struggling to unite its many ethnic groups, set up a democratic government, and build a stable economy.

◀ Siberian tiger in a forest in eastern Russia

NGS ONLINE
www.nationalgeographic.com/education

Focus on:

Russia

RUSSIA IS THE WORLD'S LARGEST COUNTRY. Spanning 11 time zones, it is almost twice the size of the United States. Russia's far northern location affects its climate and the lifestyle of its people. Being somewhat isolated from the rest of the world has played a major role in the country's history, politics, and economic development.

The Land

Stretching nearly halfway around the globe, Russia sprawls from the Bering Sea in the east to the Baltic Sea in the west. Its northernmost lands lie above the Arctic Circle; its southern border winds through the middle of Asia.

Russia is a land of sweeping plains and plateaus interrupted by mountain ranges. The Ural Mountains run north to south, dividing the country into a European region and a much larger Asian region. West of the Urals is the fertile North European Plain—home to three-fourths of the country's population. East of the Urals lies Siberia, which means "sleeping land." Immense and sparsely populated, Siberia is an area of harsh, forbidding landscapes. Although rich in natural resources, much of it remains a wilderness.

The Caspian Sea—actually a saltwater lake—lies at the base of the Caucasus Mountains in Russia's southwest. It is the largest inland body of water in the world. Farther east is Lake Baikal, the world's deepest lake. Many rivers wind through Russian landscapes. The Volga flows southward to empty into the Caspian Sea. The Lena, Yenisey, and Ob Rivers all flow north to the Arctic Ocean.

The Climate

Most of Russia has a cold climate due to its northern location. In Siberia's far north, the landscape is dominated by tundra, a treeless plain. Winters on the tundra are long, dark, and fiercely cold. During the brief summers, only the top few inches of soil thaw out. Deeper down is permafrost—permanently frozen ground.

South of the tundra are immense evergreen forests. This vast woodland area, known as the taiga, is the largest continuous expanse of forest on the earth. Snow blankets the taiga for as much as eight months of the year.

Farther south, the taiga gives way to flat, grass-covered plains, or steppes. Here the climate is less harsh, and the soil quite rich. The steppes make up Russia's most productive agricultural area.

Wildflowers and wooden churches on the North European Plain, in northwestern Russia ▼

◀ Reindeer pulling sled across the tundra, Siberia

The Economy

For nearly 70 years in the twentieth century, Russia and the other republics of the Soviet Union shared a single economy that was controlled by Communist authorities. Wheat and other crops were grown on huge government-owned farms. The top economic priority was heavy industry—the manufacturing of goods such as machinery and military equipment. Russia's rich deposits of minerals, coal, and oil supplied the raw materials and energy for many of its industries. The push to industrialize, however, led to widespread pollution of the air, soil, and water. Industrial growth also was more important than the needs of the people. Shortages of consumer goods—clothing and household products, for example—were common.

In the 1990s, when Russia and the other republics of the Soviet Union became independent countries, each took charge of its own economy. Today Russia is struggling to make the transition to a free enterprise system, in which people run their own businesses and farms.

The People

Roughly 147 million people live in Russia. Most live west of the Ural Mountains, where the climate is mildest and the land most fertile. Moscow and St. Petersburg are the region's largest cities.

Although people of many ethnic groups can be found within Russia's borders, most Russians are descendants of Slavic peoples, or Slavs. Centuries ago, Slavs from northeastern Europe settled in what is now western Russia. Their settlements grew into city-states ruled by princes. By the 1400s, these states were united under the rule of a czar. For more than 400 years, Russia was governed by a series of powerful, and often ruthless, czars. Their armies conquered surrounding lands to gradually create a huge Russian Empire.

In 1917 the last czar was overthrown, and a Communist dictatorship emerged. The Russian Empire became the Union of Soviet Socialist Republics. The Communists made the Soviet Union an industrial power but denied the people basic freedoms. In 1991 the Communists fell from power, and the Soviet Union disintegrated.

1. **Why might Russia's north-flowing rivers be difficult to travel in winter?**
2. **Why would it be hard to grow crops on the tundra?**
3. **What was the top economic priority of Communist authorities?**
4. **To what ethnic group do most Russians belong?**

◀ Russian worker inspecting tractors in a factory

UNIT 5

▲ Young people strolling and singing in St. Petersburg

REGIONAL ATLAS

Russia

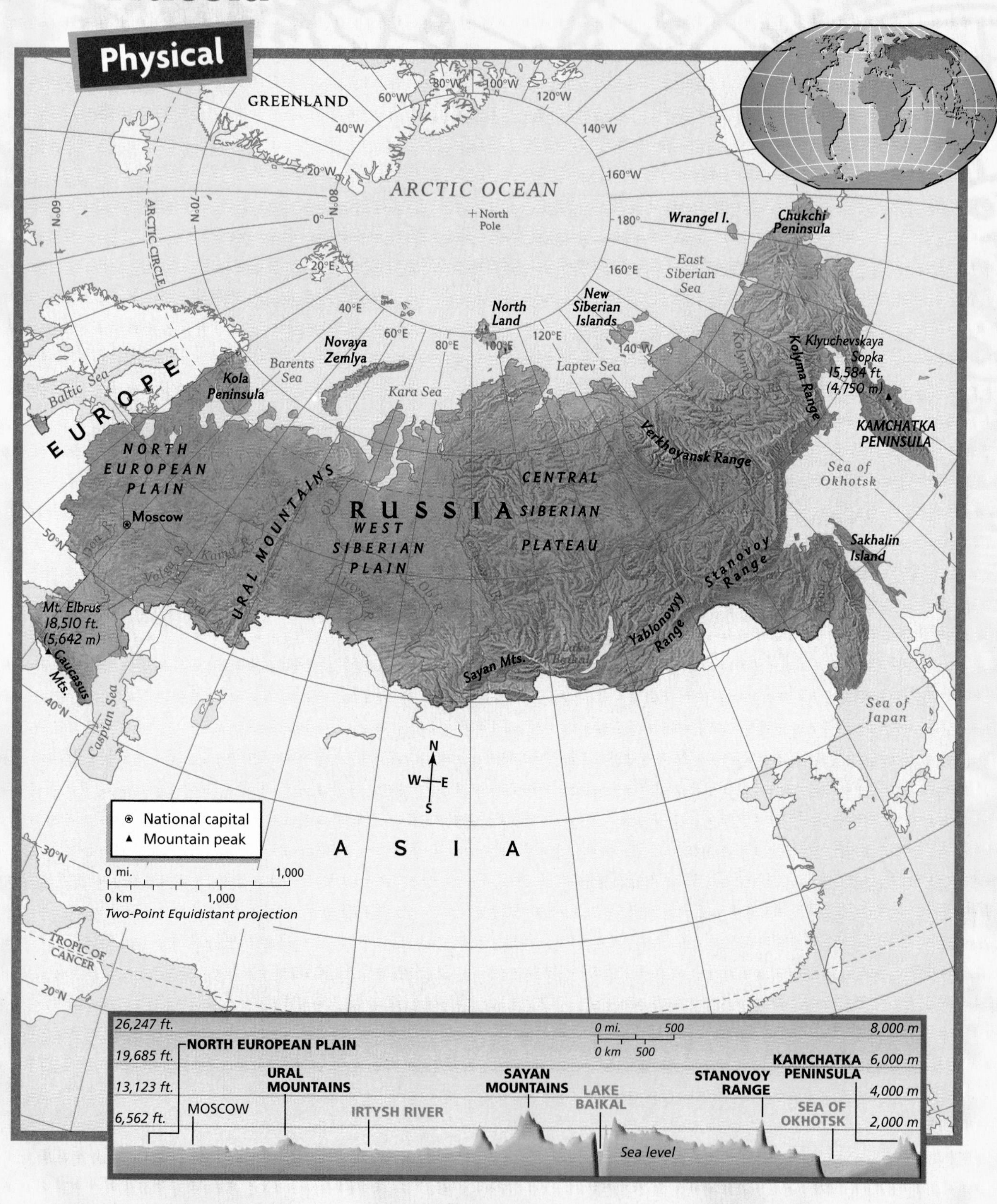

UNIT 5

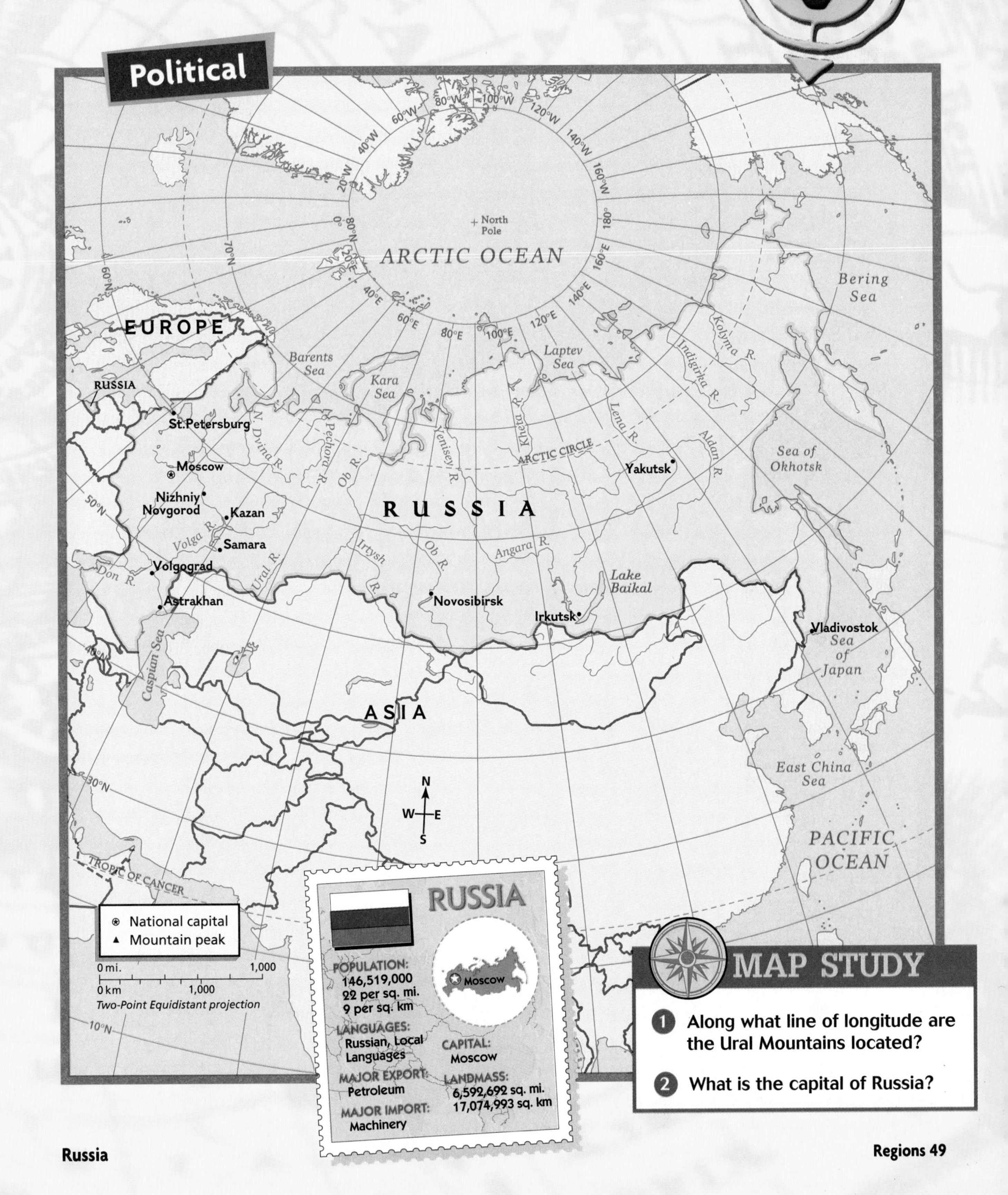

MAP STUDY

1. Along what line of longitude are the Ural Mountains located?
2. What is the capital of Russia?

REGIONAL ATLAS

Russia

The Russian Winter

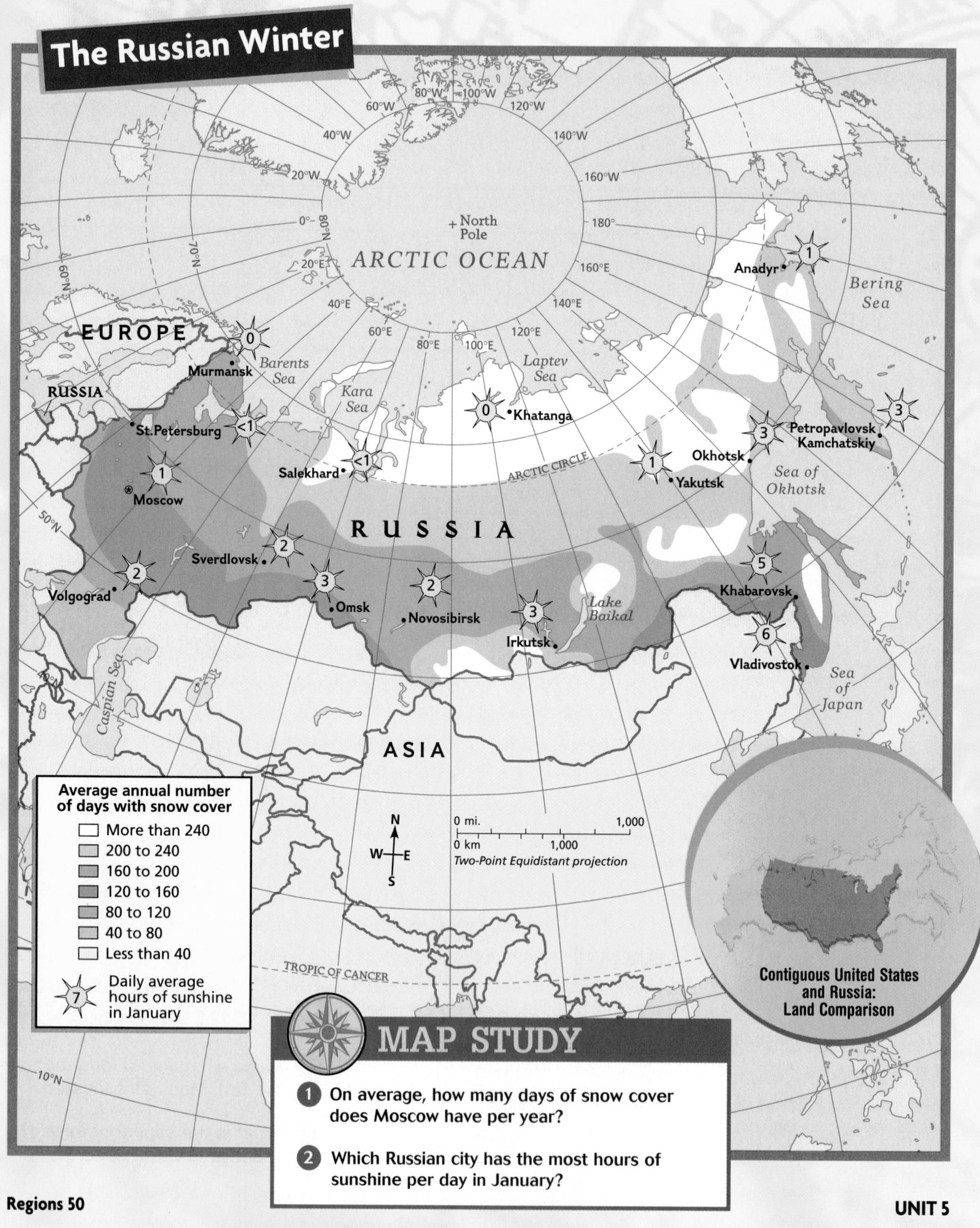

MAP STUDY

1. On average, how many days of snow cover does Moscow have per year?
2. Which Russian city has the most hours of sunshine per day in January?

Geo Extremes

① **HIGHEST POINT**
Mount Elbrus
18,510 ft.
(5,642 m) high

② **LOWEST POINT**
Caspian Sea
92 ft. (28 m)
below sea level

③ **LONGEST RIVER**
Ob-Irtysh
3,362 mi.
(5,411 km) long

④ **LARGEST LAKE**
Caspian Sea
143,244 sq. mi.
(371,000 sq. km)

⑤ **DEEPEST LAKE**
Lake Baikal
5,315 ft.
(1,620 m) deep

⑥ **LARGEST ISLAND**
Sakhalin
29,500 sq.mi.
(76,405 sq. km)

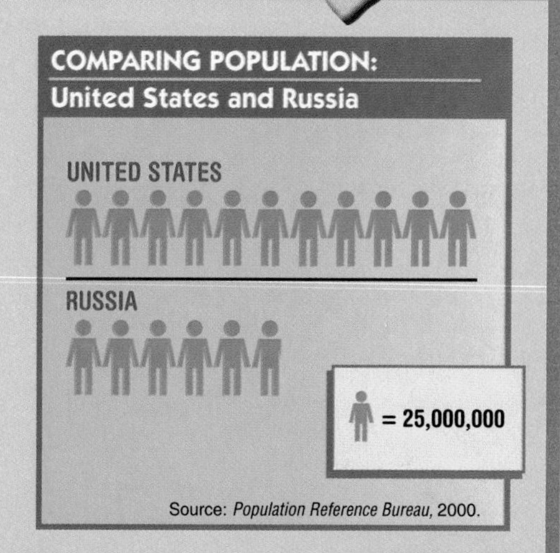

Source: *Population Reference Bureau*, 2000.

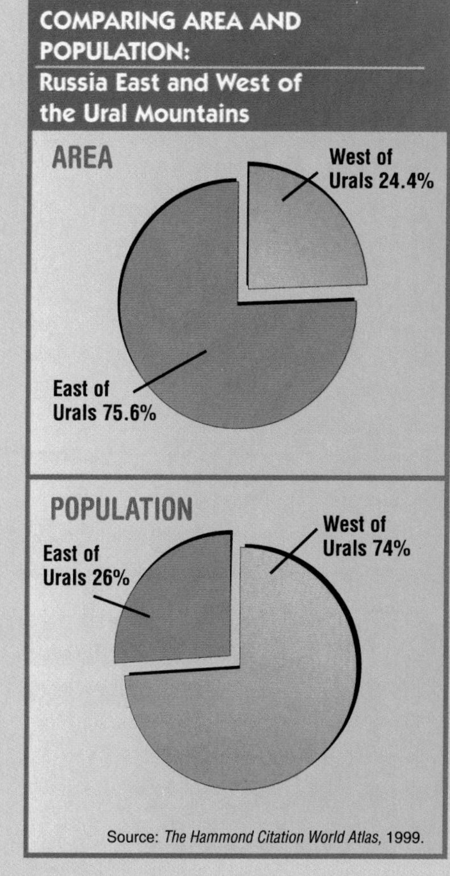

Source: *The Hammond Citation World Atlas*, 1999.

GRAPHIC STUDY

1. What two "extremes" does the Caspian Sea lay claim to?
2. What percentage of Russia's people live west of the Ural Mountains?

Unit 6

Business district at dusk, Dubai, United Arab Emirates

Young woman in national dress, Turkmenistan ▶

North Africa, Southwest Asia, and Central Asia

Ancient Egyptian pyramids overlook industrial smokestacks. Three-thousand-year-old stone temples tower over sparkling new oil derricks. Remote mountain villages and endless desert seas of sand and gravel contrast with modern beaches overrun by tourists. All of these extremes can be found within the culture region of North Africa, Southwest Asia, and Central Asia.

NGS ONLINE
www.nationalgeographic.com/education

Shepherd tending sheep, Atlas Mountains, Morocco

Focus on:

North Africa, Southwest Asia, and Central Asia

LYING AT THE INTERSECTION of Europe, Asia, and Africa, this sprawling region has long been a meeting place for diverse peoples and cultures. Troubled by bitter conflicts and plagued by a scarcity of water, the region also is extremely rich in oil and other natural resources.

The Land

Glance at a physical map of this region and you will see a jumble of mountain chains. In the west, the Atlas Mountains—Africa's longest range—run through Morocco and Algeria. The Caucasus Mountains span the land between the Black and Caspian Seas and form the northern border of Georgia and Azerbaijan. Slanting southeast through Turkey and Iran are the Zagros Mountains, where earthquakes frequently occur. Farther east are the Hindu Kush in Afghanistan and the Tian Shan, or "heavenly mountains," in Kyrgyzstan—home to some of the world's largest glaciers.

Mountains block moist winds, helping to create vast deserts across much of the region. The Sahara, in North Africa, is the world's largest hot desert. The Rub' al Khali, or Empty Quarter, covers about one-fourth of the Arabian Peninsula. The Garagum and Qizilqum lie in Turkmenistan and Uzbekistan.

Through these arid landscapes flow great rivers that bring life-giving water. The world's longest river, the Nile, runs 4,241 miles (6,825 km) through Egypt to the Mediterranean Sea. The Tigris and Euphrates Rivers flow southeast through Turkey, Syria, and Iraq.

The Climate

Water is precious in much of this region. Most areas receive a meager 10 inches (25 cm) or less of rainfall each year. In such arid lands, agriculture is only possible along rivers and canals or in places where natural springs bubble to the surface to create lush but isolated oases.

In areas with a steppe climate, where enough rain falls to support grasses, people raise livestock such as sheep, camels, and goats. Steppes cover parts of many Southwest and Central Asian countries. A narrow band of steppe runs along the northern edge of the Sahara, too.

The areas that border the Mediterranean, Black, and Caspian Seas enjoy a milder Mediterranean climate. Although summers are hot and dry, winters bring enough precipitation to turn coastal lowlands into green landscapes.

UNIT 6

Nile River flowing through Aswan, Egypt

Man gazing out across the vast expanses of the Sahara

The Economy

Like water, natural resources are distributed unevenly across the region. This helps to create great differences in living standards. The region includes some of the world's wealthiest nations—and some of its poorest. Enormous reserves of oil and natural gas lie in certain areas, including lands in central North Africa, along the Persian Gulf, and around the Caspian Sea. Countries such as Saudi Arabia and Kuwait, which export petroleum products to fuel-hungry societies, generally enjoy high standards of living.

On the other hand, those countries with economies based on agriculture have much lower standards of living. Only a small percentage of the region's land is suitable for growing crops. In river valleys and along the coasts, where there is water and fertile soil, farmers raise citrus fruits, grapes, dates, grains, and cotton. Nomadic herding is common across the large expanses of this region that are too dry for crops.

The People

Great pyramids, built as tombs for Egyptian rulers, rise above desert sands. They are a reminder that some of the world's oldest civilizations developed in this region. Roughly 5,000 years ago, the ancient Egyptians built a kingdom along the life-giving Nile River. The Sumerian civilization, an even older society, flourished in the fertile valley between the Tigris and Euphrates Rivers.

Water still dictates where people settle in this region. Most cities lie along seacoasts or rivers, or near desert oases. Among the largest cities are Cairo, Egypt; Istanbul, Turkey; and Tehran, Iran.

In North Africa and Southwest Asia, most of the people are Arabs. Turkic ethnic groups are the majority in Central Asia. Many other ethnic groups also live in the region. Despite the ethnic diversity, many who live here are united by religion. Most people practice Islam, which developed in this region centuries ago. Two other major religions, Judaism and Christianity, also began here. Georgia and Armenia are two of the world's oldest Christian countries, and the country of Israel is the Jewish national homeland.

Exploring the Region

1. **How do mountains help create deserts across much of the region?**
2. **What parts of the region receive the most rainfall?**
3. **How do most people in the region make their living?**
4. **What religion do most people in the region practice?**

◀ **Construction of an oil pipeline across a desert, Yemen**

UNIT 6

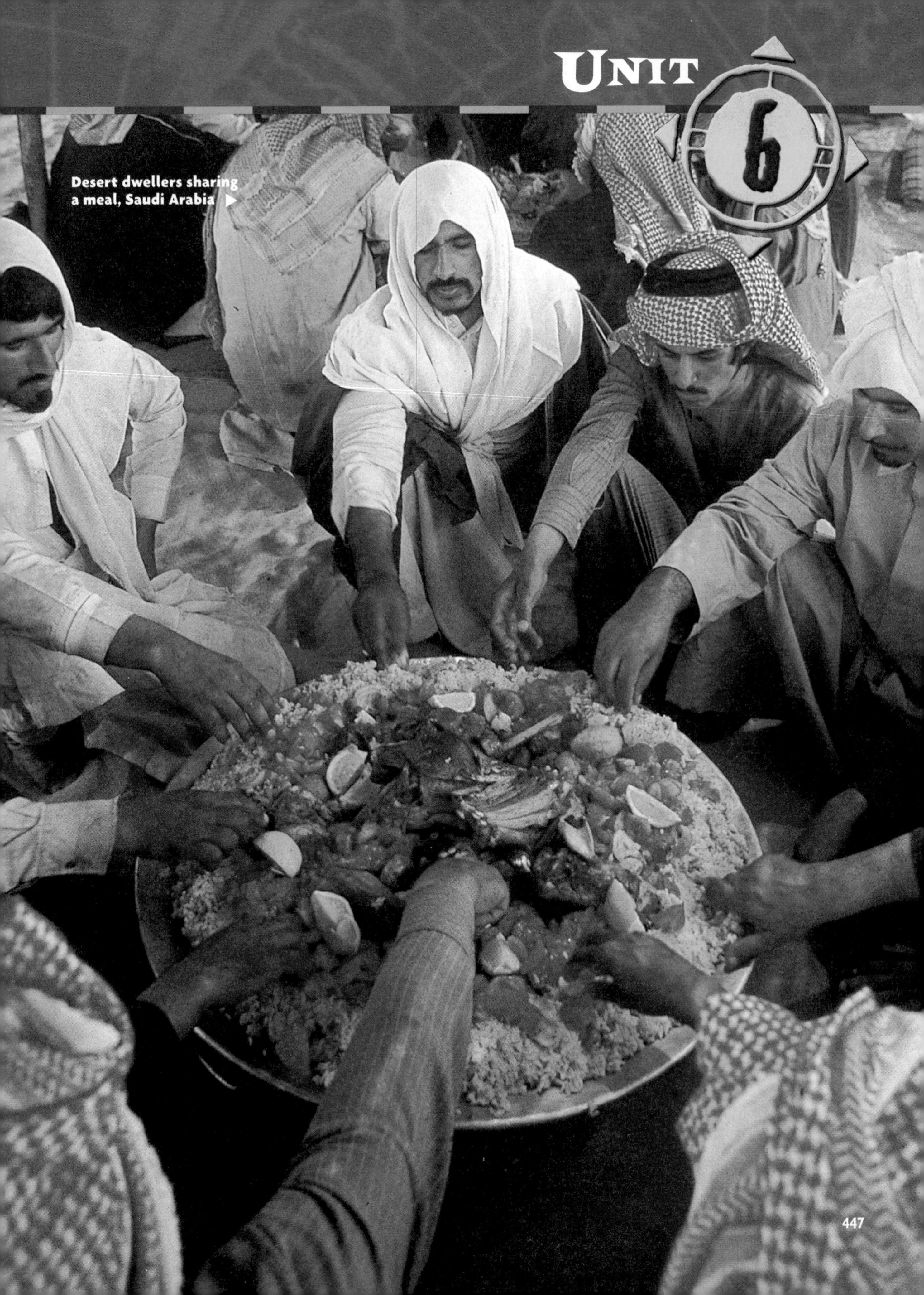

Desert dwellers sharing a meal, Saudi Arabia ▶

Regional Atlas

North Africa, Southwest Asia, and Central Asia

Physical

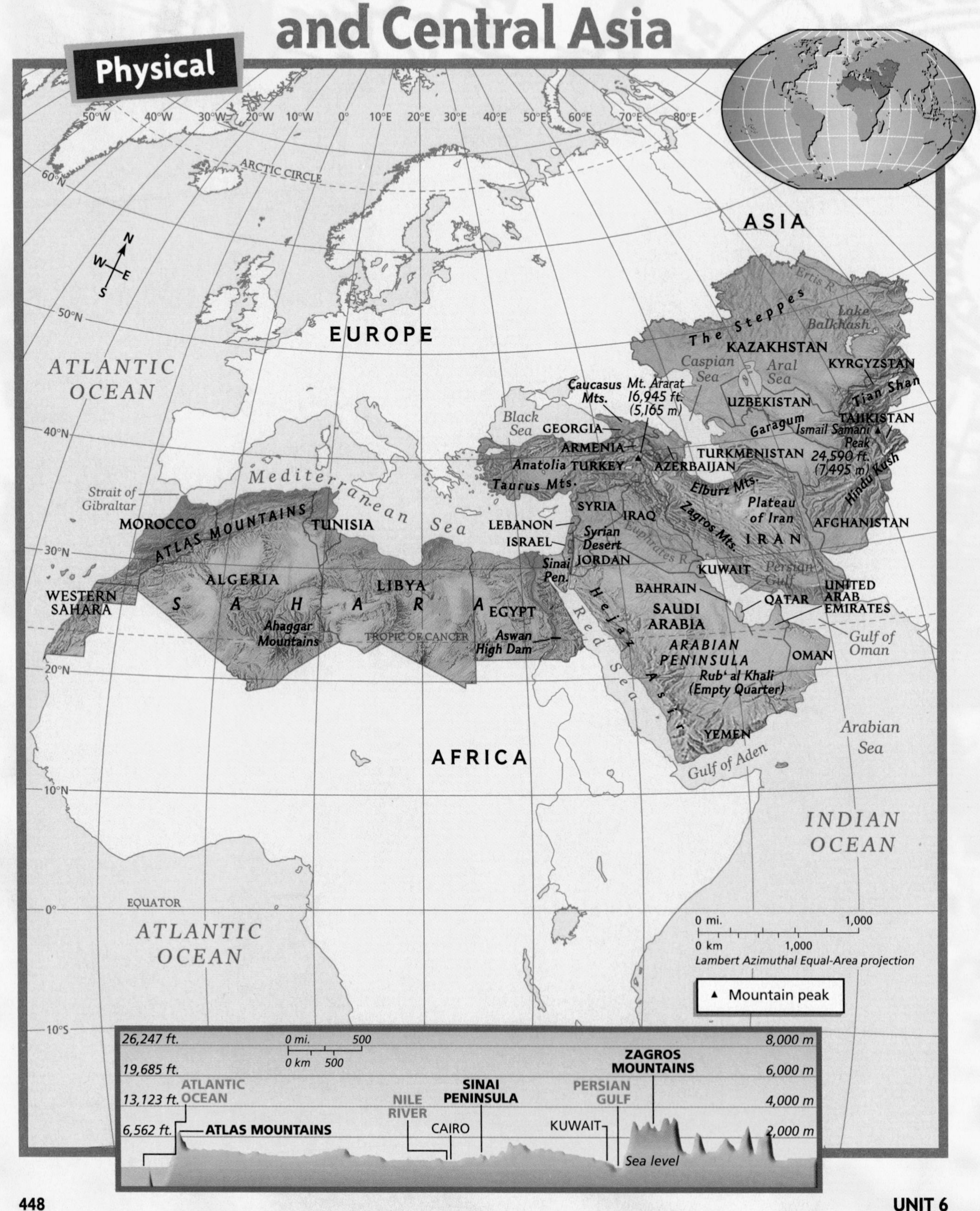

ARCTIC CIRCLE
ASIA
EUROPE
ATLANTIC OCEAN
AFRICA
INDIAN OCEAN
Mediterranean Sea
Black Sea
Caspian Sea
Aral Sea
Lake Balkhash
Ertis R.
The Steppes
KAZAKHSTAN
KYRGYZSTAN
Tian Shan
UZBEKISTAN
TAJIKISTAN
Ismail Samani Peak 24,590 ft. (7,495 m)
Hindu Kush
Garagum
TURKMENISTAN
Caucasus Mts.
Mt. Ararat 16,945 ft. (5,165 m)
GEORGIA
ARMENIA
AZERBAIJAN
Anatolia TURKEY
Taurus Mts.
Elburz Mts.
Plateau of Iran
Zagros Mts.
IRAN
AFGHANISTAN
SYRIA
IRAQ
LEBANON
ISRAEL
Syrian Desert
JORDAN
Euphrates R.
KUWAIT
Persian Gulf
BAHRAIN
QATAR
UNITED ARAB EMIRATES
Gulf of Oman
OMAN
SAUDI ARABIA
ARABIAN PENINSULA
Rub' al Khali (Empty Quarter)
Hejaz
Asir
Red Sea
YEMEN
Gulf of Aden
Arabian Sea
Sinai Pen.
EGYPT
Aswan High Dam
LIBYA
TUNISIA
ALGERIA
MOROCCO
ATLAS MOUNTAINS
Strait of Gibraltar
WESTERN SAHARA
SAHARA
Ahaggar Mountains
TROPIC OF CANCER
EQUATOR
0 mi. 1,000
0 km 1,000
Lambert Azimuthal Equal-Area projection
▲ Mountain peak
26,247 ft.
19,685 ft.
13,123 ft.
6,562 ft.
8,000 m
6,000 m
4,000 m
2,000 m
0 mi. 500
0 km 500
ATLANTIC OCEAN
ATLAS MOUNTAINS
NILE RIVER
CAIRO
SINAI PENINSULA
PERSIAN GULF
KUWAIT
ZAGROS MOUNTAINS
Sea level

UNIT 6

Political

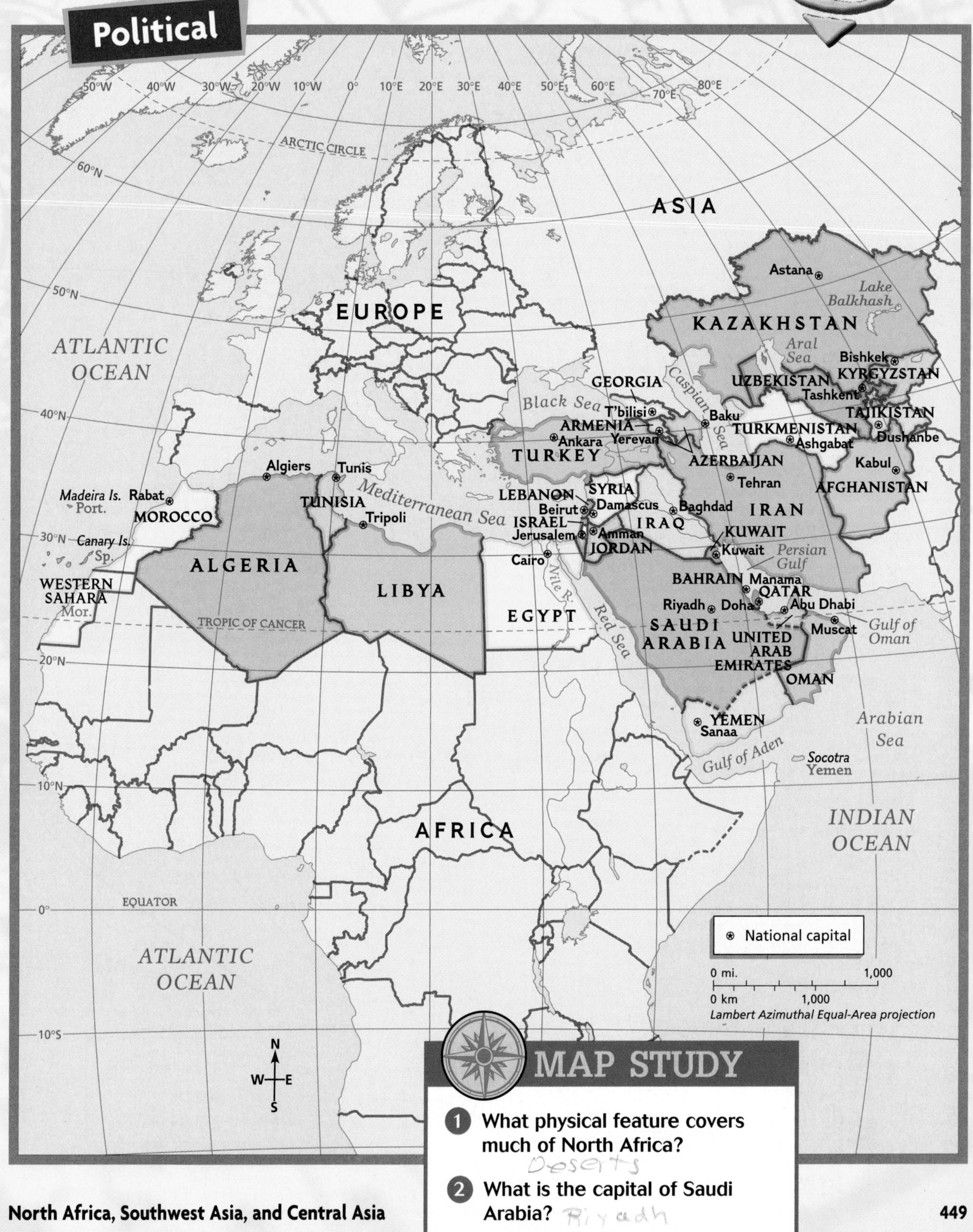

MAP STUDY

1. What physical feature covers much of North Africa? Deserts
2. What is the capital of Saudi Arabia? Riyadh

North Africa, Southwest Asia, and Central Asia

Oil and Gas Production and Distribution

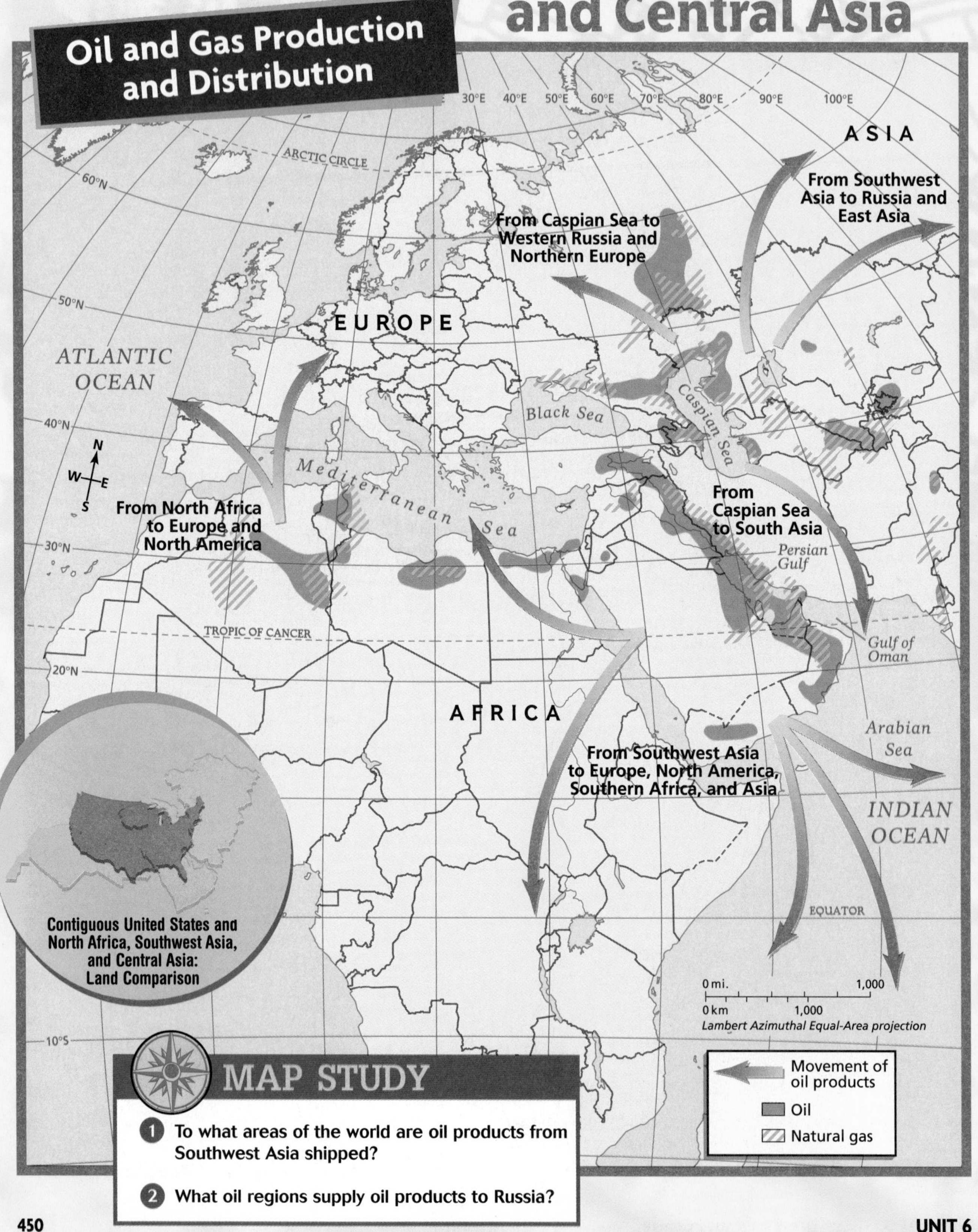

MAP STUDY

1. To what areas of the world are oil products from Southwest Asia shipped?
2. What oil regions supply oil products to Russia?

Geo Extremes

① **HIGHEST POINT**
Ismail Samani Peak
(Tajikistan)
24,590 ft. (7,495 m) high

② **LOWEST POINT**
Dead Sea
(Israel and Jordan)
1,349 ft. (411 m)
below sea level

③ **LONGEST RIVER**
Nile River
4,241 mi.
(6,825 km) long

④ **LARGEST LAKE**
Caspian Sea
143,244 sq. mi.
(371,000 sq. km)

⑤ **LARGEST DESERT**
Sahara (northern Africa)
3,475,000 sq. mi.
(9,000,208 sq. km)

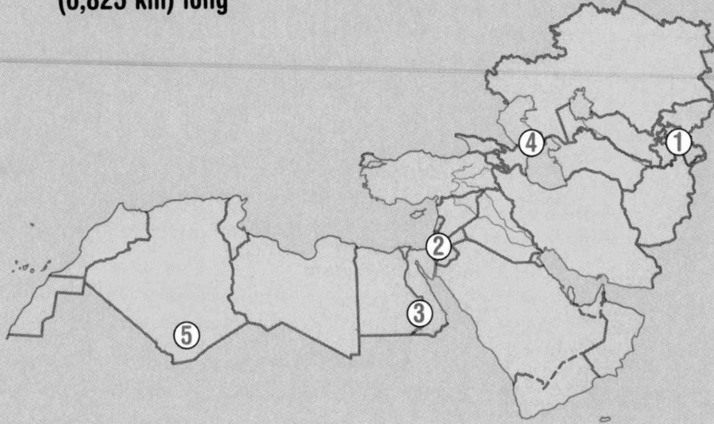

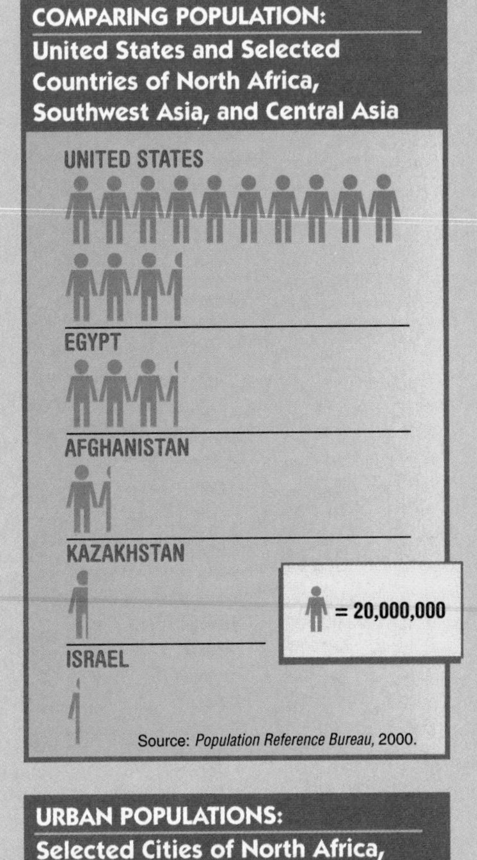

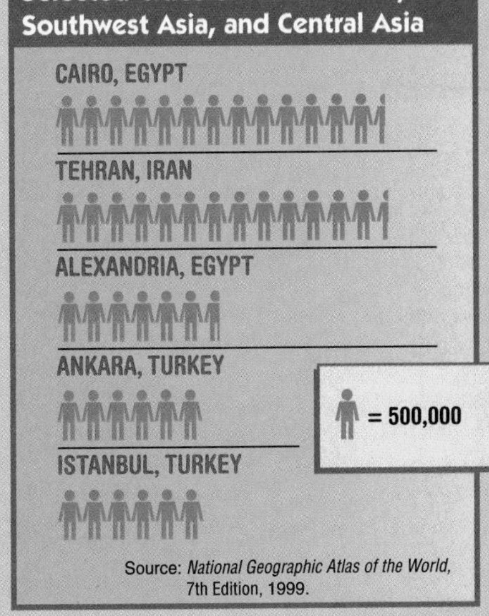

GRAPHIC STUDY

1. The lowest point on the earth is found in this region. Where is it?
   Dead Sea in Israel + Jordan
2. How does the population of Cairo compare to that of Tehran? How does the population of Ankara compare to that of Cairo?

Country Profiles

IRAQ

POPULATION: 22,450,000
133 per sq. mi.
51 per sq. km

LANGUAGES: Arabic, Kurdish

MAJOR EXPORT: Crude Oil

MAJOR IMPORT: Machinery

CAPITAL: Baghdad

LANDMASS: 169,235 sq. mi.
438,317 sq. km

ISRAEL

POPULATION: 6,135,000
765 per sq. mi.
295 per sq. km

LANGUAGES: Hebrew, Arabic

MAJOR EXPORT: Polished Diamonds

MAJOR IMPORT: Chemicals

CAPITAL: Jerusalem *

LANDMASS: 8,019 sq. mi.
20,770 sq. km

* Israel has proclaimed Jerusalem as its capital, but many countries' embassies are located in Tel Aviv.

JORDAN

Amman

POPULATION: 4,731,000
133 per sq. mi.
52 per sq. km

LANGUAGE: Arabic

MAJOR EXPORT: Phosphates

MAJOR IMPORT: Crude Oil

CAPITAL: Amman

LANDMASS: 35,467 sq. mi.
91,860 sq. km

Countries and flags not drawn to scale

KAZAKHSTAN

POPULATION: 15,417,000; 15 per sq. mi.; 6 per sq. km
LANGUAGES: Kazakh, Russian
MAJOR EXPORT: Petroleum
MAJOR IMPORT: Machinery
CAPITAL: Astana
LANDMASS: 1,049,039 sq. mi.; 2,716,998 sq. km

KYRGYZSTAN

POPULATION: 4,728,000; 62 per sq. mi.; 24 per sq. km
LANGUAGES: Kirghiz, Russian
MAJOR EXPORT: Cotton
MAJOR IMPORT: Grain
CAPITAL: Bishkek
LANDMASS: 76,834 sq. mi.; 198,999 sq. km

KUWAIT

POPULATION: 2,076,000; 302 per sq. mi.; 117 per sq. km
LANGUAGE: Arabic
MAJOR EXPORT: Petroleum
MAJOR IMPORT: Foods
CAPITAL: Kuwait
LANDMASS: 6,880 sq. mi.; 17,818 sq. km

LEBANON

POPULATION: 4,070,000; 1,014 per sq. mi.; 391 per sq. km
LANGUAGES: Arabic, French
MAJOR EXPORT: Paper
MAJOR IMPORT: Machinery
CAPITAL: Beirut
LANDMASS: 4,015 sq. mi.; 10,399 sq. km

LIBYA

POPULATION: 4,992,000; 7 per sq. mi.; 3 per sq. km
LANGUAGE: Arabic
MAJOR EXPORT: Crude Oil
MAJOR IMPORT: Machinery
CAPITAL: Tripoli
LANDMASS: 679,362 sq. mi.; 1,759,540 sq. km

MOROCCO*

POPULATION: 28,248,000; 103 per sq. mi.; 40 per sq. km
LANGUAGES: Arabic, French, Berber
MAJOR EXPORT: Foods
MAJOR IMPORT: Manufactured Goods
CAPITAL: Rabat
LANDMASS: 275,117 sq. mi.; 712,550 sq. km

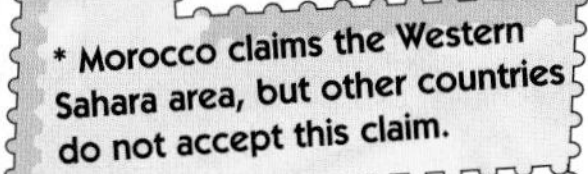

* Morocco claims the Western Sahara area, but other countries do not accept this claim.

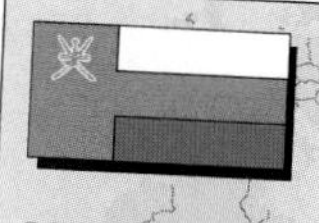

OMAN

POPULATION: 2,460,000; 30 per sq. mi.; 12 per sq. km
LANGUAGE: Arabic
MAJOR EXPORT: Petroleum
MAJOR IMPORT: Machinery
CAPITAL: Muscat
LANDMASS: 82,030 sq. mi.; 212,457 sq. km

QATAR

POPULATION: 541,000; 127 per sq. mi.; 49 per sq. km
LANGUAGE: Arabic
MAJOR EXPORT: Petroleum
MAJOR IMPORT: Machinery
CAPITAL: Doha
LANDMASS: 4,247 sq. mi.; 11,000 sq. km

SAUDI ARABIA

POPULATION: 20,899,000; 25 per sq. mi.; 10 per sq. km
LANGUAGE: Arabic
MAJOR EXPORT: Petroleum
MAJOR IMPORT: Machinery
CAPITAL: Riyadh
LANDMASS: 830,000 sq. mi.; 2,149,690 sq. km

SYRIA

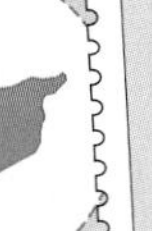

POPULATION: 16,033,000; 226 per sq. mi.; 87 per sq. km
LANGUAGES: Arabic, Kurdish, Armenian
MAJOR EXPORT: Petroleum
MAJOR IMPORT: Machinery
CAPITAL: Damascus
LANDMASS: 71,044 sq. mi.; 184,004 sq. km

TAJIKISTAN

POPULATION: 6,213,000; 113 per sq. mi.; 43 per sq. km
LANGUAGES: Tajik, Russian
MAJOR EXPORT: Aluminum
MAJOR IMPORT: Fuels
CAPITAL: Dushanbe
LANDMASS: 55,213 sq. mi.; 143,001 sq. km

Country Profiles

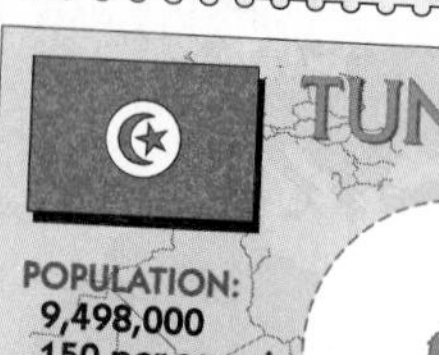

TUNISIA

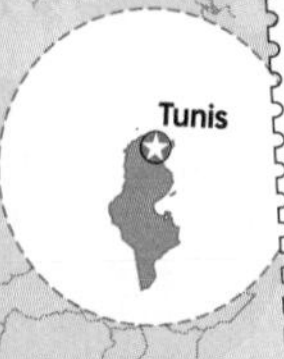

POPULATION:
9,498,000
150 per sq. mi.
58 per sq. km

LANGUAGES:
Arabic, French

MAJOR EXPORT:
Petroleum Products

MAJOR IMPORT:
Machinery

CAPITAL:
Tunis

LANDMASS:
63,170 sq. mi.
163,610 sq. km

TURKEY

POPULATION:
65,869,000
219 per sq. mi.
85 per sq. km

LANGUAGES:
Turkish, Kurdish

MAJOR EXPORTS:
Foods and Livestock

MAJOR IMPORT:
Machinery

CAPITAL:
Ankara

LANDMASS:
300,948 sq. mi.
779,452 sq. km

TURKMENISTAN

POPULATION:
4,779,000
25 per sq. mi.
10 per sq. km

LANGUAGES:
Turkmen, Russian, Uzbek

MAJOR EXPORT:
Natural Gas

MAJOR IMPORT:
Machinery

CAPITAL:
Ashgabat

LANDMASS:
188,418 sq. mi.
488,000 sq. km

UNITED ARAB EMIRATES

Abu Dhabi

POPULATION:
2,779,000
86 per sq. mi.
33 per sq. km

LANGUAGES:
Arabic, Persian

MAJOR EXPORT:
Petroleum

MAJOR IMPORT:
Manufactured Goods

CAPITAL:
Abu Dhabi

LANDMASS:
32,278 sq. mi.
83,600 sq. km

UZBEKISTAN

Tashkent

POPULATION:
24,416,000
141 per sq. mi.
55 per sq. km

LANGUAGES:
Uzbek, Russian, Tajik

MAJOR EXPORT:
Cotton

MAJOR IMPORT:
Machinery

CAPITAL:
Tashkent

LANDMASS:
172,588 sq. mi.
447,001 sq. km

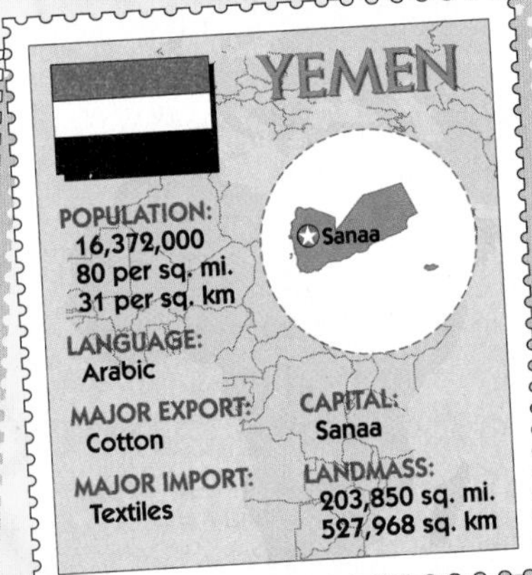

YEMEN

Sanaa

POPULATION:
16,372,000
80 per sq. mi.
31 per sq. km

LANGUAGE:
Arabic

MAJOR EXPORT:
Cotton

MAJOR IMPORT:
Textiles

CAPITAL:
Sanaa

LANDMASS:
203,850 sq. mi.
527,968 sq. km

GEO BEE: Questions From Buzz Bee!

The following questions are taken from National Geographic GeoBees. Use your textbook, the Internet, and other library resources to find the answers.

1. Which of the countries in North Africa has a king as head of state?
   Morocco
2. Which African capital city is only 150 miles (241 km) from Sicily?
   Tunisia
3. The name of a group of people who have traditionally been nomadic herders comes from an Arabic word that means "desert dwellers." Name this group. Beduoins

Anthias fish and coral reef,
Red Sea, Egypt

Chapter

16 North Africa

The World and Its People NATIONAL GEOGRAPHIC

To learn more about the people and places of North Africa, view ***The World and Its People*** **Chapter 16** video.

Chapter Overview Visit the ***Geography: The World and Its People*** Web site at gwip.glencoe.com and click on **Chapter 16–Chapter Overviews** to preview information about North Africa.

Egypt

Guide to Reading

Main Idea

Egypt's Nile River and desert landscape have shaped the lives of the Egyptian people for centuries.

Terms to Know

- delta
- silt
- oasis
- phosphate
- hieroglyphics
- republic
- fellahin
- bazaar
- mosque

Places to Locate

- Nile River
- Sinai Peninsula
- Suez Canal
- Red Sea
- Eastern Desert
- Libyan Desert
- Sahara
- Cairo
- Alexandria

Reading Strategy

Draw a chart like this. Then list five physical features of Egypt and their effects on life in Egypt.

Physical Feature	Effect on Egyptians
→	
→	
→	
→	
→	

◀ The pyramids at El Giza, Egypt

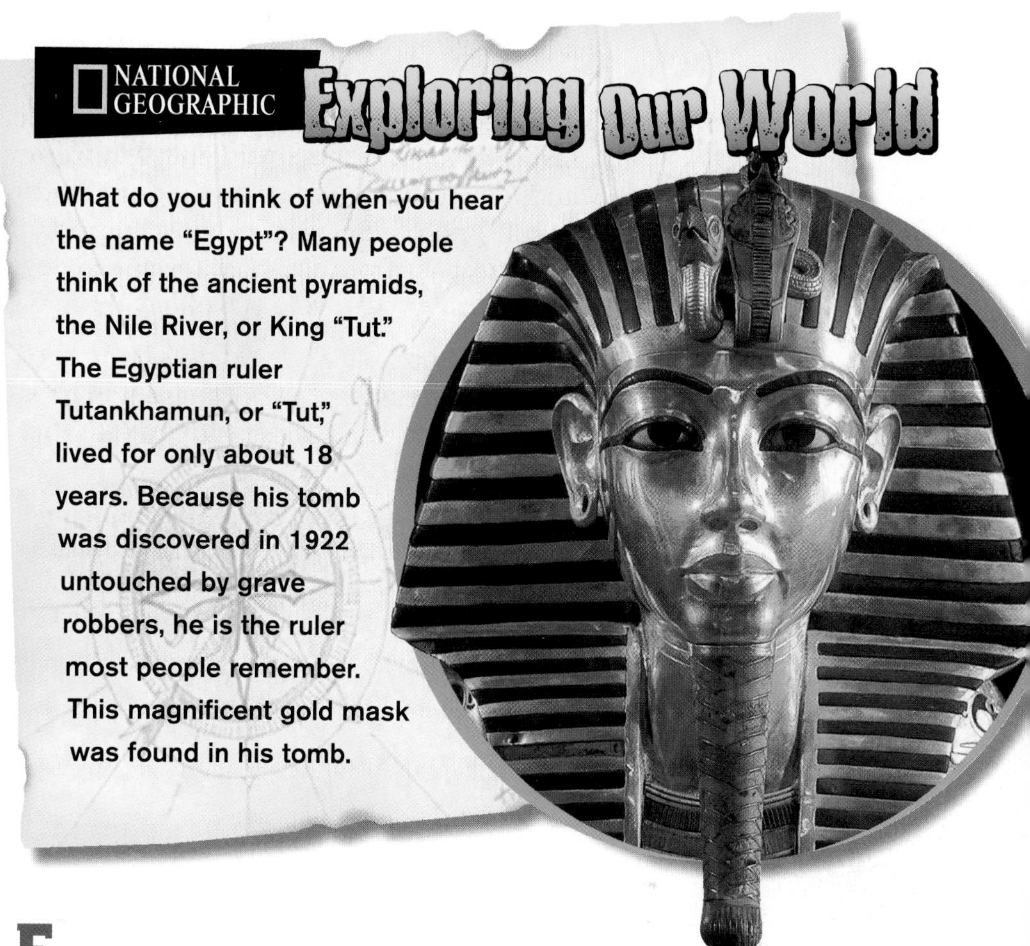

NATIONAL GEOGRAPHIC **Exploring Our World**

What do you think of when you hear the name "Egypt"? Many people think of the ancient pyramids, the Nile River, or King "Tut." The Egyptian ruler Tutankhamun, or "Tut," lived for only about 18 years. Because his tomb was discovered in 1922 untouched by grave robbers, he is the ruler most people remember. This magnificent gold mask was found in his tomb.

Egypt lies in Africa's northeast corner. Vast deserts sweep over most of the country. On the map on page 459, notice the blue line running through Egypt. This is the **Nile River.** Egypt's location, deserts, and above all, the Nile have shaped life in Egypt for thousands of years.

Egypt's Land and Climate

Egypt is a large country about the same size as Texas and New Mexico together. Yet most of it is desert. So Egypt's people crowd into less than 4 percent of the land, or an area a little larger than the size of Maryland. The lifeline of Egypt is the Nile River, which supplies 85 percent of the country's water. From its sources in East Africa, the river flows 4,241 miles (6,825 km) north to the Mediterranean Sea. This long journey makes the Nile the world's longest river. Along the Nile's banks, you can see mud-brick villages, ancient ruins, and, once in a while, a city or town of modern buildings. Where the river empties into the Mediterranean Sea, you find the Nile's delta. A **delta** is the area formed from soil deposited by a river at its mouth. The Nile delta is a fan-shaped area where the Nile splits into several smaller waterways.

For centuries, the Nile's waters would rise in the spring. The swollen river carried **silt,** or small particles of rich soil. When it reached Egypt, the Nile flooded its banks. As the floodwaters receded, the silt was left behind and made the land good for farming. Today dams and channels control the river's flow and the irrigation of farmland.

Sinai Peninsula The triangle-shaped **Sinai** (SY•NY) **Peninsula** lies southeast of the Nile delta. This area is a major crossroads between Africa and Southwest Asia. A human-made waterway called the **Suez Canal** separates the Sinai Peninsula from the rest of Egypt. Egyptians and Europeans built the canal in the mid-1860s. Today the Suez Canal is still one of the world's most important waterways. Ships use the canal to pass from the Mediterranean Sea to the **Red Sea.** In making this journey, they avoid traveling all the way around Africa.

Desert Areas Vast deserts cover most of Egypt. East of the Nile River spreads the **Eastern Desert,** also known as the Arabian Desert. West of the Nile is the much larger **Libyan** (LIH•bee•uhn) **Desert,** which covers

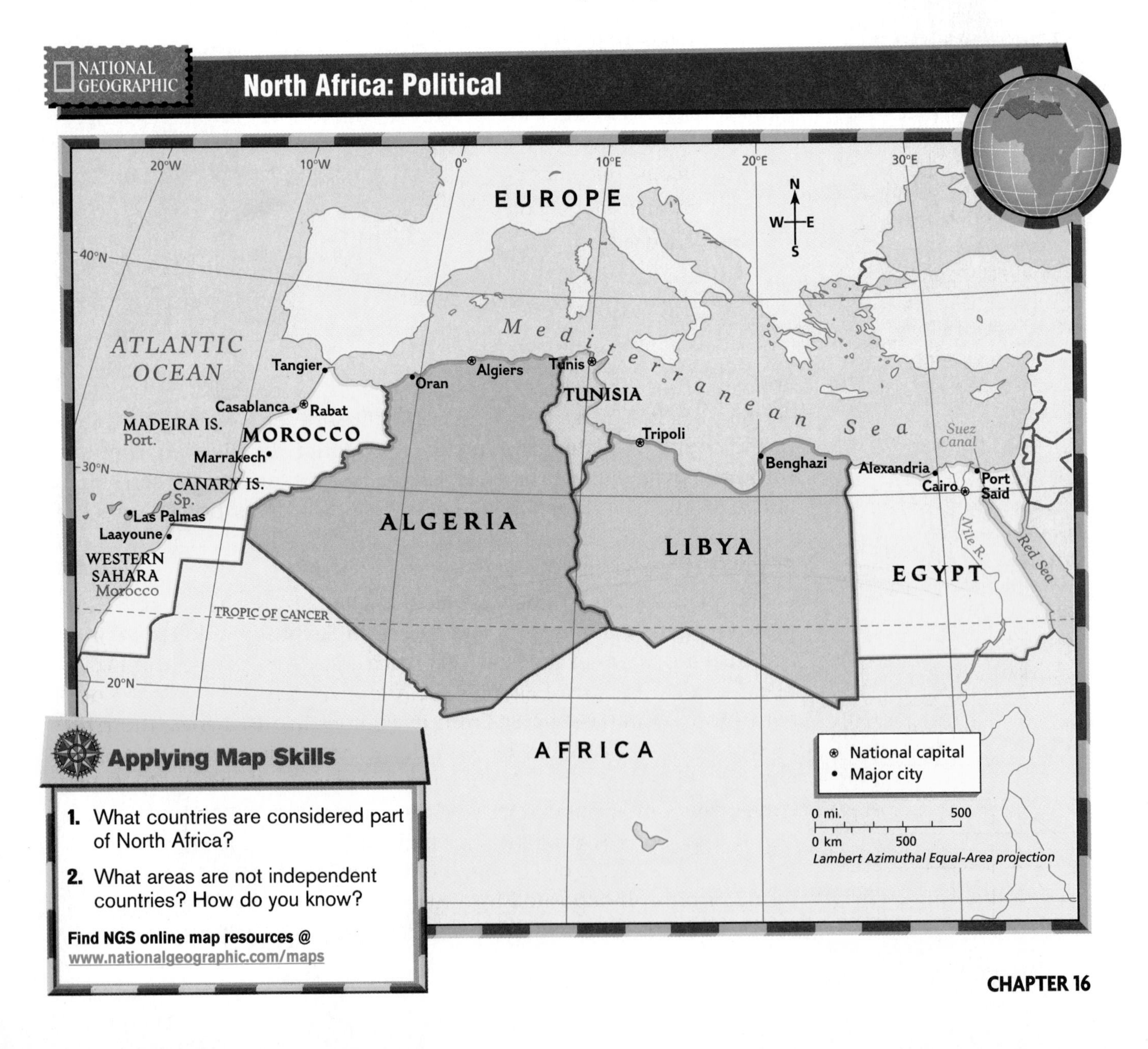

Applying Map Skills

1. What countries are considered part of North Africa?
2. What areas are not independent countries? How do you know?

Find NGS online map resources @ www.nationalgeographic.com/maps

about two-thirds of the country. Dotting both deserts are oases, or areas fed by underground water. The water allows plants to grow, giving these spots lush green growth in the midst of the hot sands.

The Eastern and Libyan Deserts are part of the **Sahara**—one of the largest desert areas in the world. *Sahara* comes from the Arabic word meaning "desert." The Sahara—about the size of the United States—stretches from Egypt westward across North Africa to the Atlantic Ocean.

Climate Wherever you go in Egypt, you find a dry desert climate of hot summers and mild winters. Egypt as a whole receives little rainfall. **Cairo,** the capital, averages only about 0.4 inch (1 cm) a year. In fact, some areas receive no rain for years at a time.

Springtime in Egypt brings hot winds instead of cooling rains. These winds move west across Egypt, reaching up to 87 miles (140 km) per hour. The powerful winds can harm crops and damage houses.

✓Reading Check **Why is the Suez Canal one of the world's most important waterways?** Ships use the canal to pass from the Med. Sea to Red Sea

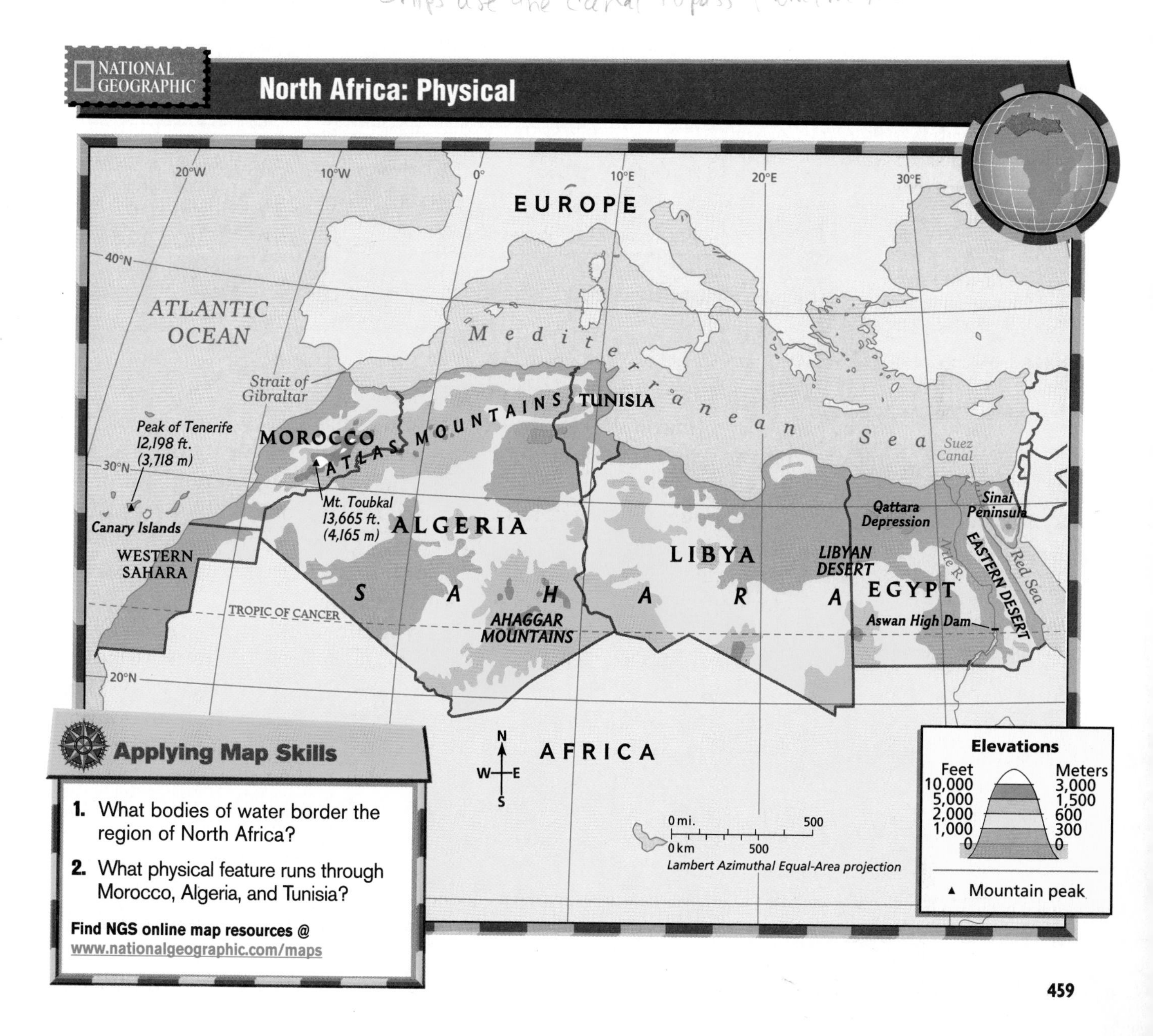

Applying Map Skills

1. What bodies of water border the region of North Africa?
2. What physical feature runs through Morocco, Algeria, and Tunisia?

Find NGS online map resources @ www.nationalgeographic.com/maps

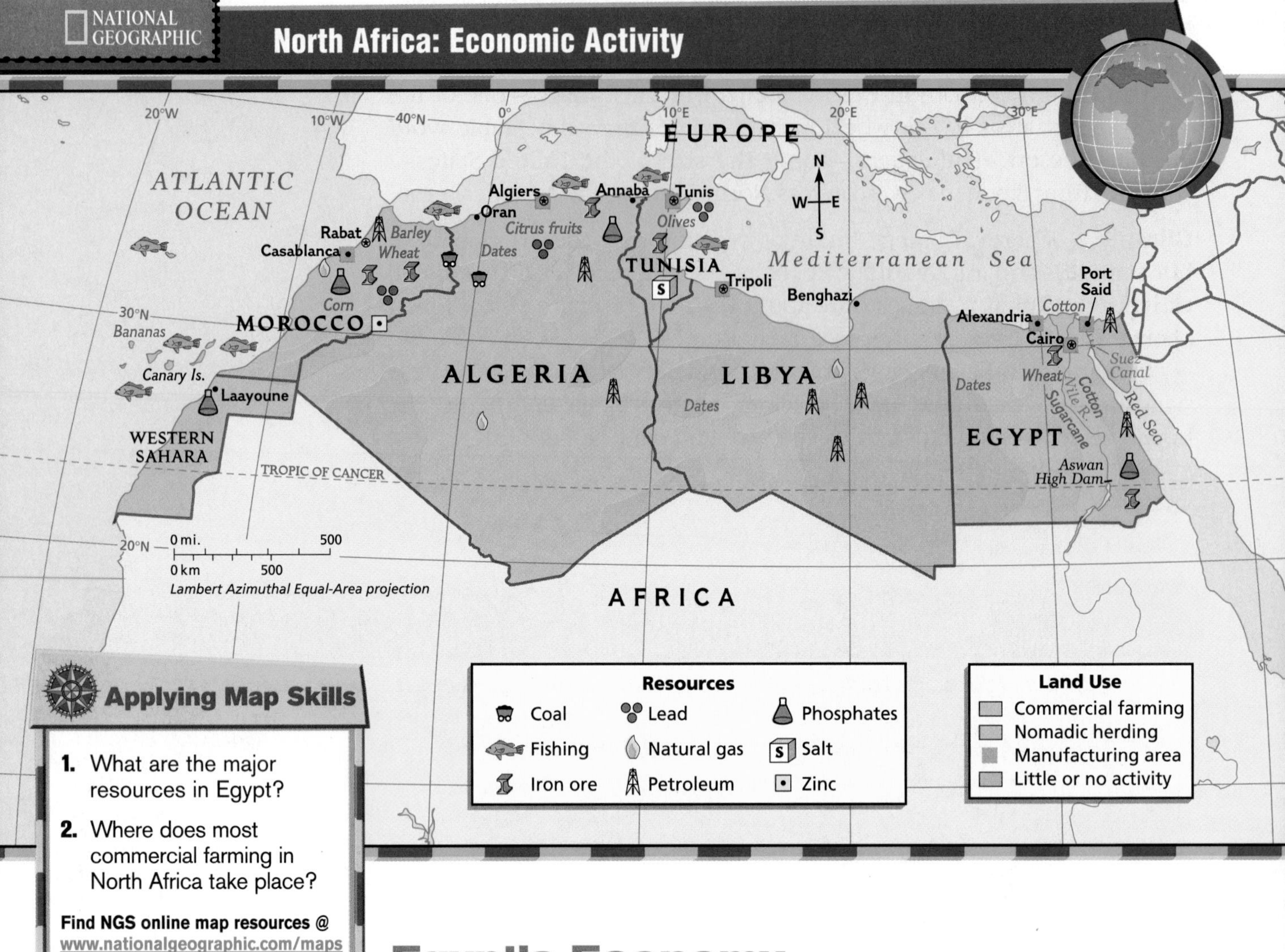

Applying Map Skills

1. What are the major resources in Egypt?
2. Where does most commercial farming in North Africa take place?

Find NGS online map resources @ www.nationalgeographic.com/maps

Egypt's Economy

Egypt has a developing economy that has grown considerably in recent years. Agriculture, however, remains the main economic activity. About 40 percent of Egypt's people work in agriculture, but only about 4 percent of Egypt's land is used for farming. The best farmland lies in the fertile Nile River valley. Egypt's major crops include sugarcane, grains, vegetables, fruits, and cotton. Raw cotton, cotton yarn, and clothing are among the country's main exports.

Some farmers still work the land with the simple practices and tools of their ancestors. Many use modern methods and machinery. All, however, rely on dams to control the water needed for their fields. The largest dam is called the Aswan High Dam. Find it on the map above. The dams give people control over the Nile's floodwaters. They can store the water for months behind the dams. Then they can release it several times during the year, rather than having just the spring floods. This allows farmers to harvest two or three crops a year.

The dams bring challenges as well as benefits. They block the flow of silt, which means farmland is becoming less fertile. Farmers now rely on chemical fertilizers to grow crops. In addition, the dams prevent quantities of freshwater from reaching the delta. So salt water from the Mediterranean Sea now flows deeper into the delta, making the land less fertile.

Industry The Aswan High Dam provides hydroelectric power, which Egypt uses to run its growing industries. The largest industrial centers are the capital city of Cairo and the seaport of **Alexandria.** Egyptian factories make food products, textiles, and consumer goods.

Egypt's main energy resource is oil, found in and around the Red Sea. Petroleum products make up almost half the value of Egypt's exports. The country also has iron ore and phosphates. **Phosphate** is a mineral salt used in fertilizers. Another important industry is tourism. Visitors come to see the magnificent ruins of ancient Egypt.

✓Reading Check **On what crop are many of Egypt's exports based?**

Egypt's History and People

Egypt's fascinating past began about 8,000 years ago, when people first settled along the Nile to farm. In about 3100 B.C., one king brought several small kingdoms together under one rule. Egypt emerged as a powerful nation. For most of the next 3,000 years, Egypt was ruled by a line of pharaohs, or kings. To show the pharaohs' wealth and power, stonecutters and laborers built huge pyramids and majestic temples. Turn to page 463 to learn how the magnificent pyramids were built.

The people of early Egypt created one of the world's most advanced civilizations. They wrote using **hieroglyphics,** or picture signs and symbols. They made paper from papyrus, a reed that grew along the banks of the Nile. Other Egyptian achievements were the invention of a calendar to keep track of the growing season and medical skills to treat injuries and diseases.

From 300 B.C. to A.D. 300, Egypt fell under the influence of Greece and Rome. You may have heard of Cleopatra, an Egyptian queen who ruled during the time of Rome's rise as a Mediterranean power. In A.D. 641, Arabs from Southwest Asia took control of Egypt. They practiced Islam, a religion based on the belief in one God known as Allah. Most of Egypt's people began to speak the Arabic language and became Muslims, as the followers of Islam are called. Today about 94 percent of Egypt's people are Muslims.

Egypt's Modern History By the end of the 1800s, all of Egypt, including the Suez Canal, had become part of the British Empire. Unhappy with British rule, the people of Egypt protested many times. Finally, in 1952 a group of army officers overthrew the British-supported king, and Egypt became independent. One of the army leaders, Gamal Abdel Nasser (guh•MAHL AHB•duhl NAH•suhr), was Egypt's president from 1954 to 1970. Nasser made Egypt one of the most powerful countries in the Muslim world.

Egypt is a **republic,** or a government headed by a president. A legislature makes the laws, but the president has

Ramses II

When the Aswan High Dam was built, it blocked the flow of the Nile River and created a reservoir 300 miles (483 km) long. This reservoir, called Lake Nasser, covered land on which 50,000 Egyptians lived. They were moved to a new location. The lake waters also would have covered four colossal statues of Ramses II built into a cliff temple. These statues, measuring 66 feet (20 m) in height, were moved as well. An international team of engineers and scientists dug away the top of the cliff and took apart the statues and temple. They rebuilt them on high ground 200 feet (61 m) above their original location. In all, some 16,000 blocks were moved.

broad powers in running the country. In recent years, various political groups have opposed the government. These groups have used armed attacks in an effort to reach their political goals. The government has tried to stop this violence.

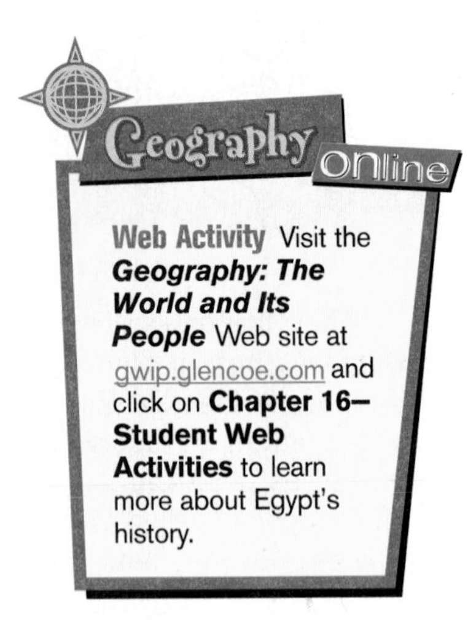

Geography online
Web Activity Visit the Geography: The World and Its People Web site at gwip.glencoe.com and click on Chapter 16–Student Web Activities to learn more about Egypt's history.

Rural and Urban Life Look at the population density map on page 467. Most of Egypt's 66.9 million people live within 20 miles (32 km) of the Nile River. More than half of Egypt's people live in rural areas. Most are farmers called **fellahin** (FEHL•uh•HEEN). They live in villages and farm small plots of land that they rent from landowners. Many fellahin raise only enough food to feed their families. Any food left over is sold in towns at a **bazaar,** or marketplace.

Life is more modern in Egypt's cities. Many city dwellers have jobs in service, manufacturing, or construction industries. In bustling ports like Alexandria and Port Said (sah•EED), people engage in trade. Cairo is a huge and rapidly growing city. Almost 7 million people are crowded into its central area, with another 6 million living in its suburbs. It is the largest city in Africa. For centuries, Cairo has been a leading center of the Muslim world. Throughout the city you see schools, universities, and **mosques,** or places of worship for followers of Islam.

Cairo's population is increasing at a rapid rate. Why? First, Egypt is a country with a high birthrate. Second, many fellahin have moved to Cairo to find work. The crowded city cannot provide enough houses, schools, and hospitals for all of its people.

Reading Check **When did Egypt become fully independent?**

Assessment

Defining Terms

1. **Define** delta, silt, oasis, phosphate, hieroglyphics, republic, fellahin, bazaar, mosque.

Recalling Facts

2. **Human/Environment Interaction** Why is the Nile River important to Egypt?

3. **History** What were three achievements of the ancient Egyptians?

4. **Culture** What are the major language and religion of Egypt?

Critical Thinking

5. **Understanding Cause and Effect** How has the Aswan High Dam affected farmers?

6. **Making Predictions** If violence in Egypt continues, what effect do you think it could have on the economy?

Graphic Organizer

7. **Organizing Information** In a chart like the one below, fill in three facts about Egypt for each category.

Agriculture	Industry
1.	1.
2.	2.
3.	3.

Applying Geography Skills

8. **Analyzing Maps** Study the political map on page 458. What direction is Alexandria from Cairo?

ART SCIENCE LITERATURE TECHNOLOGY

The Egyptian Pyramids

The ancient Egyptians viewed the pharaoh, or king, as the most important person on the earth. They believed he was a god who would continue to guide them after his death. A pyramid served as a tomb for the pharaoh and provided a place where the body would safely pass into the afterlife. Rooms inside the pyramid held food, clothing, weapons, furniture, jewels, and everything else the pharaoh might need in the afterlife.

The Great Pyramid at El Giza

The largest of Egypt's pyramids is the Great Pyramid of El Giza, built nearly 5,000 years ago. When the pyramid was new, it stood 481 feet (147 m) high—as tall as a 50-story building. The square base of the pyramid covers 13 acres (5 ha). More than 2 million limestone and granite blocks were used in building it. These are no ordinary-sized blocks, however. The huge stones weigh an average of 2.5 tons (2.3 t) each.

Construction

For thousands of years, people have wondered how the Egyptians built the pyramids without modern tools or machinery. In the fifth century B.C., a Greek historian thought it took 100,000 people to build the Great Pyramid. Today archaeologists believe a workforce of about 20,000 did the job in about 20 years. Barges carried supplies and building materials for the pyramid down the Nile River. Nearby quarries supplied most of the stone. Skilled stonecutters carved the stones into the precise size and shape so that no mortar, or cementing material, was needed to hold the stones together.

Engineers think that workers built ramps and used papyrus twine to drag the huge stones to the pyramid. They formed ramps up all four sides of the pyramid and made the ramps higher and longer as the pyramid rose. They then dragged the stones up the ramps. Once finished, the ramps were cleared away. Then stonemasons smoothed and polished the stone, and the finished pyramid towered over the surrounding desert.

Making the Connection

1. Why did the Egyptians build the pyramids?
2. How many workers did ancient historians and modern archaeologists each think it took to build the Great Pyramid?
3. **Sequencing Information** Describe the process experts think Egyptians used to build the pyramids.

◀ The Great Pyramid at El Giza, Egypt

Section 2

Libya and the Maghreb

Guide to Reading

Main Idea

The countries of Libya, Tunisia, Algeria, and Morocco share a desert environment and a mostly Arab culture.

Terms to Know

- aquifer
- dictatorship
- erg
- civil war
- secular
- casbah
- constitutional monarchy

Places to Locate

- Libya
- Tunisia
- Atlas Mountains
- Algeria
- Morocco
- Strait of Gibraltar
- Western Sahara

Reading Strategy

Draw a diagram like this one. In the four outer ovals, list facts about each country under their headings. In the center oval, write three facts that all four countries have in common.

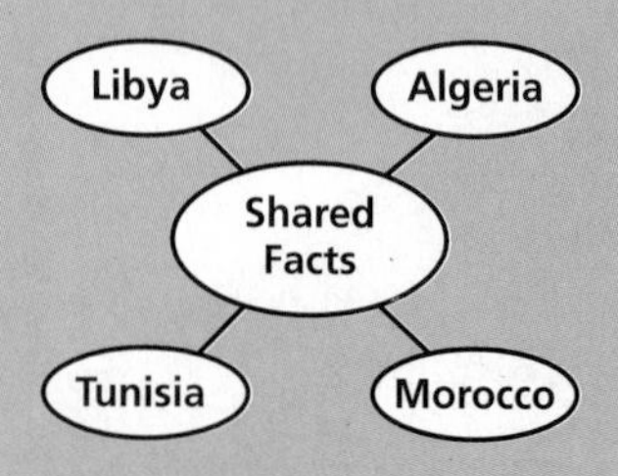

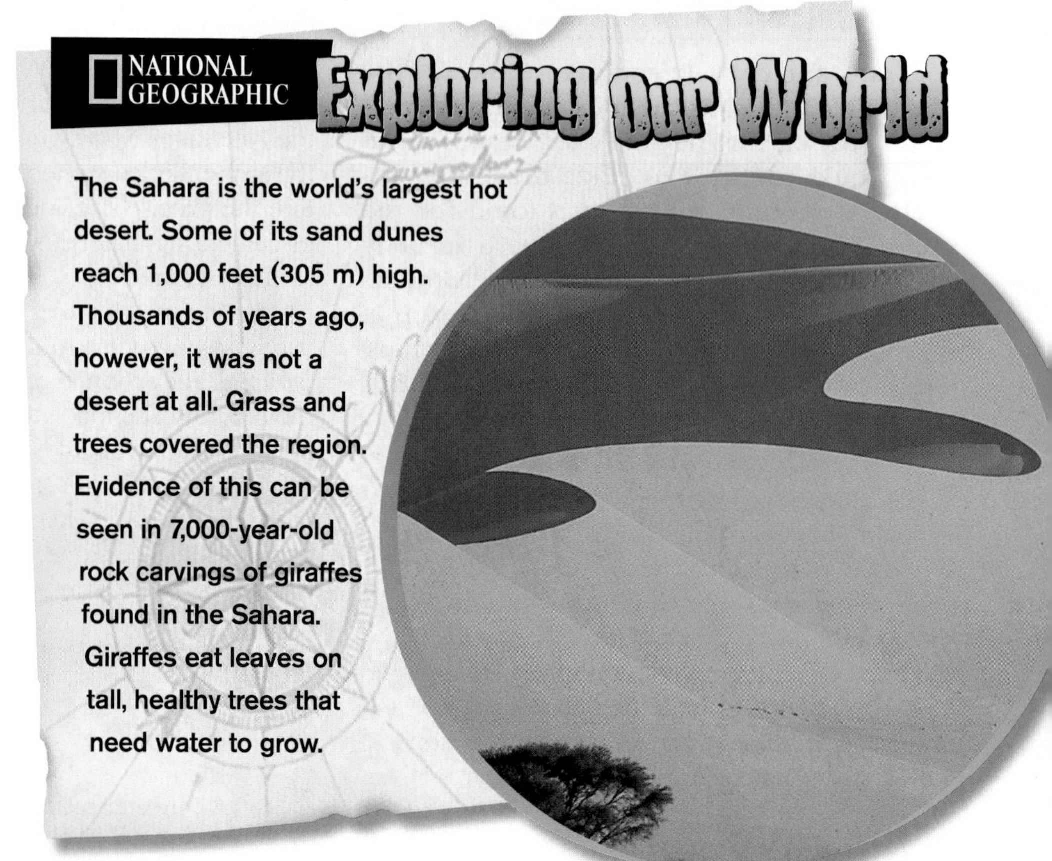

The Sahara is the world's largest hot desert. Some of its sand dunes reach 1,000 feet (305 m) high. Thousands of years ago, however, it was not a desert at all. Grass and trees covered the region. Evidence of this can be seen in 7,000-year-old rock carvings of giraffes found in the Sahara. Giraffes eat leaves on tall, healthy trees that need water to grow.

Tunisia, Algeria, and Morocco make up a region of North Africa commonly called the Maghreb (MUH•gruhb). Like Egypt, these countries plus Libya have economies based on oil and other resources in the Sahara. Unlike Egypt, however, none of these nations enjoys the benefits of a life-giving river such as the Nile.

Libya

Libya is slightly larger than Alaska. Two narrow strips of lowland stretch along Libya's Mediterranean coast. As you move about 50 miles (80 km) inland, the lowlands rise to a plateau in the center of the country. This lofty flatland climbs even higher to mountains in the south.

Except for the coastal lowlands, Libya is a desert area with only a few oases. In fact, the Sahara covers more than 90 percent of Libya. During the spring and fall, dust-heavy winds blow from the desert. When these fierce winds strike, temperatures in coastal areas can reach 110°F (43°C).

Libya has no permanent rivers, but aquifers lie beneath the vast desert. **Aquifers** are underground rock layers that store large amounts of water. In the 1990s, the government built pipelines to carry underground water from the desert to coastal areas.

Poor soil and a hot climate mean that Libya has to import about three-fourths of its food. The discovery of oil in Libya in 1959 brought the country great wealth. Libya's government uses oil money to buy food, build schools and hospitals, and maintain a strong military.

Libya's People and History Almost all of Libya's 5 million people have mixed Arab and Berber heritage. The Berbers were the first people known to live in North Africa. During the A.D. 600s, the Arabs brought Islam and the Arabic language to North Africa. Since then, Libya has been a Muslim country, and most of its people speak Arabic.

About 86 percent of Libyans live along the Mediterranean coast. Most live in two modern cities—Tripoli, the capital, and Benghazi (behn•GAH•zee). Libya became independent in 1951 under a king. In 1969 a military officer named Muammar al-Qaddhafi (kuh•DAH•fee) gained power and overthrew the king. Qaddhafi set up a **dictatorship**, or a government under the control of one all-powerful leader.

✓Reading Check **How has Libya been governed since 1969?**

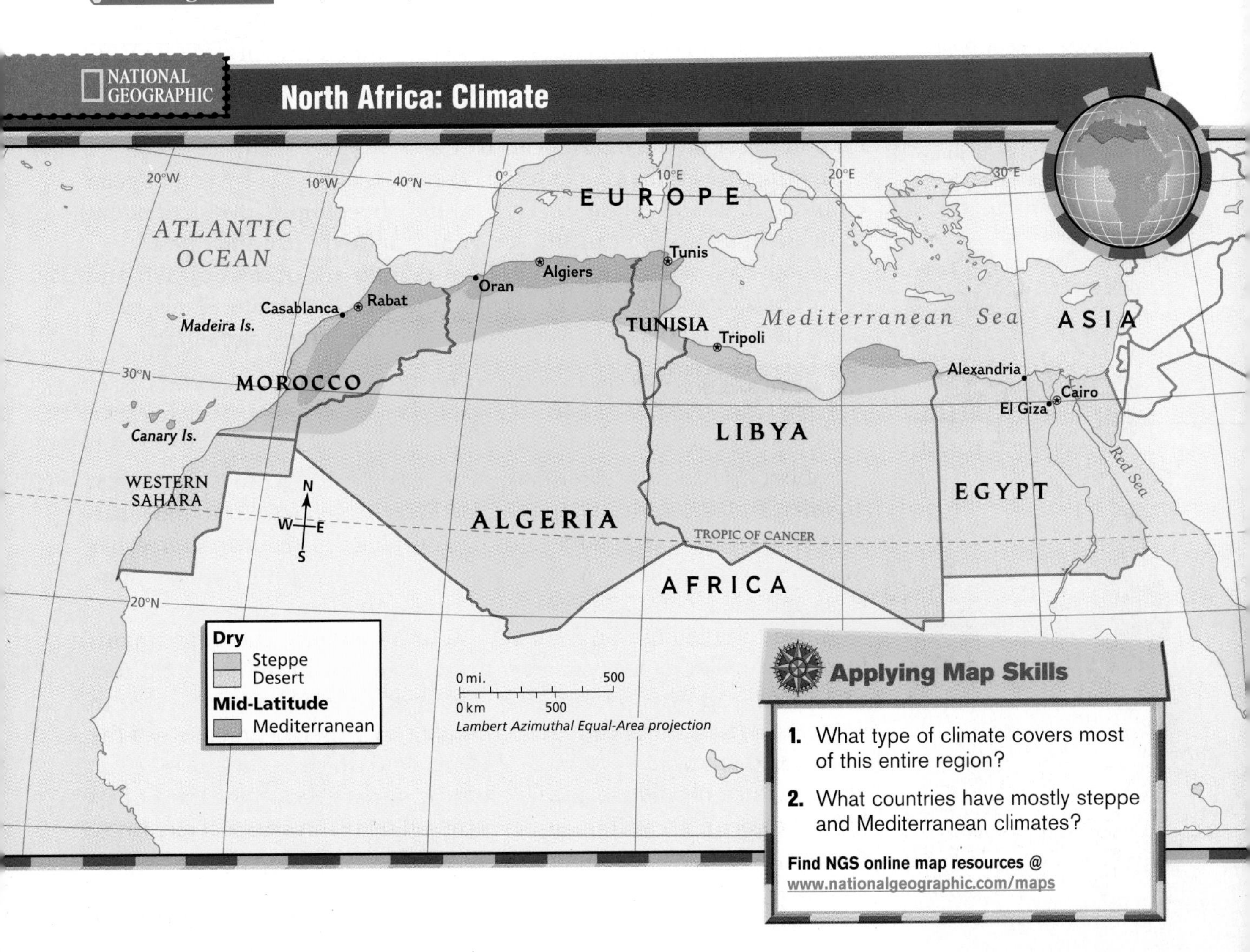

Applying Map Skills

1. What type of climate covers most of this entire region?
2. What countries have mostly steppe and Mediterranean climates?

Find NGS online map resources @ www.nationalgeographic.com/maps

Tunisia

Tunisia, Algeria, and Morocco form a region known as the Maghreb. *Maghreb* means "the land farthest west" in Arabic. These three countries were given this name because they are the westernmost part of the Arabic-speaking Muslim world.

About the size of the state of Georgia, **Tunisia** is North Africa's smallest country. Tunisia has a long Mediterranean coastline and desert areas. Low plains cover the coast and central part of the country. The **Atlas Mountains** reach into the northwest.

Farming and herding take place in much of Tunisia. The climate map on page 465 shows why this is so. Northern and central areas have Mediterranean or steppe climates, which provide some rainfall. Along the fertile eastern coast, farmers grow wheat, olives, fruits, and vegetables. Fishing is an important industry as well.

The country's major resources include phosphates, iron, oil, and natural gas. Tunisian factories produce food products, textiles, and oil products. In addition, tourism is a growing industry. Many visitors enjoy Tunisia's sunny shores and explore its Roman ruins and outdoor markets.

Bazaar!

Taha Hammam makes pottery to sell at the bazaar. "Going to the bazaar is a lot like going to an American mall," he says. "It's a big party where everyone talks and eats and buys and sells things." Taha lives in Algiers. Although Taha wears jeans and sneakers, his parents dress in traditional clothes. His mother wears a black outer dress over a bright housedress, and covers her hair with a long veil that reaches the ground. Taha's father dresses in a long robe. In school, Taha studies Arabic, religious studies, arithmetic, social studies, science, and art.

Tunisia's History and People Tunisia's coastal location has drawn people, ideas, and trade throughout the centuries. The Berbers were the first to settle the area. About 2,800 years ago, the Phoenicians (fih•NEE•shuhnz) from Southwest Asia founded the city of Carthage in the northern part of the country. This city was the center of a powerful trading empire. Carthage later fought unsuccessfully with Rome for control of the Mediterranean world.

During the following centuries, Tunisia was part of several Muslim empires. It was a colony of France until becoming an independent republic in 1956. You can still see French influence in the cities.

Almost all of Tunisia's 9.5 million people are of mixed Arab and Berber ancestry. They speak Arabic and practice Islam. Tunis, with more than 1,000,000 people, is the capital and largest urban area.

✓ Reading Check **Why can farming and herding take place in Tunisia?**

Algeria

About one and a half times the size of Alaska, **Algeria** is the largest country in North Africa. Along the Mediterranean coast, you find a narrow strip of land consisting of hills and plains. This coastal area has Algeria's best farmland. A Mediterranean climate with hot, dry summers and mild, rainy winters helps crops grow here.

Inland, the land slopes up to the Atlas Mountains. These mountains form two major ranges with an average height of about 7,000 feet (2,134 m). Another range—the Ahaggar (uh•HAH•guhr)—lies in southern Algeria. Between these mountain ranges, you find parts of the Sahara known as **ergs,** or huge shifting sand dunes.

Like neighboring Libya, Algeria must import about one-third of its food. It pays for these food imports by selling oil and natural gas. Large deposits of these resources lie in the Sahara. They have helped

Algeria's industrial growth, but widespread poverty and lack of jobs still exist. Many Algerians have moved to France and other European countries to find work.

Algeria's History and People About 30.8 million people live in Algeria. The people here—as in Libya and Tunisia—have mixed Arab and Berber heritage. Most of them are Muslim and speak Arabic. If you visited Algeria, you would discover centuries-old Muslim traditions blending with those of France. Why? From 1834 to 1962, Algeria was a French colony. In 1954, Algerian Arabs wanting freedom rose up against the French. A bloody **civil war,** or conflict between different groups inside a country, erupted. When the fighting ended in 1962, Algeria won independence. Many of the French fled to France.

Today Algeria is a republic, with a strong president and a legislature. In the early 1990s, Muslim political parties formed that opposed many of the government's **secular,** or nonreligious, policies. The Muslims gained enough support to win a national election. The government, however, rejected the election results and imprisoned many Muslim opponents. A civil war began that has taken many lives.

Algiers is the country's capital and largest city. Nearly 2.2 million people live here. Many of them live in the newer sections of the city,

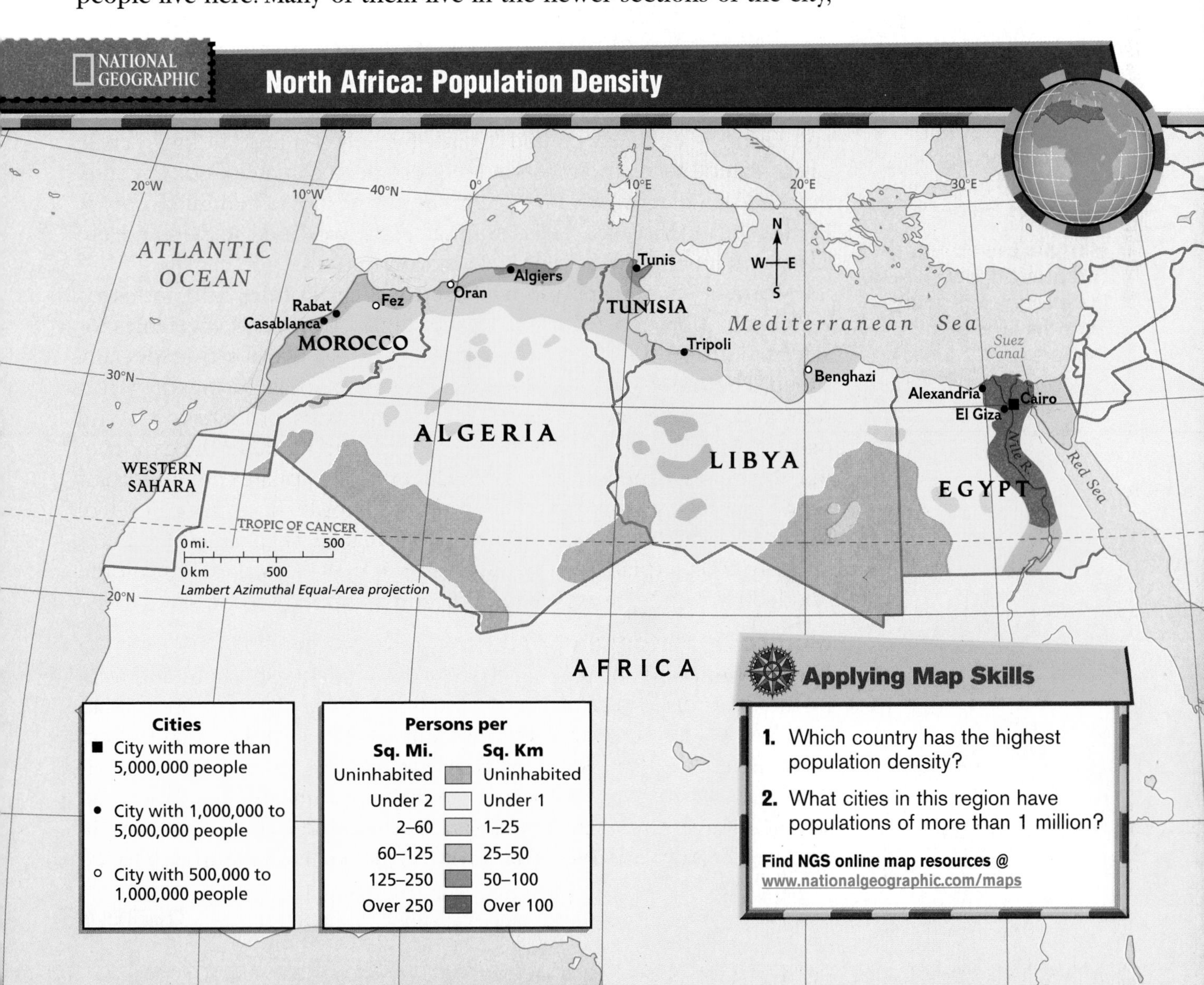

Applying Map Skills

1. Which country has the highest population density?
2. What cities in this region have populations of more than 1 million?

Find NGS online map resources @ www.nationalgeographic.com/maps

Morocco

Even though North Africa is mostly hot, snow can fall high in the Atlas Mountains where this Berber lives.

Human/Environment Interaction **How does the environment influence the lives of the Berbers?**

with modern buildings and broad streets. They enjoy visiting the older sections of the city, though, which are called **casbahs.** There they walk down narrow streets, sometimes stopping to bargain with merchants in shops and bazaars. They might step into a historic mosque for their daily prayers. Many people in Algeria's cities speak French as well as Arabic.

✓Reading Check **What conflict has affected life in Algeria since the early 1990s?**

civil war

Morocco

Slightly larger than the state of California, **Morocco** has a long coastline that borders two bodies of water—the Mediterranean Sea on the north and the Atlantic Ocean on the west. The map on page 458 shows you that Morocco's northern tip almost touches Europe. Here you will find the **Strait of Gibraltar.** It separates Africa and Europe—or Morocco and Spain—by only 9 miles (14 km). Fertile plains extend along the coasts. A Mediterranean climate of hot, dry summers and mild, rainy winters occurs in northern and central Morocco. Farther inland rise the Atlas Mountains, snowcapped in winter. Here and along much of the Atlantic coast, you find a steppe climate.

Morocco has an economy based on agriculture and industry. Farmers in Morocco grow sugar beets, grains, fruits, and vegetables for sale to Europe during the winter. Many raise livestock—especially sheep. The other half is involved in mining, manufacturing, construction, and service industries. Moroccan factories process foods and make chemicals and textiles. Morocco leads the world in the export of phosphate rock and is a leading producer of phosphates. An important service industry in Morocco is tourism. People flock to cities like Marrakech (mahr•uh•KEHSH) and Casablanca (KAH•suh•BLAHNG•kuh). In marketplaces called souks (SOOKS), sellers in traditional hooded robes offer wares made of leather, copper, and brass.

Morocco's History and People Morocco was first settled by the Berbers thousands of years ago. Their descendants still herd and farm in the foothills of the Atlas Mountains. During the A.D. 600s, Arab invaders swept into Morocco. A century later, Arabs and Berbers together crossed the Strait of Gibraltar and conquered Spain. Their descendants, called Moors, ruled parts of Spain and developed an advanced civilization until Christian Spanish rulers drove them out in the late 1400s. Many descendants of the Moors live in Morocco today.

As the Moors flourished in Spain, a Muslim kingdom prospered and ruled much of Morocco. In the early 1900s, the Moroccan kingdom weakened, and France and Spain gained control. In 1956 Morocco became independent once again. Today the country is a **constitutional monarchy.** In this form of government, a king or queen is head of state, but elected officials run the government. In Morocco, the monarch still holds many powers, however.

Beginning in the 1970s, Morocco claimed the desert region of **Western Sahara.** The discovery of phosphates there sparked a costly war between Morocco and a rebel group wanting Western Sahara to be independent. The United Nations wants to hold a vote that would allow the people of Western Sahara to decide their own future.

Morocco has about 28.2 million people. Most of them live in the coastal regions, but many also follow a traditional herding way of life in the mountains. Casablanca is Morocco's largest city, home to about 3 million people. Rabat, with a population of 1.2 million, is Morocco's capital.

Morocco's traditional culture is based on Arab, Berber, and African traditions. Moroccan music today blends the rhythms of these groups, sometimes with a dash of European pop and rock. Morocco also is known for its skilled artisans who make a variety of goods, such as carpets, pottery, jewelry, brassware, and woodwork.

✓Reading Check **Who were the Moors?** descendants of Arabs + Berbers in Morocco who conquered much of Spain in the AD 600's.

Section 2 Assessment

Defining Terms

1. **Define** aquifer, dictatorship, erg, civil war, secular, casbah, constitutional monarchy.

Recalling Facts

2. **History** Who were the Berbers?
3. **Culture** Why do many Algerians speak French?
4. **History** Why is there a dispute over control of Western Sahara?

Critical Thinking

5. **Making Generalizations** How have the physical features of North Africa affected where people live?
6. **Analyzing Information** How might the countries studied in this section improve their economies?

Graphic Organizer

7. **Organizing Information** Draw a diagram like the one below. Choose one country from this section and fill in each outer part of the diagram with a fact about that country.

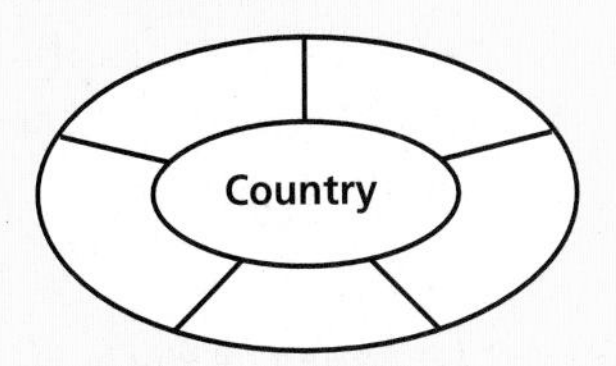

Applying Geography Skills

8. **Analyzing Maps** Study the political map on page 458. Rabat is located on the coast of which country? Along what body of water is it located?

Technology Skill

Using a Spreadsheet

A **spreadsheet** is an electronic worksheet that can manage numbers quickly and easily. Spreadsheets are powerful tools because you can change or update information, and the spreadsheet automatically performs the calculations.

Learning the Skill

All spreadsheets follow a basic design of rows and columns. Each column is assigned a letter, and each row is assigned a number. Each point where a column and a row intersect is called a *cell.* The cell's position on the spreadsheet is labeled according to its corresponding column and row—*A1* is column A, row 1; *B2* is column B, row 2, and so on.

Spreadsheets use *formulas* to calculate numbers. To create a formula, highlight the cell you want the results in. Type an equal sign (=) and then build the formula, step-by-step. If you type the formula *=B4+B5+B6* in cell B7, the numbers in these cells are added together, and the sum shows up in cell B7.

To use division, the formula would look like this: *=A5/C2.* This divides A5 by C2. An asterisk (*) signifies multiplication: *=(B2*C3)+D1* means you want to multiply B2 times C3, then add D1.

Practicing the Skill

Use these steps to create a spreadsheet.

1. In cells B1, C1, and D1, type the years 1980, 1990, and 2000. In cell E1, type the word *Total.*
2. In cells A2 through A6, type the names of North Africa's countries. In cell A7, type the word *Total.*
3. In row 2, enter the number of tons of oil produced by Algeria in 1980, 1990, and 2000.
4. Repeat step 3 in rows 3 through 6 for each country. You can find the information you need for each country in a world almanac or an encyclopedia.
5. Create a formula that tells which cells to add together so the computer can calculate the number of tons of oil for each country. For example, in cell E2, you should type *=B2+C2+D2* to find the total amount of oil Algeria produced in those years.

B2 =

	A	B	C	D	E
1		1980	1990	2000	Total
2	Algeria				
3	Egypt				
4	Libya				
5	Morocco				
6	Tunisia				
7	Total				
8					

▲ **The computer highlights the cell in which you are working.**

Applying the Skill

Use the spreadsheet you have created to answer these questions: Which country is the largest producer of oil? Has it always been number one? Are countries in North Africa together producing more oil or less oil today than they did 20 years ago?

Chapter 16 Reading Review

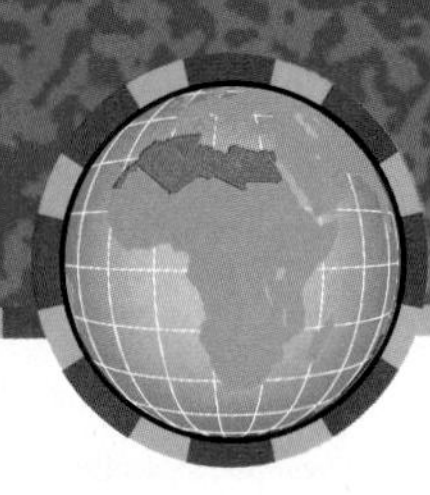

Section 1 Egypt

Terms to Know
- delta
- silt
- oasis
- phosphate
- hieroglyphics
- republic
- fellahin
- bazaar
- mosque

Main Idea

Egypt's Nile River and desert landscape have shaped the lives of the Egyptian people for centuries.

✓ **Location** Most people of Egypt live along the Nile River or in its delta.

✓ **Economics** Forty percent of Egypt's workers live by farming, but industry has grown in recent years.

✓ **History** The people of ancient Egypt built a rich and highly accomplished civilization.

✓ **Culture** Today most people in Egypt are Muslims who follow the religion of Islam.

✓ **Culture** More people in Egypt live in rural areas than in cities, but Cairo is the largest city in Africa.

Section 2 Libya and the Maghreb

Terms to Know
- aquifer
- dictatorship
- erg
- civil war
- secular
- casbah
- constitutional monarchy

Main Idea

The countries of Libya, Tunisia, Algeria, and Morocco share a desert environment and a mostly Arab culture.

✓ **Region** North Africa includes Libya and the three countries called the Maghreb—Tunisia, Algeria, and Morocco.

✓ **Location** These countries are all located on the Mediterranean Sea. Morocco also has a coast along the Atlantic Ocean.

✓ **Region** The landscape of this region is mostly desert and mountains.

✓ **Economics** Oil, natural gas, and phosphates are among the important resources in these countries.

✓ **Culture** Most of the people in these countries are Muslims and speak Arabic. Most also are of mixed Arab and Berber heritage.

◄ Desert areas begin where the fertile Nile River Valley ends.

Chapter 16 Assessment and Activities

Using Key Terms

Match the terms in Part A with their definitions in Part B.

A.

1. oasis
2. secular
3. delta
4. casbah
5. bazaar
6. dictatorship
7. silt
8. erg
9. aquifer
10. hieroglyphics

B.

a. government under a single leader
b. underground rock layer that stores water
c. old area of cities with narrow streets and small shops
d. marketplace
e. area formed by soil at a river's mouth
f. water and vegetation surrounded by desert
g. nonreligious
h. desert region of shifting sand dunes
i. particles of soil deposited by water
j. ancient form of Egyptian writing

Reviewing the Main Ideas

Section 1 Egypt

11. **Place** What is the capital of Egypt?
12. **Movement** What two bodies of water does the Suez Canal connect?
13. **Economics** Name four of Egypt's agricultural products.
14. **Culture** What huge monuments are a symbol of Egypt?
15. **Government** What type of government does Egypt have today?

Section 2 Libya and the Maghreb

16. **Human/Environment Interaction** Why must Libya depend on aquifers for water?
17. **Region** What does *maghreb* mean?
18. **History** Who founded the city of Carthage in Tunisia?
19. **History** What foreign country controlled Algeria from 1834 to 1962?
20. **Economics** What energy resource is important to almost all of North Africa's countries?

North Africa

Place Location Activity

On a separate sheet of paper, match the letters on the map with the numbered places listed below.

1. Red Sea
2. Morocco
3. Libya
4. Algeria
5. Atlas Mountains
6. Nile River
7. Cairo
8. Tunisia
9. Tripoli
10. Sinai Peninsula

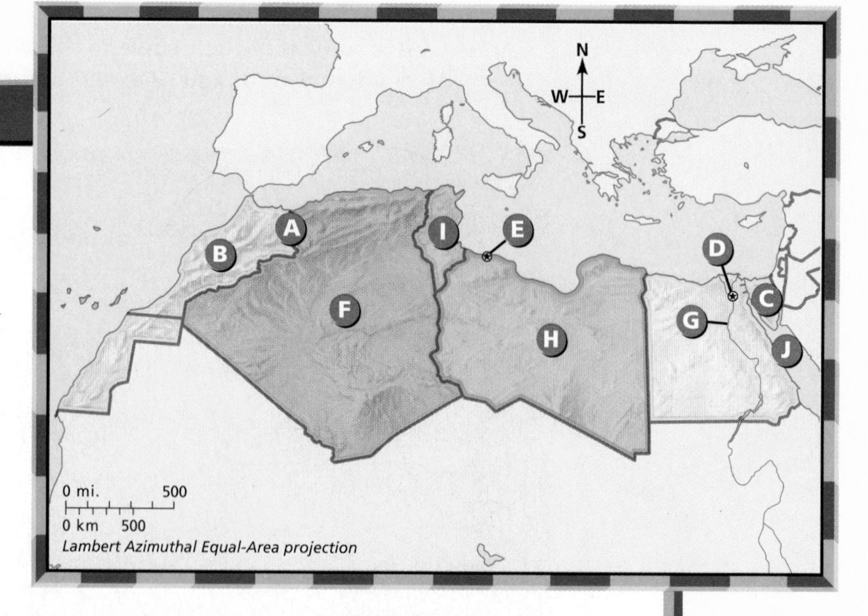

Self-Check Quiz Visit the ***Geography: The World and Its People*** Web site at gwip.glencoe.com and click on **Chapter 16—Self-Check Quizzes** to prepare for the Chapter Test.

Critical Thinking

21. **Understanding Cause and Effect** Why are the most densely populated areas of North Africa along the Mediterranean Sea and the Nile River?
22. **Sequencing Information** On a time line like the one below, label five events or eras in Egyptian history. Include their dates.

GeoJournal Activity

23. **Writing a Dialogue** Imagine that one of the pharaohs of ancient Egypt is able to visit modern times. Write down the conversation you might have as you take him on a tour of Egypt today. Describe the places you would visit. Explain the contrasts between the way people in Egypt live today and the past. What similarities between the past and present will the pharaoh find?

Mental Mapping Activity

24. **Focusing on the Region** Draw a simple outline map of North Africa, then label the following:

- Mediterranean Sea
- Red Sea
- Atlantic Ocean
- Nile River
- Atlas Mountains
- Egypt
- Libya
- Morocco
- Tunisia
- Algeria

Technology Skills Activity

25. **Using the Internet** Use the Internet to research life in the desert. Besides the Sahara, what other large deserts are there in the world? What kinds of life do deserts support? How do humans adapt to life in the desert? Are deserts changing in size and shape? Why? Use your research to create a bulletin board display on "Desert Life."

Standardized Test Practice

Directions: Study the graph, then answer the following question.

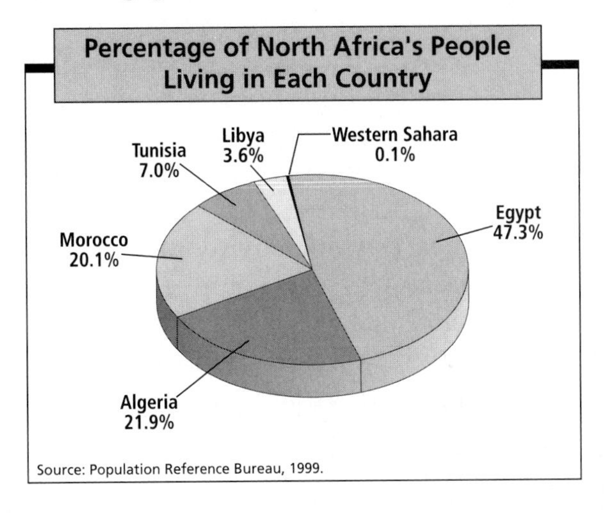

1. According to the graph above, which one of the following statements is true?

F Almost half of the people of North Africa live in Egypt.

G Almost half of the people of North Africa live in Algeria.

H Egypt's land area is much larger than Algeria's land area.

J Algeria's land area is much larger than Libya's land area.

Test-Taking Tip: When analyzing circle or pie graphs, first look at the title to see what the graph shows. Next read each section of the "pie" and compare the sections to one another. Notice that no actual population figures are given on the pie graph, only percentages. All the pie sections are different sizes, but together they add up to 100 percent.

GeoLAB ACTIVITY

Oil on the Ocean

1 Background

How did a place that is more than 90 percent desert become one of the richest countries in North Africa? The discovery of oil in the 1950s changed and helped Libya's economy, but it also created possible environmental problems. Libya's oil is shipped across the Mediterranean Sea to Europe and Asia. Whenever oil is transported by ship, there is a risk that it can leak or spill. This activity will demonstrate the far-reaching effects caused by a small amount of oil spilled in water.

2 Materials

- **1 large, shallow pan**
- **water**
- **small amount of motor oil**
- **1 medicine dropper**
- **string**
- **a piece of paper**

Oil Freighter in the Port of Tripoli, Libya

Believe It or Not!

Wealthy women living in ancient Egypt made use of oil's interesting properties by wearing cones of scented grease on top of their heads. The heat of the day caused the grease to melt and drip down their bodies, covering their skin and clothing with a fragrant, shiny gleam.

What to Do

1. Fill the pan two-thirds full with water.
2. Create an oil spill by gently adding one dropperful of motor oil to the center of the pan of water.
3. Loop the string around the edges of the "spill."
4. Remove the string. Mark, measure, and record the length of the string that circled the spill.
5. Wait three minutes, then loop the string around the spill and measure again. Repeat the measurement after another three minutes. Record a total of five different measurements after each three-minute waiting period.
6. Gently shake the pan. Record your observations of the oil spill.
7. Use the paper to gently fan the water's surface. Record your observations of what happens to the oil in the water.

LAB ACTIVITY REPORT

1. How did the oil spill's size change during the 15-minute measuring period?
2. What happened when you shook the pan? Does this occur in the ocean?
3. What did you demonstrate when you fanned the water's surface with the sheet of paper?
4. **Drawing Conclusions** If left alone, what will happen to the oil in the pan over time?

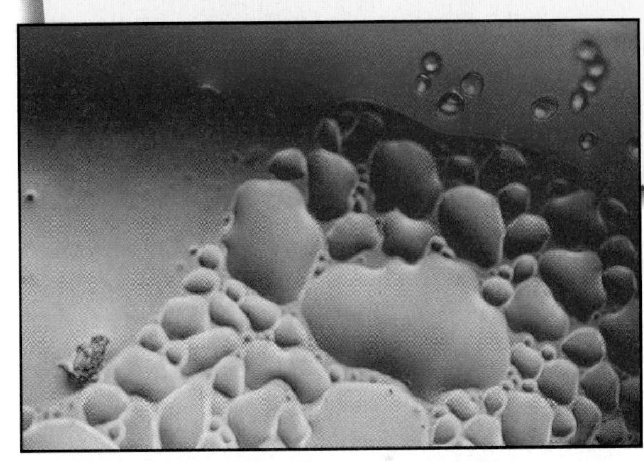

▲ **This is what oil and water look like under a microscope.**

5 Extending the Lab

Activity

Dip several feathers (which symbolize wildlife) into the oil spill. Use different cleaners (dish detergent, clothes detergent, hand soap, and so on) to clean the feathers. Are the cleaners effective? Which cleaner works best? How are the feathers affected by the cleaning? Summarize your findings in a written report.

Chapter 17

Southwest Asia

The World and Its People — NATIONAL GEOGRAPHIC

To learn more about the people and places of Southwest Asia, view ***The World and Its People*** **Chapter 17** video.

Geography online

Chapter Overview Visit the ***Geography: The World and Its People*** Web site at gwip.glencoe.com and click on **Chapter 17–Chapter Overviews** to preview information about Southwest Asia.

Section 1 Turkey

Guide to Reading

Main Idea

Rapidly modernizing Turkey is a link between Asia and Europe.

Terms to Know

- mosque
- migrate
- secular

Places to Locate

- Bosporus
- Sea of Marmara
- Dardanelles
- Anatolia
- Pontic Mountains
- Taurus Mountains
- Black Sea
- Istanbul
- Ankara

Reading Strategy

Make a chart like this one, filling in at least two key facts about Turkey in each category.

Turkey	Fact #1	Fact #2
Land		
Economy		
People		

NATIONAL GEOGRAPHIC **Exploring Our World**

There is only one city in the world that lies on two continents. The Bosporus (BAHS•puhr•uhs), a strait in Turkey, separates this city—Istanbul. The Bosporus also divides Europe from Asia. It is an important seaway that links the Black Sea to the Sea of Marmara and, eventually, to the Mediterranean Sea.

A little larger than the state of Texas, Turkey has a unique location—it bridges the continents of Asia and Europe. The large Asian part of Turkey occupies the peninsula once known as Asia Minor. The much smaller European part lies on Europe's Balkan Peninsula. Three important waterways—the **Bosporus,** the **Sea of Marmara** (MAHR•muh•ruh), and the **Dardanelles** (DAHRD•uhn•EHLZ) separate the Asian and European parts of Turkey. Together, these waterways are called the Turkish Straits. Find these bodies of water on page RA21 of the Reference Atlas.

Turkey's Land and Economy

The center of Turkey is **Anatolia** (A•nuh•TOH•lee•uh), a plateau region rimmed by mountains. The **Pontic Mountains** border the plateau on the north. The **Taurus Mountains** tower over it on the south. Earthquakes often strike the region, causing much damage and death. Lowland plains curve along Turkey's three coasts. Grassy plains cover the northern part along the **Black Sea.** On the western coast, you find broad, fertile river valleys extending inland from the Aegean Sea. In the south, coastal plains stretch along the Mediterranean Sea.

Turkey's climate varies throughout the country. If you lived on the Anatolian plateau, you would experience the hot, dry summers and cold, snowy winters of the steppe climate. People living in the coastal areas enjoy a Mediterranean climate—hot, dry summers and mild, rainy winters.

Nearly half of Turkey's people are farmers. Many live in the mild coastal areas, where they raise livestock and plant crops such as cotton, tobacco, fruits, and nuts for export. On the drier inland plateau, farmers grow mostly wheat and barley for use at home. In eastern Turkey, grains and goats are raised in mountain valleys.

Turkey has rich mineral resources of coal, copper, and iron. The most important industrial activities are oil refining and the making of textiles and clothing. Turkish factory workers also process foods and make cars, steel, and building materials. About one-third of Turkey's people work in service industries. The country's beautiful beaches and historic sites have made tourism another growing industry.

✓Reading Check **What are Turkey's most important industrial activities?**

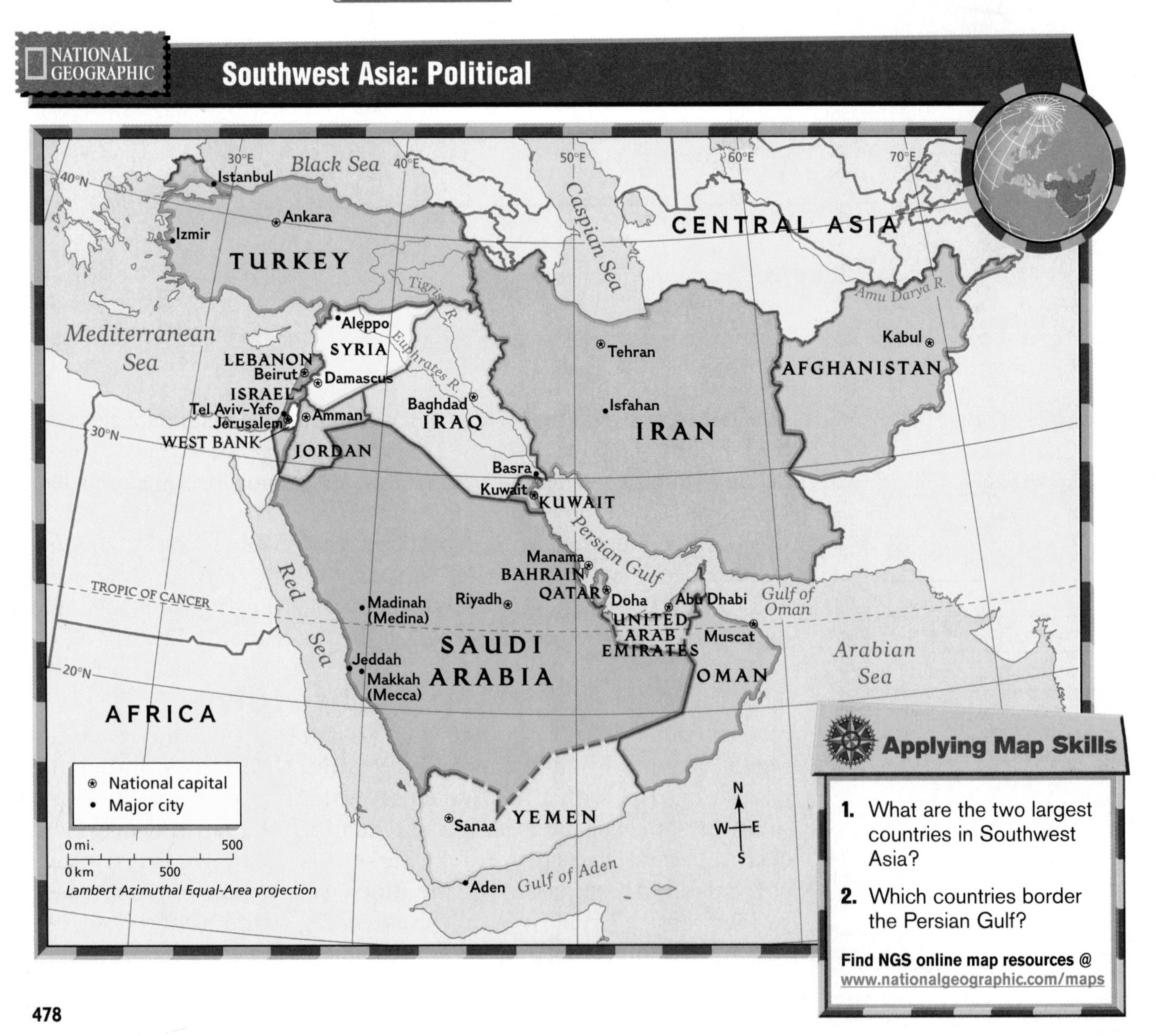

Applying Map Skills

1. What are the two largest countries in Southwest Asia?
2. Which countries border the Persian Gulf?

Find NGS online map resources @ www.nationalgeographic.com/maps

Turkey's People

Most of Turkey's 65.9 million people live in the northern part of Anatolia, on coastal plains, or in valleys. About 98 percent are Muslims, or followers of Islam. Turkish is the official language, but Kurdish and Arabic are also spoken. Kurdish is the language of the Kurds, an ethnic group who make up about 20 percent of Turkey's people. The Turkish government has tried to turn the Kurds away from Kurdish culture and language. Unwilling to abandon their identity, the Kurds have demanded their own independent state. Tensions between the two groups have resulted in violent clashes.

Almost 70 percent of Turkey's people live in cities or towns. **Istanbul** is Turkey's largest city with nearly 8 million people. It is the only city in the world located on two continents. Istanbul is known for its beautiful palaces, museums, and mosques. **Mosques** are places of worship for followers of Islam. Thanks to its location at the entrance to the Black Sea, Istanbul is a major trading center. Turkey's capital and second-largest city is **Ankara.**

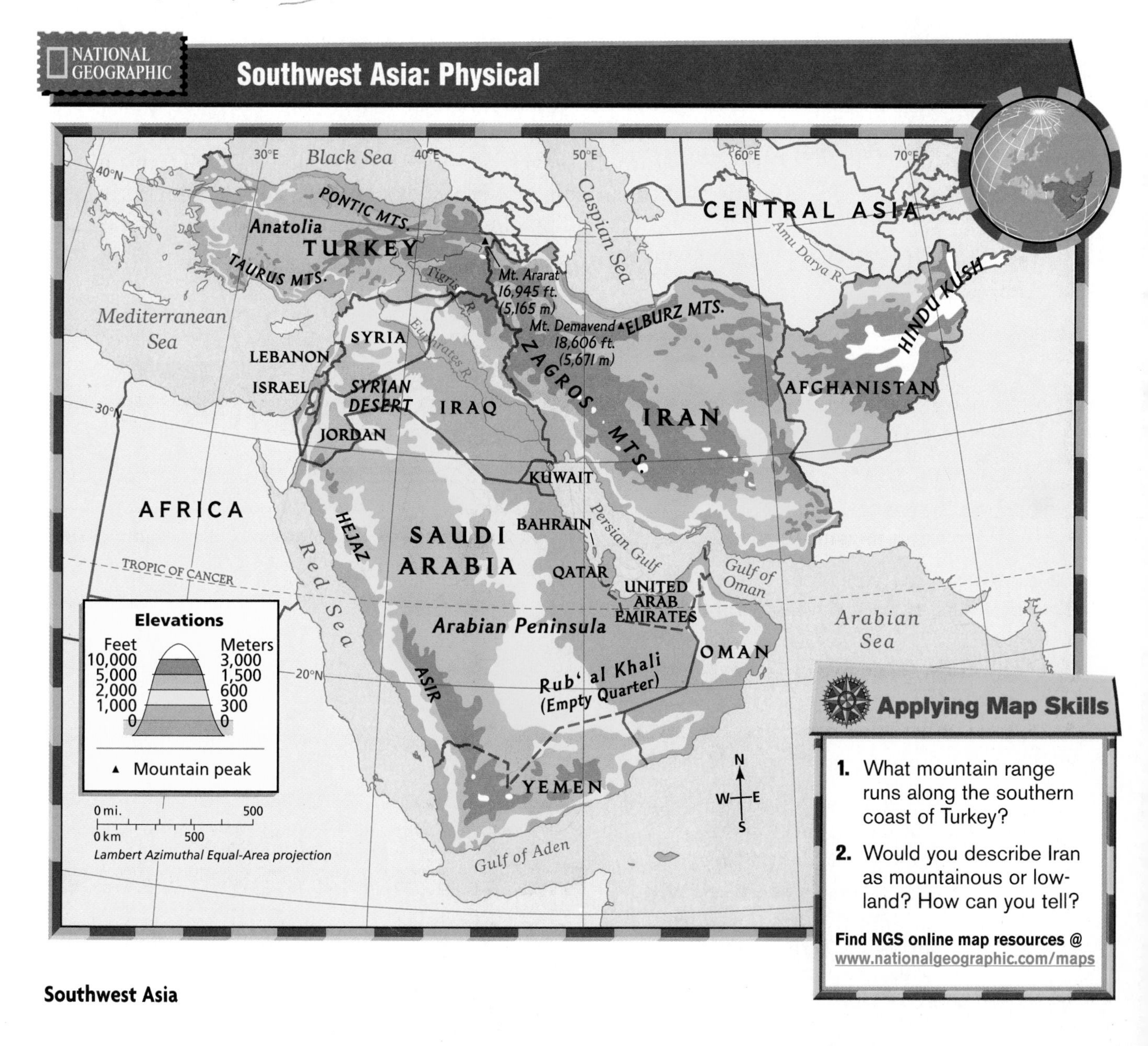

Applying Map Skills

1. What mountain range runs along the southern coast of Turkey?
2. Would you describe Iran as mountainous or lowland? How can you tell?

Find NGS online map resources @ www.nationalgeographic.com/maps

Festival Time

Kudret Özal lives in Söğüt, Turkey. Every year, she and her family attend a festival that honors a warrior ancestor. She says, "At night everyone gathers to sing, dance, and tell jokes and stories." According to custom, Kudret wears clothing that covers her head, arms, shoulders, and legs.

History and Culture Istanbul began as a Greek port called Byzantium more than 2,500 years ago. Later the Romans expanded the city and renamed it Constantinople. For hundreds of years, the city was the glittering capital of the Byzantine Empire.

Many of Turkey's people today are descendants of an Asian people called Turks. These people migrated to Anatolia during the A.D. 900s. **Migrating** means moving from one place to another. One group of Turks—the Ottomans—conquered Constantinople in the 1400s. They, too, renamed the city, calling it Istanbul. The city served as the brilliant capital of a powerful Muslim empire called the Ottoman Empire. At its height, this empire ruled much of southeastern Europe, North Africa, and Southwest Asia.

The Ottomans' defeat in World War I led to the breakup of the empire. During most of the 1920s and 1930s, Kemal Atatürk, a military hero, served as Turkey's first president. Atatürk introduced many political and social changes to modernize the country. Turkey soon began to consider itself European as well as Asian. Many Turkish people, however, continued to value the Muslim faith. During the 1990s, Muslim and **secular,** or nonreligious, political groups struggled for control of Turkey's government.

Throughout the country, you can see traditional Turkish arts: colored tiles, finely woven carpets, and beautifully decorated books. Turkish culture, however, has its modern side as well. Folk music blends traditional and modern styles. Turkey also has recently produced many outstanding films that deal with social and political issues.

✓Reading Check **What is unusual about Turkey's largest city?**

Section 1 Assessment

Defining Terms

1. **Define** mosque, migrate, secular.

Recalling Facts

2. **Place** What bodies of water form the Turkish Straits?
3. **Economics** Name five of Turkey's agricultural products.
4. **History** What other names has the city of Istanbul had?

Critical Thinking

5. **Analyzing Information** How has Istanbul's location made it a trading center?
6. **Understanding Cause and Effect** Why have violent clashes occurred between the Kurds and Turkey's government?

Graphic Organizer

7. **Organizing Information** On a diagram like this one, label an example of Turkey's culture at the end of each line.

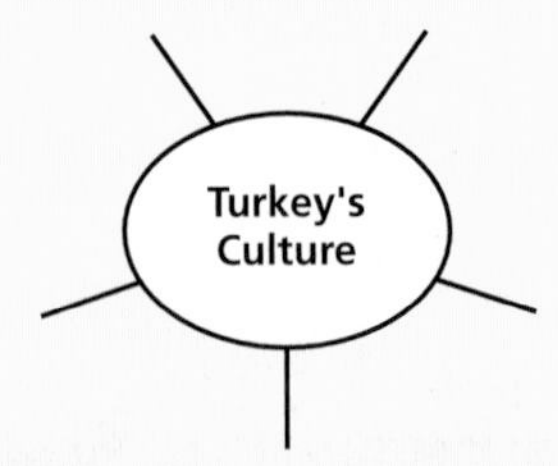

Applying Geography Skills

8. **Analyzing Maps** Study the physical map on page 479. What large body of water borders Turkey on the north? On the south?

Making Connections

ART SCIENCE  TECHNOLOGY

A Poem of Freedom

Through many of his poems, Turk Nazim Hikmet criticized actions of the Turkish government. In 1938 he was arrested and sent to prison for about 10 years. Hikmet wrote some of his finest poems there, including the following one.

The World, My Friends, My Enemies, You, and the Earth

by Nazim Hikmet (1902–1963)

I'm wonderfully happy I came into the world,
I love its earth, its light, its struggle, and its bread.
Even though I know its dimensions from pole to pole to the centimeter,
and while I'm not unaware that it's a mere toy next to the sun,
the world for me is unbelievably big.
I would have liked to go around the world
and see the fish, the fruits, and the stars that I haven't seen.
However,
I made my European trip only in books and pictures.
In all my life I never got one letter
with its blue stamp canceled in Asia.
Me and our corner grocer,
we're both mightily unknown in America.
Nevertheless,
from China to Spain, from the Cape of Good Hope to Alaska,
in every nautical mile, in every kilometer, I have friends and enemies.
Such friends that we haven't met even once—
we can die for the same bread, the same freedom, the same dream.
And such enemies that they're thirsty for my blood,
I am thirsty for their blood.
My strength
is that I'm not alone in this big world.
The world and its people are no secret in my heart,
no mystery in my science.
Calmly and openly
I took my place
in the great struggle.
And without it,
you and the earth
are not enough for me.
And yet you are astonishingly beautiful,
the earth is warm and beautiful.

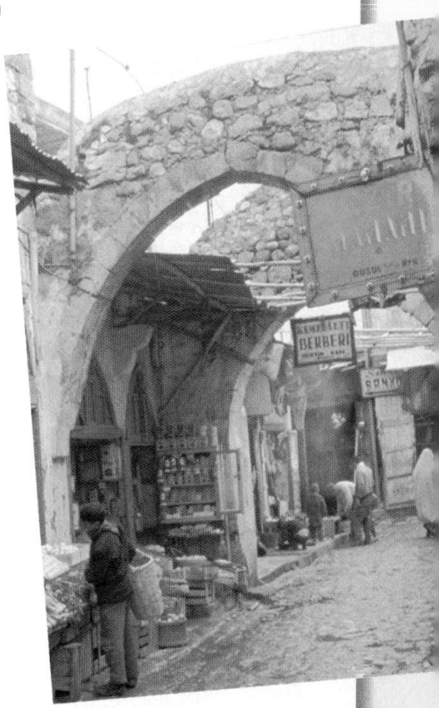

▲ **Old-fashioned outdoor market in Urfa, Turkey**

Making the Connection

1. What does Hikmet say he loves about the world?
2. How has Hikmet "traveled" to Europe?
3. **Drawing Conclusions** What feelings toward life do you get from the poem?

Section 2 Israel

Guide to Reading

Main Idea

After years of conflict, the Jewish nation of Israel and neighboring Arab countries are trying to achieve peace.

Terms to Know

- kibbutz
- moshav
- Diaspora
- Holocaust
- monotheism

Places to Locate

- Golan Heights
- Dead Sea
- Tel Aviv-Yafo
- Jerusalem
- West Bank
- Gaza Strip

Reading Strategy

Create a time line like this one to track four key events with dates in Israel since its founding in 1948.

NATIONAL GEOGRAPHIC Exploring Our World

Modern and traditional cultures exist side by side throughout Israel. Israel was established as an independent country in 1948 after many years of trying to create a homeland for Jews. Since then, Jewish people from more than 100 countries have migrated to this small country. At this market in Tel Aviv-Yafo, buyers and sellers talk about current events and sports.

Israel lies at the eastern end of the Mediterranean Sea. Slightly smaller than the state of New Jersey, it is 256 miles (412 km) long from north to south and only 68 miles (109 km) wide from east to west.

Israel's Land and Climate

The mountains of Galilee lie in Israel's far north. East of these mountains is a plateau called the **Golan Heights.** South of the Golan Heights, between Israel and Jordan, is the **Dead Sea.** At 1,349 feet (411 m) below sea level, the shores of the Dead Sea are the lowest place on the earth's surface. The Dead Sea is also the earth's saltiest body of water—about nine times saltier than ocean water. The map on page 485 shows you where the Golan Heights and the Dead Sea are located.

In southern Israel, a desert called the Negev (NEH•gehv) covers almost half the country. A fertile plain no more than 20 miles (32 km) wide lies along the country's Mediterranean coast. To the east, the Jordan River cuts through the floor of a long, narrow valley before flowing into the Dead Sea.

Northern Israel has a Mediterranean climate with hot, dry summers and mild winters. About 40 inches (102 cm) of rain fall in the north each year. Southern Israel has a desert climate. Summer temperatures soar higher than 120°F (49°C), and annual rainfall is less than 1 inch (2.5 cm).

Reading Check **What plateau is found in northeastern Israel?**

Israel's Economy

Israel's best farmland stretches along the Mediterranean coastal plain. For centuries, farmers here have grown citrus fruits, such as oranges, grapefruits, and lemons. Citrus fruits are still Israel's major agricultural export. Farther inland, you find that the desert actually blooms. This is possible because farmers add fertilizers to the soil and carefully use scarce water resources. In very dry areas, crops are grown with drip irrigation. This method uses computers to release specific amounts of water from underground tubes to the roots of plants. As a result of technology, Israel's farmers not only feed the country's people—they even export some food to other countries.

About 9 percent of Israelis live and work on farm settlements. Many join together to grow and sell crops. People in one type of settlement called a **kibbutz** (kih•BUTS) share all of the property and may also produce goods such as clothing and electronic equipment. Another

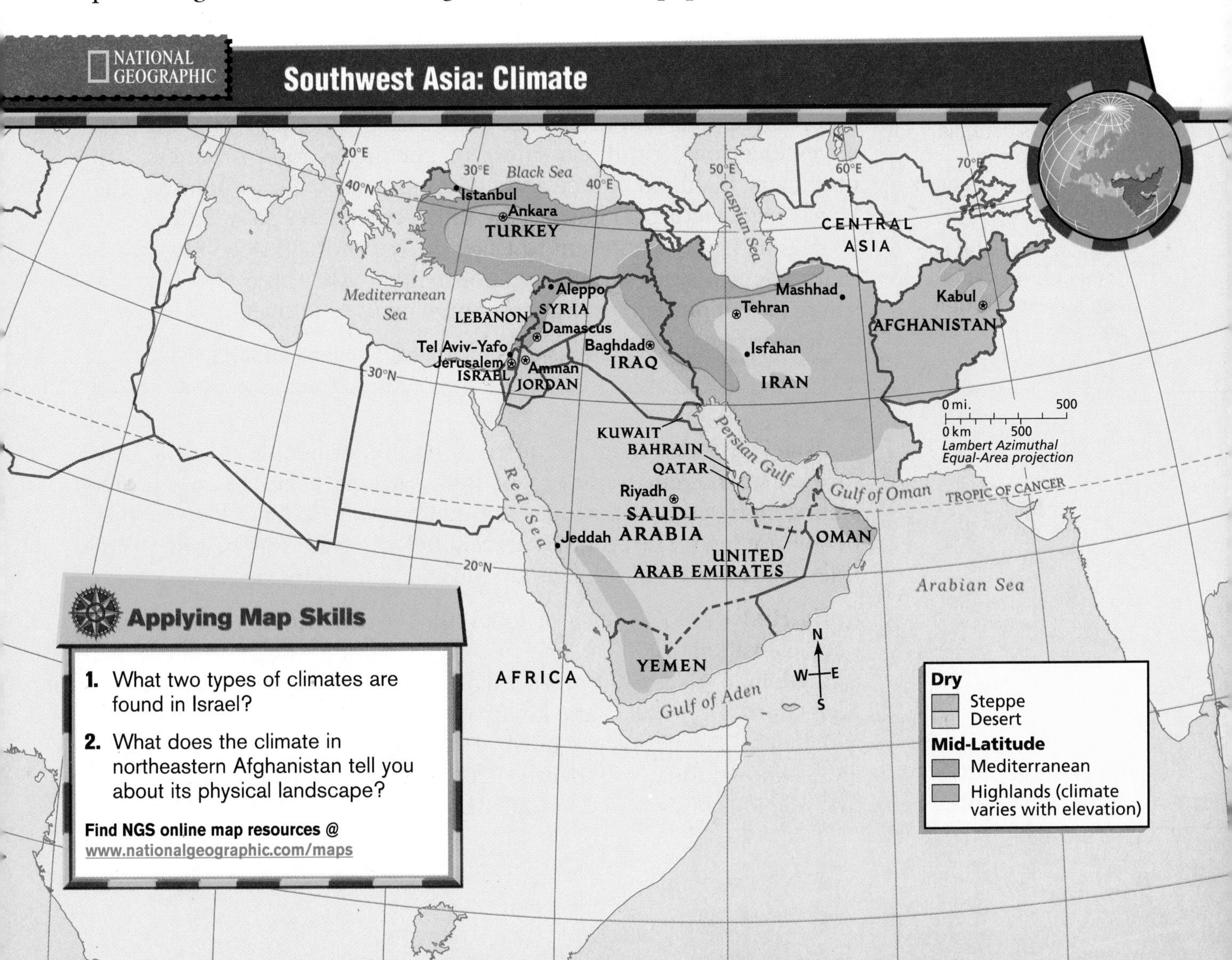

Israel's Landscapes

Salt and other minerals have crystallized on the shores of the Dead Sea (above). In other areas, Israelis have turned the desert into productive farmland (right).

Region **On what two types of farm settlements do most rural Israelis live?**

kind of settlement is called a **moshav** (moh•SHAHV). People in a moshav share in farming, production, and selling, but each person is allowed to own some private property as well.

Israel is the most industrialized country in Southwest Asia. Its economic development has been supported by large amounts of aid from European nations and the United States. Israel's skilled workforce produces electronic products, clothing, chemicals, food products, and machinery. Diamond cutting and polishing is also a major industry. The largest manufacturing center is the urban area of **Tel Aviv-Yafo.**

Mining is also important to Israel's economy. The Dead Sea area is rich in deposits of potash, a type of mineral salt. The Negev is also a source of copper and phosphate, a mineral used in making fertilizer.

Reading Check **How have Israelis improved their land for farming?**

The Israeli People

Israel has been home to different groups of people over the centuries. The ancient traditions of these groups have led to conflict among their descendants today. About 80 percent of Israel's 6.1 million people are Jews. The other 20 percent belong to an Arab people called Palestinians. Most Palestinians are Muslims, but some are Christians.

Early History The Jews of today trace their origins to an ancient people. In about 1000 B.C., the ancient Jews under King David created a kingdom in the area of present-day Israel. The kingdom's capital was **Jerusalem.** By 922 B.C., the kingdom had split into two states—Israel and Judah. The people of Judah came to be called Jews. Their religion—known today as Judaism (JOO•duh•IH•zuhm)—was one of the first religions known to emphasize faith in one God.

Over time, the region was ruled by Greeks, Romans, Byzantines, Arabs, and Ottoman Turks. Under the Romans, the area was called Palestine. The Jews twice revolted against Roman rule but failed to win their freedom. In response, the Romans ordered all Jews out of the land.

During the next 1,900 years, Jews lived in settlements scattered around the world. These settlements were known together as the **Diaspora** (dy•AS•puh•ruh), or scattering of a people. Prejudice against the Jews caused them much hardship. In the late 1800s, some European Jews began to move back to Palestine. These settlers, known as Zionists, had planned to set up a safe homeland for Jews in their ancestral land.

The Birth of Israel During World War I, the British won control of Palestine. They supported a Jewish homeland there. Most of the people living in Palestine, however, were Arabs who also claimed the area as their homeland. To keep peace with the local Arab population, the British began to limit the number of Jews entering Palestine.

During World War II, Germans killed millions of Europe's Jews and others. The mass imprisonment and slaughter of European Jews is known as the **Holocaust.** It brought worldwide attention to Jews. In 1947 the United Nations voted to divide Palestine into a Jewish and an Arab state. The Arabs in Palestine and in neighboring countries disagreed with this division. In 1948 the British left the area, and the Jews immediately declared an independent country called Israel. David Ben-Gurion (BEHN•gur•YAWN) became Israel's first leader.

War soon broke out between Israel and its Arab neighbors. The war ended with Israel's victory. It also brought about significant changes in the area's population. The fighting made many Palestinian Arabs flee to neighboring Arab countries. Jews began arriving from Europe and other areas.

After its war of independence, Israel fought three more wars with its Arab neighbors. In these conflicts, Israel won control of some of its neighbors' land. Palestinian Arabs, now left homeless, demanded their own country. During the 1970s and 1980s, many Palestinians and Israelis died fighting each other. Steps toward peace began when Israel and Egypt signed a treaty in 1979. Agreements made between Israel and Palestinian Arab leaders in 1993 and between Israel and Jordan in 1994 also moved toward peace.

In the 1993 agreement, Israel agreed to turn over two areas to the Palestinians. The **West Bank** lies on the western bank of the

Jordan River and surrounds Jerusalem. The **Gaza Strip** is located on the Mediterranean coast and shares a border with Egypt. Find these areas on the map on page 485. Palestinians now have limited control of some of these areas. Yet some Jews still live in these two regions, and tensions between the two groups remain. Many issues—particularly control of Jerusalem—need to be settled before Palestinians achieve independence. In addition, Palestinian Arabs have fewer freedoms and economic opportunities than their Jewish neighbors. In late 2000, violence erupted again because of the inability to resolve these issues.

Israel Today Israel is a democratic republic, a government headed by elected officials. A president represents the country at national events. A prime minister heads the government. The Israeli parliament, or Knesset, meets in a modern building in Jerusalem.

A single law—the Law of Return—increased Israel's population more than any other factor. Passed in 1950, the law states that Jews anywhere in the world can come to Israel to live. As a result, Jewish people have moved to Israel from many countries.

More than 90 percent of Israel's people live in urban areas. The largest cities are Jerusalem, Tel Aviv-Yafo, and Haifa (HY•fuh). Israel proclaimed Jerusalem as its capital in 1950. Turn to page 504 to see why Jerusalem is a holy city for Christians and Muslims as well as Jews. All three religions began in Southwest Asia centuries ago. They all practice **monotheism,** or the belief in one God.

Reading Check Over what two areas do Palestinians have limited control?

Assessment

Defining Terms

1. **Define** kibbutz, moshav, Diaspora, Holocaust, monotheism.

Recalling Facts

2. **Location** What is the lowest place on the earth's surface? What is its elevation?
3. **History** What was a Zionist?
4. **Place** Why is Jerusalem important?

Critical Thinking

5. **Analyzing Information** What is the major disagreement between Israel and the Palestinians?
6. **Drawing Conclusions** Why do you think Israel has worked so hard to develop its agricultural and manufacturing industries?

Graphic Organizer

7. **Organizing Information** On a diagram like this one, list three things that helped Israel's agricultural success.

Applying Geography Skills

8. **Analyzing Maps** Study the political map on page 478. What country borders Israel to the north? To the northeast? To the east?

Geography Skill

Reading a Time Zones Map

The earth rotates 360° in 24 hours. The earth's surface has been divided into 24 time zones. Each time zone represents 15° longitude, or the distance that the earth rotates in 1 hour.

Learning the Skill

The Prime Meridian, or 0° longitude, is the starting point for figuring out time around the world. Traveling west from 0° longitude, it becomes 1 hour earlier for each time zone crossed. Traveling east, it becomes 1 hour later for each time zone crossed. The international date line is set at the 180° line of longitude. Traveling west across this imaginary line, you add a day. Traveling east, you subtract a day. To read a time zones map:

- Choose a place for which you already know the time and locate it on the map.
- Locate another place and determine if it is east or west of the first place.
- Count the time zones between the two.
- Calculate the time by either adding (going east) or subtracting (going west) an hour for each time zone.
- Determine whether you have crossed the date line, and identify the day of the week.

Practicing the Skill

1. On the map below, if it is 4 P.M. in Rio de Janeiro, what time is it in Honolulu?
2. If it is 10:00 A.M. in Tokyo on Tuesday, what day and time is it in Moscow?

Applying the Skill

Imagine you have a friend living in Rome, Italy. What time (your time) would you call if you wanted to talk to your friend after 7:00 P.M.?

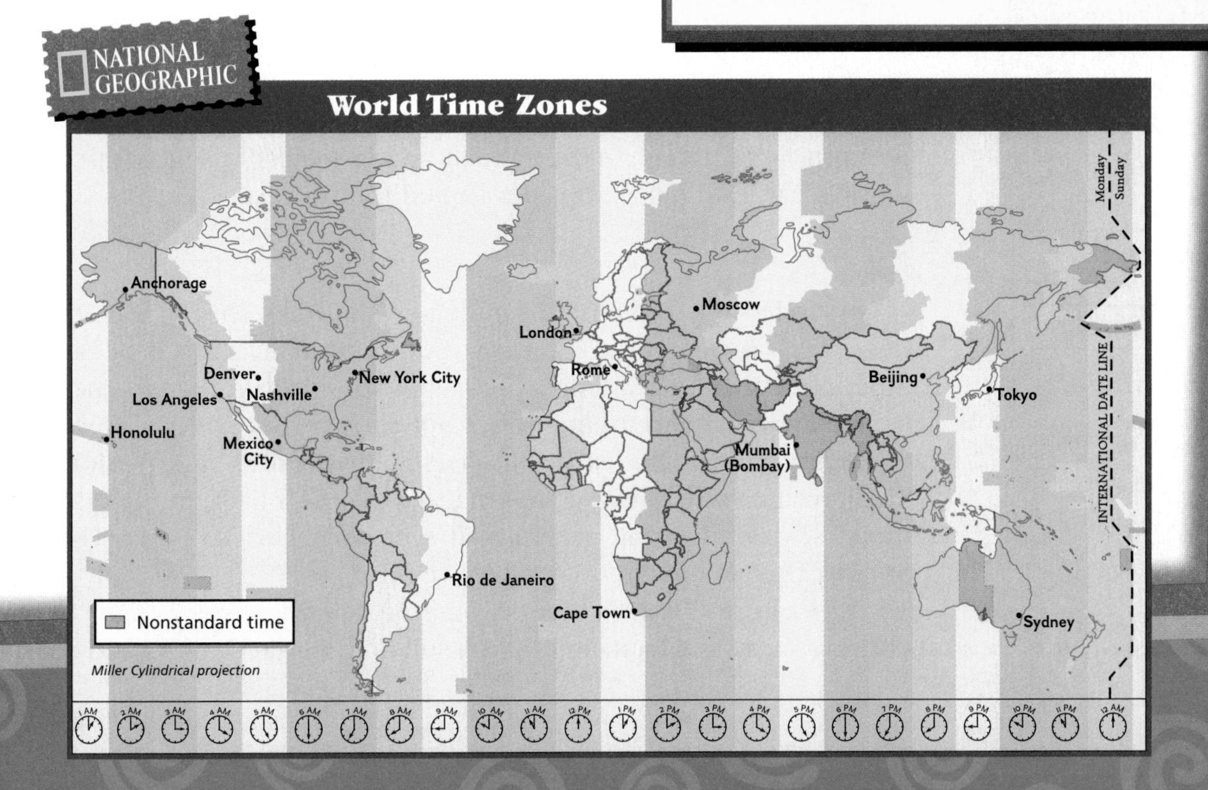

Section 3

Syria, Lebanon, and Jordan

Guide to Reading

Main Idea

Syria, Lebanon, and Jordan have largely Arab populations but have different economies and forms of government.

Terms to Know

- Bedouins
- civil war
- constitutional monarchy

Places to Locate

- Syria
- Syrian Desert
- Euphrates River
- Damascus
- Lebanon
- Beirut
- Jordan
- Jordan River
- Amman

Reading Strategy

Create a chart like this one, listing three key economic activities for each country.

Country	Economic Activities
Syria	
Lebanon	
Jordan	

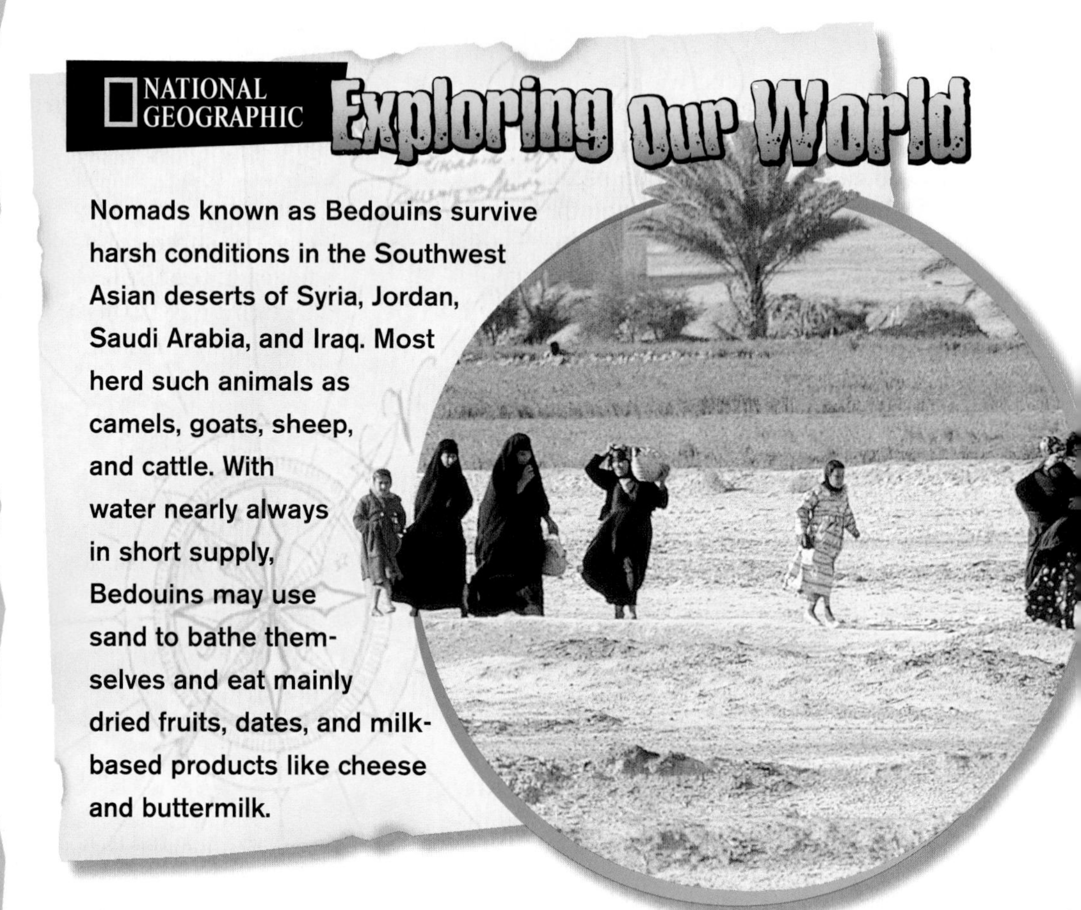

NATIONAL GEOGRAPHIC Exploring Our World

Nomads known as Bedouins survive harsh conditions in the Southwest Asian deserts of Syria, Jordan, Saudi Arabia, and Iraq. Most herd such animals as camels, goats, sheep, and cattle. With water nearly always in short supply, Bedouins may use sand to bathe themselves and eat mainly dried fruits, dates, and milk-based products like cheese and buttermilk.

Much of Jordan and large parts of Syria are desert. Lebanon and western Syria receive more rainfall and have fertile soil. Some of the world's earliest civilizations developed in this area.

Syria

Syria has been a center of trade for centuries. Throughout its early history, Syria was a part of many empires, but in 1946 it became an independent country. Since the 1960s, one political party has controlled Syria's government. It does not allow many political freedoms.

Syria's land includes fertile coastal plains and valleys along the Mediterranean Sea. Inland mountains running north and south keep moist sea winds from reaching the eastern part of Syria. The vast **Syrian Desert** covers this eastern region.

Agriculture is Syria's main economic activity. Farmers raise mostly cotton, wheat, and fruits in the rich soil of mountain valleys and coastal plains. In the dry northeast, farmers rely on irrigation. The Syrian government has built dams on the **Euphrates River,** which flows through the country. These dams provide water for irrigation as well as hydroelectric power for cities and industries. Future conflict over Euphrates water is a possibility. Turkey, Syria's upstream neighbor, is building a huge dam that will reduce the flow of water to Syria. Iraq, Syria's downstream neighbor, is concerned about the reduced flow of water that will make its way there.

Like many other countries of Southwest Asia, Syria has reserves of oil—the country's main export. Other industries are food processing and textiles. Syrian fabrics have been highly valued since ancient times.

Syria's People Almost half of Syria's 16 million people live in rural areas. A few are **Bedouins**—nomadic desert peoples who follow a traditional way of life. The word *bedouin* means "desert dweller" in Arabic. Most other Syrians live in cities. The two largest cities are Aleppo, in the north, and **Damascus,** the capital, in the south. Damascus is one of the oldest continuously inhabited cities in the world. It was founded as a trading center more than 5,000 years ago.

Like many other Southwest Asians, the people of Syria are mostly Arab Muslims. Islam has deeply influenced Syria's traditional arts and buildings. In many Syrian cities, you can see spectacular mosques and palaces. As in other Arab countries, hospitality is a major part of life in Syria. Group meals are a popular way of strengthening family ties and friendships. The most favored foods are lamb, flat bread, and bean dishes flavored with garlic and lemon.

Reading Check **On what river has Syria built dams?**

Music

The most common stringed instrument of Southwest Asia is the oud. Often pear-shaped, its neck bends sharply backward. Music from this region uses semitones that are not heard in Western music. Semitones are the "invisible" notes that lie between the black and white keys of a piano. Legend says that the oud owes its special tones to the birdsongs absorbed by the wood from which the oud was made.

Looking Closer **What instrument in our culture do you think came from the oud?**

GO TO **World Music: A Cultural Legacy**
Hear music of this region on Disc 1, Track 25.

Lebanon

Lebanon is about half the size of New Jersey. Sandy beaches and a narrow plain run along its Mediterranean coast. Rugged mountains rise east of the plain and along Lebanon's eastern border. A fertile valley lies between the two mountain ranges. Because the country is so small, you can swim in the warm Mediterranean Sea, then throw snowballs in the mountains—all in the same day.

Cedar trees once covered Lebanon. Now only a few lonely groves survive on mountaintops. Still, Lebanon is the most densely wooded of all the Southwest Asian countries. Many kinds of pine trees thrive on the mountains. Fruit trees flourish on the coastal land.

More than 60 percent of Lebanon's people work in service industries such as banking and insurance. Manufactured products include food, cement, textiles, chemicals, and metal products. Lebanese farmers grow citrus fruits, vegetables, grains, olives, and grapes. Shrimp is harvested from the Mediterranean.

Petra

One of Jordan's major tourist attractions, Petra was built by an Arab ruling family during the 300s B.C. The city's temples and monuments were carved out of cliffs in the Valley of Moses in Jordan. It was a major center of the spice trade that reached as far as China, Egypt, Greece, and India. When the trade route bypassed Petra, the city gradually faded in importance. It became unknown to the Western world until it was rediscovered in 1812. Archaeologists have found dams, rock-carved channels, and ceramic pipes that brought water to the 30,000 people who once lived here.

The Lebanese People More than 80 percent of Lebanon's nearly 4.1 million people live in coastal urban areas. **Beirut** (bay•ROOT), the capital and largest city, was once a major banking and business center. European tourists called Beirut "the Paris of the East" because of its elegant shops and sidewalk cafés. Today, however, Beirut is still rebuilding after a civil war that lasted from 1975 to 1991.

A **civil war** is a fight among different groups within a country. Lebanon's civil war arose between groups of Muslims and Christians. About 70 percent of the Lebanese are Arab Muslims and most of the rest are Arab Christians. Many lives were lost in the war, and Lebanon's economy was almost destroyed. Israel invaded Lebanon during the war, finally withdrawing its troops from the south in 2000.

Arabic is the most widely spoken language in Lebanon. French is also an official language. Why? France ruled Lebanon before the country became independent in the 1940s. Local foods reflect a blend of Arab, Turkish, and French influences.

Reading Check Why is Beirut in the process of rebuilding?

Jordan

Slightly larger than Kentucky, **Jordan** lies south of Syria and east of Israel. From 1948 to 1967, Jordan's borders included land west of the **Jordan River.** This area—the West Bank—was occupied by Israel during the 1967 war between Israel and its neighbors.

A land of contrasts, Jordan stretches from the fertile Jordan River valley in the west to dry, rugged country in the east. Jordan lacks water resources. Small amounts of irrigated farmland lie in the Jordan River valley, however. Here farmers grow wheat, fruits, and vegetables. Jordan's desert is home to tent-dwelling Bedouins, who raise livestock. In recent years, many Bedouins have settled in towns to find work.

Jordan also lacks energy resources. The majority of its people work in service and manufacturing industries. The leading manufactured goods are phosphates, potash, pottery, chemicals, and processed foods. Phosphate and potash are mined near the Dead Sea and made into fertilizer at a plant in Aqaba (AH•kah•buh), Jordan's only port.

Jordan's People Most of Jordan's 4.7 million people are Arab Muslims. They include more than 1 million Palestinian Arabs who fled Israel or the West Bank. **Amman** is the capital and largest city. At least 5,000 years old, Amman is sprinkled with Roman ruins. Petra, another ancient city, was carved completely from a cliff face.

During the early 1900s, the Ottoman Empire ruled this area. After the Ottoman defeat in World War I, the British set up a territory under the Hashemites, a ruling family from the Arabian Peninsula. This territory, later known as Jordan, became independent in 1946. The Hashemite family still rules Jordan today. The country has a **constitutional monarchy.** Elected leaders govern, but a king or queen is the official head of state. From 1952 to 1999, King Hussein (hoo•SAYN) I ruled Jordan. He worked to blend the country's traditions with modern ways of life. The present ruler of Jordan is Hussein's son, King Abdullah (uhb•dul•LAH) II.

Reading Check What are Jordan's leading manufactured goods?

Assessment

Defining Terms

1. **Define** Bedouins, civil war, constitutional monarchy.

Recalling Facts

2. **Location** What landform covers the eastern part of Syria?
3. **Place** What is the capital of Lebanon?
4. **History** What is Petra?

Critical Thinking

5. **Understanding Cause and Effect** How could a dam on the Euphrates River cause a conflict among Turkey, Syria, and Iraq?
6. **Analyzing Information** How and why did Jordan's borders change in 1967?

Graphic Organizer

7. **Organizing Information** Draw a diagram like this one. Inside the large oval, list characteristics that Syria, Lebanon, and Jordan share.

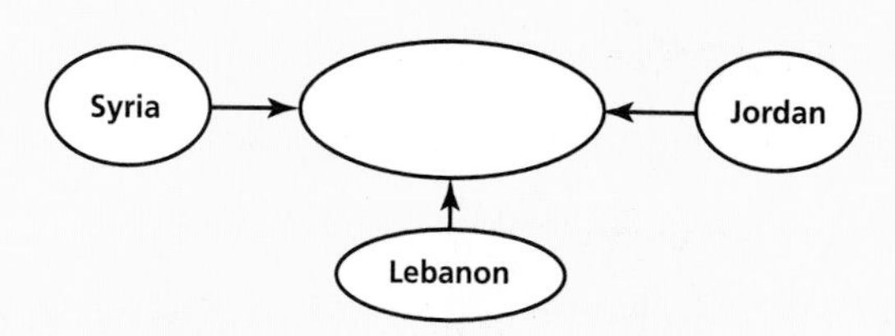

Applying Geography Skills

8. **Analyzing Maps** Study the maps on pages 485 and 479. Jordan's only port, Aqaba, is located on the Gulf of Aqaba. This gulf is part of what larger body of water?

Section 4

The Arabian Peninsula

Guide to Reading

Main Idea

Money from oil exports has boosted standards of living in most countries of the Arabian Peninsula.

Terms to Know

- wadi
- oasis
- desalinization
- hajj

Places to Locate

- Persian Gulf
- Saudi Arabia
- Riyadh
- Makkah
- Kuwait
- Bahrain
- Qatar
- United Arab Emirates
- Oman
- Yemen
- Arabian Sea

Reading Strategy

Create a diagram like this one and give four examples of how oil has benefited the Arabian Peninsula.

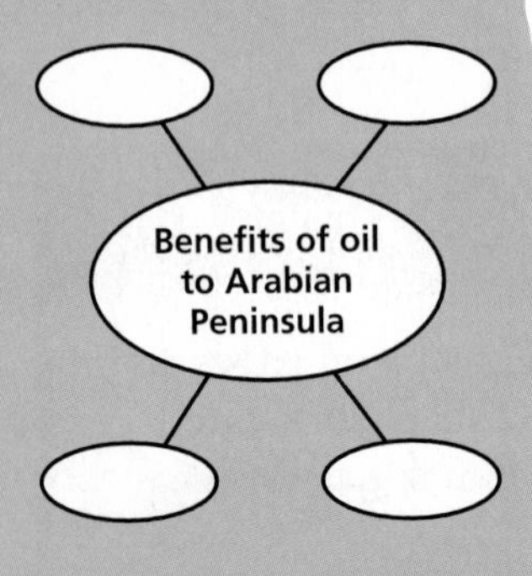

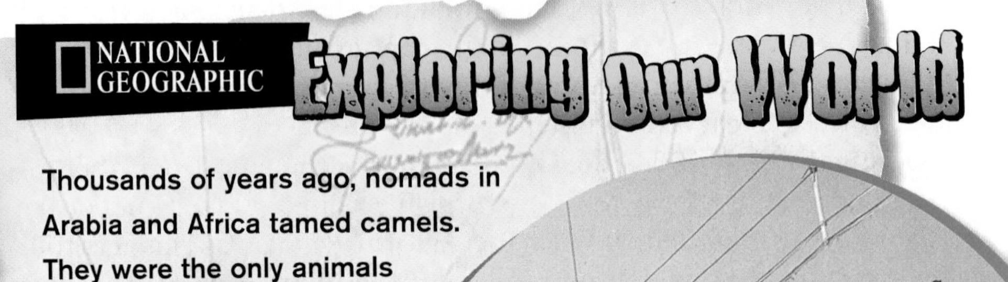

Thousands of years ago, nomads in Arabia and Africa tamed camels. They were the only animals that could make the long journey across the desert, thanks to their ability to go for days without food or water. For centuries, camels were the main source of transport, milk, and meat in the desert. Today they are also valued for their racing speed.

Find the Arabian Peninsula on the map on page 479. Notice that its highest elevations are in the south. The mostly desert land in the north borders Iraq, then it slopes toward the **Persian Gulf.** The country of Saudi Arabia takes up about 80 percent of the quite large Arabian Peninsula.

Saudi Arabia

Saudi Arabia, the largest country in Southwest Asia, is about the size of the eastern half of the United States. Vast deserts cover this region. The largest and harshest desert is the Rub' al Khali, or Empty Quarter, in the southeast. The Empty Quarter has mountains of sand that reach heights of more than 1,000 feet (305 m).

Because of the generally dry, desert climate, Saudi Arabia has no rivers or permanent bodies of water. Highlands dominate the southwest, however, and rainfall there irrigates fertile croplands in the valleys. Water for farming sometimes comes from seasonal **wadis,** or dry riverbeds filled by rainwater from rare downpours. The desert also has **oases**—green areas in the desert fed by underground water.

An Oil-Based Economy Saudi Arabia holds a major share of the world's oil. This entire region is by far the world's leading producer of oil. The graph on page 494 compares the amount of oil reserves in Southwest Asia with those of other regions.

Since 1960 Saudi Arabia and some other oil producers have formed the Organization of Petroleum Exporting Countries (OPEC). Together they work to increase income from the sale of oil. Today OPEC countries supply more than 40 percent of the world's oil. By increasing or reducing supply, they are able to influence world oil prices.

Oil has helped Saudi Arabia boost its standard of living. Money earned by selling oil has built schools, hospitals, roads, and airports. Aware that someday its oil will run out, Saudi Arabia's government has been trying to broaden its economy. In recent years, it has given more emphasis to industry and agriculture. Agriculture in Saudi Arabia has not developed greatly because of the limited amount of water and farmland. To get more water and grow more food, the government of Saudi Arabia has spent much money on irrigation and another process. This process—**desalinization**—takes salt out of seawater.

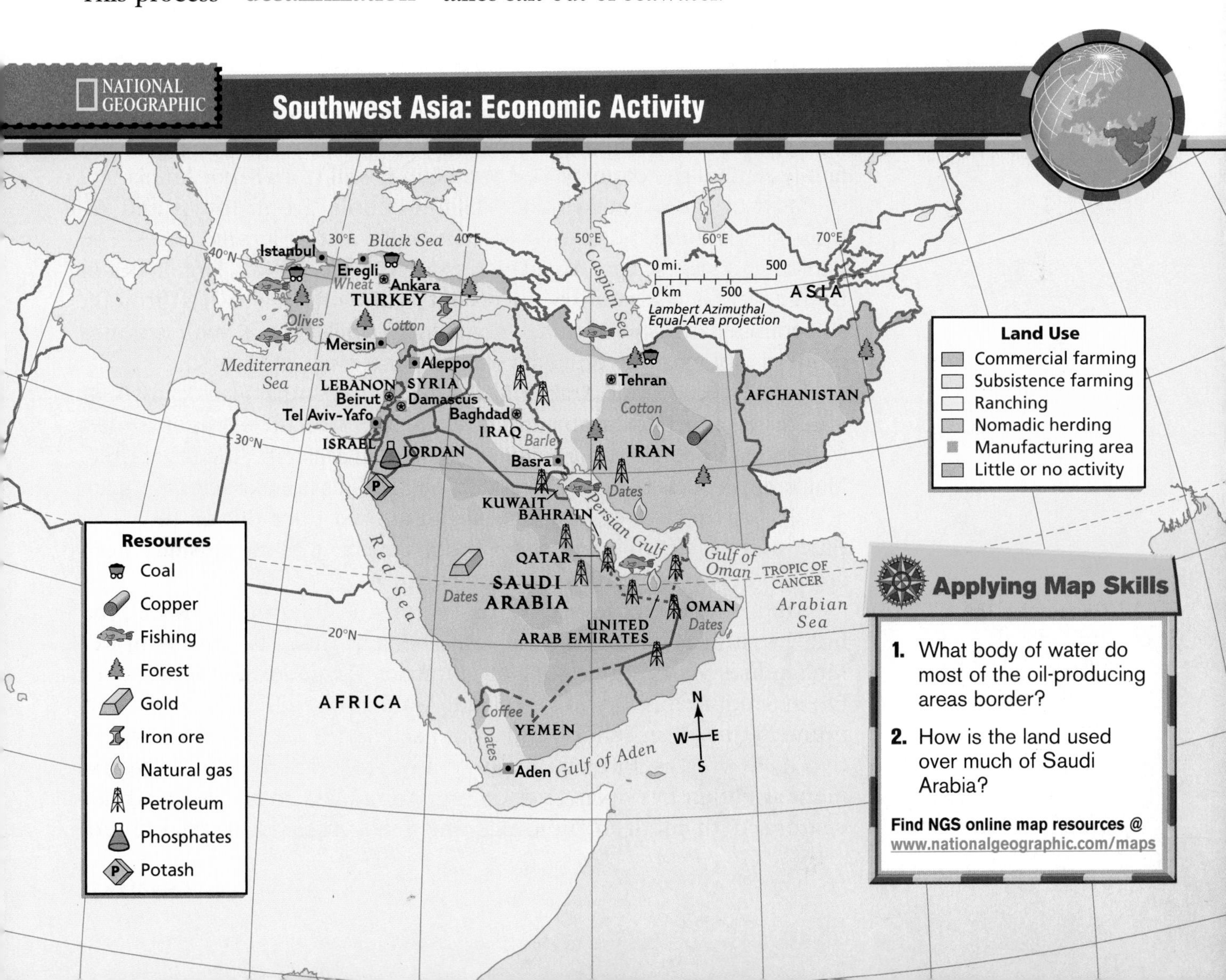

World Oil Production

Analyzing the Graph

Southwest Asia has more known oil than all other regions of the world combined.

Region **What percentage of the world's known oil reserves does Southwest Asia hold?**

Visit gwip.glencoe.com and click on **Chapter 17–Textbook Updates.**

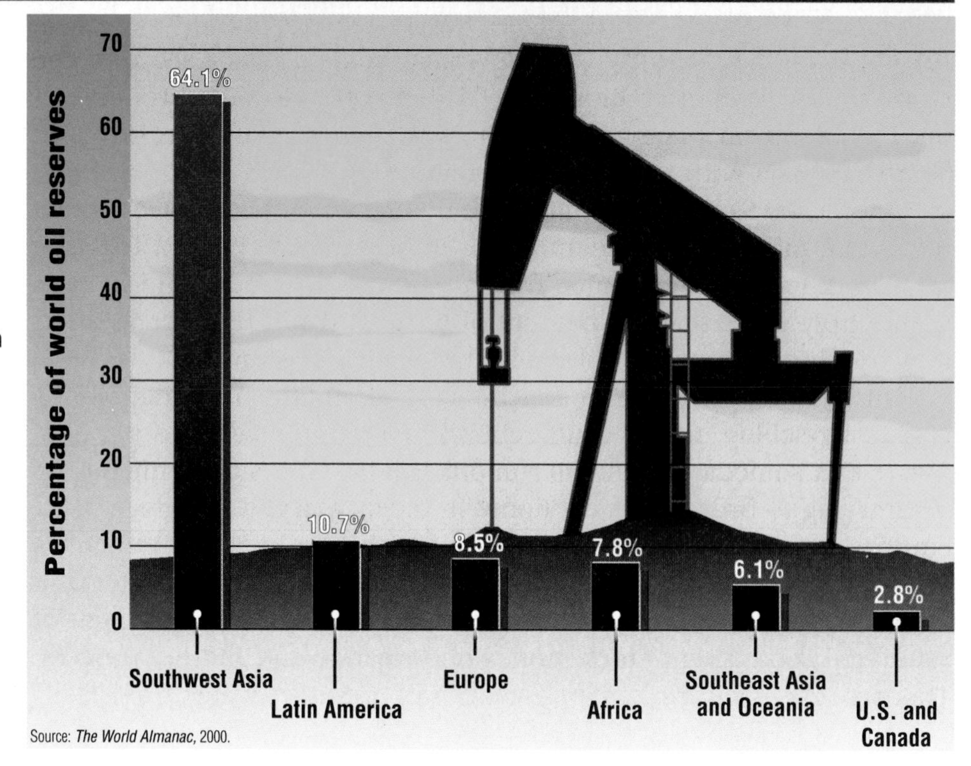

Source: *The World Almanac*, 2000.

History and People The people of Saudi Arabia were once divided into many different family groups. In 1932 a monarchy led by the Saud family unified the country. The Saud family still rules Saudi Arabia.

Most of Saudi Arabia's 20.9 million people live in towns and villages either along the oil-rich Persian Gulf coast or around oases. The capital and largest city, **Riyadh** (ree•YAHD), sits amid a large oasis in the center of the country. In recent years, oil wealth has brought sweeping changes to Riyadh. Once a small rural town, Riyadh is now a modern city with towering skyscrapers and busy highways.

In western Saudi Arabia **Makkah** (MAH•kuh), also known as Mecca, is another important city. In the A.D. 600s, the prophet Muhammad preached the religion of Islam in Makkah. Since that time, Makkah has been Islam's holiest city. All Muslims are expected to make a **hajj,** or religious journey, to Makkah at least once during their lifetime, if they are able to do so. Today several million Muslims from around the world visit Makkah each year.

Saudi Arabia's customs are based on the Quran (kuh•RAN), the holy writings of Islam, and on the example of the prophet Muhammad. As in other Muslim countries, Islam strongly influences life in Saudi Arabia. If you visit Saudi Arabia, you will notice how government, business, school, and home schedules are timed to Islam's five daily prayers and two major yearly celebrations. Much government attention has been given to preparing Makkah and Madinah for visitors, with modern buildings and beautiful architecture. Saudi

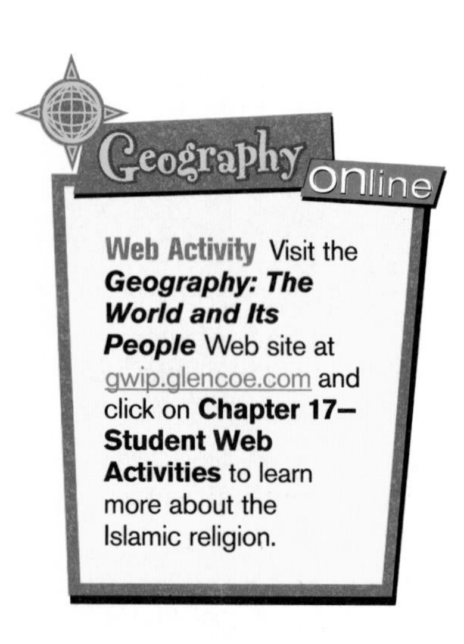

Web Activity Visit the ***Geography: The World and Its People*** Web site at gwip.glencoe.com and click on **Chapter 17–Student Web Activities** to learn more about the Islamic religion.

customs concerning the roles of men and women in public life are stricter than in most other Muslim countries. Some Saudi women do not drive cars. They may work outside the home but only in jobs in which they avoid close contact with men.

Reading Check What influences almost every part of Saudi Arabian culture?

The Persian Gulf States

Kuwait (ku•WAYT), **Bahrain** (bah•RAYN), **Qatar** (KAH•tuhr), and the **United Arab Emirates** are located along the Persian Gulf. Beneath their flat deserts and offshore areas lie vast deposits of oil. The Persian Gulf states have used profits from oil exports to build prosperous economies. Political and business leaders, however, are aware that oil revenues depend on constantly changing world oil prices. As a result, they have encouraged the growth of other industries. Their goal is to build a more varied economy.

The people of the Persian Gulf states once made a living from activities such as pearl diving, fishing, and camel herding. Now they have modern jobs in the oil and natural gas industries. They also enjoy a high standard of living. Using income from oil, their governments provide free education, health care, and other services. Many workers from other countries have settled in these countries to work in the modern cities and oil fields and to benefit from the economic boom.

Reading Check How have the economies of the Persian Gulf states changed since the discovery of oil?

Makkah, Saudi Arabia

The Kaaba, the most sacred Islamic shrine, is in the courtyard of Makkah's Grand Mosque.

Place On what are Saudi Arabia's customs based?

Oman and Yemen

At the southeastern and southern ends of the Arabian Peninsula are the countries of **Oman** and **Yemen.** Oman is largely desert, but its bare land yields oil—the basis of the country's economy. Until recently, most of the people of Oman lived in rural villages. The oil industry has drawn many of these people—and foreigners—to Muscat, the country's capital. Other natural resources include natural gas, copper, marble, and limestone. Agricultural products include dates, bananas, camels, and cattle.

Oman's location also has made the country important to world oil markets. The northern part of Oman guards the strategic Strait of Hormuz. Oil-bearing tankers have to go through this narrow waterway to pass from the Persian Gulf into the **Arabian Sea.**

Southwest of Oman lies Yemen, which is made up of a narrow coastal plain and inland mountains. In ancient times, Yemen was famous for its rich trade in fragrant tree resins such as myrrh (MUHR) and frankincense. Yemen's capital, the walled city of Sanaa (sahn•AH), was once a crossroads for camel caravans that carried goods from as far away as China.

Today Yemen is the only country of the Arabian Peninsula that does not have large deposits of oil. Most of the people are farmers or herd sheep and cattle. They live in the high fertile interior where Sanaa is located. Farther south lies Aden (AH•duhn), a major port for ships traveling between the Arabian Sea and the Red Sea.

Reading Check What makes Yemen different from other countries in the Arabian Peninsula?

Assessment

Defining Terms

1. **Define** wadi, oasis, desalinization, hajj.

Recalling Facts

2. **Place** What is the Empty Quarter?
3. **Government** Who rules Saudi Arabia and what is its form of government?
4. **Culture** What is the significance of the city of Makkah?

Critical Thinking

5. **Analyzing Information** Why is the Strait of Hormuz considered to be of such strategic importance?
6. **Drawing Conclusions** How do the nations of OPEC affect your life?

Graphic Organizer

7. **Organizing Information** On a diagram like this one, compare your life to the life of a woman in Saudi Arabia. Put facts about your daily life and a Saudi woman's daily life in the outer circles. In the overlapping area, put things that you have in common.

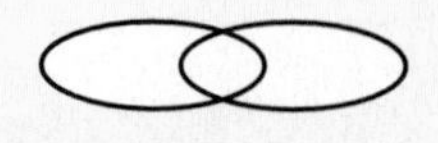

Applying Geography Skills

8. **Analyzing Maps** Study the political map on page 478. What capital city is located on the Tropic of Cancer?

Section 5

Iraq, Iran, and Afghanistan

Guide to Reading

Main Idea

Iraq, Iran, and Afghanistan have recently fought in wars and undergone sweeping political changes.

Terms to Know

- alluvial plain
- embargo
- shah
- Islamic republic

Places to Locate

- Tigris River
- Euphrates River
- Iraq
- Persian Gulf
- Baghdad
- Iran
- Elburz Mountains
- Zagros Mountains
- Caspian Sea
- Tehran
- Afghanistan
- Hindu Kush
- Kabul

Reading Strategy

Create a chart like this one and list the main landforms in each country.

Country	Landforms
Iraq	
Iran	
Afghanistan	

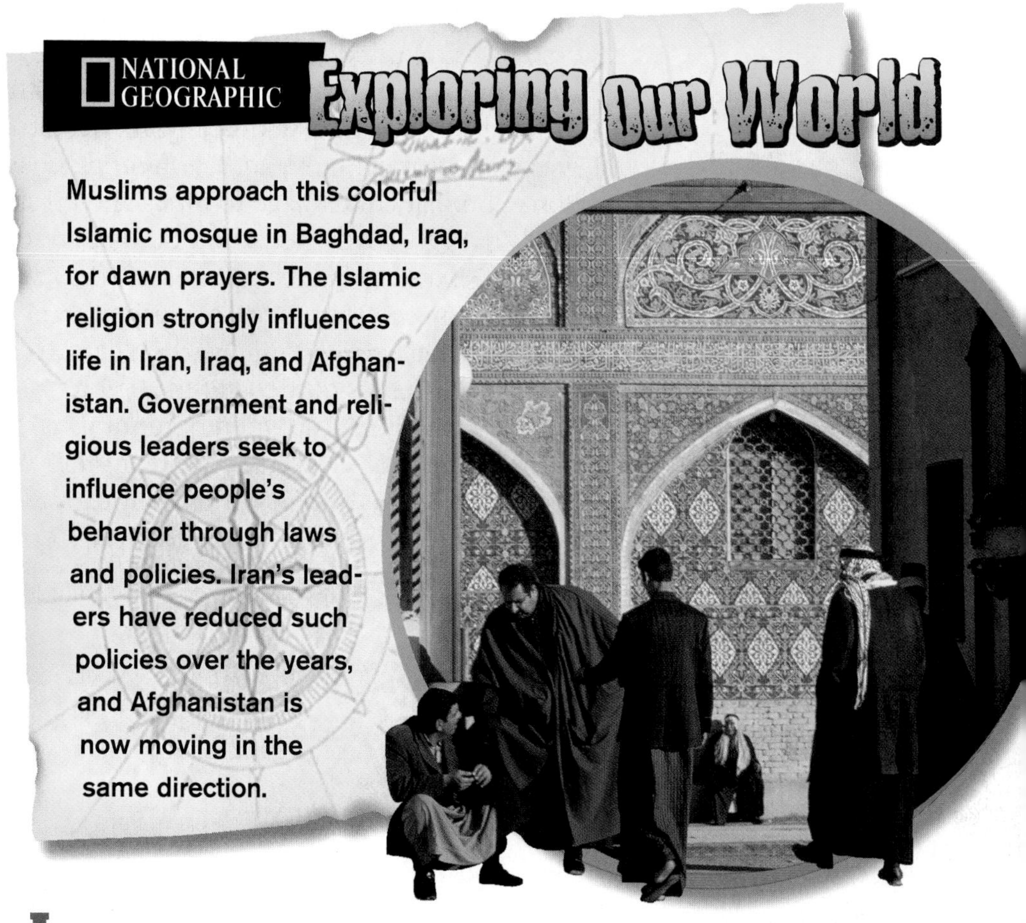

Muslims approach this colorful Islamic mosque in Baghdad, Iraq, for dawn prayers. The Islamic religion strongly influences life in Iran, Iraq, and Afghanistan. Government and religious leaders seek to influence people's behavior through laws and policies. Iran's leaders have reduced such policies over the years, and Afghanistan is now moving in the same direction.

Iraq, Iran, and Afghanistan are located in a region where some of the world's oldest civilizations developed. This region has experienced turmoil throughout history and even today.

Iraq

More than 4,000 years ago, a people known as the Sumerians built the world's first known cities between the **Tigris** and **Euphrates Rivers.** These rivers are the major geographic features of **Iraq.** They flow through Iraq's northern highlands before entering the **Persian Gulf.** Between the two rivers is an alluvial plain—an area that is built up by rich fertile soil left by river floods. Most farming takes place here. Farmers grow wheat, barley, dates, cotton, and rice.

Oil is the country's major export. Manufacturing has also developed in the past few years. Iraq's factories process foods and make textiles, chemicals, and construction materials.

The Iraqi People About 70 percent of Iraq's 22.5 million people live in urban areas. **Baghdad,** the capital, is the largest city. From the A.D. 700s to 1200s, Baghdad was the center of a large Muslim empire that made many advances in the arts and sciences. Muslim Arabs make up the largest group in Iraq's population. The second-largest group consists of another Muslim people, the Kurds. As you learned in Section 1, the Kurds want to form their own country.

Modern Iraq gained its independence as a kingdom in 1932. In 1958 the last king was overthrown in a revolt. Since then, military leaders have governed Iraq as a dictatorship. The current leader, Saddam Hussein (sah•DAHM hoo•SAYN), rules with an iron hand.

In the 1980s, Iraq, with aid from Western countries, fought a bloody war with its neighbor Iran. The fighting cost thousands of lives and billions of dollars in damage to cities and oil-shipping ports in the Persian Gulf. In 1990, partly because of a dispute over oil, Iraq invaded neighboring Kuwait. A year later, in the Persian Gulf War, a United Nations force led by the United States pushed Iraqi troops out of Kuwait.

After the Persian Gulf War, Saddam Hussein refused to cooperate with the demands of the United Nations. In response, the United States and other nations put an embargo on trade with Iraq. An **embargo** is an order that restricts trade with another country. Since then, Iraq has not exported as much oil as before and could not import certain goods. This has severely damaged Iraq's economy and cost many lives.

Reading Check **What two rivers have influenced the history of Iraq?**

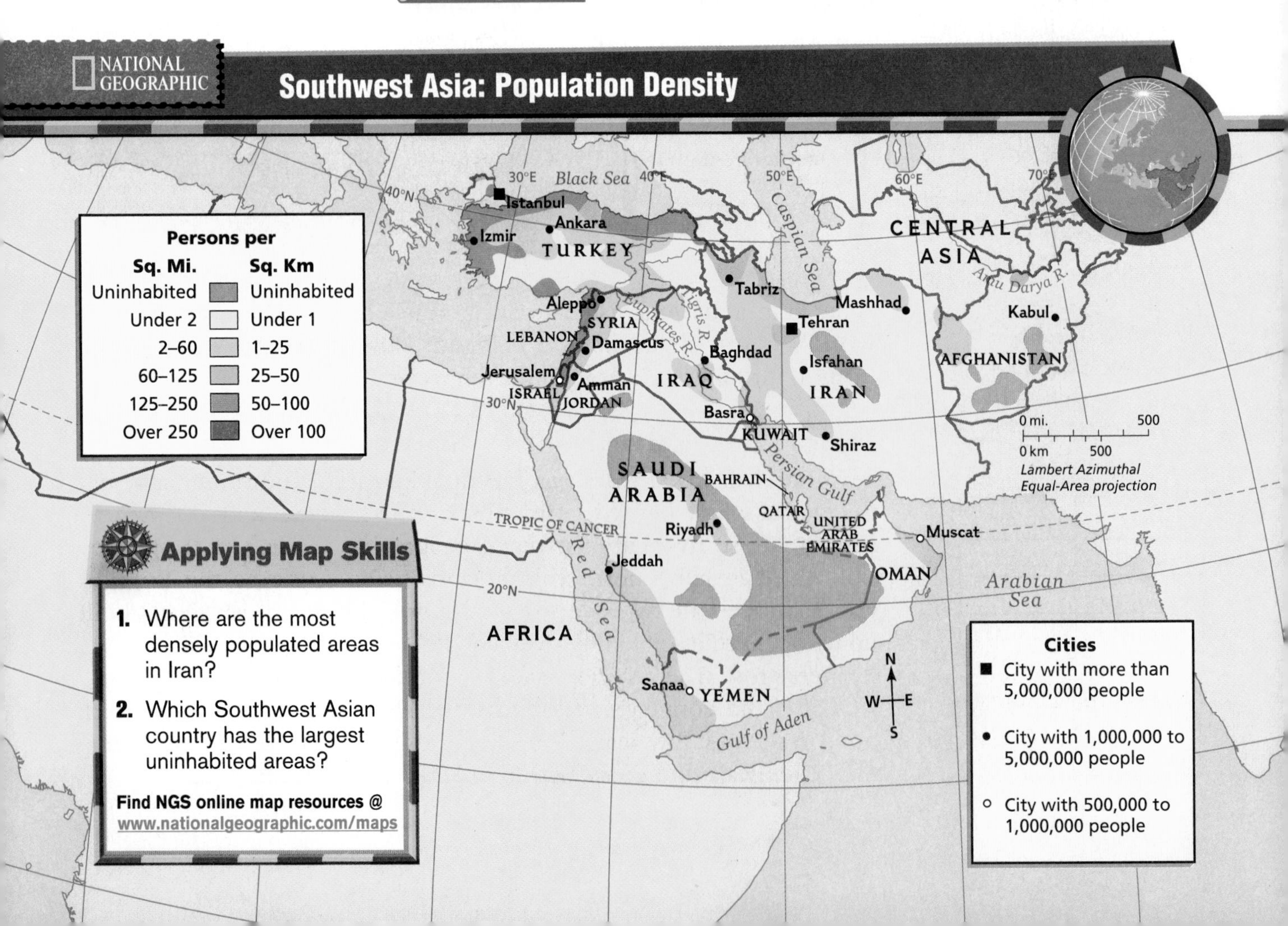

Iran and Afghanistan

An Iranian family picnics in the hills above Tehran (left). Under the Taliban, women in Afghanistan were rarely allowed in public (above).

Place **What form of government does Iran have? What group led Afghanistan in the 1990s?**

Iran

Once known as Persia, **Iran** is about the size of Alaska. In the center of the country you find a high plateau of deserts and salt flats. Two vast ranges—the **Elburz Mountains** in the north and the **Zagros Mountains** in the west—surround the plateau. Coastal plains wind along the **Caspian Sea** in the north and the Persian Gulf in the south.

Iran is an oil-rich nation where the first oil wells in Southwest Asia were drilled in 1908. The country is trying to promote other industries to make Iran depend less on oil earnings. Major Iranian industries produce textiles, metal goods, and construction materials. The beautiful carpets woven in Iran are valued around the world.

With limited supplies of water, less than 12 percent of Iran's land can be farmed. Some farmers use ancient underground channels to bring water to their fields. They grow wheat, rice, sugar beets, and cotton. Iran is also the world's largest producer of pistachio nuts.

The Iranian People Iran's 66.2 million people differ from those of other Southwest Asian countries. More than one-half are Persians, not Arabs or Turks. The Persians' ancestors migrated from Central Asia centuries ago. They speak Farsi, or Persian, the official language of Iran. Other languages include Kurdish, Arabic, and Turkish. About 60 percent of Iranians live in urban areas. **Tehran,** located in northern Iran, is the largest city and the capital. Iran is also home to about 2 million people from Iraq and Afghanistan who have fled recent wars. Nearly 98 percent of Iran's people practice some form of Islam.

About 2,000 years ago, Iran was the center of the powerful Persian Empire ruled by kings known as **shahs.** In 1979 Muslim religious leaders led a movement that successfully overthrew the last monarchy. Iran has an **Islamic republic,** a government run by Muslim religious leaders.

The government has introduced laws based on its understanding of the Quran, the Muslim sacred writings. Many Western customs seen as a threat to Islam are now forbidden.

✓Reading Check How do Iranians differ from most other Southwest Asians?

Afghanistan

A landlocked nation, **Afghanistan** (af•GA•nuh•STAN) is mostly covered with the rugged peaks of the **Hindu Kush** mountain range. The Khyber (KY•buhr) Pass cuts through the Hindu Kush. For centuries, this passageway has been a major trade route linking Southwest Asia with other parts of Asia. Grasslands and deserts make up the other areas.

Almost 70 percent of the people farm. They grow wheat, fruits, and nuts as well as herd sheep and goats. Afghanistan is rich in minerals but has little industry. A few mills produce textiles, and skilled craftspeople produce beautiful jewelry and handwoven carpets.

Most Afghans live in the fertile valleys of the Hindu Kush. The capital city, **Kabul** (KAH•buhl), lies in one of these valleys. The country's 25.8 million people are divided into about 20 different ethnic groups. The two largest groups are the Pashtuns (PUHSH•TOONS) and the Tajiks (tah•JIHKS).

For centuries, foreign invaders and rival Afghan groups have fought over Afghanistan. During the 1990s, a group known as the Taliban ruled the country harshly. In 2001, U.S.-led forces overthrew the Taliban in a war against terrorists in the country. **Terrorists** are people who illegally use violence against people or property to make a point.

✓Reading Check What do most Afghans do for a living?

Assessment

Defining Terms

1. **Define** alluvial plain, embargo, shah, Islamic republic.

Recalling Facts

2. **Economics** What is Iraq's major export?
3. **Place** Describe the physical features of Iran.
4. **History** What has been the significance of the Khyber Pass?

Critical Thinking

5. **Understanding Cause and Effect** Why have Iraq and the United States continued to treat each other with hostility?
6. **Drawing Conclusions** Why would Iran's leaders see Western customs as a threat?

Graphic Organizer

7. **Organizing Information** Draw a chart like this one, then write one fact about Afghanistan under each heading.

Afghanistan		
Capital	Landforms	Agriculture
Crafts	People	Government

Applying Geography Skills

8. **Analyzing Maps** Study the physical map on page 479. Between what two bodies of water is Iran located?

Reading Review

Section 1 Turkey

Terms to Know
- mosque
- migrate
- secular

Main Idea

Rapidly modernizing Turkey is a link between Asia and Europe.

✓ Location Turkey lies in both Europe and Asia.

✓ Economics Turkey is becoming more industrialized, with textiles and clothing as major industries.

✓ Culture Most of Turkey's people now live in cities or towns.

Section 2 Israel

Terms to Know
- kibbutz
- moshav
- Diaspora
- Holocaust
- monotheism

Main Idea

After years of conflict, the Jewish nation of Israel and neighboring Arab countries are trying to achieve peace.

✓ Culture About 80 percent of Israel's population are Jews. They have moved to Israel from many countries.

✓ History Israel and its Arab neighbors continue to work toward a peaceful settlement of issues that divide them.

Section 3 Syria, Lebanon, and Jordan

Terms to Know
- Bedouins
- civil war
- constitutional monarchy

Main Idea

Syria, Lebanon, and Jordan have largely Arab populations but have different economies and forms of government.

✓ Economics Farming is the main economic activity in Syria.

✓ History Lebanon is rebuilding and recovering after a civil war.

✓ Place Water shortages in Jordan restrict the land available for farming.

Section 4 The Arabian Peninsula

Terms to Know
- wadi
- oasis
- desalinization
- hajj

Main Idea

Money from oil exports has boosted standards of living in most countries of the Arabian Peninsula.

✓ Economics Saudi Arabia is the world's leading oil producer.

✓ Culture The Islamic religion affects almost all aspects of life in Saudi Arabia.

✓ Economics The Persian Gulf states have strong economies based on oil.

Section 5 Iraq, Iran, and Afghanistan

Terms to Know
- alluvial plain
- embargo
- shah
- Islamic republic

Main Idea

Iraq, Iran, and Afghanistan have recently fought in wars and undergone sweeping political changes.

✓ Economics Iraq is suffering because of an international trade embargo.

✓ Culture Oil-rich Iran is ruled by Muslim religious leaders.

✓ Place Afghanistan is mountainous and relatively undeveloped.

Chapter 17

Assessment and Activities

Using Key Terms

Match the terms in Part A with their definitions in Part B.

A.

1. monotheism
2. desalinization
3. alluvial plain
4. mosque
5. hajj
6. Bedouins
7. Holocaust
8. secular
9. embargo
10. Diaspora

B.

a. taking salt out of seawater
b. scattering of a people
c. mass slaughter of European Jews
d. nomadic, desert people
e. belief in one god
f. plain built up from soil deposited by a river
g. nonreligious
h. restriction on trade
i. Islamic place of worship
j. religious journey to Makkah

Reviewing the Main Ideas

Section 1 Turkey

11. **Place** What is Turkey's largest city?
12. **Economics** Name five of Turkey's products.

Section 2 Israel

13. **History** When was the modern nation of Israel created?
14. **Government** What is the Law of Return?

Section 3 Syria, Lebanon, and Jordan

15. **Place** What makes Damascus an important city?
16. **History** What area of land did Jordan lose to Israel during the 1967 war?

Section 4 The Arabian Peninsula

17. **Economics** What is OPEC?
18. **Place** What is the capital of Saudi Arabia?

Section 5 Iraq, Iran, and Afghanistan

19. **Economics** Where does most of the farming in Iraq take place?
20. **Culture** How do the people of Iran differ from other Southwest Asian peoples?
21. **Place** What landform makes up Afghanistan?

Southwest Asia

Place Location Activity

On a separate sheet of paper, match the letters on the map with the numbered places listed below.

1. Persian Gulf
2. Zagros Mountains
3. Euphrates River
4. Turkey
5. Iran
6. Israel
7. Iraq
8. Saudi Arabia
9. Makkah
10. Jerusalem

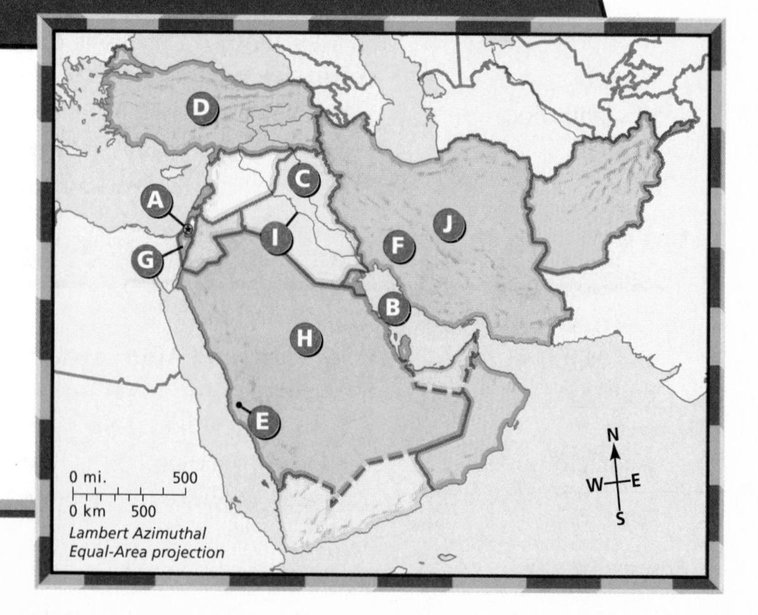

Self-Check Quiz Visit the ***Geography: The World and Its People*** Web site at gwip.glencoe.com and click on **Chapter 17–Self-Check Quizzes** to prepare for the Chapter Test.

Critical Thinking

22. **Evaluating Information** Goods were moved from Southwest Asia to other parts of the world through several routes. How are goods brought to your community? Make a list of all the routes a product would take to get from Southwest Asia to your town.
23. **Analyzing Information** On a chart like this, list a reason for the importance of oil and water to Southwest Asia and one result of their abundance or scarcity.

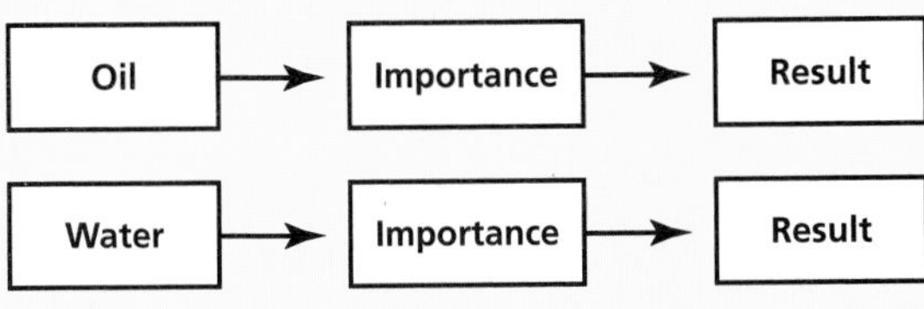

GeoJournal Activity

24. **Writing a Report** Jerusalem is a city holy to three of the world's major religions. Research these religions and identify at least three holy sites you might visit to learn more about them.

Mental Mapping Activity

25. **Focusing on the Region** Draw a simple outline map of Southwest Asia, then label the following:
    - Turkey
    - Persian Gulf
    - Red Sea
    - Israel
    - Mediterranean Sea
    - Iran
    - Saudi Arabia
    - Yemen
    - Iraq
    - Afghanistan

Technology Skills Activity

26. **Using the Internet** Search the Internet and find several newspapers that publish current events online. Research an event that took place in one of the countries of Southwest Asia. Create a poster that will inform your class about the event you researched.

Standardized Test Practice

Directions: Study the graph, then answer the question that follows.

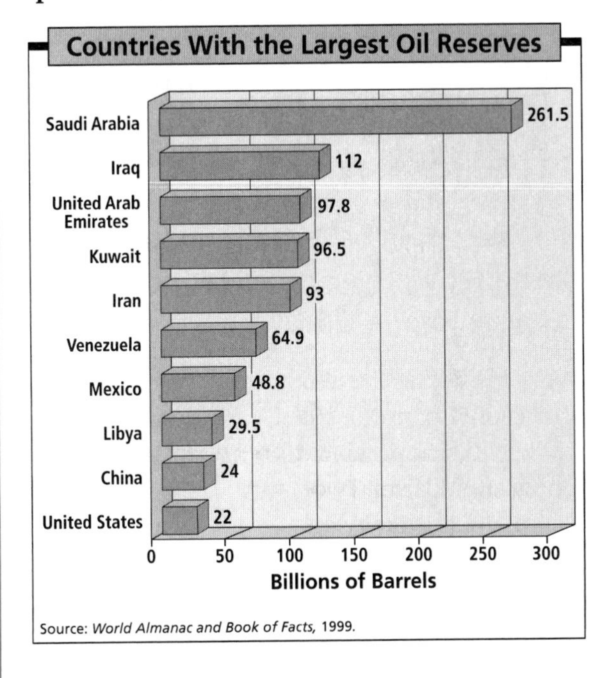

1. **How many of the 10 countries with the largest oil reserves are located in Southwest Asia?**
   A one
   B three
   C five
   D seven

Test-Taking Tip: You need to rely on your memory as well as analyze the graph to answer this question. Look at each country, then think back to the countries you studied in Chapter 17. Which of those listed on the graph did you just learn about?

Chapter 18

The Caucasus and Central Asia

The World and Its People NATIONAL GEOGRAPHIC

To learn more about the people and places of the Caucasus and Central Asia, view ***The World and Its People*** **Chapter 18** video.

Geography Online

Chapter Overview Visit the ***Geography: The World and Its People*** Web site at gwip.glencoe.com and click on **Chapter 18–Chapter Overviews** to preview information about the countries of the Caucasus and Central Asia.

Section 1

Republics of the Caucasus

Guide to Reading

Main Idea

The Caucasus countries have developed their own cultures despite long periods of rule by powerful neighbors.

Terms to Know

- fault
- landlocked
- enclave
- cash crop

Places to Locate

- Armenia
- Georgia
- Azerbaijan
- Caspian Sea
- Black Sea
- Caucasus Mountains
- Yerevan
- T'bilisi
- Baku

Reading Strategy

Make a diagram like this one. Complete the diagram by writing facts about each country's people under the headings.

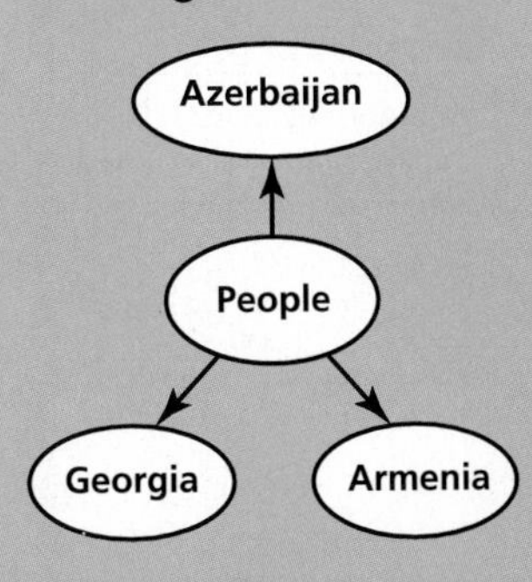

NATIONAL GEOGRAPHIC **Exploring Our World**

Look at the flag of Azerbaijan shown on page 452. Notice that a crescent—a symbol of Islam—appears on the flag. When the Soviet Union invaded Azerbaijan in 1922, most of the country's mosques, or Muslim houses of worship, were forcibly closed. With the Soviet breakup in 1991, mosques like the one shown here have reopened.

The republics of the Caucasus are **Armenia, Georgia,** and **Azerbaijan** (A•zuhr•by•JAHN). They all lie south of Russia between the **Caspian** and **Black Seas.** The towering **Caucasus Mountains,** which give the region its name, run through all three countries. Mt. Kazbek, in Georgia, stands 15,386 feet (4,690 m) high.

The Caucasus Mountains have not been a barrier to the many armies, traders, and settlers who have swept through its valleys over the centuries. Arabs, Turks, Persians, and Russians all have ruled this region at one time or another. Many of these people stayed, making up the different groups living here today. Disagreements among some of these groups have sparked violent conflicts across the region.

The Caucasus republics once were part of the Soviet Union. When the Soviet Union collapsed in 1991, the republics became independent for the first time in centuries. Since then, the region has struggled to move to a free market economy.

◀ Oygaing River and Tian Shan mountain range in Kazakhstan

Armenia

Armenia, about the size of Maryland, sits uneasily on top of many **faults,** or cracks in the earth's crust. It often suffers serious earthquakes. In a 1988 earthquake, about 25,000 people died and another 500,000 lost their homes.

Armenia's 3.8 million people are mostly ethnic Armenians who share a unique language and culture. Nearly 70 percent of the people live in cities. Founded in 782 B.C., **Yerevan** (YEHR•uh•VAHN), the capital, is one of the world's most ancient cities. Armenians are proud of its wide streets, attractive fountains, and colorful buildings made of volcanic stone. An Armenian king made Christianity the official religion in A.D. 301—the first country to do so. About 94 percent of the nation's people belong to the Armenian Orthodox Church.

About 2,000 years ago, Armenia ruled a large empire from the Caspian Sea to the Mediterranean Sea. Today Armenia is not only

Applying Map Skills

1. What country is split into two parts by Armenia?
2. What is the capital of Georgia?

Find NGS online map resources @ www.nationalgeographic.com/maps

smaller but also **landlocked.** It has no land touching a sea or an ocean. One group of Armenians lives in an enclave in neighboring Azerbaijan. An **enclave** is a small territory entirely surrounded by another territory. Cut off from their homeland, these Armenians wish to rejoin Armenia. In the 1990s, newly independent Armenia and Azerbaijan began to fight over the enclave. The shooting has stopped, but the dispute continues.

Armenia paid a high price during the war with Azerbaijan. Officials in Azerbaijan halted the flow of fuel and other resources across their country to landlocked Armenia. This action seriously hurt Armenia's economy and environment. Without any fuel, Armenians cut down trees for firewood. This has led to a massive loss of forests. The drop in industry made agriculture more important to Armenia's economy. Farmers grow wheat, fruits, and vegetables in river valleys. Herders tend sheep and goats that feed on grassy mountain slopes.

Reading Check **How has being landlocked recently affected Armenia?**

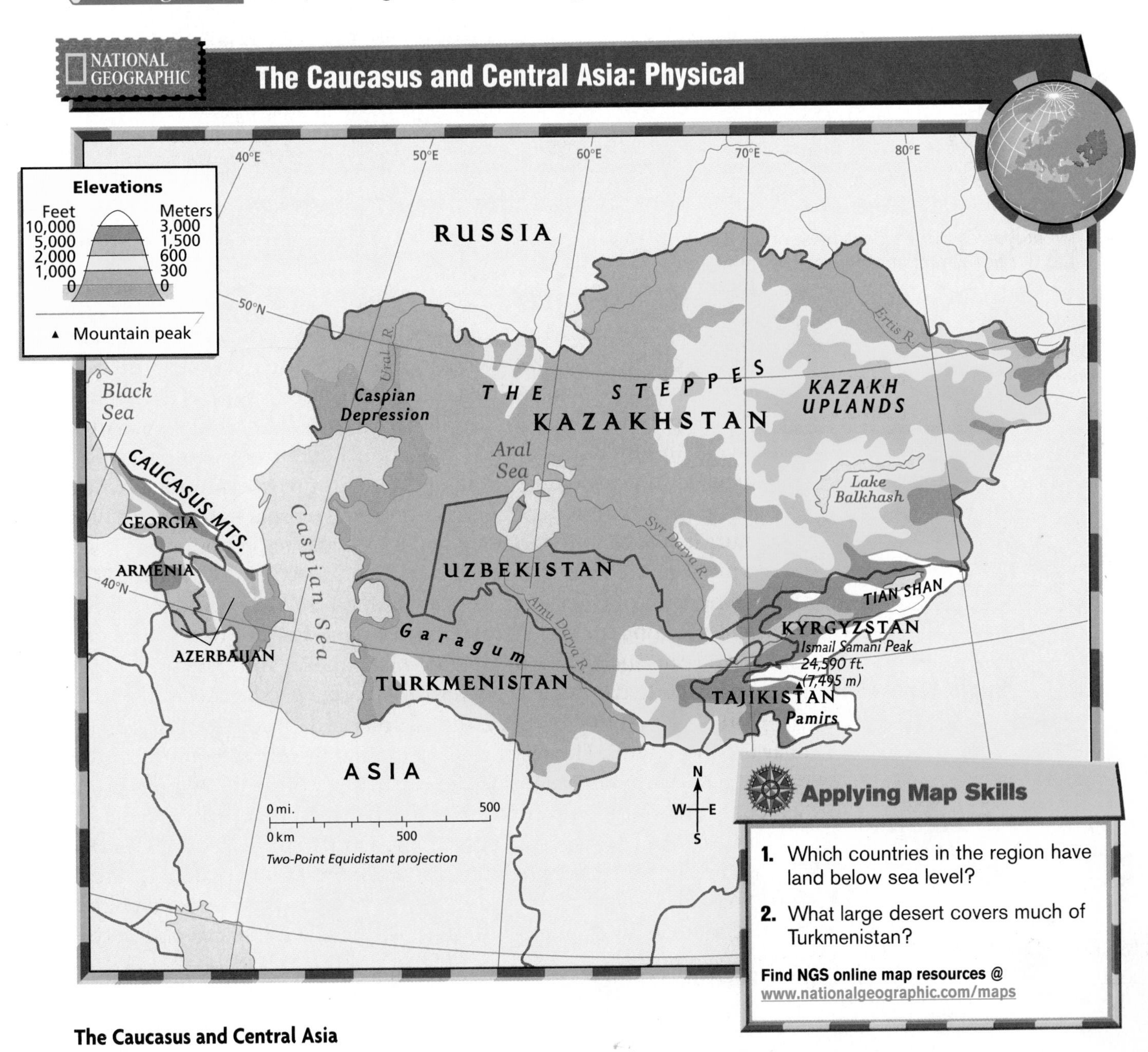

Applying Map Skills

1. Which countries in the region have land below sea level?
2. What large desert covers much of Turkmenistan?

Find NGS online map resources @ www.nationalgeographic.com/maps

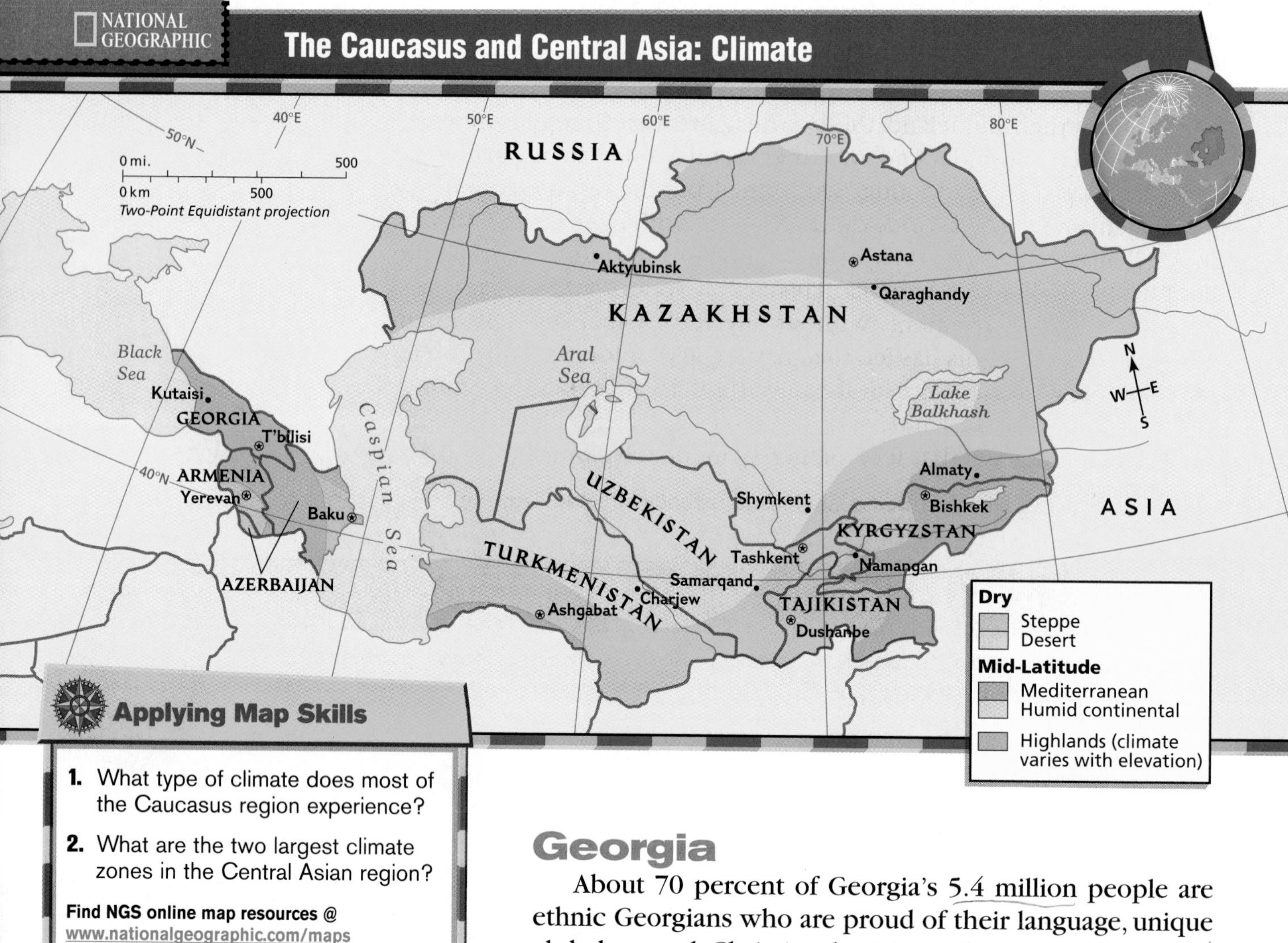

Applying Map Skills

1. What type of climate does most of the Caucasus region experience?
2. What are the two largest climate zones in the Central Asian region?

Find NGS online map resources @ www.nationalgeographic.com/maps

Georgia

About 70 percent of Georgia's 5.4 million people are ethnic Georgians who are proud of their language, unique alphabet, and Christian heritage. The rest of Georgia's people belong to a number of other ethnic groups. After independence in 1991, conflict broke out between Georgians and other ethnic groups. These groups wanted to separate and set up their own independent countries. Georgia's ethnic troubles have hurt its move to democracy and a free market economy.

Farmers in Georgia are few—but they are very productive. Less than 10 percent of Georgians work in agriculture. Farm products, however, make up one-third of all the country's goods. Georgia also has natural resources, such as copper, coal, manganese, and some oil. Swift rivers provide hydroelectric power for Georgia's industries.

T'bilisi (tuh•bih•LEE•see), the capital and largest city, is located near the mountains. The city has warm mineral springs—water heated by the high temperatures inside the earth. Resorts along the mild Black Sea coast draw thousands of tourists each year.

Like Armenia, Georgia accepted Christianity in the A.D. 300s. About 65 percent of Georgians belong to the Georgian Orthodox Church. More than half live in cities. The Georgians are known as skilled cooks. Each region has its own types of food flavored with spices and herbs.

Reading Check What has hurt Georgia's attempts to become a democracy?

Azerbaijan

Azerbaijan is split in two by Armenian territory. Most people belong to a group called Azeris. They speak the Azeri language, which is related to Turkish. They also follow the religion of Islam.

About one-third of Azerbaijan's people work in agriculture. Farmers in dry steppe areas use irrigation to grow cotton and tobacco as **cash crops,** or products grown for sale as exports. The humid climate near the Caspian shore helps farmers grow citrus fruits, grapes, and tea. Oil and natural gas deposits under the Caspian Sea are the most promising part of Azerbaijan's economy. The country has made agreements with foreign companies to develop these resources.

More than half of Azerbaijan's 7.7 million people live in cities. The capital, **Baku** (bah•KOO), is a port on the Caspian Sea. The center of the country's oil industry and manufacturing, Baku is known for the strong winds that blow through the city. In fact, its name comes from the Persian word for "windy town."

Azerbaijan's culture is a blend of Southwest Asian and European influences. The people of Azerbaijan like to eat richly spiced lamb and chicken along with rice and vegetables. The Caspian Sea yields catches of sturgeon, a fish whose eggs are prized as caviar. Craftspeople in Azerbaijan are known for their embroidered textiles. Weavers use colorful threads to create ornate patterns on wool fabric.

Reading Check **How does Azerbaijan's religion differ from that in Armenia and Georgia?**

Assessment

Defining Terms

1. **Define** fault, landlocked, enclave, cash crop.

Recalling Facts

2. **Culture** What is the official religion of Armenia?
3. **Economics** Name four of Georgia's natural resources.
4. **Place** What is the most promising part of Azerbaijan's economy?

Critical Thinking

5. **Making Comparisons** What common characteristics make the countries of Armenia, Georgia, and Azerbaijan a region?
6. **Making Predictions** How might the loss of so many trees affect the country of Armenia?

Graphic Organizer

7. **Organizing Information** Make a diagram like the one below. Then fill in details for each category for the Caucasus republics in the ovals.

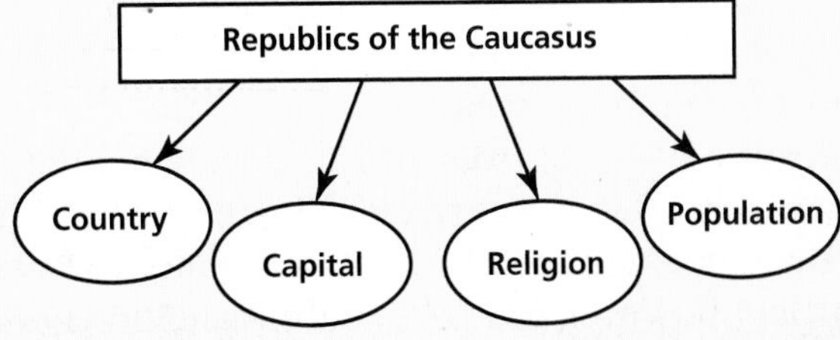

Applying Geography Skills

8. **Analyzing Maps** Study the political map on page 510. What capital city is located on the Caspian Sea? Now turn to the physical map on page 511. At what elevation does this capital city lie?

Making Connections

ART | SCIENCE | LITERATURE | TECHNOLOGY

Carpet Weaving

For thousands of years, people have been making the hand-knotted floor coverings sometimes called Oriental rugs. Valued for their rich color and intricate design, these handmade rugs are unique works of art.

History

Most experts think that the nomadic peoples of Central Asia were among the first to make hand-knotted carpets. They used their carpets as wall coverings, curtains, and saddlebags, as well as to cover the bare ground in their tents. The soft, thick rugs blocked out the cold and could also be used as a bed or blanket.

As the nomads moved from place to place, they spread the art of carpet-making to new lands and peoples. Throughout the years, the greatest carpet-producing areas have included Turkey, the republics of the Caucasus, Persia, and Turkmenistan. People in other countries, including Afghanistan, Pakistan, Nepal, India, and China, also became skilled carpet weavers.

Weaving and Knotting

Early nomads wove their carpets from sheep's wool on simple wooden looms that could be rolled up for traveling. Each carpet was woven with two sets of threads. The *warp* threads run from top to bottom, and the *weft* threads are woven from side to side. Hand-tied knots form the carpet's colorful pattern. A skillful weaver can tie about 15 knots a minute. The best carpets, however, can have more than 500 knots per square inch!

Color and Design

The beauty of woven carpets comes from the endless combination of colors and designs. Over the years, various regions developed their own carpet patterns. These were passed down from generation to generation. Often the images hold special meanings. For instance, the palm and coconut often symbolize happiness and blessings.

▲ Women weave silk carpets in Uzbekistan.

The very first rugs were colored gray, white, brown, or black—the natural color of the wool. Then people learned to make dyes from plants and animals. The root of the madder plant, as well as certain insects, provided red and pink dye. Turmeric root and saffron supplied shades of yellow, while the indigo plant provided blue.

Making the Connection

1. How did the art of carpet weaving spread from one place to another?
2. What creates the pattern in an Oriental carpet?
3. **Drawing Conclusions** In what way do hand-knotted carpets combine art with usefulness?

Section 2

Central Asian Republics

Guide to Reading

Main Idea

The Muslim countries of Central Asia are trying to build new economies and governments.

Terms to Know

- steppe
- delta
- nomad
- elevation
- clan
- bilingual
- oasis

Places to Locate

- Kazakhstan
- Kyrgyzstan
- Tajikistan
- Uzbekistan
- Turkmenistan
- Aral Sea
- Garagum

Reading Strategy

In a chart like this, write two facts about each country in Central Asia.

Country	Facts
Kazakhstan	
Kyrgyzstan	
Tajikistan	
Uzbekistan	
Turkmenistan	

NATIONAL GEOGRAPHIC **Exploring Our World**

For centuries, the city of Bukhara was a stop along an ancient trading route called the Silk Road, which stretched from China to Europe. In the past, precious Chinese silk was carried on the backs of camels. Silk is still sold in Bukhara's markets. Today, however, international trade occurs through a system of fiber-optic cables laid along the ancient tracks of the Silk Road camels.

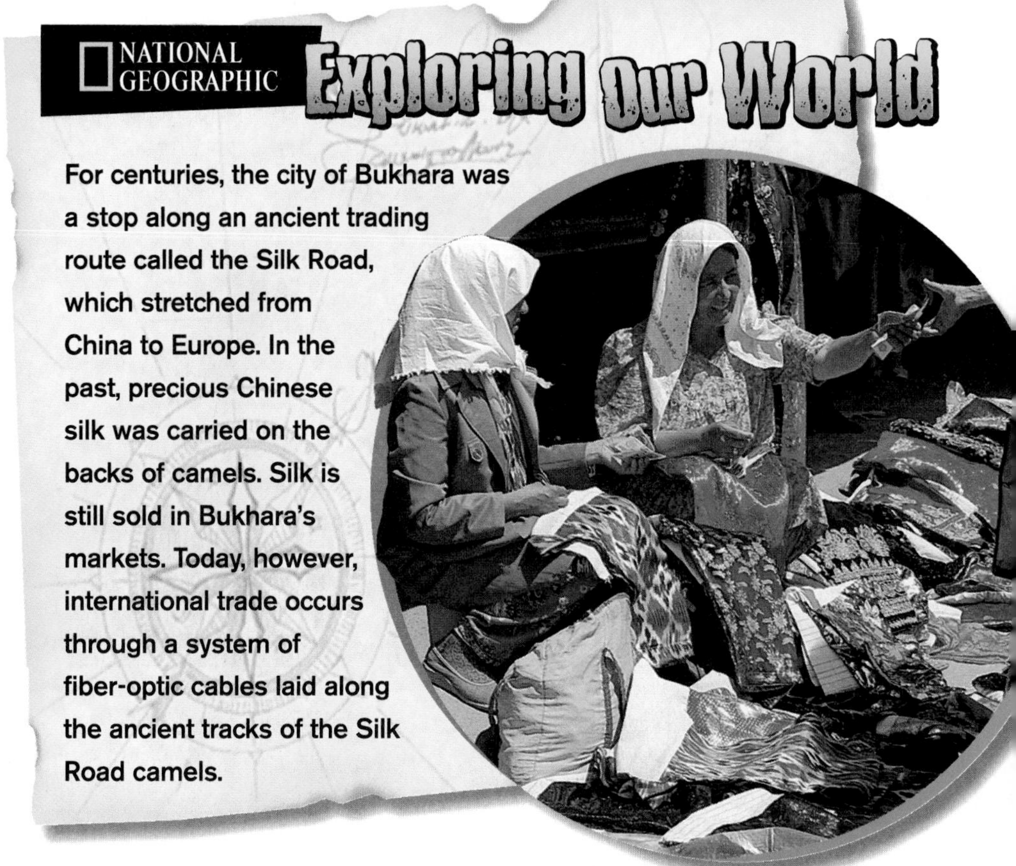

Five countries make up the Central Asian republics. They are **Kazakhstan** (kuh•ZAHK•STAHN), **Kyrgyzstan** (KIHR•gih•STAN), **Tajikistan** (tah•JIH•kih•STAN), **Uzbekistan** (UZ•BEH•kih•STAN), and **Turkmenistan** (tuhrk•MEH•nuh•STAN). They were part of the Soviet Union. When that nation fell apart in 1991, the republics became independent. In addition, these republics all follow the Islamic religion.

Kazakhstan

Almost four times the size of Texas, Kazakhstan dominates the northern part of Central Asia. The map on page 511 shows you that the Kazakh Uplands make up the eastern part. In this rugged area, you will find the unusual Lake Balkhash. It is so long that there is salt water in its eastern half and freshwater in its western half. Toward the center of Kazakhstan lie the Steppes. A **steppe** is a dry, treeless plain often found on the edges of a desert. In the west, the Caspian Depression lies

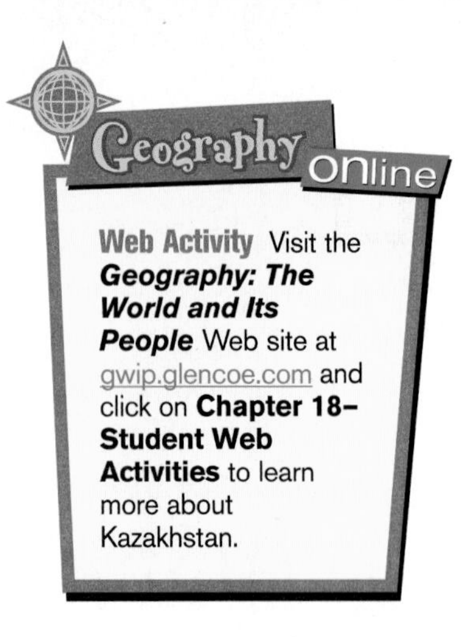

below sea level. Part of this region is the delta of the Ural River. A **delta** is the area where soil has been deposited by a river at its mouth. Two other rivers are important to Kazakhstan. The Ertis (ehr•TIHS) River, called the Irtysh when it flows through Russia, provides water for farmland in the northeast. The Syr Darya (sihr duhr•YAH) River flows across south-central deserts to reach the **Aral Sea.**

Kazakhstan's mineral resources include copper, manganese, gold, zinc, and petroleum. Kazakhstan's factories are very productive. They make machinery and chemicals and process foods. Farming is difficult in the harsh climate, but raising livestock on ranches is an important industry.

Many of Kazakhstan's people are ethnic Kazakhs, whose ancestors were horse-riding warriors called the Mongols. Like the Mongols, the Kazakhs were mostly **nomads,** or people who move from place to place with herds of animals. Under Soviet Communist rule, farms were placed under government control. Kazakh nomads were forced to settle in one place. During this time—in the 1930s—about 1 million Kazakhs died of starvation.

After World War II, the Soviet Union rapidly set up factories, and Russian workers poured into the country. Even today, Kazakhs make up less than one-half of the country's people. Russians form the second-largest group.

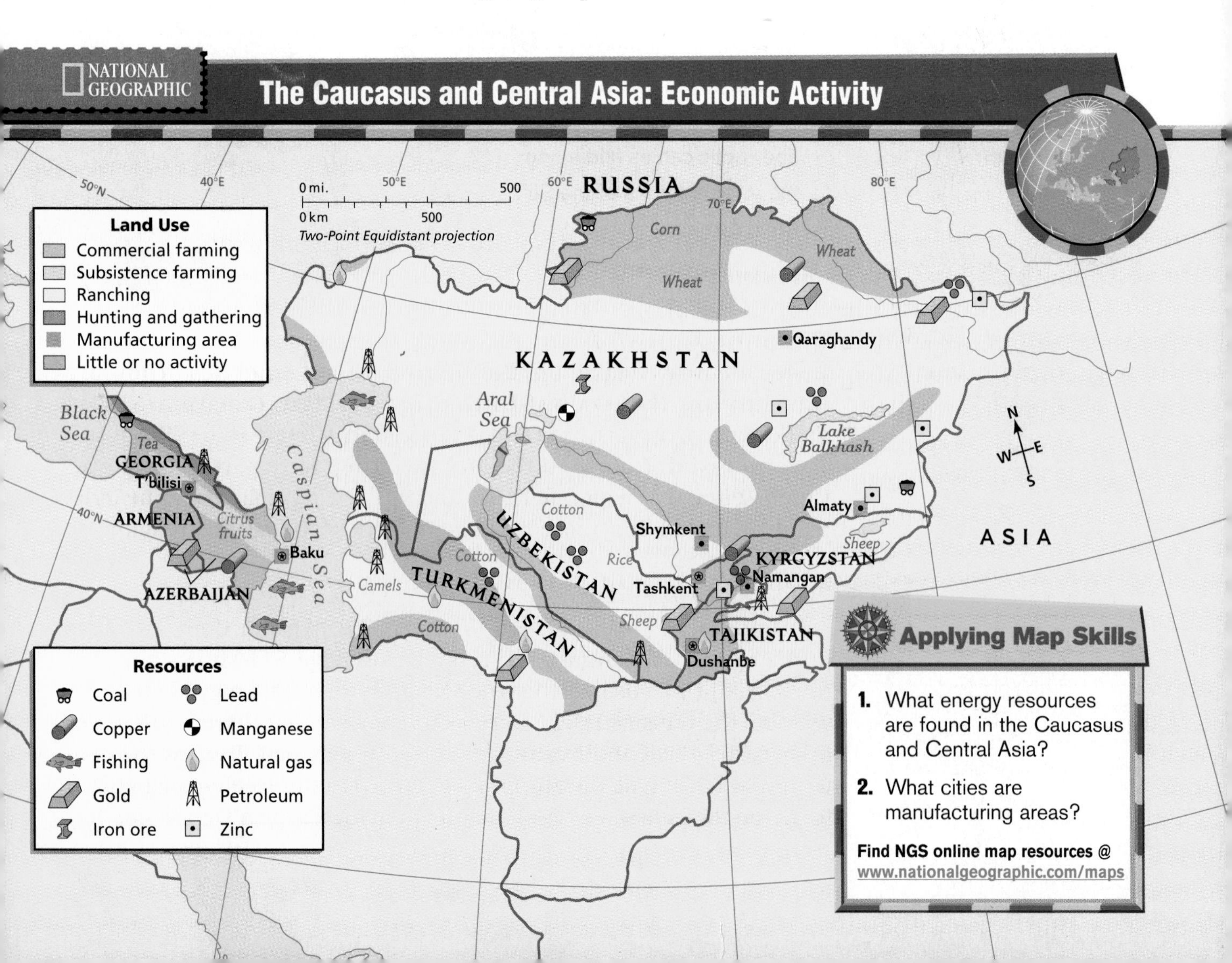

Rapid industrialization has ruined the environment. Diverting water for irrigation has drained rivers, and chemical fertilizers have polluted the soil. Environmental damage also occurred during Soviet testing of nuclear weapons in Kazakhstan. The new Kazakh government has faced the enormous challenge of cleaning up the environment.

Reading Check What people are ancestors of the Kazakhs?

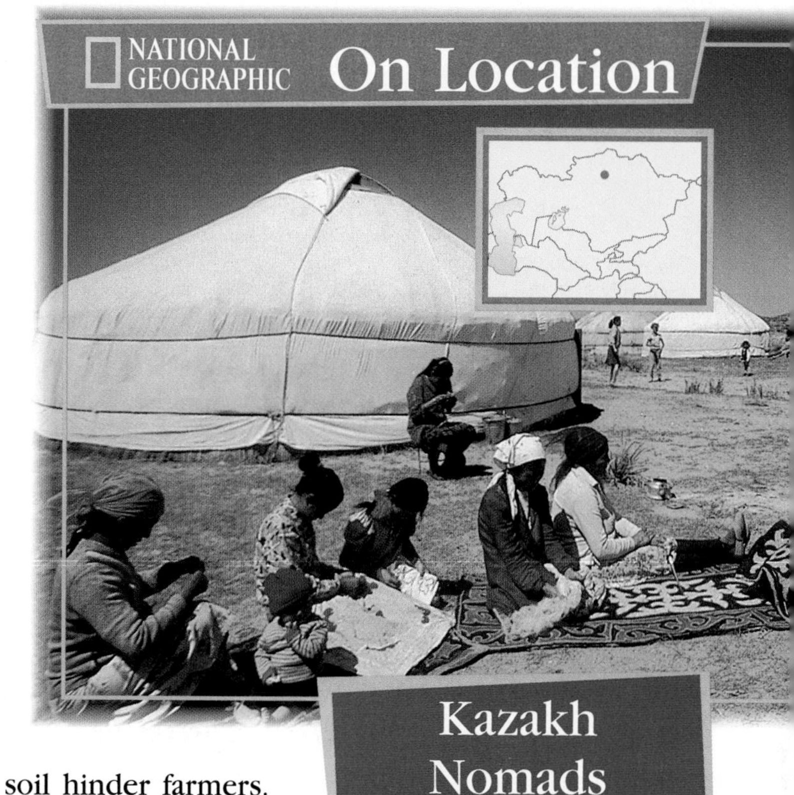

Kazakh Nomads

The traditional home of Kazakhs—called a yurt—can be easily taken apart and moved.

Movement Why would these features be important to the Mongols and early Kazakh people?

Kyrgyzstan

The lofty Tian Shan (tee•AHN SHAHN) mountain range makes up most of Kyrgyzstan. The climate depends on an area's height above sea level, or **elevation.** Lower valleys and plains have warm, dry summers and chilly winters. Higher areas have cool summers and bitterly cold winters. The harsh climate and lack of fertile soil hinder farmers. They manage to grow cotton, vegetables, and fruits, though. Many also raise sheep or cattle. Kyrgyzstan has few industries, but it does have valuable deposits of mercury and gold.

More than half of the people belong to the Kyrgyz ethnic group. Differences among **clans,** or family groups, often separate one part of the country from another. Kyrgyzstan is a **bilingual** country—one that has two official languages. These are Kirghiz, related to Turkish, and Russian. About 40 percent of the people live in cities, such as the capital, Bishkek.

Reading Check What are Kyrgyzstan's official languages?

Tajikistan

Tajikistan is even more mountainous than Kyrgyzstan. Find the mountains known as the Pamirs (puh•MIHRZ) on the map on page 511. Now look for Ismail Samani Peak. Once called Communism Peak, it is the highest mountain in Central Asia.

Agriculture is the most important activity in Tajikistan. Farmers grow cotton, rice, and fruits in fertile river valleys. Mountain streams provide water for irrigation as well as for hydroelectric power.

The largest city is Dushanbe (doo•SHAM•buh), the capital. Most of Tajikistan's people are Tajiks, who are related to the Persians. Another 25 percent are Uzbeks, a group related to the Turks. In 1992 a bitter civil war broke out between rival clans. Many people were killed, and the economy was severely damaged. Despite a peace agreement in 1997, tensions still remain high.

Reading Check What has recently damaged the economy of Tajikistan?

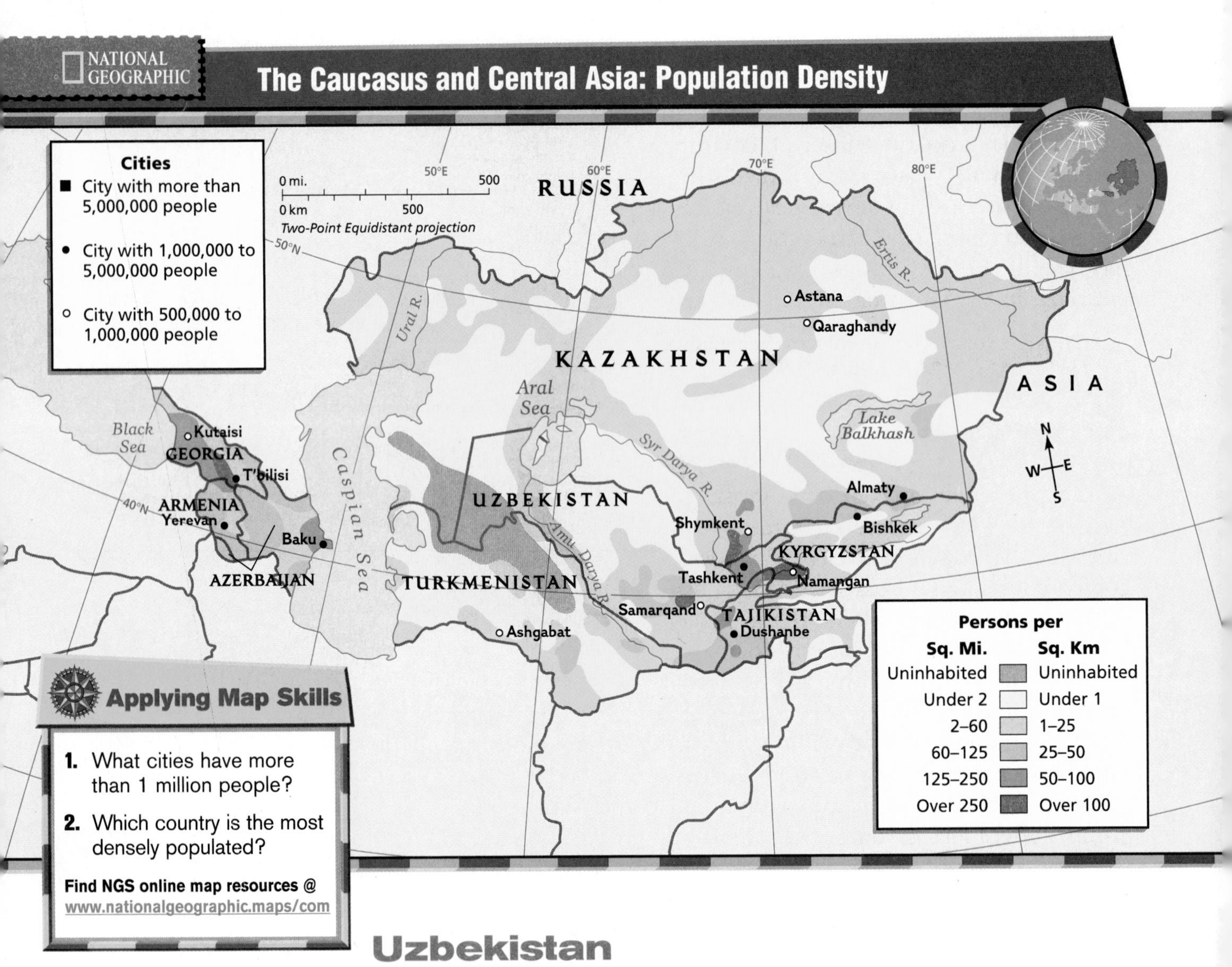

Uzbekistan

Uzbekistan, slightly larger than California, lies between the Amu Darya and the Syr Darya Rivers. Its economy relies on crops grown in fertile river valleys. Uzbekistan is one of the world's largest cotton producers. This boom in cotton, unfortunately, has had disastrous effects on the environment. Large farms needing irrigation have nearly drained away the rivers flowing into the Aral Sea. Receiving less freshwater, the sea has steadily shrunk, and its salt level has increased. Fish and wildlife have disappeared, and salt particles have polluted the air and soil. To create prosperity, Uzbek leaders are now trying to add more variety to the economy. They also want to use newly discovered deposits of oil, gas, and gold.

Most of Uzbekistan's 24.4 million people are Uzbeks who generally live in fertile valleys and oases. An **oasis,** as you recall, is a green area in a desert watered by an underground spring. Tashkent, the capital, is the largest city and industrial center in Central Asia. About 2,000 years ago, the oases of Tashkent, Bukhara, and Samarqand were part of the busy trade route—the Silk Road—that linked China and Europe.

Reading Check **Why has the Aral Sea shrunk in size?**

Turkmenistan

Turkmenistan is larger than neighboring Uzbekistan, but it has far fewer people. Why? Most of this vast land—about 85 percent—is part of a huge desert called the **Garagum** (GAHR•uh•GOOM). Making up the northern and central region, *Garagum* means "black sand." Hidden beneath it are still-to-be-tapped oil and natural gas deposits.

Despite the harshness of the land and climate, agriculture is the leading economic activity. Raising livestock is another important activity. However, not enough produce is available to feed everyone in the country, and much food has to be imported.

More than 75 percent of Turkmenistan's 4.8 million people belong to the Turkmen ethnic group. Ashgabat, the capital, is the country's largest city and leading economic and cultural center. More than one-half of Turkmenistan's people live in rural areas, though. Turkmen villages usually are located near oases formed by mountain streams along Turkmenistan's southern border.

The Turkmen people were nomads who raised camels and other livestock in the desert. Under Soviet rule, the nomads were forced to settle on farms. In the 1950s, Soviet engineers built a large irrigation and shipping canal. This technological feat greatly increased the land area used for growing cotton. However, as in Uzbekistan, cotton growing has helped to dry up the Aral Sea. Turkmenistan is hoping that its oil and natural gas resources will give it a brighter future.

The Aral Sea

This ship once moved along the waters of the Aral Sea. The sea was huge—the fourth-largest inland body of water in the world. To irrigate fields of cotton, Soviet leaders took water from the rivers that flowed into the Aral Sea. The sea shrank to one-half its former size in just 40 years. Now camels walk where fish once swam.

Reading Check Why do most Turkmen live along the southern border?

Section 2 Assessment

Defining Terms

1. **Define** steppe, delta, nomad, elevation, clan, bilingual, oasis.

Recalling Facts

2. **Economics** What products are made by factories in Kazakhstan?
3. **Location** What is the highest mountain in Central Asia?
4. **Place** What is unusual about Lake Balkhash?

Critical Thinking

5. **Understanding Cause and Effect** How does a trade route like the Silk Road contribute to the development of civilization?
6. **Drawing Conclusions** How would adding variety to an economy create prosperity?

Graphic Organizer

7. **Organizing Information** Draw a diagram like this and list the results of diverting the rivers that empty into the Aral Sea.

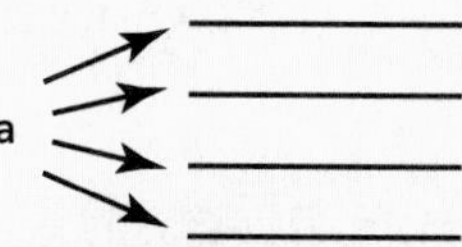

Applying Geography Skills

8. **Analyzing Maps** Study the physical map on page 511. What river forms part of the border between Turkmenistan and Uzbekistan?

Critical Thinking Skill

Analyzing Information

So much information comes our way in today's world. How can you analyze it to decide what is truly useful and accurate?

Learning the Skill

There are two basic types of information sources. *Primary sources* are original records of events made by the people who witnessed them. They include letters, photographs, and artifacts. *Secondary sources* are documents created after an event occurred. They report an event.

When reading sources, try to learn more about the person who wrote the information. Most people have a point of view, or bias. This bias influences the way they write about events.

To analyze information, follow these steps:

- Determine whether the information is a primary or secondary source.
- Identify who created the document and when it was created.
- Read the document. Who and what is it about? What are its purpose and main ideas?
- Determine how the author's point of view, or bias, is reflected in the work.

▼ A polluted playground in Azerbaijan

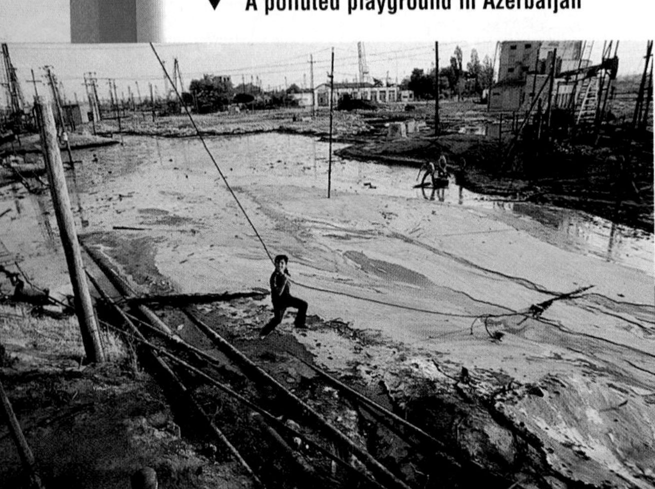

Practicing the Skill

Read the passage below, then answer the questions that follow.

> I went south to Kazakhstan and, at 4:45 A.M., stumbled off a train in Aral and went to the hospital. Beginning in the 1970s the people became ill with hepatitis, typhus, and other diseases. They drank from the rivers, as always, but now the shrunken rivers ran with sewage, industrial metals, and poisons such as DDT. "It wasn't possible to mix infant formula with that water," said a doctor. "It made goo, like soft cheese."
>
> Dust storms often blow for days, sweeping up tons of salts and fertilizers. Doctors brace then to receive children with breathing problems. Kazakhstan, declared a Kazakh writer, was the Soviet Union's "junk heap."

Adapted from "The U.S.S.R.'s Lethal Legacy" by Mike Edwards, *National Geographic,* August 1994.

1. Is this a primary or secondary source?
2. Who is the author of this passage?
3. What is the document about?
4. Where does it take place?
5. What is the purpose of this passage?
6. What, if any, evidence of bias do you find?

Applying the Skill

Analyze one of the letters to the editor in your local newspaper. Summarize the main idea, the writer's purpose, and any primary sources the writer may refer to.

GO TO

Practice key skills with **Glencoe Skillbuilder Interactive Workbook, Level 1.**

Reading Review

Section 1 Republics of the Caucasus

Terms to Know
- fault
- landlocked
- enclave
- cash crop

Main Idea

The Caucasus countries have developed their own cultures despite long periods of rule by powerful neighbors.

✓ **Region** The tall Caucasus Mountains run through Armenia, Georgia, and Azerbaijan. All three countries were part of the Soviet Union until its breakup in 1991.

✓ **Economics** Farming is important in all three countries. Georgia and Armenia have industries, including tourism.

✓ **History** Armenia and Georgia were ancient kingdoms that adopted Christianity many centuries ago.

✓ **Economics** Azerbaijan, a Muslim country, has an important oil industry.

Section 2 Central Asian Republics

Terms to Know
- steppe
- delta
- nomad
- elevation
- clan
- bilingual
- oasis

Main Idea

The Muslim countries of Central Asia are trying to build new economies and governments.

✓ **Region** Much of the land in the Central Asian republics is covered by desert or grassy steppe. Tall mountains rise in the southeast.

✓ **Culture** Almost all of the people in the five Central Asian republics are Muslims.

✓ **Economics** Kazakhstan has rich deposits of oil, copper, and gold.

✓ **Economics** Most of these countries are poor, and most people engage in farming.

✓ **Human/Environment Interaction** Uzbekistan and Turkmenistan have thriving cotton production, but the water taken to irrigate their fields is making the Aral Sea shrink.

Republic Square in Yerevan, Armenia ▶

Chapter 18 Assessment and Activities

Using Key Terms

Match the terms in Part A with their definitions in Part B.

A.

1. landlocked
2. oasis
3. bilingual
4. elevation
5. enclave
6. steppe
7. delta
8. fault
9. nomads
10. cash crop

B.

a. dry, treeless grasslands
b. green vegetation surrounded by desert
c. people who move from place to place
d. crack in the earth's crust
e. small region located inside a larger country
f. height above sea level
g. country that has no land on a sea or ocean
h. soil deposited by water at a river's mouth
i. products grown for sale as exports
j. having or speaking two languages

Reviewing the Main Ideas

Section 1 Republics of the Caucasus

11. **Location** Between what two bodies of water are Armenia, Georgia, and Azerbaijan located?
12. **Region** Why do earthquakes occur in Armenia?
13. **Government** What caused Armenia and Azerbaijan to go to war in the 1990s?
14. **Place** What is the capital of Azerbaijan?

Section 2 Central Asian Republics

15. **Region** What are the five countries that make up the Central Asian republics?
16. **History** Why did 1 million Kazakhs die during the 1930s?
17. **Place** What mountain range makes up most of Kyrgyzstan?
18. **Economics** What is the most important economic activity in Tajikistan?
19. **Place** What desert occupies most of Turkmenistan?
20. **Culture** What religion do the people of the Central Asian republics follow?

The Caucasus and Central Asia

Place Location Activity

On a separate sheet of paper, match the letters on the map with the numbered places listed below.

1. Kazakhstan
2. Aral Sea
3. Caspian Sea
4. Turkmenistan
5. Azerbaijan
6. Armenia
7. Tajikistan
8. Baku

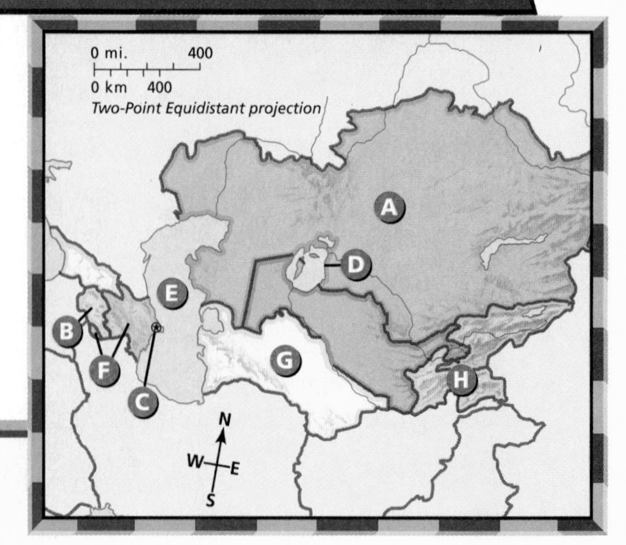

Self-Check Quiz Visit the ***Geography: The World and Its People*** Web site at gwip.glencoe.com and click on **Chapter 18–Self-Check Quizzes** to prepare for the Chapter Test.

Critical Thinking

21. **Understanding Cause and Effect** Why are there so many Russians living in the countries of Central Asia?
22. **Organizing Information** Make a chart like this one. Then list the similarities and differences between two Central Asian republics' economies.

Country	Similarities	Differences

GeoJournal Activity

23. **Writing a Story** Most of the Central Asian republics have a rich tradition of songs, poems, and stories told by word of mouth. Write a story that explains some custom or aspect of your culture. Practice telling your story aloud. Then share it with the class.

Mental Mapping Activity

24. **Focusing on the Region** Draw a simple outline map of the Caucasus and Central Asia, then label the following:

- Black Sea
- Caspian Sea
- Aral Sea
- Kazakhstan
- Caucasus Mountains
- Georgia
- Garagum
- Turkmenistan
- Lake Balkhash
- Tashkent

Technology Skills Activity

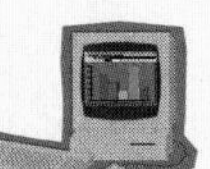

25. **Developing a Multimedia Presentation** Research the Silk Road and create a presentation about it. Your presentation should include a map marked with the route and important cities. Also include examples of the goods that were traded. Try to find the names of any important travelers who may have used the Silk Road.

Standardized Test Practice

Directions: Read the paragraph below, then answer the question that follows.

You may be surprised to know that Kazakhstan was—and still is—important to the exploration of outer space. The Russian space center Baikonur (BY•kuh•NOOR) lies in south-central Kazakhstan. During the Soviet period, Baikonur was used for many space launches. Several historic "firsts in space" occurred here. For example, the first satellite was launched in 1957. The first manned flight took place when cosmonaut Yuri Gagarin orbited the earth in 1961. In addition, the flight of the first woman in space, Valentina Tereshkova, was launched in 1963. After the Soviet collapse, the Russian-owned center remained on independent Kazakh territory.

1. The Soviet space program at Baikonur holds great importance, mostly because

F it is located in south-central Kazakhstan.
G it provides jobs for the people who live near the launch site.
H many "first in space" flights were launched from it.
J Valentina Tereshkova was the first woman in space.

Test-Taking Tip: When a question uses the word *most* or *mostly,* it means that more than one answer may be correct. Your job is to pick the *best* answer. For example, Baikonur's location in Kazakhstan may be important to the people who live near it, which is answer G. Another answer, however, provides a more general reason for Baikonur's importance.

Waterfront of Cape Town, South Africa

Woman making butter in Chad

Giraffe on a plain in Kenya ▼

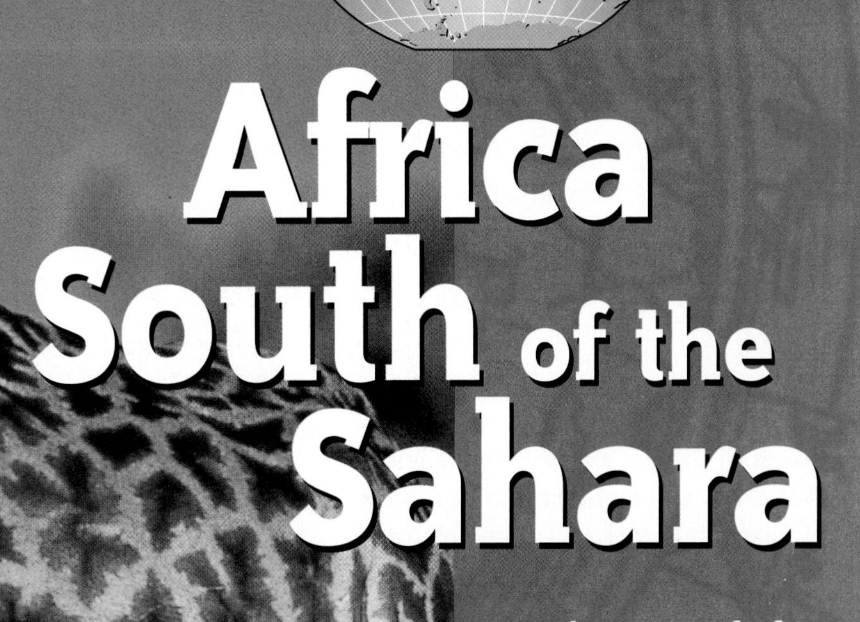

Africa South of the Sahara

The region of Africa south of the Sahara is home to more than 2,000 ethnic groups. Its hot, humid forests and dry grasslands support a variety of wild animals. Both people and animals face tough challenges in this region. The people are struggling to build stable governments and economies. Some animals are threatened with extinction as human activities destroy natural habitats.

NGS ONLINE
www.nationalgeographic.com/education

Focus on:

Africa South of the Sahara

STRADDLING THE EQUATOR, Africa south of the Sahara lies almost entirely within the tropics. Famous for its remarkable wildlife, this region also has the world's fastest-growing human population. Settling ethnic rivalries and improving low standards of living are just two of the challenges facing the people in Africa south of the Sahara.

The Land

Africa south of the Sahara has the highest overall elevation of any world region. A narrow band of low plains hugs the coastline. Inland, the land rises from west to east in a series of steplike plateaus. Separating the plateaus are steep cliffs. The region has no long mountain ranges and few towering peaks, although Mt. Kenya and Kilimanjaro are exceptions. At 19,340 feet (5,895 m), Kilimanjaro's summit is the highest point on the African continent.

Great rivers arise in this region's interior highlands. As rivers spill from one plateau to the next, they create thundering waterfalls, such as the spectacular Victoria Falls (right). It is known locally as *Mosi oa Tunya*—"smoke that thunders." Although the Nile River is Africa's longest river, the Congo River is a giant in its own right, winding 2,715 miles (4,370 km) through Africa's heart, near the Equator.

The Great Rift Valley slices through eastern Africa like a steep-walled gash in the continent. Formed by movements of the earth's crust, the valley extends from Southwest Asia southward to the Zambezi River in Mozambique. It cradles a chain of deep lakes.

The Climate

If you started at the Equator and traveled north or south from there, you would pass through four major climate regions, one after the other. Tropical rain forests lie along the Equator and fill the great basin of the Congo River in central and western Africa. Head away from the Equator and you see rain forests give way to tropical savannas. These vast grasslands are home to some of the continent's most famous large mammals, including elephants, lions, rhinoceroses, and giraffes.

As you move farther from the Equator, rainfall becomes scarce, and savannas give way to tropical steppes. Finally you encounter very dry areas where deserts dominate the landscape. Deserts cover more of Africa than any other continent. The largest deserts south of the Sahara are the Namib and the Kalahari.

UNIT 7

Victoria Falls, on the Zambezi River

Elephants browsing near Kilimanjaro, Tanzania

The Economy

Africa south of the Sahara is rich in mineral resources, but these resources are not evenly distributed. Nigeria has huge reserves of oil. South Africa has fabulous deposits of gold and diamonds, making it the wealthiest country in the region. Overall, however, Africa south of the Sahara has the lowest standard of living of any world region.

Manufacturing plays only a small role in the region's economy. In the past, colonial rulers used Africa as a source of raw materials and left the continent largely undeveloped. Today the nations south of the Sahara are struggling to industrialize.

Most people in Africa south of the Sahara still depend on small-scale farming or livestock herding for their livelihoods. They usually are able to raise only enough food to feed their families. Some farmers work on plantations that grow crops for export to other countries. Such crops include coffee, cacao, cotton, peanuts, tea, bananas, and sisal (a fiber). Drought is a constant problem for the region's farmers.

The People

Thousands of years ago, great kingdoms and empires developed in Africa south of the Sahara. In the 1400s and 1500s, Europeans began trading with African societies, carrying away gold, spices, ivory, and enslaved people.

By the late 1800s, European nations had claimed almost all of Africa. For profit and political advantage, they carved up the continent into colonies. In the process, they ripped apart once-unified regions and threw together ethnic groups that did not get along.

Most African nations won their independence in the mid-1900s. Most emerged from colonial rule politically unstable and with crippled economies.

Today more than 625 million people inhabit Africa south of the Sahara. They represent some 2,000 ethnic groups and speak 800 different languages. Nearly three-fourths of the population lives in rural areas. Although Africa is the least urbanized continent, its cities are growing. Lured by the promise of better living conditions, people are flocking to African cities, which are among the fastest-growing urban areas in the world.

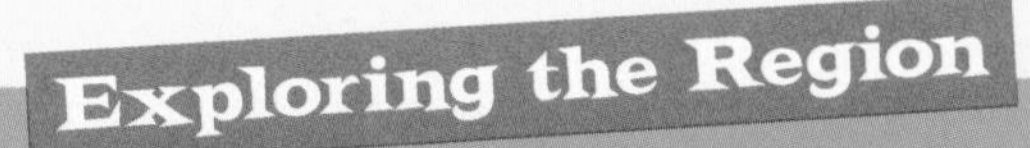

Exploring the Region

1. **What happens when Africa's rivers flow from one plateau to another?**
2. **What climate zone is centered on the Equator?**
3. **What makes South Africa the region's most prosperous country?**
4. **How did colonial rule affect Africa south of the Sahara?**

◀ **Woman fertilizing crops in Zimbabwe**

UNIT 7

◀ Crowded market in Lagos, Nigeria

REGIONAL ATLAS

Africa South of the Sahara

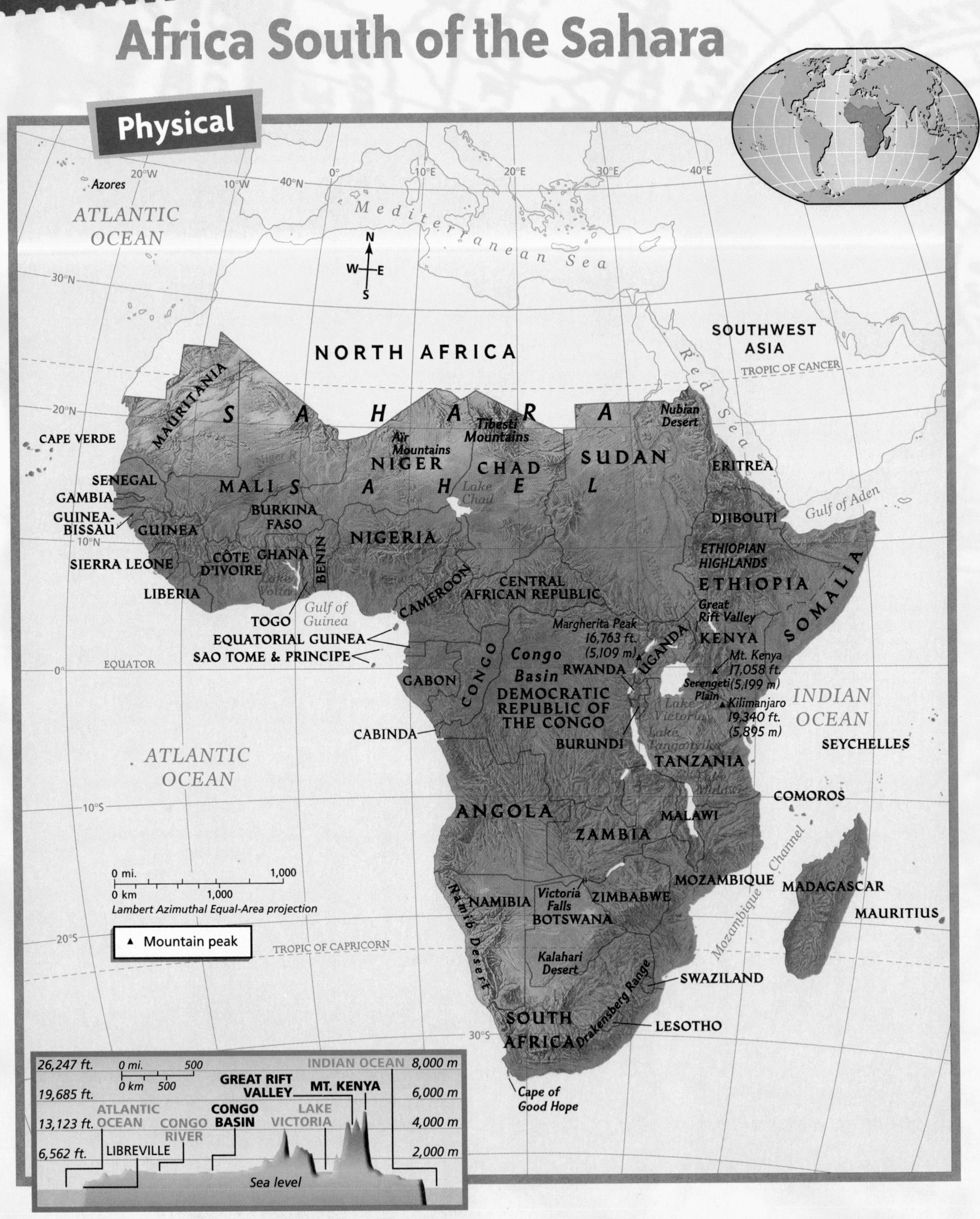

UNIT 7

Political

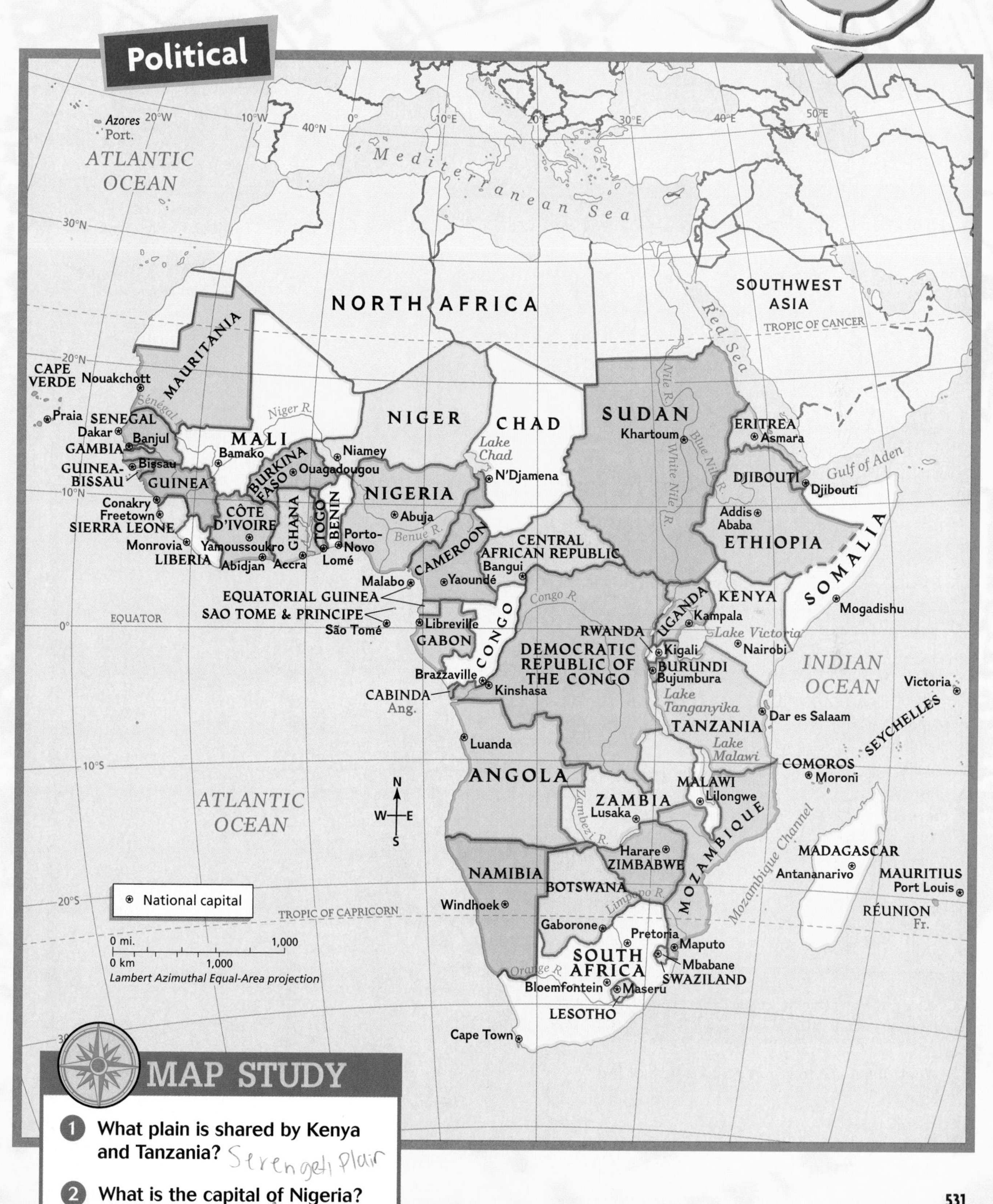

MAP STUDY

1. What plain is shared by Kenya and Tanzania? Serengeti Plain
2. What is the capital of Nigeria? Abuja

Africa South of the Sahara

Gems and Minerals

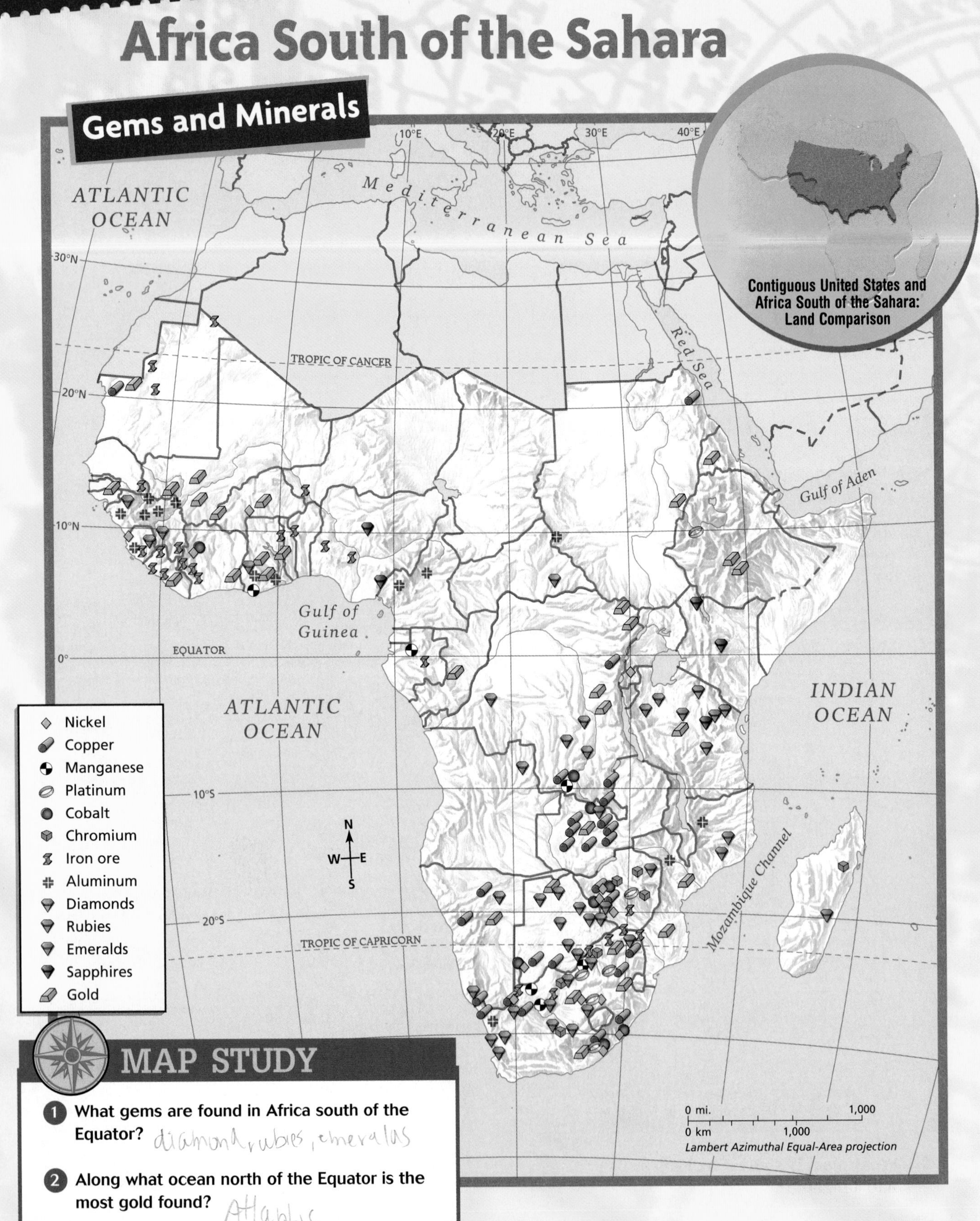

MAP STUDY

1. What gems are found in Africa south of the Equator? diamond, rubies, emeralds
2. Along what ocean north of the Equator is the most gold found? Atlantic

Geo Extremes

① **HIGHEST POINT**
Kilimanjaro (Tanzania)
19,340 ft. (5,895 m) high

② **LOWEST POINT**
Lake Assal (Djibouti)
512 ft. (156 m)
below sea level

③ **LONGEST RIVER**
Nile River
4,241 mi.
(6,825 km) long

④ **LARGEST LAKE**
Lake Victoria (Kenya, Uganda, and Tanzania)
26,834 sq. mi.
(69,500 sq. km)

⑤ **LARGEST ISLAND**
Madagascar
226,642 sq. mi.
(587,000 sq. km)

⑥ **HOTTEST PLACE**
Dalol, Denakil Depression (Ethiopia)
93°F (34°C) annual average temperature

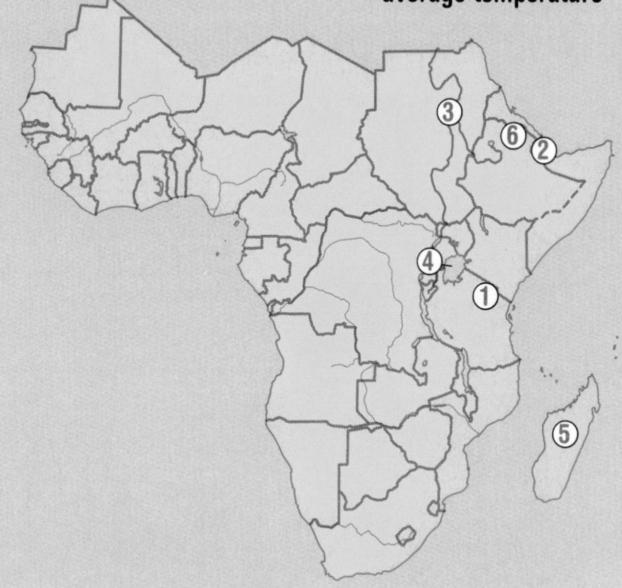

COMPARING POPULATION: United States and Selected Countries of Africa South of the Sahara

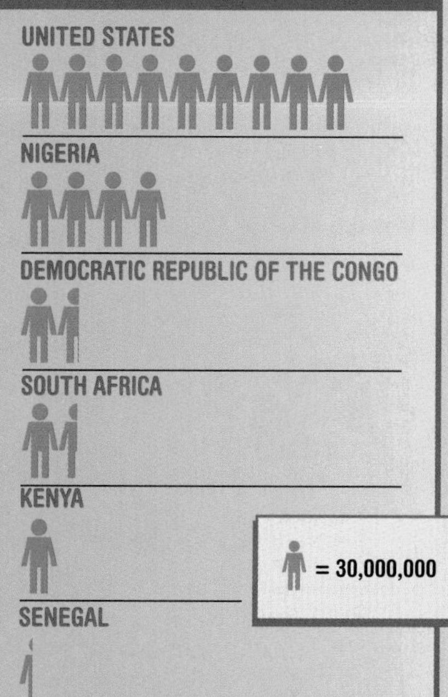

Source: *Population Reference Bureau*, 2000.

SELECTED RURAL AND URBAN POPULATIONS: Africa South of the Sahara

	Rural	Urban
WEST AFRICA		
Niger	83%	17%
Cape Verde	56%	44%
CENTRAL AFRICA		
Angola	68%	32%
Central African Republic	61%	39%
EAST AFRICA		
Rwanda	95%	5%
Djibouti	17%	83%
SOUTHERN AFRICA		
Lesotho	84%	16%
South Africa	55%	45%

Source: *Population Reference Bureau*, 2000.

GRAPHIC STUDY

1. What is the longest river in Africa?
2. Of the African countries shown in the chart at lower right, which is least urbanized? Which is most urbanized?

Country Profiles

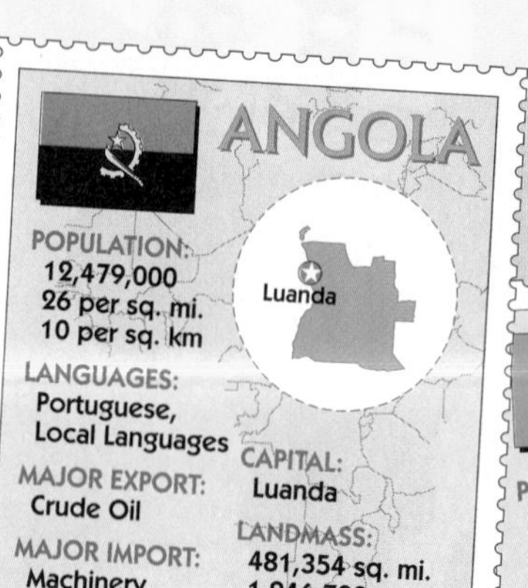

ANGOLA

POPULATION: 12,479,000; 26 per sq. mi.; 10 per sq. km
LANGUAGES: Portuguese, Local Languages
MAJOR EXPORT: Crude Oil
MAJOR IMPORT: Machinery
CAPITAL: Luanda
LANDMASS: 481,354 sq. mi.; 1,246,700 sq. km

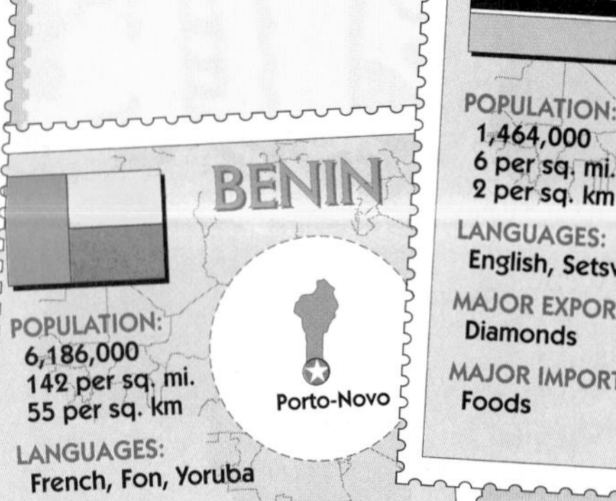

BENIN

POPULATION: 6,186,000; 142 per sq. mi.; 55 per sq. km
LANGUAGES: French, Fon, Yoruba
MAJOR EXPORT: Cotton
MAJOR IMPORT: Foods
CAPITAL: Porto-Novo
LANDMASS: 43,484 sq. mi.; 112,622 sq. km

BOTSWANA

POPULATION: 1,464,000; 6 per sq. mi.; 2 per sq. km
LANGUAGES: English, Setswana
MAJOR EXPORT: Diamonds
MAJOR IMPORT: Foods
CAPITAL: Gaborone
LANDMASS: 231,805 sq. mi.; 600,372 sq. km

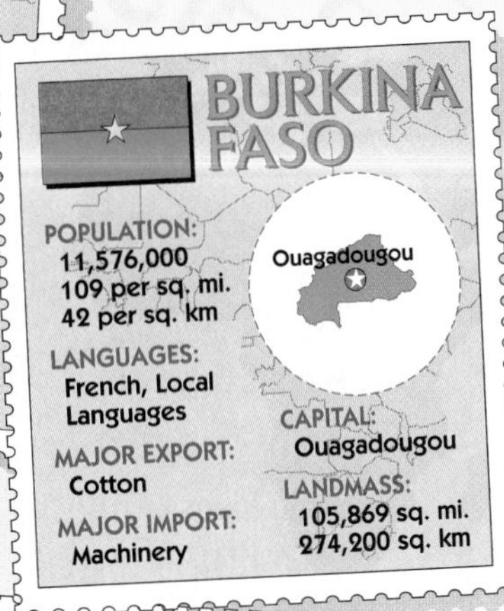

BURKINA FASO

POPULATION: 11,576,000; 109 per sq. mi.; 42 per sq. km
LANGUAGES: French, Local Languages
MAJOR EXPORT: Cotton
MAJOR IMPORT: Machinery
CAPITAL: Ouagadougou
LANDMASS: 105,869 sq. mi.; 274,200 sq. km

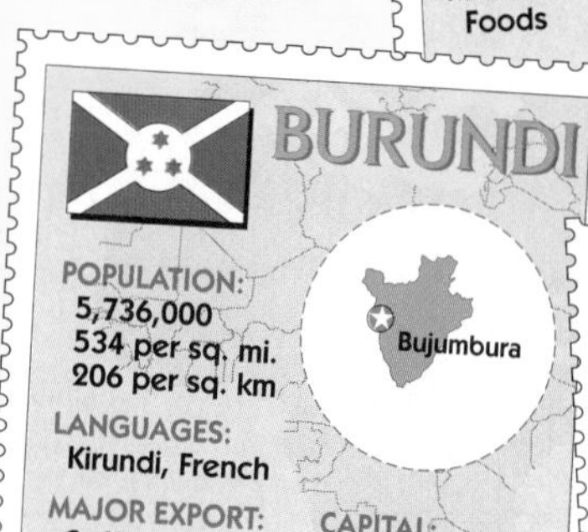
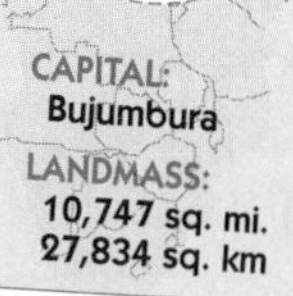

BURUNDI

POPULATION: 5,736,000; 534 per sq. mi.; 206 per sq. km
LANGUAGES: Kirundi, French
MAJOR EXPORT: Coffee
MAJOR IMPORT: Machinery
CAPITAL: Bujumbura
LANDMASS: 10,747 sq. mi.; 27,834 sq. km

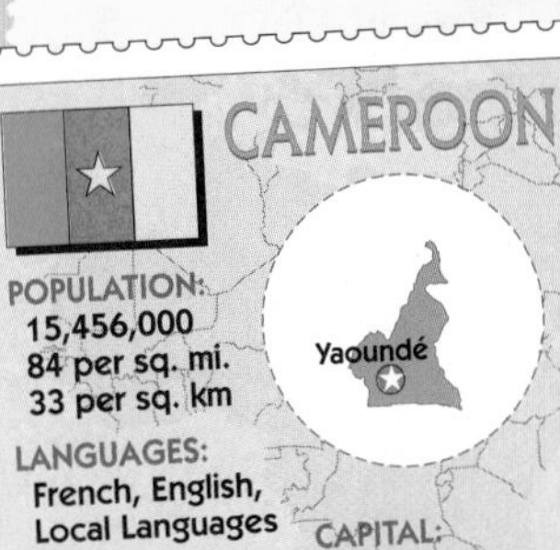
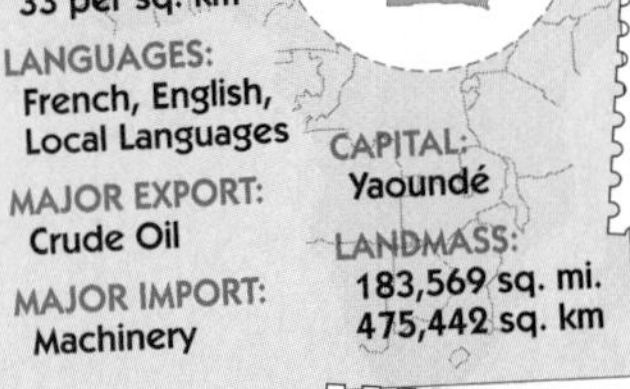

CAMEROON

POPULATION: 15,456,000; 84 per sq. mi.; 33 per sq. km
LANGUAGES: French, English, Local Languages
MAJOR EXPORT: Crude Oil
MAJOR IMPORT: Machinery
CAPITAL: Yaoundé
LANDMASS: 183,569 sq. mi.; 475,442 sq. km

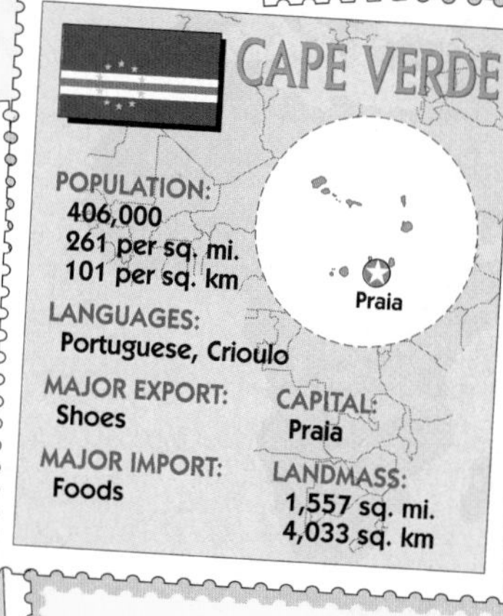

CAPE VERDE

POPULATION: 406,000; 261 per sq. mi.; 101 per sq. km
LANGUAGES: Portuguese, Crioulo
MAJOR EXPORT: Shoes
MAJOR IMPORT: Foods
CAPITAL: Praia
LANDMASS: 1,557 sq. mi.; 4,033 sq. km

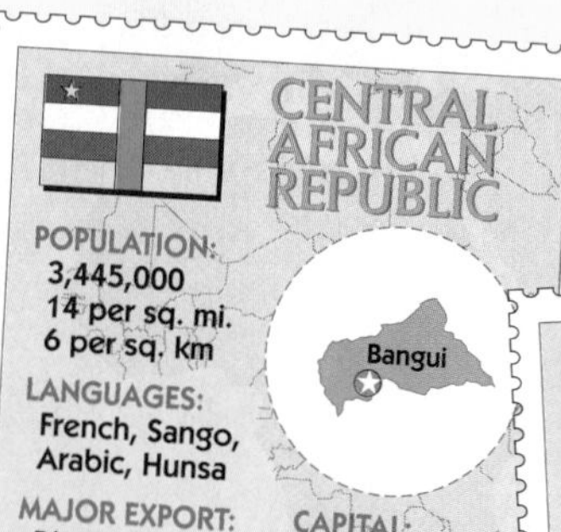
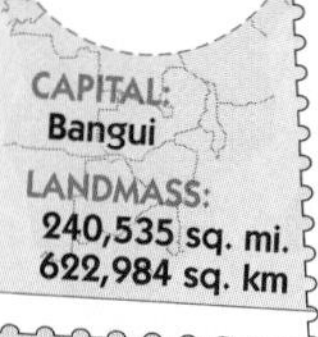

CENTRAL AFRICAN REPUBLIC

POPULATION: 3,445,000; 14 per sq. mi.; 6 per sq. km
LANGUAGES: French, Sango, Arabic, Hunsa
MAJOR EXPORT: Diamonds
MAJOR IMPORT: Foods
CAPITAL: Bangui
LANDMASS: 240,535 sq. mi.; 622,984 sq. km

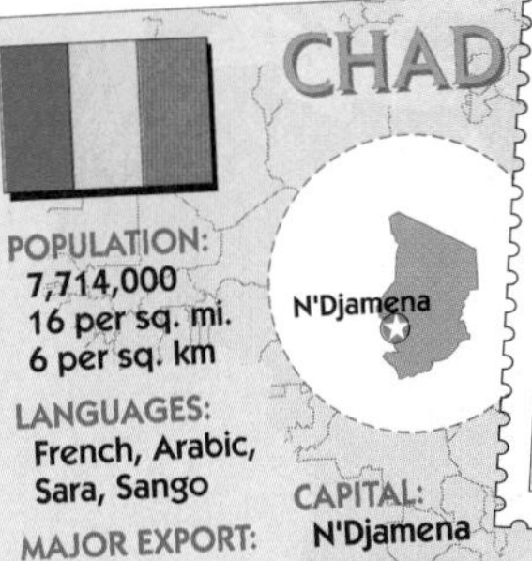
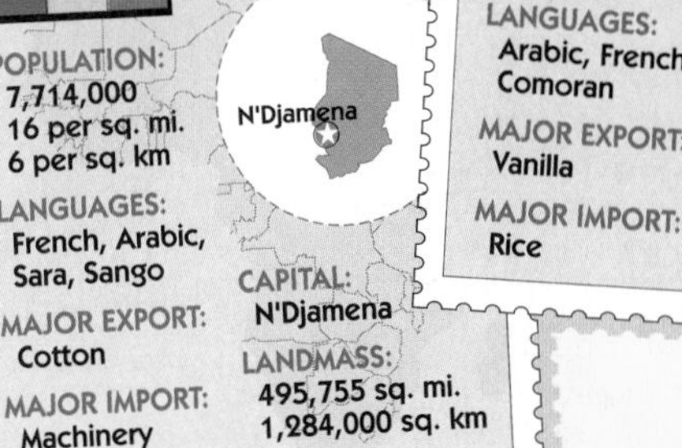

CHAD

POPULATION: 7,714,000; 16 per sq. mi.; 6 per sq. km
LANGUAGES: French, Arabic, Sara, Sango
MAJOR EXPORT: Cotton
MAJOR IMPORT: Machinery
CAPITAL: N'Djamena
LANDMASS: 495,755 sq. mi.; 1,284,000 sq. km

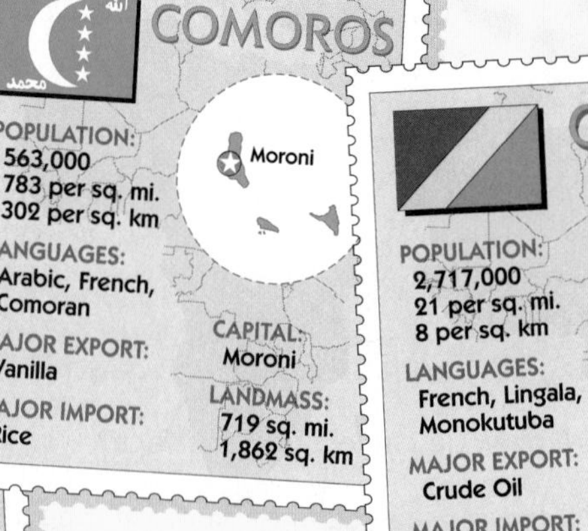

COMOROS

POPULATION: 563,000; 783 per sq. mi.; 302 per sq. km
LANGUAGES: Arabic, French, Comoran
MAJOR EXPORT: Vanilla
MAJOR IMPORT: Rice
CAPITAL: Moroni
LANDMASS: 719 sq. mi.; 1,862 sq. km

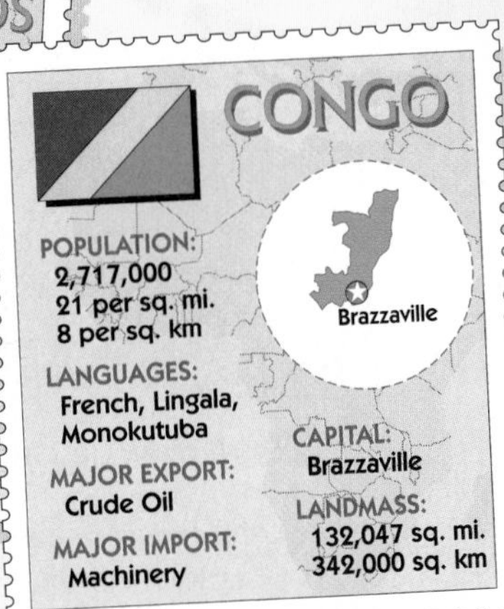

CONGO

POPULATION: 2,717,000; 21 per sq. mi.; 8 per sq. km
LANGUAGES: French, Lingala, Monokutuba
MAJOR EXPORT: Crude Oil
MAJOR IMPORT: Machinery
CAPITAL: Brazzaville
LANDMASS: 132,047 sq. mi.; 342,000 sq. km

Countries and flags not drawn to scale

CONGO, Democratic Republic of the

Kinshasa

POPULATION:
50,481,000
56 per sq. mi.
22 per sq. km

LANGUAGES:
French, Lingala, Kingwana

MAJOR EXPORT:
Diamonds

MAJOR IMPORT:
Manufactured Goods

CAPITAL:
Kinshasa

LANDMASS:
905,568 sq. mi.
2,345,409 sq. km

CÔTE D'IVOIRE

Yamoussoukro
Abidjan

POPULATION:
15,818,000
127 per sq. mi.
49 per sq. km

LANGUAGES:
French, Dioula

MAJOR EXPORT:
Cocoa

MAJOR IMPORT:
Foods

CAPITALS:
Yamoussoukro, Abidjan

LANDMASS:
124,504 sq. mi.
322,463 sq. km

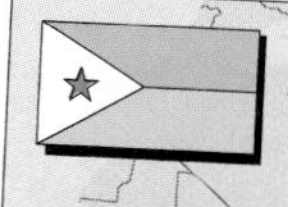

DJIBOUTI

Djibouti

POPULATION:
629,000
70 per sq. mi.
27 per sq. km

LANGUAGES:
French, Arabic

MAJOR EXPORTS:
Hides and Skins

MAJOR IMPORT:
Foods

CAPITAL:
Djibouti

LANDMASS:
8,958 sq. mi.
23,200 sq. km

EQUATORIAL GUINEA

Malabo

POPULATION:
442,000
41 per sq. mi.
16 per sq. km

LANGUAGES:
Spanish, French, Fang, Bubi, Ibo

MAJOR EXPORT:
Petroleum

MAJOR IMPORT:
Machinery

CAPITAL:
Malabo

LANDMASS:
10,831 sq. mi.
28,051 sq. km

ERITREA

Asmara

POPULATION:
3,985,000
85 per sq. mi.
33 per sq. km

LANGUAGES:
Afar, Amharic, Arabic, Tigre

MAJOR EXPORT:
Livestock

MAJOR IMPORT:
Processed Foods

CAPITAL:
Asmara

LANDMASS:
46,842 sq. mi.
121,320 sq. km

ETHIOPIA

Addis Ababa

POPULATION:
59,680,000
140 per sq. mi.
54 per sq. km

LANGUAGES:
Amharic, Tigrinya, Orominga

MAJOR EXPORT:
Coffee

MAJOR IMPORTS:
Foods and Livestock

CAPITAL:
Addis Ababa

LANDMASS:
424,934 sq. mi.
1,100,574 sq. km

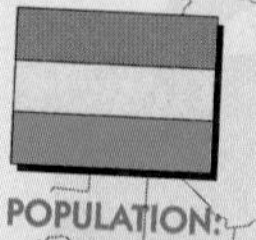

GABON

POPULATION:
1,197,000
12 per sq. mi.
4 per sq. km

LANGUAGES:
French, Local Languages

MAJOR EXPORT:
Crude Oil

MAJOR IMPORT:
Machinery

CAPITAL:
Libreville

LANDMASS:
103,347 sq. mi.
267,667 sq. km

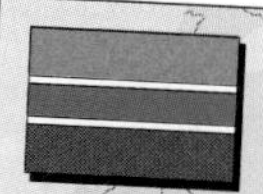

GAMBIA

Banjul

POPULATION:
1,268,000
291 per sq. mi.
112 per sq. km

LANGUAGES:
English, Mandinka, Fula, Wolof

MAJOR EXPORT:
Peanuts

MAJOR IMPORT:
Foods

CAPITAL:
Banjul

LANDMASS:
4,361 sq. mi.
11,295 sq. km

GHANA

Accra

POPULATION:
19,678,000
214 per sq. mi.
82 per sq. km

LANGUAGES:
English, Local Languages

MAJOR EXPORT:
Gold

MAJOR IMPORT:
Machinery

CAPITAL:
Accra

LANDMASS:
92,100 sq. mi.
238,537 sq. km

GUINEA

Conakry

POPULATION:
7,539,000
79 per sq. mi.
31 per sq. km

LANGUAGES:
French, Local Languages

MAJOR EXPORT:
Bauxite

MAJOR IMPORT:
Petroleum Products

CAPITAL:
Conakry

LANDMASS:
94,926 sq. mi.
245,857 sq. km

GUINEA-BISSAU

POPULATION:
1,187,000
85 per sq. mi.
33 per sq. km

LANGUAGES:
Portuguese, Crioulo, Local Languages

MAJOR EXPORT:
Cashews

MAJOR IMPORT:
Foods

CAPITAL:
Bissau

LANDMASS:
13,948 sq. mi.
36,125 sq. km

Country Profiles

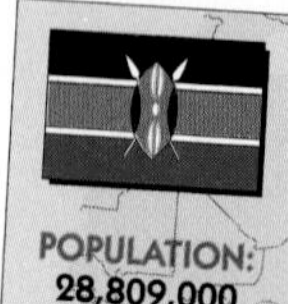

KENYA

Nairobi

POPULATION:
28,809,000
126 per sq. mi.
49 per sq. km

LANGUAGES:
English, Swahili

MAJOR EXPORT:
Tea

MAJOR IMPORT:
Machinery

CAPITAL:
Nairobi

LANDMASS:
228,861 sq. mi.
592,747 sq. km

LESOTHO

Maseru

POPULATION:
2,129,000
182 per sq. mi.
70 per sq. km

LANGUAGES:
English, Sesotho, Zulu, Xhosa

MAJOR EXPORT:
Clothing

MAJOR IMPORT:
Corn

CAPITAL:
Maseru

LANDMASS:
11,720 sq. mi.
30,355 sq. km

LIBERIA

Monrovia

POPULATION:
2,924,000
68 per sq. mi.
26 per sq. km

LANGUAGES:
English, Local Languages

MAJOR EXPORT:
Diamonds

MAJOR IMPORT:
Natural Gas

CAPITAL:
Monrovia

LANDMASS:
43,000 sq. mi.
111,369 sq. km

MADAGASCAR

Antananarivo

POPULATION:
14,417,000
64 per sq. mi.
25 per sq. km

LANGUAGES:
French, Malagasy

MAJOR EXPORT:
Coffee

MAJOR IMPORT:
Machinery

CAPITAL:
Antananarivo

LANDMASS:
226,658 sq. mi.
587,041 sq. km

MALAWI

Lilongwe

POPULATION:
10,000,000
219 per sq. mi.
84 per sq. km

LANGUAGES:
Chewa, English

MAJOR EXPORT:
Tobacco

MAJOR IMPORT:
Foods

CAPITAL:
Lilongwe

LANDMASS:
45,747 sq. mi.
118,484 sq. km

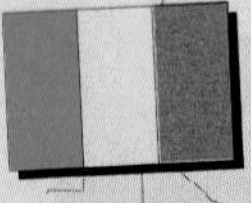

MALI

POPULATION:
10,960,000
23 per sq. mi.
9 per sq. km

LANGUAGES:
French, Bambara

MAJOR EXPORT:
Cotton

MAJOR IMPORT:
Machinery

CAPITAL:
Bamako

LANDMASS:
478,841 sq. mi.
1,240,192 sq. km

MAURITANIA

Nouakchott

POPULATION:
2,598,000
7 per sq. mi.
3 per sq. km

LANGUAGES:
Hasaniya Arabic, Wolof

MAJOR EXPORT:
Fish

MAJOR IMPORT:
Foods

CAPITAL:
Nouakchott

LANDMASS:
397,955 sq. mi.
1,030,700 sq. km

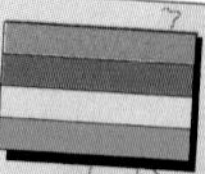

MAURITIUS

Port Louis

POPULATION:
1,172,000
1,487 per sq. mi.
575 per sq. km

LANGUAGES:
English, Creole, Bhojpuri, French

MAJOR EXPORT:
Sugar

MAJOR IMPORT:
Foods

CAPITAL:
Port Louis

LANDMASS:
788 sq. mi.
2,040 sq. km

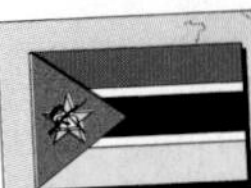

MOZAMBIQUE

Maputo

POPULATION:
19,124,000
62 per sq. mi.
24 per sq. km

LANGUAGES:
Portuguese, Local Languages

MAJOR EXPORT:
Cashews

MAJOR IMPORT:
Foods

CAPITAL:
Maputo

LANDMASS:
308,642 sq. mi.
799,380 sq. km

NAMIBIA

Windhoek

POPULATION:
1,648,000
5 per sq. mi.
2 per sq. km

LANGUAGES:
English, Local Languages

MAJOR EXPORT:
Diamonds

MAJOR IMPORT:
Construction Materials

CAPITAL:
Windhoek

LANDMASS:
318,261 sq. mi.
824,292 sq. km

NIGER

Niamey

POPULATION:
9,962,000
20 per sq. mi.
8 per sq. km

LANGUAGES:
French, Hausa, Djerma

MAJOR EXPORT:
Uranium Ore

MAJOR IMPORT:
Manufactured Goods

CAPITAL:
Niamey

LANDMASS:
489,191 sq. mi.
1,267,000 sq. km

Countries and flags not drawn to scale

UNIT 7

NIGERIA

POPULATION: 113,829,000; 319 per sq. mi.; 123 per sq. km
LANGUAGES: English, Hausa, Yoruba, Igbo
MAJOR EXPORT: Petroleum
MAJOR IMPORT: Machinery
CAPITAL: Abuja
LANDMASS: 356,669 sq. mi.; 923,768 sq. km

RWANDA

POPULATION: 8,155,000; 802 per sq. mi.; 310 per sq. km
LANGUAGES: Kinyarwanda, French, English
MAJOR EXPORT: Coffee
MAJOR IMPORT: Foods
CAPITAL: Kigali
LANDMASS: 10,169 sq. mi.; 26,338 sq. km

SAO TOME and PRINCIPE

POPULATION: 155,000; 417 per sq. mi.; 161 per sq. km
LANGUAGES: Portuguese, Crioulo
MAJOR EXPORT: Cocoa
MAJOR IMPORT: Textiles
CAPITAL: São Tomé
LANDMASS: 372 sq. mi.; 964 sq. km

SENEGAL

POPULATION: 9,240,000; 122 per sq. mi.; 47 per sq. km
LANGUAGES: French, Wolof, Pulaar, Diola
MAJOR EXPORT: Fish
MAJOR IMPORT: Foods
CAPITAL: Dakar
LANDMASS: 75,955 sq. mi.; 196,722 sq. km

SEYCHELLES

POPULATION: 80,000; 457 per sq. mi.; 177 per sq. km
LANGUAGES: English, French, Creole
MAJOR EXPORT: Fish
MAJOR IMPORT: Foods
CAPITAL: Victoria
LANDMASS: 175 sq. mi.; 453 sq. km

SIERRA LEONE

POPULATION: 5,297,000; 191 per sq. mi.; 74 per sq. km
LANGUAGES: English, Mende, Temne, Krio
MAJOR EXPORT: Diamonds
MAJOR IMPORT: Foods
CAPITAL: Freetown
LANDMASS: 27,699 sq. mi.; 71,740 sq. km

SOMALIA

POPULATION: 7,141,000; 29 per sq. mi.; 11 per sq. km
LANGUAGES: Somali, Arabic
MAJOR EXPORT: Livestock
MAJOR IMPORT: Textiles
CAPITAL: Mogadishu
LANDMASS: 246,201 sq. mi.; 637,657 sq. km

SOUTH AFRICA

POPULATION: 42,579,000; 90 per sq. mi.; 35 per sq. km
LANGUAGES: Afrikaans, English, Local Languages
MAJOR EXPORT: Gold
MAJOR IMPORT: Transport Equip.
CAPITALS: Pretoria, Cape Town, Bloemfontein
LANDMASS: 471,445 sq. mi.; 1,221,037 sq. km

SUDAN

POPULATION: 28,883,000; 30 per sq. mi.; 12 per sq. km
LANGUAGES: Arabic, Nubian, Ta Bedawie
MAJOR EXPORT: Cotton
MAJOR IMPORT: Petroleum Products
CAPITAL: Khartoum
LANDMASS: 963,600 sq. mi.; 2,495,712 sq. km

SWAZILAND

POPULATION: 985,000; 147 per sq. mi.; 57 per sq. km
LANGUAGES: English, Swazi
MAJOR EXPORT: Soft Drink Concentrates
MAJOR IMPORT: Machinery
CAPITAL: Mbabane
LANDMASS: 6,704 sq. mi.; 17,364 sq. km

TANZANIA

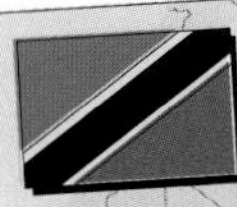

POPULATION: 31,271,000; 86 per sq. mi.; 33 per sq. km
LANGUAGES: Swahili, English
MAJOR EXPORT: Coffee
MAJOR IMPORT: Machinery
CAPITAL: Dar es Salaam
LANDMASS: 364,900 sq. mi.; 945,087 sq. km

REGIONAL ATLAS

Country Profiles

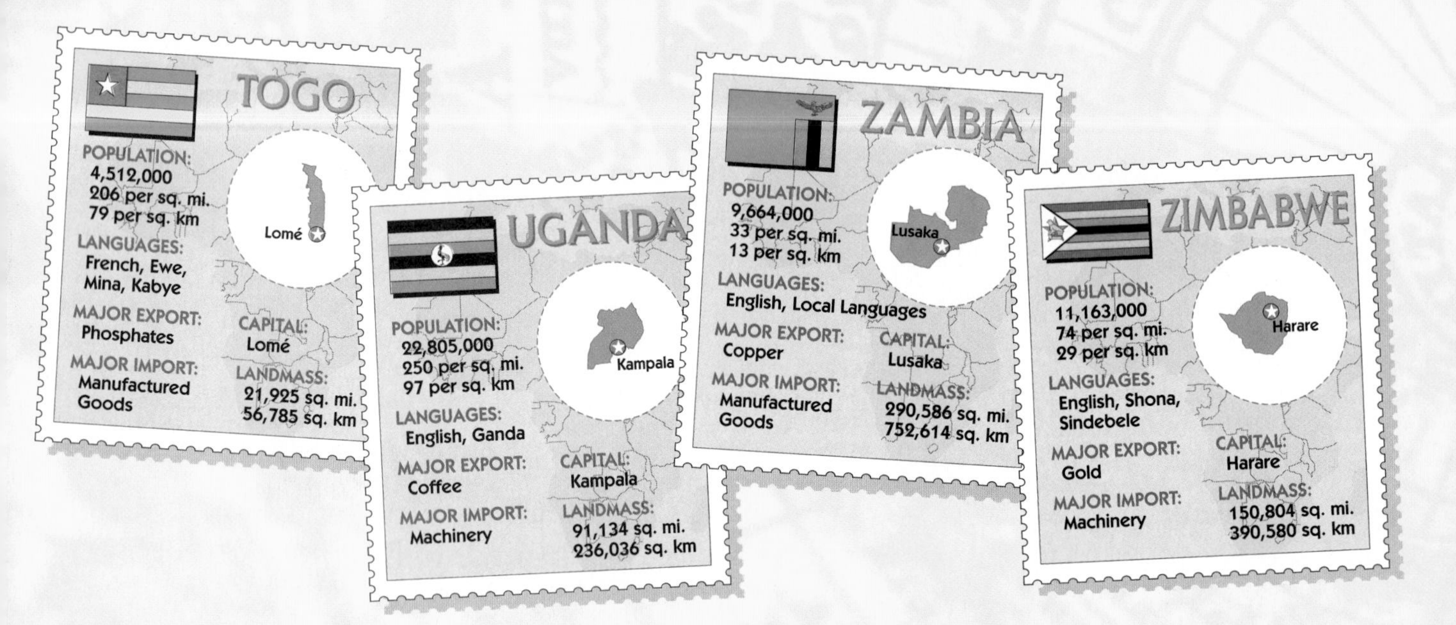

TOGO
POPULATION:
4,512,000
206 per sq. mi.
79 per sq. km
LANGUAGES:
French, Ewe,
Mina, Kabye
MAJOR EXPORT:
Phosphates
MAJOR IMPORT:
Manufactured
Goods
CAPITAL:
Lomé
LANDMASS:
21,925 sq. mi.
56,785 sq. km
Lomé
UGANDA
POPULATION:
22,805,000
250 per sq. mi.
97 per sq. km
LANGUAGES:
English, Ganda
MAJOR EXPORT:
Coffee
MAJOR IMPORT:
Machinery
CAPITAL:
Kampala
LANDMASS:
91,134 sq. mi.
236,036 sq. km
Kampala
ZAMBIA
POPULATION:
9,664,000
33 per sq. mi.
13 per sq. km
LANGUAGES:
English, Local Languages
MAJOR EXPORT:
Copper
MAJOR IMPORT:
Manufactured
Goods
CAPITAL:
Lusaka
LANDMASS:
290,586 sq. mi.
752,614 sq. km
Lusaka
ZIMBABWE
POPULATION:
11,163,000
74 per sq. mi.
29 per sq. km
LANGUAGES:
English, Shona,
Sindebele
MAJOR EXPORT:
Gold
MAJOR IMPORT:
Machinery
CAPITAL:
Harare
LANDMASS:
150,804 sq. mi.
390,580 sq. km
Harare

Countries and flags not drawn to scale

Questions From Buzz Bee!

The following questions are taken from National Geographic GeoBees. Use your textbook, the Internet, and other library resources to find the answers.

1. The Dutch who settled in what is now South Africa became known by a Dutch word that means "farmer." By what name were these settlers known? Boers
2. Volcanic activity created the highest peak along the Great Rift Valley. Name the peak.
3. The Hutu and the Tutsi are ethnic groups that have been fighting for control of two countries. Name these countries.

UNIT 7

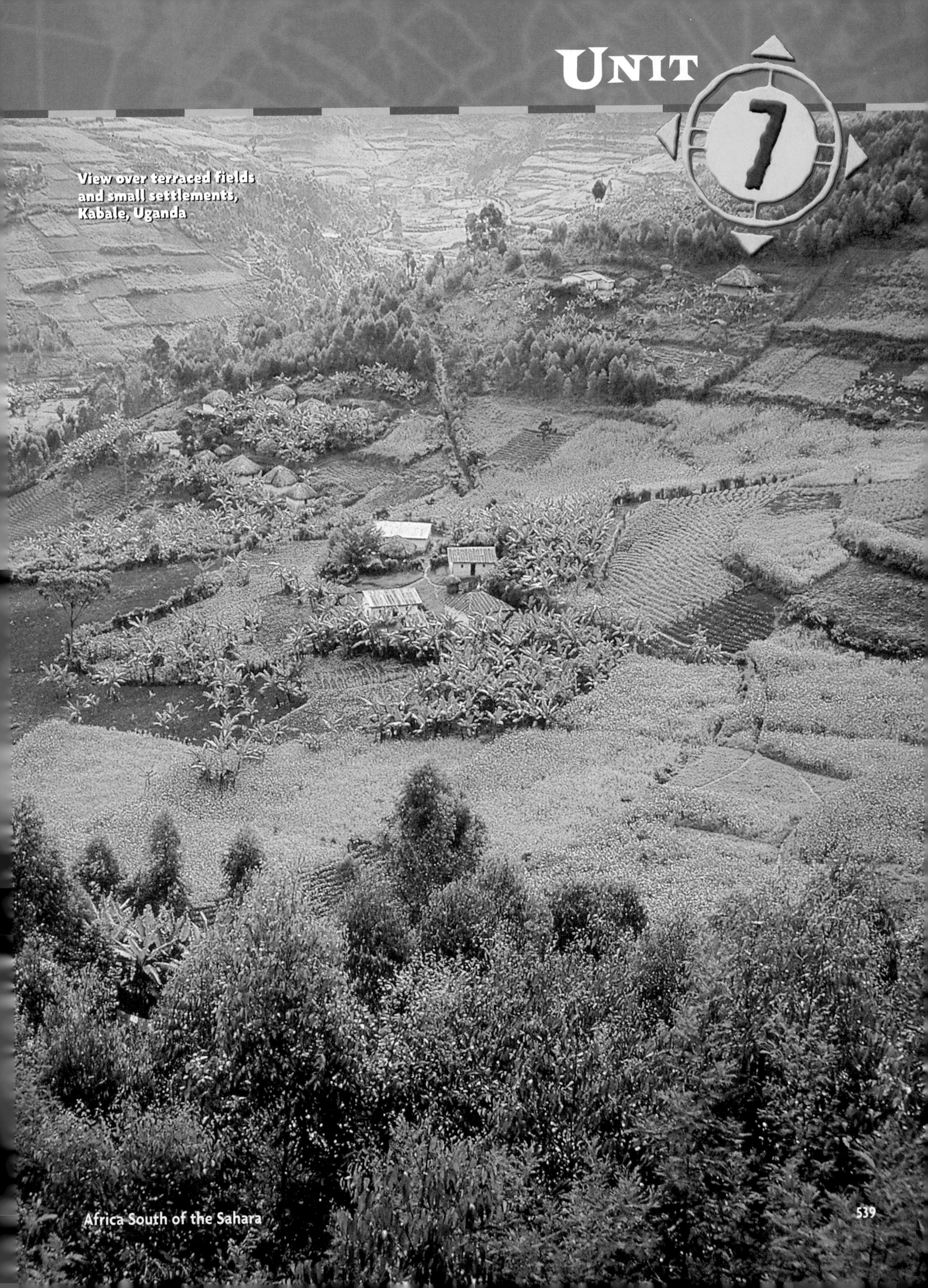

View over terraced fields and small settlements, Kabale, Uganda

Chapter
19
West Africa
The World and Its People
NATIONAL GEOGRAPHIC
To learn more about the people and places of West Africa, view **The World and Its People** Chapter 19 video.
Geography online
Chapter Overview Visit the ***Geography: The World and Its People*** Web site at gwip.glencoe.com and click on **Chapter 19—Chapter Overviews** to preview information about West Africa.

Section 1 Nigeria

Guide to Reading

Main Idea

A large, oil-rich country, Nigeria has more people than any other African nation.

Terms to Know

- mangrove
- savanna
- harmattan
- subsistence farm
- cacao
- compound
- civil war

Places to Locate

- Nigeria
- Niger River
- Gulf of Guinea
- Kano
- Lagos
- Abuja

Reading Strategy

Create a chart like the one below. Then list two facts about Nigeria in each category.

Nigeria	Fact #1	Fact #2
Land		
Economy		
People		

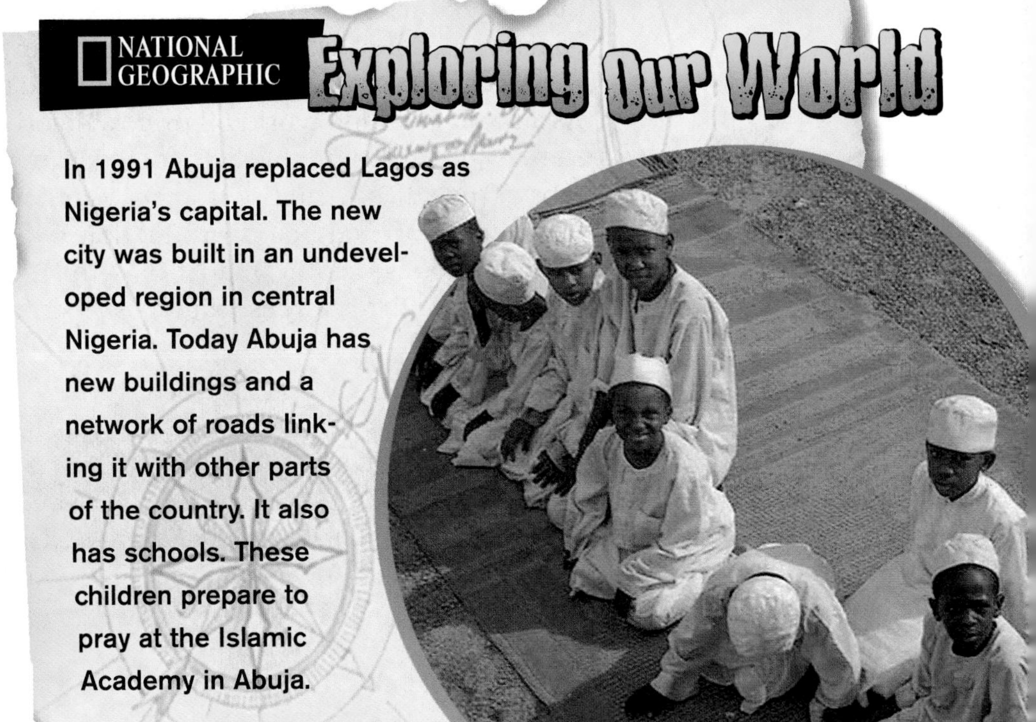

NATIONAL GEOGRAPHIC **Exploring Our World**

In 1991 Abuja replaced Lagos as Nigeria's capital. The new city was built in an undeveloped region in central Nigeria. Today Abuja has new buildings and a network of roads linking it with other parts of the country. It also has schools. These children prepare to pray at the Islamic Academy in Abuja.

The West African country of **Nigeria** takes its name from the **Niger River,** which flows through western and central Nigeria. One of the largest nations in Africa, Nigeria is more than twice the size of California.

From Tropics to Savanna

Nigeria has a long coastline on the **Gulf of Guinea,** an arm of the Atlantic Ocean. Along Nigeria's coast, the land is covered with mangrove swamps. A **mangrove** is a tropical tree with roots that extend both above and beneath the water.

As you travel inland, the land becomes vast tropical rain forests. Small villages appear in only a few clearings. The forests gradually thin into savannas in central Nigeria. **Savannas** are tropical grasslands with only a few trees. Highlands and plateaus also make up this area. Most of the country has a tropical savanna climate with high average temperatures and seasonal rains. The grasslands of the far north have a dry steppe climate. In the winter months, a dusty wind called the **harmattan** blows south from the Sahara.

Reading Check What kinds of vegetation are found in Nigeria?

◀ Colorful side street in Dakar, Senegal

Economic Challenges

Nigeria is one of the world's major oil-producing countries. More than 90 percent of the country's income comes from oil exports. The government has used money from oil to build highways, schools, skyscrapers, and factories. These factories make food products, textiles, chemicals, machinery, and vehicles. Still, more than one-third of Nigeria's people lack jobs and live in poverty.

Nigeria began to experience economic troubles during the 1980s. As a result of falling world oil prices, Nigeria's income dropped. At the same time, many people left their farms in search of better-paying jobs in the cities. In addition, a few years of low rainfall meant smaller harvests. As a result, food production fell. Nigeria—which had once exported food—had to import food to feed its people.

Despite oil resources, Nigeria's people mainly work as farmers. Most have **subsistence farms,** or small plots that grow just enough

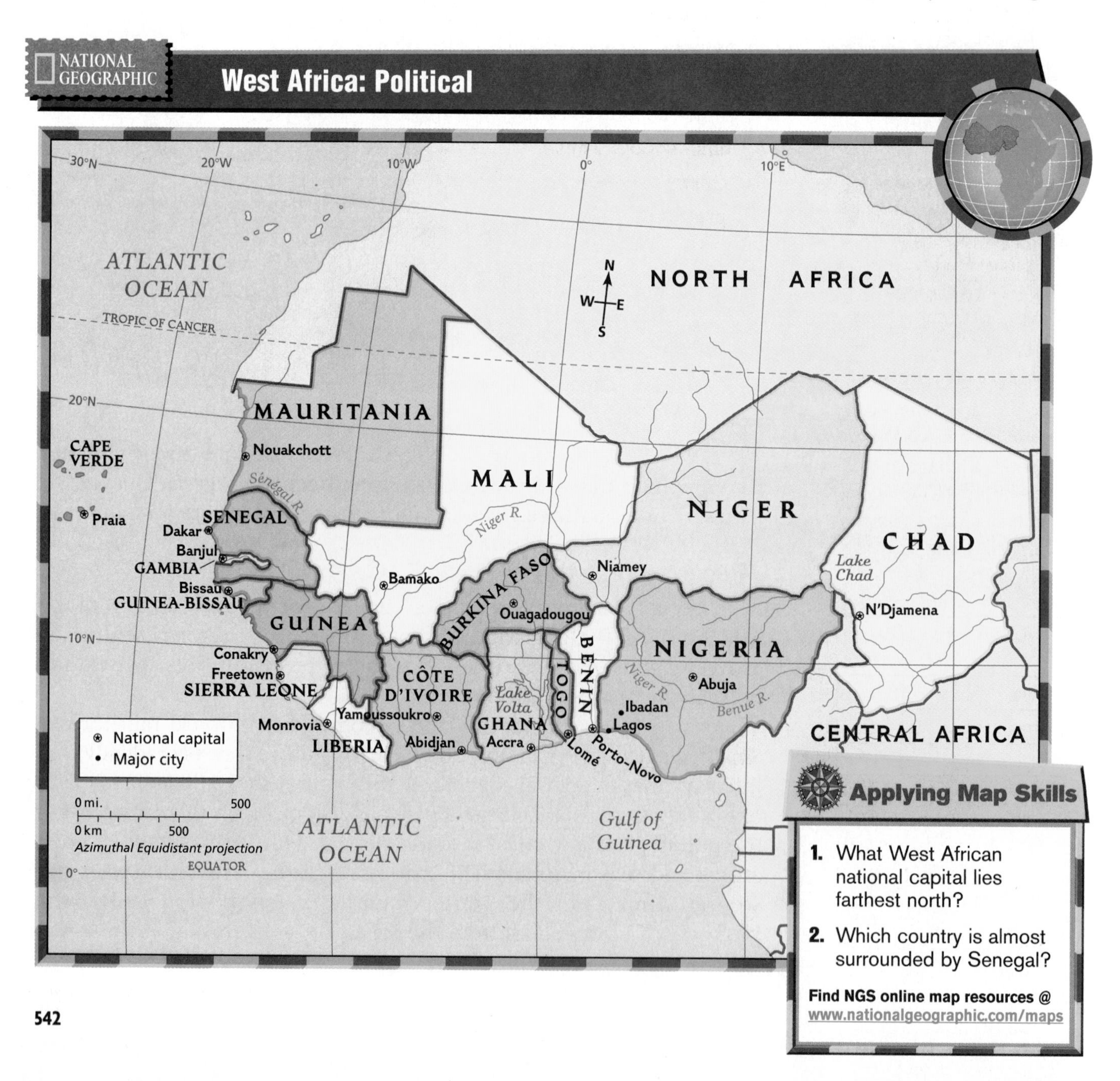

Applying Map Skills

1. What West African national capital lies farthest north?
2. Which country is almost surrounded by Senegal?

Find NGS online map resources @ www.nationalgeographic.com/maps

to feed their families. Some work on larger farms that produce such cash crops as rubber, peanuts, palm oil, and cacao. The **cacao** is a tropical tree whose seeds are used to make chocolate and cocoa. Nigeria is a leading producer of cacao beans.

✓ Reading Check **How has Nigeria's government used money from oil exports?**

Nigeria's People

About 113.8 million people live in Nigeria—more people than in any other country in Africa. The map on page 554 shows that most of the people live along the coast and around the city of **Kano** in the north.

One of the strongest bonds that Africans have is a sense of belonging to a group or family. Nigeria has more than 300 ethnic groups. The four largest are the Hausa (HOW•suh), Fulani (foo•LAH•nee), Yoruba (YAWR•uh•buh), and Ibo (EE•boh). Nigerians speak many different

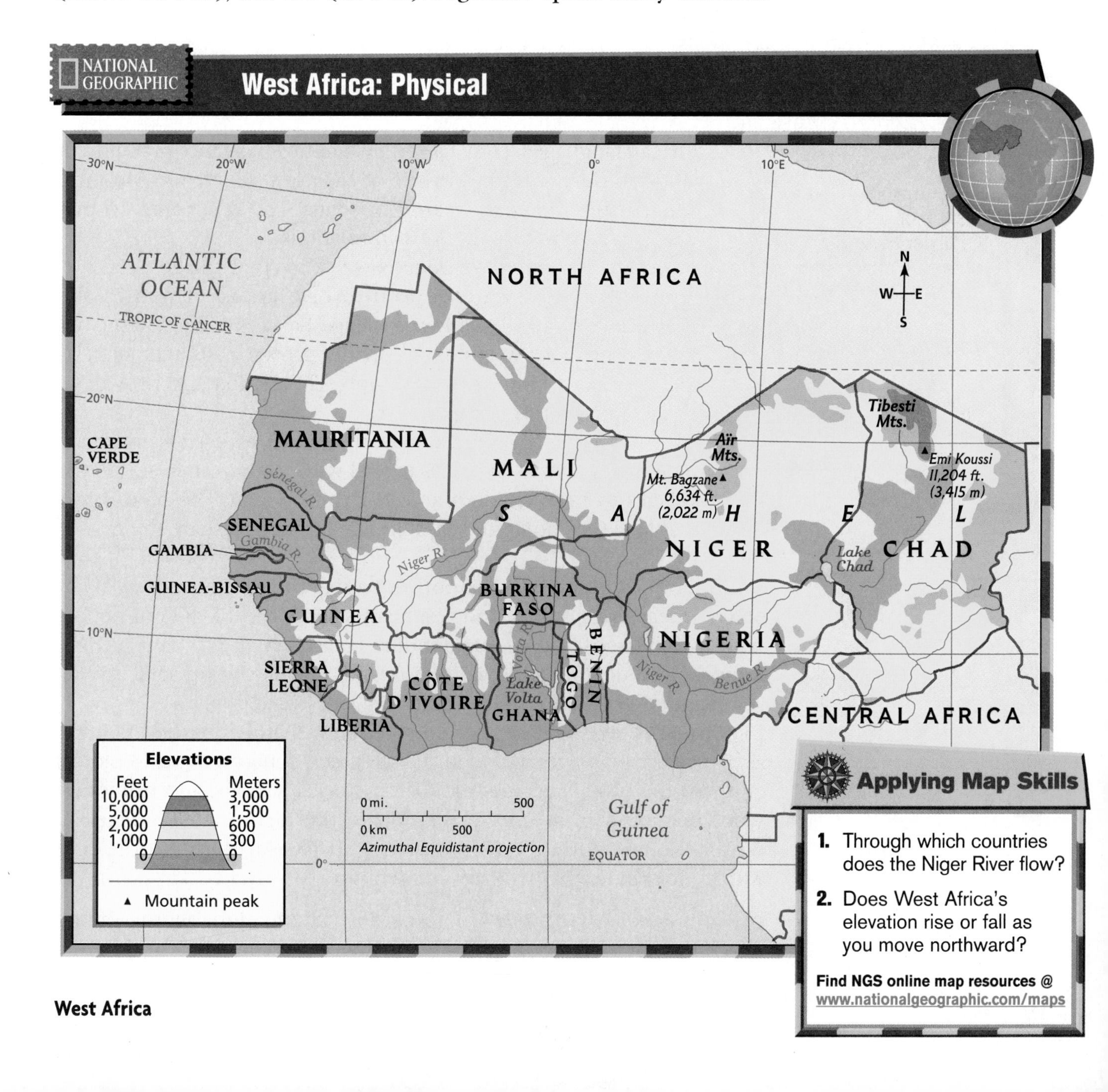

Nigeria's Economy

Nigerian women canoe past an oil refinery in the Niger River delta (above). Cacao pods are harvested in Nigeria (right).

Human/Environment Interaction **What are Nigeria's important cash crops?**

African languages. They use English in business and government affairs, though. About one-half of Nigeria's people are Muslims, and another 40 percent are Christians. The remaining 10 percent practice traditional African religions.

About 60 percent of Nigerians live in rural villages. The typical family lives in a **compound,** or a group of houses surrounded by walls. Usually the village has a weekly market run by women. The women sell locally grown products such as meat, palm oil, cloth, yams, and nuts. The market also provides a chance for friends to meet.

Long-standing rural ways are changing, however. Many young men now move to the cities to find work. The women stay in the villages to raise children and to farm the land. The men, when they are able, return home to see their families and to share the money they have made.

Nigeria's largest city is the port of **Lagos,** the former capital. Major banks, department stores, and restaurants serve the 11 million people who live in Lagos and its surrounding areas. Ibadan (EE•bah•DAHN), Kano, and Abuja (ah•BOO•jah) lie inland. **Abuja,** the present capital, is a planned city that was begun during the 1980s.

Nigerians take pride in both old and new features of their culture. Artists make elaborate wooden masks, metal sculptures, and colorful cloth. In the past, Nigerians passed on stories, sayings, and riddles by word of mouth from one generation to the next. In 1986 Nigerian writer Wole Soyinka (WAW•lay shaw•YIHNG•ka) became the first African to win the Nobel Prize in literature.

History and Government The earliest known inhabitants of the area were the Nok people. They lived between the Niger and Benue

Rivers between 300 B.C. and A.D. 200. The Nok were known as skilled metalworkers and traders.

Over the centuries, powerful city-states and kingdoms became centers of trade and the arts. People in the north came in contact with Muslim cultures and adopted the religion of Islam. People in the south developed cultures based on traditional African religions.

During the 1400s, Europeans arrived in Africa looking for gold and Africans to take overseas as enslaved laborers. In 1884 European leaders divided most of Africa into colonies. The borders of these colonies, however, often sliced through ethnic lands. As a result, many ethnic groups found their members living in two or more separate territories. By the early 1900s, the British had taken control of Nigeria.

In 1960 Nigeria finally became an independent country. Ethnic, religious, and political disputes soon tore it apart, however. One ethnic group, the Ibo, tried to set up their own country. A **civil war**—a fight among different groups within a country—resulted. In this bloody war, starvation and conflict led to 2 million deaths. The Ibo were defeated, and their region remained part of Nigeria.

Nigeria has faced the challenge of building a stable government. Military leaders have often ruled the country. In 1999 Nigerians were able to vote for a president in free elections. Nigeria's democratic government is working to build greater national unity.

Reading Check What are the four largest ethnic groups in Nigeria?

Ken Saro-Wiwa

Ken Saro-Wiwa was a Nigerian environmentalist. In the 1990s, he fought to protect farmlands and fisheries on the Niger River from damage caused by oil spills and polluted air. After his death, an international group set up a fund named after Saro-Wiwa. The fund will help people who work to save the environment.

Assessment

Defining Terms

1. **Define** mangrove, savanna, harmattan, subsistence farm, cacao, compound, civil war.

Recalling Facts

2. **Place** Describe the changes in Nigeria's physical geography as you move from the coast inland.
3. **Place** What is the capital of Nigeria?
4. **Culture** How many ethnic groups are represented by the people of Nigeria?

Critical Thinking

5. **Drawing Conclusions** Why do you think government leaders moved the capital city to a new location in the interior of Nigeria?
6. **Understanding Cause and Effect** Why did a drop in oil prices cause an economic depression in Nigeria in the 1980s?

Graphic Organizer

7. **Organizing Information** On a time line like the one below, place the following events and their dates in order: Nigeria becomes independent; Nok people work in metal and trade for goods; Free elections are held; Europeans create colonies in Nigeria.

Applying Geography Skills

8. **Analyzing Maps** Study the physical map on page 543. Into what larger body of water does the Niger River empty?

Critical Thinking Skill

Drawing Inferences and Conclusions

Suppose your teacher brought to class a colorful wooden mask, and a classmate said, "Wow. That's from Nigeria." You might infer that your classmate has an interest in African art and, therefore, recognizes the mask as coming from Nigeria.

Yoruba wood masks ▲

Learning the Skill

To *infer* means to evaluate information and arrive at a conclusion. When you make inferences, you "read between the lines" or draw conclusions that are not stated directly in the text. You must use the available facts *and* your own knowledge and experience to form a judgment or opinion about the material.

Use the following steps to help draw inferences and make conclusions:

- Read carefully for stated facts and ideas.
- Summarize the information and list the important facts.
- Apply related information that you may already know to make inferences.
- Use your knowledge and insight to develop some conclusions about these facts.

Practicing the Skill

Read the passage below, then answer the questions that follow.

> Nigerian art forms reflect the people's beliefs in spirits and nature. Yoruba masks are carved out of wood, reflecting the forces of nature and gods. The masks are used in ceremonies to help connect with the spirit of their ancestors. The masks also appear at funerals in order to please the spirits of the dead. Of all the Yoruba masks, the helmet masks of the Epa cult are the most spectacular.

1. What topic is the writer describing?
2. What facts are presented?
3. What can you infer about the role of masks in Nigerian life?
4. What do you already know about religious ceremonies?
5. What conclusion can you make about traditional religions in Nigeria?

Applying the Skill

Study the photos of Nigerians on page 544. What can you infer about life in Nigeria from the photographs? What evidence supports this inference or conclusion?

GO TO Practice key skills with **Glencoe Skillbuilder Interactive Workbook, Level 1.**

Section 2

The Sahel Countries

Guide to Reading

Main Idea

The Sahel countries face a continuing struggle to keep grasslands from turning into desert.

Terms to Know

- overgraze
- drought
- desertification

Places to Locate

- Sahel
- Mauritania
- Mali
- Burkina Faso
- Niger
- Chad

Reading Strategy

Create a diagram like this one. Then list two of the factors that have led to the loss of vegetation in the Sahel countries.

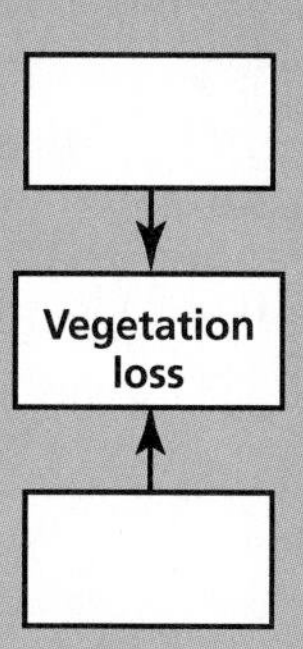

NATIONAL GEOGRAPHIC **Exploring Our World**

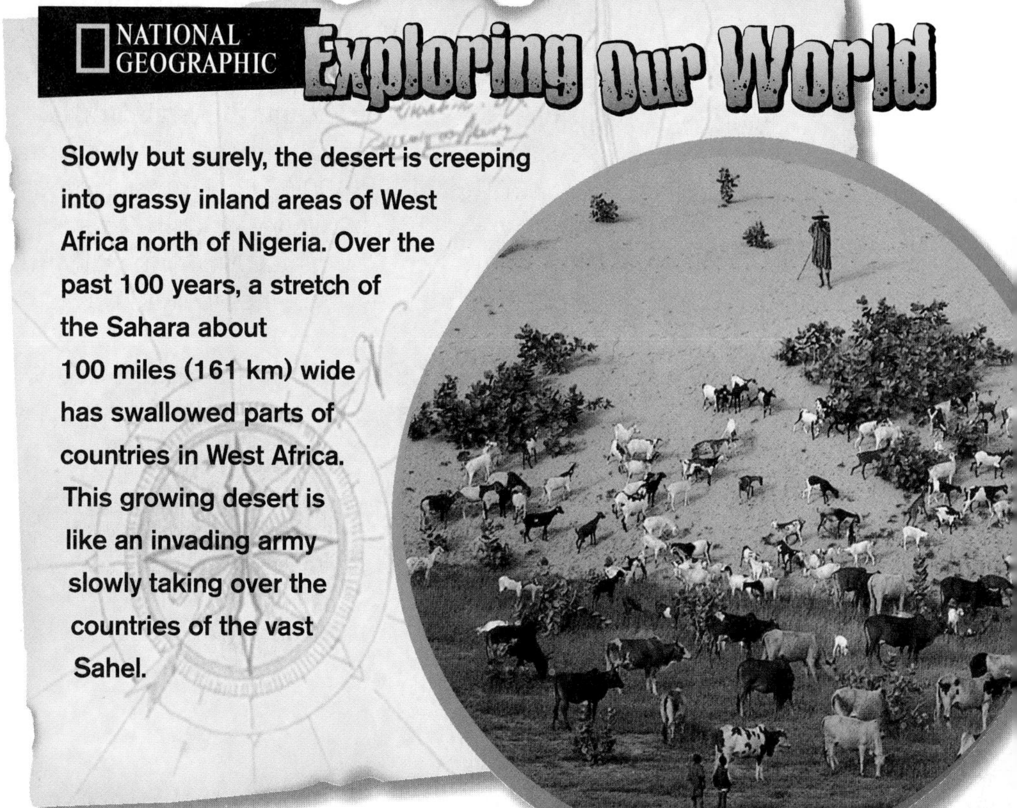

Slowly but surely, the desert is creeping into grassy inland areas of West Africa north of Nigeria. Over the past 100 years, a stretch of the Sahara about 100 miles (161 km) wide has swallowed parts of countries in West Africa. This growing desert is like an invading army slowly taking over the countries of the vast Sahel.

Five countries—Mauritania (MAWR•uh•TAY•nee•uh), Mali (MAH•lee), Burkina Faso (bur•KEE•nuh FAH•soh), Niger (NY•juhr), and Chad—are located in an area known as the **Sahel.** The word *Sahel* comes from an Arabic word that means "border." The Sahel in West Africa forms the border between the Sahara to the north and the fertile lands to the south.

The Land and History

The Sahel receives little rainfall, so only short grasses and small trees can support grazing animals. Most people have traditionally herded livestock. Their flocks, unfortunately, have overgrazed the land in some places. When animals **overgraze** land, they strip areas so bare that plants cannot grow back. Then bare soil is blown away by winds.

In the Sahel, dry and wet periods usually follow each other. When the seasonal rains do not fall, drought takes hold. A **drought** is a long period of extreme dryness and water shortages. The latest drought occurred in the 1980s. Rivers dried up, crops failed, and millions of animals died. Thousands of people died of hunger. Millions of others fled to more productive southern areas.

Over the years, both overgrazing and drought have ruined once-productive areas of the Sahel. Many grassland areas have become desert—a process called **desertification.** Scientists believe that the increasing human use of land in the Sahel will only increase damage to the land. More of the Sahel will become desert.

History From the A.D. 500s to 1500s, three great African empires—Ghana (GAH•nuh), Mali, and Songhai (SAWNG•HY)—arose in the Sahel. These empires controlled the trade in gold, salt, and other goods between West Africa and the Arab lands of North Africa and Southwest Asia. To learn more about the salt trade, turn to page 560.

In the early 1300s, Mali's most famous ruler, Mansa Musa, made a journey in grand style to Makkah. This is the holy city of Islam located in the Arabian Peninsula. A faithful Muslim, Mansa Musa made his capital, Tombouctou (TOH•book•TOO), a leading center of Islamic learning. People came from all over the Muslim world to study there.

Invaders from North Africa defeated Songhai—the last of the great empires—in the late 1500s. During the 1800s, the Sahel region came

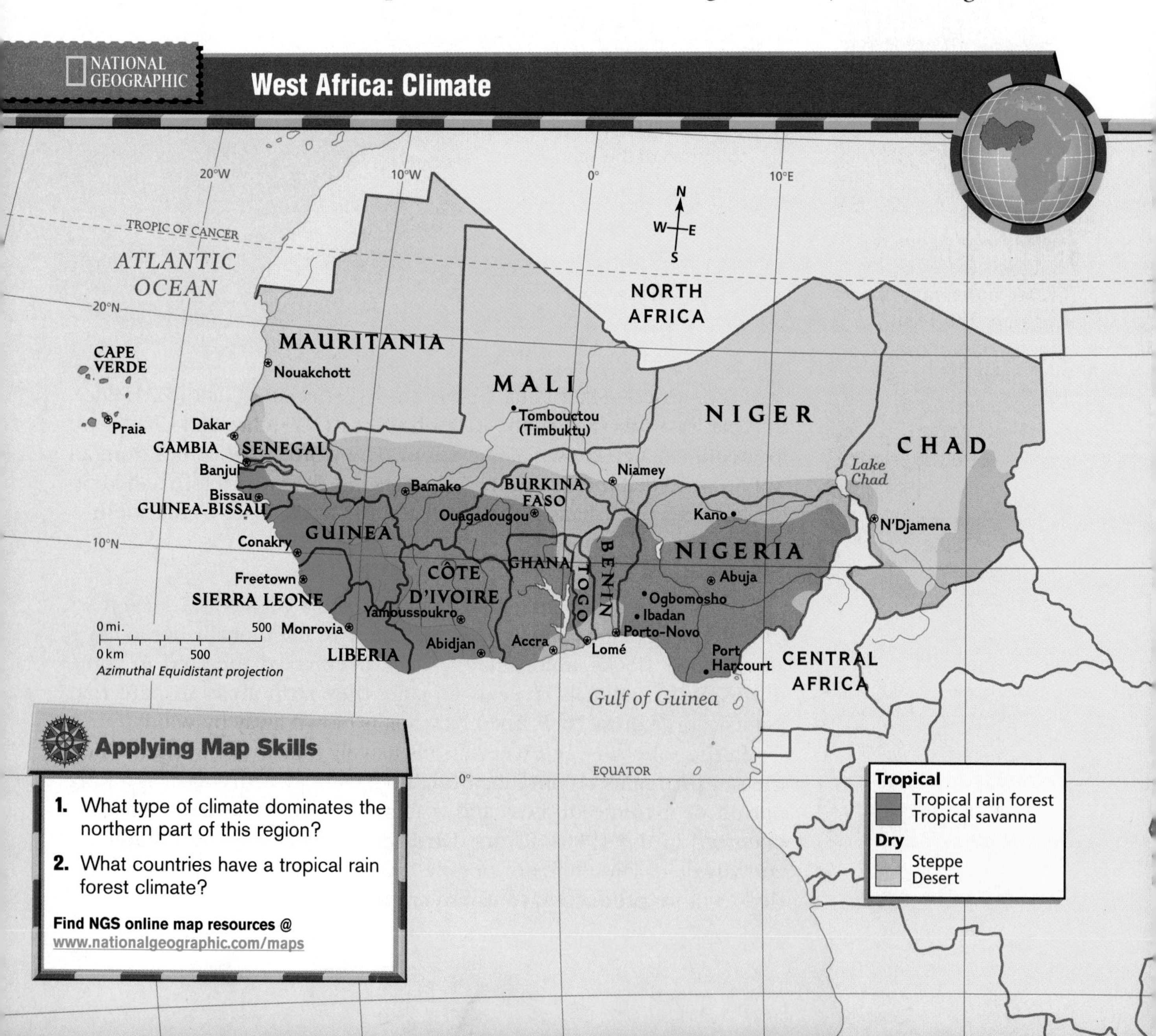

Applying Map Skills

1. What type of climate dominates the northern part of this region?
2. What countries have a tropical rain forest climate?

Find NGS online map resources @ www.nationalgeographic.com/maps

Clothing

To protect themselves from the hot Saharan sun, the Tuareg people wear layers of clothing under their long flowing robes. These loose cotton clothes help slow the evaporation of sweat and conserve body moisture. As a sign of respect for their superiors, Tuareg men cover their mouths and faces with veils. Women usually wear veils only for weddings. The veils are made of blue cloth dyed from crushed indigo. The blue dye easily rubs off onto the skin, earning the Tuareg men the nickname the Blue Men of the Desert.

Looking Closer **How is the clothing of the Tuareg appropriate for the land in which they live?**

under French rule. The French created five colonies in the area. In 1960 these five colonies became the independent nations of Mauritania, Mali, Upper Volta (now Burkina Faso), Niger, and Chad.

✓Reading Check **What has caused the desertification of the Sahel?**

The People of the Sahel

The Sahel countries are large in size but have small populations. If you look at the map on page 554, you will see that most people live in the southern areas of the Sahel. Rivers flow here, and the land can be farmed or grazed. Yet even these areas do not have enough water and fertile land to support large numbers of people.

Today most people in the Sahel live in small towns. They are subsistence farmers who grow grains, such as millet and sorghum (SAWR•guhm). For years, many people were nomads. Groups such as the Tuareg (TWAH•REHG) and the Fulani, for example, would cross the desert with herds of camels, cattle, goats, and sheep. The recent droughts forced many of them to give up their traditional way of life and move to the towns. Here they often live in crowded camps of tents.

The people of the Sahel practice a mix of African, Arab, and European traditions. Most are Muslims and follow the Islamic religion. They speak Arabic as well as a variety of African languages. In many of the larger cities, French is also spoken.

Mauritania The westernmost Sahel country, **Mauritania** borders the Atlantic Ocean. The waters off the coast are a rich fishing ground, but fishing ships from other countries have overfished these waters. As a result, fewer people can make a living from the sea than in the past. Mauritania's first port suitable for large ships opened in 1986 near the

Section 3

Coastal Countries

Guide to Reading

Main Idea

West Africa's coastal countries have a favorable climate for agriculture.

Terms to Know

- cassava
- bauxite
- phosphate

Places to Locate

- Gambia
- Senegal
- Guinea
- Guinea-Bissau
- Cape Verde
- Liberia
- Sierra Leone
- Côte d'Ivoire
- Ghana

Reading Strategy

Create five charts like this one, filling in at least one key fact about five coastal countries for each category.

Country	
Land	
Economy	
Culture	

NATIONAL GEOGRAPHIC **Exploring Our World**

A royal parade in Ghana is a proud display of wealth. The Ashanti people—who live in this country—have been mining gold since the 1300s. The twenty-fifth anniversary of the reign of their king was a perfect opportunity to celebrate the country's rich history. Skilled goldsmiths created the bracelets, rings, and other jewelry worn by the king and his bearers.

In addition to Nigeria and the Sahel countries, West Africa includes 11 coastal countries. One country—Cape Verde—is a group of islands in the Atlantic Ocean. The other countries, including Togo and Benin, stretch along the Gulf of Guinea and the Atlantic coast.

The Land and Economy

Sandy beaches, thick mangrove swamps, and rain forests cover the shores of West Africa's coastal countries. Highland areas with grasses and trees lie inland. Several major rivers flow from these highlands to the coast. They include the Sénégal, Gambia, Volta, and Niger Rivers. Rapids and shallow waters prevent large ships from traveling far inland.

Because they border the ocean, the coastal countries receive plenty of rainfall. Warm currents in the Gulf of Guinea create a moist, tropical rain forest climate in most coastal lowlands year-round. The highland areas have a tropical savanna climate with dry and wet seasons. Some areas have drier steppe and desert climates.

Most people in the region are subsistence farmers. They grow yams, corn, rice, cassava, and other foods for their families. **Cassava** is a plant whose roots are ground into flour to make bread. Coffee, rubber, cacao, palm oil, and kola nuts are cash crops grown on plantations. Kola nuts help give colas their special flavor.

Despite these rich agricultural resources, coastal West African countries import more in industrial goods than they export in natural products. Why? Agricultural products often rise and fall in price suddenly, and their value is not equal to finished goods. To meet their countries' needs, governments have to borrow money from other countries or international organizations.

Reading Check **What types of climates do the coastal countries have?**

History and People

In early times, the powerful and wealthy kingdoms of Ashanti and Abomey ruled West Africa's coastal region. These kingdoms were centers of trade, learning, and the arts. From the late 1400s to the early 1800s, Europeans set up trading posts along the West African coast. From these posts they traded with Africans for gold, ivory, and other goods that people in Europe wanted.

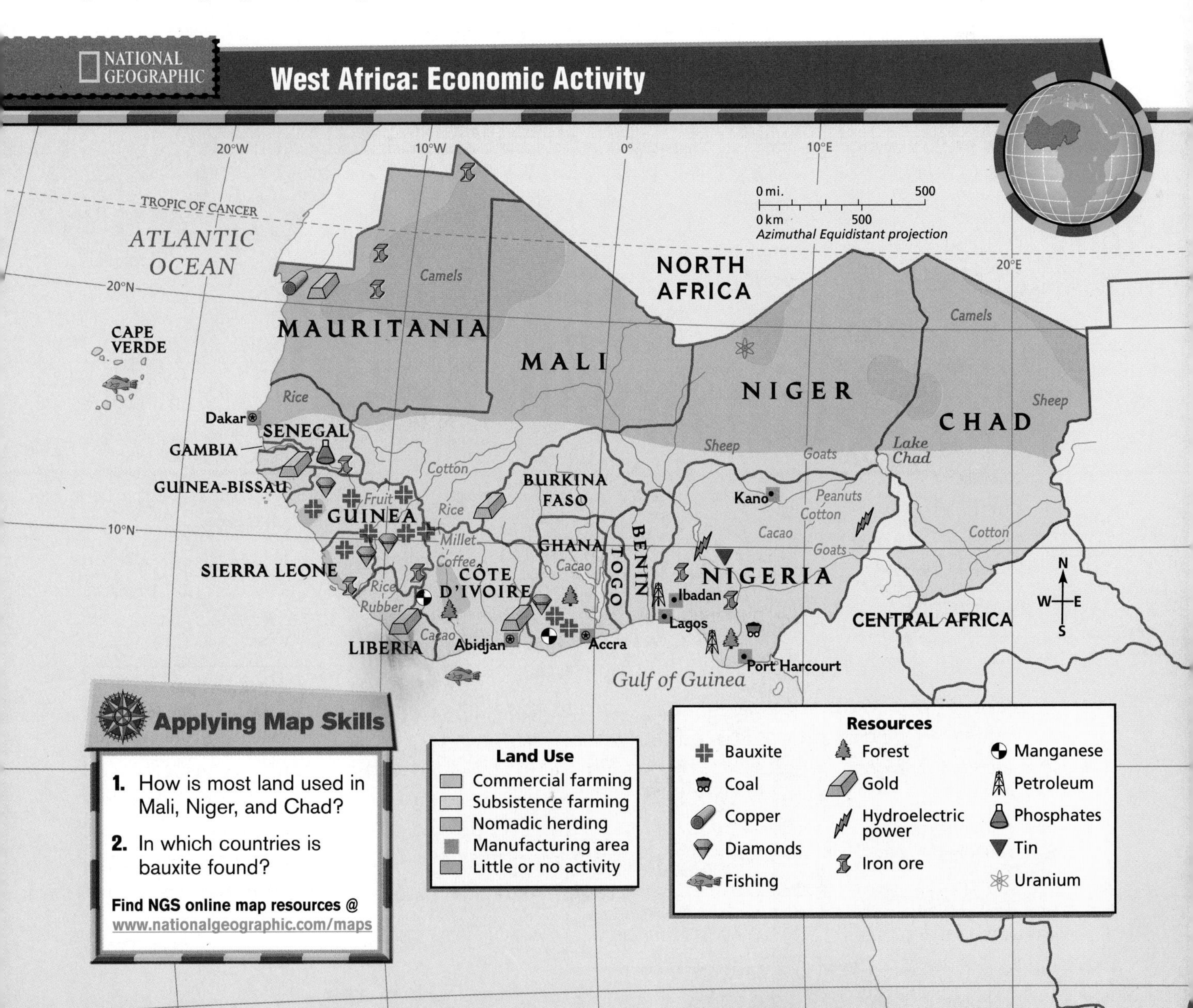

Geography Online
Web Activity Visit the Geography: The World and Its People Web site at gwip.glencoe.com and click on Chapter 19—Student Web Activities to learn more about Senegal.

The Europeans also enslaved and forced millions of Africans to migrate to the Americas to work on plantations and in mines. This trade in human beings, which also took place among African countries, was a disaster for West Africa. The removal of so many young and skilled people devastated West African families, villages, and economies.

The French, British, and Portuguese eventually divided up the coastal region and set up colonies to obtain the region's rich resources. In 1957 Ghana became the first country in West Africa to become independent. By the late 1970s, no West African country was under European rule.

People in coastal West Africa cherish family ties. Some practice traditional African religions, whereas others are Christians or Muslims. Local African languages are spoken in everyday conversation. Languages such as French, English, and Portuguese are used in business or government.

Cities in coastal West Africa are modern and growing. If you were to visit cities such as Ghana's capital—Accra (AH•kruh)—or Côte d'Ivoire's capital—Abidjan (AH•bee•JAHN)—you would find busy downtown areas with modern government and office buildings. You would also see some people dressed in Western-style business clothes and others in traditional African clothing.

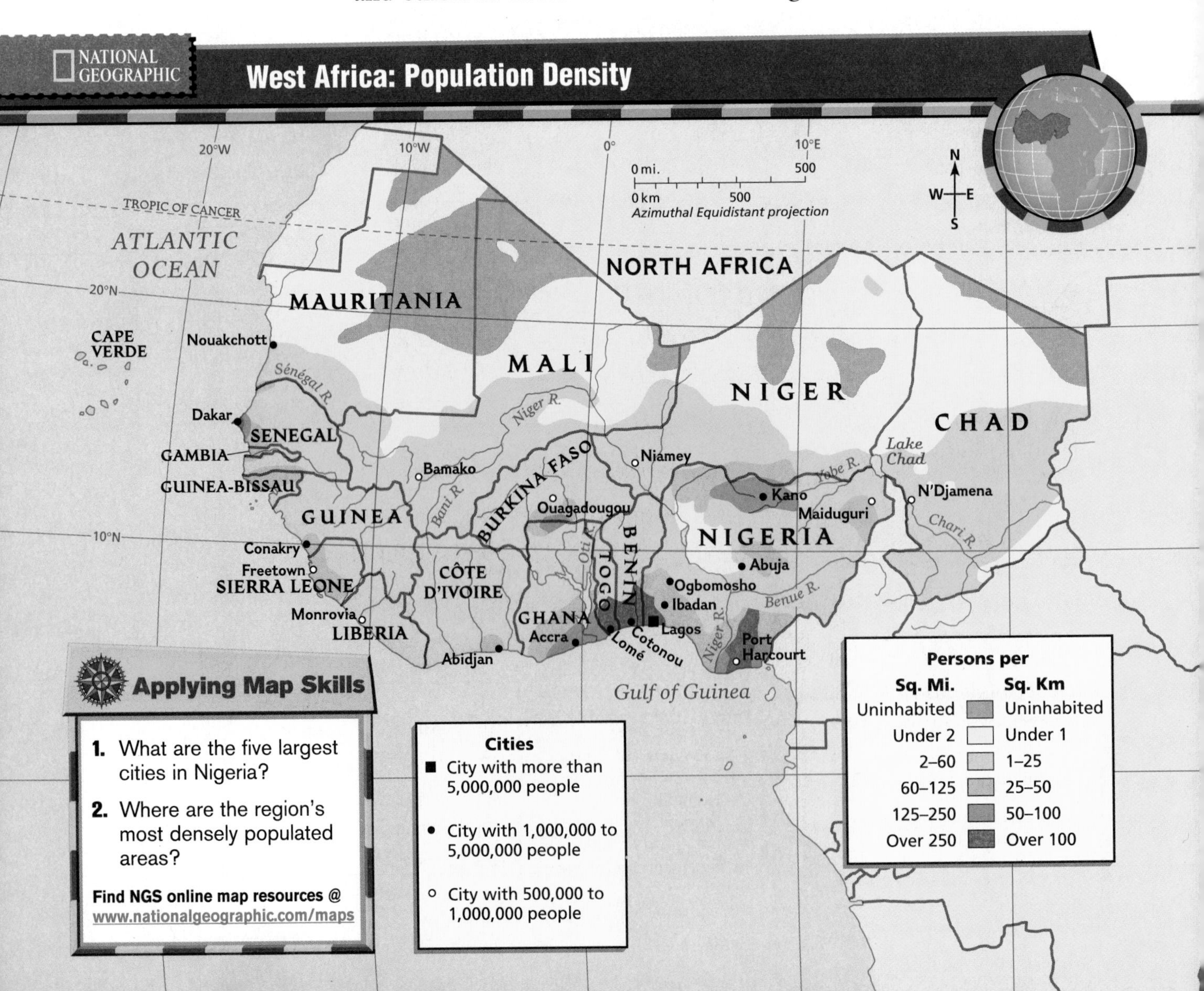

Applying Map Skills

1. What are the five largest cities in Nigeria?
2. Where are the region's most densely populated areas?

Find NGS online map resources @ www.nationalgeographic.com/maps

Gambia, Senegal, and Guinea Most of the people in **Gambia, Senegal,** and **Guinea** work in agriculture. Guinea is also rich in bauxite and diamonds. **Bauxite** is a mineral used to make aluminum. Guinea has about 25 percent of the world's reserves of bauxite. Senegal is an important source of phosphate. **Phosphate** is mineral salt that has phosphorus, which is used in fertilizers.

Gambia proclaimed independence from the British in 1965. Senegal and Guinea were French colonies until becoming independent in 1958 (Guinea) and 1960 (Senegal). Senegal has more people—9.2 million—than the other two countries. About 42 percent live in cities. Dakar (dah•KAHR), Senegal's capital, is a coastal city known for its tree-lined streets, cafés, and markets.

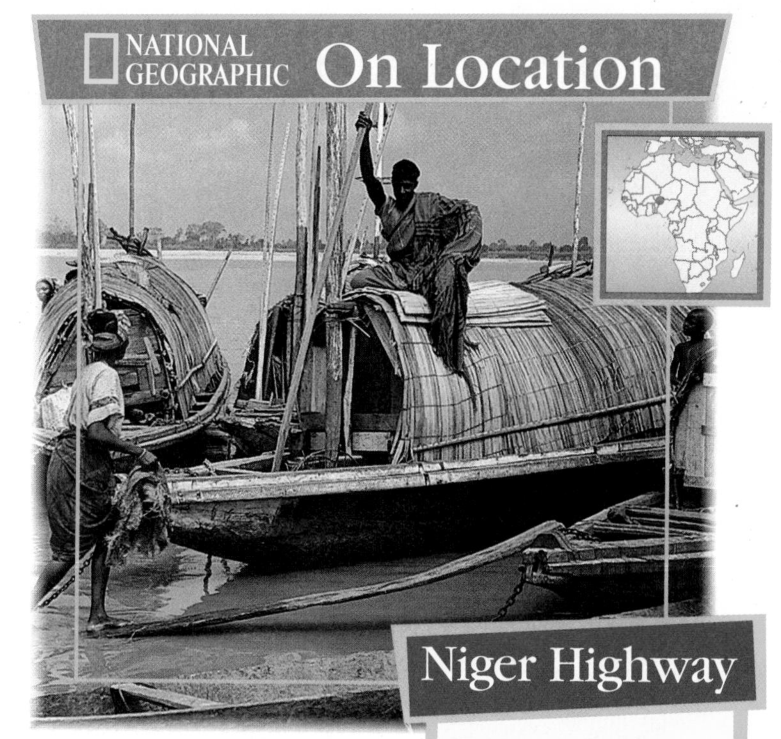

Niger Highway

Rivers of West Africa provide not only water but transportation. Here, freight boats on the Niger River deliver goods to Benin's people.

Place **What prevents large ships from traveling far inland on West Africa's rivers?**

Guinea-Bissau and Cape Verde **Guinea-Bissau** and **Cape Verde** were Portuguese colonies until they won independence in 1975. The 15 volcanic islands of Cape Verde lie about 375 miles (604 km) offshore in the Atlantic Ocean. You will notice a mix of African and Portuguese influences in the languages and cultures of these two countries.

Most of the 1.2 million people in Guinea-Bissau make their living by farming or fishing. Rice—the major crop—is also the basic food item. Do you like to eat cashew nuts? Guinea-Bissau is one of the world's top producers of cashews. Poor soil and low rainfall make the Cape Verde islands unsuitable for farming. Most of the people here work in service industries related to trade, government, and transportation. The islands' 400,000 people must import about 90 percent of their food.

Liberia and Sierra Leone **Liberia** is the only West African nation that was never a colony. African Americans freed from slavery founded Liberia in 1822. Monrovia, Liberia's capital, was named for James Monroe—the president of the United States when Liberia was founded. From 1989 to 1996, a civil war cost many lives and destroyed much of the country's economy.

Like Liberia, **Sierra Leone** was founded as a home for people freed from slavery. The British ruled Sierra Leone from 1787 until 1961. Most of the land is used for farming, but the country also has mineral resources, especially diamonds. As with Liberia, civil war has destroyed recent economic development. Sierra Leone's population of 5.3 million is nearly twice Liberia's 2.9 million.

Côte d'Ivoire **Côte d'Ivoire** has a French name that means "ivory coast." From the late 1400s to the early 1900s, a trade in elephant ivory tusks in Côte d'Ivoire brought profits to European traders. Today the ivory trade is illegal, and the country protects its elephants.

Côte d'Ivoire was a French colony before winning its independence in 1960. It is the world's top producer of cacao beans and also a major producer of coffee and palm oil. The port of Abidjan is the largest urban area and economic center. This busy city has towering office buildings and wide avenues. Abidjan is the official seat of government, but Yamoussoukro (YAH•moo•SOO•kroh), some 137 miles (220 km) inland, has been named the new capital.

Ghana Like Côte d'Ivoire, **Ghana** produces cacao beans for export. Gold and timber are also major sources of income. The country is rich in natural resources, but many people subsistence-farm.

Ghana won its independence from the British in 1957. For a long time, dictators or military leaders ruled the country. Unwise government decisions slowed Ghana's economic growth and brought hardships to its people. In the 1990s, Ghana moved toward democracy.

Ghana's people belong to about 100 ethnic groups. The Ashanti and the Fante are the largest. Many groups still keep their local kings, but these rulers have no political power. The people respect these ceremonial rulers and look to them to keep traditions alive.

About 35 percent of Ghana's people live in cities. Accra, on the coast, is the capital and largest city. A giant dam on the Volta River provides hydroelectric power to urban areas. The dam also has created Lake Volta, the world's largest artificial lake.

Reading Check What are the capitals of Ghana and Côte d'Ivoire?

Assessment

Defining Terms

1. **Define** cassava, bauxite, phosphate.

Recalling Facts

2. **Economics** What types of goods must West African countries import?
3. **History** Which West African country was never a colony?
4. **Government** How much political power do the local kings in Ghana have?

Critical Thinking

5. **Drawing Conclusions** Why is coastal West Africa divided into so many countries?
6. **Analyzing Information** What West African products do you use?

Graphic Organizer

7. **Organizing Information** On a diagram like this one, record when each of the coastal countries discussed in this section gained their independence and from whom.

Country	Date of Independence	Independence From Whom?

Applying Geography Skills

8. **Analyzing Maps** Study the physical map on page 543. Which West African nation has the highest overall elevation? What is that elevation in feet and meters?

Chapter 19 Reading Review

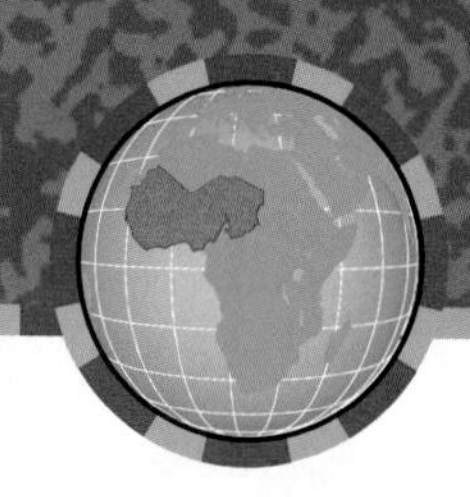

Section 1 Nigeria

Terms to Know
- mangrove
- savanna
- harmattan
- subsistence farm
- cacao
- compound
- civil war

Main Idea

A large, oil-rich country, Nigeria has more people than any other African nation.

✓ **Place** Nigeria's major landforms are coastal lowlands, savanna highlands, and partly dry grasslands.

✓ **Economics** More than 90 percent of Nigeria's income comes from oil exports.

✓ **Culture** Nigeria has more than 300 ethnic groups. The government is taking steps to reduce ethnic conflicts.

Section 2 The Sahel Countries

Terms to Know
- overgraze
- drought
- desertification

Main Idea

The Sahel countries face a continuing struggle to keep grasslands from turning into desert.

✓ **Region** The Sahel countries are Mauritania, Mali, Niger, Chad, and Burkina Faso.

✓ **Region** The Sahel forms a border between the Sahara to the north and fertile lands to the south.

✓ **Human/Environment Interaction** Overgrazing and drought have caused many grassland areas in this region to become desert.

✓ **Economics** Most people in the Sahel are subsistence farmers or livestock herders.

Section 3 Coastal Countries

Terms to Know
- cassava
- bauxite
- phosphate

Main Idea

West Africa's coastal countries have a favorable climate for agriculture.

✓ **Region** The 11 countries that make up coastal West Africa are Senegal, Gambia, Guinea, Guinea-Bissau, Cape Verde, Liberia, Sierra Leone, Côte d'Ivoire, Ghana, Togo, and Benin.

✓ **Economics** West Africa's coastal countries have a good climate for agriculture, and most people are farmers.

✓ **History** With the exception of Liberia, the coastal countries were all European colonies. All had gained their independence by the late 1970s.

The port of Abidjan, Côte d'Ivoire ▶

Chapter 19

Assessment and Activities

Using Key Terms

Match the terms in Part A with their definitions in Part B.

A.

1. overgraze
2. cassava
3. drought
4. mangrove
5. compound
6. phosphate
7. desertification
8. cacao
9. subsistence farm
10. savanna

B.

- **a.** process in which deserts expand
- **b.** a group of houses surrounded by a wall
- **c.** plant whose roots are ground into flour and eaten
- **d.** mineral salt used in fertilizers
- **e.** tropical tree whose seeds are used to make cocoa and chocolate
- **f.** tropical grassland with scattered trees
- **g.** produces enough to support a family's needs
- **h.** extended period of extreme dryness
- **i.** when animals strip the land so bare that plants cannot grow
- **j.** tropical tree with roots above and beneath the water

Reviewing the Main Idea

Section 1 Nigeria

11. **Economics** What is Nigeria's major export?
12. **History** Why have there been so many conflicts in Nigeria since 1960?
13. **Culture** Who was the first African to win the Nobel Prize in literature?
14. **Culture** What are the four largest ethnic groups in Nigeria?

Section 2 The Sahel Countries

15. **Region** The Sahel forms the boundary between what two regions in West Africa?
16. **History** Who was Mansa Musa?
17. **Culture** What religion do most people of the Sahel follow?
18. **Economics** Why has fishing declined in Mauritania?

Section 3 Coastal Countries

19. **Movement** Why are ships unable to sail very far inland in coastal West Africa?
20. **Economics** How are kola nuts used?
21. **History** What two kingdoms controlled the coastal area prior to the 1400s?
22. **Economics** What country is one of the world's top producers of cashews?

West Africa

Place Location Activity

On a separate sheet of paper, match the letters on the map with the numbered places listed below.

1. Gulf of Guinea
2. Nigeria
3. Niger River
4. Liberia
5. Cape Verde
6. Lagos
7. Mali
8. Ghana
9. Chad
10. Monrovia

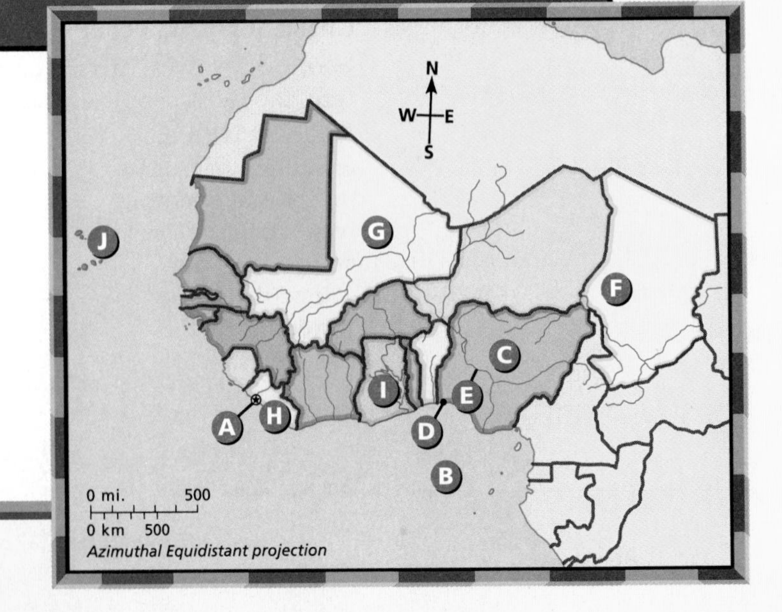

Self-Check Quiz Visit the ***Geography: The World and Its People*** Web site at gwip.glencoe.com and click on **Chapter 19—Self-Check Quizzes** to prepare for the Chapter Test.

Critical Thinking

23. **Evaluating Information** What do you feel is the major challenge facing the countries of West Africa today? Explain your answer.
24. **Sequencing Information** After reviewing the entire chapter, choose what you feel are five of the most important events in the history of West Africa. Place those events and their dates on a time line like this one.

GeoJournal Activity

25. **Writing a Poem** Find out more about one of the countries of West Africa. Imagine that you are there and write an "I am . . ." poem. Begin each line with the words "I am . . ." and then complete it with a description, action, or emotion that you feel relates to the subject. Share your poem with the rest of the class, and ask classmates to identify your poem's subject.

Mental Mapping Activity

26. **Focusing on the Region** Draw a simple outline map of West Africa, then label the following:

- Niger River
- Atlantic Ocean
- Gulf of Guinea
- Tropic of Cancer
- Nigeria
- Niger
- Senegal
- Côte d'Ivoire
- Chad
- Mali
- Mauritania
- Liberia

Technology Skills Activity

27. **Using the Internet** Conduct a search for information about one of the ancient empires or kingdoms of West Africa. Look for maps, pictures, and descriptions of the important rulers. Then write a report using the information you found. Share your report with the rest of the class.

Standardized Test Practice

Directions: Study the graph, then answer the question that follows.

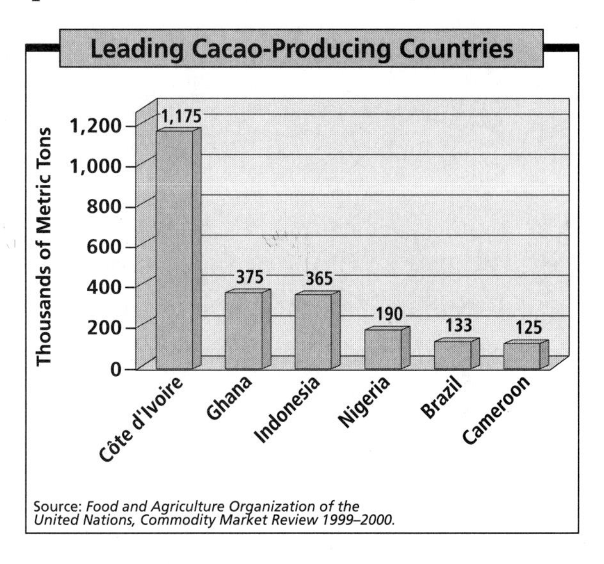

Source: *Food and Agriculture Organization of the United Nations, Commodity Market Review 1999–2000.*

1. What countries on the graph are top cacao-producing countries from West Africa?

A Ghana, Indonesia, and Nigeria
B Côte d'Ivoire, Ghana, and Indonesia
C Côte d'Ivoire, Nigeria, and Cameroon
D Côte d'Ivoire, Ghana, and Nigeria

Test-Taking Tip: The important words in this question are "from West Africa." You need to use information on the graph as well as information you learned in Chapter 19 to answer this question. As with any graph, read the title bar and information along the side and bottom of the graph first. Then analyze and compare the sizes of the bars to one another.

Chapter 20

Central Africa

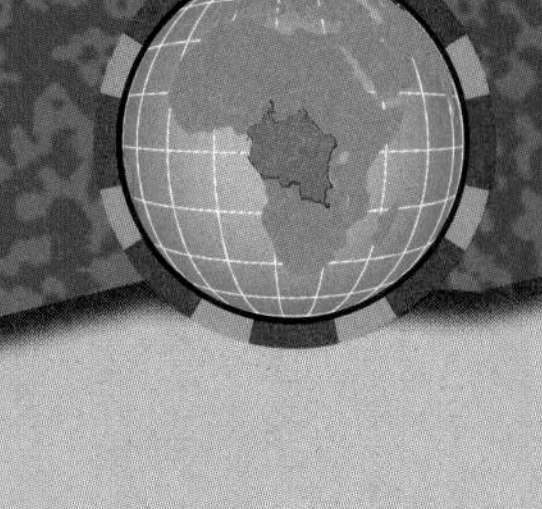

The World and Its People NATIONAL GEOGRAPHIC

To learn more about the people and places of Central Africa, view ***The World and Its People*** **Chapter 20** video.

Geography Online

Chapter Overview Visit the ***Geography: The World and Its People*** Web site at gwip.glencoe.com and click on **Chapter 20—Chapter Overviews** to preview information about Central Africa.

Section 1

Democratic Republic of the Congo

Guide to Reading

Main Idea

The Democratic Republic of the Congo has rich natural resources that are largely undeveloped because of civil war and poor government decisions.

Terms to Know

- savanna
- basin
- canopy
- hydroelectric power
- dictator
- refugee

Places to Locate

- Congo River
- Lake Albert
- Lake Edward
- Lake Kivu
- Lake Tanganyika
- Kinshasa

Reading Strategy

Make a chart like this one. Then list two facts about the Democratic Republic of the Congo for each category.

	Fact #1	Fact #2
Land		
Economy		
People		

NATIONAL GEOGRAPHIC Exploring Our World

Barges act as floating malls on the Congo River. They provide goods to areas too remote for overland travel. At each village along the river, local people come to the barge to sell captured animals, crops, or baskets they have woven. Then they buy goods brought upriver from the cities—anything from bread to clothes to cassette tapes.

Africa's second-longest river—the **Congo River**—flows through the heart of the Democratic Republic of the Congo. This large country is located on the Equator in the very heart of Africa.

The Land and Climate

One-fourth the size of the United States, the Democratic Republic of the Congo has only about 23 miles (37 km) of coastline. Most of its land borders other African countries—nine in all.

High, rugged mountains rise in the eastern part of the country. Here you will find four large lakes—**Lake Albert, Lake Edward, Lake Kivu,** and **Lake Tanganyika** (TAN•guhn•YEE•kuh). Lake Tanganyika is the longest freshwater lake in the world. It is also the second deepest, after Russia's Lake Baikal. Savannas, or tropical grasslands with few trees, cover the highlands in the far north and south of the country. In these areas, lions and leopards stalk antelopes and zebras for food.

◀ Fishers on the Congo River near Kisangani, Democratic Republic of the Congo

One of the world's largest rain forests covers the central basin of the Democratic Republic of the Congo. A **basin** is a broad flat valley. The treetops form a **canopy,** or an umbrella-like forest covering so thick that sunlight rarely reaches the forest floor. More than 750 different kinds of trees grow here. The rain forests are being destroyed at a rapid rate, however, as they are cleared for timber and farmland.

The mighty Congo River—about 2,800 miles (4,506 km) long—weaves its way through the Congo Basin on its journey to the Atlantic Ocean. The river current is so strong that it carries water

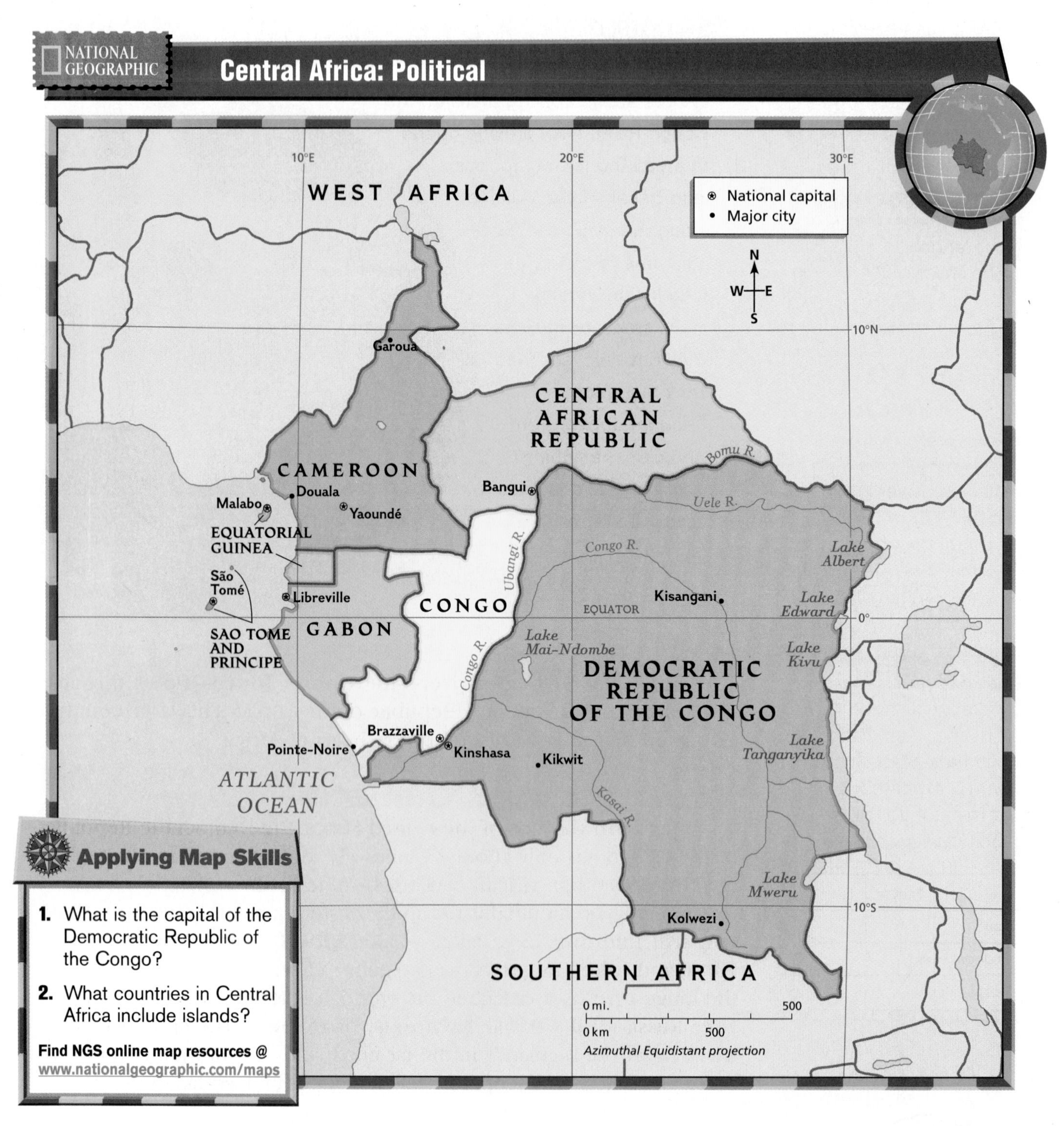

Applying Map Skills

1. What is the capital of the Democratic Republic of the Congo?
2. What countries in Central Africa include islands?

Find NGS online map resources @ www.nationalgeographic.com/maps

about 100 miles (161 km) into the ocean. The Congo River and its tributaries, such as the Kasai River, provide **hydroelectric power,** or electricity generated by flowing water. In fact, these rivers produce more than 10 percent of all the world's hydroelectric power. The Congo is also the country's highway for trade and travel.

A Tropical Climate Because of its location on the Equator, the Democratic Republic of the Congo has a tropical climate. Along the Congo River and in the rain forests, the climate is hot and humid.

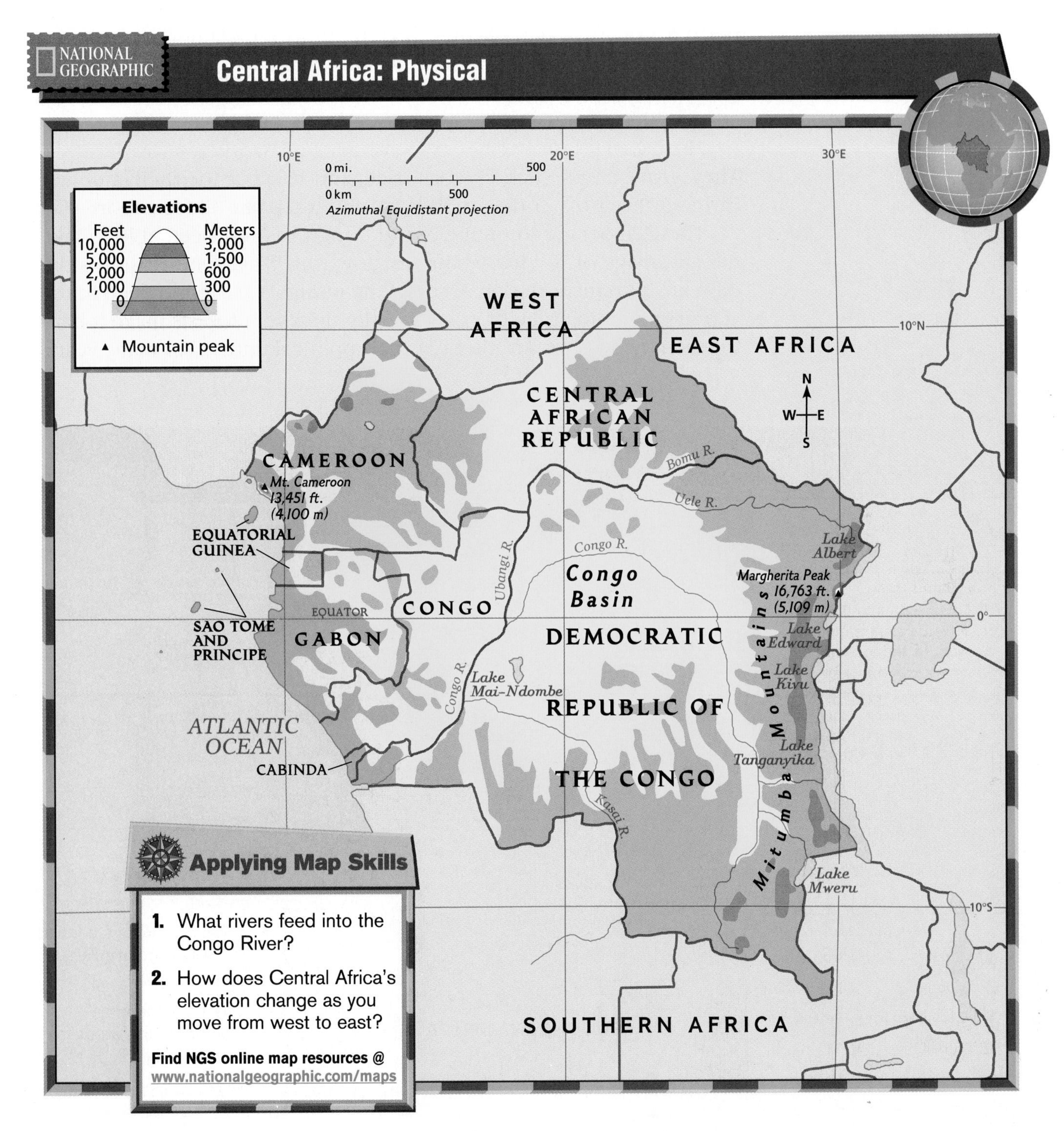

Applying Map Skills

1. What rivers feed into the Congo River?
2. How does Central Africa's elevation change as you move from west to east?

Find NGS online map resources @ www.nationalgeographic.com/maps

Heavy rainstorms bring 80 inches (203 cm) or more of rain each year. In the southern and northern grasslands, rain tends to fall just a few months of the year. The highlands in the east are cooler and drier.

✓Reading Check **Why does the Democratic Republic of the Congo generally have a warm, tropical climate?**

The Economy

The Democratic Republic of the Congo has the opportunity to be a wealthy nation. The map below shows you its many valuable mineral resources. The country exports gold, petroleum, diamonds, and copper. It is Central Africa's main source of industrial diamonds, as shown on the graph on page 567. These diamonds are used in making strong industrial tools that cut metal. The country's factories make steel, cement, tires, shoes, textiles, processed foods, and beverages.

Most Congolese—about two-thirds of all workers—are farmers. They grow crops such as corn, rice, and cassava for their families. Farmers also grow coffee, rubber, cotton, and palm oil for export.

The Democratic Republic of the Congo has not been able to take full advantage of its rich resources, however. Why? One reason is the difficulty of transportation. Many of the minerals are found deep in the country's interior. Lack of roads and the thick rain forests make it hard to reach these areas. Another reason is political unrest. For many years,

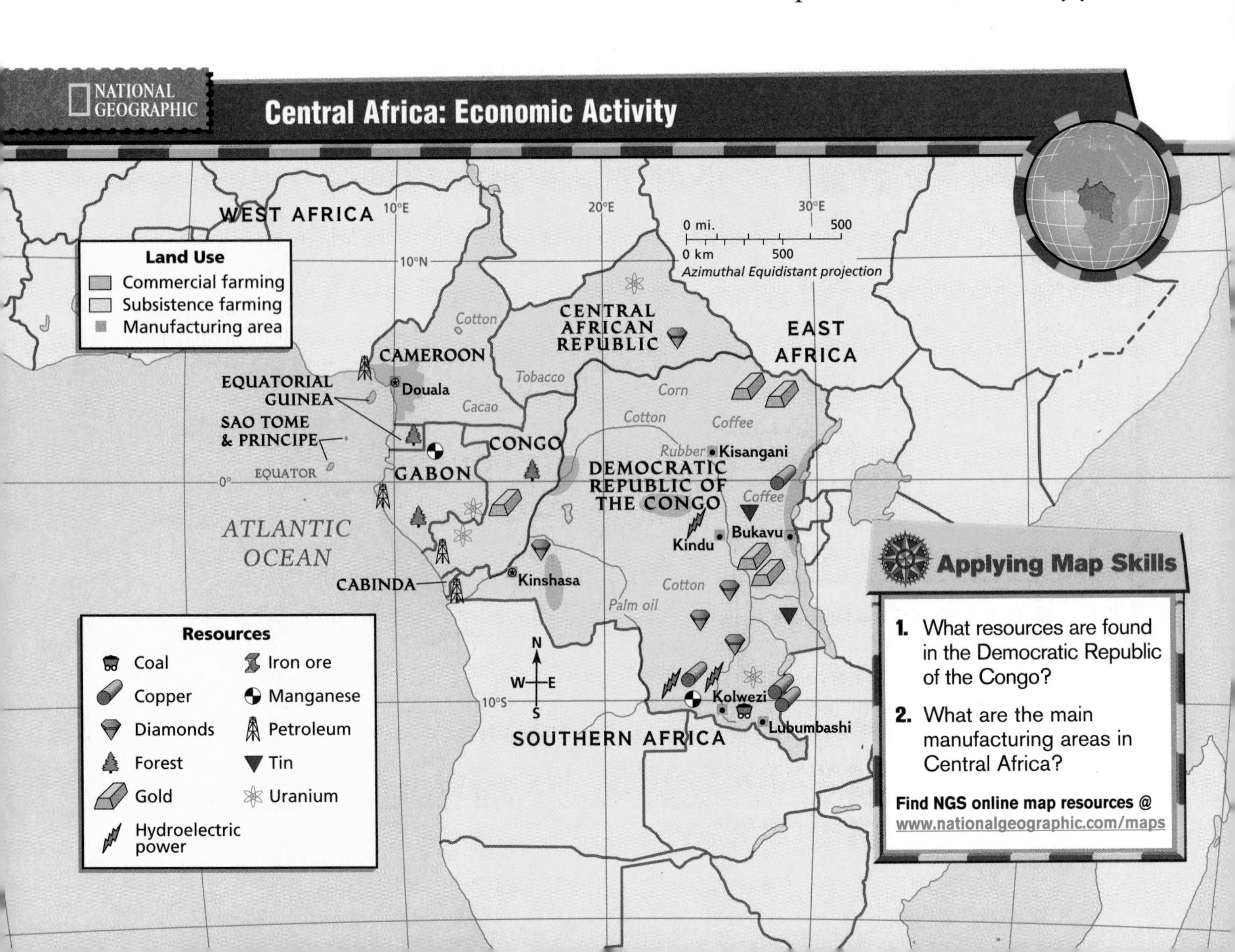

Leading Diamond-Producing Countries

Analyzing the Graph

Three of the world's top diamond-producing countries are in Africa south of the Sahara.

Place **What two countries produce the most diamonds in Africa?**

Visit gwip.glencoe.com and click on **Chapter 20—Textbook Updates.**

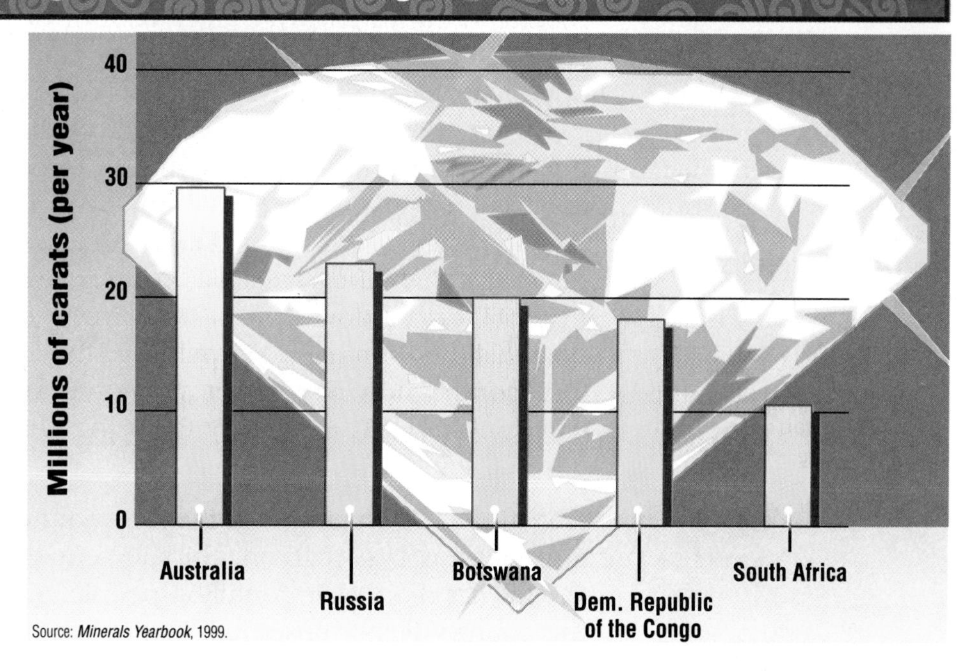

Source: *Minerals Yearbook*, 1999.

power-hungry leaders kept the nation's wealth for themselves. Then a civil war broke out in the late 1990s. This war has hurt efforts to develop the country's economy.

✓Reading Check **What mineral resources does the Democratic Republic of the Congo export?**

The People

The Democratic Republic of the Congo's 50.5 million people consists of as many as 250 different ethnic groups. One of these groups is the Kongo people, after whom the country is named. The official language is French, but many people speak local languages, such as Lingala or Kingwana. More than 75 percent of Congolese are Christians. Most of these are Roman Catholic.

History and Government The Congo region was first settled about 10,000 years ago. The Bantu people—ancestors of most of the Congolese people today—moved here from Nigeria around the A.D. 600s and 700s. Several powerful kingdoms arose in the savannas south of the rain forests. The largest of these kingdoms was the Kongo.

In the late 1400s, Portuguese and other European traders arrived in Central Africa. During the next 300 years, they enslaved many people from the Congo region. Most of these Africans were shipped to the Americas. In the late 1800s, King Leopold II of Belgium made the region his personal plantation. Belgium's government took over the area in 1908 and called the colony the Belgian Congo.

After World War II, Africans pushed European governments to end the practice of ruling Africa as colonies. In 1960 the Belgian Congo

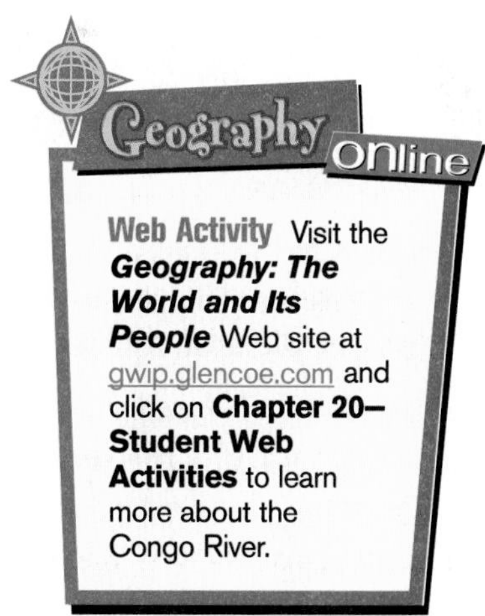

Web Activity Visit the ***Geography: The World and Its People*** Web site at gwip.glencoe.com and click on **Chapter 20—Student Web Activities** to learn more about the Congo River.

The Okapi

The Democratic Republic of the Congo provides the only home for the okapi (oh•KAH•pee). With its long, tough tongue, the okapi pulls leaves off the branches of young trees. Its tongue is so long that the okapi can use it to clean its eyes.

became independent. The country's name was changed to Zaire. From 1965 to 1997, a dictator named Mobutu Sese Seko ruled Zaire. A **dictator** is a leader who takes control of a government and directs affairs as he or she wishes. The corrupt government treated critics and opponents harshly and weakened Zaire's economy.

In the 1990s, civil wars erupted in the neighboring countries of Rwanda and Burundi. These conflicts forced thousands of refugees to enter Zaire. **Refugees** are people who flee to another country to escape danger or disaster. These refugees fought each other at the same time that rebels waged a civil war against Zaire's corrupt leaders. In 1997 Zaire's government was finally overthrown, and Zaire was renamed the Democratic Republic of the Congo. After the rebel leader took control, however, he set up another dictatorship and was assassinated in 2001. The country faces the challenge of building a stable government.

Daily Life Most people in the Democratic Republic of the Congo live in rural areas. Less than one-third are city dwellers. Still, **Kinshasa,** the capital, has more than 5 million people. After years of civil war, life in the country is still unsettled. The economy has nearly collapsed, and many people in the cities are without work.

In rural areas people follow traditional ways of life. They plant seeds, tend fields, and harvest crops. Most of the harvest goes to feeding the family. Any extra goes to the local market—or to the boats moving along the rivers—to sell or trade for goods the people need.

✓Reading Check **What kind of government does the Democratic Republic of the Congo now have?**

Assessment

Defining Terms

1. **Define** savanna, basin, canopy, hydroelectric power, dictator, refugee.

Recalling Facts

2. **Place** What is the second-longest river in Africa?
3. **Economics** Why has the Democratic Republic of the Congo not been able to take full advantage of its resources?
4. **History** When did the Belgian Congo become an independent country?

Critical Thinking

5. **Analyzing Information** Although the Democratic Republic of the Congo is located on the Equator, it does not have the highest temperatures in Africa. Why?
6. **Evaluating Information** Using the Congo River as an example, explain why rivers are so important to economic development.

Graphic Organizer

7. **Organizing Information** On a time line like the one below, label five important events and their dates in the history of the Democratic Republic of the Congo.

Applying Geography Skills

8. **Analyzing Maps** Study the physical map on page 565. What is the highest point in the Democratic Republic of the Congo?

Geography Skill

Mental Mapping

Think about how you get from place to place each day. In your mind you have a picture—or **mental map**—of your route. If necessary, you could probably draw sketch maps like the one below of many familiar places.

Learning the Skill

To develop your mental mapping skills, follow these steps.

- When a country or city name is mentioned, find it on a map to get an idea of where it is and what is near it.
- Draw a sketch map of it and include a compass rose to determine the cardinal directions.
- As you read or hear information about the place, try to picture where on your sketch you would fill in this information.
- Compare your sketch to an actual map of the place. Change your sketch if you need to, thus changing your mental map.

Practicing the Skill

Study the sketch map at right. Picture yourself standing *in* the map, then answer the following questions.

1. If you were facing north, looking at the Chicago Public Library, what route would you take to reach the Chicago Harbor?
2. You are at the Sears Tower, one of the tallest buildings in the world. About how many miles would you have to walk to get to the Medinah Temple?
3. If you met your friend at the public library, would it be too far to walk to the Art Institute? Should you take a taxi? Explain.

Applying the Skill

Think about your own neighborhood. Draw a sketch map of it from your mental map. Which neighborhood streets or roads did you include? What are the three most important features on your map?

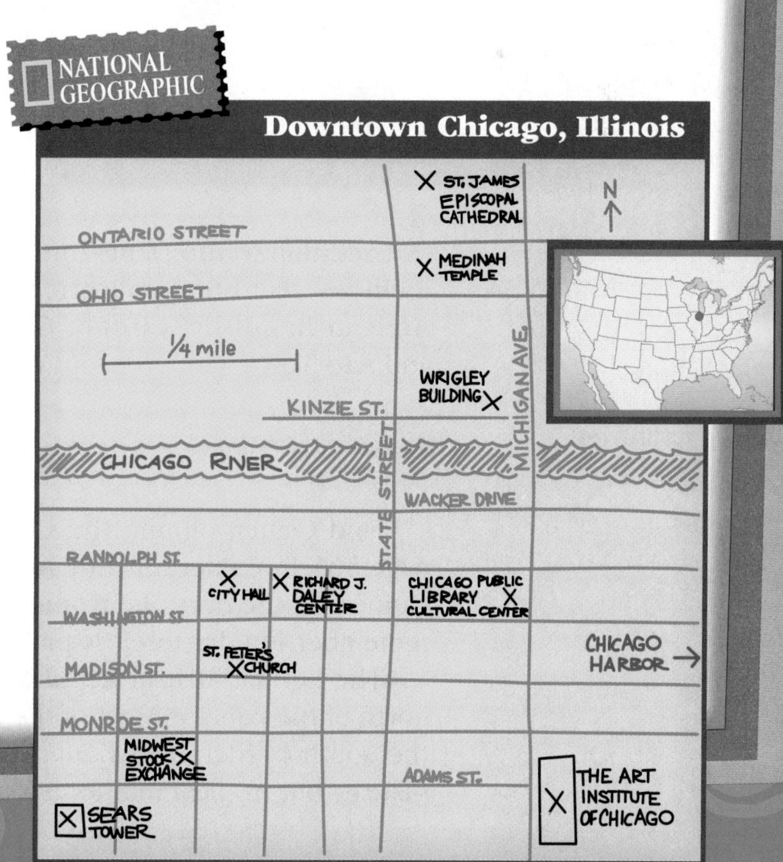

Section 2

Other Countries of Central Africa

Guide to Reading

Main Idea

Other countries in Central Africa have fairly small populations, with most people farming the land.

Terms to Know

- steppe
- tsetse fly
- industrialize
- deforestation

Places to Locate

- Cameroon
- Central African Republic
- Congo
- Gabon
- Equatorial Guinea
- Sao Tome and Principe
- Ubangi River

Reading Strategy

List facts about two countries in Central Africa in the outer parts of ovals like these. Where the ovals overlap, list facts that are true of both countries.

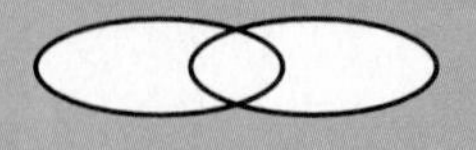

In the 2000 Olympic Summer Games, the gold medal in men's soccer went to Cameroon's team, the Indomitable Lions. The streets of Cameroon's capital, Yaoundé, and other cities were jammed with wildly excited fans screaming with joy. Cameroon's president even declared the following Monday a national holiday to celebrate the victory.

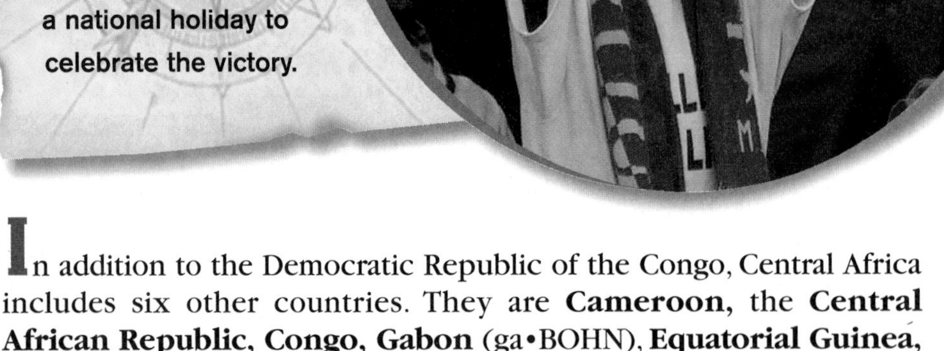

In addition to the Democratic Republic of the Congo, Central Africa includes six other countries. They are **Cameroon,** the **Central African Republic, Congo, Gabon** (ga•BOHN), **Equatorial Guinea,** and **Sao Tome** (sow too•MAY) **and Principe** (PREEN•see•pee).

Cameroon and the Central African Republic

Find Cameroon and the Central African Republic on the map on page 565. In Cameroon, hot, humid lowlands stretch along the Gulf of Guinea. To the north lie tropical savannas and steppes. A **steppe,** you remember, is a dry treeless plain often found on the edges of a desert.

The Central African Republic lies deep in the middle of Africa, just north of the Equator. Most of the country lies on a flat plateau, although the south borders the Congo Basin. Savannas make up most of the plateau, but tropical rain forests are found in the south.

The Economy and People Most people in the Central African Republic and Cameroon farm for a living. A few large plantations raise cacao, cotton, tobacco, and rubber for export. Some people herd livestock in areas that are safe from tsetse flies. The bite of the **tsetse** (SEET•see) **fly** causes a deadly disease called sleeping sickness. Turn to page 574 to find out more about sleeping sickness.

These two countries are only beginning to **industrialize,** or base their economies more on manufacturing and less on farming. Cameroon has had greater success in this effort. It has coastal ports and forest products, petroleum, and bauxite. The Central African Republic can claim only diamond mining as an important industry.

A colony of France until 1960, the Central African Republic recognizes French as its official language. Yet most of its people speak Sango, the national African language, to ease communication among the many ethnic groups. Cameroon, divided between the British and the French until 1960, uses both English and French as its official languages. The largest cities are Douala (doo•AH•lah), Yaoundé (yown•DAY), Cameroon's capital, and Bangui (bahng•GEE), the capital of the Central African Republic.

Reading Check **Why has Cameroon had greater success than the Central African Republic in industrializing?**

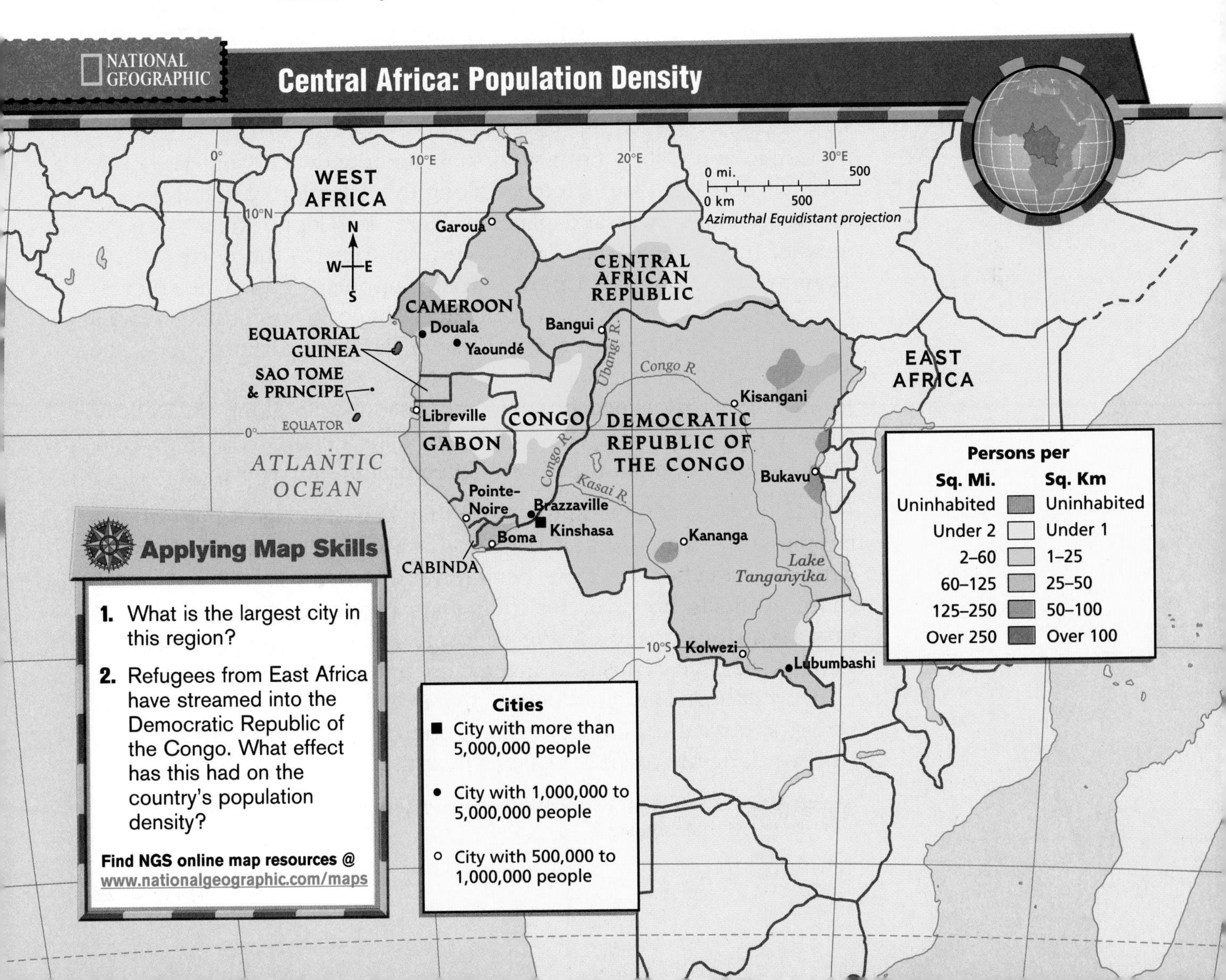

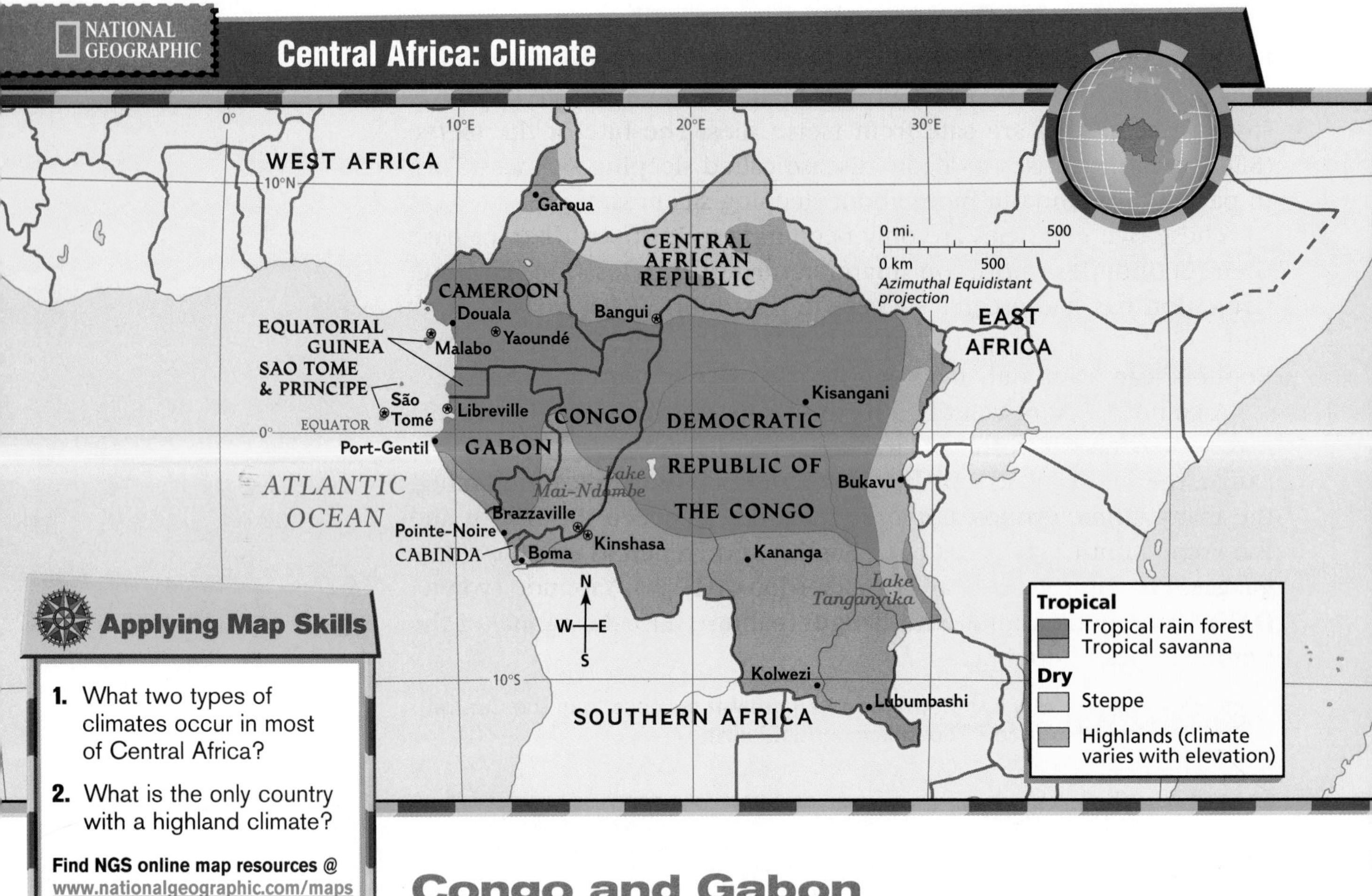

Applying Map Skills

1. What two types of climates occur in most of Central Africa?
2. What is the only country with a highland climate?

Find NGS online map resources @ www.nationalgeographic.com/maps

Congo and Gabon

Congo and Gabon both won their independence from France in 1960. In Congo, a plain stretches along the Atlantic coast and rises to low mountain ranges and plateaus. Here the Congo River supports most of the country's farmlands and industries. To the north, a large swampy area along the **Ubangi River** supports dense vine thickets and tropical trees. Both the Ubangi and Congo Rivers provide Congo with hydroelectric power. They also make Congo the door to the Atlantic Ocean for trade and transport.

In Gabon, palm-lined beaches and swamps run along the Gulf of Guinea. Dense rain forests cover most of the country. This lush landscape is home to more than 3,000 different kinds of vegetation.

The Economy and People More than half of Congo's and Gabon's people farm small plots of land. Both countries' economies rely on exports of lumber. They are beginning to depend more on rich offshore oil fields, however, for their main export. Gabon suffers from **deforestation,** or the cutting of too many trees too quickly. Gabon also has valuable deposits of manganese and uranium.

Only about 1.2 million people live in Gabon—mainly along rivers or in the coastal capital, Libreville. Congo's 2.7 million people generally live along the Atlantic coast or near the capital, Brazzaville.

Reading Check **What two exports are most important to Congo and Gabon?**

Island Countries

The map on page 564 shows you that Equatorial Guinea and Sao Tome and Principe are both island countries. Equatorial Guinea includes land on the mainland of Africa and five islands. Sao Tome and Principe consists of two main islands and several smaller ones.

Equatorial Guinea Once a Spanish colony, Equatorial Guinea won its independence in 1968. Today the country is home to about 400,000 people. Most live on the mainland, although the capital and largest city—Malabo (mah•LAH•boh)—is on the country's largest island.

Farming, fishing, and harvesting wood are the country's main economic activities. For many years, timber and cacao grown in the islands' rich volcanic soil were the main exports. Oil was recently discovered and now leads all other exports.

Sao Tome and Principe The island country of Sao Tome and Principe gained its independence from Portugal in 1975. The Portuguese had first settled here about 300 years earlier. At that time, no people lived on the islands. Today about 200,000 people live here, with almost all living on the main island of Sao Tome.

Sao Tome and Principe are volcanic islands. As a result, the soil is rich and productive. Farmworkers on the islands grow various crops, including coconuts and bananas for export. The biggest export crop, however, is cacao, which makes cocoa.

✓Reading Check **Which of these island countries is also located on the African mainland?**

Assessment

Defining Terms

1. **Define** steppe, tsetse fly, industrialize, deforestation.

Recalling Facts

2. **Place** What is the main landform of the Central African Republic?
3. **Economics** Name three natural resources of Cameroon.
4. **Economics** What natural resource was recently discovered in Equatorial Guinea?

Critical Thinking

5. **Evaluating Information** Why do you think Europeans wanted to colonize parts of Africa, like the Congo?
6. **Understanding Cause and Effect** How could furniture buyers in the United States affect the countries of Central Africa?

Graphic Organizer

7. **Organizing Information** Complete a chart like this with one fact about each country.

Country	Fact
Cameroon	
Central African Republic	
Congo	
Gabon	
Equatorial Guinea	
Sao Tome & Principe	

Applying Geography Skills

8. **Analyzing Maps** Study the physical map on page 565. The Ubangi River forms part of the boundaries of what countries?

Making Connections

ART SCIENCE LITERATURE TECHNOLOGY

Battling Sleeping Sickness

Since the 1300s, people in Africa south of the Sahara have battled a disease now commonly called sleeping sickness. Yet it was not until the early 1900s that scientists began to understand the disease and that it was transmitted through the bite of an infected tsetse fly.

The Tsetse Fly

Found only in parts of Africa, the tsetse fly is the common name for any of about 21 species of flies that can transmit sleeping sickness. The flies are larger than the houseflies common to the United States. Tsetse flies thrive in forests and in areas of thick shrubbery and trees near lakes, ponds, and rivers.

Although the bite of a tsetse fly is painful, the bite itself is not necessarily harmful. What gives the tsetse fly its dreadful reputation is the disease-causing parasite it may carry.

Sleeping Sickness

The World Health Organization (WHO) estimates that more than 60 million people in Africa are at risk of being infected with sleeping sickness. As many as 500,000 people carry the disease. If left untreated, the disease leads to a slow breakdown of bodily functions and, eventually, death. Sleeping sickness need not be fatal. When the disease is treated in its early stages, most people recover. Treatment is expensive, however, and many of those infected lack medical care. Even if they are cured, they may become infected again.

Disease Prevention

Stopping sleeping sickness is a major challenge for poor countries struggling to meet the many needs of their people. There is no known vaccine. Controlling the disease involves isolating and treating infected humans. Wild animals with the disease must be removed so that tsetse flies cannot get the parasite in the first place. In addition, woodlands and brush near cities must be cleared. Because the disease is such a threat to human life, it is important that governments and private companies support scientists' efforts to wipe out the disease.

Making the Connection

1. Where do tsetse flies live?
2. What causes sleeping sickness?
3. **Drawing Conclusions** Why is treatment of infected humans only part of the solution to eliminating sleeping sickness?

◀ **Children in Central Africa have learned to report bites of the tsetse fly.**

Reading Review

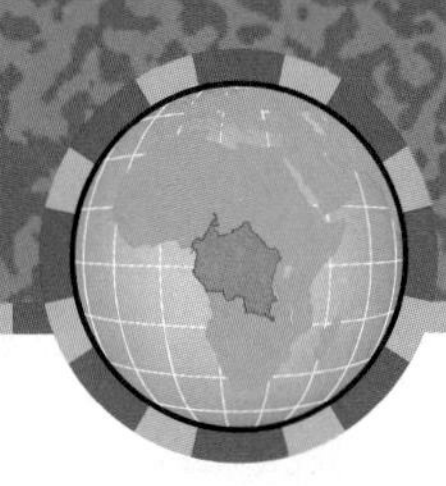

Section 1 Democratic Republic of the Congo

Terms to Know
savanna
basin
canopy
hydroelectric power
dictator
refugee

Main Idea

The Democratic Republic of the Congo has rich natural resources that are largely undeveloped because of civil war and poor government decisions.

- ✓ **Place** Large areas of rain forests and savannas cover the Democratic Republic of the Congo. It has a warm climate because of its location on the Equator.
- ✓ **Movement** The Congo River—the second-largest river in Africa—provides transportation and hydroelectric power.
- ✓ **Place** The Democratic Republic of the Congo has many resources, but most people live in the countryside and farm.
- ✓ **History** Many years under a corrupt and harsh ruler have prevented the economy from fully developing.
- ✓ **Government** A recent civil war overthrew a harsh ruler, but an elected government is not yet in place.

Section 2 Other Countries of Central Africa

Terms to Know
steppe
tsetse fly
industrialize
deforestation

Main Idea

Other countries in Central Africa have fairly small populations, with most people farming the land.

- ✓ **Region** Rain forests cover much of the other countries of Central Africa. The Central African Republic, which receives less rainfall, has only a small area of rain forests.
- ✓ **Culture** Most people in these countries make their living by farming.
- ✓ **Economics** Cameroon earns money by exporting oil and bauxite.
- ✓ **Economics** Congo uses river water to generate hydroelectric power.
- ✓ **Economics** Gabon exports lumber from its rain forests, but now most of its export earnings come from selling oil.
- ✓ **Place** Equatorial Guinea—once a Spanish colony—includes land on the African mainland and some islands.

Market in Kinshasa, Democratic Republic of the Congo ▶

Chapter 20

Assessment and Activities

Using Key Terms

Match the terms in Part A with their definitions in Part B.

A.

1. industrialize
2. basin
3. hydroelectric power
4. canopy
5. deforestation
6. dictator
7. tsetse fly
8. savanna
9. steppe
10. refugee

B.

a. insect whose bite causes sleeping sickness
b. economy based on manufacturing
c. dry, treeless grasslands
d. tropical grassland with scattered trees
e. person who flees to another country for safety
f. electric power generated by flowing water
g. a single all-powerful leader
h. loss of forests due to widespread cutting of trees
i. topmost layer of a rain forest
j. broad, flat lowland area surrounded by higher areas

Reviewing the Main Ideas

Section 1 Democratic Republic of the Congo

11. **Location** What important latitude line runs across the Democratic Republic of the Congo?
12. **Human/Environment Interaction** Why are the rain forests in the Democratic Republic of the Congo being destroyed at a rapid rate?
13. **Place** What is the longest freshwater lake in the world?
14. **Economics** What is the major use of the diamonds found in the Democratic Republic of the Congo?
15. **Government** What is the current government of the Democratic Republic of the Congo?

Section 2 Other Countries of Central Africa

16. **Place** Why must Cameroon's people watch where they herd livestock?
17. **Culture** What is the official language of the Central African Republic? Why?
18. **Place** What is the capital of the Congo?
19. **Economics** Name three of Gabon's natural resources.
20. **History** Who originally settled Sao Tome and Principe?

Central Africa

Place Location Activity

On a separate sheet of paper, match the letters on the map with the numbered places listed below.

1. Democratic Republic of the Congo
2. Cameroon
3. Congo River
4. Kinshasa
5. Lake Tanganyika
6. Gabon
7. Central African Republic
8. Equatorial Guinea
9. Kasai River
10. Ubangi River

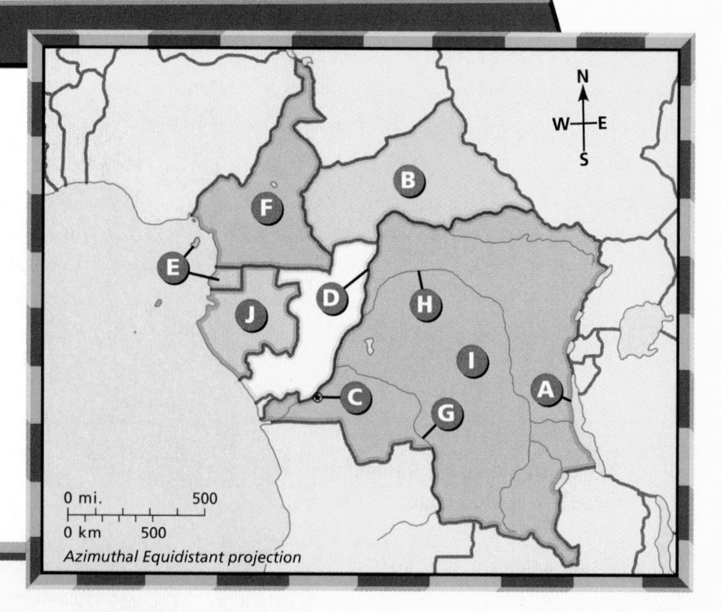

Self-Check Quiz Visit the ***Geography: The World and Its People*** Web site at gwip.glencoe.com and click on **Chapter 20–Self-Check Quizzes** to prepare for the Chapter Test.

Critical Thinking

21. **Synthesizing Information** If you were a government leader in the Democratic Republic of the Congo, what steps would you take to develop the country's resources?
22. **Comparing Information** Compare the economies of the Central African Republic and Cameroon. At the ends of the arrows on a diagram like the one below, list the factors that show which country's economy is the strongest.

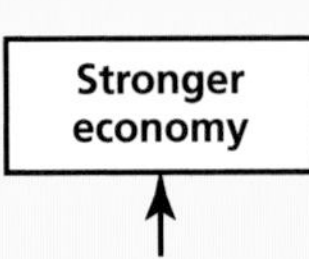

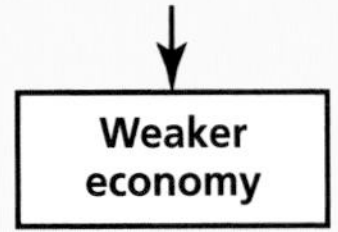

GeoJournal Activity

23. **Writing a Diary** Research the events leading to the overthrow of Mobutu Sese Seko, the former dictator of Zaire. Write at least five diary entries that an eyewitness might have recorded during that turbulent period.

Mental Mapping Activity

24. **Focusing on the Region** Draw a simple outline map of Africa, then label the following:
    - Kinshasa
    - Congo River
    - Cameroon
    - Democratic Republic of the Congo
    - Lake Tanganyika
    - Gabon
    - Congo

Technology Skills Activity

25. **Using a Spreadsheet** Use an almanac or an online source to find the value of the exports for each country of Central Africa. Enter this information and each country's name into a spreadsheet. Then find each country's per capita gross domestic product (GDP) and enter this into the spreadsheet. Create graphs that compare the exports and GDP.

Standardized Test Practice

Directions: Read the paragraph below, then answer the questions that follow.

Gabon is known for its tropical rain forests and savanna grasslands. Both the floor and the canopy of the rain forests are alive with animals. Squirrels, monkeys, baboons, lemurs, and parrots often live their entire lives in the dense branches of the canopy. Pythons, vipers, porcupines, and tortoises move along the rain forest floor. Crocodiles and hippopotamuses live along the riverbanks, while elephants and antelopes roam the savanna. Gorillas, which are endangered in most other places of Africa, are very numerous in Gabon.

1. **Which of the following animals are least likely to live in a tropical rain forest?**
   - **A** elephants
   - **B** baboons
   - **C** pythons
   - **D** parrots
2. **Which of the following would most likely harm Gabon's rain forest habitats?**
   - **F** building a port on the coast
   - **G** discovering oil in the Gulf of Guinea
   - **H** farming small plots of land
   - **J** exporting lumber to furniture makers

Test-Taking Tip: Look in the paragraph to find clues to support your answer. Plus, think about the words *rain forests* and *savanna*. Which has a wet climate and which has a dry climate? As you study for an exam, make note of important words, such as *habitat*. The glossary of your textbook can help you define these words. You may also want to make your own word list as you read new chapters.

Chapter 21

East Africa

The World and Its People NATIONAL GEOGRAPHIC

To learn more about the people and places of East Africa, view ***The World and Its People*** **Chapter 21** video.

Geography Online

Chapter Overview Visit the ***Geography: The World and Its People*** Web site at gwip.glencoe.com and click on **Chapter 21–Chapter Overviews** to preview information about East Africa.

Climate Because of its location on the Equator, Kenya has temperatures that tend to be warm year-round. Remember that **altitude,** or height above sea level, also affects climate. The highland regions tend to be cooler than lowlands. The highlands, as well as the coastal regions, also receive more rain than do the central plains. Kenya's best farmland is found in the highlands.

✓Reading Check **What huge fault runs through East Africa?**

Kenya's Economy

Kenya has a developing economy based on a **free enterprise system.** In this economic system, people can start and run businesses with limited government involvement. Kenya's capital,

Applying Map Skills

1. What highland plain does Kenya share with Tanzania?
2. What is the average elevation of Rwanda and Burundi?

Find NGS online map resources @ www.nationalgeographic.com/maps

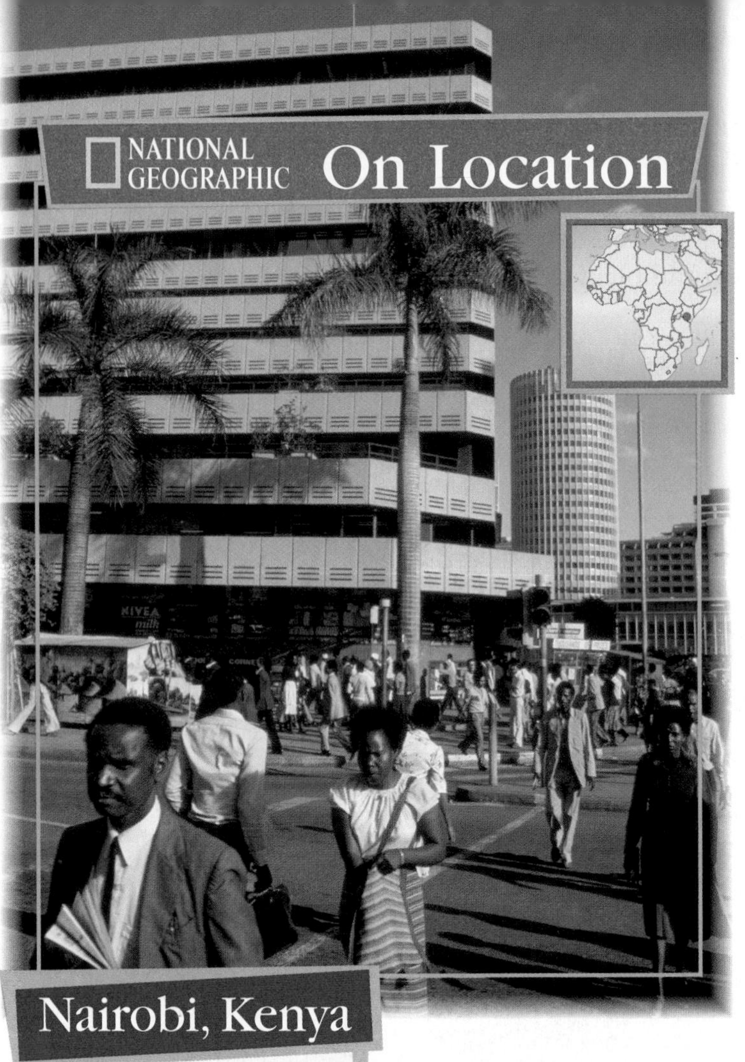

Nairobi, Kenya

Like most cities, Kenya's capital has crowded markets, high-rise office buildings, and elegant mansions. Many city workers maintain close ties to relatives in the countryside.

Place **About how many people live in Nairobi?**

Nairobi (ny•ROH•bee), has become a center of business and commerce for all of East Africa. Foreign companies have set up regional headquarters in this city. Nairobi offers them good transportation and communications systems.

Many Kenyans remain poor, however. Nomadic herding and subsistence farming are still the main economic activities. The main crops are corn, cassava, potatoes, sweet potatoes, and bananas. **Cassava** is a plant whose roots are ground into flour to make bread. Some larger farms raise coffee and tea for export. In recent years, the weather has not been good for crops. Also, corrupt practices of government officials have hurt the economy.

Manufacturing forms only a small part of Kenya's economy. Factories process foods and make vehicles, chemicals, metal, and paper. One of the fastest-growing industries is tourism. Thousands of tourists visit Kenya each year to take trips called safaris. On safaris, visitors tour the nature preserves in jeeps and buses to see the country's wildlife in its natural surroundings.

Reading Check **What problems have hurt Kenya's economy?**

Kenya's People

The people of Kenya believe in *harambee,* which means "pulling together." Even though they come from many different ethnic groups, they have worked together to try to strengthen their country. The spirit of *harambee* has led people to build schools and clinics in their communities. They have raised money to send good students to universities.

History and Government Many scientists believe that people may have first lived in Kenya about 2 million years ago. Over the centuries, many groups in Africa came to the region. During the A.D. 700s, Arab traders from Southwest Asia settled along the coast.

Africans and Arabs lived and worked together. As a result, a blending of cultures took place. The Swahili language came about from this blending. Swahili developed in coastal areas of Kenya. The name *Swahili* comes from an Arabic word meaning "of the coast." The language includes features of several African languages as well as Arabic. Today Swahili is one of Kenya's two official languages. English is the other.

Kenya

Guide to Reading

Main Idea

Kenya is a country of diverse landscapes and peoples.

Terms to Know

- coral reef
- nature preserve
- poaching
- fault
- escarpment
- altitude
- free enterprise system
- cassava

Places to Locate

- Kenya
- Indian Ocean
- Great Rift Valley
- Mt. Kenya
- Nairobi
- Mombasa

Reading Strategy

Draw a diagram like this one. Under each heading in the outer ovals, write two examples of the diversity in Kenya's land, economy, and people.

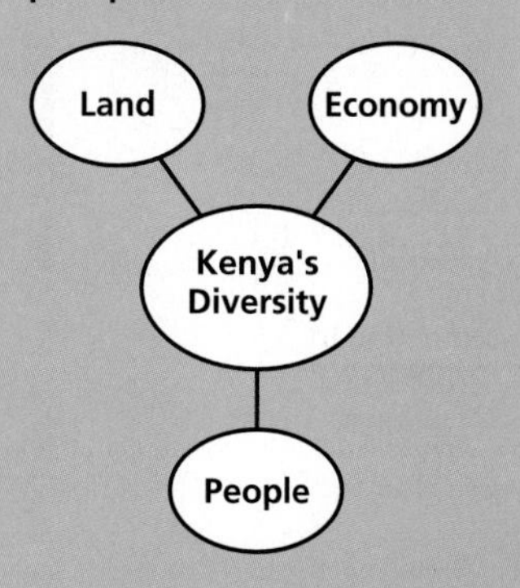

The Masai (mah•SY) is one of Kenya's many ethnic groups. Rituals have shaped their lives for hundreds of years. Young men take part in an important four-day ceremony. When it ends, they become elders and help make group decisions. In the ceremony, elders tell them, "Drop your weapons and use your head and wisdom instead."

Both traditional and modern cultures meet in **Kenya.** Groups like the Masai follow ways of life similar to their ancestors. People in cities live in apartments and work in offices.

Kenya's Land and Climate

About two times the size of Nevada, Kenya straddles the middle of East Africa. The country's **Indian Ocean** coastline has stretches of white beaches lined with palm trees. Offshore lies a **coral reef,** a natural formation at or near the water's surface that is made of the skeletons of small sea animals. The beautiful beaches and the colorful fish of the coral reef attract tourists from around the world.

West of Kenya's coastal plain, the land rises gently to the central part of the country. Much of this upland plain is dry, but enough rain falls on some hilly areas to support farming. Lions, elephants, rhinoceroses, and other wildlife roam this plain. Millions of acres are designated **nature preserves**—land set aside by a government to protect plants and wildlife. Still, in recent years there has been heavy **poaching,** the illegal hunting of protected animals.

◀ A view of Kilimanjaro in Tanzania, from Kenya

Dominating the western part of the country are highlands and the **Great Rift Valley.** This valley is really a fault—a crack in the earth's crust. The Great Rift Valley begins in southeastern Africa and stretches about 3,500 miles (5,633 km) north to the Red Sea.

The Great Rift Valley actually has two branches. The western branch plows through the western borders of Tanzania, Burundi, Rwanda, and Uganda. The eastern, or Kenyan, branch ranges from 30 to 80 miles (48 to 129 km) wide. Steep cliffs called escarpments tower on both sides of the valley. In many places water has flooded the valley to form lakes. Volcanoes also dot the area. One of them—**Mt. Kenya**—rises 17,058 feet (5,199 m) high.

The British gained influence in Kenya during the late 1800s and made it a colony after World War I. Attracted by the mild climate and fertile soil, many British people moved to the highlands. They took land from the Africans and set up farms to grow coffee and tea for export.

By the 1940s, Kenya's African groups organized and fought to end British rule. Kenya finally won its independence in 1963 and became a republic. The country's first president, Jomo Kenyatta (JOH•moh kehn•YAHT•uh), won respect as an early leader in Africa's movement for freedom. Under Kenyatta, Kenya enjoyed economic prosperity and had a stable government. In recent years, the economy has weakened under the president who followed Kenyatta. In response, many Kenyans have demanded democratic changes.

Kenya Today Kenya's 28.8 million people are divided among 40 different ethnic groups. The Kikuyu (kee•KOO•yoo) people are Kenya's main group, making up less than one-fourth of the population. If you visited Kenya, you would discover that most Kenyans live in rural areas. With the constant threat of drought, many people struggle to grow crops. In recent years, large numbers of people have moved to cities in search of a better life.

About one-third of Kenya's people live in cities. Nairobi is the largest city, with 1 million people. **Mombasa** (mohm•BAH•sah) is Kenya's chief port on the Indian Ocean. This city has the best harbor in East Africa, making it an ideal site for oceangoing trade.

✓Reading Check **What city is Kenya's chief port?**

I Am a Samburu

Nimfa Lekuuk is a member of the Samburu of northern Kenya. The word *Samburu* means "the people with the white goats." Nimfa wears the traditional clothes of Samburu women. She is in standard 7 now. "Standard" is the Kenyans' term for *grade.* She studies language, math, history, geography, science, arts and crafts, and religions.

Assessment

Defining Terms

1. Define coral reef, nature preserve, poaching, fault, escarpment, altitude, free enterprise system, cassava.

Recalling Facts

2. Place Describe the Great Rift Valley.

3. Culture What are Kenya's official languages?

4. History Who was Jomo Kenyatta?

Critical Thinking

5. Making Predictions How might a rapidly growing population create problems for Kenya?

6. Evaluating Information Why do you think many African people move from rural areas to urban areas?

Graphic Organizer

7. Organizing Information *Harambee* means "pulling together." On the lines of a diagram like the one below, show examples of *harambee* in Kenya.

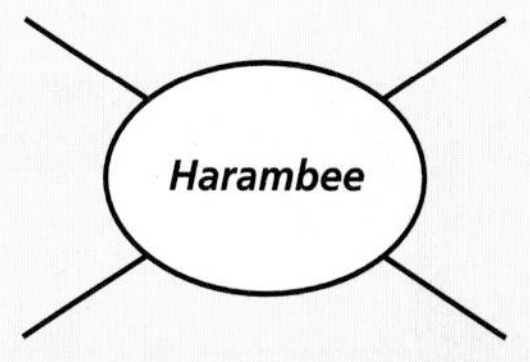

Applying Geography Skills

8. Analyzing Maps Study the political map on page 582. What is Nairobi's latitude and longitude?

Critical Thinking Skill

Making Predictions

Predicting consequences is obviously difficult and sometimes risky. The more information you have, however, the more accurate your predictions will be.

Learning the Skill

Follow these steps to learn how to better predict consequences:

- Gather information about the decision or action you are considering.
- Use your knowledge of history and human behavior to identify what consequences could result.
- Analyze each of the consequences by asking yourself: How likely is it that this will occur?

Practicing the Skill

Study the graph below, then answer these questions:

1. What is measured on this graph? Over what time period?
2. In what year did the fewest tourists visit Kenya?
3. What trend does the graph show?
4. Do you think this trend is likely to continue?
5. On what do you base this prediction?
6. What are three possible consequences of this trend?

Applying the Skill

Analyze three articles in your local newspaper. Predict three consequences of the actions in each of the articles. On what do you base your predictions?

GO TO Practice key skills with **Glencoe Skillbuilder Interactive Workbook, Level 1.**

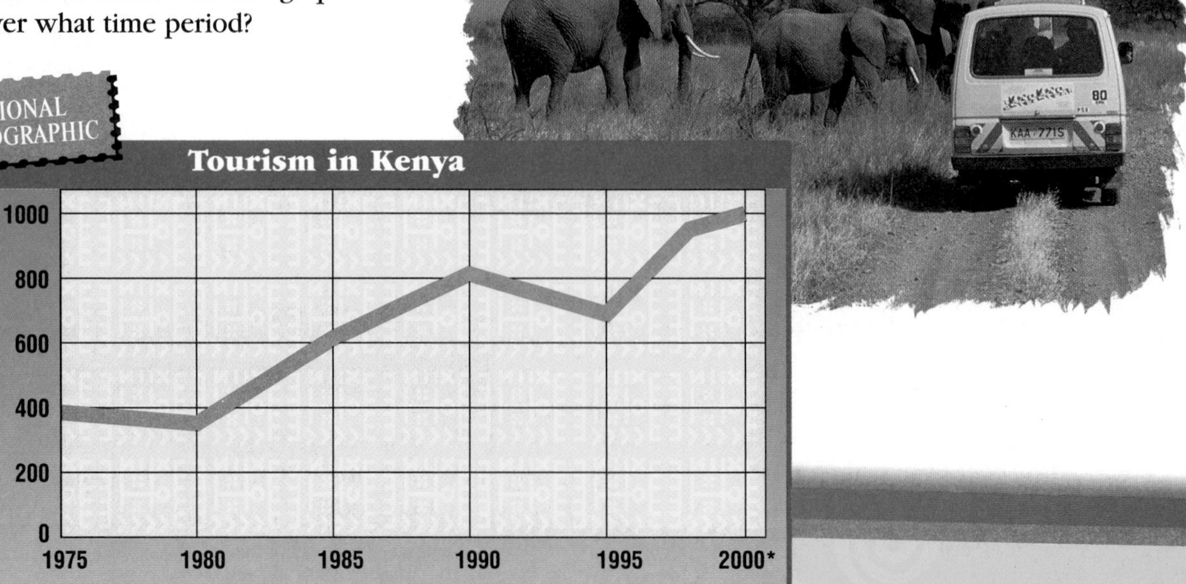

Section 2

Tanzania

Guide to Reading

Main Idea

Tanzania is located on the Indian Ocean and relies on agriculture and tourism.

Terms to Know

- sisal
- habitat
- eco-tourist

Places to Locate

- Tanzania
- Serengeti Plain
- Kilimanjaro
- Great Rift Valley
- Lake Tanganyika
- Lake Victoria
- Dar es Salaam

Reading Strategy

Make a chart like this one. Then list two facts about the land, economy, and people of Tanzania.

Tanzania	Fact #1	Fact #2
Land		
Economy		
People		

This red colobus monkey in Tanzania is tired! You would be too if you had spent the day like she did. She usually leaps about 25 feet (8 m) from tree to tree—with a baby hanging on to her. If she were human size, it would be the same as jumping 50 feet (15 m)—without a running start—while carrying an extra 25 pounds (11 kg) of weight.

If you step out into the vast open plains of **Tanzania,** you may suddenly feel very small. You are in the territory of one of the largest wild animal populations in the world. Tanzania includes a large mainland area—once called Tanganyika (TAN•guhn•YEE•kuh)—and three islands—once called Zanzibar. In 1964 these two areas united, forming the country of Tanzania.

Tanzania's Land

About two times the size of California, Tanzania has landforms and climates similar to those in Kenya. The Indian Ocean coastline boasts white beaches and palm trees, bordered by a thin band of humid lowlands. Farther inland, the land gradually slopes upward to a plateau. The **Serengeti** (SEHR•uhn•GEH•tee) **Plain,** with its huge grasslands and patches of trees and shrubs, dominates this plateau. To the north, near the Kenyan border, a snowcapped mountain called **Kilimanjaro** towers over this region. It is the highest point in Africa.

The **Great Rift Valley** cuts two gashes through Tanzania, one in the center of the country and the other along the western border. As in Kenya, parts of the western valley are covered by lakes. Unusual fish swim in the deep, dark waters of **Lake Tanganyika.** You can see another body of water in the northwest corner of Tanzania—**Lake Victoria.** This body of water is Africa's largest lake and one of the sources of the Nile River.

Just east of the mainland lie Tanzania's three islands—Pemba, Zanzibar, and Mafia Island. Find these islands on the map on page 583. They were all formed from coral.

Reading Check **What mountain in Tanzania is the highest point in Africa?**

Tanzania's Economy

More than 80 percent of all Tanzanians work in farming or herding. Most of them grow food on subsistence farms. Some farmers grow coffee, cotton, tea, cashews, and tobacco for export. Another export crop is sisal, a plant fiber used to make rope and twine. Do you enjoy eating baked ham? If so, you might have tasted the spice called cloves, often used to flavor ham. The islands of Zanzibar and Pemba produce more cloves than any other places in the world.

The country's few factories process foods or make cement, soap, or textiles. Tanzania has deposits of gold and precious gems like emeralds and diamonds. Yet it does not have enough money to set up large-scale mining operations to get to these resources.

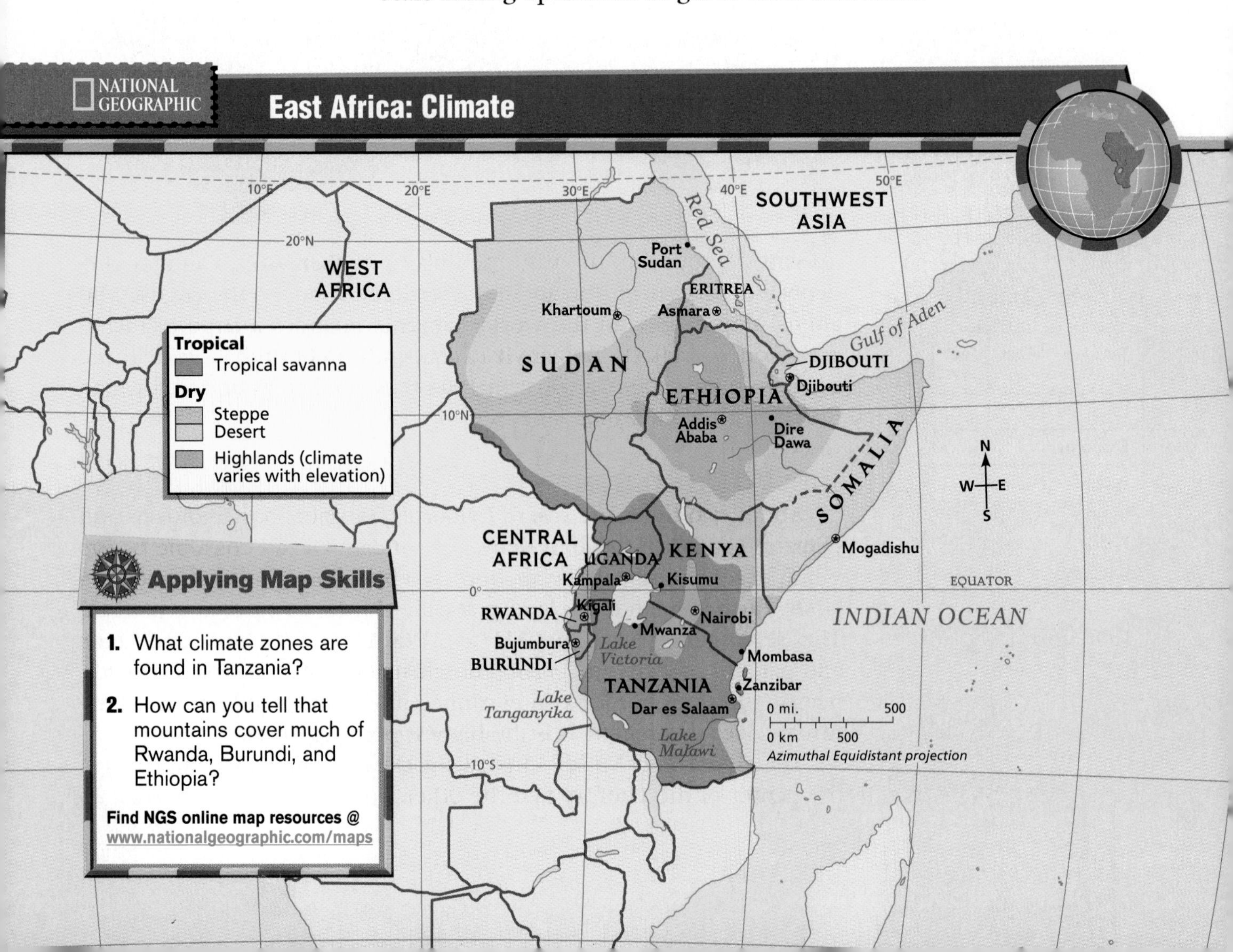

Applying Map Skills

1. What climate zones are found in Tanzania?
2. How can you tell that mountains cover much of Rwanda, Burundi, and Ethiopia?

Find NGS online map resources @ www.nationalgeographic.com/maps

Two Views of Tanzania

Village markets in Tanzania offer both food products and a chance to socialize (above). Zanzibar, once called Spice Island, has a growing tourist industry (left).

Place Zanzibar leads the world in the production of what spice?

Tourism is a fast-growing industry in Tanzania. The government has set aside several national parks to protect the habitats of the country's wildlife. A **habitat** is the type of environment in which a particular animal species lives. Serengeti National Park covers about 5,600 square miles (14,504 sq. km). Lions and wild dogs hunt among thousands of zebras, wildebeests, and antelopes. The park attracts many **eco-tourists,** or people who travel to another country to view its natural wonders.

Tanzania's leaders are also taking steps to preserve farmland. In recent years, many trees have been cut down. Without trees, the land cannot hold soil or rainwater in place. As a result, the land dries out, and soil blows away. To prevent the land from becoming desert, the government of Tanzania has announced a new policy. For every tree that is cut down, five new trees should be planted.

✓Reading Check **What is Tanzania doing to try to stop the land from becoming a desert?**

Tanzania's History and Government

Scientists have found what they believe are the remains of some of the earliest human settlements in Tanzania. By about A.D. 500, groups from other parts of Africa had settled in the area. About 200 years later, Arabs from Southwest Asia set up trading centers on Zanzibar and coastal areas. As in Kenya, the Arabs along the coast mixed with the local Africans to form a new culture in which Swahili was spoken.

In the early 1500s, the Portuguese set up a trading post in Zanzibar. Later, in the late 1800s, Germany won control of the southern part of East Africa and set up a colony here. By the early 1900s, tens of thousands of East Africans rebelled against German rule. Many died in fighting, but the rebellion launched a movement for self-rule. After World War I, the British took charge of Zanzibar and mainland Tanganyika. The mainland became independent in 1961, followed by Zanzibar two years later. In 1964 the two countries united as Tanzania. Since then, Tanzania has been one of Africa's most politically stable republics.

During the 1960s, Tanzania's government controlled the economy. By the 1990s, however, it had moved the country toward a free market system. In taking this step, Tanzania's leaders hoped to improve the economy and reduce poverty. Meanwhile, the country's government also became more democratic with more than one political party.

✓Reading Check **How did Tanzania's economy and government change in the 1990s?**

Tanzania's People

Tanzania's 31.3 million people include more than 120 different ethnic groups. Each group has its own language, but most people also speak Swahili. The two main religions are Christianity and Islam.

About three-fourths of Tanzania's people live in rural areas. **Dar es Salaam,** on the Indian Ocean, is Tanzania's capital and chief port. With nearly 1.4 million people, it is also the country's largest city. The central area of Tanzania has few people. In an effort to encourage people to move there, the Tanzanian government plans to eventually move the capital inland to the city of Dodoma.

Tanzanian music and dance dominate much of East Africa's culture. In Dar es Salaam, you can sway to the music's strong rhythms and Swahili-based sounds performed by local dance bands. In Tanzania, as well as in Kenya, the most popular food is barbecued meat.

✓Reading Check **Where do three-fourths of Tanzania's people live?**

Assessment

Defining Terms

1. **Define** sisal, habitat, eco-tourist.

Recalling Facts

2. **Place** What is the highest mountain in Africa?
3. **Economics** Why are tourists drawn to Tanzania?
4. **Culture** What are the two major religions of Tanzania?

Graphic Organizer

5. **Organizing Information** On a time line like the one below, label five important events and their dates in Tanzania's history.

Critical Thinking

6. **Evaluating Information** Why do you think two countries such as Tanganyika and Zanzibar would unite?
7. **Analyzing Information** Why would the government of Tanzania put so much effort into preserving its national parks?

Applying Geography Skills

8. **Analyzing Maps** Study the physical and political maps on pages 582 and 583. Name the four bodies of water that border Tanzania. On what body of water is Dar es Salaam located?

Making Connections

ART | SCIENCE | LITERATURE | TECHNOLOGY

A Changing Kenya

As developing countries modernize, traditional ways of life often change. In her poem, Kenyan poet and playwright Micere Githae Mugo expresses the challenge of living in a changing world.

Where are those Songs?

by Micere Githae Mugo

Where are those songs
my mother and yours
always sang
fitting rhythms
to the whole
vast span of life?

What was it again
they sang
harvesting maize, threshing millet, storing
the grain . . .

What did they sing
bathing us, rocking us to sleep . . .
and the one they sang
stirring the pot
(swallowed in parts by choking smoke)?

What was it
the woods echoed
as in long file
my mother and yours and all the women on
our ridge
beat out the rhythms
trudging gaily
as they carried
piles of wood
through those forests
miles from home

What song was it?

. . . .

▲ A mother takes care of her children in Kenya.

Sing
I have forgotten
my mother's song
my children
will never know.
This I remember:
Mother always said
sing child sing
make a song
and sing
beat out your own rhythms
the rhythms of your life
but make the song soulful
and make life
sing

. . . .

Making the Connection

1. Based on the poem, how do you think the poet's mother felt about her responsibilities?
2. How do you think the poet feels about her mother?
3. **Analyzing Information** What do you think the poet's mother meant when she said to "beat out your own rhythms"?

Section 3

Inland East Africa

Guide to Reading

Main Idea

Rwanda, Burundi, and Uganda have suffered much conflict in recent years.

Terms to Know

- plantains
- autonomy
- watershed
- endangered species
- refugee

Places to Locate

- Uganda
- Rwanda
- Burundi
- Lake Victoria
- Nile River
- Kampala

Reading Strategy

Make a chart like this one. On the left, write the cause of conflict in each country under that country's name. Then write the effects of that conflict.

Cause of conflict in:	Effects of conflict
Uganda	
Rwanda	
Burundi	

NATIONAL GEOGRAPHIC **Exploring Our World**

If you walk through the mountain rain forests of Rwanda, you might feel you are being watched. Who's watching you? It could be one of the world's 600 remaining gorillas—the rarest and largest of the great apes. Every day these gorillas face the threat of death from poachers, loss of their habitat, disease, and civil war.

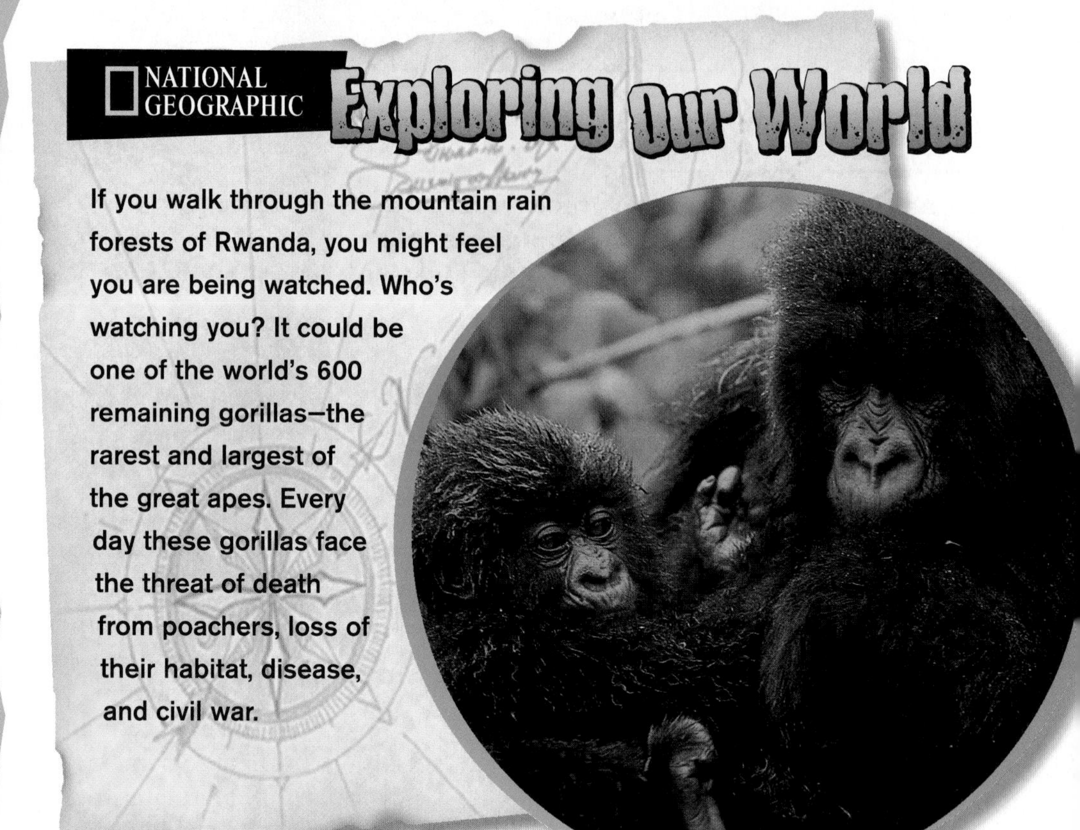

West of Kenya and Tanzania lie **Uganda, Rwanda,** and **Burundi.** All three are landlocked—they have no land touching a sea or an ocean. Instead, they use three large lakes for transportation and trade.

Uganda

Once called "the pearl of Africa," Uganda is a fertile, green land of mountains, lakes, and wild animals. About the size of Oregon, Uganda sits astride the western branch of the Great Rift Valley. Find Uganda on the map on page 583. The country consists mainly of a central plateau. South of the plateau you find **Lake Victoria.** From here, part of the **Nile River** flows through the central plateau to lakes in the west. Although Uganda lies on the Equator, temperatures are mild because of the country's high elevation.

Uganda's rich soil and plentiful rain make the land good for farming. More than 80 percent of Uganda's workers are employed in agriculture. Most farmers work on subsistence farms. They grow **plantains**—a kind of banana—cassava, potatoes, corn, and grains. Some plantations grow coffee, cotton, and tea for export. Coffee makes up nearly three-fourths of the country's exports. Uganda's few factories make cement, soap, sugar, metal, and shoes.

Uganda's People Uganda's 22.8 million people live mainly in rural villages in the southern part of the country. **Kampala,** the capital, lies on the shores of Lake Victoria, making it a port city for local trade.

About two-thirds of Ugandans are Christians. The remaining one-third practice Islam or traditional African religions. At one time there were large numbers of Hindus and Sikhs from South Asia living in the country. A dictator drove them out in 1972. Recently, the Ugandan government has invited them back, and many are now returning.

The people of Uganda belong to more than 40 different ethnic groups. They have a rich cultural heritage of songs, folktales, and poems. In the past, these were passed by word of mouth from one generation to the next. Today they are now in print. For the most part, Ugandans have a very basic diet. Their meals often include beans, beef, goat, mutton, cornmeal, and a variety of tropical fruits.

History and Government For much of the 1900s, the British ruled Uganda. After Uganda won its freedom in 1962, fighting broke out among ethnic groups. These ethnic groups had enjoyed **autonomy,** or self-government, in their local territories under their kings. These kings lost power in 1967, and the ethnic regions were tightly bound to the central government. A period of rule by Idi Amin, a cruel dictator, hurt

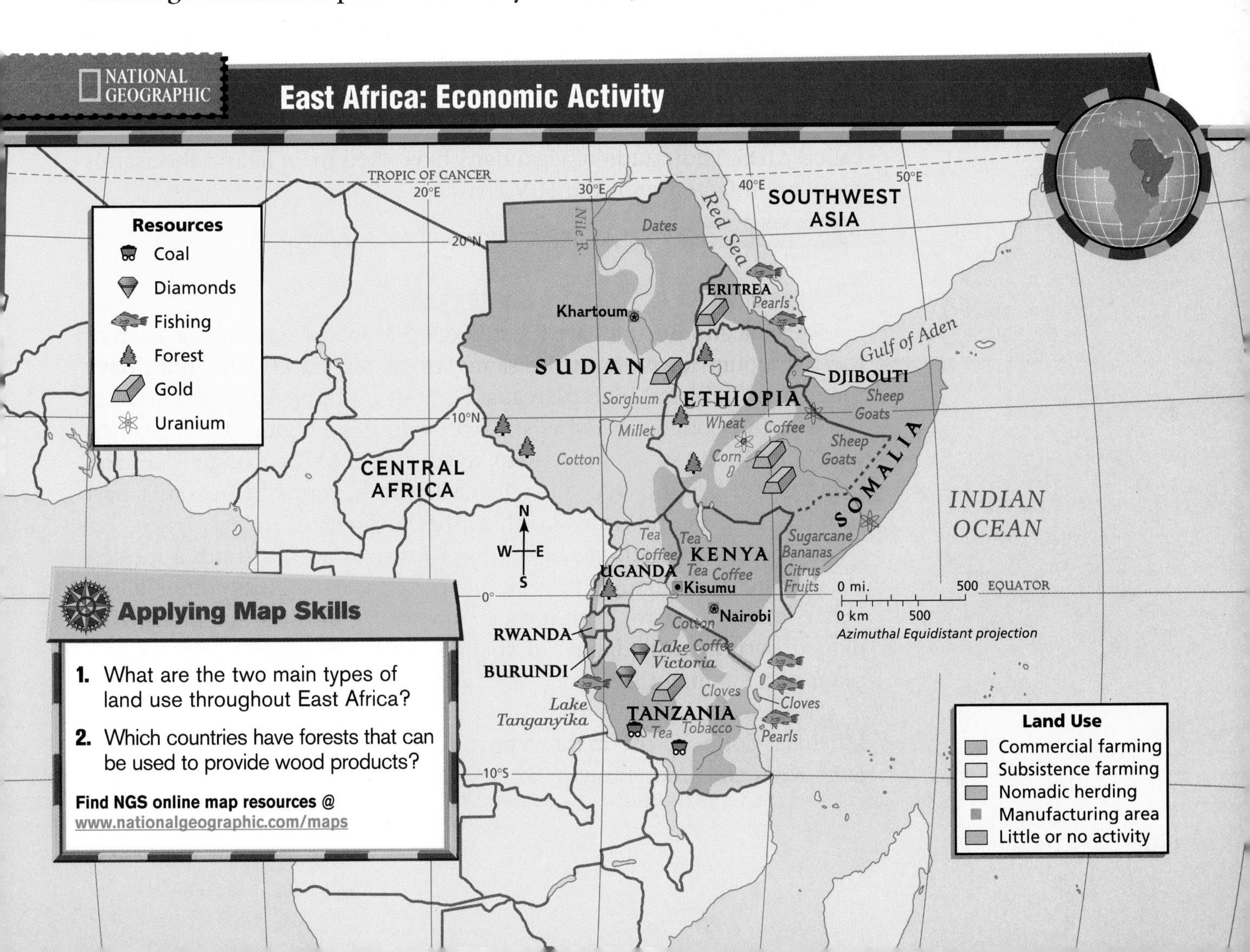

Kampala, Uganda

Public transportation in Kampala includes passenger boats on Lake Victoria.

Place **Part of what major river flows from Lake Victoria?**

Uganda throughout much of the 1980s. Since the mid-1990s, the national government has allowed ethnic groups to once again have kings, but only as local ceremonial leaders.

Ugandans now have one of the fastest-growing economies in Africa. Uganda also enjoys a stable government. It is a republic with an elected president and legislature. Still, the future is clouded. Uganda, along with other African countries, faces a new threat: the disease called AIDS. Thousands of Ugandans have died from it, and thousands more are infected with the HIV virus.

✓Reading Check **What kind of government does Uganda have today?**

Rwanda and Burundi

Rwanda and Burundi are located deep in inland East Africa. Each of the two countries is about the same size as Maryland. They both have mountains, hills, and high plateaus. They sit on the ridge that separates the Nile and Congo watersheds. A **watershed** is a region that is drained by a river. To the west of the ridge, water runs into the Congo River and flows to the Atlantic Ocean. To the east, water eventually becomes part of the Nile River and flows north to the Mediterranean Sea.

As in Uganda, high elevation gives Rwanda and Burundi a moderate climate even though they lie near the Equator. Heavy rains allow dense forests to grow. Within these forests live gorillas. Scientists have named gorillas an endangered species. An **endangered species** is a plant or an animal threatened with extinction.

Farmers in Burundi and Rwanda work small plots that dot the hillsides. Coffee is the main export crop. The people who live along Lake Kivu and Lake Tanganyika also fish. Because both countries are landlocked, they have trouble getting their goods to foreign buyers. Few paved roads and no railroads exist. Most goods must be transported by road to Lake Tanganyika, where boats take them to Tanzania or the Democratic Republic of the Congo. Another route is by dirt road to Tanzania and then by rail to Dar es Salaam.

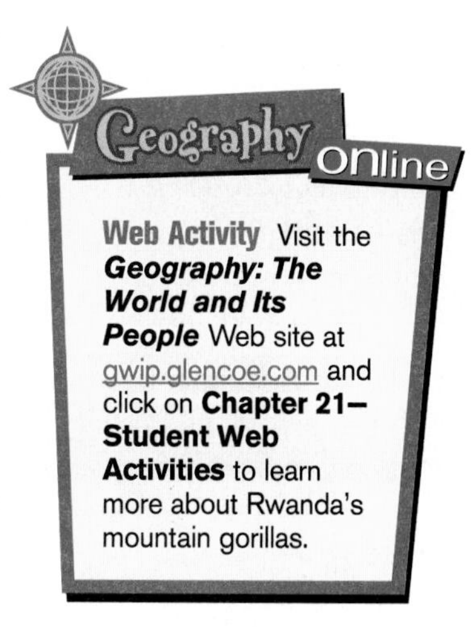

Geography online
Web Activity Visit the Geography: The World and Its People Web site at gwip.glencoe.com and click on Chapter 21—Student Web Activities to learn more about Rwanda's mountain gorillas.

Ethnic Conflict Rwanda and Burundi have large populations and small areas. As a result, they are among the most densely populated countries in Africa. Rwanda, for example, has an average of 802 people per square mile (310 per sq. km). Yet fewer than 10 percent of the people live in cities.

Two ethnic groups form most of the population of both countries—the Hutu and the Tutsi. The Hutu make up 80 percent or more of the population in both Rwanda and Burundi. Over the years, however, the Tutsi have controlled the governments and economies of these countries. Since the countries became independent in 1962, the Hutu have tried to gain some of this power.

That effort led to a terrible civil war in the 1990s. Hundreds of thousands of people were killed. Two million more became **refugees,** or people who flee to another country to escape persecution or disaster. The fighting between the Hutu and Tutsi has lessened, but both countries face many challenges as they try to rebuild.

✓Reading Check **Which ethnic group makes up the majority of the population in Rwanda and Burundi?**

Assessment

Defining Terms

1. **Define** plantains, autonomy, watershed, endangered species, refugee.

Recalling Facts

2. **Location** Explain the factors that affect Uganda's climate.
3. **Place** What is the capital of Uganda?
4. **Region** What endangered species lives in the forests of Rwanda and Burundi?

Critical Thinking

5. **Evaluating Information** How could a deadly epidemic, such as AIDS, affect a country's economy?
6. **Analyzing Information** How do ethnic differences create problems for the countries of East Africa?

Graphic Organizer

7. **Organizing Information** Draw a diagram like the one below. Then write two facts about Uganda under each of the category headings in the outer ovals.

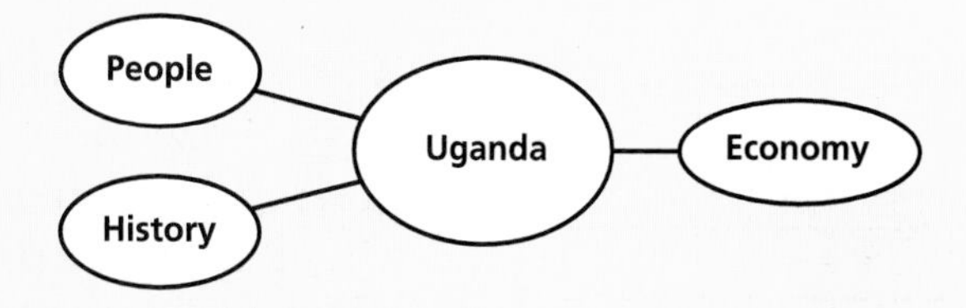

Applying Geography Skills

8. **Analyzing Maps** Study the political map on page 582. What are the four landlocked countries of East Africa?

Section 4

The Horn of Africa

Guide to Reading

Main Idea

The countries of the Horn of Africa have all been scarred by conflict in recent years.

Terms to Know

- drought
- plate
- clan

Places to Locate

- Sudan
- Ethiopia
- Eritrea
- Djibouti
- Somalia
- Blue Nile River
- White Nile River
- Khartoum
- Addis Ababa

Reading Strategy

Make a chart like this one. Then fill in two facts that are true of each country.

Country	Fact #1	Fact #2
Sudan		
Ethiopia		
Eritrea		
Djibouti		
Somalia		

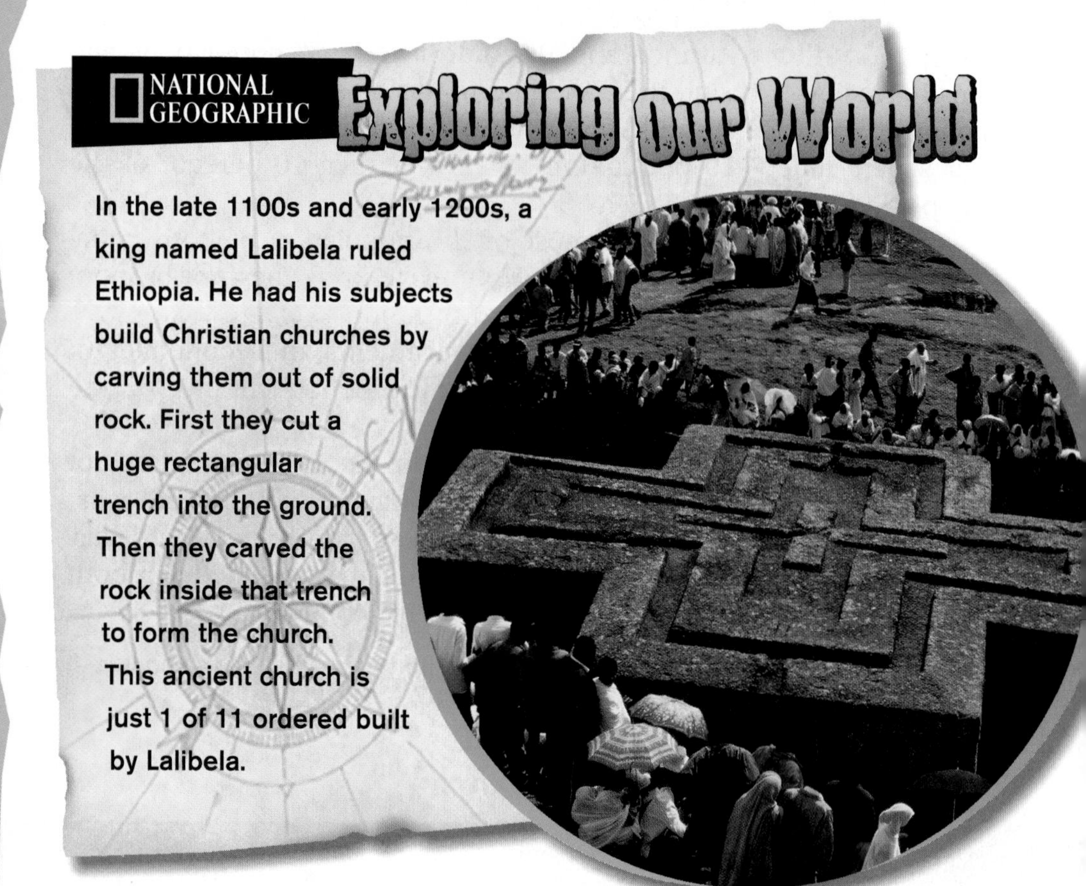

In the late 1100s and early 1200s, a king named Lalibela ruled Ethiopia. He had his subjects build Christian churches by carving them out of solid rock. First they cut a huge rectangular trench into the ground. Then they carved the rock inside that trench to form the church. This ancient church is just 1 of 11 ordered built by Lalibela.

The northern part of East Africa is a region called the Horn of Africa. This region got its name because it is shaped like a horn that juts out into the Indian Ocean. The countries here are **Sudan, Ethiopia, Eritrea** (EHR•uh•TREE•uh), **Djibouti** (jih•BOO•tee), and **Somalia.**

Sudan

Sudan is the largest country in Africa—about one-third the size of the continental United States. The northern part is covered by the sand dunes of the Sahara and Nubian Desert. Nomads raise camels and goats here. The most fertile part of the country is the central region. In this area of grassy plains, the two main tributaries of the Nile River—the **Blue Nile River** and the **White Nile River**—join together at **Khartoum** (kahr•TOOM), Sudan's capital. The southern part of Sudan receives plenty of rain and has some fertile soil. It also holds one of the world's largest swamps, which drains into the White Nile.

Most of Sudan's people live along the Nile River or one of its tributaries. They use water from the Nile to irrigate their fields. Farmers grow sugarcane, grains, nuts, dates, and cotton—the country's leading export. Sheep and gold are other important exports. Recently discovered oil fields in the south offer another possibility of income.

Sudan's People and History Most people in the northern two-thirds of the country are Muslim Arabs. People in the southern one-third come from many different African groups. Most of these southern groups practice Christianity or traditional African religions.

In ancient times, Sudan was the center of a powerful civilization called Kush. The people of Kush had close cultural and trade ties with the Egyptians to the north. During the A.D. 500s, missionaries from Egypt brought Christianity to the area. About 900 years later, Muslim Arabs entered northern Sudan and converted its people to Islam.

From the late 1800s to the 1950s, the British and the Egyptians together ruled the entire country. Sudan became an independent nation in 1956. Since then, military leaders generally have ruled Sudan. Its people have faced an uncertain future. In the 1980s, a fierce civil war broke out between the northern and southern peoples. The fighting has disrupted the economy and caused widespread hunger, especially in the south. A recent **drought**—a long period of extreme

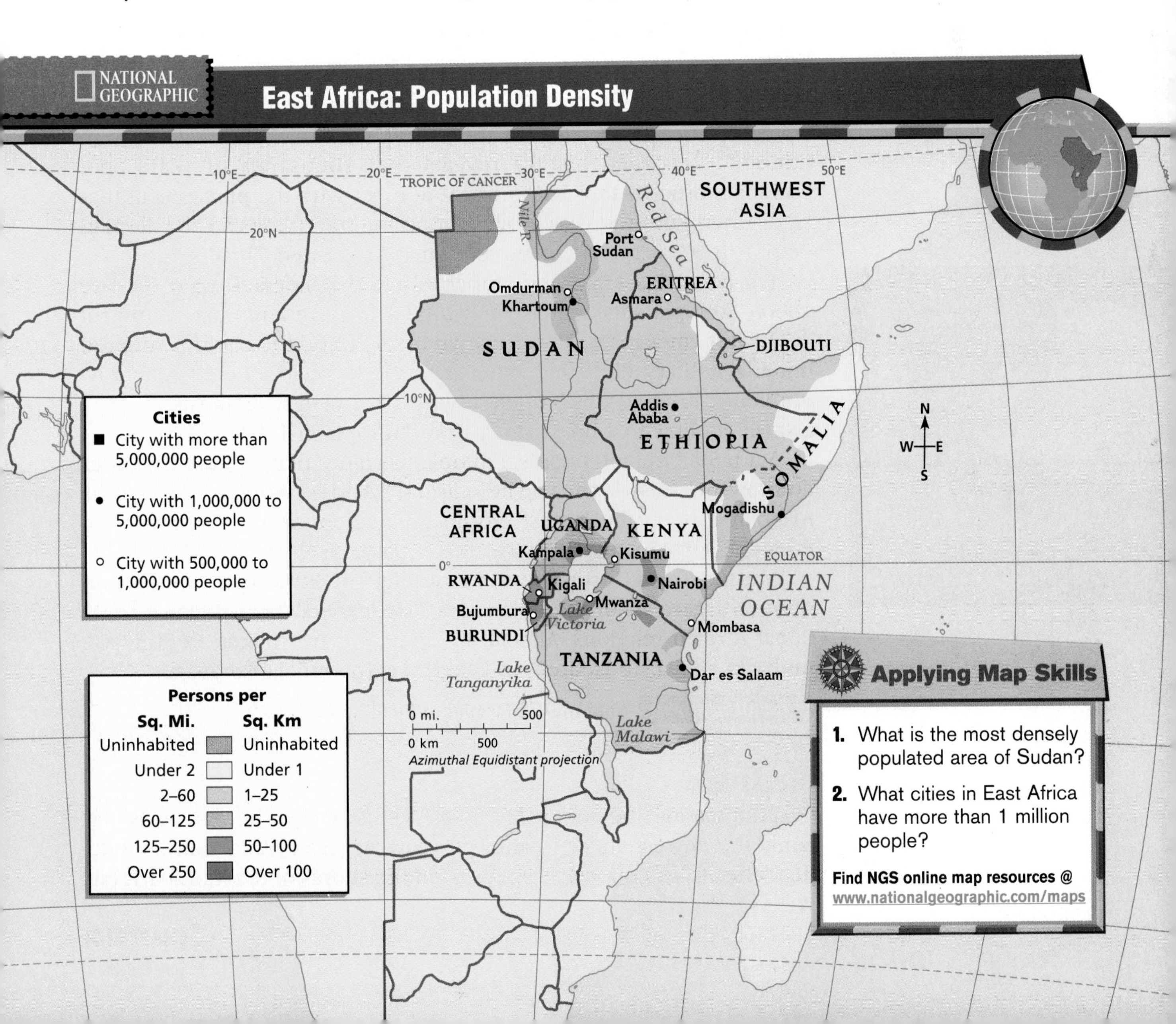

dryness and water shortages—made the situation worse. Millions of people have starved to death, and major outbreaks of diseases have swept through the country. To end the war, the government has announced that it might allow the south to become independent.

Reading Check What is the main export of Sudan?

Casting Votes

In May 2000, Ethiopians voted in their second-ever democratic election. This woman puts her two ballot papers into the ballot box. One ballot was her vote for her region's lawmakers. The other was a vote for the national parliament. People stood in line for hours to exercise their right to vote in a country that has long suffered under military dictators.

Ethiopia

Landlocked Ethiopia is almost twice the size of Texas. Ethiopia's landscape varies from hot lowlands to rugged mountains. The central part of Ethiopia is a highland plateau sliced through by the Great Rift Valley. The valley forms deep river gorges and sparkling waterfalls. Mild temperatures and good soils make the highlands Ethiopia's best farming region. Farmers raise grains, sugarcane, potatoes, and coffee. Coffee is a major export crop. The southern highlands are believed to be the world's original home of coffee.

Rain is not consistent in many parts of Ethiopia, however. Low rainfall in some years brings on drought, and Ethiopia's people suffer. Famine brought Ethiopia to the world's attention in the 1980s. At that time, a drought turned fields once rich in crops into seas of dust. Despite food aid, more than 1 million Ethiopians died.

Ethiopia's History and People Scientists have found what they believe to be the remains of the oldest known human ancestors in Ethiopia. Recorded history reveals that, thousands of years ago, Ethiopian officials traveled to Egypt to meet with the pharaohs of that land. Later, Ethiopia developed important trade links to the Roman Empire. In the A.D. 300s, many Ethiopians accepted Christianity.

For centuries, kings and emperors ruled Ethiopia. During the late 1800s, Ethiopia successfully withstood European attempts to control it. The last emperor was overthrown in 1974, and the country suffered under a military dictator. Now it is trying to build a democratic government. This goal was hindered by warfare with neighboring Eritrea, a small country that broke away from Ethiopia in 1993.

With 59.7 million people, Ethiopia has more people than any other country in East Africa. The capital, **Addis Ababa** (AHD•dihs AH•bah•BAH), is the largest city in the region. About 85 percent of Ethiopians live in rural areas.

Muslims now form about 45 percent of Ethiopia's population. About 40 percent of Ethiopians are Christians. Others practice traditional African religions. Almost 80 languages are spoken in Ethiopia. Amharic, similar to Hebrew and Arabic, is the official language.

Reading Check What makes farming difficult in Ethiopia?

Eritrea

Ethiopia may be one of Africa's oldest countries, but Eritrea is certainly the newest. In 1993, after 30 years of war, Eritrea won its independence from Ethiopia. Eritrea sits on the shores of the Red Sea. It has

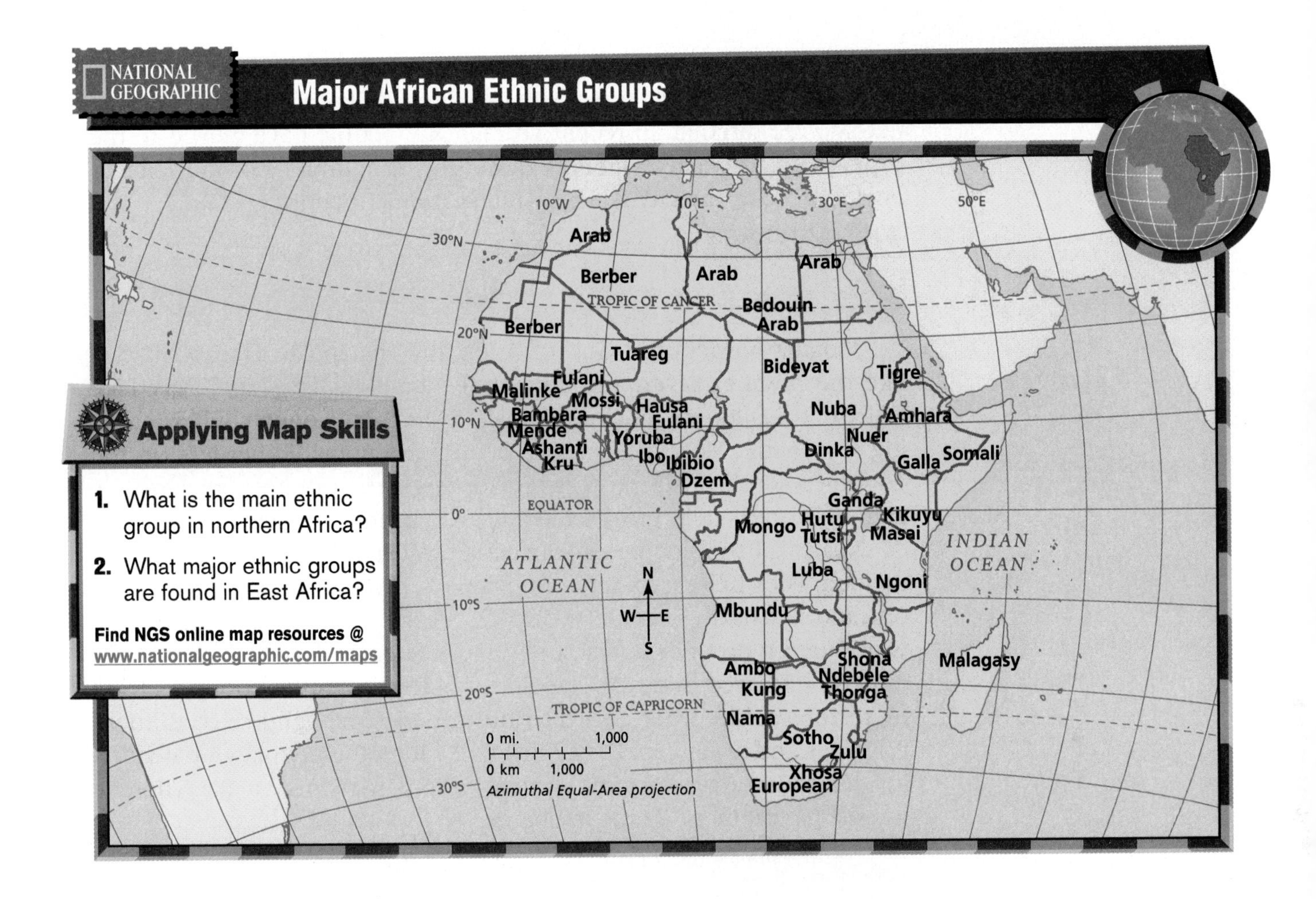

a narrow plain that stretches about 600 miles (966 km) along the coast. When Eritrea became a country, Ethiopia became landlocked.

Most of Eritrea's 4 million people farm the land. Farming is uncertain work because the climate is dry. The long war with Ethiopia also hurt farming. The war did have a positive effect on some of Eritrea's people, however. Women formed about one-third of the army that won the war. After the war ended, the new government passed laws that gave women more rights than they had ever had before.

✓Reading Check **When and from what country did Eritrea win independence?**

Djibouti

To see the earth undergoing change, visit Djibouti. This country lies at the northern tip of the Great Rift Valley, where three of the earth's plates join. **Plates** are huge slabs of rock that make up the earth's crust. In Djibouti, two of these plates are pulling away from each other. As they separate, fiery hot rock rises to the earth's surface, causing volcanic activity.

Djibouti wraps around a natural harbor at the point where the Red Sea meets the Gulf of Aden. This tiny country is one of the hottest, driest places on the earth. Its landscape is covered by rocky desert. Here and there, you will find the desert interrupted by salt lakes and rare patches of grassland.

Djibouti's 600,000 people are mostly Muslims. In the past, they lived a nomadic life of herding. Because of Djibouti's dry climate, farming and herding are difficult. In recent years, many people have moved to the capital city, also called Djibouti. Here they have found jobs in the city's docks, because the city is a busy international seaport.

Reading Check Why does Djibouti experience volcanic activity?

Somalia

Somalia borders the Gulf of Aden and the Indian Ocean. Shaped like the number seven, the country is almost as large as Texas. Like Eritrea and Djibouti, much of Somalia is hot, dry country where farming is difficult. Most of Somalia's people are nomadic herders on the country's scrubby plateaus. In the south, rivers provide water for irrigation. Farmers here grow fruits, sugarcane, and bananas.

Nearly all the people of Somalia are Muslims, but they are deeply divided. They belong to different **clans,** or groups of people related to one another. In the late 1980s, disputes between these clans led to civil war. When a drought struck a few years later, hundreds of thousands of people starved to death. The United States and other countries tried to restore some order and distribute food. The fighting continued, however, and often kept the aid from reaching the people who needed it. Even today, armed groups control various parts of Somalia. There is no real government that is in charge.

Reading Check What kind of conflict led to civil war in Somalia?

Assessment

Defining Terms

1. **Define** drought, plate, clan.

Recalling Facts

2. **Place** What is the capital of Sudan?
3. **History** What is the only country of East Africa that was never colonized by Europeans?
4. **Government** Describe the current political situation in Somalia.

Critical Thinking

5. **Analyzing Information** What factors do you think might have led to the settlement of Khartoum, Sudan's capital?
6. **Understanding Cause and Effect** How did a war bring increased rights to women in Eritrea?

Graphic Organizer

7. **Organizing Information** On a diagram like the one below, list the major religions practiced by countries of the Horn of Africa. Write the religions under each country's name.

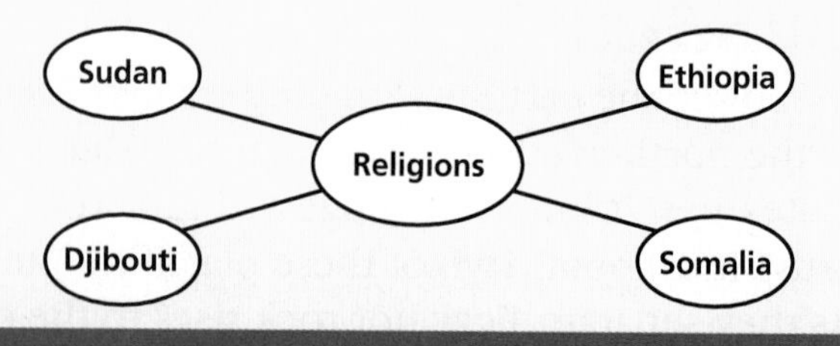

Applying Geography Skills

8. **Analyzing Maps** Study the climate map on page 588. What four types of climates are found in Sudan?

Chapter 21 Reading Review

Section 1 Kenya

Terms to Know
- coral reef
- nature preserve
- poaching
- fault
- escarpment
- altitude
- free enterprise system
- cassava

Main Idea

Kenya is a country of diverse landscapes and peoples.

✓ **Place** Eastern Kenya is covered by lowlands. The western half is marked by highlands and the wide Great Rift Valley.

✓ **Economics** Most people in Kenya carry out subsistence farming. Coffee and tea are grown for export. Tourism is also a major industry in Kenya.

✓ **Culture** Kenya's people come from many different ethnic groups. They speak Swahili, a blended language.

✓ **Government** Since becoming independent in 1963, Kenya has had a mostly stable government.

Section 2 Tanzania

Terms to Know
- sisal
- habitat
- eco-tourist

Main Idea

Tanzania is located on the Indian Ocean and relies on agriculture and tourism.

✓ **Economics** Farming and tourism are the main activities in Tanzania. The country is too poor to develop its mineral resources.

✓ **Government** Tanzania's government has been stable and democratic.

Section 3 Inland East Africa

Terms to Know
- plantains
- autonomy
- watershed
- endangered species
- refugee

Main Idea

Rwanda, Burundi, and Uganda have suffered much conflict in recent years.

✓ **Place** Uganda, Rwanda, and Burundi are landlocked countries with high elevation and rainy, moderate climates.

✓ **Economics** Most people in all three countries practice subsistence farming. Major exports include coffee, tea, and cotton.

✓ **History** Rwanda and Burundi suffered a brutal civil war in the 1990s between the Hutu and the Tutsi ethnic groups.

Section 4 The Horn of Africa

Terms to Know
- drought
- plate
- clan

Main Idea

The countries of the Horn of Africa have all been scarred by conflict in recent years.

✓ **History** Sudan has been torn by a civil war between the northern Muslim Arabs and the southern African peoples.

✓ **Human/Environment Interaction** Ethiopia has good farmland, but scarce rainfall can cause drought.

✓ **Government** Eritrea recently won its independence from Ethiopia.

✓ **History** Fighting between rival clans and drought have caused suffering in Somalia.

EYE on the Environment

Endangered Spaces

Shrinking Habitats When you think of Africa, what images come to mind? Roaring lions? Sprinting cheetahs? Lumbering elephants? Unless conditions change, some wild African animals may soon be only memories. Many are endangered, primarily because their habitats—their grassland and forest homes—are being destroyed in many ways.

 Population growth — Africa south of the Sahara has the world's highest population growth rate. Farmers and ranchers turn wild lands into fields and pastures to raise food. Urban sprawl also takes over habitats.

 Logging — Logging companies build roads and cut valuable trees, destroying forest habitats.

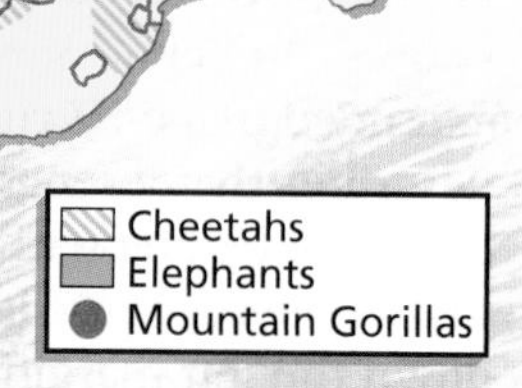

Mining — Open pit mines scar the land, pollute waters, and destroy trees.

As habitats shrink, so do populations of African animals.

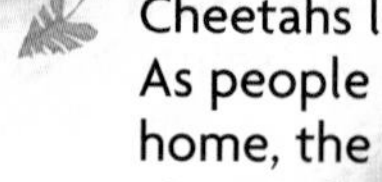

Cheetahs live in Africa's grasslands. As people move into the cheetahs' home, the big cats struggle to survive. Only about 12,000 cheetahs are left in the wild.

 Mountain gorillas live in the misty mountain forests of Central and East Africa. Logging and mining are destroying these forests. Only about 650 mountain gorillas remain.

These and other endangered African animals will survive only if their habitats are saved.

Loggers destroy a forest in the Democratic Republic of the Congo.

Cheetahs are running out of room in Africa.

Making a Difference

The Cheetah Conservation Fund Cheetahs in Africa are getting a helping hand from the Cheetah Conservation Fund (CCF). This organization is based in Namibia, which is home to about 2,500 cheetahs. Namibian ranchers often trap and shoot cheetahs to protect their livestock. The CCF has donated nearly 80 special herding dogs to ranchers. The dogs protect the livestock and keep cheetahs out of harm's way at the same time. The CCF also teaches villagers and schoolchildren about cheetahs and about why it is important to save these big cats and their habitats.

Namibian children learn about cheetahs.

Protecting Gorillas For nearly 18 years, Dian Fossey studied mountain gorillas in Rwanda. Through her book, *Gorillas in the Mist*, which was made into a movie, Fossey told others about mountain gorillas and how their survival was threatened by habitat destruction and poaching. Fossey established the Karisoke Research Center and an international fund to support gorilla conservation.

Dian Fossey fought fiercely to end gorilla poaching. Although Fossey was murdered at Karisoke in 1985, the Dian Fossey Gorilla Fund International continues its work protecting mountain gorillas and their habitat.

A mountain gorilla

What Can You Do?

Adopt a Cheetah

You and your classmates can help save cheetahs in the wild by adopting one. To learn more, contact the Cheetah Conservation Fund at www.cheetah.org

Find Out More

What animal habitats are endangered where you live? Work with a partner to investigate endangered spaces in your area. Summarize your findings in a report to the class.

Use the Internet

How can people help save vanishing habitats? Check The Nature Conservancy's Web site at www.tnc.org to learn how this group works to preserve habitats worldwide.

South Africa and Its Neighbors

The World and Its People NATIONAL GEOGRAPHIC

To learn more about the people and places of South Africa and its neighbors, view ***The World and Its People*** **Chapter 22** video.

Geography Online

Chapter Overview Visit the ***Geography: The World and Its People*** Web site at gwip.glencoe.com and click on **Chapter 22–Chapter Overviews** to preview information about South Africa and its neighbors.

Section 1

Republic of South Africa

Guide to Reading

Main Idea

Rich in resources, South Africa has recently seen major social and political changes.

Terms to Know

- high veld
- escarpment
- developed country
- Boer
- apartheid
- township
- enclave

Places to Locate

- South Africa
- Namib Desert
- Cape of Good Hope
- Drakensberg Range
- Cape Town
- Johannesburg
- Durban
- Pretoria
- Lesotho
- Swaziland

Reading Strategy

Create a time line like this one. Then list five key events and their dates in South Africa's history.

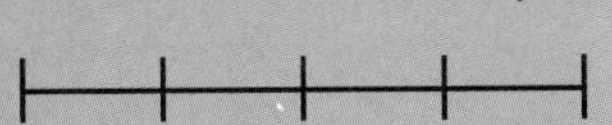

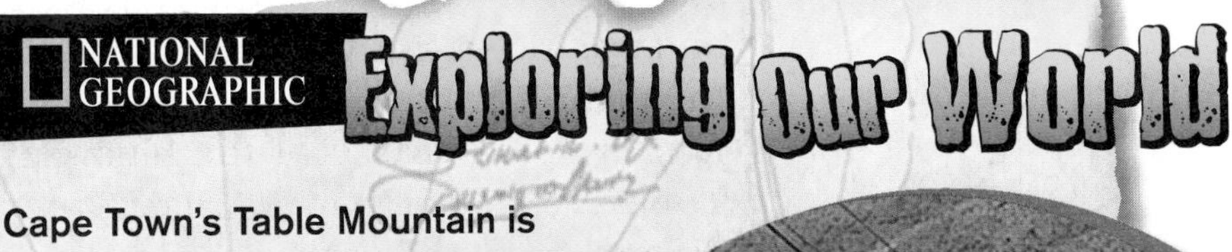

Cape Town's Table Mountain is famous for its flat top. A cable car takes people to the top. If you are physically fit and have four hours to spare, you can also climb the mountain. From the top, you can see Cape Town—South Africa's legislative capital. The country's executive capital is Pretoria, while the judicial capital is Bloemfontein.

South Africa (officially called the Republic of South Africa) spreads across the southern end of the African continent. It is a land of beautiful scenery and great mineral wealth. It is also a land of great change.

A Land of Variety

The Republic of South Africa covers an area almost twice the size of Texas. Here you will find the continent's biggest mammal, the African elephant, and smallest mammal, the miniature shrew. To protect these creatures, the government has set aside land as national parks.

South Africa borders the Atlantic Ocean on the west and the Indian Ocean on the south and east. The vast **Namib Desert** reaches into the northwest. Farther south you find the **Cape of Good Hope,** the southernmost point of Africa. Two small plateaus rise above the coastal plain east of the cape. These dry plateaus are the Great Karroo and the Little Karroo. *Karroo* means "land of thirst" in a local African language.

◄ Table Mountain overlooks Cape Town, South Africa.

A large plateau spreads through the center of South Africa. Part of this plateau is made up of flat, grass-covered plains called the **high veld.** The high veld has soil too poor for farming, but the grasslands can support grazing livestock. The plateau is separated from coastal areas by the Great Escarpment. An **escarpment** is a steep cliff between higher and lower land. The Great Escarpment reaches its highest elevations in the **Drakensberg Range** in the east.

Climate South Africa lies south of the Equator. As a result, its seasons are opposite to those in the Northern Hemisphere. The climate map on page 615 shows you that South Africa has a variety of climates. On the plateau, winters are cool and sunny with some rainfall. Summers remain mild because of the high elevation. The area around **Cape Town** has a Mediterranean climate of cool, rainy winters and hot, often dry summers. The farther east you go along the coast, the handier an umbrella becomes. Warm winds from the Indian Ocean bring a humid subtropical climate and rain.

Reading Check **Which part of South Africa has a Mediterranean climate?**

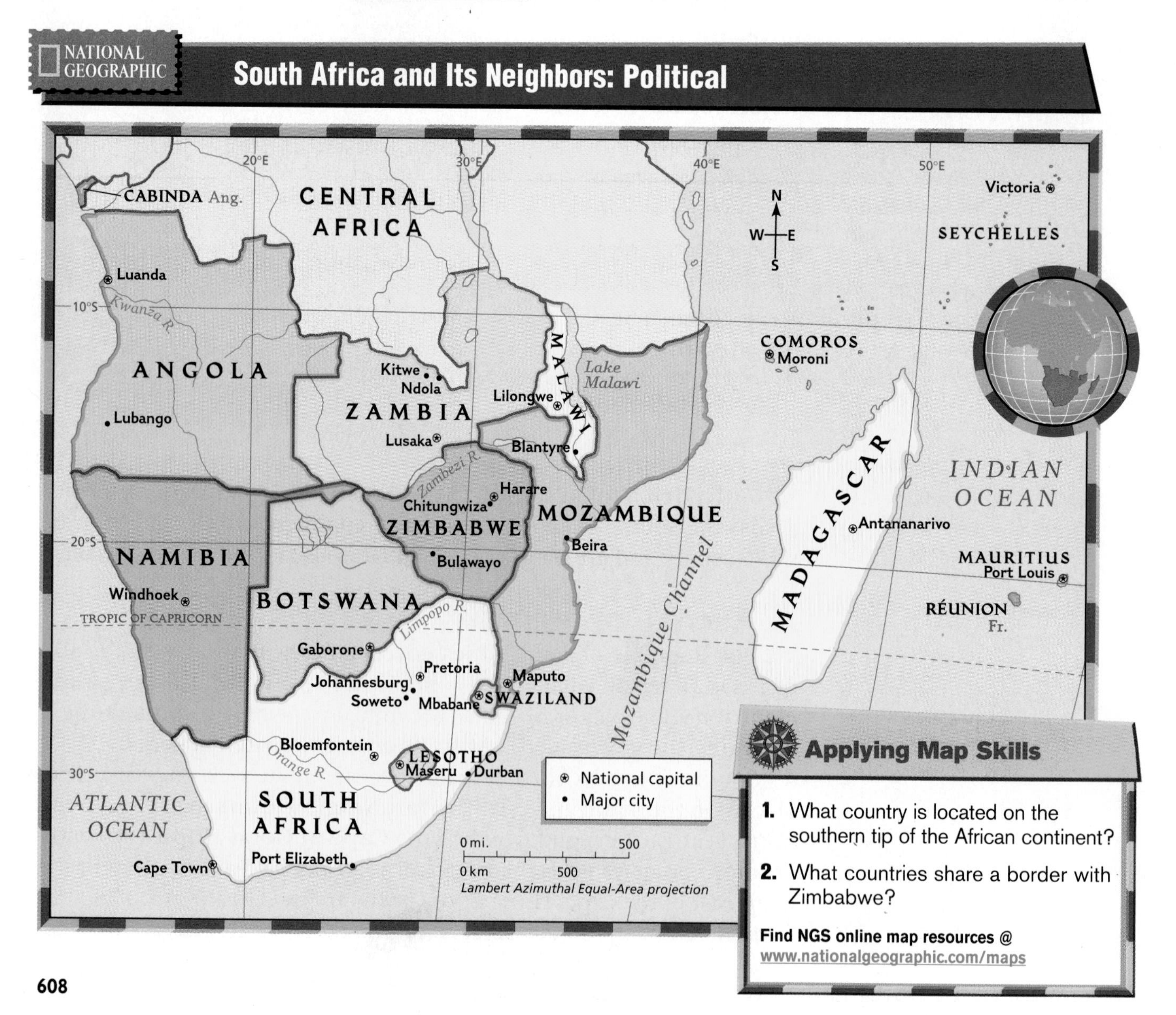

Applying Map Skills

1. What country is located on the southern tip of the African continent?
2. What countries share a border with Zimbabwe?

Find NGS online map resources @ www.nationalgeographic.com/maps

A Developed Economy

South Africa is the most developed country in Africa. As you learned in Chapter 3, a **developed country** is one in which a great deal of manufacturing occurs. Not all South Africans benefit from this prosperous economy, however. In rural areas, many people continue to depend on subsistence farming and live in poverty.

In terms of mineral resources, South Africa is one of the richest countries in the world. It is the world's largest producer and exporter of gold. The Witwatersrand (WIHT•WAW•tuhrz•RAHND)—an area around the city of **Johannesburg**—is the site of the world's largest and richest gold field. South Africa also has large deposits of diamonds, chromite, platinum, and coal. The income earned from these resources has led to the growth of cities, industrial centers, huge cattle and sheep ranches, and high-technology farms.

Factories produce many of the manufactured goods used by South Africans themselves. The country also exports machinery, chemicals, clothing, and processed foods.

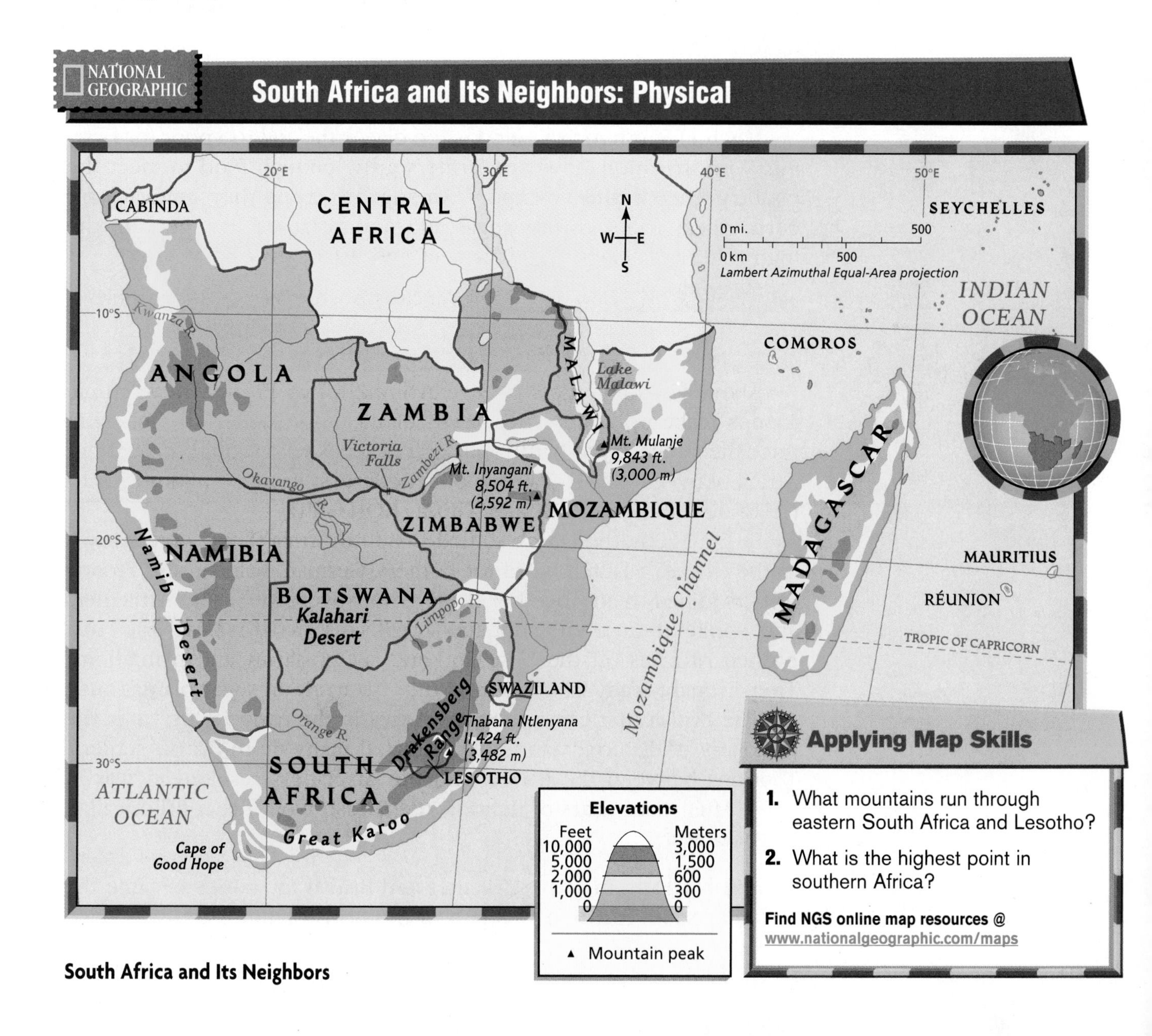

Applying Map Skills

1. What mountains run through eastern South Africa and Lesotho?
2. What is the highest point in southern Africa?

Find NGS online map resources @ www.nationalgeographic.com/maps

Music

The talking drum is a popular instrument in Africa south of the Sahara. Animal skins cover both ends and are held together by strings. While holding the drum under the arm, the musician strikes one skin with a curved wooden mallet. By squeezing down on the strings with the arm, the skins are stretched tighter and the pitch of the drum becomes higher. The loosening and tightening of the strings give the drum its characteristic "talking" sound.

Looking Closer **How is the talking drum different from the traditional drums in the United States?**

GO TO
World Music: A Cultural Legacy
Hear music of this region on Disk 2, Track 1.

Much of South Africa's land is too dry or the soil too poor for farming. With irrigation, however, farmers grow enough food to meet the country's needs and to export. Among the crops they cultivate are corn, wheat, fruits, cotton, sugarcane, and potatoes. Ranchers on the high veld raise sheep, cattle for beef, and dairy cows.

✓Reading Check **How have South Africa's resources helped its economy?**

South Africa's History and People

About 42.6 million people live in South Africa. Black African ethnic groups make up about 78 percent of the population. Most of them trace their ancestry to Bantu-speaking peoples who settled throughout Africa between A.D. 100 and 1000. The largest groups in South Africa today are the Sotho, Zulu, and Xhosa (KOH•suh).

In the 1600s, the Dutch settled in South Africa. They were known as the **Boers,** a Dutch word for farmers. German, Belgian, and French settlers joined them. Together these groups were known as Afrikaners and spoke their own language—Afrikaans (A•frih•KAHNS). They pushed Africans off the best land and set up farms and plantations. They brought many laborers from India to work on sugar plantations.

The British first came to South Africa in the early 1800s. Later, the discovery of diamonds and gold attracted many more British settlers. Tensions between the British and the Afrikaners resulted in war in 1899. After three years of fighting, the British won this conflict called the Boer War.

A New Nation In 1910 Afrikaner and British territories became the Union of South Africa. It was part of the British Empire and was ruled by whites. Black South Africans founded the African National Congress (ANC) in 1912 in hopes of gaining power.

In the 1930s, the whites set up a system of apartheid, or "apartness." **Apartheid** (uh•PAHR•TAYT) involved laws that separated racial and ethnic groups and limited the rights of blacks. For example, laws forced black South Africans to live in separate areas, called "homelands." People of non-European background were not even allowed to vote.

For more than 40 years, people inside and outside South Africa protested against the practice of apartheid. Many black Africans were jailed for voicing their opposition to apartheid. The United Nations declared that apartheid was "a crime against humanity." Many countries cut off trade with South Africa.

Finally, in 1991 the South African government agreed to end apartheid. In April 1994, South Africa held its first democratic election that allowed all people to vote. South Africans elected their first black president, Nelson Mandela.

The People About 55 percent of South Africans live in urban areas. The largest cities are Cape Town, Johannesburg, **Durban,** and **Pretoria.** South Africa has 11 official languages, including Afrikaans, English, Zulu, Xhosa, and other African ethnic languages. About two-thirds of South Africans are Christians. Almost one-third practice traditional African religions.

One of the challenges facing South Africa today is to develop a better standard of living for its poor people. Most European South Africans live in modern homes and enjoy a high standard of living. Most black African, Asian, and mixed-group South Africans live in rural areas and crowded **townships,** or neighborhoods outside cities. The government has introduced measures to improve education and basic services for these people. It is also working to provide more jobs for the poor.

Another challenge facing South Africa is the AIDS epidemic. Millions of people throughout Africa have been infected with the virus that causes AIDS. South Africa is one of the hardest-hit countries. The government is looking for ways to prevent the spread of the disease and to treat those who have it.

Lesotho

Although altitude, soil, and climate make farming difficult in Lesotho, most people are subsistence farmers.

Place **At what elevation does most of Lesotho lie?**

Nelson Mandela

As a young man, Nelson Mandela spoke out against apartheid. He was arrested and jailed. In 1990 he was released after spending a total of 27 years in jail. After becoming president in 1994, Mandela worked hard to maintain peace between the blacks and whites of South Africa.

The country's arts reflect the long struggle for justice and equality. In recent years, musicians have combined traditional African dance rhythms and modern rock music. Since the 1980s, groups such as Ladysmith Black Mambazo have made South African sounds popular worldwide. Other groups preserve traditional instruments and songs.

✓Reading Check **What has the South African government done to improve the lives of its poor people?**

Lesotho and Swaziland

Look at the political map on page 608. Within South Africa lie two other African nations—**Lesotho** (luh•SOH•toh) and **Swaziland.** They are enclaves—small countries located inside a larger country. Both are poor countries that depend heavily on South Africa.

A mountainous kingdom about the size of Maryland, Lesotho's only important natural resource is water. Lesotho sells some of this water to South Africa. Most people subsistence-farm and herd livestock. About one-third of Lesotho's male workers labor in the mines of South Africa. Their wages help support their families in Lesotho.

Swaziland is another tiny kingdom almost completely surrounded by South Africa. It shares a short border with Mozambique. About 60 percent of its people are engaged in subsistence farming. Others work in Swaziland's coal and asbestos mines or travel to South Africa to work in the mines there. Swaziland is ruled by a royal family that has been in power for more than 400 years.

✓Reading Check **How do Lesotho and Swaziland earn money from South Africa?**

Assessment

Defining Terms

1. Define high veld, escarpment, developed country, Boer, apartheid, township, enclave.

Recalling Facts

2. Place Name the largest and the smallest mammal in Africa.

3. Government Who is Nelson Mandela?

4. Culture What challenges face South Africa?

Critical Thinking

5. Drawing Conclusions How did the rest of the world view apartheid?

6. Analyzing Information Why do you think many of the workers in Lesotho and Swaziland travel to South Africa to find jobs?

Graphic Organizer

7. Organizing Information In a chart like the one below, write the resources and products of South Africa in the two boxes.

South Africa	
Resources	Products

Applying Geography Skills

8. Analyzing Maps Study the physical map on page 609. Into what body of water does the Orange River empty?

Making Connections

ART  SCIENCE  LITERATURE TECHNOLOGY

Mining and Cutting Diamonds

A diamond is a mineral made entirely of carbon. It is the hardest-known substance on the earth and the most popular gemstone. Most diamonds formed billions of years ago deep inside the earth's mantle. There, intense pressure and heat transformed carbon into diamond crystal.

Mining

There are two major techniques used for mining diamonds: open pit and underground mining. In open pit mining, the earth is dug out in layers, creating a series of steps or roads that circle down into a pit. After drills and explosives loosen the rock containing diamonds, shovels and trucks remove it. When the pit becomes too deep to reach easily, underground mining may begin.

Underground mining requires sinking a shaft into the ground and tunneling to the rock. Explosives blast the rock loose, and the resulting rubble is crushed and carried to the surface for further processing.

To remove the diamonds, the crushed rock is mixed with water and placed in a washing pan. Heavier minerals, such as diamonds, settle to the bottom, while lighter wastes rise to the top and overflow. Next, the heavier mixture travels to a grease table. Diamonds cling to the grease while other wetted minerals flow past. Workers continue the sorting and separating by hand.

Cutting

The newly mined diamond resembles a piece of glass, not a sparkling jewel. To enhance their brilliance and sparkle, gem-quality diamonds are precisely cut and polished. The cutter uses high-speed diamond-tipped tools to cut facets, or small flat surfaces, into the stone. One of the most popular diamond cuts is the brilliant cut, which has 58 facets. The job of the cutter requires extreme skill, because the angles of the facets must be exactly right to bring out the diamond's beauty.

Making the Connection

1. Of what are diamonds made?
2. Why are gemstone diamonds cut and polished?
3. **Making Comparisons** How are open pit and underground diamond mining techniques alike? How are they different?

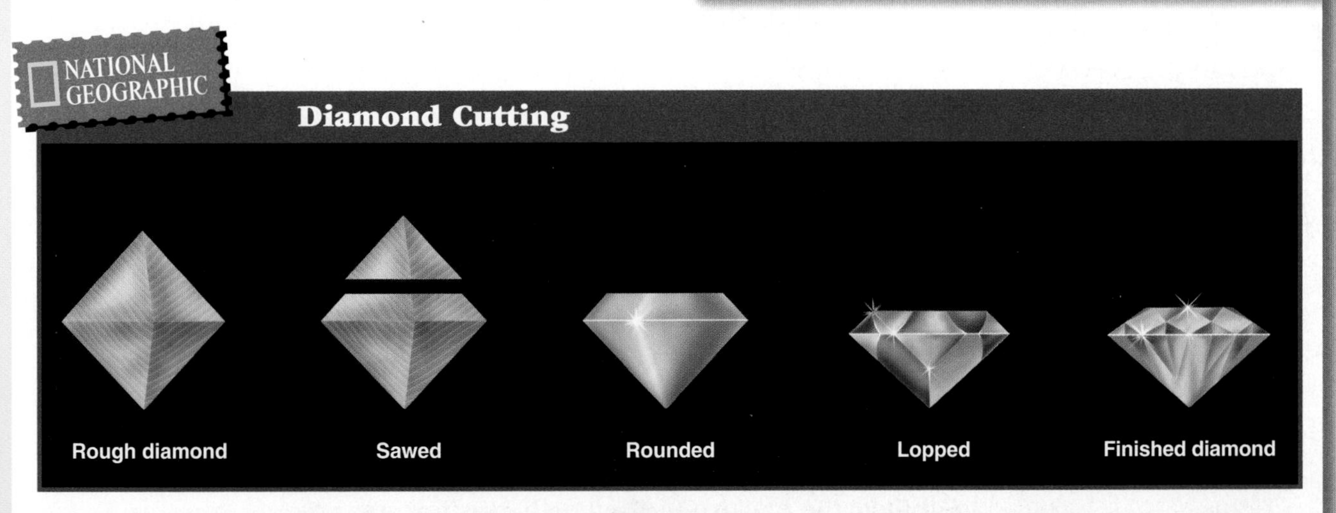

Section 2

Atlantic Countries

Guide to Reading

Main Idea

Angola and Namibia, although rich in resources, are struggling to develop their economies.

Terms to Know

- exclave

Places to Locate

- Angola
- Namibia
- Namib Desert
- Cabinda
- Kalahari Desert

Reading Strategy

Create a diagram like this one. Write facts about Angola and Namibia in the outer ovals under each heading. Where the ovals overlap, write statements that are true of both countries.

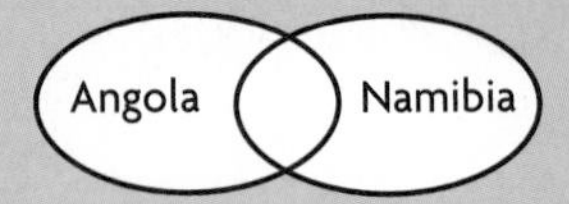

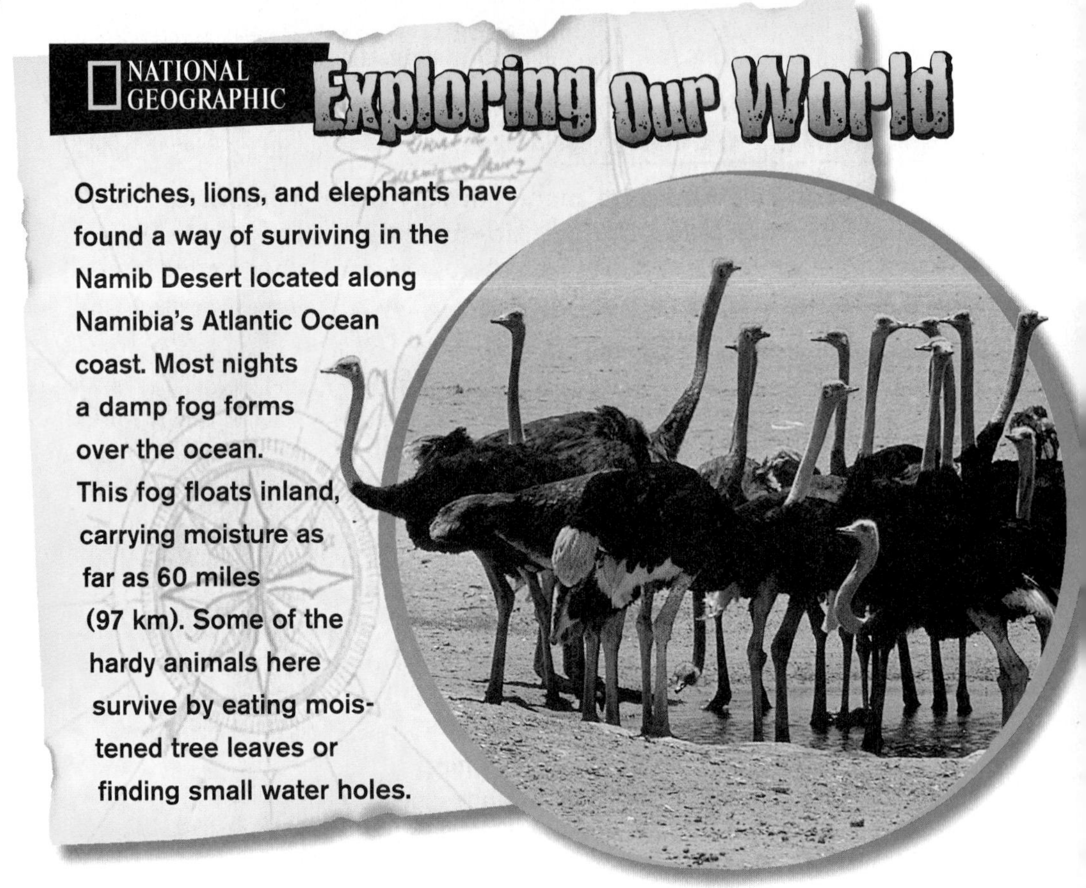

Ostriches, lions, and elephants have found a way of surviving in the Namib Desert located along Namibia's Atlantic Ocean coast. Most nights a damp fog forms over the ocean. This fog floats inland, carrying moisture as far as 60 miles (97 km). Some of the hardy animals here survive by eating moistened tree leaves or finding small water holes.

Angola and **Namibia** have long coastlines on the Atlantic Ocean. For this reason, they are known as southern Africa's Atlantic countries. They share the **Namib Desert**—a vast expanse of sand and rock that stretches north to south about 1,200 miles (1,931 km). It begins in Angola, extends through Namibia, and stretches into South Africa.

Angola

Angola is almost twice the size of Texas. Angola also includes a tiny exclave called **Cabinda.** An exclave is a small part of a country that is separated from the main part. Look at the map on page 608. You will see that Cabinda lies just to the north of Angola.

Most of Angola is part of the same inland plateau that sweeps through South Africa. Many rivers cross the country. Some flow into the Congo River in the north. Others flow into the Atlantic Ocean.

Hilly grasslands cover northern Angola. The southern part of the country is a rocky desert. A low strip of land winds along the Atlantic coastline. Most of this lowland area has little natural vegetation. In Cabinda, the exclave to the north, rain forests thrive.

The climate map below shows you that Angola's coastal areas have a desert climate. Inland from the coast, you pass through a narrow stretch of land with a steppe climate. The large inland plateau has a tropical savanna climate of wet and dry seasons. This area receives enough rainfall for farming.

Angola's Economy Angola's main economic activity is agriculture. About 85 percent of the people make their living from subsistence farming. Some farmers grow coffee and cotton for export. Angola's main source of income, however, is oil. Oil deposits off the coast of Cabinda account for 90 percent of Angola's export earnings. Other important industries include diamond mining, fish processing, and textiles. Still, Angola is not a wealthy country. Different groups have struggled for control of the country, which has hurt the economy.

Angola's History Most of Angola's people belong to several African ethnic groups. They trace their ancestry to the Bantu-speaking peoples who spread across much of Africa many centuries ago. In the 1400s, the Kongo kingdom ruled a large part of northern Angola.

From the 1500s until its independence in 1975, Angola was a colony of Portugal. Portugal is still an important trading partner, and Portuguese is the official language. Bantu and other African languages are also widely spoken. Almost 50 percent of Angolans practice the Roman Catholic faith brought to Angola by the Portuguese.

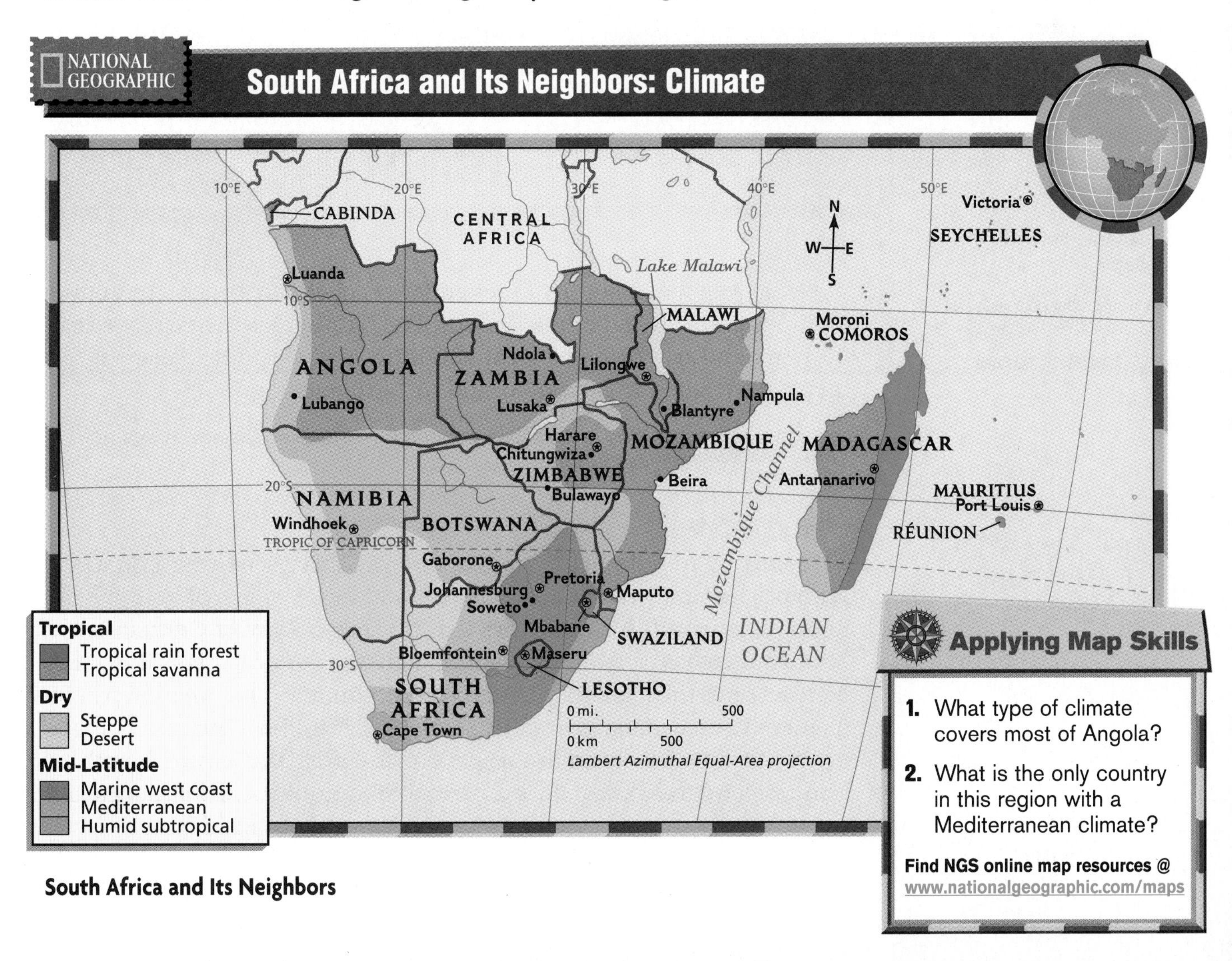

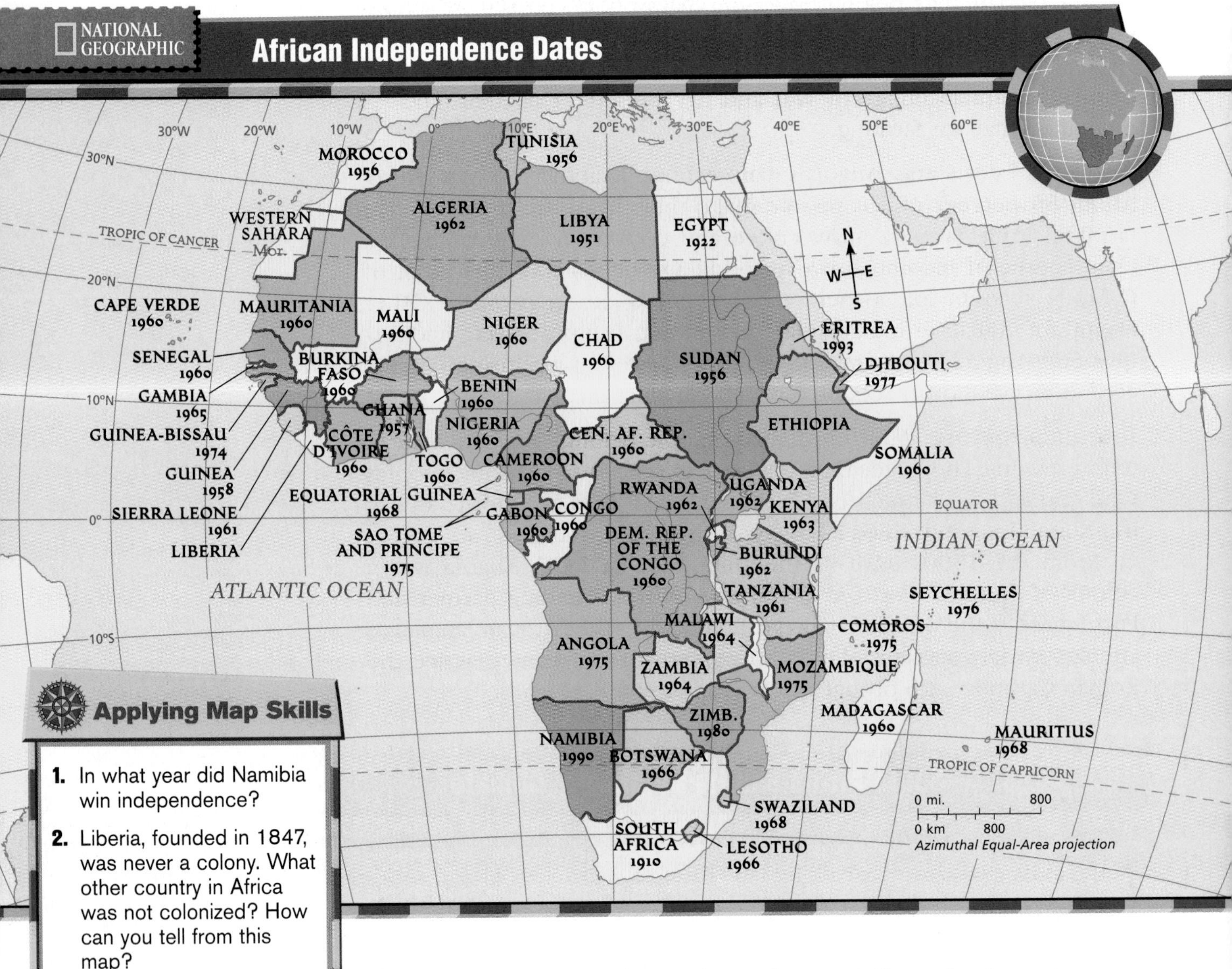

Applying Map Skills

1. In what year did Namibia win independence?
2. Liberia, founded in 1847, was never a colony. What other country in Africa was not colonized? How can you tell from this map?

Find NGS online map resources @ www.nationalgeographic.com/maps

After Angola gained its independence, civil war broke out among different political and ethnic groups. The fighting has lasted more than 25 years and has brought great suffering to the people. Peace agreements have failed to take hold, and the struggle continues.

Reading Check **Why is Angola's economy weak, even though it has rich resources?**

Namibia

South of Angola lies Namibia, one of Africa's newest countries. Namibia became independent in 1990 after 75 years of rule by the Republic of South Africa. Before that it was a colony of Germany.

Namibia has a land area about half the size of Alaska. A large plateau runs through the center of the country. This area of patchy grassland is the most populous section of Namibia. The rest is made up of deserts. The Namib Desert runs almost the entire length of Namibia's Atlantic coast. It is a narrow ribbon of towering dunes and rocks. Tourists come from all over the world to "sand-board" down these dunes. Another desert—the **Kalahari Desert**—stretches across the southeastern part of the country. It mainly consists of sand and scrub. As you might guess, most of Namibia has a hot, dry climate.

Namibia's Economy Namibia has rich deposits of diamonds, copper, gold, zinc, silver, and lead. It is the world's fifth-largest producer of uranium, a substance used for making nuclear fuels. The country's economy depends on the mining, processing, and exporting of these minerals. Many people work in Namibia's mines.

Despite this mineral wealth, most of Namibia's people live in poverty. The income from mineral exports stays in the hands of a small group of people. Large amounts also go to the foreign companies that have invested in Namibia's mineral resources. As a result, half of Namibia's people depend on subsistence farming. They also herd cattle, sheep, and goats. Some work in food industries, packing meat and processing fish and dairy products.

Namibia's People Only 1.6 million people live in Namibia. It is one of the most sparsely populated countries in Africa. In fact, in the language of Namibia's Nama ethnic group, *namib* means "the land without people." Most Namibians belong to African ethnic groups. A small number are of European ancestry. Namibians speak African languages, while most of the white population speaks Afrikaans and English.

✓ Reading Check **When did Namibia become an independent country?**

Assessment

Defining Terms

1. **Define** exclave.

Recalling Facts

2. **Location** Where is the Namib Desert located?
3. **Economics** What is Angola's main source of income?
4. **Place** What two deserts can be found in Namibia?

Critical Thinking

5. **Synthesizing Information** Reread "Exploring Our World" on page 614. Describe another example of animals adapting to a harsh environment.
6. **Understanding Cause and Effect** Why is Namibia one of the most sparsely populated countries in Africa?

Graphic Organizer

7. **Organizing Information** On a chart like the one below, list examples of Namibia's economy in the three categories.

Namibia's Economy		
Resources	Agriculture	Industries

Applying Geography Skills

8. **Analyzing Maps** Study the map of African independence dates on page 616. Which southern African country first achieved independence?

Section 3

Inland Southern Africa

Guide to Reading

Main Idea

Most of inland southern Africa is rich in resources and home to a wide variety of ethnic groups.

Terms to Know

- copper belt
- sorghum

Places to Locate

- Zambia
- Malawi
- Zimbabwe
- Botswana
- Zambezi River
- Victoria Falls
- Lake Malawi
- Harare
- Kalahari Desert
- Okavango River
- Gaborone

Reading Strategy

Create a chart like this one. Then list the main economic activities of each country.

Country	Economic Activities
Zambia	
Malawi	
Zimbabwe	
Botswana	

NATIONAL GEOGRAPHIC **Exploring Our World**

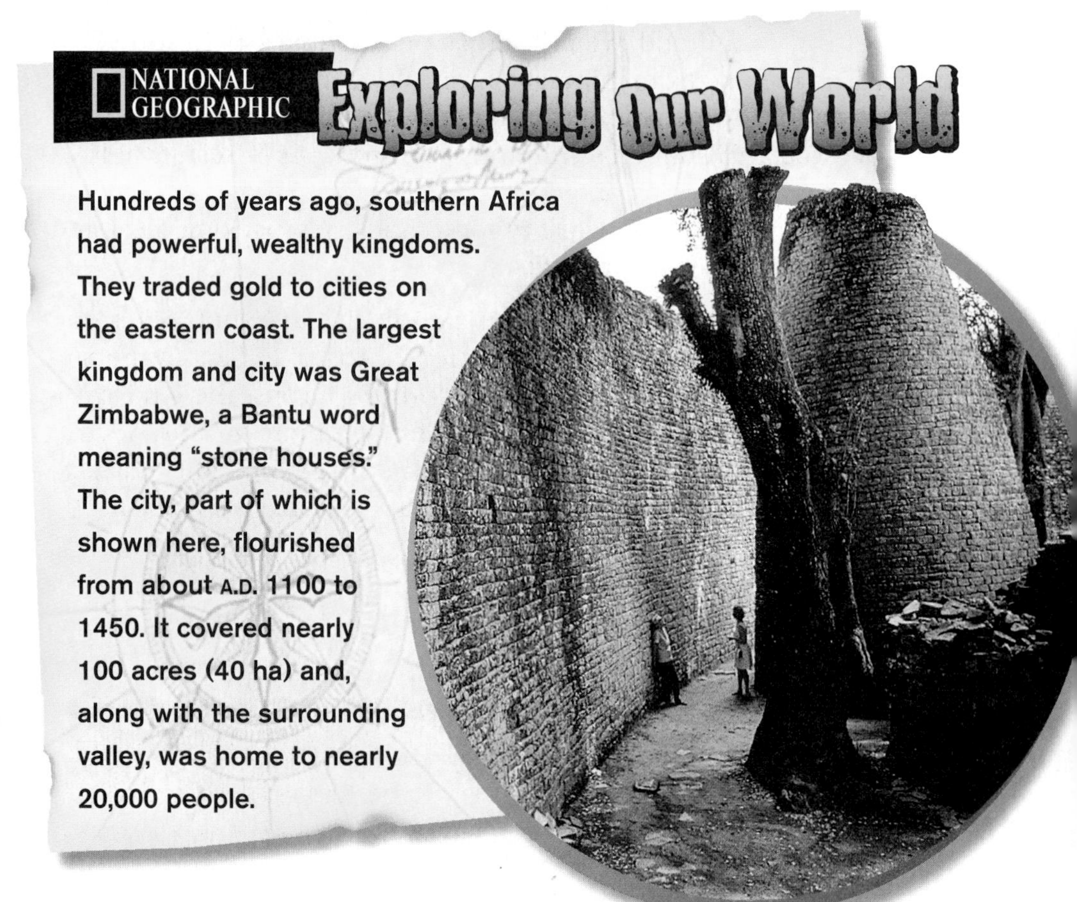

Hundreds of years ago, southern Africa had powerful, wealthy kingdoms. They traded gold to cities on the eastern coast. The largest kingdom and city was Great Zimbabwe, a Bantu word meaning "stone houses." The city, part of which is shown here, flourished from about A.D. 1100 to 1450. It covered nearly 100 acres (40 ha) and, along with the surrounding valley, was home to nearly 20,000 people.

The four countries of inland southern Africa include **Zambia, Malawi** (mah•LAH•wee), **Zimbabwe,** and **Botswana** (baht•SWAH•nah). They have several things in common. First, they all are landlocked. A high plateau dominates much of their landscape and gives them a mild climate. In addition, about 70 percent of the people practice subsistence farming in rural villages. Thousands move to cities each year to look for work.

Zambia

Zambia is slightly larger than Texas. The **Zambezi** (zam•BEE•zee) **River**—one of southern Africa's longest rivers—crosses the country. The Kariba Dam—one of Africa's largest hydroelectric projects—spans the Zambezi River. Also along the Zambezi River are the spectacular **Victoria Falls,** named in honor of British Queen Victoria, who ruled in the 1800s. The falls are known locally as *Mosi oa Tunya*—or "smoke that thunders."

A large area of copper mines, known as a **copper belt,** stretches across northern Zambia. One of the world's major producers of copper, Zambia relies on it for more than 80 percent of its income. As a result, when world copper prices go down, Zambia's income goes down too. As copper reserves dwindle, the government has encouraged city dwellers to return to farming. Zambia must import much of its food.

Once a British colony, Zambia gained its independence in 1964. The country's 9.7 million people belong to more than 70 ethnic groups and speak many languages. English is the official language. Those who live in urban areas such as Lusaka, the capital, work in mining and service industries. Villagers grow corn, rice, and other crops to support their families. Their main food is porridge made from corn.

✓Reading Check **What happens to Zambia when copper prices go down?**

Malawi

If you travel through narrow Malawi, you see green plains and savanna grasslands in western areas. Vast herds of elephants, zebras, and antelope roam national parks and animal reserves here.

The Great Rift Valley runs through eastern Malawi. In the middle of it lies beautiful **Lake Malawi.** This lake holds about 500 fish species, more than any other inland body of water in the world. Malawi is also famous for its more than 400 orchid species.

Malawi has few mineral resources and little industry. Tobacco, tea, sugar, coffee, and peanuts are exported. Small farmers also grow **sorghum,** a tall grass whose seeds are used as grain and to make syrup. Donations from international organizations support Malawi's people.

Bantu-speaking people arrived in the area about 2,000 years ago, bringing with them knowledge of iron working. The most famous European explorer to reach Malawi was the Scottish missionary David Livingstone during the mid-1800s. Today most people in Malawi are Protestant Christians as a result of missionaries.

In 1964 the British colony became independent. Malawi has recently returned to democratic government after a long period of rule by a dictator. After years of harsh government, the works of many modern writers emphasize themes such as human rights and abuse of power.

Malawi is one of the most densely populated countries in Africa. It has 219 people per square mile (84 people per sq. km). Jobs are scarce, so thousands of men seek work in South Africa and Zambia.

✓Reading Check **What types of landforms cover western Malawi?**

What's for Dinner?

Kabemba Mwape hurries home from the market in Chavuma, Zambia. He carries a live pig on the back of his bike. Although Kabemba's family will enjoy the pig for dinner, they usually eat porridge. Kabemba's family is relatively wealthy. They can afford to pay for meat, bicycles, and the high price of school uniforms and books for Kabemba's education. He knows English, but he speaks his native language of Lozi while at the market.

Zimbabwe

Crossing Zimbabwe, you might think you were in the western United States. The vast plateau is studded with large outcrops of rock. The Limpopo River winds through southern lowlands. The Zambezi River crosses the north.

Mining gold, copper, iron ore, and asbestos provides most of the country's income. Some large plantations grow coffee, cotton, and tobacco. Europeans own many of the large plantations, while many

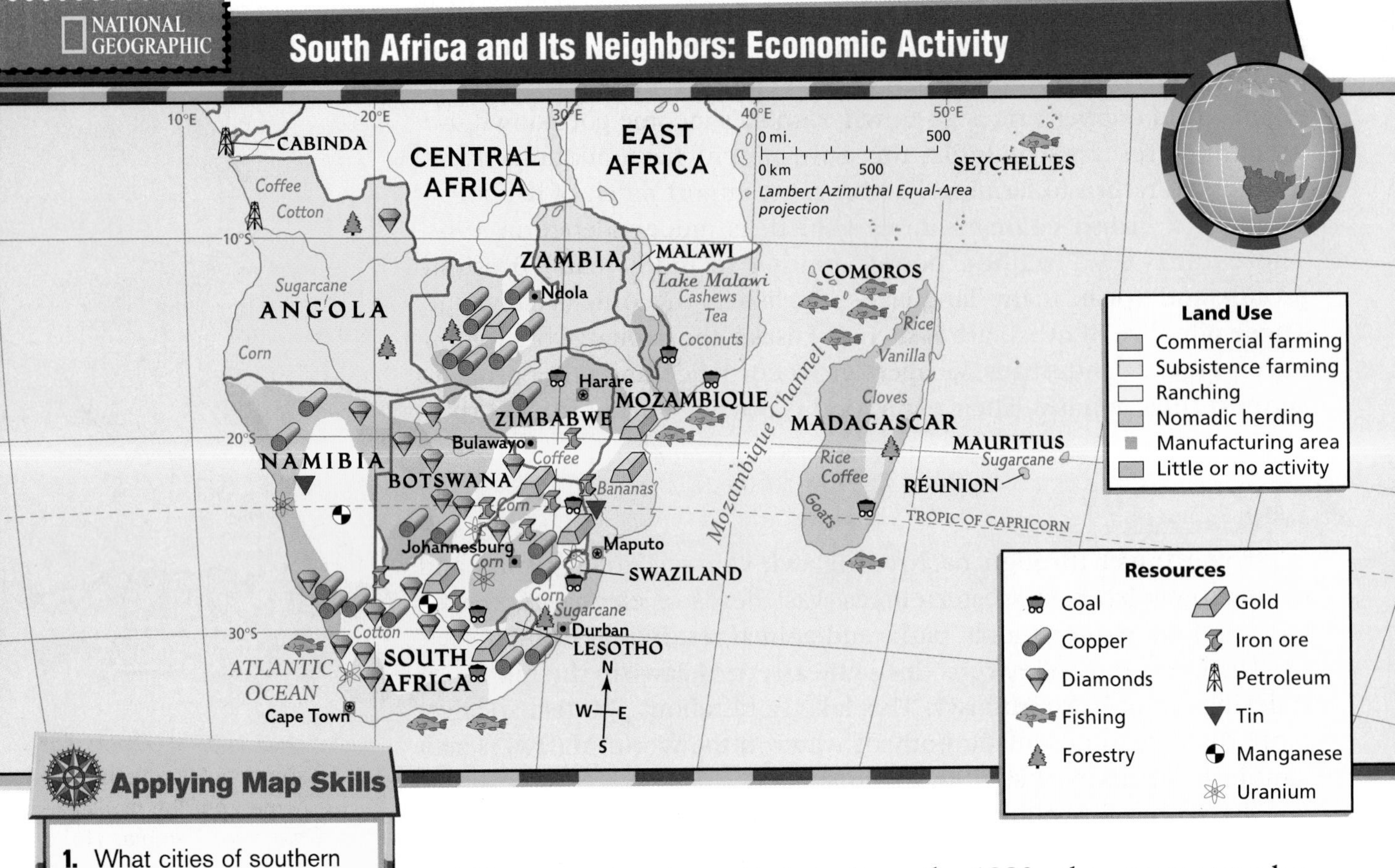

Applying Map Skills

1. What cities of southern Africa are manufacturing areas?
2. What resources are mined in southern Africa?

Find NGS online map resources @ www.nationalgeographic.com/maps

Africans work only small plots. Since the 1980s, the government has tried to give some of the plantation land to Africans. Progress has been slow, and protesters recently took over some European-owned farms to force changes.

Another serious challenge to Zimbabwe's economy comes from the spread of AIDS. People who have the disease often cannot work to support their families. Many children have been orphaned by AIDS. Also, the nation's health care costs have risen sharply.

Zimbabwe takes its name from an ancient African city and trading center—Great Zimbabwe. This remarkable stone fortress was built by an ethnic group called the Shona in the A.D. 1100s to 1400s. The Shona and the Ndebele (ehn•duh•BEH•leh) ruled large stretches of south-central Africa until the late 1800s. In the 1890s, the British controlled the area and called it Rhodesia. They named it after Cecil Rhodes, a British businessman who expanded British rule in Africa.

Europeans ran Rhodesia and owned all the best farmland. In response, the Africans organized into political groups and fought European rule. In 1980 free elections brought an independent African government to power. The country was renamed Zimbabwe.

Today Zimbabwe has about 11.2 million people. Most of them belong to the Shona and Ndebele ethnic groups. About half of the population is Christian. The other half practices traditional African religions. The largest city is **Harare** (hah•RAH•ray), the capital.

Reading Check **How has AIDS affected Zimbabwe's economy?**

Botswana

Botswana lies in the center of southern Africa. The map on page 609 shows you that the vast **Kalahari Desert** spreads over southwestern Botswana. This hot, dry area has rolling red sands and low thorny shrubs. The **Okavango River** in the northwest forms one of the largest swamp areas in the world. This area of shifting streams has much wildlife.

Botswana's national emblem has a one-word motto, *Pula,* meaning "rain." The people of Botswana all agree that there is never much of it. From May to October, the sun bakes the land. Droughts strike often, and many years can pass before the rains fall again.

Geography Online
Web Activity Visit the ***Geography: The World and Its People*** Web site at gwip.glencoe.com and click on **Chapter 22–Student Web Activities** to learn more about Botswana.

Botswana is rich in mineral resources, especially diamonds and copper. Diamonds account for more than 75 percent of Botswana's export income. Thousands of tourists visit Botswana's game preserves every year. Farming is difficult, and the country grows only about 50 percent of its food needs. It must import the rest. To earn a living, many people work in South Africa for several months a year.

During most of the 1800s, the Tswana ethnic group ruled most of what is today Botswana. Afrikaners from South Africa tried to seize the territory. The Tswana appealed for help to the British, who then took over the land for themselves. After nearly 80 years of British rule, Botswana became independent in 1966. Today it has one of Africa's strongest democracies. Most of Botswana's people are Christians, although a large number practice traditional African religions. The official language is English, but 90 percent of the population speaks an African language called Setswana. **Gaborone** is the capital and largest city.

✓Reading Check **What is Botswana's biggest source of export income?**

Section 3 Assessment

Defining Terms

1. **Define** copper belt, sorghum.

Recalling Facts

2. **Economics** What is Zambia's most important export?
3. **Place** What makes Lake Malawi unique?
4. **Culture** Where did Zimbabwe get its name?

Graphic Organizer

5. **Organizing Information** Choose two of the countries in this section. Put the name and facts about each country in the outer ovals. Where the ovals overlap, put facts that are true of both countries.

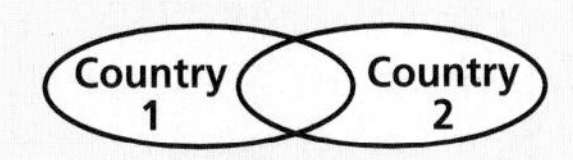

Critical Thinking

6. **Synthesizing Information** Imagine that someone from Great Zimbabwe traveled to that country today. What do you think he or she would describe as the greatest difference between then and now?
7. **Analyzing Information** Why do you think the people of Botswana chose *Pula,* or "rain," as their motto?

Applying Geography Skills

8. **Analyzing Maps** Study the political map on page 608. What five African nations does the Tropic of Capricorn cross?

Technology Skill

Developing Multimedia Presentations

Your geography homework is to make a presentation about Botswana. You want to make your presentation informative but also interesting and fun. How can you do this? One way is to combine several types of media into a **multimedia presentation.**

Learning the Skill

A multimedia presentation involves using several types of media, including photographs, videos, or sound recordings. The equipment can range from simple cassette players to overhead projectors to VCRs to computers and beyond. In your presentation on Botswana, for example, you might show photographs of cheetahs in the Okavango River delta or women fishing. You could also play a recording of local music or find a video of people working in diamond mines. If you have the proper equipment, you can then combine all these items on a computer.

Computer multimedia programs allow you to combine text, video, audio, art or graphics, and animation. The tools you need include computer graphic and drawing programs, animation programs that make certain images move, and systems that tie everything together. Your computer manual will tell you which tools your computer can support.

Practicing the Skill

Use the following questions as a guide when planning your presentation:

1. Which forms of media do I want to include? Video? Sound? Animation? Photographs? Graphics?
2. Which of the media forms does my computer support?
3. Which kinds of media equipment are available at my school or local library?
4. What types of media can I create to enhance my presentation?

Applying the Skill

Plan and create a multimedia presentation on a country discussed in this unit. List three ideas you would like to cover. Use as many multimedia materials as possible and share your presentation with the class.

▼ **Various equipment is needed to make multimedia presentations. For example, photographs and videos of cheetahs will make your report on Botswana more interesting.**

Section 4

Indian Ocean Countries

Guide to Reading

Main Idea

Africa's Indian Ocean countries are mostly farming nations that are struggling to develop more varied economies.

Terms to Know

- cyclone
- slash-and-burn farming
- deforestation

Places to Locate

- Mozambique
- Madagascar
- Comoros
- Seychelles
- Mauritius
- Zambezi River
- Maputo
- Victoria

Reading Strategy

Create a chart like the one below. Then fill in two key facts about each of the Indian Ocean countries.

Country	Fact #1	Fact #2
Mozambique		
Madagascar		
Comoros		
Seychelles		
Mauritius		

The island country of Madagascar has wildlife that appears nowhere else on the earth. Lemurs, chameleons, baobabs, geckos, and octopus trees are just a few animals and plants that exist on the island. Ring-tailed lemurs like this one spend their lives in the trees and on the ground, eating fruits, leaves, and seeds.

Mozambique in southern Africa borders the Indian Ocean. Four island countries—**Madagascar** (MA•duh•GAS•kuhr), **Comoros** (KAH•muh•ROHZ), **Seychelles** (say•SHEHL), and **Mauritius** (maw•RIH•shuhs)—also form part of southern Africa's Indian Ocean region.

Mozambique

Sand dunes, swamps, and fine natural harbors line Mozambique's long Indian Ocean coastline. In the center of this Y-shaped country stretches a flat plain covered with grasses and tropical forests. High plateaus and mountains lie along the northwestern border with Malawi. The **Zambezi River** splits Mozambique in two. The Cabora Bassa Dam on this river provides irrigation and electric power.

Most of Mozambique has a tropical savanna climate with wet and dry seasons. Mozambique also experiences deadly cyclones. A **cyclone** is an intense storm system with heavy rain and high winds.

Maputo, Mozambique

A high-rise building is being constructed in Maputo. Hotels and industrial projects are helping the city's economy to grow.

Place What slowed industrial growth in Maputo in the 1980s and 1990s?

Most people in Mozambique are farmers. Some practice **slash-and-burn farming**—a method of clearing land for planting by cutting and burning forest. One result of slash-and-burn farming is **deforestation,** or cutting down of forests. Deforestation can, in turn, lead to flooding during the rainy season. Such floods drove millions of people from their homes in early 2000.

Mozambique's major crops are cashews, cotton, sugarcane, tea, coconuts, and tropical fruits. Fishing provides an income for people who live along the coast. The main source of income, however, comes from its seaports. South Africa, Zimbabwe, Swaziland, and Malawi all pay to use the docks at **Maputo,** the capital, and other ports.

During the 1980s and early 1990s, a fierce civil war slowed industrial growth. In recent years, however, foreign companies have begun to invest in metal production, natural gas, fishing, and transportation services.

About 19.1 million people live here. Nearly all belong to 16 major African ethnic groups. Portuguese is Mozambique's official language, but most people speak African languages. About half of the people practice traditional African religions. Most of the rest are Muslim or Christian.

Reading Check What is a negative result of slash-and-burn farming?

Madagascar

Madagascar is the world's fourth-largest island. Madagascar broke away from the African mainland about 160 million years ago. The island's location kept it isolated from other parts of the world. As a result, it has many plants and animals that are not found elsewhere.

If you like vanilla ice cream, thank Madagascar. It produces most of the world's vanilla beans. The main cash crop is coffee, grown on the humid eastern plains. Rice is grown on the central plateau. About 80 percent of the island has been slashed and burned by people who must farm and herd to survive. Rainfall erodes the red clay soil. The government has taken steps to save what forests are left. It has increased the amount of forested land under its protection.

Only about 22 percent of the people are city dwellers. Antananarivo (AHN•tah•NAH•nah•REE•voh), the capital, lies in the central plateau. Called "Tana" for short, this city is known for it colorful street markets, where craftspeople sell a variety of products.

Music revolves around dance rhythms that reflect Madagascar's Southeast Asian and African heritage. The people are known for their rhythmic style of singing accompanied only by hand clapping. Most of the songs have themes about love, poverty, and hope for the future.

Reading Check **What percentage of Madagascar's people live in cities?**

Small Island Countries

Far from Africa in the Indian Ocean are three other island republics—Comoros, Seychelles, and Mauritius. The people of these countries have many different backgrounds.

Comoros Volcanoes formed the Comoros thousands of years ago. Dense tropical forests cover Comoros today. Most of the 600,000 people are farmers. The main crops are rice, vanilla, cloves, coconuts, and bananas. Even though agriculture employs 80 percent of the workforce, Comoros cannot grow enough food for its growing population. The government is trying to encourage industry, including tourism.

The people of Comoros are a mixture of Arabs, Africans, and people from Madagascar. They speak Arabic, French, and Comoran. Most practice Islam. Once ruled by France, the people of Comoros declared their independence in 1975. Since then, they have suffered from fighting among political groups for control of the government.

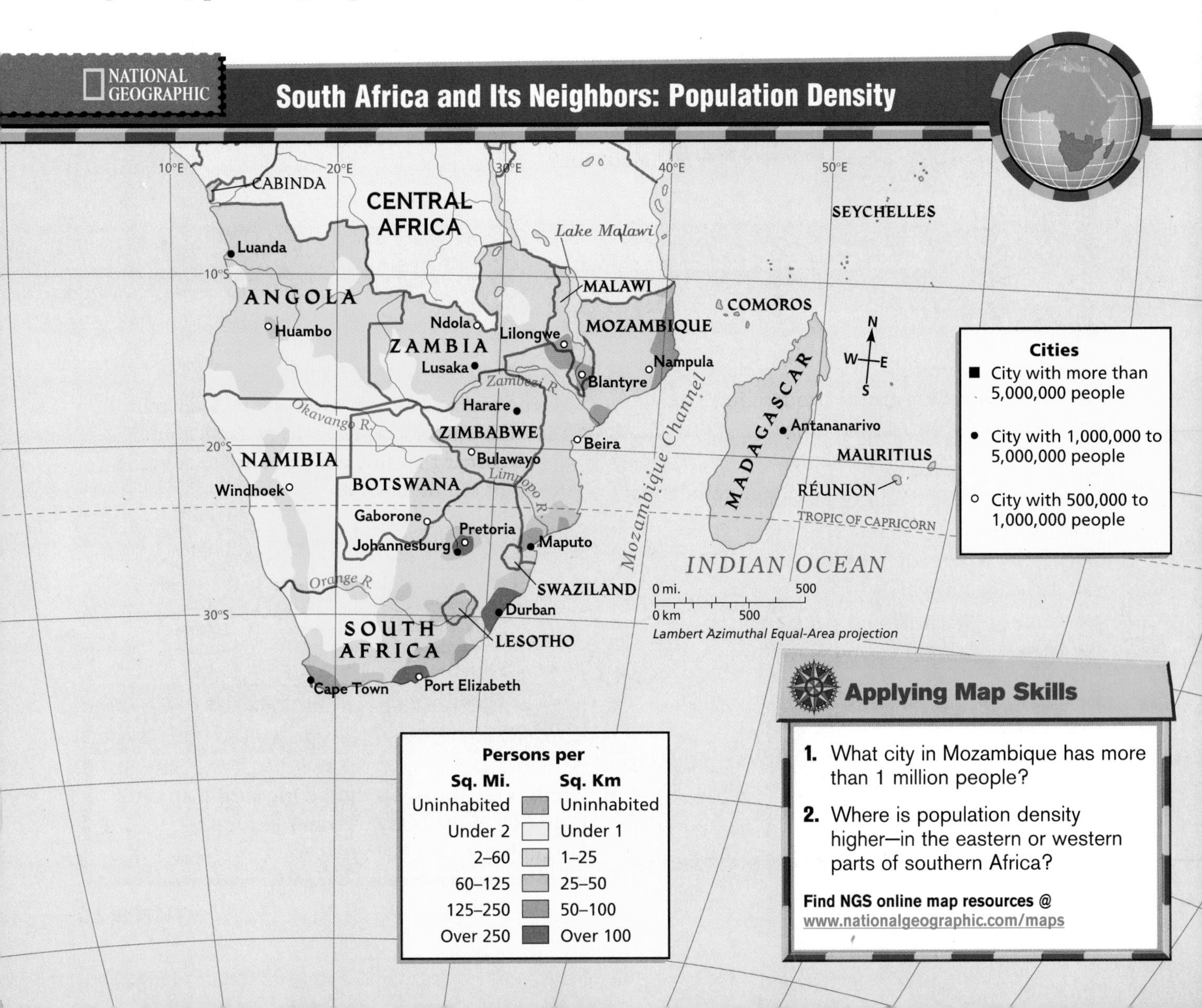

Applying Map Skills

1. What city in Mozambique has more than 1 million people?
2. Where is population density higher—in the eastern or western parts of southern Africa?

Find NGS online map resources @ www.nationalgeographic.com/maps

Seychelles About 90 islands form the country of Seychelles. About 40 of the islands are granite with high green peaks. The rest are small, flat coral islands with few people. Nearly 90 percent of the country's 100,000 people live on Mahé, the largest island. Here you will also find the capital, **Victoria.**

Seychelles was not inhabited until the 1700s. Under French and then British rule, it finally became independent in 1976. Most of the country's people are of mixed African, European, and Asian descent. They grow coconuts and cinnamon, the chief cash crops, on the larger islands. Fishing and tourism are important industries as well.

Mauritius Like Comoros, the islands of Mauritius were formed by volcanoes. Bare, black peaks rise sharply above green fields, and palm-dotted white beaches line the coasts. Mauritius has few natural resources, but it has succeeded in building a varied economy. Sugar is its main agricultural export. Major industries are located in Port Louis, the capital. Clothing and textiles account for about half of Mauritius's export earnings. Tourism is an important industry, too.

Mauritius has about 1.2 million people who come from many different backgrounds. About 70 percent are descendants of settlers from India. The rest are of African, European, or Chinese ancestry. Because of this varied ethnic heritage, the foods of Mauritius have quite a mix of ingredients. You can sample Indian chicken curry, Chinese pork, African-made roast beef, and French-style vegetables.

✓Reading Check **What are the capitals and populations of Seychelles and Mauritius?**

Assessment

Defining Terms

1. **Define** cyclone, slash-and-burn farming, deforestation.

Recalling Facts

2. **Economics** Name four of Mozambique's major crops.
3. **Location** Where are most of the world's vanilla beans grown?
4. **Movement** What natural force created the islands of Comoros and Mauritius?

Critical Thinking

5. **Analyzing Information** Why are Mozambique's seaports so important?
6. **Evaluating Information** How do the foods of Mauritius show its heritage?

Graphic Organizer

7. **Organizing Information** Draw a diagram like the one below. Then write facts about Madagascar that fit each category heading in each of the outer ovals.

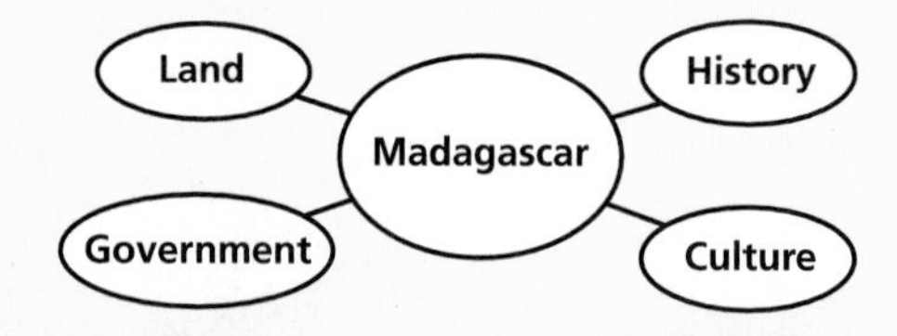

Applying Geography Skills

8. **Analyzing Maps** Study the physical map on page 609. What body of water separates Mozambique and Madagascar?

Chapter 22

Reading Review

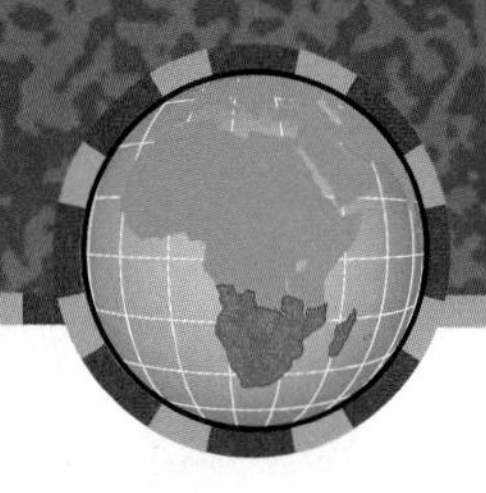

Section 1 Republic of South Africa

Terms to Know
high veld
escarpment
developed country
Boer
apartheid
township
enclave

Main Idea

Rich in resources, South Africa has recently seen major social and political changes.

✓ Economics Because of its abundant mineral resources, South Africa has the most developed economy in Africa.

✓ Government South Africa held its first democratic election in which people from all ethnic groups could vote in 1994.

✓ Culture South Africa is working to improve the lives of its poorer citizens.

Section 2 Atlantic Countries

Terms to Know
exclave

Main Idea

Angola and Namibia, although rich in resources, are struggling to develop their economies.

✓ Economics Angola's main source of income is oil.

✓ Culture Few Namibians benefit from the country's rich mineral wealth. The majority live in poverty.

Section 3 Inland Southern Africa

Terms to Know
copper belt
sorghum

Main Idea

Most of inland southern Africa is rich in resources and home to a wide variety of ethnic groups.

✓ Economics Zambia is one of the world's largest producers of copper.

✓ Place Zimbabwe has many mineral resources and good farmland.

✓ Economics Mining and tourism earn money for Botswana, but most of its people are farmers.

Section 4 Indian Ocean Countries

Terms to Know
cyclone
slash-and-burn farming
deforestation

Main Idea

Africa's Indian Ocean countries are mostly farming nations that are struggling to develop more varied economies.

✓ Human/Environment Interaction Slash-and-burn farming in Mozambique has led to deforestation and flooding. Neighboring countries pay fees for the use of its ports.

✓ Location Madagascar's island location has resulted in many plants and animals found nowhere else in the world.

✓ Economics Comoros continues to be a mainly agricultural economy, but Mauritius has succeeded in developing a variety of industries.

✓ Place Seychelles's beaches and tropical climate draw many tourists.

Chapter 22 Assessment and Activities

Using Key Terms

Match the terms in Part A with their definitions in Part B.

A.

1. escarpment
2. cyclone
3. exclave
4. slash-and-burn farming
5. township
6. apartheid
7. Boer
8. deforestation
9. sorghum
10. enclave

B.

a. separating racial and ethnic groups
b. storm with high circular winds
c. widespread cutting of trees
d. small nation located inside a larger country
e. steep cliff separating two fairly flat surfaces
f. small part of a nation separated from the main part of the country
g. areas of forest are cleared by burning
h. tall grass used as grain and to make syrup
i. settlements outside cities in South Africa
j. Dutch farmer

Reviewing Main Ideas

Section 1 Republic of South Africa

11. **Location** What is the southernmost point of Africa?
12. **History** When was South Africa's first election allowing all people to vote?
13. **Economics** What is Lesotho's only important natural resource?

Section 2 Atlantic Countries

14. **Region** What desert extends along the Atlantic coast of Angola and Namibia?
15. **History** What European country colonized Angola?
16. **Culture** What does *namib* mean?

Section 3 Inland Southern Africa

17. **Place** What river crosses Zambia?
18. **Economics** Where is the copper belt?
19. **History** What was Great Zimbabwe?

Section 4 Indian Ocean Countries

20. **Economics** What industries does Mozambique hope will help its economy?
21. **Place** What is the capital of Mozambique?
22. **Economics** What is Madagascar's main cash crop?

South Africa and Its Neighbors

Place Location Activity

On a separate sheet of paper, match the letters on the map with the numbered places listed below.

1. Madagascar
2. Lake Malawi
3. Zambezi River
4. Kalahari Desert
5. Angola
6. Zimbabwe
7. Pretoria
8. Mozambique
9. South Africa
10. Namibia

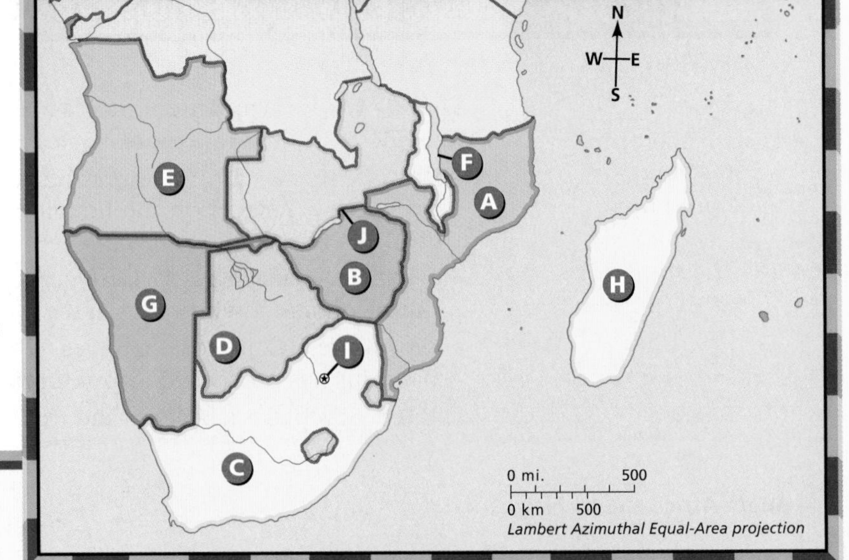

Self-Check Quiz Visit the ***Geography: The World and Its People*** Web site at gwip.glencoe.com and click on **Chapter 22–Self-Check Quizzes** to prepare for the Chapter Test.

Critical Thinking

23. **Supporting Generalizations** What facts support the statement "South Africa has the most developed economy in Africa"?
24. **Evaluating Information** Many countries of southern Africa are hoping to build and improve their industries. On a chart like the one below, list the positive and negative aspects of industrialization under the correct headings.

Industrialization	
Positives	Negatives

GeoJournal Activity

25. **Writing a Myth** Throughout history, different cultures have created myths or stories to explain events in nature, such as thunder and lightning or an eclipse. Write a story that might explain some aspect of African life. Story ideas include how deserts formed, why a zebra has stripes, or why a waterfall has "smoke that thunders."

Mental Mapping Activity

26. **Focusing on the Region** Draw a simple outline map of Africa, then label the following:
    - Atlantic Ocean
    - Madagascar
    - South Africa
    - Namib Desert
    - Indian Ocean
    - Pretoria
    - Angola
    - Cape Town
    - Mozambique
    - Kalahari Desert

Technology Skills Activity

27. **Using the Internet** The Zulu are a well-known ethnic group in Africa. Research this group on the Internet. Write a speech answering these questions: Who are the Zulu? How have they affected the history of southern Africa? Where do they live today?

Standardized Test Practice

Directions: Study the graph below, then answer the question that follows.

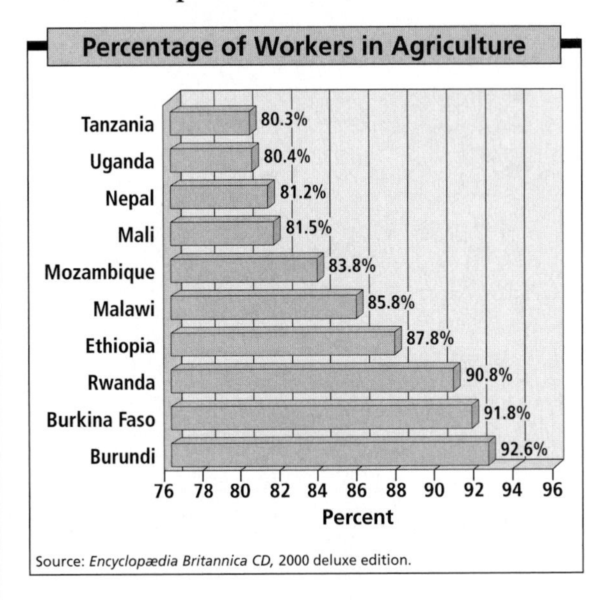

Source: *Encyclopædia Britannica CD,* 2000 deluxe edition.

1. **What percentage of Malawi's people do NOT work in agriculture?**
   - **F** 14.2%
   - **G** 85.8%
   - **H** 81.5%
   - **J** 19.7%

Test-Taking Tip: Check your answer by asking yourself, "Did I read the question carefully?" The question is referring to Malawi, so double-check Malawi's percentage of agricultural workers on the graph. Then take another look at the question. Overlooking the word NOT in a question is a common error. If you know the percentage of agricultural workers, how do you find those NOT involved in agriculture?

Unit 8

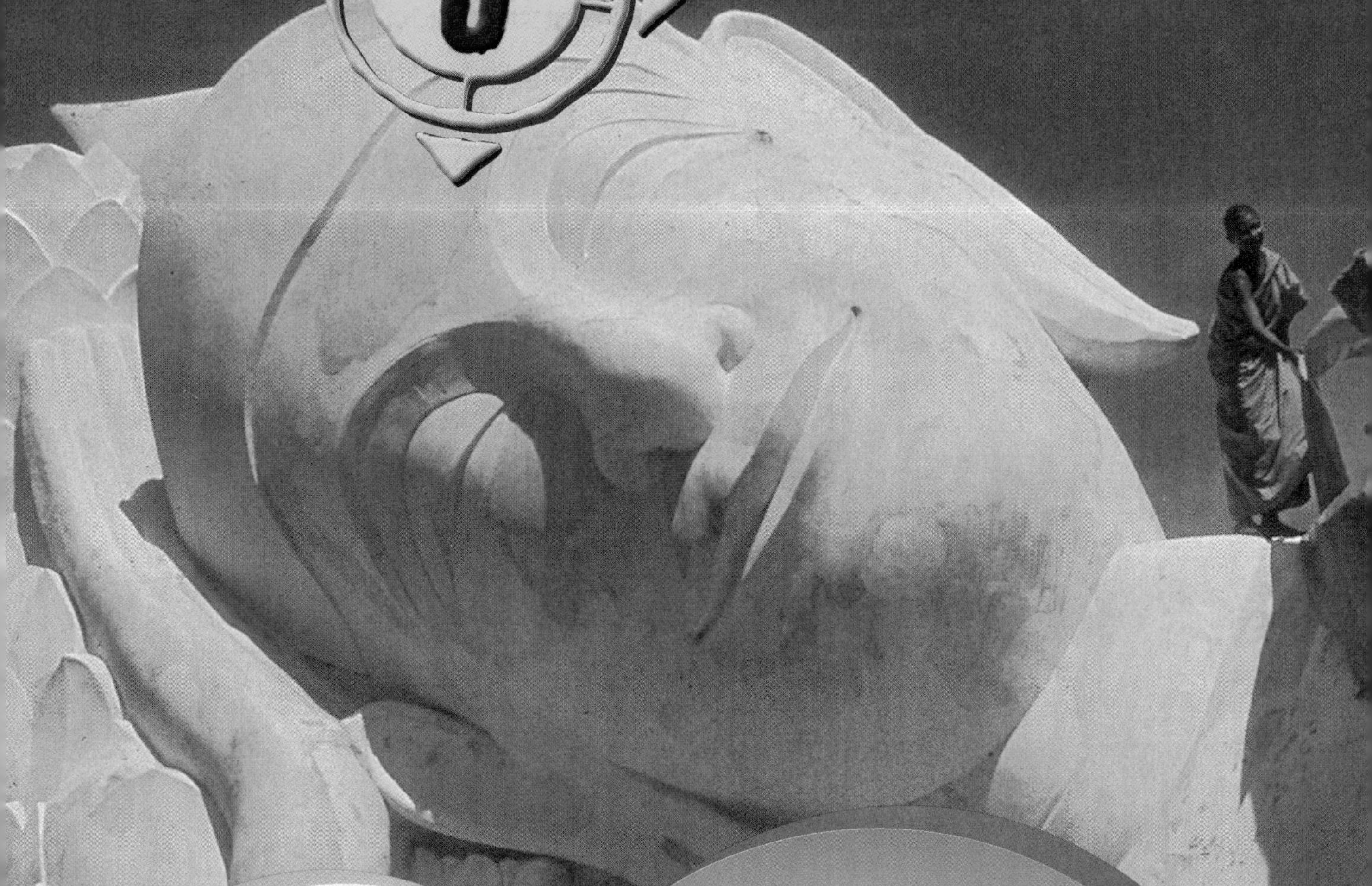

Taj Mahal, Agra, India

Macaques in a hot spring, Japan

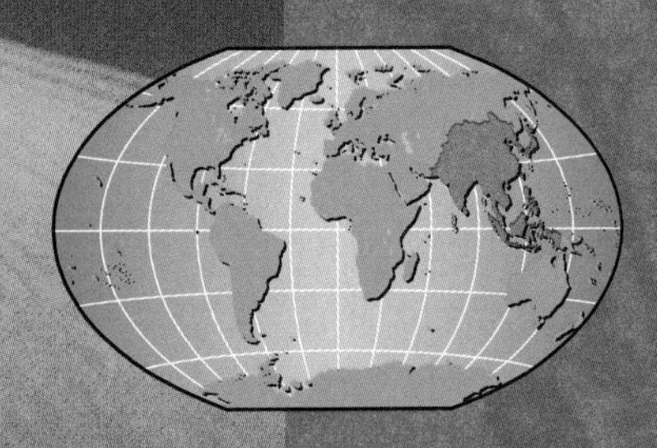

Asia

For many people in the Western Hemisphere, the region of Asia—in the Eastern Hemisphere—brings to mind exotic images. Ancient temples stand in dense rain forests. Farmers work in flooded rice fields. Pandas nibble bamboo shoots. Yet bustling cities, gleaming skyscrapers, and high-technology industries also can be found here. Turn the page to learn more about this region and its 3 billion people.

Monks wrapping statue of Buddha in yellow cloth, Thailand

NGS ONLINE
www.nationalgeographic.com/education

Focus on:

Asia

THE REGION OF ASIA is made up of surprisingly diverse landscapes. It includes a large chunk of the Asian continent, together with island groups that fringe its southern and eastern shores. Some of the world's oldest civilizations and religions had their beginnings in Asia. Now more than 3 billion people call this region home.

The Land

Covering roughly 7.8 million square miles (20.2 sq. km), the Asian region stretches from the mountains of western Pakistan to the eastern shores of Japan. It reaches from the highlands of northeastern China to the tropical islands of Indonesia. The region's long, winding coastlines are washed by two major oceans—the Indian and the Pacific—as well as many seas.

Several mountain ranges slice through central Asia. Most famous are the towering Himalaya, the site of Mt. Everest—the earth's tallest peak. North of the Himalaya lies the vast Plateau of Tibet, so high it has been called the Roof of the World. Beyond the plateau are two immense deserts: the Taklimakan and the Gobi.

Other mountain ranges cut across northeastern China, slant down the Korean Peninsula, and sweep through the peninsulas of Southeast Asia. Offshore, Japan, Indonesia, and other mountainous islands lie along the Ring of Fire. This is an area where adjoining plates of the earth's crust slip and buckle, setting off earthquakes and volcanic eruptions.

Great rivers begin in Asia's lofty center. On their journey to the sea, they flow through fertile plains in several countries. The most important rivers include the Indus in Pakistan, the Ganges and Brahmaputra in India and Bangladesh, the Yangtze and Yellow in China, and the Mekong in Southeast Asia.

The Climate

A person traveling across Asia would need clothes to suit almost every imaginable climate. The snowcapped mountains and high, windswept plateaus of northern and central Asia can be bitterly cold. The deserts can shimmer with heat by day, yet be frosty at night. Lowlands and coastal plains enjoy milder climates. The peninsulas of Southeast Asia and the islands straddling the Equator have mostly tropical climates. They are cloaked in dense rain forests. Seasonal winds called monsoons blow across much of Asia, bringing dry weather in winter and drenching rains in summer.

Unit 8

Terraced rice fields, Bali, Indonesia

◀ Street flooded by monsoon rain, Tamil Nadu, India

The Economy

Agriculture is the major economic activity across most of Asia. The region's rugged mountains and vast deserts mean that only a small amount of the land is suitable for growing crops, however. For example, only about 10 percent of China's land can be used for agriculture. To feed the region's huge population, Asian farmers must make the most of every possible bit of farmland. Terraces allow farmers to grow rice even on steep hillsides. Rice, which grows well in places with warm temperatures and plenty of water, is the most important food crop in Asia. China, India, Indonesia, and Bangladesh are the leading rice producers in the world.

Most of Asia's manufacturing takes place in Japan, South Korea, Taiwan, China, and India. China and India both are rich in coal, iron ore, and other natural resources. Japan, however, has few mineral resources and must import fuel and nearly all the raw materials it needs. Still, Japan has become one of the world's leading manufacturers of cars, electronic products, and other goods. In some of the region's other countries, such as Laos, Vietnam, and Bhutan, industry is less well developed.

The People

Nestled in fertile river valleys, some of the world's oldest civilizations arose in Asia thousands of years ago. Ancient religions took root here as well. Both Hinduism and Buddhism, for example, originated in India. Hindus remain concentrated in India, but over time Buddhism spread throughout the region. The region's most widespread faith—Islam—began in Southwest Asia.

Europeans arrived in the region around 1500, bringing Christianity to some of the people. By the early 1800s, many Asian countries had become European colonies. Most of these countries became independent in the mid-1900s. In many cases, however, independence was followed by political turmoil and conflict between Communist and non-Communist forces. Today China, Vietnam, and North Korea have Communist governments; Nepal is ruled by a monarch; and military leaders control Myanmar. Japan, India, and the Philippines are democracies.

About 3.3 billion people live in Asia, but this huge population is very unevenly distributed. Most people live near the coasts and in the fertile river valleys. As a result, some parts of Asia are among the most crowded places in the world.

Exploring the Region

1. **Why is the Plateau of Tibet called the Roof of the World?**
2. **How do monsoons affect the region?**
3. **What is the most important food crop in Asia?**
4. **What are two religions that originated in the region?**

◀ **Robot welding car bodies in a factory, Japan**

UNIT 8

Three generations of a Chinese family

REGIONAL ATLAS

Asia

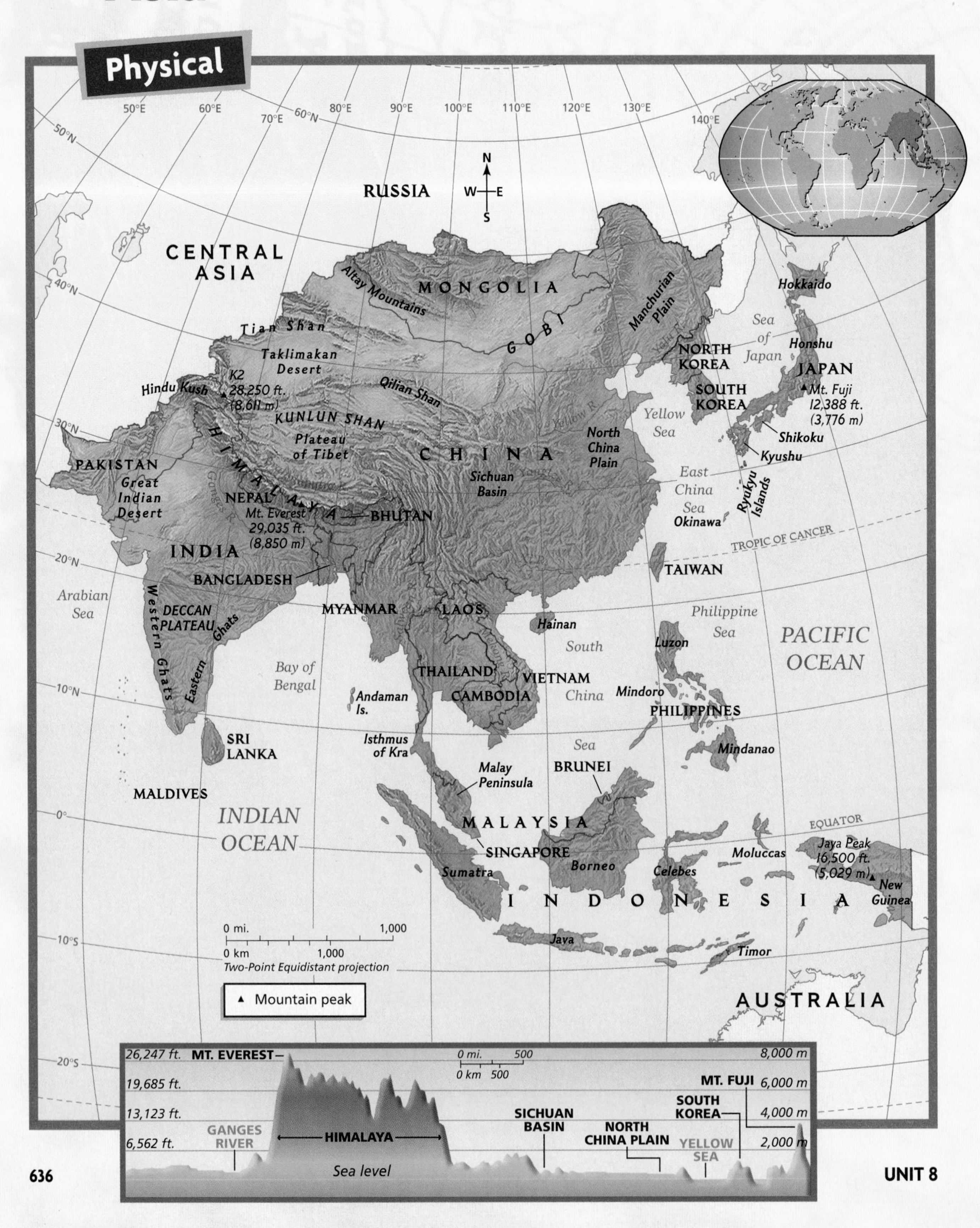

Physical
RUSSIA
CENTRAL ASIA
MONGOLIA
Altay Mountains
GOBI
Manchurian Plain
Tian Shan
Taklimakan Desert
Qilian Shan
Hindu Kush
K2 28,250 ft. (8,611 m)
KUNLUN SHAN
Plateau of Tibet
HIMALAYA
PAKISTAN
Great Indian Desert
NEPAL
Mt. Everest 29,035 ft. (8,850 m)
BHUTAN
CHINA
Sichuan Basin
North China Plain
Yellow Sea
NORTH KOREA
SOUTH KOREA
Sea of Japan
Hokkaido
Honshu
JAPAN
Mt. Fuji 12,388 ft. (3,776 m)
Shikoku
Kyushu
East China Sea
Ryukyu Islands
Okinawa
TROPIC OF CANCER
TAIWAN
INDIA
BANGLADESH
Arabian Sea
Western Ghats
DECCAN PLATEAU
Eastern Ghats
MYANMAR
LAOS
Hainan
Philippine Sea
PACIFIC OCEAN
Luzon
South China Sea
Bay of Bengal
THAILAND
VIETNAM
CAMBODIA
Andaman Is.
Mindoro
PHILIPPINES
SRI LANKA
Isthmus of Kra
Mindanao
Malay Peninsula
BRUNEI
MALDIVES
INDIAN OCEAN
MALAYSIA
SINGAPORE
EQUATOR
Jaya Peak 16,500 ft. (5,029 m)
Moluccas
Borneo
Celebes
Sumatra
New Guinea
INDONESIA
Java
Timor
AUSTRALIA
0 mi. 1,000
0 km 1,000
Two-Point Equidistant projection
Mountain peak
26,247 ft. MT. EVEREST
19,685 ft.
13,123 ft.
6,562 ft.
GANGES RIVER
HIMALAYA
Sea level
SICHUAN BASIN
NORTH CHINA PLAIN
YELLOW SEA
SOUTH KOREA
MT. FUJI
8,000 m
6,000 m
4,000 m
2,000 m
0 mi. 500
0 km 500

UNIT 8

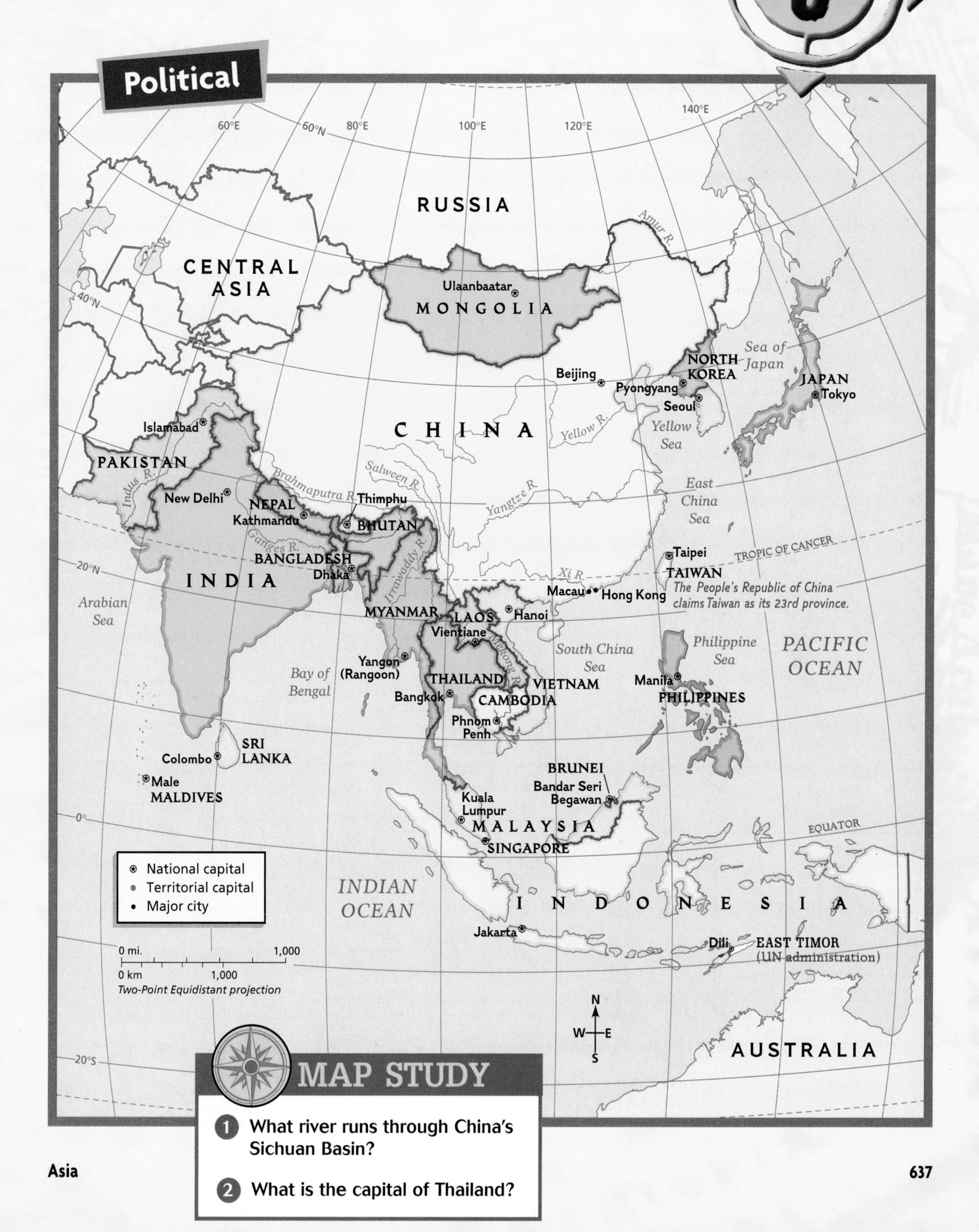
Political
RUSSIA
CENTRAL ASIA
MONGOLIA
Ulaanbaatar
CHINA
Beijing
NORTH KOREA
Pyongyang
Seoul
JAPAN
Tokyo
Sea of Japan
Yellow Sea
East China Sea
Amur R.
Yellow R.
Yangtze R.
Salween R.
Brahmaputra R.
Xi R.
Islamabad
PAKISTAN
Indus R.
New Delhi
NEPAL
Kathmandu
Thimphu
BHUTAN
Ganges R.
BANGLADESH
Dhaka
INDIA
Irrawaddy R.
Arabian Sea
Bay of Bengal
MYANMAR
Yangon (Rangoon)
LAOS
Vientiane
Hanoi
Mekong R.
THAILAND
Bangkok
VIETNAM
CAMBODIA
Phnom Penh
Macau
Hong Kong
Taipei
TAIWAN
The People's Republic of China claims Taiwan as its 23rd province.
TROPIC OF CANCER
South China Sea
Philippine Sea
PACIFIC OCEAN
Manila
PHILIPPINES
SRI LANKA
Colombo
Male
MALDIVES
BRUNEI
Bandar Seri Begawan
Kuala Lumpur
MALAYSIA
SINGAPORE
EQUATOR
INDIAN OCEAN
INDONESIA
Jakarta
Dili
EAST TIMOR (UN administration)
AUSTRALIA
60°E
80°E
100°E
120°E
140°E
60°N
40°N
20°N
0°
20°S
National capital
Territorial capital
Major city
0 mi. 1,000
0 km 1,000
Two-Point Equidistant projection
N
S
E
W

MAP STUDY

1. What river runs through China's Sichuan Basin?
2. What is the capital of Thailand?

Asia

Monsoons

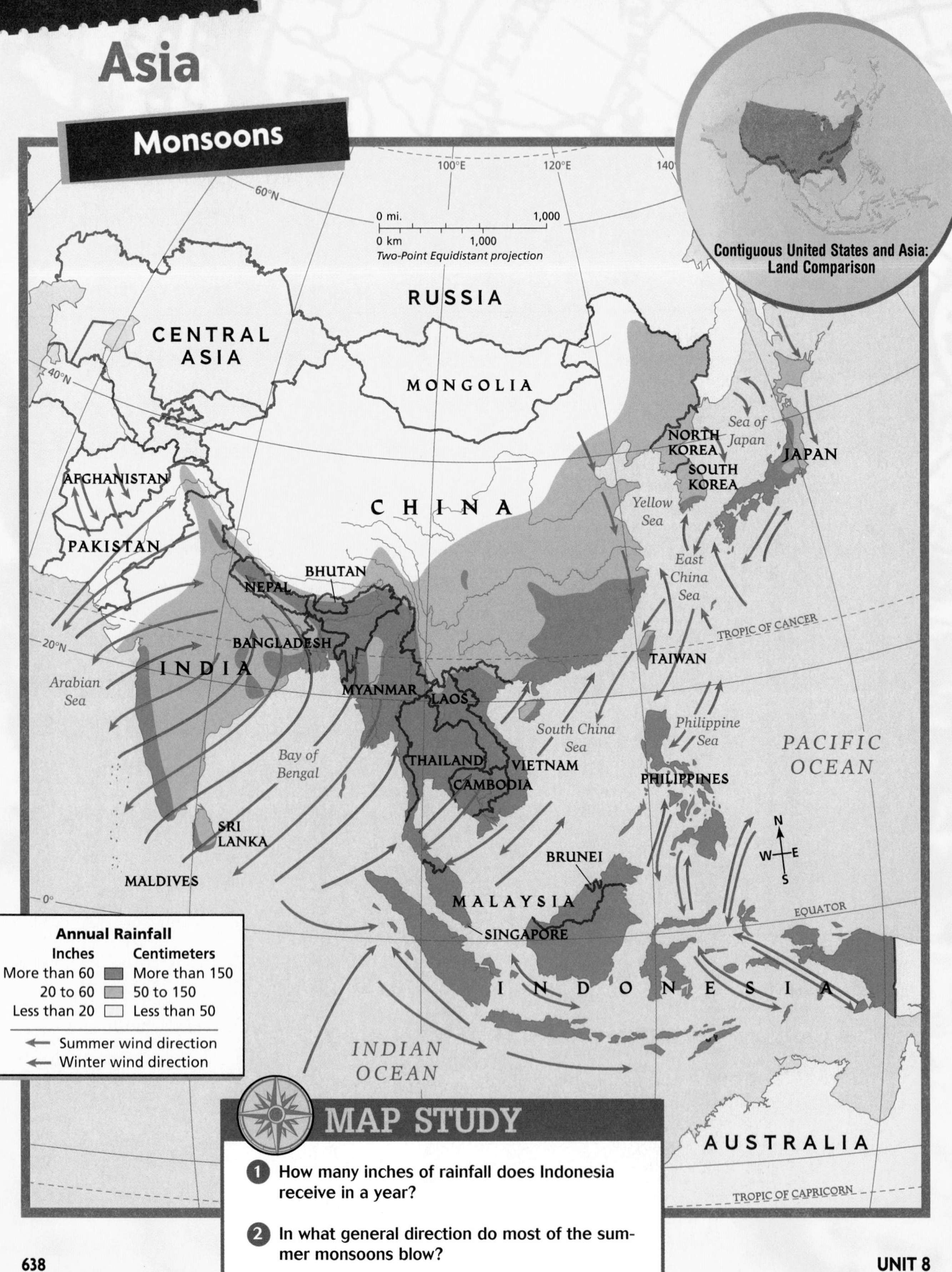

MAP STUDY

1. How many inches of rainfall does Indonesia receive in a year?

2. In what general direction do most of the summer monsoons blow?

Unit 8

Geo Extremes

① **HIGHEST POINT**
Mt. Everest
(Nepal and Tibet)
29,035 ft. (8,850 m) high

② **LOWEST POINT**
Turpan Depression (China)
505 ft. (154 m)
below sea level

③ **LONGEST RIVER**
Yangtze (China)
3,964 mi.
(6,380 km) long

④ **LARGEST DESERT**
Gobi (Mongolia and China)
500,000 sq. mi.
(1,295,000 sq. km)

⑤ **HIGHEST WATERFALL**
Mawsmai (India)
1,148 ft. (350 m) high

⑥ **LARGEST ISLAND**
New Guinea (Indonesia and Papua New Guinea)
306,000 sq. mi.
(792,536 sq. km)

⑦ **WETTEST PLACE**
Mawsynram (India)
467 in. (1,186 cm)
average annual rainfall

COMPARING POPULATION: United States and Selected Countries of Asia

UNITED STATES

CHINA

INDIA

INDONESIA

JAPAN

(one figure) = 70,000,000

Source: *Population Reference Bureau*, 2000.

WORLD POPULATION: Asia's Share of the World's People

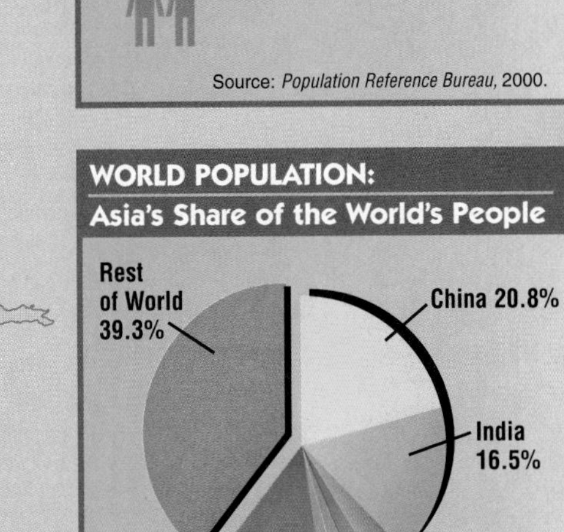

Source: *Population Reference Bureau*, 2000.

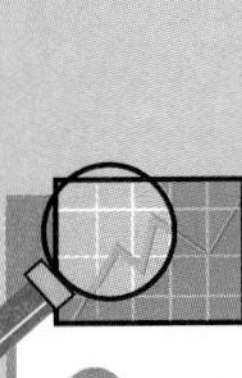

GRAPHIC STUDY

1. The highest point in Asia is also the highest point in the world. What is it?
2. What percentage of the world's population lives in Asia?

Country Profiles

BANGLADESH

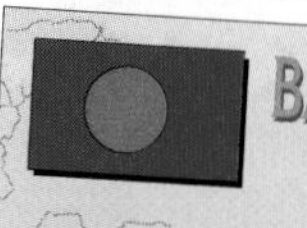

POPULATION:
125,721,000
2,261 per sq. mi.
873 per sq. km

LANGUAGE:
Bengali

MAJOR EXPORT:
Clothing

MAJOR IMPORT:
Machinery

CAPITAL:
Dhaka

LANDMASS:
55,598 sq. mi.
143,998 sq. km

BHUTAN

POPULATION:
800,000
44 per sq. mi.
17 per sq. km

LANGUAGES:
Dzonkha, Local Languages

MAJOR EXPORT:
Cardamom

MAJOR IMPORT:
Fuels

CAPITAL:
Thimphu

LANDMASS:
18,147 sq. mi.
47,001 sq. km

BRUNEI

POPULATION:
323,000
145 per sq. mi.
56 per sq. km

LANGUAGES:
Malay, English, Chinese

MAJOR EXPORT:
Crude Oil

MAJOR IMPORT:
Machinery

CAPITAL:
Bandar Seri Begawan

LANDMASS:
2,226 sq. mi.
5,765 sq. km

CAMBODIA

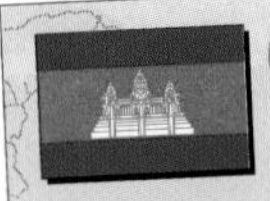

Phnom Penh

POPULATION:
11,851,000
170 per sq. mi.
65 per sq. km

LANGUAGES:
Khmer, French

MAJOR EXPORT:
Timber

MAJOR IMPORT:
Construction Materials

CAPITAL:
Phnom Penh

LANDMASS:
69,898 sq. mi.
181,035 sq. km

CHINA

POPULATION:
1,254,062,000
339 per sq. mi.
131 per sq. km

LANGUAGE:
Mandarin Chinese

MAJOR EXPORT:
Machinery

MAJOR IMPORT:
Machinery

CAPITAL:
Beijing

LANDMASS:
3,705,820 sq. mi.
9,598,032 sq. km

INDIA

POPULATION:
1,000,849,000
788 per sq. mi.
304 per sq. km

LANGUAGES:
Hindi, English, Local Languages

MAJOR EXPORTS:
Gems and Jewelry

MAJOR IMPORT:
Crude Oil

CAPITAL:
New Delhi

LANDMASS:
1,269,346 sq. mi.
3,287,591 sq. km

INDONESIA

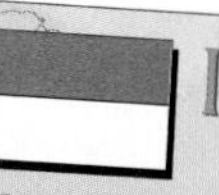

Jakarta

POPULATION:
211,806,000
286 per sq. mi.
110 per sq. km

LANGUAGES:
Bahasa Indonesia, Local Languages

MAJOR EXPORT:
Textiles

MAJOR IMPORT:
Manufactured Goods

CAPITAL:
Jakarta

LAND MASS:
741,101 sq. mi.
1,919,443 sq. km

JAPAN

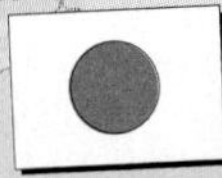

Tokyo

POPULATION:
126,745,000
869 per sq. mi.
335 per sq. km

LANGUAGE:
Japanese

MAJOR EXPORT:
Machinery

MAJOR IMPORT:
Manufactured Goods

CAPITAL:
Tokyo

LANDMASS:
145,875 sq. mi.
377,815 sq. km

LAOS

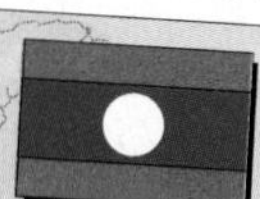

POPULATION:
5,000,000
55 per sq. mi.
21 per sq. km

LANGUAGES:
Lao, French

MAJOR EXPORT:
Wood Products

MAJOR IMPORT:
Machinery

CAPITAL:
Vientiane

LANDMASS:
91,429 sq. mi.
236,800 sq. km

Countries and flags not drawn to scale

Unit 8

Malaysia

POPULATION:
22,710,000
178 per sq. mi.
69 per sq. km

LANGUAGES:
Malay, English, Chinese

MAJOR EXPORT:
Electronic Equipment

MAJOR IMPORT:
Machinery

CAPITAL:
Kuala Lumpur

LANDMASS:
127,317 sq. mi.
329,749 sq. km

Maldives

POPULATION:
278,000
2,417 per sq. mi.
933 per sq. km

LANGUAGES:
Maldivian Divehi, English

MAJOR EXPORT:
Fish

MAJOR IMPORT:
Machinery

CAPITAL:
Male

LANDMASS:
115 sq. mi.
298 sq. km

Mongolia

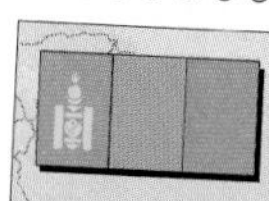

POPULATION:
2,438,000
4 per sq. mi.
2 per sq. km

LANGUAGE:
Khalkha Mongol

MAJOR EXPORT:
Copper

MAJOR IMPORT:
Fuels

CAPITAL:
Ulaanbaatar

LANDMASS:
604,250 sq. mi.
1,565,000 sq. km

Myanmar

POPULATION:
48,081,000
184 per sq. mi.
71 per sq. km

LANGUAGES:
Burmese, Local Languages

MAJOR EXPORT:
Beans

MAJOR IMPORT:
Machinery

CAPITAL:
Yangon (Rangoon)

LANDMASS:
261,218 sq. mi.
676,552 sq. km

Nepal

POPULATION:
24,303,000
447 per sq. mi.
173 per sq. km

LANGUAGE:
Nepali

MAJOR EXPORT:
Clothing

MAJOR IMPORT:
Petroleum Products

CAPITAL:
Kathmandu

LANDMASS:
54,362 sq. mi.
140,797 sq. km

North Korea

POPULATION:
21,386,000
460 per sq. mi.
177 per sq. km

LANGUAGE:
Korean

MAJOR EXPORT:
Minerals

MAJOR IMPORT:
Petroleum

CAPITAL:
Pyongyang

LANDMASS:
46,540 sq. mi.
120,538 sq. km

Pakistan

POPULATION:
146,488,000
477 per sq. mi.
184 per sq. km

LANGUAGES:
Urdu, English, Punjabi, Sindhi

MAJOR EXPORT:
Cotton

MAJOR IMPORT:
Petroleum

CAPITAL:
Islamabad

LANDMASS:
307,374 sq. mi.
796,095 sq. km

Philippines

POPULATION:
74,655,000
645 per sq. mi.
249 per sq. km

LANGUAGES:
Tagalog, English

MAJOR EXPORT:
Electronic Equipment

MAJOR IMPORT:
Raw Materials

CAPITAL:
Manila

LANDMASS:
115,831 sq. mi.
300,001 sq. km

Singapore

POPULATION:
3,999,000
16,732 per sq. mi.
6,471 per sq. km

LANGUAGES:
Chinese, Malay, Tamil, English

MAJOR EXPORT:
Computer Equipment

MAJOR IMPORT:
Aircraft

CAPITAL:
Singapore

LANDMASS:
239 sq. mi.
618 sq. km

Chapter 23 South Asia

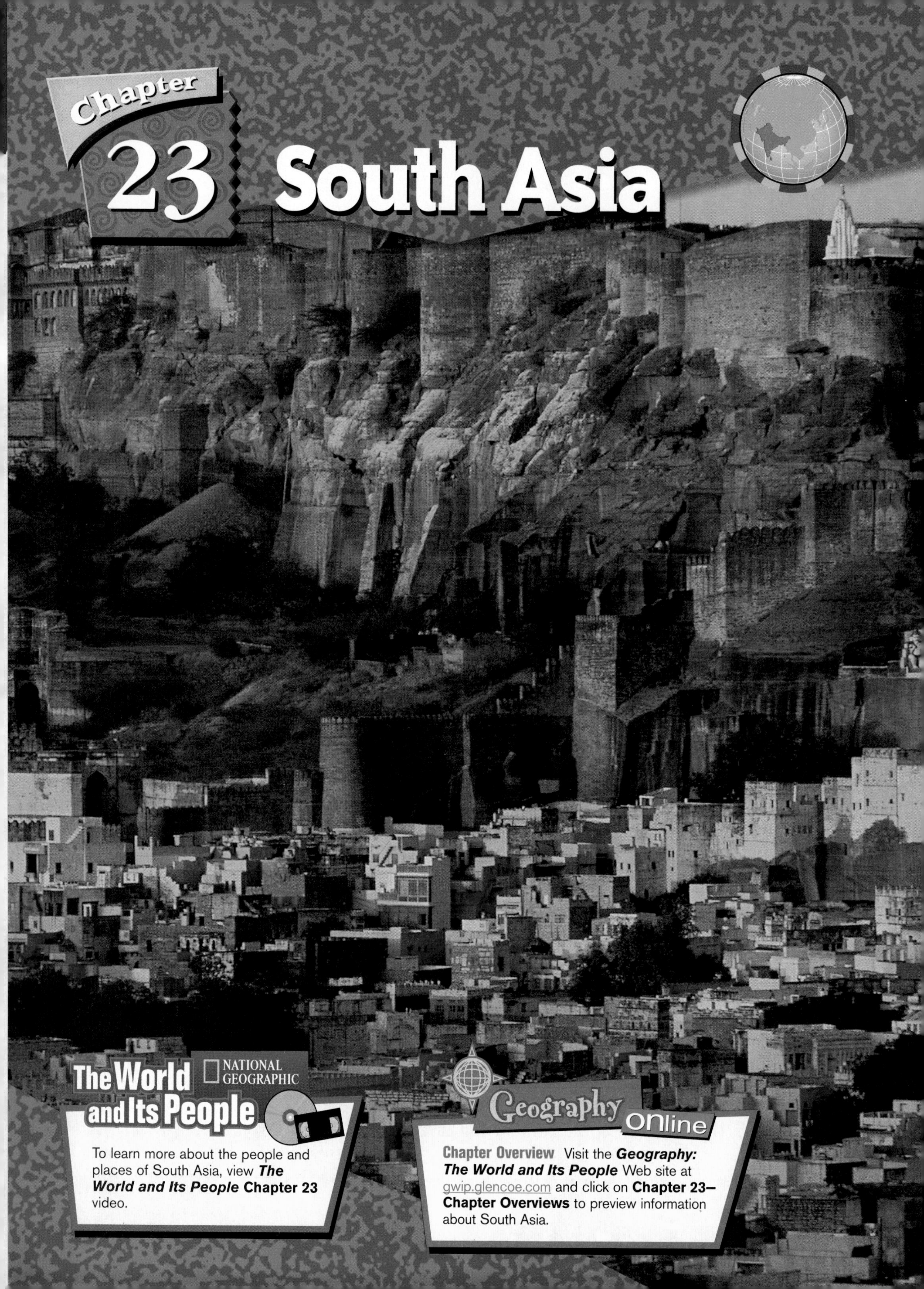

The World and Its People NATIONAL GEOGRAPHIC

To learn more about the people and places of South Asia, view ***The World and Its People*** **Chapter 23** video.

Geography Online

Chapter Overview Visit the ***Geography: The World and Its People*** Web site at gwip.glencoe.com and click on **Chapter 23–Chapter Overviews** to preview information about South Asia.

India

Guide to Reading

Main Idea

India—the world's most populous democracy—is trying to develop its resources and meet the needs of its rapidly growing population.

Terms to Know

- subcontinent
- monsoon
- jute
- cottage industry
- pesticide
- caste
- coalition government

Places to Locate

- India
- Karakoram Range
- Himalaya
- Satpura Range
- Ganges River
- Great Indian Desert
- Deccan Plateau
- New Delhi
- Mumbai
- Delhi
- Calcutta
- Chennai

Reading Strategy

Create a chart like this one. Then fill in at least two key facts about India under each category.

India	
Land	Economy
History	Religion

◀ Jodhpur, India

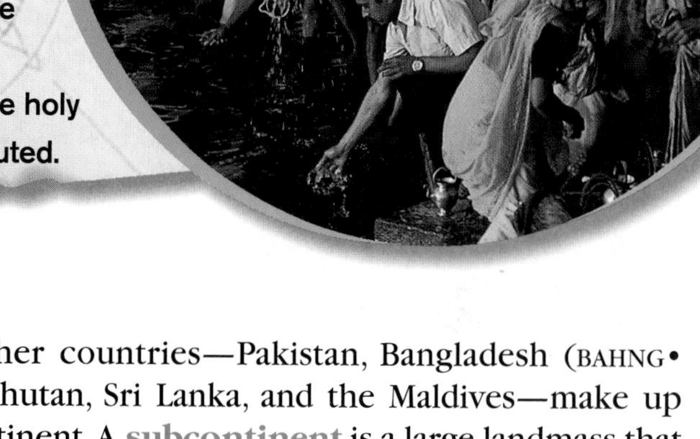

Stone steps lead to the Ganges River at this village in India. India's Hindus consider the Ganges River to be a holy river. About 1 million Hindus from all areas of India come to pray and bathe in its waters every year. People also use the river to do their laundry. In addition, industries dump waste into it. Today there is grave concern that the holy river is seriously polluted.

India and several other countries—Pakistan, Bangladesh (BAHNG•gluh•DEHSH), Nepal, Bhutan, Sri Lanka, and the Maldives—make up the South Asian subcontinent. A **subcontinent** is a large landmass that is part of another continent but distinct from it.

India's Land and Economy

Two huge walls of mountains—the **Karakoram** (KAH•rah•KOHR•ahm) **Range** and the **Himalaya** (HIH•muh•LAY•uh)—form India's northern border and separate South Asia from the rest of Asia. The tallest mountains in the world, the Himalaya's snowcapped peaks average more than 5 miles (8 km) in height. Edging India's southern coasts are the Eastern Ghats and the Western Ghats. These mountains lie just inland from the Bay of Bengal and the Arabian Sea. In central India, the **Satpura** (SAHT•puh•ruh) **Range** divides the country.

North of the Satpura lies the vast Ganges Plain. It boasts some of the most fertile soil in the country and holds about 40 percent of India's people. The **Ganges River** flows through the Ganges Plain to the Bay of Bengal. To the west lies the **Great Indian Desert.** South of the Satpura Range is the **Deccan Plateau.** Forests, farmland, and rich deposits of minerals make the Deccan Plateau a valuable region.

Most of India is warm or hot all year. The Himalaya block cold northern air from sweeping south into the country. **Monsoons,** or seasonal winds that blow steadily from the same direction for months, also influence the climate. Indians experience three seasons—cool, hot, and rainy. During the cool season (November through February) and the hot season (March through April), monsoon winds from the north bring dry air. During the rainy season (May through October), southern monsoon winds bring moist air from the Indian Ocean.

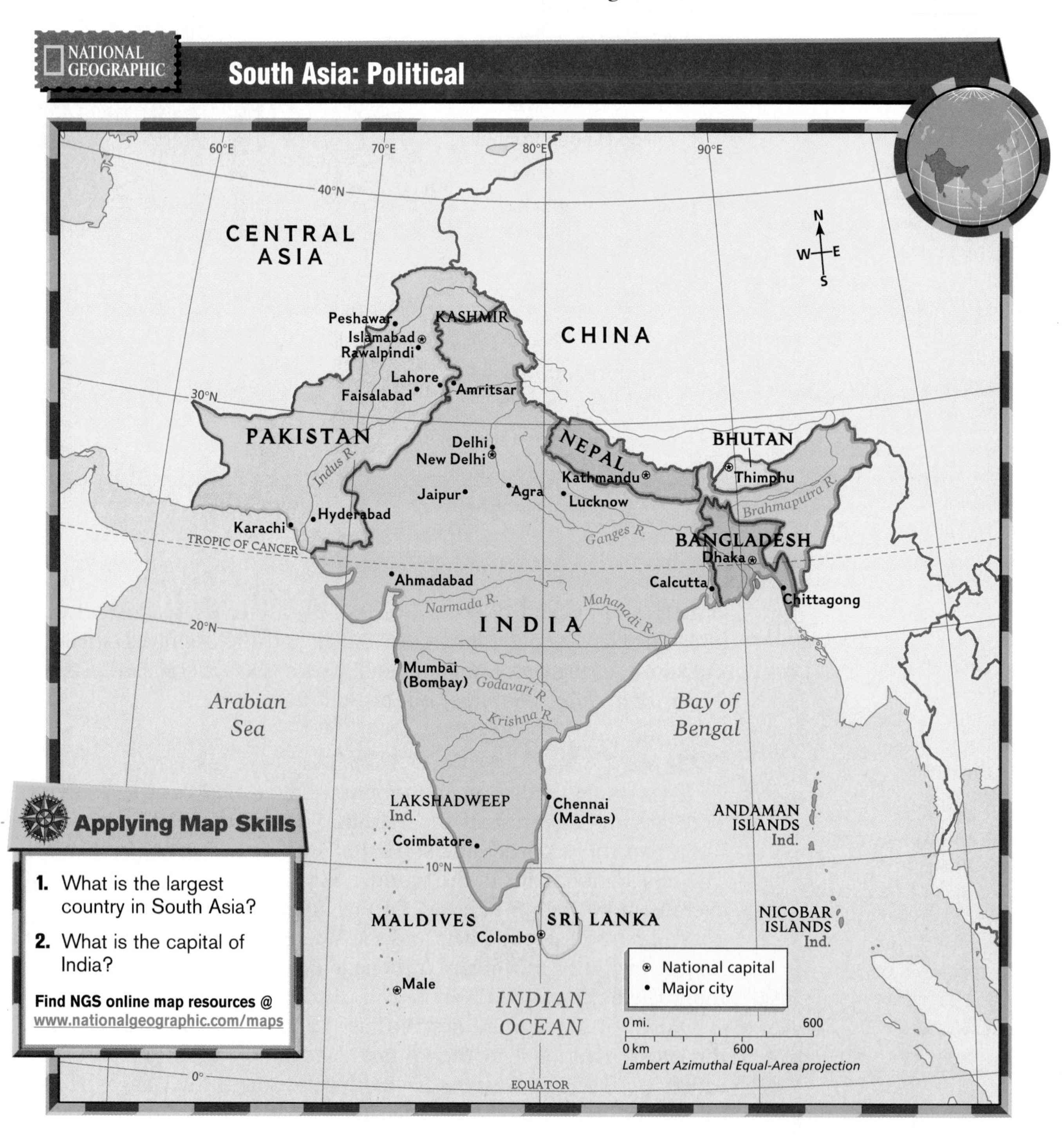

South Asia: Political

Applying Map Skills

1. What is the largest country in South Asia?
2. What is the capital of India?

Find NGS online map resources @ www.nationalgeographic.com/maps

India's Economy Agriculture and industry are equally important to India's economy. The best farmland lies in the Ganges Plain and on the Deccan Plateau. More than two-thirds of India's labor force works in agriculture, mostly on small farms. Many of them cannot afford fertilizer, good seed, or machinery. Still, India's farmers raise a variety of crops, including rice, wheat, cotton, tea, sugarcane, and jute. **Jute** is a plant fiber used for making rope, burlap bags, and carpet backing. India is the world's second-largest rice producer, after China.

Huge factories in India's cities turn out cotton textiles and produce iron and steel. Oil and sugar refineries loom over many urban skylines.

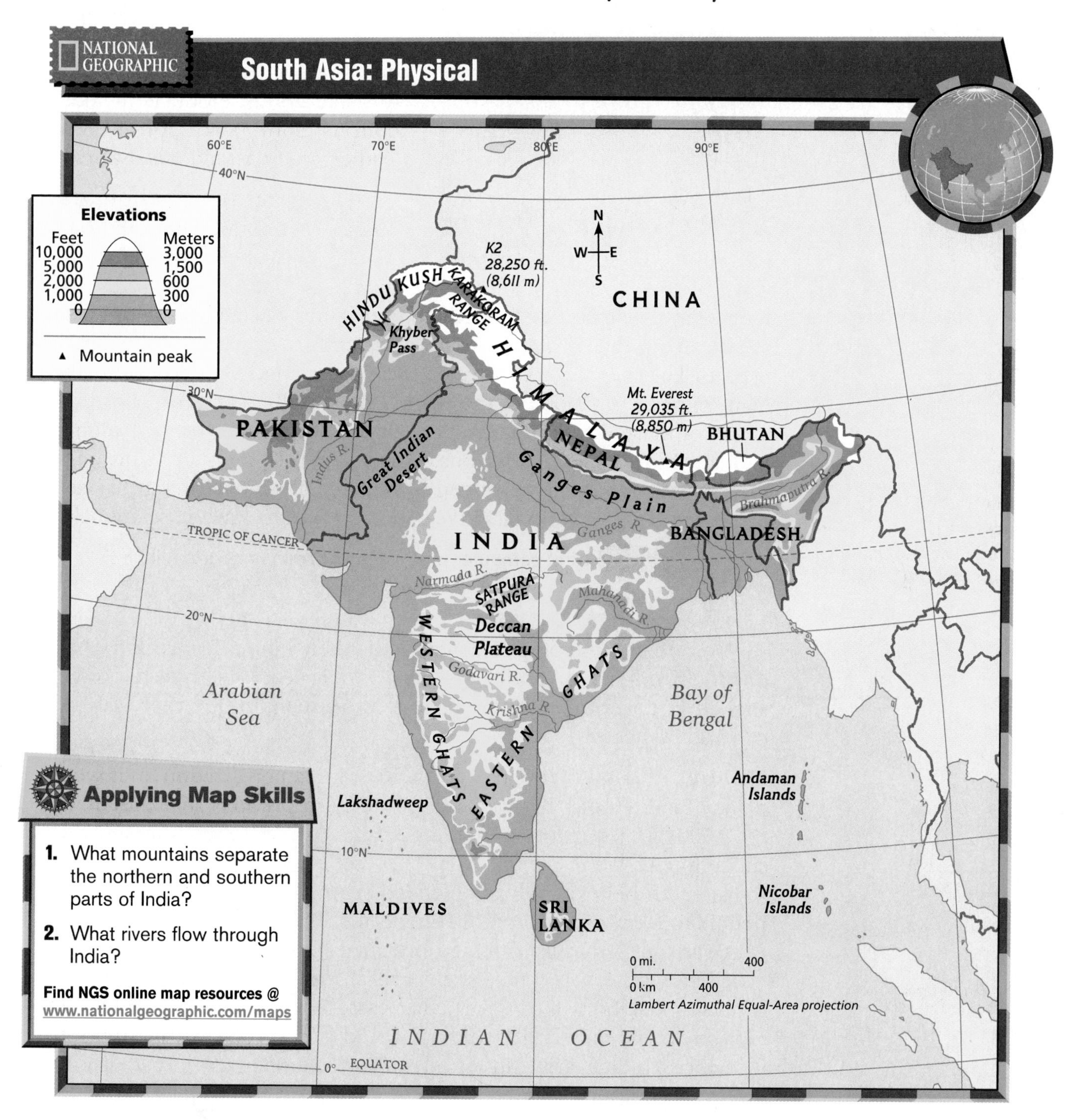

Two Views of India

The growing middle class live comfortable lives in India's suburbs (above), but the poor in India's cities must struggle to survive (right).

Movement **Why have many villagers in India moved from the countryside into the cities?**

Recently, American computer companies have opened offices in India, making it an important source of computer software. Mining is another major industry. India has rich deposits of coal, iron ore, manganese, and bauxite. Its major exports are gems and jewelry.

Many Indian products are manufactured in cottage industries. A **cottage industry** is a home- or village-based industry in which family members, including children, supply their own equipment to make goods. Products include cotton cloth, silk cloth, rugs, leather products, and metalware.

Environmental Challenges India's economic growth has brought challenges to its environment. Thousands of acres of forests have been cleared for farming. Both water and land have been polluted from burning coal, industrial wastes, and **pesticides,** or chemicals used to kill insects. India's major river, the Ganges, is considered by many experts to be one of the world's most polluted rivers.

All of these developments have played a part in destroying animal habitats. India's elephants, lions, tigers, leopards, monkeys, and panthers have been greatly reduced in number. The government has set up more than 350 national parks and preserves to save these animals.

✓Reading Check **What computer product is becoming important in India?**

India's People

More than 1 billion people call India their home. The country has 18 official languages. Hindi is the most widely used, but English is often spoken in government and business. More than 1,000 other dialects can also be heard.

History About 4,000 years ago, the first Indian civilization built well-planned cities along the Indus River valley, in present-day Pakistan. In the 1500s B.C., warriors known as Aryans (AR•ee•uhns) entered the subcontinent from Central Asia. They set up kingdoms in northern India. Aryan beliefs gradually blended with the practices of the local people to form the religion of Hinduism.

Over time, Hinduism helped organize India's society into groups called castes. A **caste** was a social class based on a person's ancestry. A person was born into a particular caste. That caste determined the jobs one could hold and whom one could marry. The caste system still influences Indian life, although laws forbid unfair treatment of one group by another.

The religion of Islam also influenced India's history. In the A.D. 700s, Muslims from Southwest Asia brought Islam to India. In the 1500s, they founded the Mogul (MOH•guhl) Empire and ruled India for 200 years.

The British were the last of India's conquerors, ruling from the 1700s to the mid-1900s. They built roads, railroads, and seaports. They also made large profits from the plantations, mines, and factories they set up. An Indian leader named Mohandas Gandhi (moh•HAHN•duhs GAHN•dee) led a nonviolent resistance movement. His efforts brought India independence from the United Kingdom in 1947.

Today India is a federal republic, or a government divided between national and state powers. India has 25 states and 7 territories. The head of state is a president, whose duties are mainly ceremonial. The real power lies with the prime minister. Voters choose from more than 20 major political parties. As a result, the prime minister often leads a coalition government. A **coalition government** is one in which two or more political parties work together to run the country. **New Delhi** was built specifically to be the country's capital.

Religion About 80 percent of India's people are Hindus, or followers of Hinduism. Hindus worship many deities, or gods and goddesses. They believe that after the body dies, the soul is reborn, often in a different form. This process is repeated until the soul reaches perfection. For this reason, many Hindus believe it is wrong to kill any living creature. Cows are particularly sacred. Indians allow them to roam freely.

Islam also has many followers. India's 140 million Muslims form one of the world's largest Muslim populations. Other religions include Christianity, Sikhism (SEE•KIH•zuhm), Buddhism, and

Music

The tabla is a pair of connected drums from India. The drums are made of wood in the shape of a cylinder. Wooden pegs and leather straps hold the skin tightly onto the right-hand drum. The skin on the left-hand drum is kept slightly loose so that players can push down into it. This creates lower and higher pitches. Although the tabla emerged in India 500 years ago, it is now heard in modern pop and jazz music all over the world.

Looking Closer **Which drum do you think has more variation in sound? Why?**

GO TO

World Music: A Cultural Legacy

Hear music of this region on Disc 2, Track 13.

Jainism (JY•NIH•zuhm). Conflict sometimes occurs between Hindus and followers of the other religions. The Sikhs, who practice Sikhism, believe in one God as Christians and Muslims do, yet Sikhs have other beliefs similar to Hindus. Today many Sikhs would like to form their own independent state.

Religion has influenced the arts of India. Ancient Hindu builders constructed temples with hundreds of statues. Hindu writers composed stories about deities. Among Muslim achievements are large mosques, palaces, and forts. One of the finest Muslim buildings in India is the Taj Mahal. Turn to page 651 to learn more about it.

Daily Life About 70 percent of the people live in farming villages. The government has been working to provide villagers with electricity, drinking water, better schools, and paved roads. Many villagers stream to cities to find jobs and a better standard of living.

India's cities are very crowded. **Mumbai** (formerly Bombay), **Delhi, Calcutta,** and **Chennai** (formerly Madras) each have more than 5 million people and are growing rapidly. Modern high-rise buildings tower over slum areas where many live in deep poverty. Bicycles, carts, animals, and people fill the streets.

Many Indians enjoy sports and celebrations. Rugby and soccer are the major sports in India. One of the most popular holidays is Diwali—the Festival of Lights. It is a Hindu celebration marking the coming of winter and the victory of good over evil. Indians also like watching movies. India's movie industry turns out more than 400 movies a year.

✓Reading Check **What percentage of India's people live in rural villages?**

Section 1 Assessment

Defining Terms

1. **Define** subcontinent, monsoon, jute, cottage industry, pesticide, caste, coalition government.

Recalling Facts

2. **Location** What two mountain ranges form India's northern border?
3. **Culture** What is the most widely used language in India?
4. **History** What Indian leader led a movement that brought India its independence in 1947?

Critical Thinking

5. **Understanding Cause and Effect** How do monsoon winds affect India's climate?
6. **Drawing Conclusions** What challenges do you think having so many languages would present to India?

Graphic Organizer

7. **Organizing Information** Draw a chart like this one. Then list both modern and traditional aspects of India.

Modern Aspects	Traditional Aspects

Applying Geography Skills

8. **Analyzing Maps** Look at the population density map on page 661. What are the most densely populated areas of India?

Making Connections

ART SCIENCE LITERATURE TECHNOLOGY

The Taj Mahal

Considered one of the world's most beautiful buildings, the Taj Mahal was built by the Muslim emperor Shah Jahan of India. He had it built to house the grave of his beloved wife, Mumtaz Mahal. She died in 1631 shortly after giving birth to their fourteenth child.

The Taj Mahal ▲

Background

While they were married, Mumtaz Mahal and Shah Jahan were constant companions. The empress went everywhere with her husband, even on military expeditions. She encouraged her husband to perform great acts of charity toward the poor. This earned her the love and admiration of the Indian people.

After her death, Shah Jahan ordered the construction of the finest monument ever. A team of architects, sculptors, calligraphers, and master builders participated in the design. More than 20,000 laborers and skilled craft workers from India, Persia, the Ottoman Empire, and Europe worked together to build the monument. For 22 years they worked to complete the Taj Mahal, which holds a tomb, mosque, rest house, elaborate garden, and arched gateway.

The Mausoleum

The central part of the Taj Mahal is the domed marble mausoleum, or tomb, built on a square marble platform. The central dome is 213 feet (65 m) tall, and four smaller domed chambers surround it. A high minaret, or tower, marks each corner of the platform.

Inside the central chamber, delicately carved marble screens enclose the caskets of Mumtaz Mahal and Shah Jahan. He was buried next to his wife after his death in 1666. Following Islamic tradition, the caskets face east toward Makkah, the religious capital of Islam.

The white marble from which the mausoleum is built seems to change color throughout the day as it reflects light from the sun and moon. Detailed flower patterns are carved into the marble walls and inlaid with colorful gemstones. Verses from Islamic religious writings are etched in calligraphy into the stone archways.

Making the Connection

1. Who is buried in the Taj Mahal?
2. Who built the Taj Mahal and how long did it take?
3. **Understanding Cause and Effect** How did Shah Jahan's feelings for his wife affect the grave site he built for her?

Section 2

Pakistan and Bangladesh

Guide to Reading

Main Idea

Once a single nation, Pakistan and Bangladesh today are separate countries that border India on the west and east.

Terms to Know

- tributary
- delta
- cyclone

Places to Locate

- Pakistan
- Bangladesh
- Kashmir
- Hindu Kush
- Khyber Pass
- Indus River
- Great Indian Desert
- Karachi
- Islamabad
- Brahmaputra River
- Ganges River
- Dhaka

Reading Strategy

Draw a diagram like this one. In the outer ovals, write statements that are true of each country under the headings. Where the ovals overlap, write statements that are true of both countries.

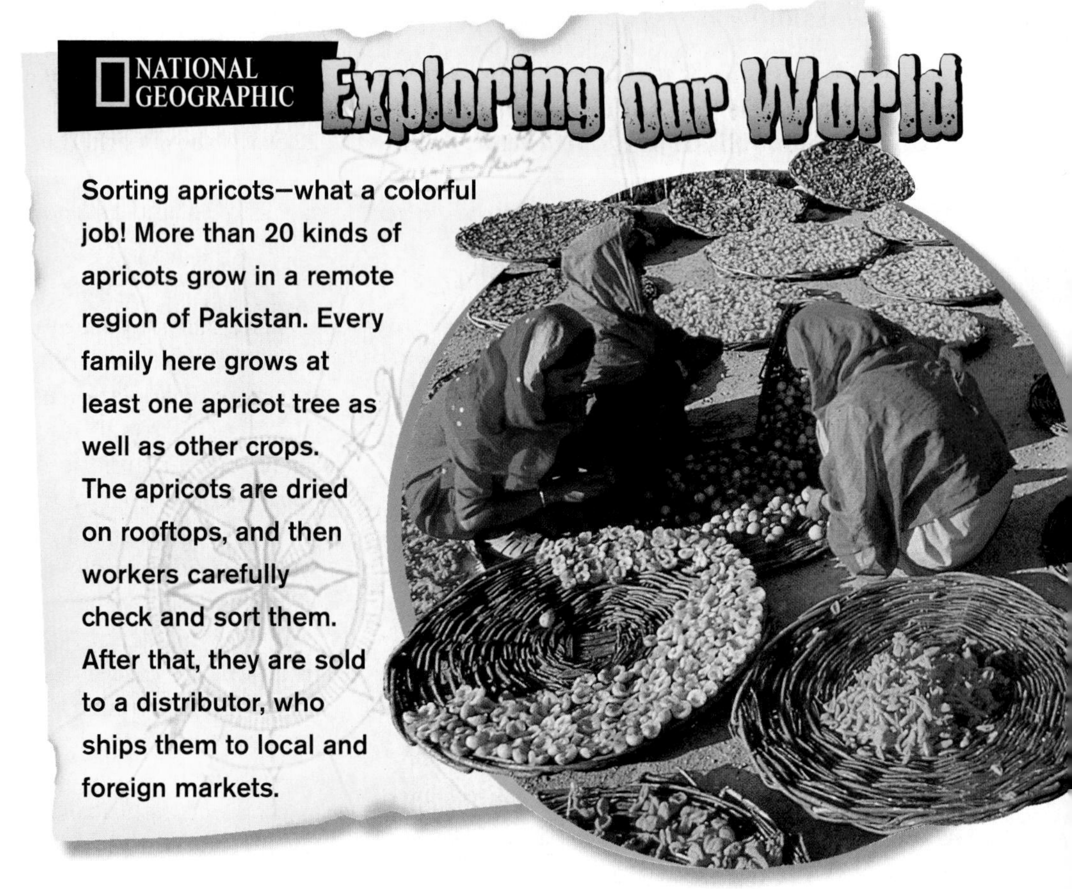

Sorting apricots—what a colorful job! More than 20 kinds of apricots grow in a remote region of Pakistan. Every family here grows at least one apricot tree as well as other crops. The apricots are dried on rooftops, and then workers carefully check and sort them. After that, they are sold to a distributor, who ships them to local and foreign markets.

Two countries in South Asia—**Pakistan** and **Bangladesh**—are largely Muslim. Although they share the same religion, the two countries have very different cultures and languages.

For many centuries, both areas were part of India. In 1947 they both separated from largely Hindu India and together formed one Muslim country called Pakistan. The western area was called West Pakistan, and the eastern area, East Pakistan. Cultural and political differences between the two parts led to a violent conflict in 1971. When the war ended, West Pakistan kept the name of Pakistan. East Pakistan became a separate new country called Bangladesh.

Pakistan

Pakistan is about twice the size of California. The country also claims **Kashmir,** a mostly Muslim territory on the northern border of India and Pakistan. Kashmir is currently divided between Pakistan and

India. Both countries want to control the entire region, mainly for its vast water resources. This dispute over Kashmir has sparked three wars between Pakistan and India. In fact, it threatens all of South Asia because both Pakistan and India have nuclear weapons.

Towering mountains occupy most of northern and western Pakistan. The world's second-highest peak, K2, rises 28,250 feet (8,611 m) in the Karakoram Range. Another range, the **Hindu Kush,** lies in the far north. Several passes cut through its rugged peaks. The best known is the **Khyber Pass.** For centuries, it has been used by people traveling through South Asia from the north.

Plains in eastern Pakistan are rich in fertile soil deposited by rivers. The major river system running through these plains is the **Indus River** and its tributaries. A **tributary** is a small river that flows into a larger one. West of the Indus River valley, the land rises to form a mostly dry plateau. Another vast barren area—the **Great Indian Desert**—lies east of the Indus River valley and reaches into India.

Pakistan has mostly desert and steppe climates, with hot summers and cool winters. Rainfall in most areas is less than 10 inches (25 cm) a year. Mountains in the north block cold air from Central Asia.

Pakistan's Economy Pakistan has fertile land and enough energy resources to meet its needs. About half of the people are farmers. A large irrigation system helps them grow crops such as sugarcane, wheat, rice, and cotton. Cotton and textiles are the country's main

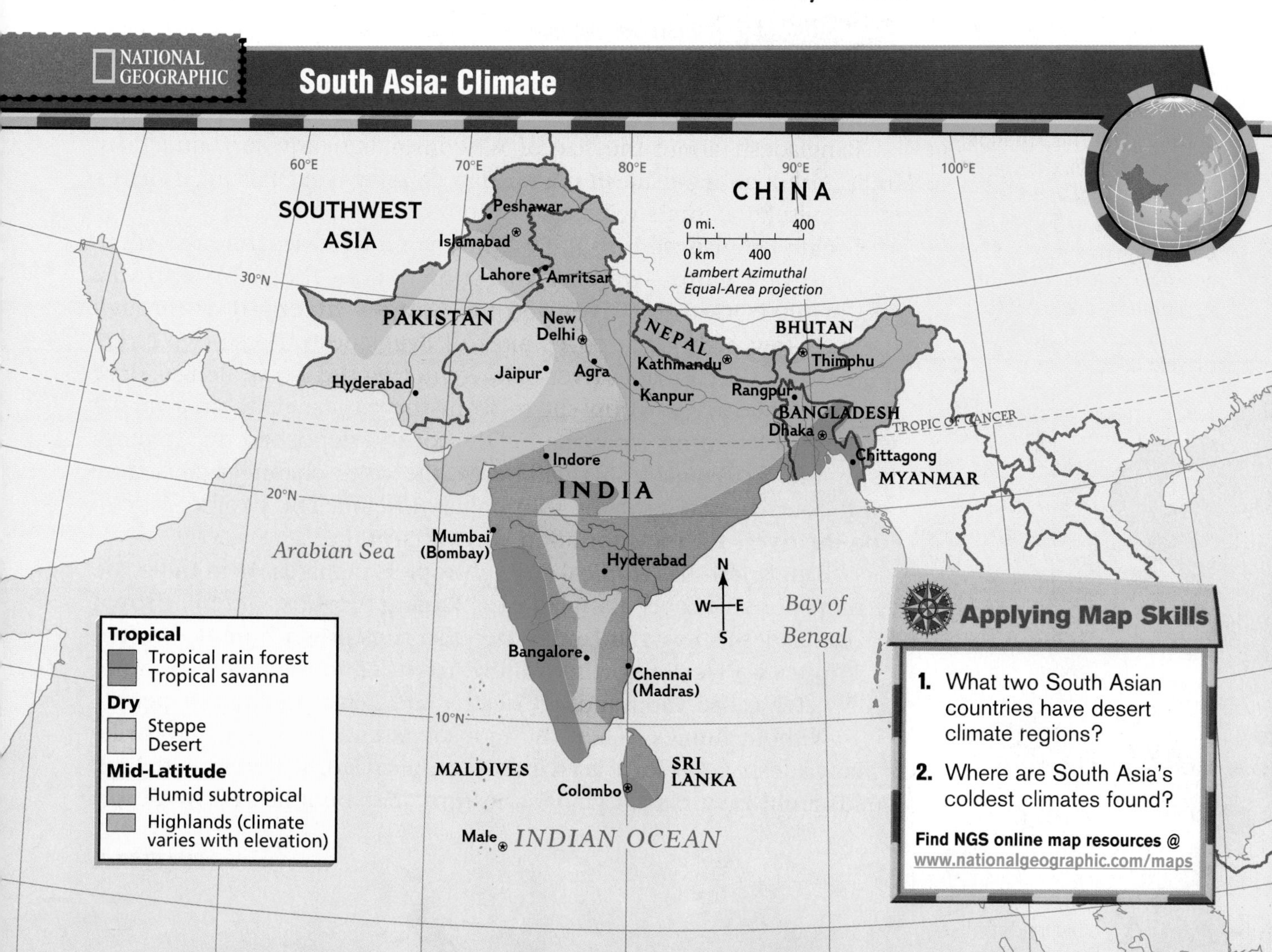

exports. Other important industries include cement, fertilizer, food processing, and chemicals. Many people make metalware, pottery, and carpets in cottage industries. Pakistan's economy is struggling, however, because of frequent changes of government.

Pakistan's People Since independence, Pakistan has had many changes of government. Some of these governments were elected, including a female prime minister, Benazir Bhutto. In other cases, the army seized power from an elected government. The most recent army takeover occurred in 1999, and military leaders still control the country.

About 97 percent of Pakistanis are Muslims. The influence of Islam is seen in large, domed mosques and people bowed in prayer at certain times of the day. Among the major languages are Punjabi and Sindhi. The official language, Urdu, is the first language of only 9 percent of the people. English is widely spoken in government.

Almost 70 percent of Pakistan's people live in rural villages. Most follow traditional customs and live in small homes of clay or sun-dried mud. Pakistanis live in large cities as well. **Karachi,** a seaport on the Arabian Sea, is a sprawling urban area. It has traditional outdoor markets, modern shops, and hotels. In the far north lies **Islamabad,** the capital. The government built this well-planned, modern city to draw people inland from crowded coastal areas. Most people in Pakistan's cities are factory workers, shopkeepers, and craft workers who live in crowded neighborhoods. Wealthier city dwellers live in modern homes.

✓Reading Check **What are Pakistan's main exports?**

Bangladesh

Bangladesh, about the size of Wisconsin, is nearly surrounded by India. Although Bangladesh is a Muslim country like Pakistan, it shares many cultural features with eastern India.

Seeing Bangladesh for the first time, you might describe the country with one word—water. Two major rivers—the **Brahmaputra** (BRAHM•uh•POO•truh) **River** and the **Ganges River**—flow through the lush, low plains that cover most of Bangladesh. These two rivers unite with a third, smaller river before entering the Bay of Bengal. Here the combined rivers drop silt to form the largest delta area in the world. A **delta** forms from the buildup of soil deposited by a river at its mouth. In Bangladesh's delta area, the river channels constantly shift course, creating many thin fingers of land. The people depend on the rivers for transportation and for farming.

Bangladesh has tropical and subtropical climates. As in India, the monsoons affect Bangladesh. Raging floods often drown Bangladesh's low, flat land. Water also runs down from deforested slopes upriver in northern India. Together, these violent flows of water cause thousands of deaths and leave millions of people without homes. When the monsoons end, cyclones may strike Bangladesh. A **cyclone** is an intense tropical storm system with high winds and heavy rains. Cyclones, in turn, may be followed by deadly

School's Out!

Adil Husain is on his way home from middle school. In Pakistan, schooling only goes to grade 10. Adil has only three more years of school before he must decide whether to go to intermediate college (grades 11 and 12) and then the university. Like most Pakistanis, Adil is Muslim. His parents have taught him to pray when he hears the call from the mosque. Afterward, he wants to start a game of cricket with his friends, Kiran and Malik. "It's a lot like American baseball where teams of 11 players bat in innings and try to score runs. Our rules and equipment are different, though. You should try playing it!"

tidal waves that surge up from the Bay of Bengal. As deadly as the monsoons and cyclones may be, it is worse if the rains come too late. When this happens, crops often fail and there is widespread hunger.

A Farming Economy Most people of Bangladesh earn their living by farming. Rice is the most important crop. The fertile soil and plentiful water make it possible for rice to be grown and harvested three times a year. Other crops include sugarcane, jute, and wheat. Cash crops of tea grow in hilly regions in the east. Despite good growing conditions, Bangladesh cannot grow enough food for its people. Its farmers have few modern tools and use outdated farming methods. In addition, the disastrous floods can drown crops and cause food shortages.

Bangladesh has an important clothing industry. It exports large amounts of manufactured clothing to other countries. *You* may even be wearing clothes made in Bangladesh.

The People With about 125.7 million people, Bangladesh is one of the most densely populated countries in the world. It is also one of the poorest countries. More than 80 percent of the people live in rural areas. Because of floods, people in rural Bangladesh have to build their houses on platforms. Many people have moved to crowded urban areas to find work in factories. Their most common choice is **Dhaka** (DA•kuh), Bangladesh's capital and major port.

Most of Bangladesh's people speak Bengali. About 88 percent of the people are Muslim, and most of the rest are Hindus. Muslim influences are strong in the country's art, literature, and music.

✓Reading Check What is an important industry in Bangladesh?

Ship Breakers

On a beach near Karachi, Pakistan, ship breakers haul an old cargo vessel to shore. Their next task? The men will use hammers, crowbars, and wrenches to pull the ship apart. They will then sell the pipes, chains, portholes, steel plates, and other reusable parts. The work is exhausting, but in this poor country it is a way of making a living.

Section 2 Assessment

Defining Terms

1. **Define** tributary, delta, cyclone.

Recalling Facts

2. **Region** What region has been the source of conflict between Pakistan and India?
3. **History** Why has the Khyber Pass been important?
4. **Movement** Why was Islamabad built inland?

Critical Thinking

5. **Analyzing Information** Why can rice be grown three times a year in Bangladesh?
6. **Drawing Conclusions** Why are Pakistan's and Bangladesh's economies struggling?

Graphic Organizer

7. **Organizing Information** Draw a diagram like this one. At the ends of the arrows, list three effects on Bangladesh caused by summer monsoon rains.

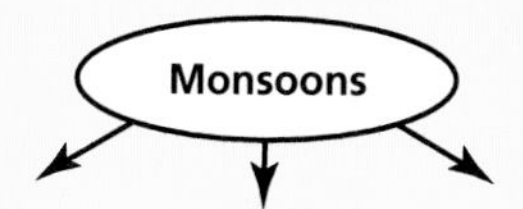

Applying Geography Skills

8. **Analyzing Maps** Look at the physical map on page 647. What rivers have deltas in Bangladesh?

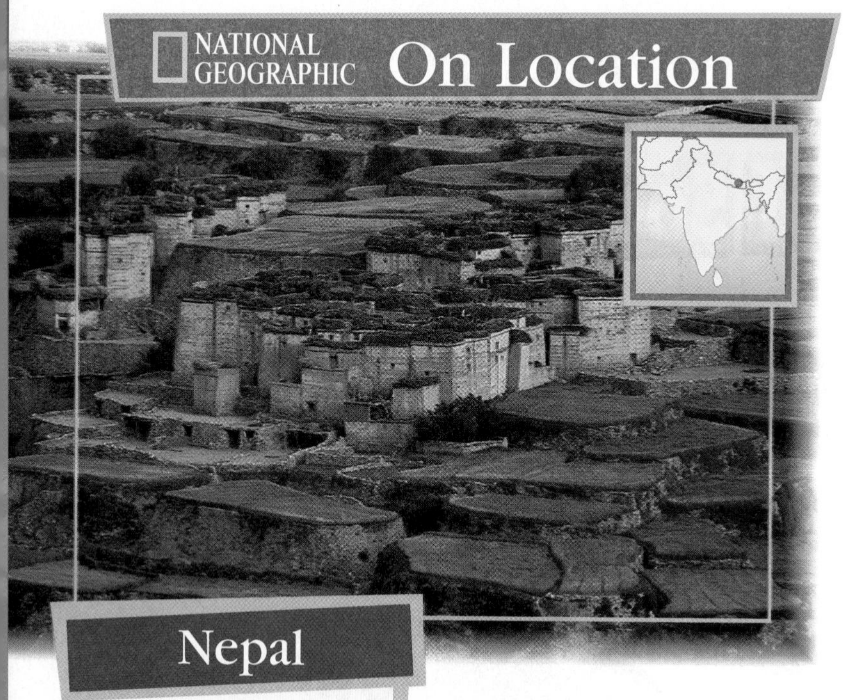

Nepal

Years of hard labor turned mountain slopes into terraced fields in Nepal. Farmers also built channels to bring melting snow from the mountains to the fields.

Place **What mountains dominate Nepal?**

Nepal has a humid subtropical climate in the south and a highland climate in the north. Monsoon rains often flood the southern plains area.

Nepal's economy depends almost entirely on farming. Farmers grow rice, sugarcane, wheat, corn, and potatoes to feed their families. Most fields are located on the southern plains or on the lower mountain slopes.

As the population has increased, Nepalese farmers have moved higher up the slopes. There they clear the forests for new fields and use the cut trees for fuel. Stripped of trees, however, the slopes erode very easily. Valleys often are flooded, fields destroyed, and rivers filled with mud.

With few roads or railroads, Nepal carries on limited trade with the outside world. Herbs, jute, rice, and wheat are exported to India. In return, Nepal imports gasoline, fertilizer, and machinery. Clothing and carpets now make up the country's most valuable exports. Nepal's rugged mountains attract thousands of climbers and hikers each year, creating a growing tourist industry.

Nepal's People Nepal has 24.3 million people. Most are related to peoples in northern India and Tibet. One group—the Sherpa—is known for its skill in guiding mountain climbers. About 85 percent of Nepal's people live in rural villages. A growing number live in **Kathmandu,** Nepal's capital and largest city. Nepal is a parliamentary democracy ruled by a prime minister, who is appointed by Nepal's king.

The founder of Buddhism, Siddartha Gautama (sihd•DAHR•tuh GOW•tuh•muh), was born in the Kathmandu region about 563 B.C. Raised as a prince, Gautama gave up his wealth and became a holy man in India. Known as the Buddha, or "Enlightened One," he taught that people could find peace from life's troubles by living simply, doing good deeds, and praying. Buddhism later spread to other parts of Asia.

Today Hindu is the official religion, but Buddhism is practiced as well. If you visit Nepal, you will find temples and monuments of both religions scattered throughout the country.

Reading Check **Why does Nepal have limited trade with other countries?**

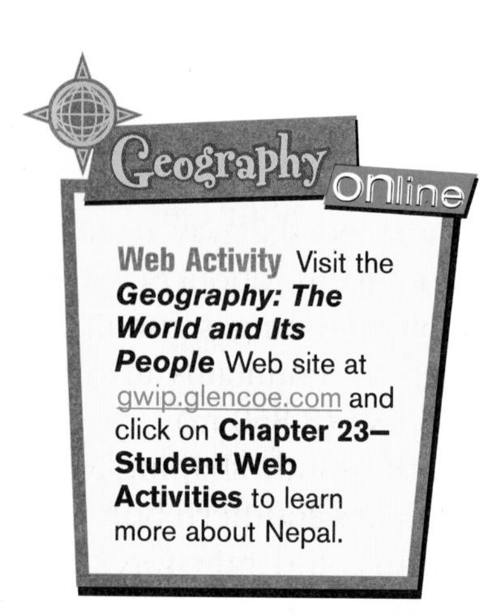

Web Activity Visit the ***Geography: The World and Its People*** Web site at gwip.glencoe.com and click on **Chapter 23—Student Web Activities** to learn more about Nepal.

Bhutan

East of Nepal lies an even smaller kingdom—Bhutan. Bhutan is about half the size of Indiana. The map on page 659 shows you that a small part of India separates Bhutan from Nepal.

As in Nepal, the Himalaya are the major landform of Bhutan. Violent mountain storms are common and are the basis of Bhutan's name, which means "land of the thunder dragon." In the foothills of the Himalaya, the climate is mild. Thick forests cover much of this area. To the south—along Bhutan's border with India—lies an area of subtropical plains and river valleys.

More than 90 percent of Bhutan's people are subsistence farmers. They live in the fertile mountain valleys and grow the spice cardamom, oranges, rice, corn, and potatoes. People also herd cattle and yaks, which are a type of oxen. Bhutan is trying to develop its economy, but the mountains slow progress. Building roads is difficult, and there are no railroads. Bhutan has built hydroelectric plants to create electricity from rushing mountain waters. It now exports electricity to India. Tourism is a new industry. However, the government limits the number of tourists in order to protect Bhutan's cultural traditions.

Bhutan's People Bhutan has about 800,000 people. Most speak the Dzonkha dialect and live in rural villages that dot southern valleys and plains. **Thimphu,** the capital, is located in the southern area.

Bhutan was once called the Hidden Holy Land because of its isolation and its Buddhist religion. In the 1960s, new roads and other connections opened Bhutan to the outside world. Most people remain deeply loyal to Buddhism. In Bhutan, Buddhist centers of prayer and study are called ***dzongs.*** They have shaped the country's art and culture.

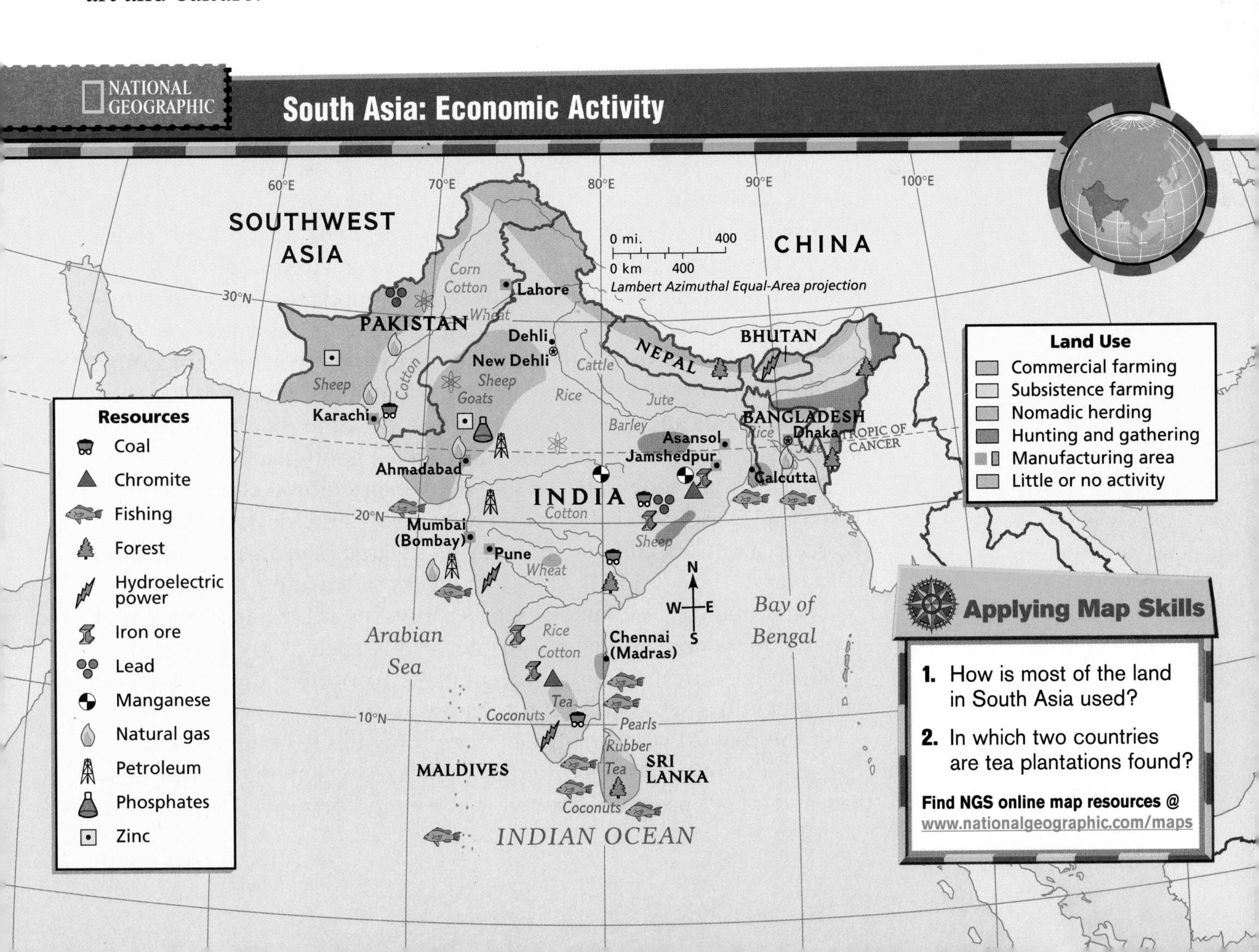

Highest Mountain on Each Continent

Analyzing the Graph

Mount Everest is the tallest mountain on the earth.

Place **What is the tallest mountain in North America?**

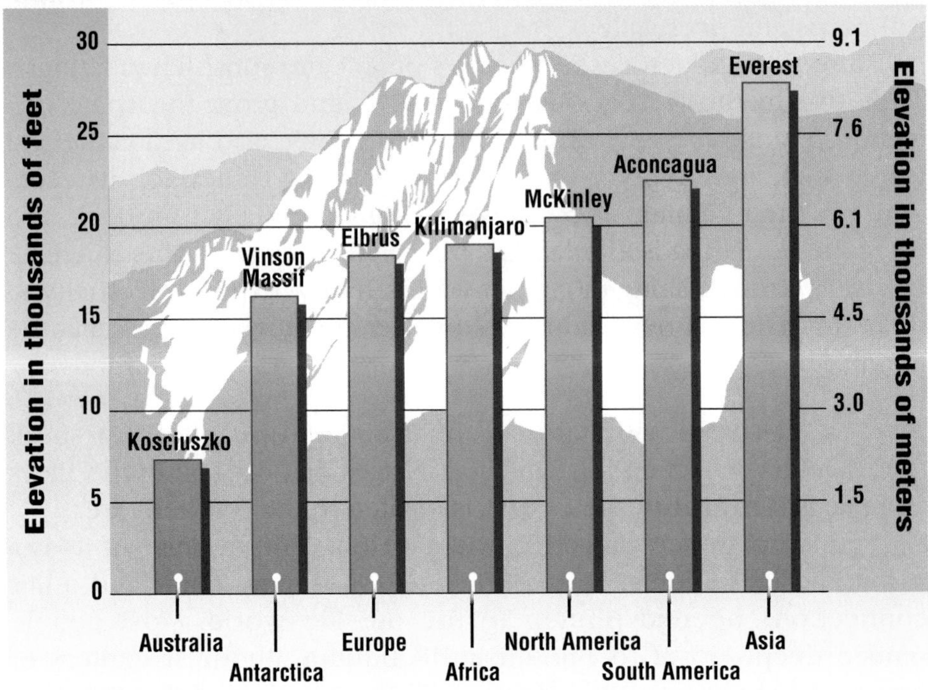

Source: *The World Almanac,* 2000.

For many years, Bhutan was ruled by strong kings. In 1998 the country began to move toward democracy. At that time, the ruling king agreed to share his power with an elected legislature.

Reading Check **What is the main religion in Bhutan?**

Sri Lanka

Pear-shaped Sri Lanka lies about 20 miles (32 km) off the southeastern coast of India. A little larger than West Virginia, Sri Lanka is a land of white beaches, dense forests, and abundant wildlife. Much of the country along the coast is rolling lowlands. Highlands cover the center. Rivers flow from the highlands, providing irrigation for crops.

The country has tropical climates with dry and wet seasons. Monsoon winds and heavy rains combine with the island's warm temperatures and fertile soil to make Sri Lanka a good place to farm.

Sri Lanka has long been known for its agricultural economy. Many farmers grow rice and other food crops in lowland areas. In higher elevations, tea, rubber, and coconuts grow on large plantations. The country is one of the world's leading producers of tea and rubber.

The country is also famous for its sapphires, rubies, and other gemstones. Forests contain many valuable woods, such as ebony and satinwood, and a variety of birds and animals. To protect the wildlife, the government has set aside land for national parks.

In the past 20 years, Sri Lanka's economy has become more industrialized. Factories produce textiles, fertilizers, cement, leather products, and wood products for export. New and growing industries are telecommunications, insurance, and banking. **Colombo,** the capital, is a bustling port on the country's western coast.

Sri Lanka's People For centuries, Sri Lanka prospered because of its location on an important ocean route between Africa and Asia. It was a natural stopping place for seagoing traders. Beginning in the 1500s, Sri Lanka—then known as Ceylon—came under the control of European countries. The British ruled the island from 1802 to 1948, when it became independent. In 1972 Ceylon took the name of Sri Lanka, an ancient term meaning "brilliant land." Today Sri Lanka is a republic with a president who carries out ceremonial duties. Real power is held by a prime minister, who is the head of government.

About 19 million people live here. They belong to two major ethnic groups: the Sinhalese (SIHNG•guh•LEEZ) and the Tamils (TA•muhlz). Forming about 74 percent of the population, the Sinhalese live in the southern and western parts of the island. They speak Sinhalese and are mostly Buddhist. The Tamils make up about 18 percent of the population. They live in the north and east, speak Tamil, and are Hindus.

Since 1983 the Tamils and the Sinhalese have fought a violent civil war. The minority Tamils claim they have not been treated justly by the

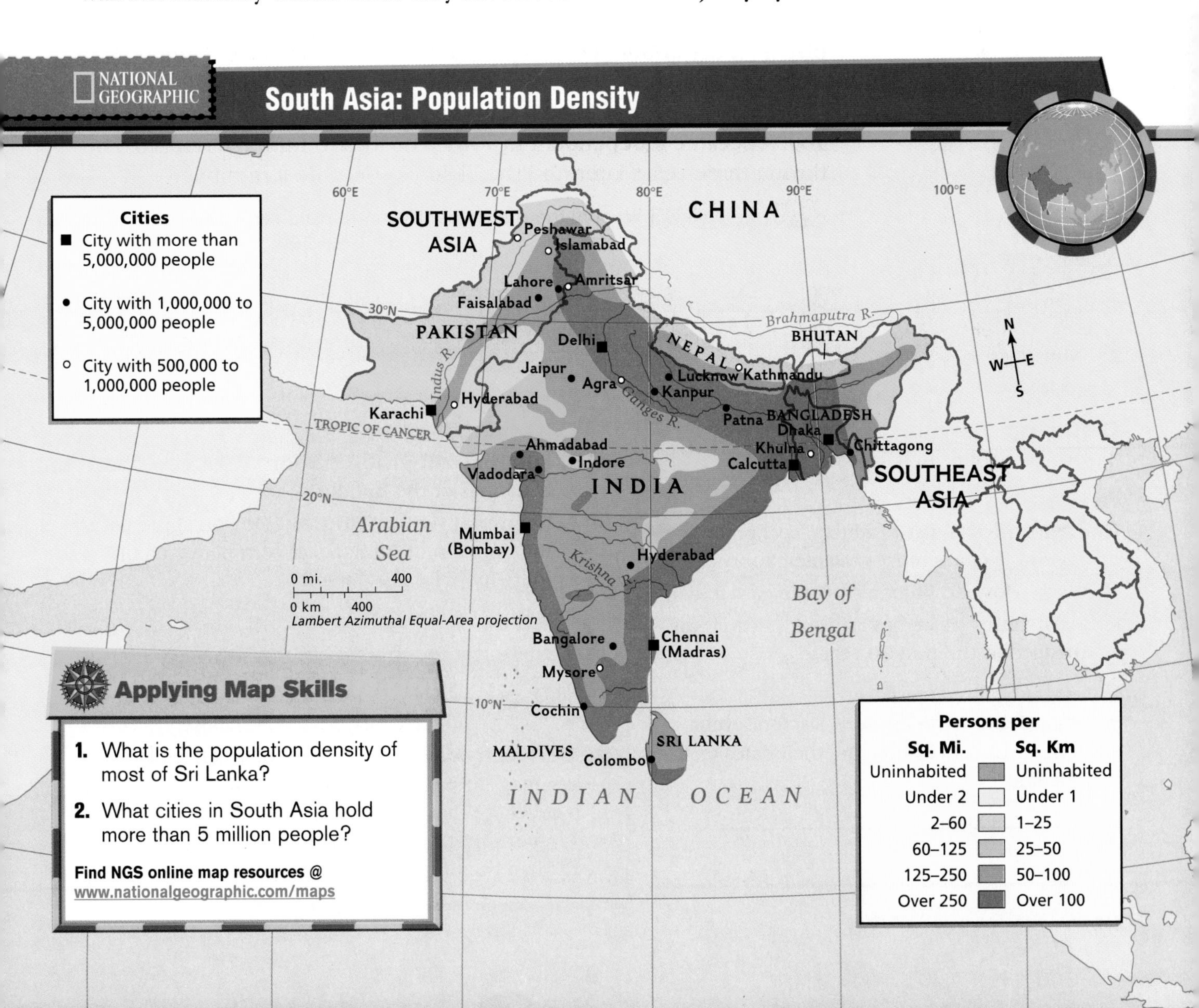

majority Sinhalese. They want to set up a separate Tamil nation in northern Sri Lanka. Thousands of Sri Lankans have lost their lives in the fighting.

Reading Check **What are the two main ethnic groups in Sri Lanka?**

The Maldives

About 370 miles (595 km) south of India lie the Maldives, made up of about 1,200 coral islands. Many of the islands are atolls. An **atoll** is a low-lying, ring-shaped island that surrounds a lagoon. A **lagoon** is a shallow pool of water surrounded by reefs, sandbars, or atolls. Only 200 of the islands are inhabited. The climate of the Maldives is warm and humid throughout the year. Monsoons bring plenty of rain.

Most of the Maldives have poor, sandy soil. Only a limited number of crops can grow, including sweet potatoes, grains, and watermelon. In recent years, the Maldives's palm-lined sandy beaches and coral formations have attracted many tourists. As a result, tourism is now the largest industry. Fishing is the second-largest industry.

The first people to arrive in the Maldives came from southern India and Sri Lanka several thousand years ago. Over the years, the islands' position near major sea routes brought traders from many other places. Today about 300,000 people live in the Maldives. Some 60,000 of them make their home in **Male** (MAH•lay), the capital. Most are Muslims. The islands, which came under British rule during the late 1890s, became independent in 1965. The local traditional ruler lost his throne three years later, and the Maldives became a republic.

Reading Check **What is the main industry in the Maldives?**

Assessment

Defining Terms

1. **Define** *dzong,* atoll, lagoon.

Recalling Facts

2. **Economics** What products have recently become Nepal's most valuable exports?
3. **Place** How do Bhutan's people earn a living?
4. **Economics** How has Sri Lanka's economy changed in the past 20 years?

Graphic Organizer

5. **Organizing Information** List four events from Sri Lanka's history and their dates on a time line.

Critical Thinking

6. **Summarizing Information** What were the teachings of the Buddha?
7. **Formulating an Opinion** Do you agree with the decision of Bhutan's government to limit tourism? Why or why not?

Applying Geography Skills

8. **Analyzing Maps** Look at the population density map on page 661 and the physical map on page 647. What is the population density of the southern part of Nepal? The northern part? Explain the difference.

Chapter 23

Reading Review

Section 1 India

Terms to Know
- subcontinent
- monsoon
- jute
- cottage industry
- pesticide
- caste
- coalition government

Main Idea

India—the world's most populous democracy—is trying to develop its resources and meet the needs of its rapidly growing population.

✓ Place India is the largest country in South Asia in size and population.

✓ Place The Himalaya and the monsoons affect India's climate.

✓ Economics India's economy is based on both farming and industry.

✓ Culture India has many languages and religions, but the majority of Indians are Hindus.

✓ Government India has a democratic government with many political parties.

Section 2 Pakistan and Bangladesh

Terms to Know
- tributary
- delta
- cyclone

Main Idea

Once a single nation, Pakistan and Bangladesh today are separate countries that border India on the west and east.

✓ History Cultural and political differences between Pakistan and Bangladesh led to war and separation in 1971.

✓ Economics Pakistan has fertile land and energy resources, but its economy is not well developed because of a history of unstable governments.

✓ Location The Ganges and Brahmaputra Rivers form deltas in Bangladesh.

✓ Place Bangladesh is a densely populated and poor country.

Section 3 Other Countries of South Asia

Terms to Know
- *dzong*
- atoll
- lagoon

Main Idea

The other countries of South Asia include mountainous Nepal and Bhutan and the island countries of Sri Lanka and the Maldives.

✓ Region The Himalaya are the major landform of Nepal and Bhutan.

✓ Economics Most people in Nepal are farmers, but the production of textiles and carpets has gained importance in recent years.

✓ Culture The Buddhist religion has shaped the art and culture of Bhutan.

✓ Economics Sri Lanka has industrialized, but agriculture is still important.

✓ Economics Tourism is the biggest industry in the Maldives.

◀ A teacher and his students have classes outdoors on a pleasant day in Bhutan.

Chapter 23 Assessment and Activities

Using Key Terms

Match the terms in Part A with their definitions in Part B.

A.

1. monsoon
2. cyclone
3. atoll
4. jute
5. subcontinent
6. delta
7. lagoon
8. caste
9. *dzong*
10. cottage industry

B.

a. social class based on a person's ancestry
b. seasonal wind
c. family members supply their own equipment to make goods
d. large landmass that is part of another continent but distinct from it
e. Buddhist center for prayer and study
f. shallow pool of water surrounded by reefs
g. intense storm system with high winds
h. ring-shaped island that surrounds a lagoon
i. area formed by soil deposited at the mouth of a river
j. plant fiber used for making rope, burlap bags, and carpet backing

Reviewing the Main Ideas

Section 1 India

11. **Place** What forms a barrier between South Asia and the rest of Asia?
12. **Place** How do the Himalaya affect India's climate?
13. **Economics** What kinds of goods are produced by India's cottage industries?
14. **Culture** What religion do most Indians practice?

Section 2 Pakistan and Bangladesh

15. **Place** What river flows through Pakistan?
16. **Human/Environment Interaction** What kind of damage can a cyclone cause?
17. **Economics** What do most of the people of Bangladesh do for a living?

Section 3 Other Countries of South Asia

18. **History** Why was Bhutan once called the Hidden Holy Land?
19. **Human/Environment Interaction** How do mountains hinder economic development in Bhutan?
20. **History** What is the basis of the civil war in Sri Lanka?
21. **Economics** What is the major industry in the Maldives?

NATIONAL GEOGRAPHIC South Asia

Place Location Activity

On a separate sheet of paper, match the letters on the map with the numbered places listed below.

1. Ganges River
2. New Delhi
3. Brahmaputra River
4. Indus River
5. Sri Lanka
6. Himalaya
7. Bangladesh
8. Mumbai
9. Western Ghats
10. Deccan Plateau

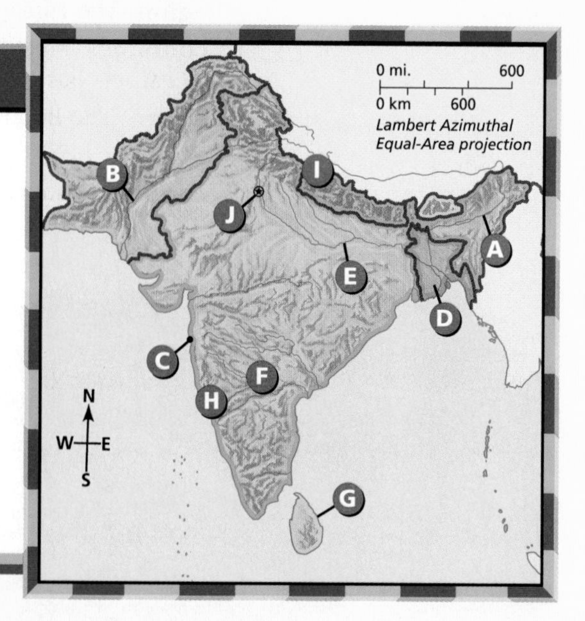

Self-Check Quiz Visit the ***Geography: The World and Its People*** Web site at gwip.glencoe.com and click on **Chapter 23–Self-Check Quizzes** to prepare for the Chapter Test.

Critical Thinking

22. **Identifying Alternatives** In this chapter you read about South Asia, a region with much poverty. What problems do you think a country faces when it has so many poor people? What are some solutions to this poverty?
23. **Understanding Cause and Effect** Draw a diagram like this one. List a physical feature of South Asia in the left-hand box. In the right-hand box, explain how that feature affects people's lives.

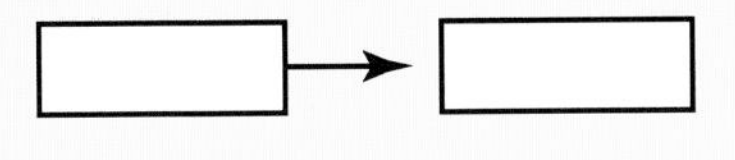

GeoJournal Activity

24. **Writing About Religion** Choose Hinduism, Islam, or Buddhism and research its main beliefs and places of worship. Find out how the religion influences family life. After your research is complete, create a chart or poster that presents your findings.

Mental Mapping Activity

25. **Focusing on the Region** Draw a simple outline map of South Asia, then label the following:

- Bay of Bengal
- Kashmir
- Nepal
- Bhutan
- Deccan Plateau
- Sri Lanka
- Pakistan
- Indian Ocean
- Bangladesh
- New Delhi

Technology Skills Activity

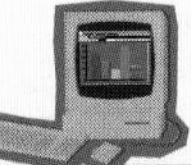

26. **Using the Internet** Use the Internet to research tourism in one of the following countries: Nepal, the Maldives, India, or Sri Lanka. Find information on the equipment and clothing that is needed, the availability of guides, costs, and so on. Publish your information in a brochure.

Standardized Test Practice

Directions: Study the graph below, then answer the following questions.

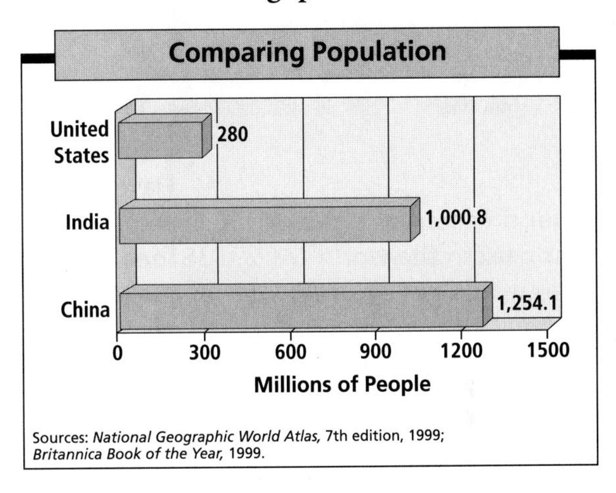

1. **How many people live in India?**
   - **A** 1,000.8
   - **B** 1,000,000.8
   - **C** 1,000,000,000.8
   - **D** 1,000,000,000,000.8
2. **About how many more people live in India than in the United States?**
   - **F** twice as many
   - **G** three times as many
   - **H** four times as many
   - **J** five times as many

Test-Taking Tip: You often need to use math skills in order to understand graphs. Look at the information along the sides and bottom of the graph to find out what the bars on the graph mean. Notice that on the graph above, the numbers represent *millions* of people. Therefore, you need to multiply the number on each bar by 1,000,000 to get the correct answer.

Chapter

24 China

The World and Its People NATIONAL GEOGRAPHIC

To learn more about the people and places of China, view ***The World and Its People*** **Chapter 24** video.

Geography Online

Chapter Overview Visit the ***Geography: The World and Its People*** Web site at gwip.glencoe.com and click on **Chapter 24—Chapter Overviews** to preview information about China.

Section 1

China's Land and Climate

Guide to Reading

Main Idea

China—the third-largest country in the world—has very diverse landforms and climates.

Terms to Know

- loess
- dike
- fault
- typhoon

Places to Locate

- China
- Himalaya
- Kunlun Shan
- Plateau of Tibet
- Taklimakan Desert
- Gobi
- Manchurian Plain
- Yangtze River
- Yellow River
- Xi River

Reading Strategy

Draw a diagram like this one. Then list two facts under each heading in the outer ovals.

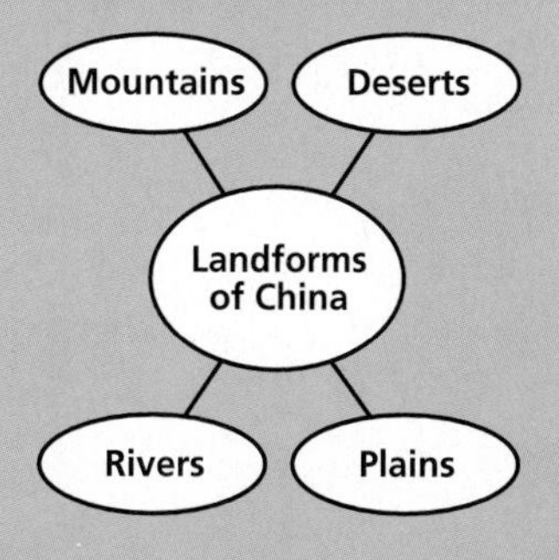

NATIONAL GEOGRAPHIC **Exploring Our World**

Giant pandas look cute and cuddly, but actually they are somewhat hot-tempered. You would be hot-tempered, too, if your habitat were dwindling in size. Fewer than 1,000 pandas live in the wild, and about 140 live in zoos. The wild pandas make their home on the eastern edge of the Plateau of Tibet. They eat mainly bamboo stems and leaves.

China (officially called the People's Republic of China) lies in the central part of eastern Asia. It is the third-largest country in area, after Russia and Canada. China is just slightly larger than the United States.

China's Land

The physical map on page 671 shows you the many landforms found within China's vast area. Rugged mountains cover about one-third of the country. The mighty **Himalaya** mountain ranges sweep along China's border with India and Nepal. Another towering range—the **Kunlun Shan**—twists and turns from Afghanistan into central China. Farther north you find still more awesome ranges—the Tian Shan and the Altay Mountains.

Between the Himalaya and the Kunlun Shan lies the **Plateau of Tibet.** The world's largest plateau, this high flat land is called the Roof of the World. Its height averages about 13,000 feet (3,962 m) above sea

◀ Part of the Great Wall of China

level. Scattered shrubs and grasses cover the plateau's harsh landscape. Pandas, golden monkeys, and other rare animals roam the thick forests found at the eastern end of this plateau.

In addition to very high elevations, western China has some extremely low areas. The Turpan Depression, east of the Tian Shan, lies about 505 feet (154 m) *below* sea level. It is partly filled with salt lakes. It also is the hottest area of China. Daytime temperatures can reach as high as 122°F (50°C).

In the north of China, mountain ranges circle desert areas. One of these areas is the **Taklimakan Desert.** It is an isolated region with very high temperatures. Sandstorms here may last for days and create huge, drifting sand dunes. Farther east lies another desert, the **Gobi.** Instead of sand, the Gobi has rocks and stones. Temperatures here can be extremely high—and extremely low. On summer days, the thermometer can rise to 110°F (43°C). Because the Gobi is far north and sits at a high elevation, the temperature can plunge as low as −30°F (−34°C) on winter nights.

Applying Map Skills

1. What is the capital of China?
2. What major cities are located along the Yangtze River?

Find NGS online map resources @ www.nationalgeographic.com/maps

The **Manchurian Plain** lies in the center of Manchuria, a region in northeastern China. To the east of the plain lies a heavily forested, hilly area near China's border with North Korea. To the south is the wide, flat North China Plain. The map below shows you that plains also run along the coasts of the South China and East China Seas. These fertile plains are rich in mineral resources. Almost 90 percent of China's people live here.

In the southeast, the land changes from plains to green highlands as you move inland. This region is one of the most scenic areas in China. Tourists come from around the world to see its numerous limestone hills, waterfalls, underground caves, and steep gorges.

Rivers Three of China's major waterways—the **Yangtze** (YANG• SEE), **Yellow,** and **Xi** (SHEE) **Rivers**—flow through the plains and southern highlands. They serve as important transportation routes and also as a source of soil. How? For centuries, these rivers have flooded their banks in the spring. The floodwaters have deposited rich soil to form flat river basins that can be farmed.

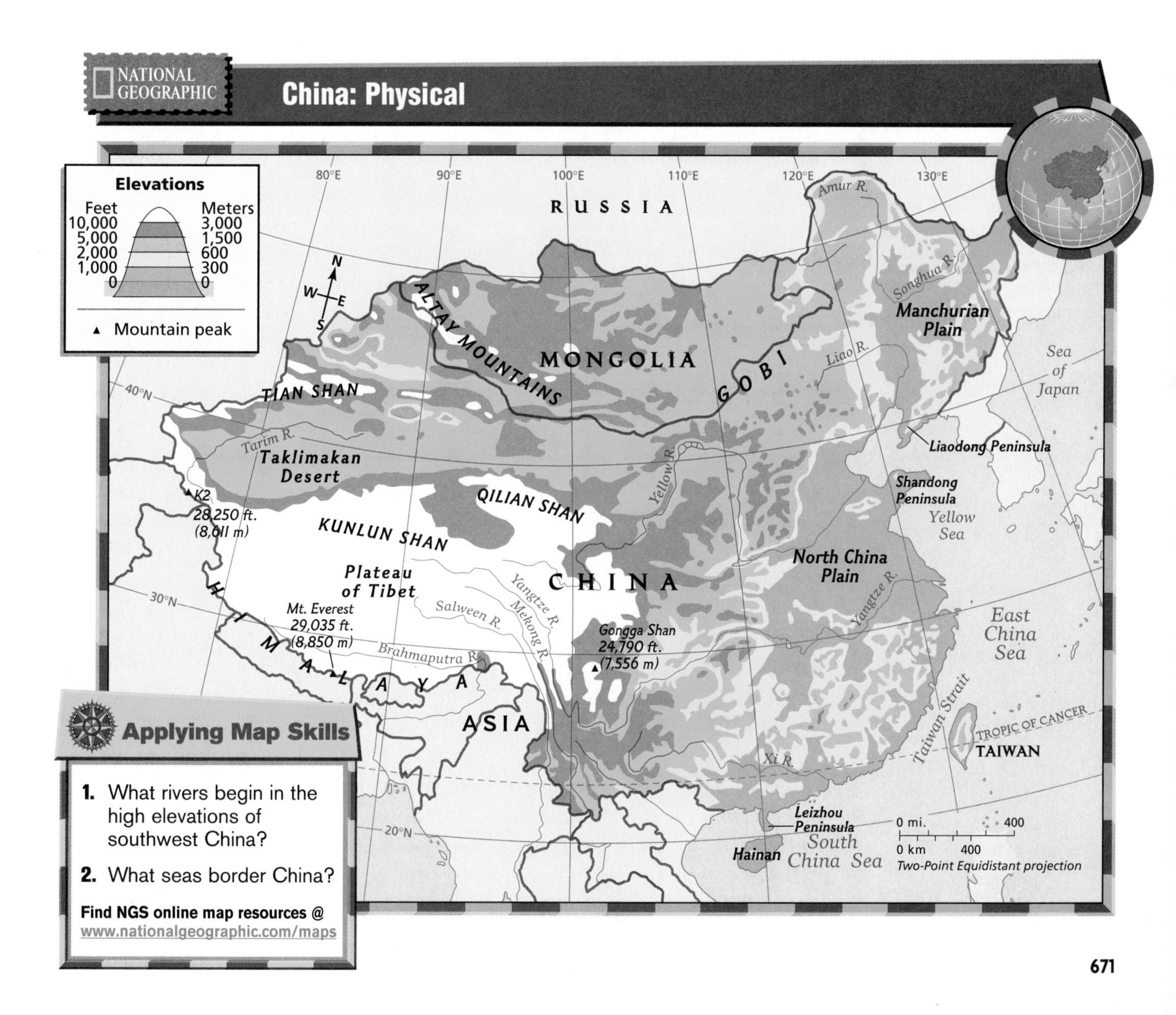

Applying Map Skills

1. What rivers begin in the high elevations of southwest China?
2. What seas border China?

Find NGS online map resources @ www.nationalgeographic.com/maps

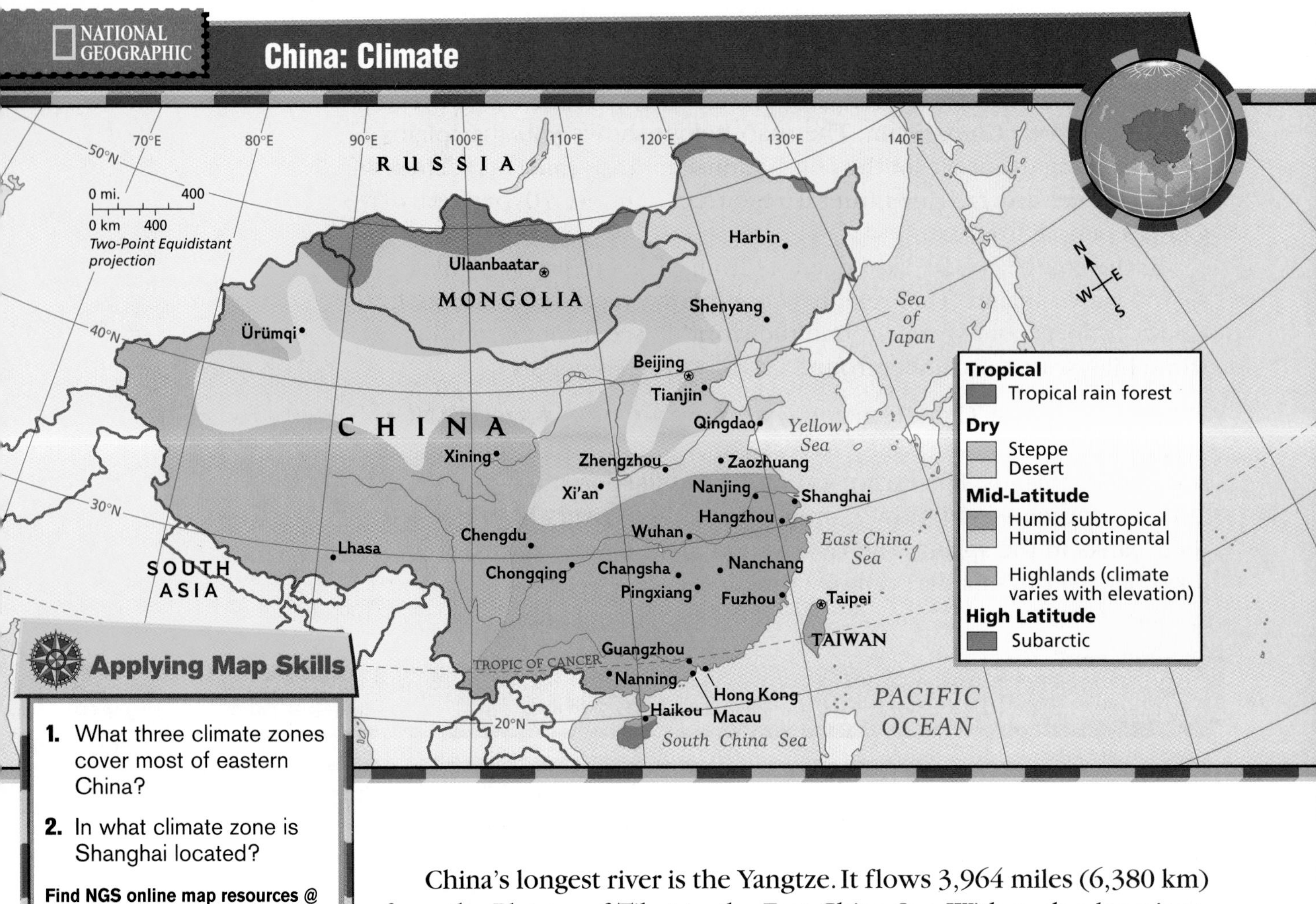

Applying Map Skills

1. What three climate zones cover most of eastern China?
2. In what climate zone is Shanghai located?

Find NGS online map resources @ www.nationalgeographic.com/maps

China's longest river is the Yangtze. It flows 3,964 miles (6,380 km) from the Plateau of Tibet to the East China Sea. With such a long journey, it is no surprise that its Chinese name, *Chang Jiang,* means "long river." The valley of the Yangtze has rich farmland and numerous industrial centers.

China's second-longest waterway, the Yellow River, also flows from the Plateau of Tibet and crosses the North China Plain. Centuries ago, Chinese civilization began in the Yellow River valley. Today the valley is an important farming area. It is thickly covered with **loess** (LEHS), a fertile, yellow-gray soil deposited by wind and water. The Chinese name of the river, *Huang He,* means "yellow river," referring to the large amounts of loess it carries.

Far to the south, you find the shortest of the three rivers—the Xi. It flows from China's southeastern highlands. The river at first passes through steep winding gorges. It later forms a fertile delta area before entering the South China Sea. Numerous farms, cities, and factories sprawl across the delta.

Despite their benefits, the rivers of China also have brought much suffering. The Chinese call the Yellow River "China's sorrow." In the past, its flooding cost hundreds of thousands of lives and much damage. To control floods, the Chinese have built dams and **dikes,** or high banks of soil, along the rivers. Turn to page 674 to learn more about the Three Gorges Dam, a project underway on the Yangtze River.

An Unsteady Land In addition to floods, people in eastern China face another danger—earthquakes. Their part of the country stretches along the Ring of Fire, a name that describes Pacific coastal areas with volcanoes and frequent earthquakes. Eastern China lies along a **fault,** or crack in the earth's crust. As a result, earthquakes in this region are common—and can be very violent. Because so many people live in eastern China, these earthquakes can bring great suffering.

✓Reading Check **What are China's three major rivers?**

China's Climate

Like the United States, China has many different climates. The map on page 672 shows you the seven climate zones of China. Location, elevation, and wind currents affect the type of climate found in any particular area of the country. Southeastern China has a humid subtropical climate with hot, humid summers. The northeast has a humid continental climate with cold winters. In the deserts of the northwest, summers are hot and winters are cold—but rain hardly ever falls. In the southwestern part of China, the high Plateau of Tibet has cool summers and bitterly cold winters.

Monsoons greatly affect China's climates. Cold, dry air blows from central Asia in winter. In summer the monsoons blow in from the sea, bringing warm, moist air. The summer monsoons often bring typhoons to coastal areas in the south. **Typhoons**—called hurricanes in the Atlantic Ocean—are tropical storms with strong winds and heavy rains.

✓Reading Check **How many different climate zones does China have?**

Assessment

Defining Terms

1. **Define** loess, dike, fault, typhoon.

Recalling Facts

2. **Place** Why is the Plateau of Tibet called the Roof of the World?

3. **Place** What are China's two large deserts?

4. **Region** What lowland region is drained by the Yangtze and Yellow Rivers?

Critical Thinking

5. **Drawing Conclusions** Why do most people live in the plains of eastern China?

6. **Analyzing Information** How are China's rivers both a blessing and a disaster?

Graphic Organizer

7. **Organizing Information** Draw a diagram like this one. In the proper places on the circle, fill in the physical features you would encounter if you traveled completely around China.

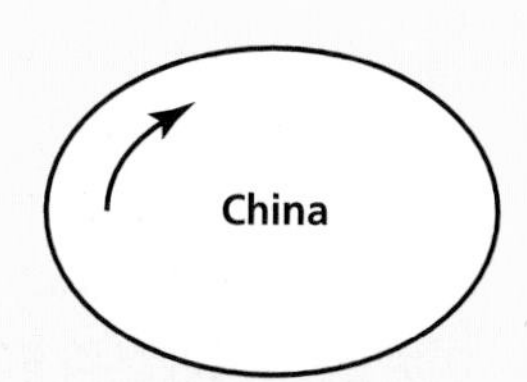

Applying Geography Skills

8. **Analyzing Maps** Look at the climate map on page 672. In what climate zone is Beijing located?

Making Connections

ART SCIENCE LITERATURE TECHNOLOGY

The Three Gorges Dam

Since 1919, Chinese officials have dreamed of building a dam across the Yangtze, the third-longest river in the world. Curving through the heart of China, the river provides an important highway for moving people and products from town to town. Yet the Yangtze is unpredictable. For thousands of years, floods have harmed the millions of people who live along its banks. Now construction is underway to build the dam.

The Dam

In 1994 the Chinese government began a 17-year-long project to build a massive dam. It will eventually be 1.5 miles (2.4 km) wide and more than 600 feet (183 m) high. The dam, called the Three Gorges Dam, will benefit China in several ways. First, it will control water flow and stop floods. Second, its system of locks will allow large ships to travel inland. This will reduce trade and transportation costs for the millions of people who live inland. Third, the dam will create electricity using water-driven engines. Engineers believe that the dam will one day create one-ninth of China's electric power.

Controversy

Even with all the proposed benefits, many people within China and elsewhere have questioned the wisdom of building the dam. When completed, the dam will create a deep reservoir nearly 400 miles (644 km) long. This reservoir will flood more than 100 towns and force nearly 2 million people to move. Many of these people must leave the farms that their families have worked for centuries. Historians point out that the reservoir will also wash away more than 1,000 important historical sites, including the homeland of the first people to settle the region about 4,000 years ago.

Environmentalists caution that the dam may create pollution and health risks. Industrial sites, once they lie underwater, may leak hazardous chemicals. Sewage from communities surrounding the dam could flow directly into the reservoir and into the Yangtze River. In the past, this problem was less serious because the fast-moving waters of the Yangtze carried waste quickly out to sea.

Making the Connection

1. How have the unpredictable waters of the Yangtze River affected the Chinese?
2. How might pollution become a more significant problem after the completion of the dam?
3. **Interpreting Points of View** List three reasons in support of constructing Three Gorges Dam and three reasons against it.

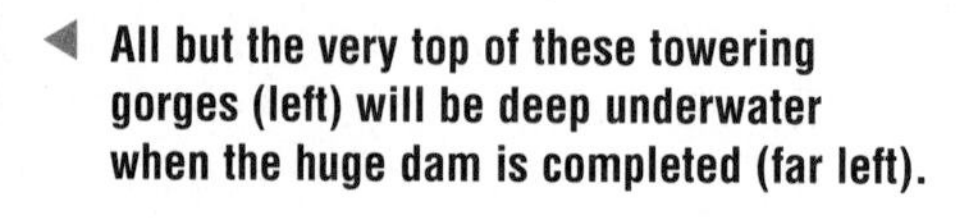

All but the very top of these towering gorges (left) will be deep underwater when the huge dam is completed (far left).

Section 2

China's New Economy

Guide to Reading

Main Idea

China's rapidly growing economy has changed in recent years.

Terms to Know

- communist state
- invest
- consumer goods
- tungsten
- terraced field

Places to Locate

- Beijing
- Shanghai
- Hong Kong
- Macau

Reading Strategy

Draw a diagram like this one. Write two statements about the Chinese economy under the headings in each oval.

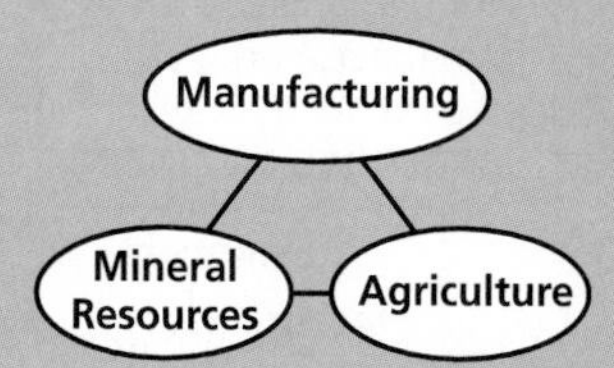

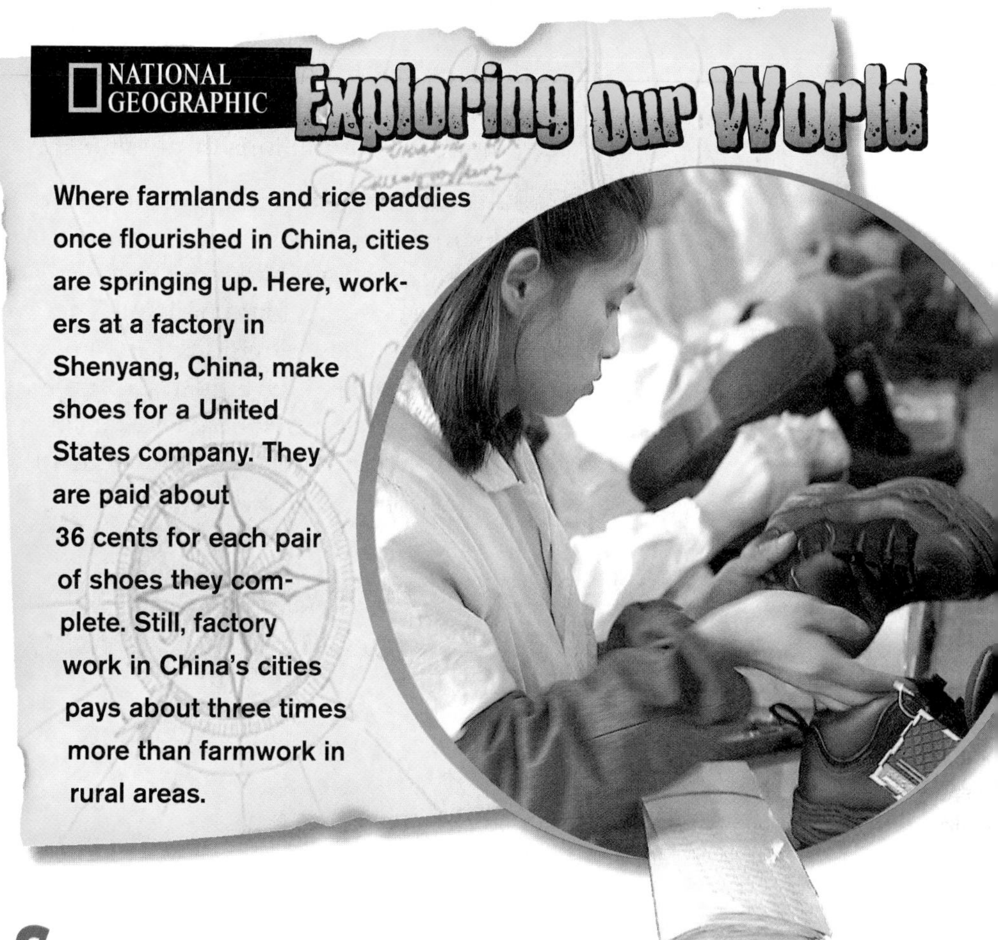

Where farmlands and rice paddies once flourished in China, cities are springing up. Here, workers at a factory in Shenyang, China, make shoes for a United States company. They are paid about 36 cents for each pair of shoes they complete. Still, factory work in China's cities pays about three times more than farmwork in rural areas.

Since 1949, China has been a **communist state,** in which the government has strong control over the economy and society as a whole. Government officials—not individuals or businesses—decide what crops are grown, what products are made, and what prices are charged. China discovered that the communist system created many problems. China fell behind other countries in technology, and manufactured goods were of poor quality.

A New Economy

In recent years, China's leaders have begun many changes to make the economy stronger. Without completely giving up communism, the government has allowed many features of the free enterprise system to take hold. The government now wants individuals to choose what jobs they want and where to start their own businesses. Workers can keep the profits they make. Farmers can grow and sell what they wish.

As a result of these and other changes, China's economy has boomed. The total value of goods and services produced in China increased four times from 1978 to 1999. Farm output also rose rapidly. Because of mountains and deserts, only 10 percent of China's land is farmed. Yet China is now a world leader in producing various agricultural products.

Foreign Trade Eager to learn about new business methods, China has asked other countries to **invest,** or put money, in Chinese businesses. Many companies in China are now jointly owned by Chinese and foreign businesspeople. Foreign companies expect two benefits from investing in China. First, they can pay Chinese workers less than they pay workers in their own countries. Second, companies in China have hundreds of millions of possible customers for their goods.

Results of Growth Because of economic growth, more of China's people are able to get jobs in manufacturing and service industries. Wages have increased, and more goods are available to buy. Some Chinese now enjoy a good standard of living. They can afford **consumer goods,** or products such as televisions, cars, and motorcycles.

Not everyone has adjusted well to the new economy. Many Chinese find that prices have risen faster than their incomes. Some Chinese have become very rich, while others remain poor.

Leading Rice-Producing Countries

Analyzing the Graph

The most important food crop in Asia is rice.

Place **How many millions of tons of rice does China produce in a year?**

Textbook update

Visit gwip.glencoe.com and click on **Chapter 24—Textbook Updates.**

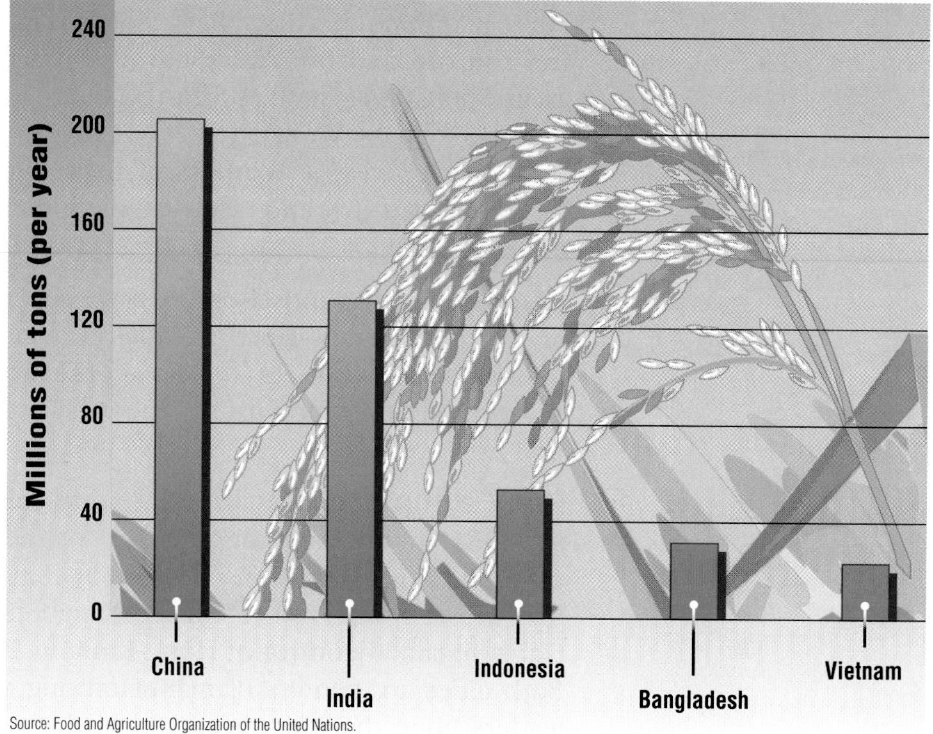

Source: Food and Agriculture Organization of the United Nations.

China's economic growth has also hurt the environment. Many factories dump poisonous chemicals into rivers. Others burn coal, which gives off smoke that pollutes the air. This pollution leads to lung disease, which causes one-fourth of all deaths in China.

Reading Check **What two benefits do foreign-owned businesses expect from operating in China?**

China's Economic Regions

The physical geography of China influences its economy. Climate zones and the availability of resources also affect economic activity. China has three economic regions: the north, south, and west.

The North The climate in the north is partly dry and often cold. As a result, farmers grow hardy crops like wheat, corn, and soybeans. The map on page 676 shows you that the north is rich in natural resources. China is a world leader in mining coal and iron ore. Other minerals include petroleum, copper, and tungsten. **Tungsten** is a metal used in electrical equipment. Trees in this region provide lumber and wood products.

Factory workers produce textiles, chemicals, electronic equipment, airplanes, and other metal goods. **Beijing** (BAY•JIHNG), China's capital and a major industrial city, is located in the north.

The South The south region has fertile soil, a humid climate, and a long growing season. In hilly areas, farmers grow crops on terraced

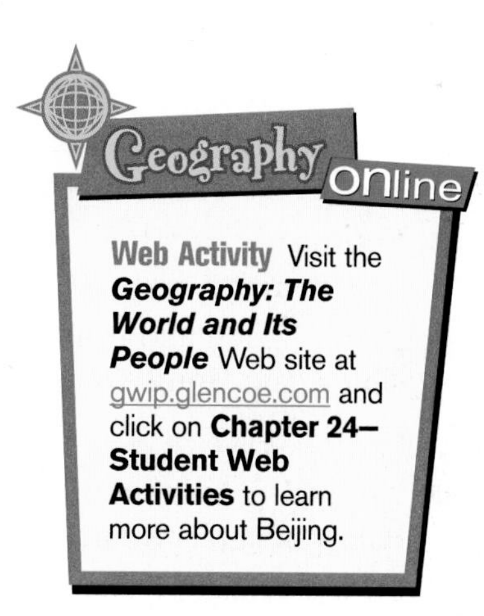

Web Activity Visit the ***Geography: The World and Its People*** Web site at gwip.glencoe.com and click on **Chapter 24—Student Web Activities** to learn more about Beijing.

fields. A terraced field has strips of land cut out of a hillside like stair steps. Rice is the south's major crop. Farmers also grow tea, fruits, and vegetables and raise silkworms, which produce silk thread.

In addition to fertile soil, the south is rich in minerals such as bauxite, iron ore, and tin. The south also has many urban manufacturing areas. Some cities, such as **Shanghai,** are located on or near the coast. Others, such as Wuhan (WOO•HAHN) or Guangzhou (GWAHNG•JOH), are on rivers. Workers in these industrial cities make ships, machinery, textiles, and electrical equipment.

The West China's western region includes large areas of mountains, deserts, and grasslands. Herders graze sheep on the grasslands. The dry and cold Plateau of Tibet provides only limited grazing land for hardy animals such as yaks. In low-lying fertile areas, farmers can grow corn, wheat, and other food crops. The west is also rich in petroleum, coal, and iron ore.

Hong Kong and Macau The cities of **Hong Kong** and **Macau** (muh•KOW) are an important part of the economic changes taking place in China. Both cities were once controlled by a European country—Hong Kong by the United Kingdom, and Macau by Portugal. China regained control of Hong Kong in 1997 and of Macau in 1999. Both cities are centers of manufacturing, trade, and finance. Chinese leaders hope that the successful businesses in these cities will help spur economic growth in the rest of the country.

Reading Check Which economic region has the longest growing season? Why?

Assessment

Defining Terms

1. Define communist state, invest, consumer goods, tungsten, terraced field.

Recalling Facts

2. Economics Why did China's leaders begin to change the country's economy in recent years?
3. Economics By how much did China's economy increase from 1978 to 1999?
4. Place When and from what countries did China gain Hong Kong and Macau?

Critical Thinking

5. Analyzing Information What benefits does China receive from foreign investment?
6. Summarizing Information What are three results of China's economic growth?

Graphic Organizer

7. Organizing Information Draw a chart like this one. Then list the agricultural and manufactured products of China's economic regions.

Region	North	South	West
Economic Products			

Applying Geography Skills

8. Analyzing Maps Look at the economic activity map on page 676. Which economic region of China has the fewest manufacturing centers?

Critical Thinking Skill

Distinguishing Fact From Opinion

Distinguishing fact from opinion can help you make reasonable judgments about what others say and write. Facts can be proved by evidence such as records, documents, or historical sources. Opinions are based on people's differing values and beliefs.

Learning the Skill

The following steps will help you identify facts and opinions:

- Read or listen to the information carefully. Identify the facts. Ask: Can these statements be proved? Where would I find information to prove them?
- If a statement can be proved, it is factual. Check the sources for the facts. Often statistics sound impressive, but they may come from an unreliable source.
- Identify opinions by looking for statements of feelings or beliefs. The statements may contain words like *should, would, could, best, greatest, all, every,* or *always.*

Practicing the Skill

Read the paragraph below, then answer the questions that follow.

> Anyone who thinks the Internet is not used in China has been asleep at the mouse. China's government-owned factories and political system may seem old-fashioned. When it comes to cyberspace, however, China is moving at Net speed. Internet use is growing explosively. Two years ago, only 640,000 Chinese were using the Internet. Now more than 4 million are. International Data Corp. estimates that by 2001, the online population should hit 27 million. China will become the greatest market for computer sales in history.

Adapted from *Business Week,* August 2, 1999.

1. Identify facts. Can you prove that Chinese Internet use is increasing?
2. Note opinions. What phrases alert you that these are opinions?
3. What is the purpose of this paragraph?

Applying the Skill

Watch a television commercial. List one fact and one opinion that are stated. Does the fact seem reliable? How can you prove the fact?

GO TO Practice key skills with **Glencoe Skillbuilder Interactive Workbook, Level 1.**

◀ Chinese students attend an Internet exhibit in Beijing.

Section 3

China's People and Culture

Guide to Reading

Main Idea

The arts and ideas of ancient times still influence China today.

Terms to Know

- dynasty
- human rights
- exile
- calligraphy
- pagoda

Places to Locate

- Beijing
- Taiwan
- Tibet

Reading Strategy

Make a chart like this one. Then list two key facts in the right column for each item in the left column.

China	
History	
Government	
Urban and Rural Life	
Arts	

NATIONAL GEOGRAPHIC **Exploring Our World**

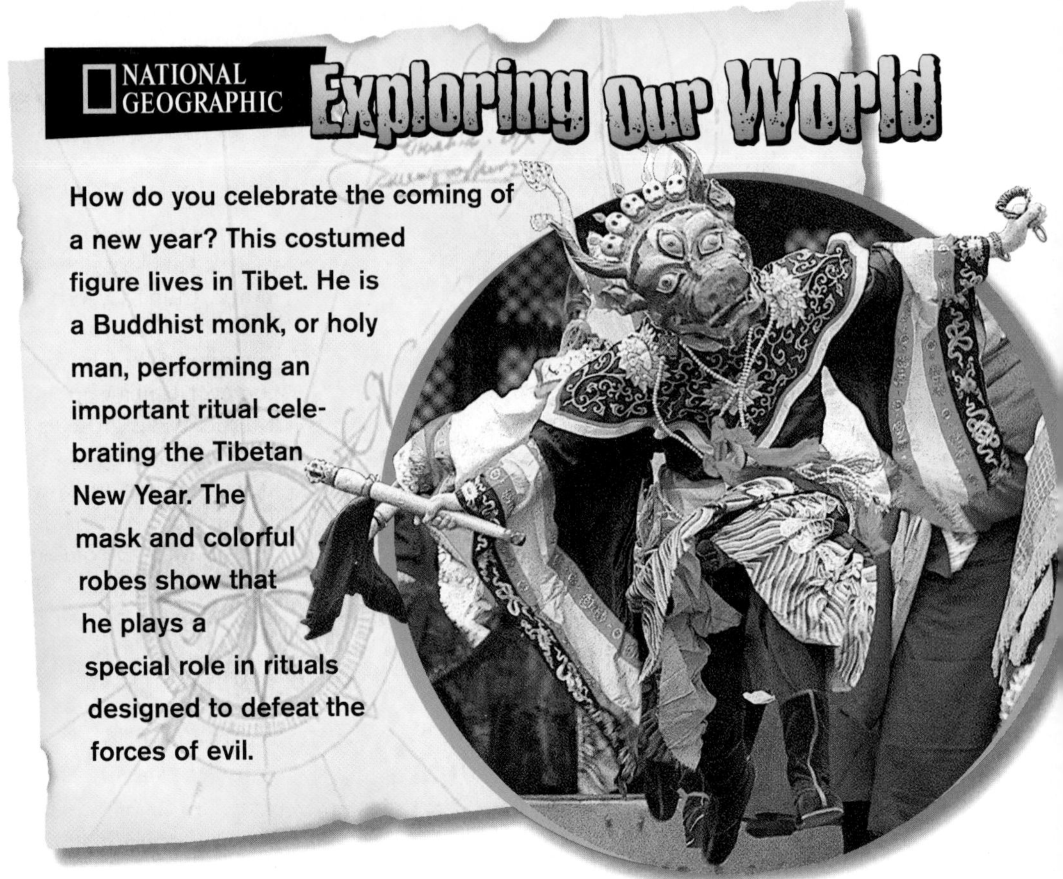

How do you celebrate the coming of a new year? This costumed figure lives in Tibet. He is a Buddhist monk, or holy man, performing an important ritual celebrating the Tibetan New Year. The mask and colorful robes show that he plays a special role in rituals designed to defeat the forces of evil.

China's population of 1.25 billion is about one-fifth of the world's people. About 92 percent of these people belong to the ethnic group called Han Chinese. They have a unique culture. The remaining 8 percent belong to 55 other ethnic groups. Most of these groups, such as the Tibetans, live in the western part of China. They have struggled to protect their traditions from Han Chinese influences.

China's History

China's civilization is more than 4,000 years old. For centuries—in fact, until the early 1900s—rulers known as emperors or empresses governed China. Many lived in the Imperial Palace, located in the heart of **Beijing.** A **dynasty,** or a line of rulers from a single family, would hold power until it was overthrown. Then a new leader would start a new dynasty. Under the dynasties, China built a highly developed culture and conquered neighboring lands.

As their civilization developed, the Chinese tried to keep out foreign invaders. In many ways, this was easy. On most of China's borders, natural barriers such as seas, mountains, and deserts already provided protection. Still, invaders threatened from the north. To defend this area, the Chinese began building the Great Wall of China about 2,200 years ago. Over the centuries, the wall was continually rebuilt and lengthened. In time, it snaked more than 4,000 miles (6,437 km) from the Yellow Sea in the east to the deserts of the west. It still stands today.

Chinese thinkers believed that learning was a key to good behavior. About 500 B.C., a thinker named Kongfuzi (KOONG•FOO•DZUH), or Confucius, taught that people should be polite, honest, brave, and wise. Children were to obey their parents, and every person was to respect the elderly and obey the country's rulers. Kongfuzi's teachings became the foundation of Chinese life. The teachings shaped China's government and society until the early 1900s.

During Kongfuzi's time, another thinker named Laozi arose. His teachings, called Daoism (DAHW•ehzm), stated that people should live simply and in harmony with nature. While Kongfuzi's ideas appealed to government leaders, Laozi's beliefs attracted artists and writers.

Beginning about A.D. 100, another religion, Buddhism, won followers among the Chinese. This faith came to China from South Asia. Buddhism taught that prayer, right thoughts, and good deeds could help people find relief from life's problems. Over time, the Chinese mixed Buddhism, Daoism, and the ideas of Kongfuzi. This mixed spiritual heritage still influences many Chinese people today.

The early Chinese were inventors as well as thinkers. Did you know that they were using paper and ink before people in other parts of the world? Other Chinese inventions included silk, the clock, the magnetic compass, printed books, gunpowder, and fireworks. For hundreds of years, China was the most advanced civilization in the world.

Communist China Foreign influences increasingly entered China during the 1700s and 1800s. Europeans especially wanted to get such fine Chinese goods as silk, tea, and pottery. The United Kingdom and other countries used military power to force China to trade.

In 1911 a Chinese uprising overthrew the last emperor. China became a republic, or a country governed by elected leaders. Disorder followed until the Nationalist political party took over. A Communist party gained power as well. After World War II, the Nationalists and the Communists fought for control of China. General Chiang Kai-shek (jee•AHNG KY•SHEHK) led the Nationalists. Mao Zedong (MOW DZUH•DOONG) led the Communists.

In 1949 the Communists won and set up the People's Republic of China under Mao Zedong. The Nationalists under Chiang Kai-shek fled to the offshore island of **Taiwan.** There they set up a rival government.

✓ Reading Check **Why was the Great Wall of China built?**

Clay Warriors

One of the most fascinating archaeological finds in China was the clay army buried to guard the tomb of China's first emperor. The huge vault, covering 20 square miles (52 sq. km), was discovered in 1974. The clay warriors stand in four separate underground pits. In pit one are 6,000 life-size figures in military formation. Pit two contains 1,400 chariots and men. The third pit has an elite command force, and the fourth pit is empty, possibly abandoned before the work was completed. Each of the nearly 7,500 foot soldiers, horsemen, archers, and chariot riders were individually crafted more than 2,200 years ago.

China's Government and Society

After 1949 the Communists completely changed the mainland of China. All land and factories were taken over by the government. Farmers were organized onto large government farms, and women joined the industrial workforce. Dams and improved agricultural methods brought some economic benefits. Yet many government plans went wrong, and individual freedoms were lost. Many people were killed because they opposed communism.

After Mao Zedong died in 1976, a new Communist leader, Deng Xiaoping (DUHNG SHOW•PIHNG), decided to take a new direction. He wanted to make China a more open country. One way to do this was to give people more economic freedom. The government kept tight control over all political activities, however. It continued to deny individual freedoms and acted harshly against any Chinese who criticized its actions. In 1989 thousands of students gathered in Beijing's Tiananmen (TEE•EHN•AH•MEHN) Square. The students called for more democracy in China. The government answered by sending in tanks and troops. These forces killed thousands of protesters and arrested many more.

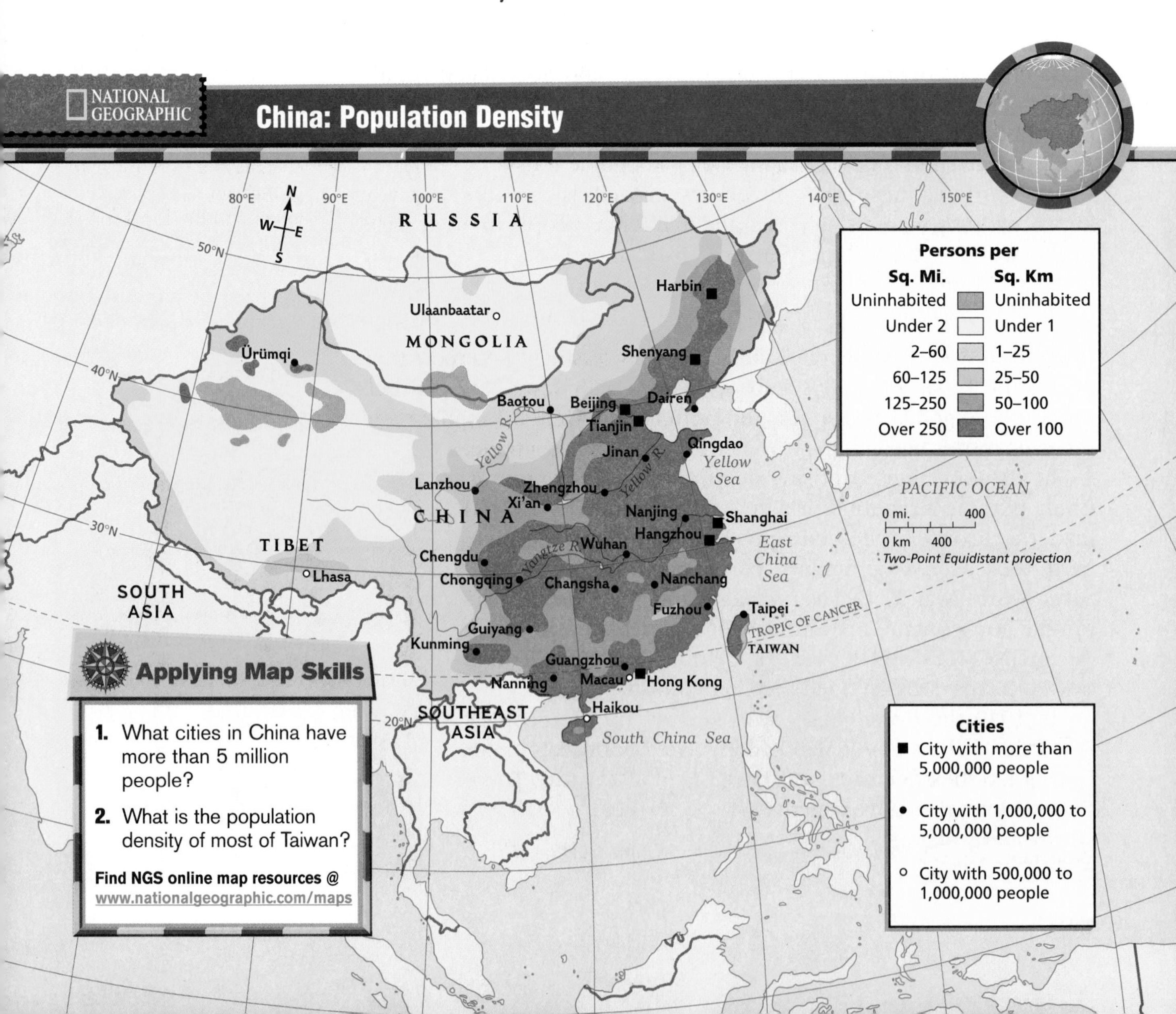

Countries around the world have protested the Chinese government's continued harsh treatment of people who criticize it. They say that Chinese leaders have no respect for **human rights**—the basic freedoms and rights, such as freedom of speech, that all people should enjoy. Because of China's actions, some people say that other countries should not trade with China.

China's leaders have also been criticized for their actions in **Tibet.** Tibet was once a separate Buddhist kingdom. China took control of the area in 1950 and crushed a rebellion there about nine years later. The Tibetan people have demanded independence since then. The Dalai Lama (DAH•LY LAH•muh), the Buddhist leader of Tibet, now lives in exile in India. Someone in **exile** is unable to live in his or her own country because of political beliefs. The Dalai Lama travels around the world trying to win support for his people.

Rural and Urban

Hundreds of thousands of people use bicycles—not cars—to get around Beijing and other cities. Still, about 70 percent of China's people live on small plots of land in rural areas.

Place **About how many people live in China's cities?**

Rural Life About 70 percent of China's people live in rural areas. The map on page 682 shows that most Chinese are crowded into the fertile river valleys of eastern China. Families work hard in their fields. They often use hand tools because mechanical equipment is too expensive.

Village life has improved in recent years. Most rural families now live in three- or four-room houses. They have enough food and some modern appliances. Many villages have community centers. People gather there to watch movies and play table tennis and basketball.

Urban Life More than 360 million Chinese people live in cities. China's cities are growing rapidly as people leave farms in the hopes of finding better-paying jobs. Living conditions in the cities are crowded, but most homes and apartments have heat, electricity, and running water. Many people now earn enough money to buy extra clothes and televisions. They also have more leisure time to attend concerts or Chinese operas, walk in parks, or visit zoos.

✓Reading Check **Why have people in other countries criticized China's government?**

China's Culture

China is famous for its traditional arts. Chinese craft workers make bronze bowls, jade jewelry, decorated silk, glazed pottery, and fine porcelain. The Chinese are also known for their painting, sculpture, and architecture.

The Chinese love of nature has influenced painting and poetry. Chinese artists paint on long panels of paper or silk. Artwork often shows scenes of mountains, rivers, and forests. Artists attempt to portray the harmony between people and nature.

Many Chinese paintings include a poem written in **calligraphy,** the art of beautiful writing. Chinese writing is different from the print you are reading right now. It uses characters that represent words or ideas instead of letters that represent sounds. There are more than 50,000 Chinese characters, but the average person recognizes only about 8,000. It takes many years to learn to write Chinese.

The Chinese developed the first porcelain centuries ago. Porcelain is made from coal dust and fine, white clay. Painted porcelain vases from early China are considered priceless today.

Most buildings in China's cities are modern. Yet traditional buildings still stand. Some have large tiled roofs with edges that curve gracefully upward. Others are Buddhist temples with many-storied towers called **pagodas.** These buildings hold large statues of the Buddha.

Foods Cooking differs greatly from region to region. In coastal areas, people enjoy fish, crab, and shrimp dishes. Central China is famous for its spicy dishes made with hot peppers. Most Chinese eat very simply. A typical Chinese meal includes vegetables with bits of meat or seafood, soup, and rice or noodles. Often the meat and vegetables are cooked quickly in a small amount of oil over very high heat. This method—called stir-frying—allows the vegetables to stay crunchy.

✓Reading Check **Why does it take many years to learn to read and write Chinese?**

Assessment

Defining Terms

1. **Define** dynasty, human rights, exile, calligraphy, pagoda.

Recalling Facts

2. **History** Who are two thinkers who influenced life in China?
3. **History** Who led the Nationalists after World War II? Who led the Communists after World War II? Who won control of China?
4. **Culture** What scenes are commonly found in Chinese paintings?

Critical Thinking

5. **Making Predictions** How would the teachings of Kongfuzi prevent rebellions in China?
6. **Summarizing Information** Why did Europeans want to force China to trade with them?

Graphic Organizer

7. **Organizing Information** Draw a time line like this one. Then list at least five dates and their events in China's history.

Applying Geography Skills

8. **Analyzing Maps** Look at the population density map on page 682. How does the population density in western China differ from that in eastern China?

China's Neighbors

Guide to Reading

Main Idea

Taiwan and Mongolia have been influenced by Chinese ways and traditions.

Terms to Know

- high-technology industry
- steppe
- nomad
- empire
- yurt

Places to Locate

- Taiwan
- Mongolia
- Taipei
- Gobi
- Ulaanbaatar

Reading Strategy

Draw a diagram like this one. Then write statements that are true of each country under their headings in the outer ovals. Where the ovals overlap, write statements that are true of both countries.

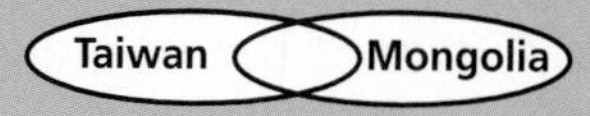

NATIONAL GEOGRAPHIC **Exploring Our World**

In the remote, harsh land of western Mongolia, a centuries-old tradition continues. Hunters train eagles to bring their kill back to the human hunter. The people say that female eagles make the best hunters. Because they weigh more than males, they can capture larger prey. Like all eagles, they have superb vision—eight times better than a human's.

Taiwan is an island close to China's mainland, and **Mongolia** borders China on the north. Throughout history, Taiwan and Mongolia have had close ties to their larger neighbor.

Taiwan

About 100 miles (161 km) off the southeastern coast of China lies the island country of Taiwan. It is slightly larger than the states of Connecticut and Massachusetts put together. Through Taiwan's center runs a ridge of steep, forested mountains. On the east, the mountains descend to a rocky coastline. On the west, they fall away to a narrow, fertile plain. This flat area is home to 90 percent of the island's people. Like southeastern China, Taiwan has a humid subtropical climate, with mild winters and hot, rainy summers.

Taiwan's Economy Taiwan has one of the world's most prosperous economies. Taiwan's wealth comes largely from high-technology industries, manufacturing, and trade with other countries. **High-technology industries** produce computers and other kinds of

Many electronic industries have headquarters in Taiwan.

Place **What kinds of products do high-technology factories in Taiwan produce?**

electronic equipment. Workers in Taiwan's factories make many different products, including computers, calculators, radios, televisions, and telephones. You have probably seen goods from Taiwan sold in stores in your community.

Taiwan has a growing economic influence on its Asian neighbors. Many powerful companies based in Taiwan have recently built factories in the People's Republic of China and Thailand. Despite their political differences, Taiwan and mainland China have increased their economic ties since the 1990s.

Agriculture also contributes to Taiwan's booming economy. The island's mountainous landscape limits the amount of land that can be farmed. Still, some farmers have built terraces on mountainsides to grow rice. Other major crops include sugarcane and fruits. In fact, Taiwan's farmers produce enough food not only to feed their own people but also to export.

Taiwan's History and People For centuries, Taiwan generally was part of China's empire. Then in 1895, Japan took the island after defeating China in war. The Japanese developed the economy of Taiwan but treated the people very harshly. After Japan's loss in World War II, Taiwan returned to China.

In 1949 the Nationalists under Chiang Kai-shek arrived in Taiwan from the mainland. Along with them came more than 1 million refugees fleeing Communist rule. Fearing a Communist invasion, the Nationalists kept a large army in the hope of someday retaking the mainland. They also blocked other political groups from sharing in the government.

By the early 1990s, the Nationalists felt secure enough to make changes. Local Taiwanese were allowed more opportunities in government. The one-party system ended, and Taiwan became a democracy. Taiwan still claims to be a Chinese country, but many people would like to declare Taiwan independent. China claims Taiwan as its twenty-third province and believes that it should be under China's control. China has threatened to use force against Taiwan if the island declares its independence.

About 75 percent of Taiwan's 22 million people live in urban areas. The most populous city—with 2.6 million people—is the capital, **Taipei.** This bustling center of trade and commerce has tall skyscrapers and modern stores. If you stroll through the city, however, you will see Chinese traditions. Buddhist temples, for example, still reflect traditional Chinese architecture.

✓Reading Check **What is the capital of Taiwan?**

Mongolia

Landlocked Mongolia is a large country about the size of Alaska. Rugged mountains and high plateaus rise in the west and central regions. The bleak desert landscape of the **Gobi** spreads over the southeast. The rest of the country is covered by **steppes,** the dry treeless plains often found on the edges of a desert.

Known as the Land of the Blue Sky, Mongolia boasts more than 260 days of sunshine. Yet its climate has extremes. Rainfall is scarce, and fierce dust storms sometimes sweep across the landscape. Temperatures are very hot in the summer. In the winter, they fall below freezing at night.

For centuries, most of Mongolia's people were nomads. **Nomads** are people who move from place to place with herds of animals. Even today, many Mongolians tend sheep, goats, cattle, or camels on the country's vast steppes. Important industries in Mongolia use products from these animals. Some factories use wool to make textiles and clothing. Others use the hides of cattle to make leather and shoes. Some farmers grow wheat and other grains. Mongolia also has deposits of copper and gold.

Mongolia's History and People

Mongolia's people are famous for their skills in raising and riding horses. In the past, they also were known as fierce fighters. In the 1200s, many groups of Mongols joined together under one leader, Genghis Khan (JEHNG•guhs KAHN). He led Mongol armies on a series of conquests. The Mongols eventually carved out the largest land empire in history. An **empire** is a collection of different territories under one ruler. The Mongol Empire stretched from China all the way to eastern Europe.

During the 1300s, the Mongol Empire weakened and fell apart. China ruled the area that is now Mongolia from the 1700s to the early 1900s. In 1924 Mongolia gained independence and created a strict Communist government. The country finally became a democracy in 1990. Since then, the Mongolian economy has moved slowly from government control to a free market system.

Ulaanbaatar began as a Buddhist community in the early 1600s. Today it is a modern cultural and industrial center.

Place **Why is Mongolia known as the Land of the Blue Sky?**

Chapter 25 Japan and the Koreas

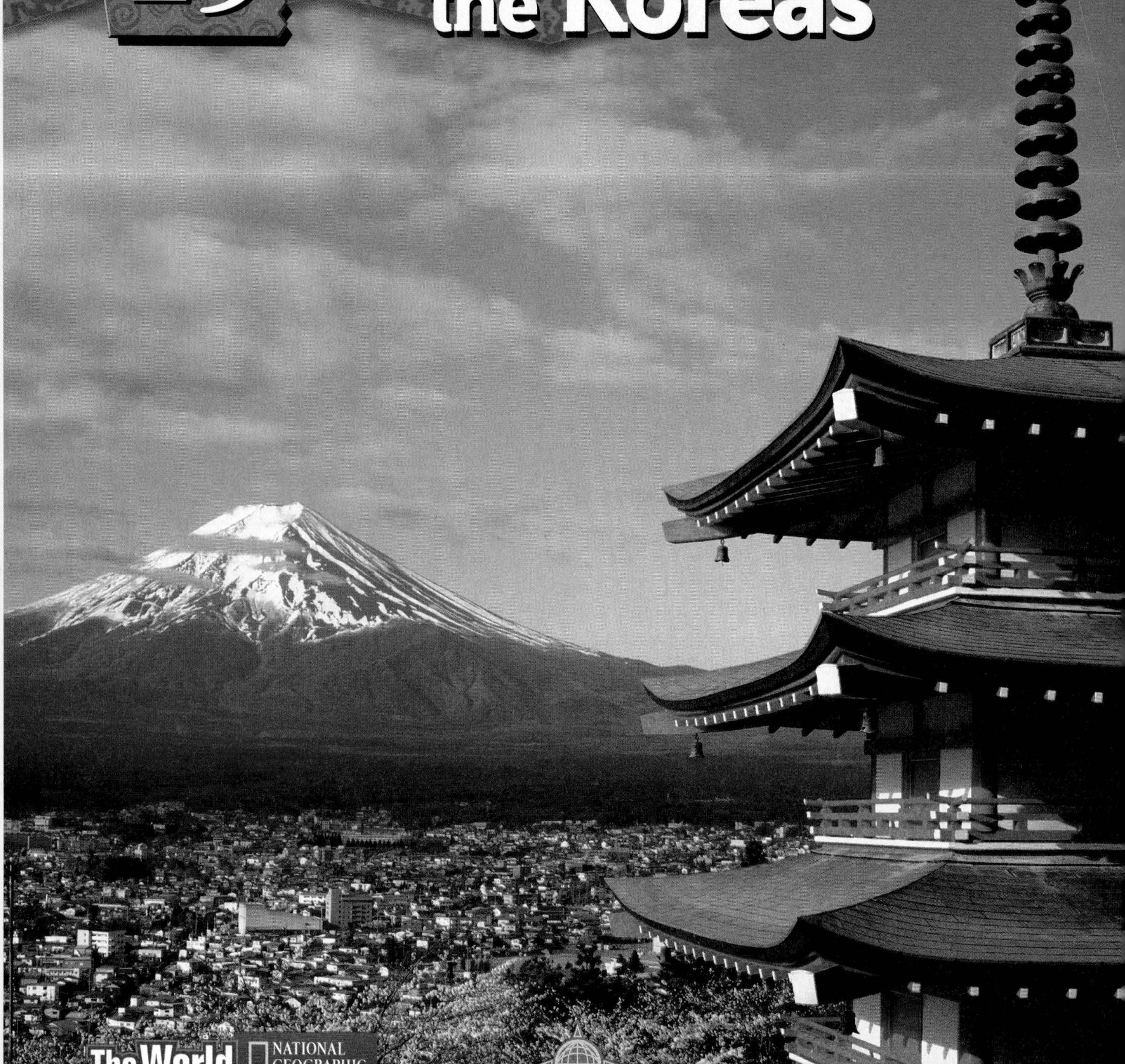

The World and Its People NATIONAL GEOGRAPHIC

To learn more about the people and places of Japan and the Koreas, view ***The World and Its People*** **Chapter 25** video.

Geography Online

Chapter Overview Visit the ***Geography: The World and Its People*** Web site at gwip.glencoe.com and click on **Chapter 25—Chapter Overviews** to preview information about Japan and the Koreas.

Japan

Guide to Reading

Main Idea

Although they have few mineral resources, Japan's people have built a prosperous country.

Terms to Know

- tsunami
- archipelago
- intensive cultivation
- clan
- shogun
- samurai
- constitutional monarchy
- megalopolis

Places to Locate

- Sea of Japan
- Hokkaido
- Honshu
- Shikoku
- Kyushu
- Mount Fuji
- Kanto Plain
- Tokyo
- Inland Sea

Reading Strategy

Make a chart like this one. In the right column, write a fact about Japan for each topic in the left column.

Japan	Fact
Land	
Economy	
History	
People	

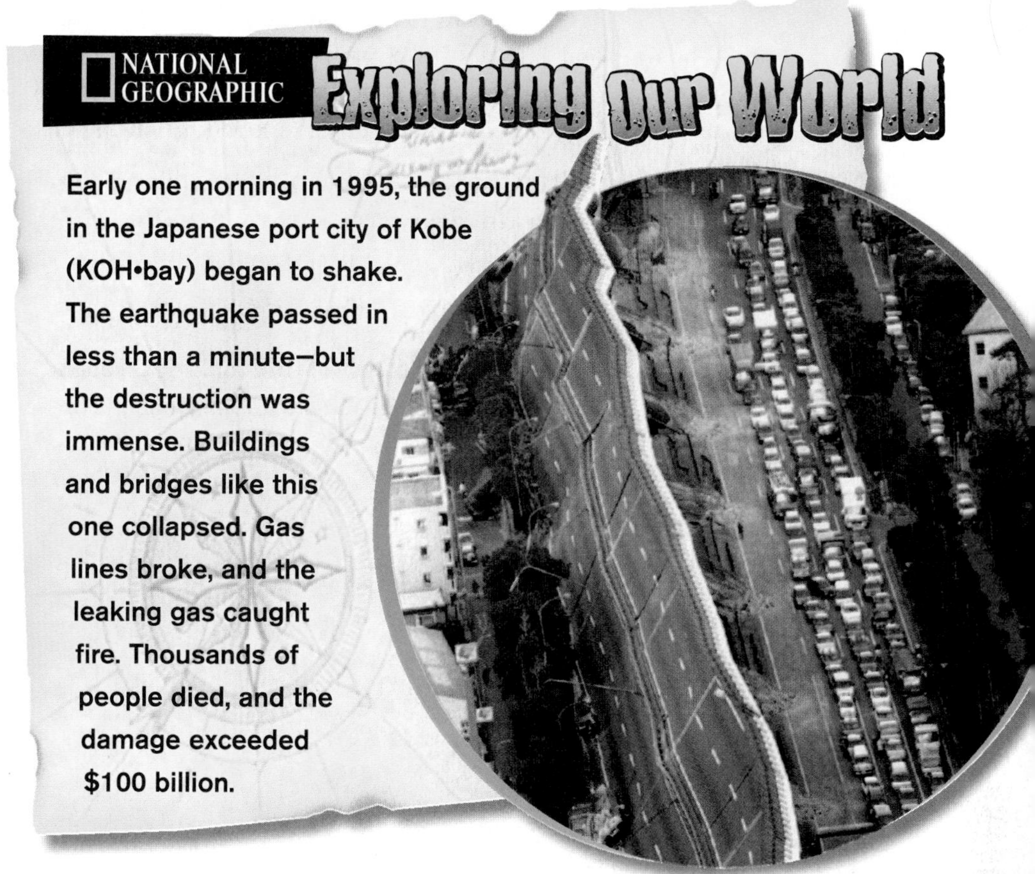

Early one morning in 1995, the ground in the Japanese port city of Kobe (KOH•bay) began to shake. The earthquake passed in less than a minute—but the destruction was immense. Buildings and bridges like this one collapsed. Gas lines broke, and the leaking gas caught fire. Thousands of people died, and the damage exceeded $100 billion.

The city of Kobe suffered an earthquake because Japan lies on the Ring of Fire. This name refers to an area surrounding the Pacific Ocean where the earth's crust often shifts. Japan experiences thousands of earthquakes a year. People in Japan also have to deal with **tsunamis** (tsu•NAH•mees). These huge sea waves caused by undersea earthquakes are very destructive along Japan's Pacific coast.

Japan's Land and Climate

Japan is an **archipelago,** or a group of islands, off the coast of eastern Asia between the **Sea of Japan** and the Pacific Ocean. Four main islands and thousands of smaller ones make up Japan's land area. The four largest islands are **Hokkaido** (hoh•KY•doh), **Honshu, Shikoku** (shee•KOH•koo), and **Kyushu** (KYOO•SHOO).

These islands are actually the peaks of mountains that rise from the floor of the Pacific Ocean. The mountains are volcanic, but many are no longer active. The most famous peak is **Mount Fuji,** Japan's highest mountain and national symbol. Rugged mountains and steep, forested hills dominate most of Japan. Narrowly squeezed between the seacoast

◀ Mount Fuji overlooks Yamanashi, Japan.

and the mountains are plains. The **Kanto Plain** in eastern Honshu is Japan's largest plain. It holds **Tokyo,** the capital, and Yokohama, one of Asia's major port cities. You will find most of Japan's cities, farms, and industries on the coastal plains.

No part of Japan is more than 70 miles (113 km) from the sea. In bay areas along the jagged coasts lie many fine harbors and ports. One of Japan's most important seacoasts is located along the **Inland Sea.** This sea winds its way among the islands of Honshu, Shikoku, and Kyushu. It provides an excellent transportation route.

The Climate Ocean currents and winds affect Japan's climate. The map on page 706 shows that the climate differs in the north and

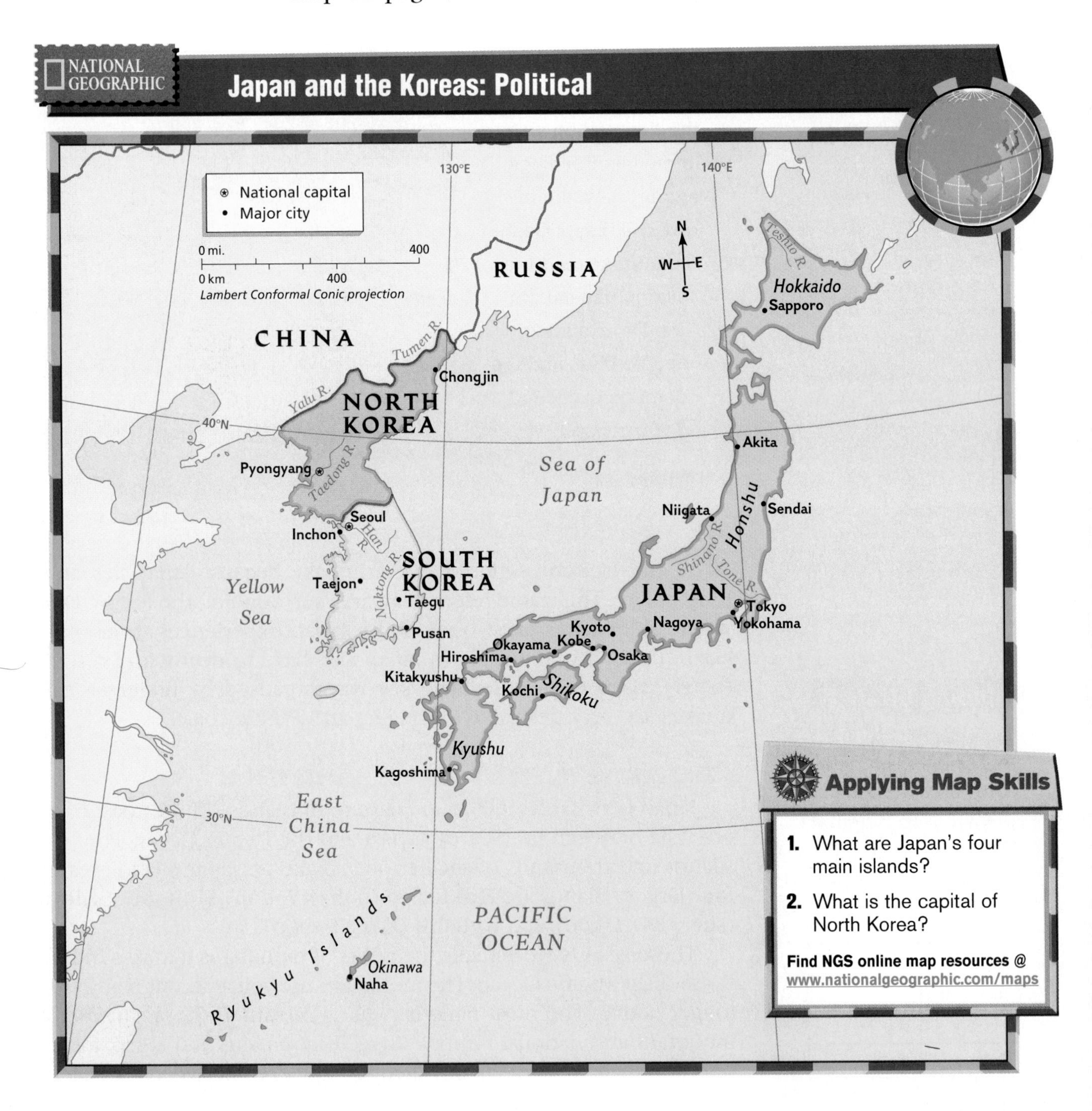

south. Cold winds and ocean currents from the Arctic bring cold, snowy winters to Hokkaido and northern Honshu. Warm ocean currents from the Pacific give the southern part of Honshu and the southern islands a humid subtropical climate.

Reading Check **What two factors affect Japan's climate?**

Japan's Economy

Japan has few mineral resources, so it must import raw materials like iron ore, coal, and oil. However, Japan is an industrial giant known around the world for the variety and quality of its manufactured goods. Japan's modern factories use new technology and robots to make their

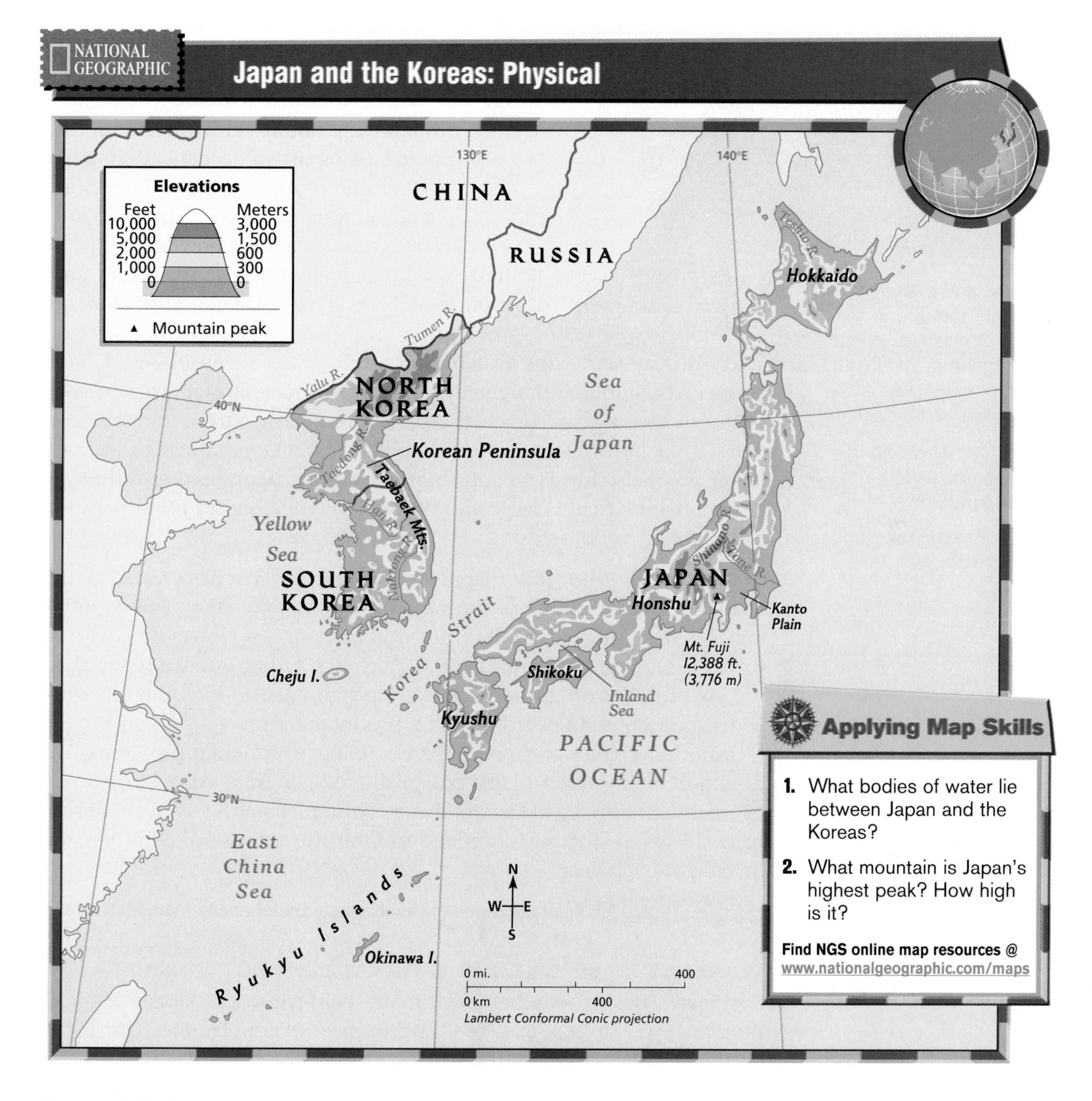

Past and Present

Past and present come together in Japan. Here, a priest of the ancient Shinto religion blesses a family's shiny new car.

Place When did Japan begin to modernize?

products quickly and carefully. These products include automobiles and other vehicles. The graph on page 16 in the **Geography Handbook** shows you that Japan leads the world in automobile production. Japan's factories also produce consumer goods like electronic equipment, watches, small appliances, and calculators. You may own a piece of electronic equipment—a television, VCR, or camera—made in Japan. Other factories produce industrial goods like steel, cement, fertilizer, plastics, and fabrics.

Japan's industries have benefited from having highly skilled workers. The people of Japan value education, hard work, and cooperation. After high school graduation, many Japanese students go on to a local university.

Farmland is very limited. Yet Japan's farmers use fertilizers and modern machinery to produce high crop yields. They also practice **intensive cultivation**—they grow crops on every available piece of land. You can see crops growing on terraces cut in hillsides and even between buildings and highways. In warmer areas, farmers harvest two or three crops a year. The chief crop is rice, a basic part of the Japanese diet. Other important crops include sugar beets, potatoes, fruits, and tea. Seafood also forms an important part of the people's diet. Although Japan's fishing fleet is large and productive, the country imports more fish than any other nation.

Economic Challenges Japan's economic success has created some challenges. Japan is one of the world's leading exporters. The country imports few finished goods from other countries, however. This has led to disagreements with trading partners. Other countries say that the government of Japan, by setting up trade restrictions, unfairly prevents their companies from selling products there.

Another challenge facing Japan is its environmental problems. Air pollution from power plants has produced acid rain. Also, because of overfishing, supplies of seafood have dropped. Japan's government has passed laws to stop pollution and to limit the amount of fish that can be caught each year.

Reading Check What are some products made by Japanese manufacturers?

Japan's History and Government

Japan's history reaches back many centuries. The Japanese trace their ancestry to various **clans**, or groups of related families, that lived on the islands as early as the late A.D. 400s.

The Japanese developed close ties with China on the Asian mainland. Ruled by emperors, Japan modeled its society on the Chinese way of life. The Japanese also borrowed the Chinese system of writing and accepted the Buddhist religion brought by Chinese missionaries. Today most Japanese practice Buddhism along with Shinto, Japan's own traditional religion.

In the 790s, the power of Japanese emperors began to decline. From the late 1100s to the 1860s, Japan was ruled by **shoguns,** or military leaders, and powerful land-owning warriors known as the **samurai.** Like China, Japan did not want to trade with foreign countries. In 1853 the United States government sent a fleet headed by Commodore Matthew Perry to Japan to demand trading privileges. In response to this action and other outside pressures, the Japanese started trading with other countries.

In the late 1800s, Japanese leaders began to use Western ideas to modernize the country, improve education, and set up industries. By the early 1900s, Japan was the leading military power in Asia.

In the 1930s, Japan needed more resources for its growing population. It took land in China and spread its influence to Southeast Asia. In 1941 Japanese forces attacked the American naval base at Pearl Harbor in Hawaii. This attack caused the United States to enter World War II. After four years of fighting, Japan surrendered when the United States dropped atomic bombs on two of its cities. By then, many of Japan's cities lay in ruins, and the economy had collapsed. With help from the United States, Japan became a democracy and rebuilt its economy.

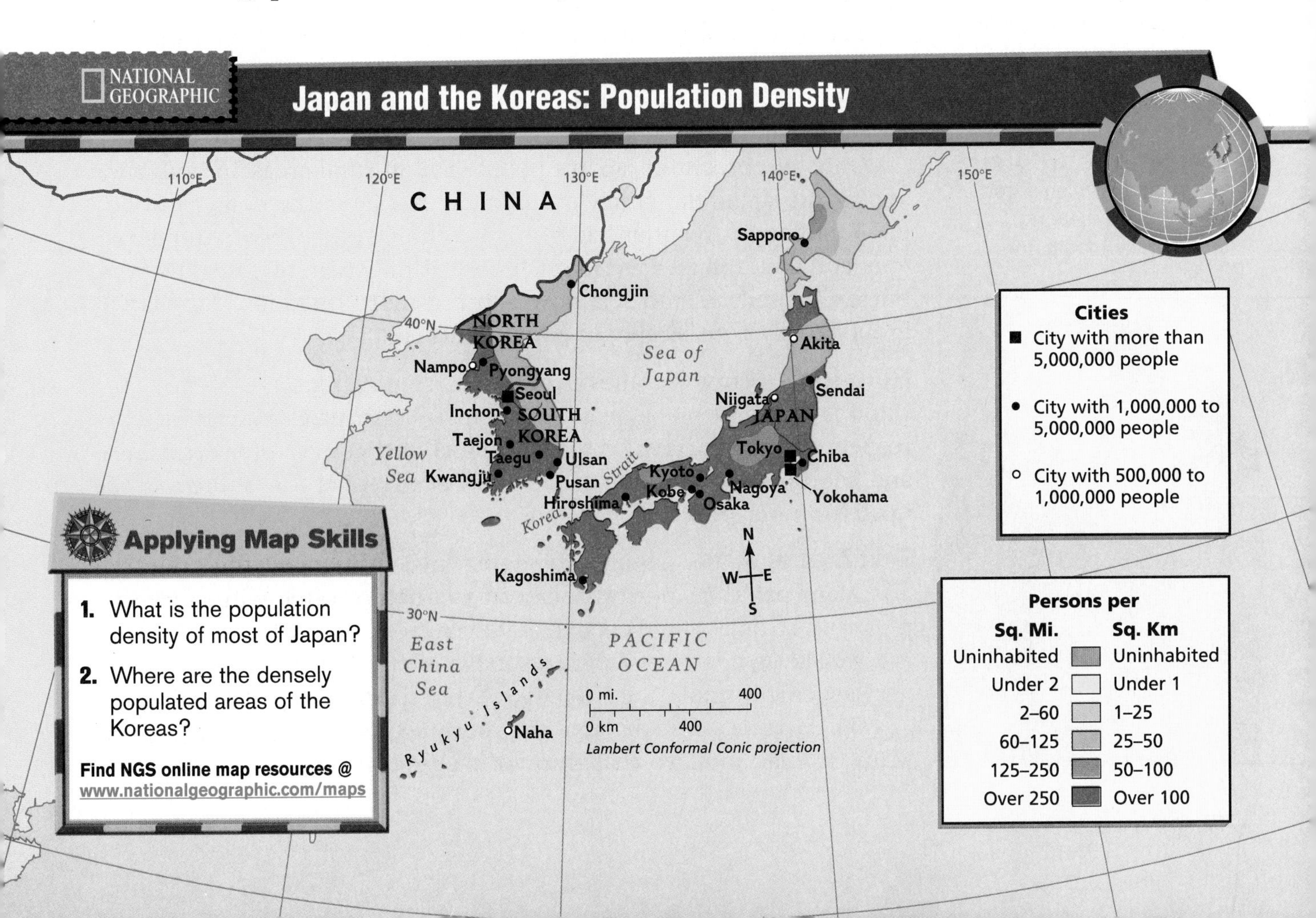

Applying Map Skills

1. What is the population density of most of Japan?
2. Where are the densely populated areas of the Koreas?

Find NGS online map resources @ www.nationalgeographic.com/maps

Japan's democracy is in the form of a **constitutional monarchy.** The emperor is the official head of state, but elected officials run the government. Voters elect representatives to the national legislature. The political party with the most members chooses a prime minister to lead the government.

Japan has great influence as a world economic power. In addition, it gives large amounts of money to poorer countries. Japan is not a military power, though. Because of the suffering that World War II caused, the Japanese have chosen to keep Japan's military small.

Reading Check What kind of government does Japan have?

Hard Hats to School?

Okajima Yukiko and Sataka Aya walk along ash-covered sidewalks to Kurokami Junior High School. Why are they wearing hard hats? Their city is near Japan's Mount Oyama Volcano, which has just erupted. Yukiko and Aya have grown up facing the dangers of volcanic eruptions, earthquakes, and tsunamis. "Right now we need to worry about our grades," says Yukiko. They study *sansu* (math), *rika* (science), *kokugo* (Japanese), *shakai* (social studies), *ongaku* (music), and *doutoku* (moral education). Their first class starts at 8:30 A.M., and their last class ends at 3:40 P.M. Yukiko and Aya must go to school every second Saturday of the month, too.

Japan's People and Culture

Although about the size of California, Japan has 126.7 million people—nearly one-half the population of the United States. Most of Japan's people belong to the same Japanese ethnic background. Look at the map on page 699 to see where most of Japan's people live. About three-fourths are crowded into urban areas on the coastal plains. The four large cities of Tokyo, Yokohama, Nagoya, and Osaka form a **megalopolis,** or a huge urban area made up of several large cities and communities near them.

Japan's cities have tall office buildings, busy streets, and speedy highways. Homes and apartments are small and close to one another. Many city workers crowd into subway trains to get to work. Men work long hours and arrive home very late. Women often quit their jobs to raise children and return to work when the children are grown.

You still see signs of traditional life, even in the cities. Parks and gardens give people a chance to take a break from the busy day. It is common to see a person dressed in a traditional garment called a kimono walking with another person wearing a T-shirt and jeans.

Only 22 percent of Japan's people live in rural areas. In both rural and urban Japan, the family traditionally has been the center of one's life. Each family member had to obey certain rules. Grandparents, parents, and children all lived in one house. Family ties still remain strong, but each family member is now allowed more freedom. Many family groups today consist only of parents and children.

Religion Many Japanese practice two religions—Shinto and Buddhism. Shinto began in Japan many centuries ago. It teaches respect for nature, love of simple things, and concern for cleanliness and good manners. Buddhism also teaches respect for nature and the need for inner peace.

Traditional Arts Japan's religions have influenced the country's arts. Many paintings portray the beauty of nature, often with a few simple brush strokes. Some even include verses of poetry. Haiku (HY•koo) is a well-known type of Japanese poetry that is written according to a very specific formula. Turn to page 702 to learn more about haiku.

The Japanese also have a rich heritage of literature and drama. Many scholars believe that the world's first novel came from Japan.

The novel is called *The Tale of Genji* and was written by a noblewoman about A.D. 1000. Since the 1600s, Japanese theater-goers have attended the historical plays of the Kabuki theater. In Kabuki plays, actors wearing brilliantly colored costumes perform on colorful stages.

Many of Japan's sports have their origins in the past. A popular sport is sumo, an ancient Japanese form of wrestling. Two ancient martial arts—judo and karate—also developed in this area. Today martial arts are practiced both for self-defense and for exercise.

Modern Pastimes Along with these traditional arts, the people of Japan enjoy modern pastimes. Many Japanese are enthusiastic about baseball, and the professional leagues in Japan field several teams. Despite Japan's strong emphasis on education, life is not all work for Japanese young people. They enjoy rock music, modern fashions, television, and movies. Japanese cartoon shows are popular around the world.

Reading Check What two religions are found in Japan?

Baseball

A Japanese fan cheers on his team.

Place What are some other popular pastimes in Japan?

Section 1 Assessment

Defining Terms

1. **Define** tsunami, archipelago, intensive cultivation, clan, shogun, samurai, constitutional monarchy, megalopolis.

Recalling Facts

2. **Location** Why does Japan experience earthquakes?
3. **History** Who were the samurai?
4. **Culture** How have Japan's religions influenced the country's arts?

Critical Thinking

5. **Summarizing Information** Why do the Japanese not want a large military?
6. **Synthesizing Information** What three values of the Japanese people have created good workers?

Graphic Organizer

7. **Organizing Information** Draw a diagram like this one. List Japan's economic successes in the large oval and its economic challenges in each of the smaller ovals.

Applying Geography Skills

8. **Analyzing Maps** Look at the economic activity map on page 705. What mineral resources are found in Japan?

Making Connections

ART | SCIENCE | LITERATURE | TECHNOLOGY

Haiku

Haiku is a type of poetry that first became popular in Japan during the 1600s. A haiku is a three-line poem, usually about nature and human emotions. The traditional haiku requires 17 syllables—5 in the first line, 7 in the second line, and 5 in the third line. All of the haiku below, written by famous Japanese poets, concern the subject of New Year's Day.*

For this New Year's Day,
The sight we gaze upon shall be
Mount Fuji.
Sôkan

That is good, this too is good,—
New Year's Day
In my old age.
Rôyto

New Year's Day;
Whosoever's face we see,
It is care-free.
Shigyoku

New Year's Day:
My hovel,
The same as ever.
Issa

New Year's Day:
What luck! What luck!
A pale blue sky!
Issa

The dawn of New Year's Day;
Yesterday,
How far off!
Ichiku

▲ **This Japanese wood-block print shows two girls playing a New Year's game.**

The first dream of the year;
I kept it a secret,
And smiled to myself.
Shô-u

*The translations may have affected the number of syllables.

Making the Connection

1. How does the poet Shigyoku think most people react to New Year's Day?
2. From his poem, how can you tell that Ichiku sees the New Year as a new beginning?
3. **Making Comparisons** Compare the two poems by Issa. How does his mood change from one to the other?

Section 2

The Two Koreas

Guide to Reading

Main Idea

South Korea and North Korea share the same peninsula and history, but they have very different political and economic systems.

Terms to Know

- dynasty
- monsoon
- anthracite
- famine

Places to Locate

- Korean Peninsula
- North Korea
- South Korea
- Seoul
- Pyongyang

Reading Strategy

Create a time line like this one to record four important dates and their events in Korean history.

NATIONAL GEOGRAPHIC **Exploring Our World**

One of Korea's most sacred places is the shrine at Sokkuram. Built in the A.D. 700s, the shrine has 40 statues, including this 11-foot (3.4-m) statue of the Buddha. The original builders created a complex system of stone passages that let air circulate in the shrine. Today air conditioning keeps the statues in good condition.

The **Korean Peninsula** juts out from northern China, between the Sea of Japan and the Yellow Sea. For centuries, this piece of land held a unified country. Today the peninsula is divided into two nations—Communist **North Korea** and non-Communist **South Korea.** For nearly 50 years after World War II, the two governments were bitter enemies. Since the 1990s, they have been drawing closer together.

A Divided Country

The Koreans trace their ancestry to people who settled on the peninsula thousands of years ago. From the 100s B.C. until the early A.D. 300s, neighboring China ruled Korea. When Chinese control ended, separate Korean kingdoms arose throughout the peninsula.

From A.D. 668 to 935, a single kingdom called Silla (SIH•luh) united much of the peninsula. During this time, Korea made many cultural and scientific advances. For example, Silla rulers built one of the world's earliest astronomical observatories in the A.D. 600s.

Other **dynasties,** or ruling families, followed the Silla. In the 1400s, scholars invented a new way of writing the Korean language. This new

NATIONAL GEOGRAPHIC **On Location**

Korean Border

Nearly 50 years after the fighting stopped in Korea, troops still patrol the border between North and South Korea (above). Seoul, South Korea's modern capital (right), is only about 25 miles (40 km) from the border.

Location **Where was the line of division drawn between the two countries?**

system—called *bangul* (HAHN•GOOL)—used only 28 symbols to write words. This is far fewer than the thousands of characters needed to write Chinese, making the Korean system much easier to learn.

The Korean Peninsula was a stepping stone between Japan and mainland Asia. Trade and ideas went back and forth. In 1910 the Japanese conquered Korea and made it part of their empire. They governed the peninsula until the end of World War II in 1945.

Division and War After World War II, troops from the Communist Soviet Union took over the northern half of Korea. American troops occupied the southern half. Korea eventually divided along the 38th parallel. A Communist state arose in what came to be called North Korea. A non-Communist government controlled South Korea.

In 1950 the armies of North Korea attacked South Korea. They hoped to unite all of Korea under Communist rule. United Nations countries, led by the United States, rushed to support South Korea. China's Communist leaders eventually sent troops across the Yalu River to help North Korea. The Korean War finally ended in 1953—without a peace treaty or a victory for either side. By the 1960s, two separate countries had developed in the Korean Peninsula.

After years of bitterness, the two Koreas in the 1990s developed closer relations. In the year 2000, the leaders of North Korea and South Korea held a meeting for the first time since the division.

✓Reading Check **Why is the Korean Peninsula divided?**

South Korea

South Korea, slightly larger than Indiana, lies at the southern end of the Korean Peninsula. The forested Taebaek Mountains dominate most of central and eastern South Korea. Plains with hills and fertile river valleys spread along the southern and western coasts. Most South Koreans live in these coastal areas.

Monsoons affect South Korea's climate. A **monsoon** is the seasonal wind that blows over Asia for months at a time. During the summer, a monsoon from the south brings hot, humid weather. In the winter, a monsoon blows in from the north, bringing cold, dry weather.

Manufacturing and trade dominate South Korea's economy. High-technology and service industries have grown tremendously. The country is a leading exporter of ships, cars, textiles, computers, and electronic appliances. The map below shows you that South Korea's mineral resources include tungsten, zinc, and anthracite. **Anthracite** is a type of hard coal. In the 1990s, South Korea faced economic difficulties, but it remains one of the economic powers of Asia.

South Korean farmers own their land, although most of their farms are very small. The major crops are rice, barley, onions, potatoes, cabbage, apples, and tangerines. Rice is the country's basic food item. One of the most popular Korean dishes is kimchi, a highly spiced blend of vegetables mixed with chili, garlic, and ginger. Many farmers also raise livestock, especially chickens. Some add to their income by fishing.

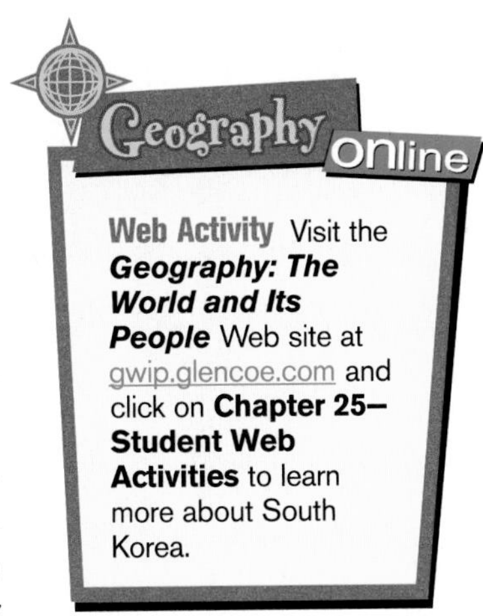

Web Activity Visit the ***Geography: The World and Its People*** Web site at gwip.glencoe.com and click on **Chapter 25—Student Web Activities** to learn more about South Korea.

South Korea's People The people of the two Koreas belong to the same Korean ethnic group. South Korea has about 46.9 million people. More than 80 percent live in cities and towns in the coastal plains. South Korea's capital of **Seoul** is the largest city.

Most city dwellers live in tall apartment buildings. Many own cars, but they also use buses, subways, and trains to travel to and from work.

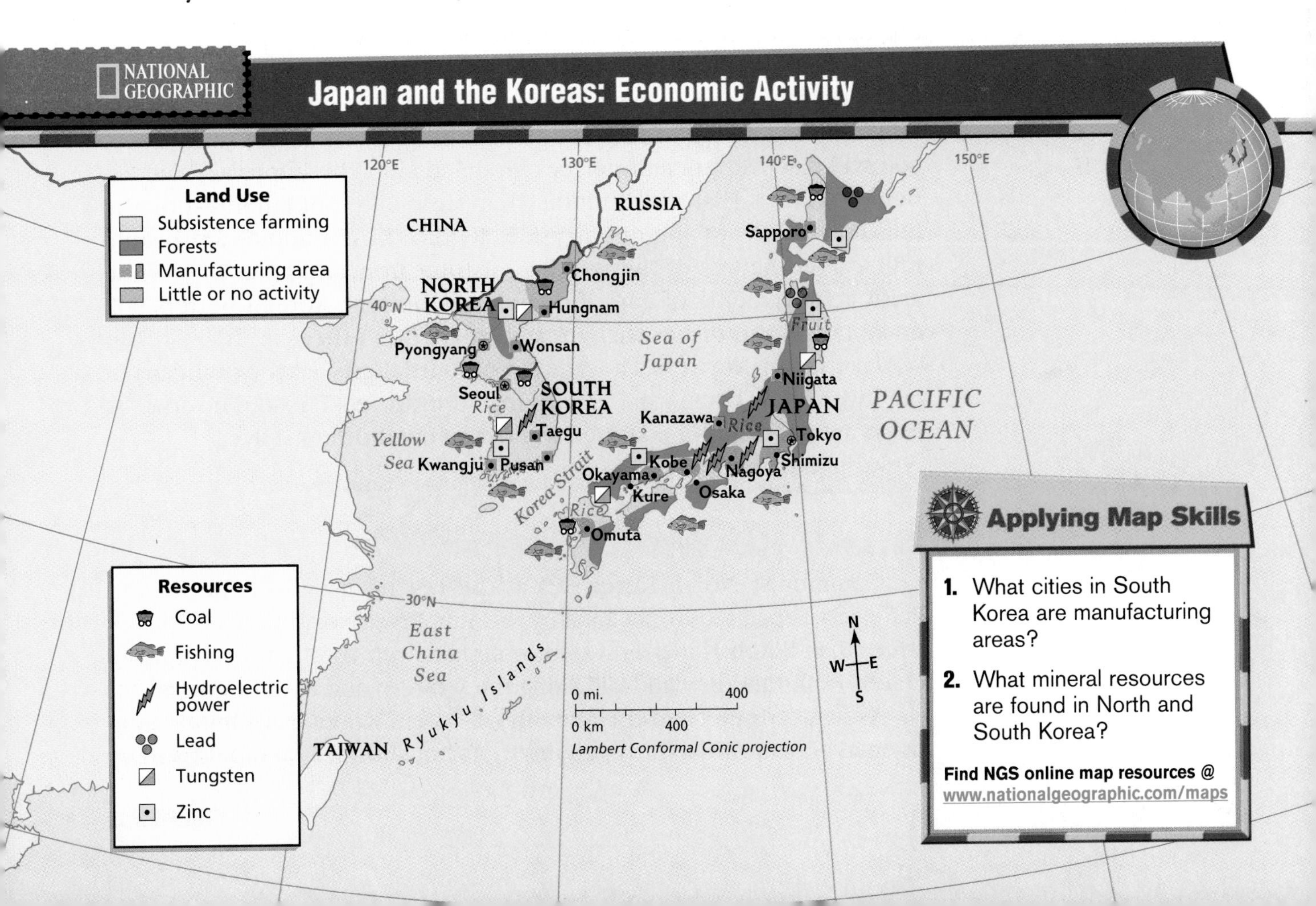

Applying Map Skills

1. What cities in South Korea are manufacturing areas?
2. What mineral resources are found in North and South Korea?

Find NGS online map resources @ www.nationalgeographic.com/maps

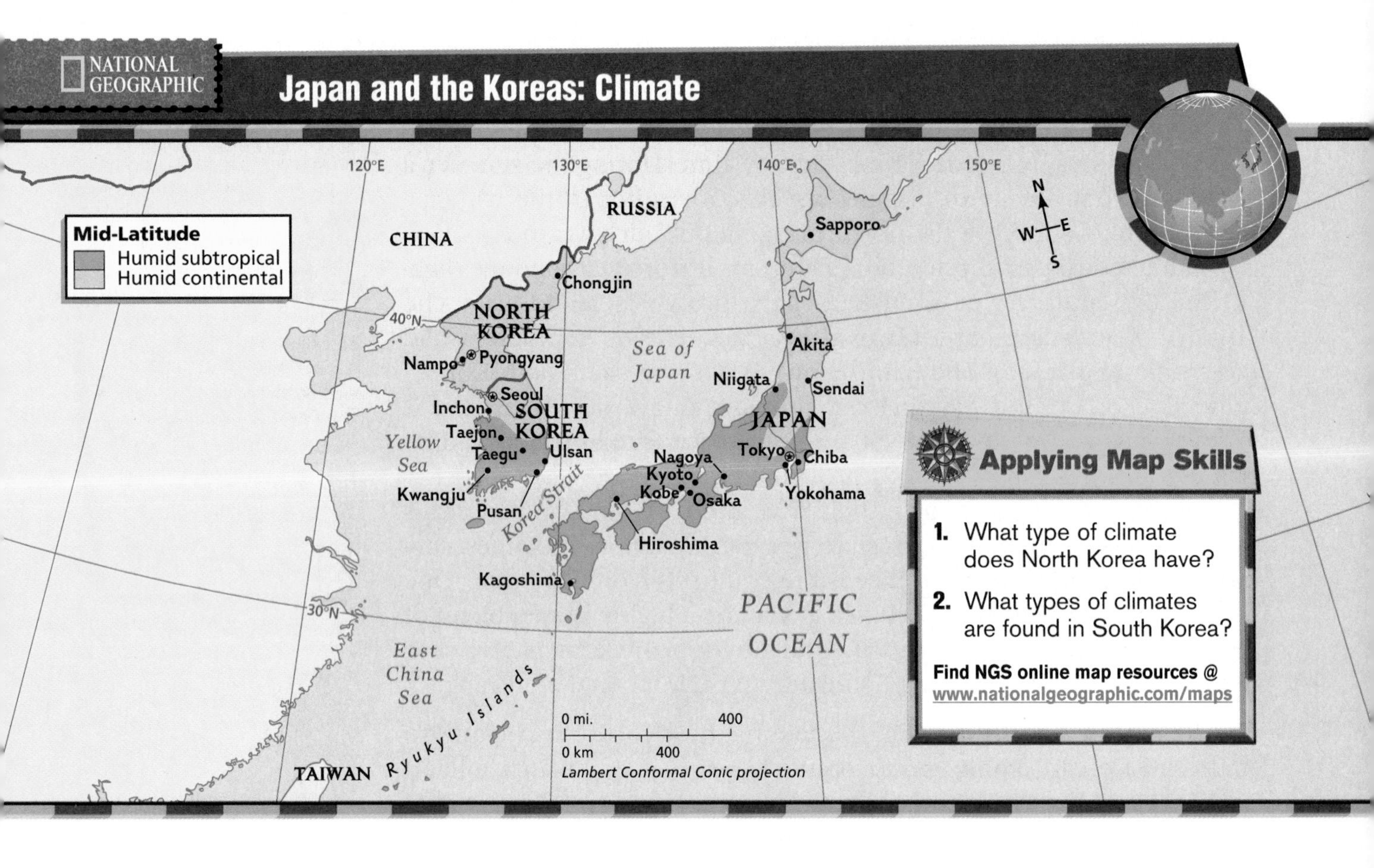

In rural areas, people live in small, one-story homes made of brick or concrete blocks. A large number of South Koreans have emigrated to the United States since the end of the Korean War.

Christianity, Buddhism, and Confucianism are South Korea's major religions. The Koreans have developed their own culture, but Chinese religion and culture influenced the traditional arts of Korea. In Seoul you will discover ancient palaces modeled after the Imperial Palace in Beijing, China. Historic Buddhist temples—like Sokkuram—dot the hills and valleys of the countryside. Within these temples are beautifully carved figures of the Buddha in stone, iron, and gold. One of the great achievements of early Koreans was pottery. Korean potters still make bowls and dishes that are admired around the world.

Like Japan, Korea has a tradition of martial arts. Have you heard of tae kwon do? This martial art originated in Korea. Those who study it learn mental discipline as well as how to defend themselves.

Reading Check **What are the major religions in South Korea?**

North Korea

Communist North Korea lies at the northern end of the Korean Peninsula. Separated from China by the Yalu River, North Korea is slightly larger than South Korea. Forested mountains run through the country's center. Plains and lowlands lie along the western and eastern coasts.

As you can see from the map above, North Korea has a humid continental climate. Most of the country has hot summers and cold, snowy winters. Monsoons affect the climate, but the central mountains block some of the winter monsoon. As a result, the eastern coast generally has warmer winters than the rest of the country.

The North Korean government owns and runs factories and farms. It spends much money on the military. Unlike prosperous South Korea, North Korea is economically poor. Coal and iron ore are plentiful, but industries suffer from old equipment and power shortages.

About 75 percent of North Korea's rugged landscape is forested. This leaves little land to farm, yet more than 40 percent of North Korea's people are farmers who work on large, government-run farms. These farms do not grow enough food to feed the country. A lack of fertilizer recently produced **famines,** or severe food shortages.

North Korea's People North Korea has about 21.4 million people. Nearly 60 percent live in urban areas along the coasts and river valleys. **Pyongyang** is the capital and largest city. Largely rebuilt since the Korean War, Pyongyang has many modern buildings and monuments to Communist leaders. Most of these monuments honor Kim Il Sung, who became North Korea's first ruler in the late 1940s. After Kim's death in 1994, his son Kim Jong Il became the ruler.

The Communist government discourages the practice of religion, although many people still hold to their traditional beliefs. The government also places the needs of the communist system over the needs of individuals and families.

Reading Check **Who controls the economy of North Korea?**

Deep Sea Diver

This Korean woman is dressed in a modern diving suit, but she is carrying on a 1,500-year-old tradition. The women of Cheju Island, off Korea's southern tip, are famous for their diving skills. At age 10, they learn to dive for shellfish and octopuses. They can go 60 feet (18 m) underwater—and hold their breath for two minutes.

Assessment

Defining Terms

1. **Define** dynasty, monsoon, anthracite, famine.

Recalling Facts

2. **Location** Where is the Korean Peninsula?
3. **History** When did Korea become divided?
4. **Economics** What products are made in South Korea?

Critical Thinking

5. **Making Comparisons** How does the standard of living in South Korea differ from that in North Korea?
6. **Summarizing Information** What country has had the greatest influence on the culture and arts of South Korea? Explain.

Graphic Organizer

7. **Organizing Information** Draw a diagram like this one. Write facts about each country under their headings in the outer ovals. Where the ovals overlap, write facts that are common to both countries.

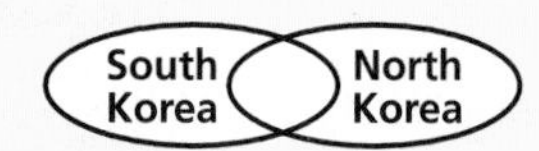

Applying Geography Skills

8. **Analyzing Maps** Turn to the political map on page 696. What is the capital of South Korea? Along what river is it located?

Critical Thinking Skill

Making Comparisons

When you make comparisons, you determine similarities and differences among ideas, objects, or events. By comparing maps and graphs, you can learn more about a region.

Learning the Skill

Follow these steps to make comparisons:

- Identify or decide what will be compared.
- Determine a common area or areas in which comparisons can be drawn.
- Look for similarities and differences within these areas.

Practicing the Skill

Use the map and graph below to make comparisons and answer these questions:

1. What is the title of the map? The graph?
2. How are the map and graph related?
3. Which country has the most exports and imports?
4. Does a country's size have any effect on the amount it exports? Explain.
5. What generalizations can you make about this map and graph?

Applying the Skill

Survey your classmates about an issue in the news. Summarize the opinions and write a paragraph comparing the different opinions.

GO TO Practice key skills with **Glencoe Skillbuilder Interactive Workbook, Level 1.**

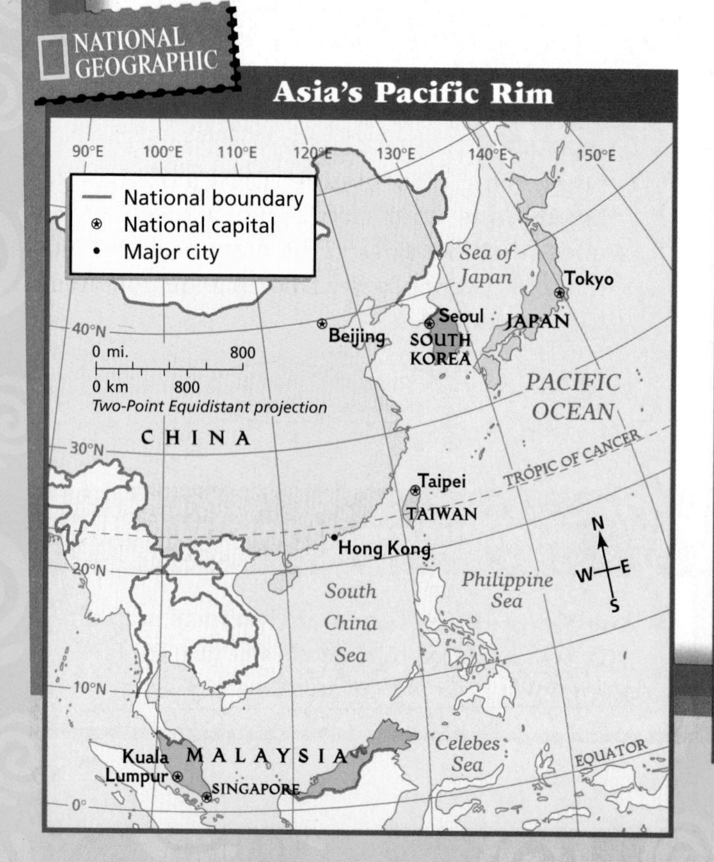

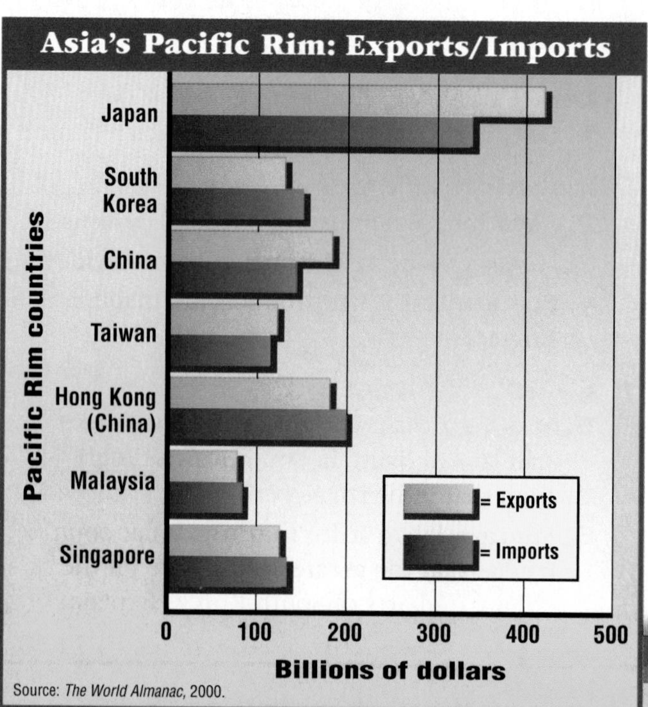

Reading Review

Section 1 Japan

Terms to Know
tsunami
archipelago
intensive cultivation
clan
shogun
samurai
constitutional monarchy
megalopolis

Main Idea

Although they have few mineral resources, Japan's people have built a prosperous country.

✓ **Location** Japan is an archipelago along the Ring of Fire in the western Pacific Ocean. Volcanoes, earthquakes, and tsunamis may strike these islands.

✓ **Place** Japan is mountainous, but its limited farmland is very productive.

✓ **Economics** Japan has few resources. Through trade, the use of advanced technology, and highly skilled workers, Japan has built a strong industrial economy.

✓ **History** The Japanese people have been strongly influenced by China and by Western countries.

✓ **Culture** Most people in Japan live in crowded cities.

✓ **Culture** Japanese religion has encouraged a love of nature and simplicity.

Section 2 The Two Koreas

Terms to Know
dynasty
monsoon
anthracite
famine

Main Idea

South Korea and North Korea share the same peninsula and history, but they have very different political and economic systems.

✓ **Location** The Korean Peninsula lies just south of northern China, and China has had a strong influence on Korean life and culture.

✓ **Government** After World War II, the peninsula became divided into two countries, with a Communist government in North Korea and a non-Communist one in South Korea.

✓ **Economics** South Korea has a strong, industrial economy.

✓ **Culture** Most South Koreans live in cities, enjoying a mix of modern and traditional life.

✓ **Government** North Korea has a Communist government that does not allow its people many freedoms and spends a great deal of money on the military.

Because of its beautiful forest-covered mountains, Korea was once known as Land of the Morning Calm. ▶

Chapter 25

Assessment and Activities

Using Key Terms

Match the terms in Part A with their definitions in Part B.

A.

1. monsoon
2. tsunami
3. intensive cultivation
4. anthracite
5. archipelago
6. dynasty
7. constitutional monarchy
8. clan
9. famine
10. megalopolis

B.

a. group of related families
b. type of hard coal
c. chain of islands
d. emperor is the official head of state, but elected officials run the government
e. seasonal wind that blows over a continent for months at a time
f. huge wave caused by an undersea earthquake
g. severe food shortage
h. huge supercity
i. ruling family
j. growing crops on every available piece of land

Reviewing the Main Ideas

Section 1 Japan

11. **Human/Environment Interaction** How do Japan's farmers achieve high crop yields?
12. **Economics** What consumer goods and industrial goods are made in Japan?
13. **History** How did Japan change in the late 1800s?
14. **Location** What four cities make up Japan's megalopolis?
15. **Culture** What are three of Japan's traditional arts?

Section 2 The Two Koreas

16. **Location** What large Asian nation lies north of the Korean Peninsula?
17. **History** Why did Korea become divided in 1945?
18. **Movement** How do summer and winter monsoons differ in Korea?
19. **Economics** What are the main economic activities in South Korea?
20. **Human/Environment Interaction** Why has North Korea suffered from famine in recent years?

Japan and the Koreas

Place Location Activity

On a separate sheet of paper, match the letters on the map with the numbered places listed below.

1. Mt. Fuji
2. Sea of Japan
3. North Korea
4. South Korea
5. Tokyo
6. Honshu
7. Yalu River
8. Seoul
9. Pyongyang
10. Hokkaido

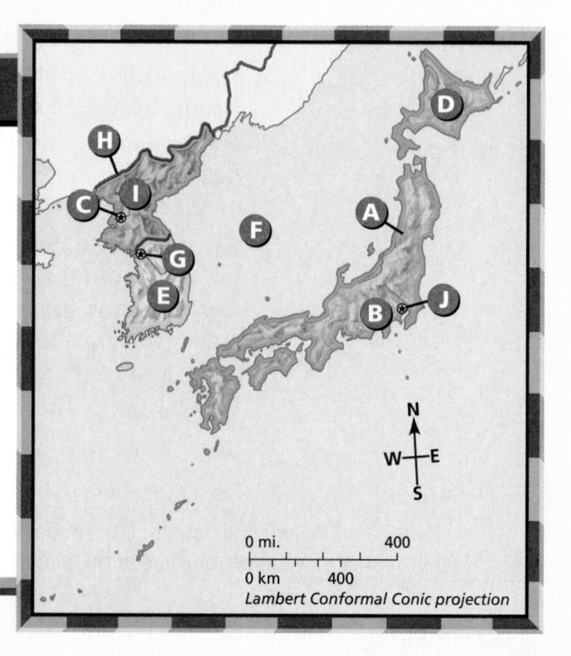

Self-Check Quiz Visit the ***Geography: The World and Its People*** Web site at gwip.glencoe.com and click on **Chapter 25–Self-Check Quizzes** to prepare for the Chapter Test.

Critical Thinking

21. **Drawing Conclusions** Why might North Korea find it difficult to change from a communist system to a non-communist system? Keep in mind the country's location.
22. **Organizing Information** Draw a chart like this one. In each column, write two main ideas about Japan, South Korea, and North Korea as they relate to the topics in the first column.

Topic	Japan	South Korea	North Korea
Land			
Economy			
History			
People			

GeoJournal Activity

23. **Writing a Poem** As you recall, a haiku is a traditional Japanese poem that requires 17 syllables—5 in the first line, 7 in the second line, and 5 in the third line. Write a haiku in which you poetically describe a scene from nature, such as a snowfall or a sunrise.

Mental Mapping Activity

24. **Focusing on the Region** Draw a map of Japan and the Koreas, then add these labels:

- Honshu
- North Korea
- Korean Peninsula
- Pacific Ocean
- Tokyo
- Sea of Japan

Technology Skills Activity

25. **Using the Internet** Use the Internet to research traditional Japanese culture. You might look at Japanese gardens, Buddhism, literature, or painting. Create a bulletin board display with pictures and write captions that explain what the images show.

Standardized Test Practice

Directions: Read the paragraph below, then answer the following questions.

In A.D. 1185 Japan's emperor gave political and military power to a shogun, or general. The shogun system proved to be quite strong. Even though the Mongol warrior Kublai Khan tried twice to invade Japan, he did not succeed. On the first invasion in 1274, Japanese warriors and the threat of a storm forced the Mongols to leave. On the second invasion in 1281, 150,000 Mongol warriors came by ship, but a typhoon arose and destroyed the fleet. The Japanese thought of the storm as the *kamikaze,* or "divine wind." They believed that their islands were indeed sacred.

1. In what century did shoguns gain political power in Japan?

A tenth century
B eleventh century
C twelfth century
D thirteenth century

2. In what century did the Mongol warrior Kublai Khan try to invade Japan?

F tenth century
G eleventh century
H twelfth century
J thirteenth century

Test-Taking Tip: Century names are a common source of error. Remember, in western societies, a baby's first year begins at birth and ends at age one. Therefore, if you are now 14 years old, you are in your *fifteenth* year. Using the same type of thinking, what century began in 1201?

GeoLAB ACTIVITY

Erosion: Saving the Soil

1 Background

Many Asian countries are affected by monsoons, which cause heavy rains or snow and strong winds. This type of weather can cause soil to *erode,* or wash away. When soil is blown or washed away, farmlands become less productive and crops are lost. What techniques do Asian farmers use to slow soil erosion?

2 Materials

- **4 aluminum pie pans**
- **potting soil (enough to fill each pan)**
- **water**
- **500 ml beaker or 2-cup measuring cup**
- **grass clippings or leaves**
- **pebbles (about 2 handfuls)**
- **watering can**
- **dishpan**
- **pencil**

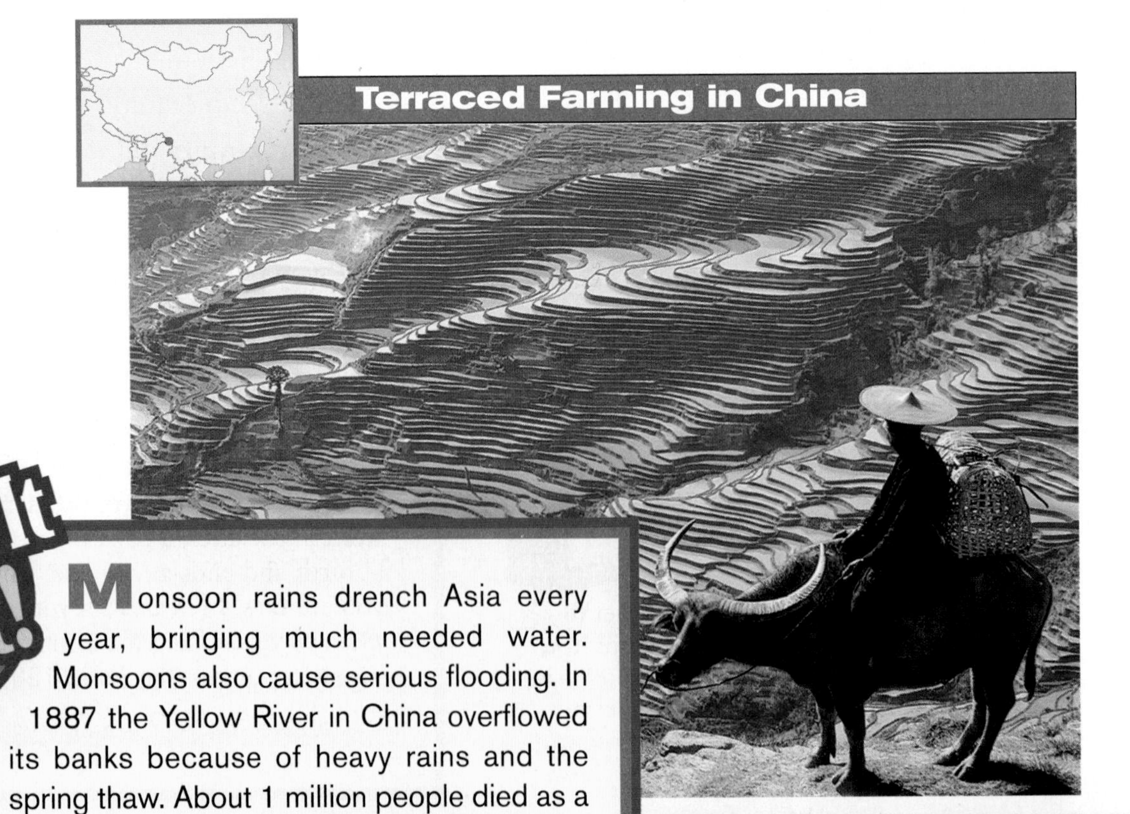

Terraced Farming in China

Believe It or Not!

Monsoon rains drench Asia every year, bringing much needed water. Monsoons also cause serious flooding. In 1887 the Yellow River in China overflowed its banks because of heavy rains and the spring thaw. About 1 million people died as a result—the greatest flood disaster in history.

What to Do

1. Fill each pan almost full of soil. Pat the soil until it is firm and flat. Label each pan: 1, 2, 3, and 4.
2. Pour 100 ml of water over the entire surface of each pan.
3. Prepare the pans for the experiment by doing the following:
   a. Pan 1: Use a pencil to make curved grooves across the surface of the soil. This represents contour plowing.
   b. Pan 2: Make several small walls of pebbles running in the same direction across the pan. This represents terracing.
   c. Pan 3: Lightly bury grass clippings or leaves into the soil surface. Leave some scattered across the top of the soil. This represents crop residue management.
   d. Pan 4: Leave this pan alone.
4. Add 200 ml of water to the watering can.
5. Hold Pan 1 at a small angle, with one end in the dishpan. *Slowly* pour the water from the watering can over the pan. Make sure the grooves are parallel to the dishpan so the water flows across them. Wait until all water stops running across the soil and collects in the dishpan.
6. Pour the water and soil that collects in the dishpan into the beaker. Record how much water is collected. Record your observations about how much soil washed away.
7. Repeat steps 4, 5, and 6 for each pan. Compare the results from each pan.

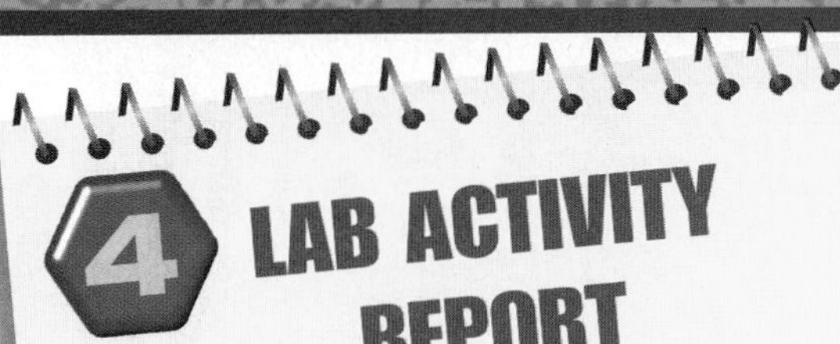

LAB ACTIVITY REPORT

1. Pans 1 and 2 represent two different ways of slowing soil erosion. Describe how these two techniques differ.
2. Which pan had the least soil erosion? Which had the most?
3. **Drawing Conclusions** Will any method used to stop soil erosion be 100 percent effective? Why or why not?

▲ **When pouring water over Pans 1 and 2, be sure that the grooves and pebble walls are parallel to the dishpan.**

Extending the Lab

Activity

Soil erosion can happen anywhere. Locate the steepest hill you can find on or near your school grounds. Make a sketch of the site. After a heavy rain, go back to the site. Do you see any evidence of erosion? Look for evidence such as gullies cut into the hillside, misplaced rocks, or excess soil that has washed to the bottom of the hill.

Chapter 26 Southeast Asia

The World and Its People NATIONAL GEOGRAPHIC

To learn more about the people and places of Southeast Asia, view the ***Geography: The World and Its People*** **Chapter 26** video.

Geography Online

Chapter Overview Visit the ***Geography: The World and Its People*** Web site at gwip.glencoe.com and click on **Chapter 26–Chapter Overviews** to preview information about Southeast Asia.

Section 1

Mainland Southeast Asia

Guide to Reading

Main Idea

The countries of mainland Southeast Asia rely on agriculture.

Terms to Know

- monsoon
- deforestation
- socialism
- delta

Places to Locate

- Malay Peninsula
- Myanmar
- Yangon
- Thailand
- Bangkok
- Laos
- Mekong River
- Vientiane
- Cambodia
- Phnom Penh
- Vietnam
- Hanoi

Reading Strategy

Make five charts like this one. Write a key fact about Myanmar, Thailand, Laos, Cambodia, and Vietnam in the right column for each topic listed in the left column.

Topic	Key Fact
Land	
Economy	
People	

NATIONAL GEOGRAPHIC **Exploring Our World**

Tattoos and high-heeled shoes in the United States are no match for the fashion statements found in Southeast Asia. This woman belongs to the Padaung ethnic group found in Myanmar and Thailand. A series of brass rings covers her neck. The rings do not stretch the woman's neck but actually push down her collarbone and ribs.

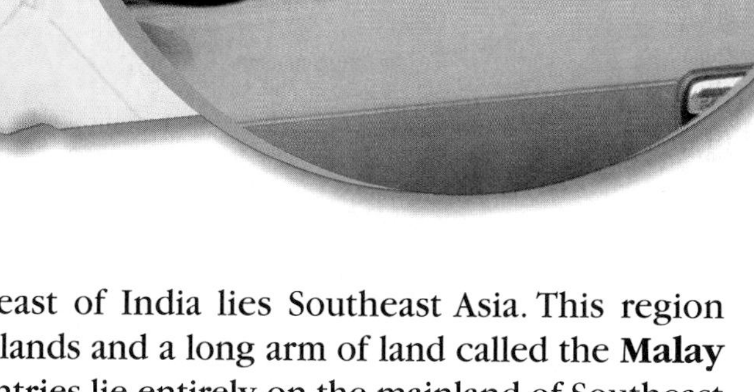

South of China and east of India lies Southeast Asia. This region includes thousands of islands and a long arm of land called the **Malay Peninsula.** Several countries lie entirely on the mainland of Southeast Asia. They are Myanmar, Thailand, Laos, Cambodia, and Vietnam.

Myanmar

Myanmar, once called Burma, is about the size of Texas. Rugged, steep mountains sweep through its western and eastern borders. Two wide rivers—the Irrawaddy (IHR•ah•WAH•dee) and the Salween—flow through vast lowland plains between these mountain ranges.

Myanmar has tropical and subtropical climates influenced by **monsoons,** or seasonal winds that blow over a continent for months at a time. Wet monsoons from the Indian Ocean bring heavy rains from June to September. During the dry monsoons of the winter, winds blowing from the north bring hardly any rain at all.

◀ Petronas Towers in Kuala Lumpur, Malaysia

About two-thirds of the country's people farm. The main crops are rice, sugarcane, beans, and nuts. Some farmers work their fields with tractors, but most rely on plows drawn by water buffalo.

Factories produce and export such goods as soap, noodles, paper, textiles, and glass bottles. Myanmar also exports precious gems like rubies, sapphires, and jade. In addition, the country provides about 75 percent of the world's teakwood. Myanmar's valuable forests are decreasing because of **deforestation,** or the widespread cutting of trees.

About 75 percent of Myanmar's 48.1 million people live in rural areas. The most densely populated part of the country is the fertile Irrawaddy River valley. Many rural dwellers build their homes on poles above the ground for protection from floods and wild animals.

The capital and largest city, **Yangon** (formerly called Rangoon), is famous both for its modern university and its gold-covered Buddhist temples. Buddhism is the main religion in Myanmar. Most people are of Burman heritage, and Burmese is the main language.

Myanmar was part of British India for many years. It became an independent republic in 1948. Since then, military leaders have turned

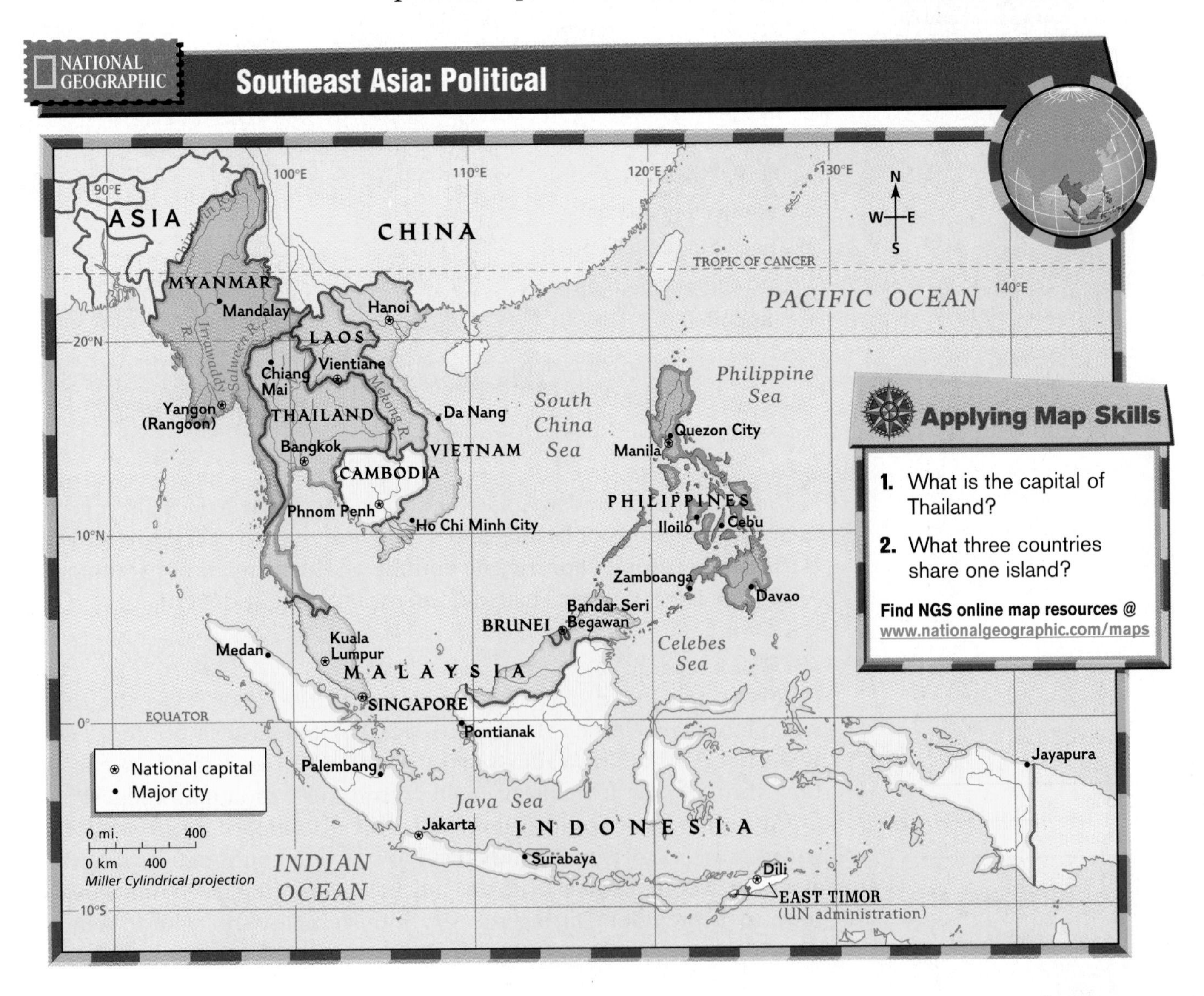

Myanmar into a socialist country. **Socialism** is an economic system in which most businesses are owned and run by the government. Some people have struggled to build a democracy in Myanmar. A woman named Aung San Suu Kyi (AWNG SAN SOO CHEE) has become a leader in this struggle. In 1991 she was awarded the Nobel Peace Prize for her efforts to bring political changes without violence.

✓ Reading Check **Where is Myanmar's most densely populated area?**

Thailand

The map below shows you that **Thailand** looks like a flower on a stem. The "flower" is the northern part, located on the mainland. The "stem" is a narrow strip on the Malay Peninsula. The country's main waterway—the Chao Phraya (chow PRY•uh) River—flows through a central plain. Thailand has wet summer monsoons and dry winter monsoons.

Once called Siam, *Thailand* means "land of the free." It is the only Southeast Asian country that has never been a European colony. The

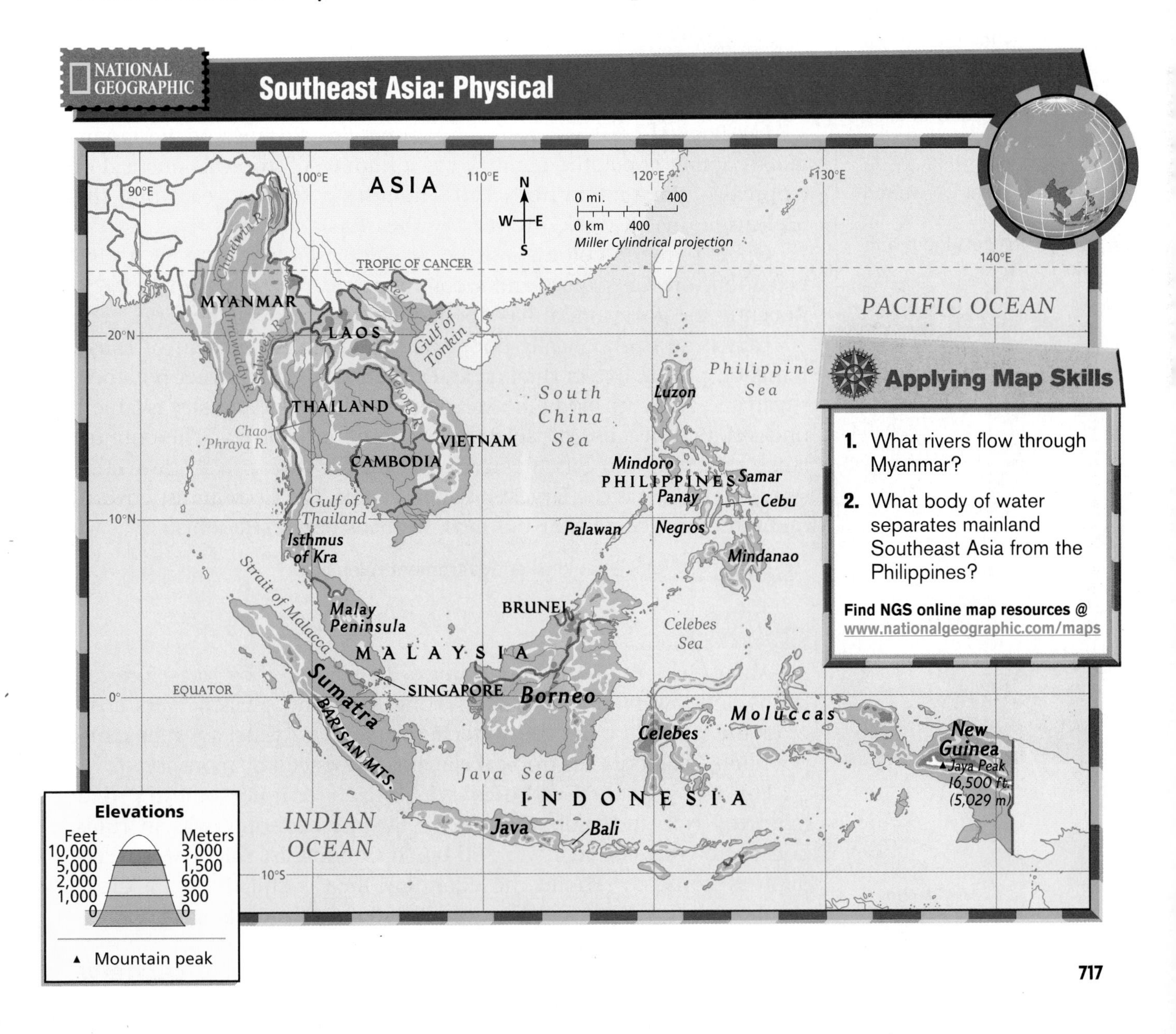

Life as a Monk

After his grandfather died, Nattawud Daoruang became a novice Buddhist monk. "You see," he says, "Thai Buddhists believe they can get to paradise by holding onto a monk's robe. So I became a monk for a month to help my grandfather get to paradise. My family and the monks shaved my hair and eyebrows, and I had to recite the monastery rules in Pali—the language of the ancient scriptures. The novice monks had to get up at 5:00 A.M. and meditate. After that, we had free time, so we read comics and played games on the monks' Play Station™. In the afternoons, we walked around the village with the monks to get food and drink."

Thai people trace their independence as a kingdom back to the A.D. 1200s. Thailand still has a king or queen and honors its royal family.

An agricultural country, Thailand's farmers grow rice, corn, fruits, and cassava as their major crops. Large land holdings in the south provide rubber for export. Teak and other woods come from Thailand's forests. The government has taken steps to limit deforestation.

Thailand is also rich in mineral resources. It is one of the world's leading exporters of tin and gemstones. Most manufacturing is located near **Bangkok,** the capital. Workers make cement, textiles, clothing, and metal products. Tourism is an important industry as well.

Most of Thailand's 61.8 million people belong to the Thai ethnic group and practice Buddhism. Hundreds of Buddhist temples called *wats* dot the cities and countryside. Buddhist monks, or holy men, carry small bowls and walk among the people to receive food offerings.

About 80 percent of Thais live in rural villages, although thousands look for jobs in Bangkok. Here, beautiful temples and royal palaces stand next to modern skyscrapers and crowded streets. Bangkok has so many cars that daily traffic jams last for hours.

✓Reading Check **What are Thailand's major crops?**

Laos

Landlocked **Laos** is covered by mountains. Because of the mountains' cooling effect, the country has a humid subtropical—instead of tropical—climate. Southern Laos includes a fertile area along the **Mekong** (MAY•KAWNG) **River,** Southeast Asia's longest river.

Once a French colony, Laos became independent in 1953. A civil war soon tore Laos apart, and a communist state was set up in 1975. Recently, the government has opened Laos to the outside world.

Laos is an economically poor country. About 80 percent of Laos's 5 million people live in rural areas. Farmers grow rice, sweet potatoes, sugarcane, and corn along the Mekong's fertile banks. Industry is largely undeveloped because of isolation and years of civil war. The country lacks railroads and has electricity in only a few cities. **Vientiane** (vyehn•TYAHN) is the largest city and capital. The Communist government discourages religion, but most Laotians remain Buddhists.

✓Reading Check **What kind of government rules Laos?**

Cambodia

About the size of Missouri, **Cambodia** is covered by lakes, forested plains, and low mountains. The Mekong River and smaller rivers crisscross the land. They provide water, fertile soil, and waterways for transportation. Cambodia has tropical climates influenced by monsoons.

For many years Cambodia was a rich farming country that exported rice and rubber. By the 1980s, its economy was in ruins because of years of civil war and harsh Communist rule. The government is trying to rebuild the economy. Rice is still the main crop. Cambodia's few factories produce food items, chemicals, and textiles.

Most of Cambodia's 11.9 million people belong to the Khmer (kuh•MEHR) ethnic group. About 80 percent live in rural villages. The rest live in cities such as the capital, **Phnom Penh** (puh•NAWM PEHN). Buddhism is Cambodia's main religion. About 1,000 years ago, Cambodia was the center of the vast Khmer Empire, during which Hinduism was also practiced. Among the achievements of the Khmer were the magnificent temples of Angkor.

In modern times, Cambodia was under French rule, finally becoming independent in 1953. Since the 1960s, it has experienced almost constant warfare among rival political groups. A Communist government took control in the mid-1970s, and the people suffered great hardships. Many people from the cities were forced to move to rural areas and work as farmers. More than 1 million Cambodians died. Some fled to other countries as refugees. In 1993 Cambodia brought back its king, but rivalry among political groups continues.

Reading Check **Why is Cambodia's economy in ruins?**

Vietnam

About the size of New Mexico, **Vietnam's** long eastern coastline borders the Gulf of Tonkin, the South China Sea, and the Gulf of Thailand. In the north lies the fertile delta of the Red River. A **delta** is an area of land formed by soil deposits at the mouth of a river. In the south you find the wide, swampy delta of the Mekong River. Monsoons bring wet and dry seasons.

Farmers grow large amounts of rice, sugarcane, cassava, sweet potatoes, corn, bananas, and coffee in river deltas. Vietnam's mountain forests provide wood, and the South China Sea yields large catches of fish.

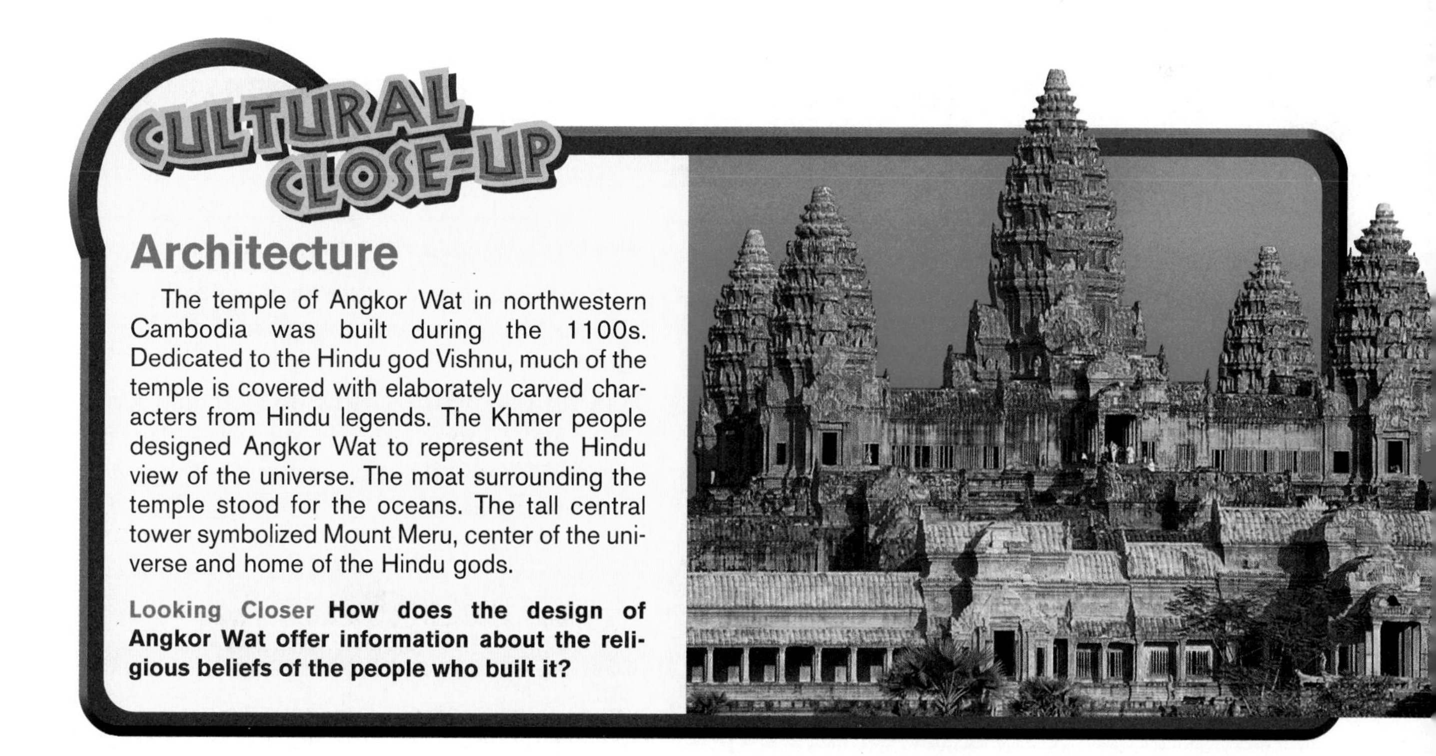

Architecture

The temple of Angkor Wat in northwestern Cambodia was built during the 1100s. Dedicated to the Hindu god Vishnu, much of the temple is covered with elaborately carved characters from Hindu legends. The Khmer people designed Angkor Wat to represent the Hindu view of the universe. The moat surrounding the temple stood for the oceans. The tall central tower symbolized Mount Meru, center of the universe and home of the Hindu gods.

Looking Closer **How does the design of Angkor Wat offer information about the religious beliefs of the people who built it?**

Northern and central Vietnam are rich in coal, natural gas, and other mineral resources. Most factories lie in the south, which is a bustling commercial area. Cement, fertilizer, steel, clothing, and bicycles are manufactured. Years of warfare and strict government controls have kept Vietnam's industries from fully developing.

With 79.5 million people, Vietnam has the largest population in mainland Southeast Asia. About 80 percent live in the countryside. The largest urban area is Ho Chi Minh (HOH CHEE MIHN) City, named for the country's first Communist leader. It used to be called Saigon (sy•GAHN). Vietnam's capital, **Hanoi,** is located in the north. Most people are Buddhists and belong to the Vietnamese ethnic group. The rest are Chinese, Cambodians, and other Asian ethnic groups. Vietnamese is the major language, but Chinese and English are also spoken.

The ancestors of Vietnam's people came from China more than 2,000 years ago. From the late 1800s to the mid-1950s, Vietnam was under French rule. Vietnamese Communists drove out the French in 1954. The Communist government controlled northern Vietnam, while an American-supported government ruled the south. In the 1960s, fighting between these two groups led to the Vietnam War. During this 10-year conflict, more than 3 million Americans helped fight against the Communists. The war was costly, and the United States eventually withdrew its forces in 1973. Within two years, the Communists had captured the south. Many thousands of people fled Vietnam, settling in the United States and other countries.

In recent years, Vietnam's Communist leaders have opened the country to Western ideas, businesses, and tourists. They also have loosened government controls on the economy. In these two ways, the Communist leaders hope to raise Vietnam's standard of living.

✓Reading Check Where are most of Vietnam's factories located?

Assessment

Defining Terms

1. Define monsoon, deforestation, socialism, delta.

Recalling Facts

2. Economics What does Myanmar export?

3. Region How do monsoons affect climate in mainland Southeast Asia?

4. Economics What has slowed the economies of Laos and Cambodia?

Graphic Organizer

5. Organizing Information Draw a time line like this one. Then list four events and their dates in Vietnam's history.

Critical Thinking

6. Summarizing Information What makes Thailand unique among the countries of Southeast Asia?

7. Making Predictions How do Vietnam's leaders hope to improve the country's standard of living? Do you think these actions will help? Why or why not?

Applying Geography Skills

8. Analyzing Maps Look at the economic activity map on page 727. What resources are found in Thailand?

Island Southeast Asia

Guide to Reading

Main Idea

The island countries of Southeast Asia have a variety of cultures and economic activities.

Terms to Know

- strait
- free port
- terraced field

Places to Locate

- Malaysia
- Strait of Malacca
- Kuala Lumpur
- Singapore
- Brunei
- Philippines
- Manila

Reading Strategy

Draw a chart like this one. As you read, list two facts about each country in the right column.

Country	Facts
Malaysia	
Singapore	
Brunei	
Philippines	

NATIONAL GEOGRAPHIC Exploring Our World

Yes, you are looking at 50,000 rubber ducks as they float down the Singapore River. Singapore, the largest port in Southeast Asia, has been described as one of the cleanest and safest cities in the world. The Great Duck Race is one way that Singapore's Red Cross Society raises money for the community.

The island countries of Southeast Asia are Malaysia, Singapore, Brunei (bru•NY), and the Philippines. Indonesia also lies in this region, but it is discussed separately in Section 3.

Malaysia

Malaysia has three parts. An area known as Malaya occupies the southern end of the Malay Peninsula. The two other parts lie on the island of Borneo. They are the territories of Sarawak (suh•RAH•wahk) and Sabah (SAH•bah). Dense rain forests and rugged mountains make up their landscape. All three parts have a tropical rain forest climate.

West of Malaya lies the **Strait of Malacca.** A strait is a narrow body of water between two pieces of land. The Strait of Malacca is an important waterway for trade between the Indian Ocean and the Java Sea.

Agriculture is an important economic activity in Malaysia, but only about one-sixth of the people farm. Some work on subsistence farms and grow rice, fruits, and vegetables. Many others work on plantations. Malaysia is one of the world's leaders in exporting rubber and palm oil. Large amounts of wood are also exported.

Malaysia is rich in minerals such as tin, iron ore, copper, and bauxite. Large amounts of oil and natural gas are found in the Sabah and Sarawak regions. Factory workers make high-technology and consumer goods. Malaysia's ports are important centers of trade.

Most of Malaysia's 22.7 million people belong to the Malay ethnic group. Their ancestors came from southern China about 4,000 years ago. In the 1800s, the British—who then ruled Malaysia—brought in Chinese and South Asian workers to mine tin and to work on rubber plantations. As a result, in marketplaces today you can hear Malay, Chinese, Tamil, and English spoken. Most Malaysians are Muslims, but there are large numbers of Hindus, Buddhists, and Christians.

In 1963 Malaysia became independent and formed a constitutional monarchy. Every five years, a council of local rulers chooses a king to serve as head of state. An elected prime minister and parliament actually run the country. **Kuala Lumpur** (KWAH•luh LUM•PUR) is the capital and largest city. The Petronas Towers—the tallest buildings in the world—soar above this city. In contrast, many rural villagers live in thatched-roof homes built on posts a few feet off the ground.

Reading Check **Where are the three parts of Malaysia located?**

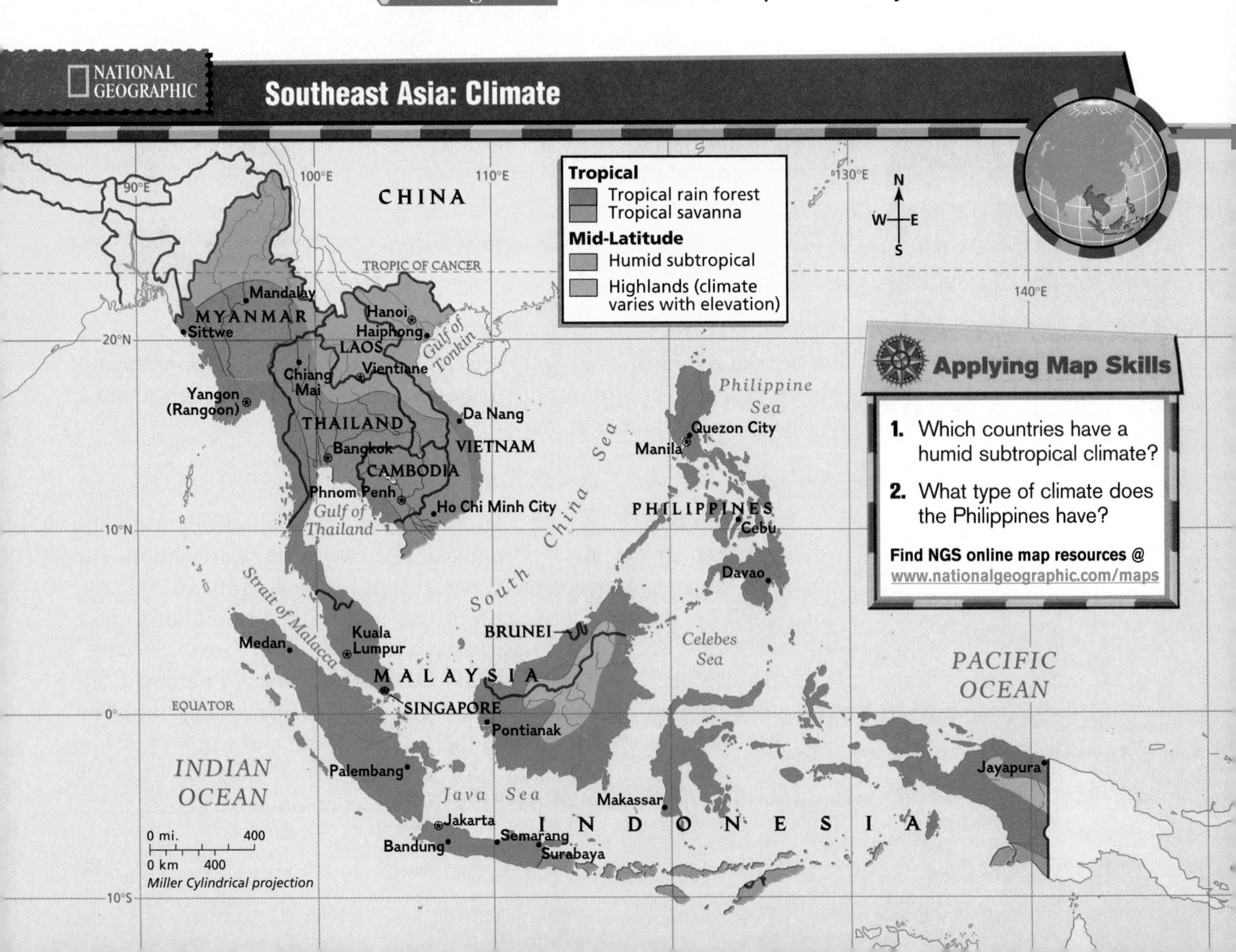

Singapore

Singapore lies off the southern tip of the Malay Peninsula. It is made up of Singapore Island and 58 smaller islands. With only 239 square miles (618 sq. km), Singapore is one of the world's smallest countries. Yet it has one of the world's most productive economies.

The city of Singapore is the capital and takes up much of Singapore Island. Once covered by rain forests, Singapore Island now has highways, factories, office buildings, and docks.

Singapore's location along a narrow sea route has resulted in an economy based mainly on trade and manufacturing. The city of Singapore has one of the world's busiest harbors. It is a **free port,** a place where goods can be loaded or unloaded, stored, and shipped again without payment of import taxes. Huge amounts of goods pass through this port. Singapore's many factories make high-tech goods, machinery, chemicals, and paper products. Because of their productive economy, the people of Singapore enjoy a high standard of living.

Founded by the British in the early 1800s, Singapore became an independent republic in 1965. Most of the country's nearly 4 million people are Chinese, but Malaysians and Indians make up about 25 percent of the population. These people practice Buddhism, Islam, Christianity, Hinduism, and traditional Chinese religions.

✓Reading Check What is a free port?

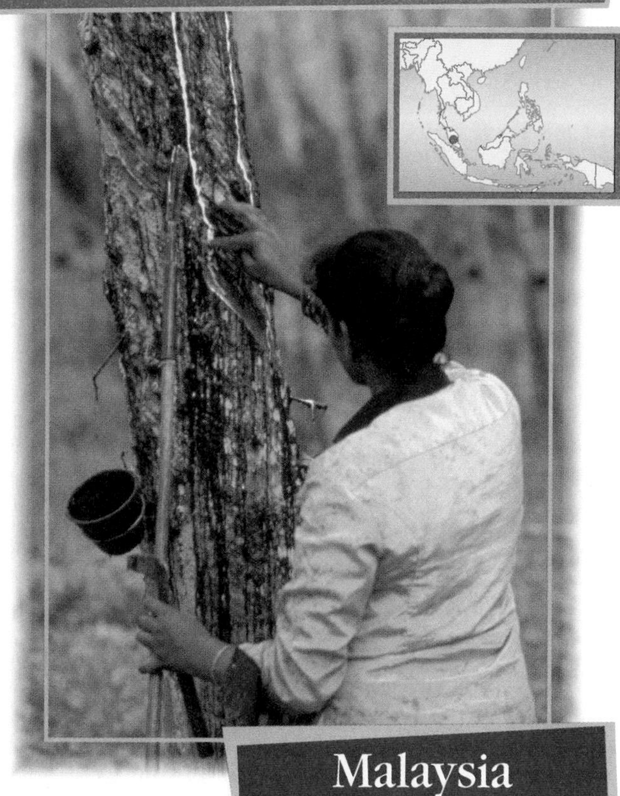

Malaysia

A Malaysian worker taps a rubber tree to get the milky liquid called latex.

Human/Environment Interaction What other Malaysian products are grown for export?

Brunei

On the northern coast of Borneo lies another small nation—**Brunei.** About the size of Delaware, Brunei is made up of two wedges of land that are separated by Malaysian territory. Low plains, coastal swamps, and mountains dominate the country.

Brunei has rich deposits of oil and natural gas. The export of these resources provides about half of the country's income. Brunei's citizens receive free education and medical care, and low-cost housing, fuel, and food. Today the government is investing in new industries to avoid too much reliance on fuels. All political and economic decisions are made by Brunei's ruler, or sultan, who governs with a firm hand.

Most of Brunei's 300,000 people are Malays or Chinese. Malay, Chinese, and English are the most commonly spoken languages. Islam is the leading religion. The largest urban area is the capital, Bandar Seri Begawan (BAN•dar SEHR•ee beh•GAH•wan).

✓Reading Check Who makes economic and political decisions in Brunei?

The Philippines

East of Vietnam in the South China Sea lie the **Philippines,** an archipelago of more than 7,000 islands. About 90 percent of the 74.7 million Filipino people live on only 11 of the islands. The largest islands are Luzon (loo•ZAWN) and Mindanao (MIHN•duh•NAH•OH).

Volcanic mountains and forests dominate the landscape. Lava from volcanoes provides fertile soil for agriculture. Farmers have built terraces on the steep slopes of the mountains. **Terraced fields** are strips of land cut out of a hillside like stair steps. The land can then hold water and be used for farming. About 40 percent of the people farm. They grow sugarcane, coconuts, and abaca—a strong fiber obtained from banana leaves.

Miners unearth coal, nickel, copper, silver, and gold. Factory workers produce high-tech goods, food products, chemicals, clothing, and shoes. **Manila,** the country's capital, is a great commercial center.

Named after King Philip II of Spain, the Philippines spent more than 300 years as a Spanish colony. As a result of the Spanish-American War, the United States controlled the islands from 1898 until World War II. In 1946 the Philippines became an independent, democratic republic.

The Philippines is the only Christian country in Southeast Asia. About 90 percent of Filipinos follow the Roman Catholic religion, brought to the islands by Spanish missionaries. The culture today blends Malay, Spanish, and American influences. American influences can be seen in the products sold along busy city streets.

Reading Check What are the two largest islands in the Philippines?

Assessment

Defining Terms

1. **Define** strait, free port, terraced field.

Recalling Facts

2. **Place** Malaysia is among the world's leading producers of what two products?
3. **Economics** On what two economic activities is Singapore's economy based?
4. **Culture** What religion do most Filipinos practice?

Critical Thinking

5. **Summarizing Information** How does Brunei's government use its fuel income?
6. **Interpreting Information** What groups influenced the culture of the Philippines?

Graphic Organizer

7. **Organizing Information** Draw a diagram like this one. In the center, list similarities of the countries listed. In the outer ovals, write two ways that the country differs from the others.

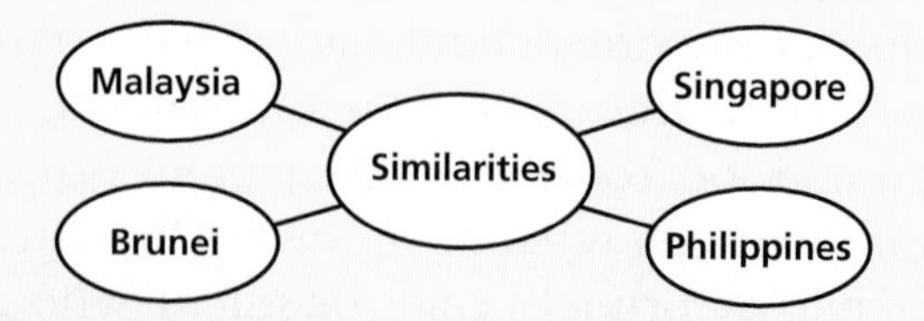

Applying Geography Skills

8. **Analyzing Maps** Look at the population density map on page 728. What cities in the Philippines have more than 1 million people?

Geography Skill

Reading a Contour Map

A trail map would show the paths you could follow if you went hiking in the mountains. How would you know if the trail follows an easy, flat route, though, or if it cuts steeply up a mountain? To find out, you need a **contour map.**

Learning the Skill

Contour maps use lines to outline the shape—or contour—of the landscape. Each contour line connects all points that are at the same elevation. This means that if you walked along one contour line, you would always be at the same height above sea level.

Where the contour lines are far apart, the land rises gradually. Where the lines are close together, the land rises steeply. For example, one contour line may be labeled 1,000 meters (3,281 ft.). Another contour line very close to the first one may be labeled 2,000 meters (6,562 ft.). This means that the land rises 1,000 meters (3,281 ft.) in just a short distance.

To read a contour map, follow these steps:

- Identify the area shown on the map.
- Read the numbers on the contour lines to determine how much the elevation increases or decreases with each line.
- Locate the highest and lowest numbers, which indicate the highest and lowest elevations.
- Notice the amount of space between the lines, which tells you whether the land is steep or flat.

Practicing the Skill

Study the contour map below, then answer the following questions.

1. What area is shown on the map?
2. What is the lowest elevation on the map?
3. What is the highest elevation on the map?
4. Where is the landscape the most flat? How can you tell?
5. How would you describe the physical geography of this island?

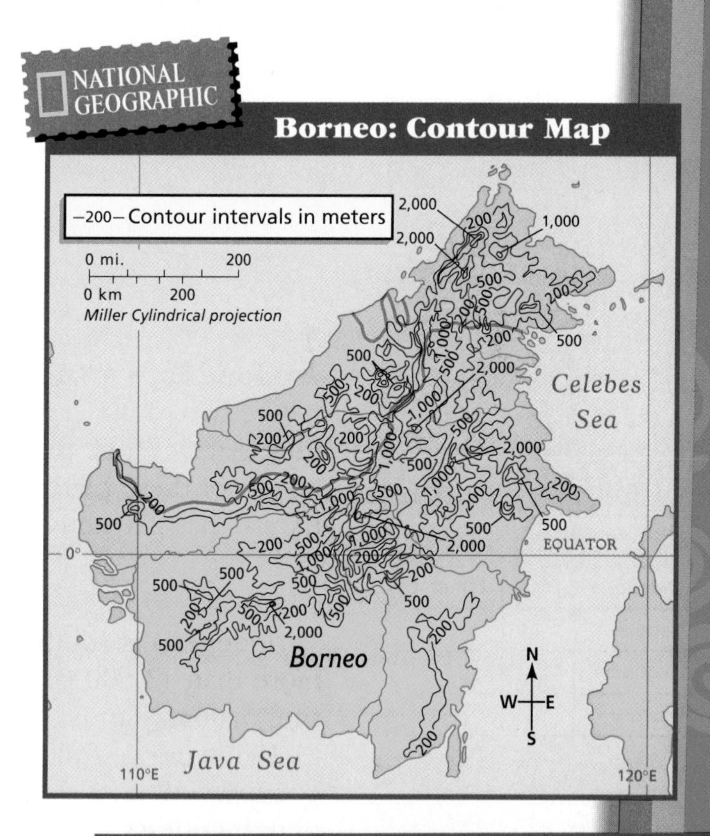

Applying the Skill

Turn to page 12 in the **Geography Handbook.** Use the contour map there to answer the five questions above.

Section 3

Indonesia

Guide to Reading

Main Idea

Indonesia has a diverse population within its many islands.

Terms to Know

- plate
- civil war
- dictatorship

Places to Locate

- Indonesia
- Sumatra
- Java
- Celebes
- Borneo
- New Guinea
- Jakarta
- Bali
- East Timor

Reading Strategy

Draw a chart like this one. Write a fact about Indonesia's land, economy, and history. Then write an effect of that fact.

Indonesia	Fact	Effect
Land		
Economy		
History		

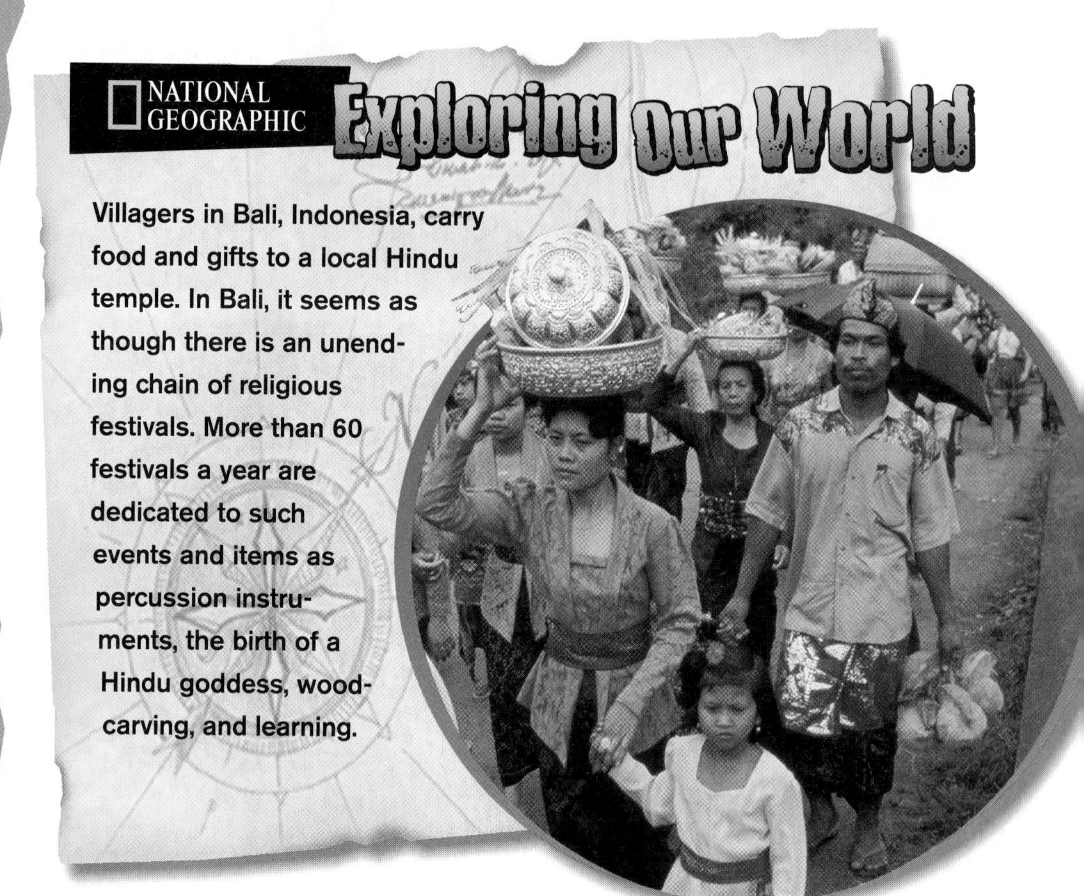

Villagers in Bali, Indonesia, carry food and gifts to a local Hindu temple. In Bali, it seems as though there is an unending chain of religious festivals. More than 60 festivals a year are dedicated to such events and items as percussion instruments, the birth of a Hindu goddess, woodcarving, and learning.

Indonesia is a large, mineral-rich country. It sprawls over an area where two of the earth's tectonic plates meet. Tectonic **plates** are the huge slabs of rock that make up the earth's crust. Indonesia's location on top of these plates causes it to experience earthquakes. The country also has 220 active volcanoes.

Indonesia's Land and Climate

Indonesia is Southeast Asia's largest country. It is an archipelago of more than 13,600 islands. The physical map on page 717 shows you the major islands of Indonesia—**Sumatra, Java,** and **Celebes** (SEH•luh•BEEZ). Indonesia also shares two large islands with other countries. Most of the island of **Borneo** belongs to Indonesia. In addition, Indonesia controls the western half of **New Guinea.** Another country—Papua New Guinea—lies on the eastern half.

Mountains rise on Indonesia's larger islands. Their rugged peaks slope down to meet coastal lowlands. The volcanoes that formed Indonesia have left a rich covering of ash that makes the soil good for farming. Because Indonesia lies on the Equator, its climate is tropical. Monsoons bring a wet season and a dry season. The tropical climate, combined with fertile soil, has allowed dense rain forests to spread.

Reading Check How have volcanoes helped farming in Indonesia?

Indonesia's Economy

Indonesia has one of the world's largest economies in regard to the amount of goods and services it produces. Agriculture provides work for nearly half of the people. Farmers grow rice, coffee, cassava, tea, coconuts, and rubber trees. Cattle and sheep are also raised.

Indonesia has large reserves of oil and natural gas. Its mines yield tin, silver, nickel, copper, bauxite, and gold. Indonesia is one of the world's leading producers of tin. Dense rain forests provide teak and other valuable woods. Some companies that own large tracts of land are cutting down the trees very quickly. The environment suffers from this deforestation. Tree roots help keep the soil in place during heavy rains. When the trees are cut down, the rich soil runs off into the sea.

Geography Online
Web Activity Visit the *Geography: The World and Its People* Web site at gwip.glencoe.com and click on **Chapter 26—Student Web Activities** to learn more about Indonesia.

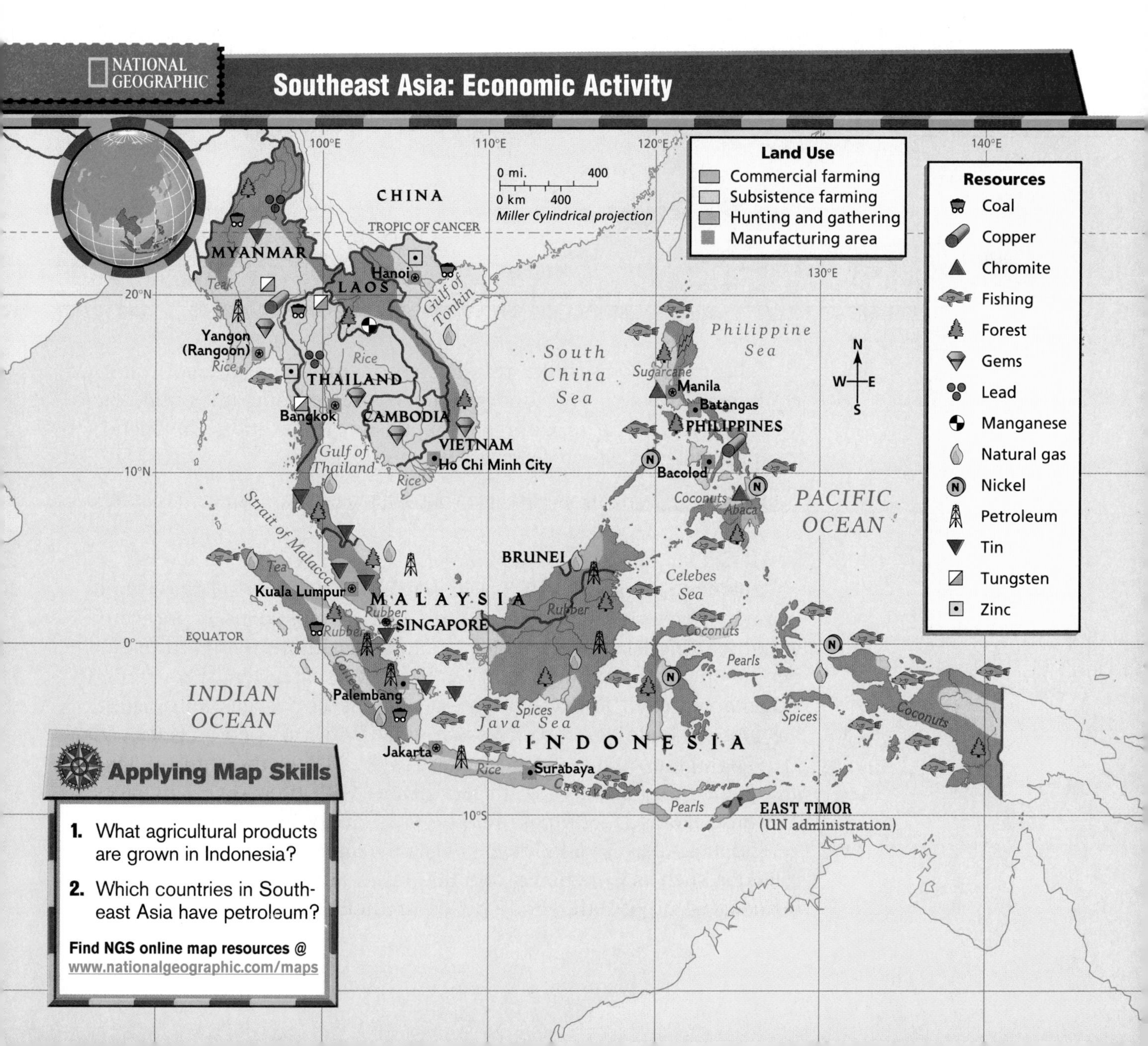

Applying Map Skills

1. What agricultural products are grown in Indonesia?
2. Which countries in Southeast Asia have petroleum?

Find NGS online map resources @ www.nationalgeographic.com/maps

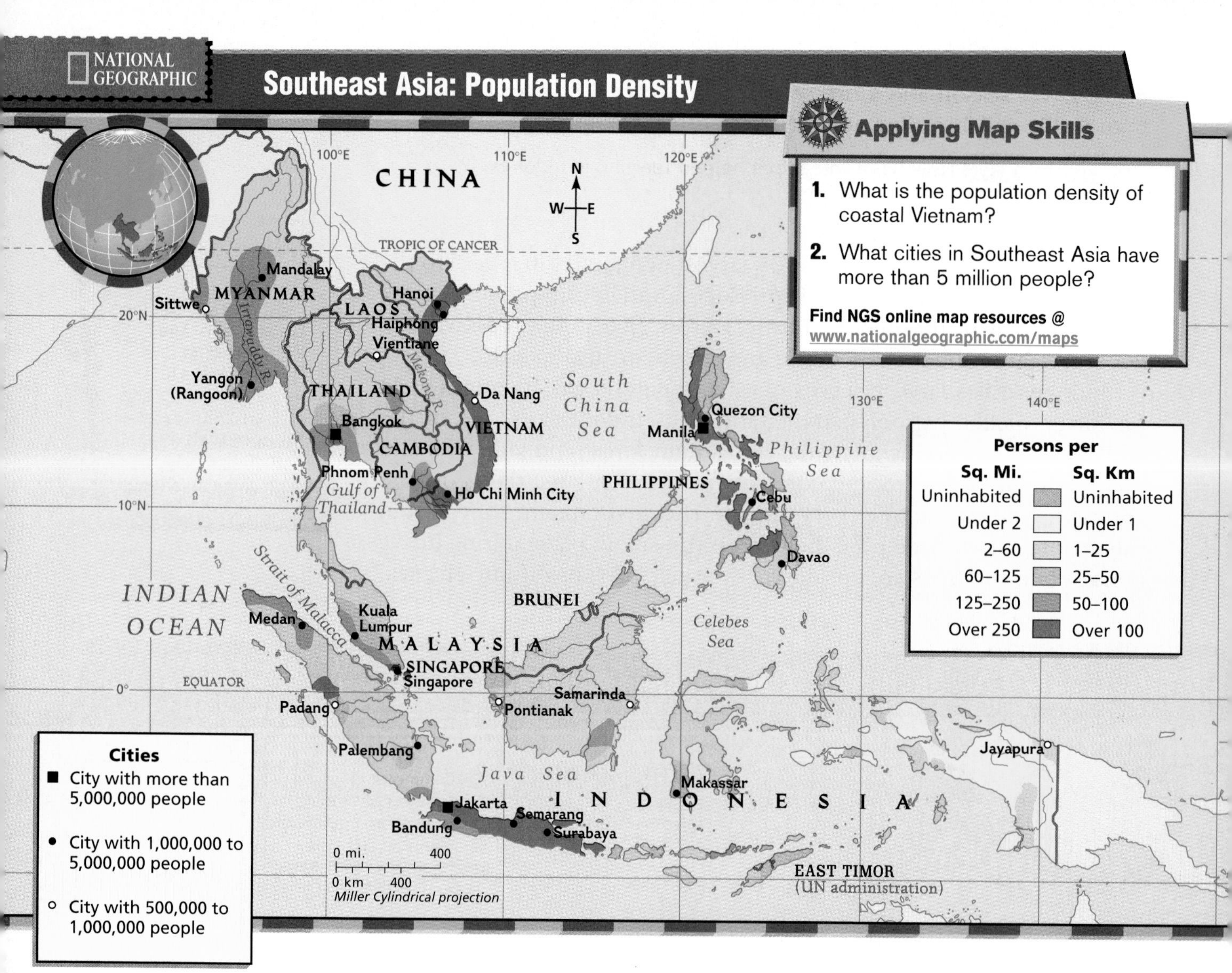

Factory workers make textiles, wood products, transportation equipment, tobacco, and food products. Foreign companies build factories on Java because labor is inexpensive. In addition, the island's location makes it easy to ship goods.

Reading Check **Of what resources does Indonesia have large reserves?**

Indonesia's People

Indonesia has about 211.8 million people—the fourth-largest population in the world. It is also one of the world's most densely populated countries. The island of Java, for example, has more than 110 million people on an area about the size of New York State. Here you will find **Jakarta** (juh•KAHR•tuh), Indonesia's capital and largest city. Jakarta has modern buildings and streets crowded with cars and bicycles.

Most of Indonesia's people belong to the Malay ethnic group. They are divided into about 300 smaller groups with their own languages. The official language, Bahasa Indonesia, is taught in schools.

Indonesia has more followers of Islam than any other country. Other religions, such as Christianity and Buddhism, are also practiced. On the beautiful island of **Bali,** Hindu beliefs are held by most of the people.

Thousands of years ago, Malays from mainland Southeast Asia settled the islands that are today Indonesia. Their descendants set up Buddhist and Hindu kingdoms. These kingdoms grew wealthy by controlling the trade that passed through the waterways between the Indian and Pacific Oceans. In the A.D. 1100s, traders from Southwest Asia brought Islam to the region. Four hundred years later, Europeans arrived to acquire the valuable spices grown here. They brought Christianity to the islands. The Dutch eventually controlled most of the islands as a colony. Independence finally came to Indonesia in 1949.

Since the 1960s, unrest and civil war have occurred on several islands. A **civil war** is a fight among different groups within a country. Most recently, the people of **East Timor,** who are largely Roman Catholic and were once ruled by Portugal, voted for independence. Indonesia has accepted the results of this election.

From the 1960s to the late 1990s, the military leader, General Suharto, ruled Indonesia. He set up a **dictatorship,** or a government under control of a single all-powerful leader. In the late 1990s, severe economic problems led to widespread unrest, and Suharto was forced to resign. Today Indonesia has a democratic government. With so many different ethnic groups, many small political parties arise. As a result, Indonesia's leaders find it difficult to form a government that is strong enough to deal with the challenges facing the country.

Reading Check When did Indonesia win its independence?

Komodo Dragon

The world's largest lizard, the Komodo dragon, is found only on a few small islands in Indonesia. Its long, forked tongue acts as a nose to "smell" nearby prey. The Komodo can grow 10 feet (3 m) long and can weigh about 300 pounds (136 kg).

Section 3 Assessment

Defining Terms

1. **Define** plate, civil war, dictatorship.

Recalling Facts

2. **Location** What five islands are the largest in Indonesia?
3. **Movement** Why do foreign companies want to build factories in Indonesia?
4. **Government** What type of government does Indonesia have today?

Critical Thinking

5. **Categorizing Information** List Indonesia's three main religions and who brought these religions to the islands.
6. **Understanding Cause and Effect** Why is it difficult for government officials to rule Indonesia?

Graphic Organizer

7. **Organizing Information** Draw a diagram like this one. Under the headings, list two strengths and two challenges of Indonesia.

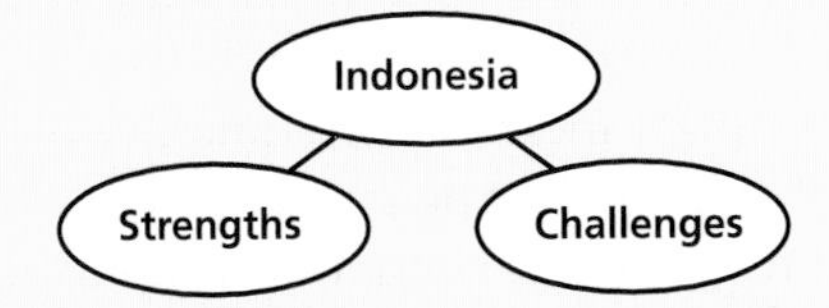

Applying Geography Skills

8. **Analyzing Maps** Look at the physical map on page 717. What are six bodies of water near Indonesia that make the country's location a good one for shipping goods?

Making Connections

ART | SCIENCE | LITERATURE | TECHNOLOGY

Shadow Puppets

Late at night, long after dark has fallen on a small stage in Java, a shadow puppet show is about to begin. The glow of a lamp shines behind a wide linen screen. Puppets stand hidden from direct view. The "good" characters are on the right. The "bad" ones are placed on the left. The audience waits anxiously on the other side of the screen. Once the story begins, the performance will continue until dawn.

The Performance

Wayang kulit, the ancient Indonesian shadow puppet theater, dates back at least 1,000 years. Today there are several thousand puppeteers. This makes shadow puppets the strongest theater tradition in Southeast Asia.

Shadow puppets are flat leather puppets, many with movable limbs and mouths, that are operated by sticks. During the show, the puppets cast their shadows onto the screen. The *dalang,* or puppeteer, sits behind the screen and manipulates the figures. He brings each to life in one of the more than 200 traditional puppet stories.

The Stories

Although Islam is now the major religion of Indonesia, much of the traditional shadow puppet theater is based on stories from two ancient Hindu epics from India. At one time the principal purpose of shadow puppetry was to provide moral and religious instruction in Hinduism. Now the stories combine Hindu themes with elements of Buddhism and Islam, as well as Indonesian history and folklore. Often the performance is given in celebration of public or religious holidays or to honor a wedding or birth.

▼ The *dalang* and his orchestra

The Puppeteer

The skill of the *dalang* is critical to the show's success. The *dalang* operates all the puppets, narrates the story, provides sound effects, and directs the gong, drum, and flute orchestra that accompanies the puppet show. The puppeteer changes his voice to create a unique sound for each character. The *dalang* performs without a script or notes, adding jokes and making small changes to suit the crowd and the occasion. Because a shadow puppet show can last as long as nine hours, the *dalang* must have both a tremendous memory and great endurance.

Many *dalang*s carve their own puppets, having learned this art from earlier generations. Each figure must appear in a specific size, body build, and costume. Even the shape of the eyes tells about the figure's character and mood.

Making the Connection

1. How do shadow puppets move?
2. What kinds of stories do shadow puppet shows present?
3. **Drawing Conclusions** In what way is the *dalang* a master of many different art forms?

Chapter 26

Reading Review

Section 1 Mainland Southeast Asia

Terms to Know
monsoon
deforestation
socialism
delta

Main Idea
The countries of mainland Southeast Asia rely on agriculture.

✓Region Mainland Southeast Asia includes the countries of Myanmar, Thailand, Laos, Cambodia, and Vietnam.

✓Place These countries have highland areas and lowland river valleys with fertile soil. Monsoons bring heavy rains in the summer.

✓History Thailand is the only country in Southeast Asia free of the influence of colonial rule.

✓Economics Conflict has hurt the economies of Laos, Cambodia, and Vietnam.

Section 2 Island Southeast Asia

Terms to Know
strait
free port
terraced field

Main Idea
The island countries of Southeast Asia have a variety of cultures and economic activities.

✓Region The island countries of Southeast Asia include Malaysia, Singapore, Brunei, and the Philippines.

✓Economics Malaysia produces palm oil and rubber, among other goods. Its capital, Kuala Lumpur, is a commercial center.

✓Economics The port of Singapore is one of the world's busiest trading centers.

✓Economics Brunei has grown wealthy from oil and gas income.

✓Culture The Philippines shows the influence of Malaysian, Spanish, and American culture.

Section 3 Indonesia

Terms to Know
plate
civil war
dictatorship

Main Idea
Indonesia has a diverse population within its many islands.

✓Place Indonesia—the world's fourth-most populous country—is an archipelago formed by volcanoes.

✓Economics Rich soil and a warm, wet climate make Indonesia good for farming.

✓Economics Indonesia has rich supplies of oil, natural gas, and minerals.

✓Government Leaders face the challenge of creating a nation out of a land with many different groups and political parties.

◄ People in Bangkok, Thailand, face traffic snarls and pollution that are among the worst in the world.

Chapter 26 Assessment and Activities

Using Key Terms

Match the terms in Part A with their definitions in Part B.

A.

1. free port
2. delta
3. plate
4. strait
5. deforestation
6. terraced field
7. civil war
8. socialism
9. dictatorship
10. monsoon

B.

a. land made from soil deposited at the mouth of a river
b. war fought between groups within a country
c. cutting strips of land out of a hillside
d. economic system in which the government owns many businesses
e. government under control of a single leader
f. place where shipped goods are not taxed
g. slabs of rock that make up the earth's crust
h. seasonal wind that blows over a continent for months at a time
i. the widespread cutting of trees
j. narrow body of water that runs between two land areas

Reviewing the Main Ideas

Section 1 Mainland Southeast Asia

11. **Economics** What resources are found in Thailand?
12. **History** What countries have poor economies because of recent conflict?
13. **Economics** How is Vietnam trying to improve its economy?

Section 2 Island Southeast Asia

14. **Place** What two kinds of farming are carried on in Malaysia?
15. **Economics** What economic activity is important in Singapore besides its shipping industry?
16. **Economics** What resources have made Brunei wealthy?
17. **Culture** How does religion show Spanish influence in the Philippines?

Section 3 Indonesia

18. **Economics** How do nearly half of the people of Indonesia make a living?
19. **Location** How does location make Indonesia a center of trade?
20. **Government** Why does Indonesia have many political parties?

Southeast Asia

Place Location Activity

On a separate sheet of paper, match the letters on the map with the numbered places listed below.

1. Mekong River
2. South China Sea
3. Gulf of Tonkin
4. Hanoi
5. Indonesia
6. Singapore
7. Thailand
8. Vietnam
9. Indian Ocean
10. Philippines

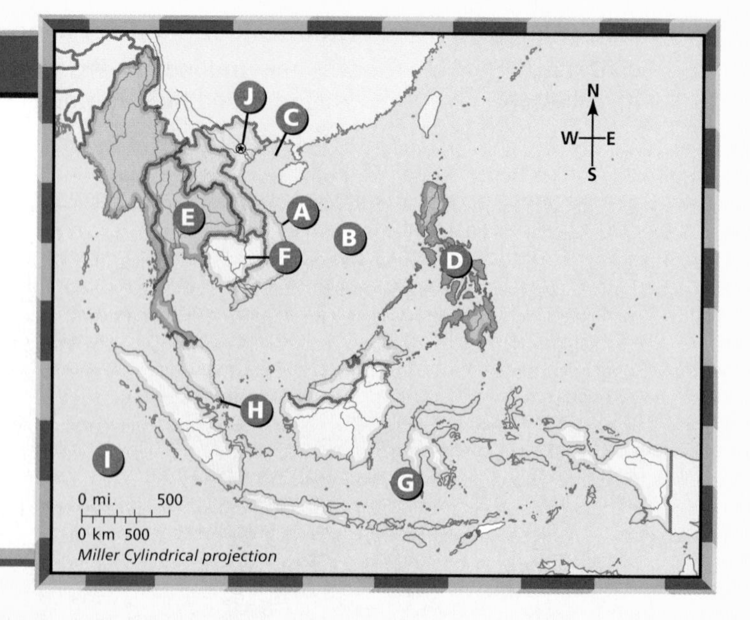

Self-Check Quiz Visit the ***Geography: The World and Its People*** Web site at gwip.glencoe.com and click on **Chapter 26–Self-Check Quizzes** to prepare for the Chapter Test.

Critical Thinking

21. **Predicting Outcomes** Experts believe that Brunei has enough oil reserves to last until 2018. What might happen to the country's economy and standard of living at that time?
22. **Organizing Information** Make a chart like this one. Under each column, write two facts about a country in Southeast Asia. Write about three countries—one from mainland Southeast Asia, one from island Southeast Asia, and Indonesia.

Country	Land	Economy	People

GeoJournal Activity

23. **Writing a Report** Learn about the culture of one of the countries in Southeast Asia. Choose one of the following topics to research: (1) the arts; (2) festivals and holidays; or (3) music and literature. Prepare a written report with illustrations or photos.

Mental Mapping Activity

24. **Focusing on the Region** Draw a map of Southeast Asia, then label the following:

- Borneo
- Irrawaddy River
- Java
- Malay Peninsula
- Philippines
- South China Sea
- Strait of Malacca
- Thailand

Technology Skills Activity

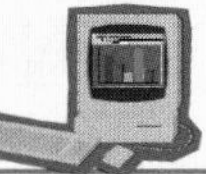

25. **Using the Internet** Use the Internet to learn about the foods in a Southeast Asian country. Find recipes and pictures. Prepare a display that shows a typical meal, or cook the meal yourself and share it with the class.

Standardized Test Practice

Directions: Study the graph below, then answer the questions that follow.

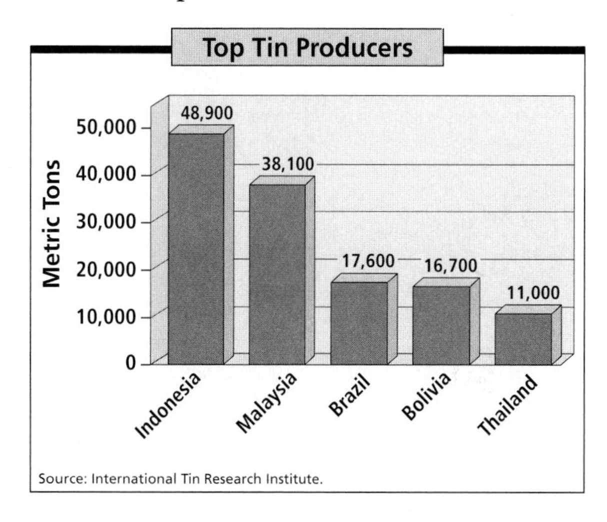

1. **About how much tin does Indonesia produce each year?**
   - **A** 48,900 metric tons
   - **B** 48,000,900 metric tons
   - **C** 48.9 million metric tons
   - **D** 48.9 billion metric tons
2. **About how much tin does Brazil produce each year?**
   - **F** 17,600 metric tons
   - **G** 17,600,000 metric tons
   - **H** 17.6 million metric tons
   - **J** 17.6 billion metric tons

Test-Taking Tip: In order to understand any type of graph, look carefully around the graph for keys that show how it is organized. On this bar graph, the numbers along the left side represent the exact number shown. You do not have to multiply by millions or billions to find the number of metric tons.

Unit 9

Fur seal on the beach, Antarctica

Boy selling fish, Samoa

Australia, Oceania, and Antarctica

Australia, Oceania, and Antarctica are grouped together more because of their nearness to one another than because of any similarities among their peoples. These lands lie mostly in the Southern Hemisphere. Australia is a dry continent that is home to unusual wildlife. Oceania's 25,000 tropical islands spread out across the Pacific Ocean. Frozen Antarctica encompasses the South Pole.

Lone tree in the outback, Australia

NGS ONLINE
www.nationalgeographic.com/education

Focus on:

Australia, Oceania, and Antarctica

LYING ALMOST ENTIRELY in the Southern Hemisphere, this region includes two continents and thousands of islands scattered across the Pacific Ocean. Covering a huge portion of the globe, the region includes landscapes ranging from polar to tropical.

The Land

Both a continent and a single country, Australia is a vast expanse of mostly flat land. The Great Dividing Range runs down the continent's eastern edge. Between this range of mountains and the Pacific Ocean lies a narrow strip of coastal land. West of the Great Dividing Range lies Australia's large—and very dry—interior. Here in the Australian "outback" are seemingly endless miles of scrubland, as well as three huge deserts.

Across the Tasman Sea from Australia lies New Zealand, made up of two main islands—North Island and South Island—and many smaller ones. Mountains and hills dominate the landscape.

North and east of New Zealand is Oceania. Its roughly 25,000 islands lie scattered across the Pacific Ocean on either side of the Equator. Some of these islands are volcanic. Others are huge formations of rock that have risen from the ocean floor. Still others are low-lying coral islands surrounded by reefs.

Antarctica, the frozen continent, covers and surrounds the South Pole. It is almost completely buried under an enormous sheet of ice.

The Climate

Australia is one of the driest continents in the world. Its eastern coast does receive rainfall from the Pacific Ocean. Mountains block this moisture from reaching inland areas, however. Much of Australia's outback has a desert climate.

No place in New Zealand is more than 80 miles (129 km) from the sea. This country has only one climate region: marine west coast. It has mild temperatures and plentiful rainfall throughout the year.

The islands of Oceania have mostly tropical climates, with warm temperatures and distinct wet and dry seasons. Rain forests cover many of the islands.

Antarctica is one of the coldest and windiest places on the earth, as well as one of the driest. It receives so little precipitation that it is considered a desert—the world's largest cold desert.

Sheep grazing near Mount Egmont, New Zealand

Emperor penguins examining the ice, Antarctica

The Economy

Mines dot the Australian landscape. Its ancient rocks and soils are rich in minerals such as uranium, bauxite, iron ore, copper, nickel, and gold. Little of Australia's land is good for growing crops. Instead, vast cattle and sheep ranches—or stations, as the Australians call them—spread across much of the country. Sheep far outnumber people in New Zealand, where pastures are lush and green almost year-round. New Zealand is one of the world's leading producers of lamb and wool.

The people of Oceania depend primarily on fishing and farming. Across much of Oceania, the soil and climate are not favorable for widespread agriculture, and islanders generally raise only enough food for themselves. Yet some larger islands have rich volcanic soil. In such places, cash crops of fruits, sugar, coffee, and coconut products are grown for export.

Antarctica is believed to be rich in mineral resources. To preserve Antarctica for research and exploration, however, many nations have agreed not to mine this mineral wealth.

The People

The first settlers in this region probably came from Asia thousands of years ago. Australia's first inhabitants, the ancestors of today's Aborigines, may have arrived as long as 40,000 years ago. Not until about A.D. 1000, however, did seafaring peoples reach the farthest islands of Oceania.

The British colonized Australia and New Zealand in the 1700s and 1800s. These two countries gained their independence in the early 1900s. Many South Pacific islands were not freed from colonial rule until after World War II. Today Australia and Oceania are a blend of European, traditional Pacific, and Asian cultures.

Despite its vast size, this is the least populous of all the world's regions. It is home to only about 30 million people. More than half of these live in Australia, where they are found mostly in coastal cities such as Sydney and Melbourne. Roughly 4 million people live in New Zealand, which also has large urban populations along its coasts. Oceania is less urbanized. Antarctica has no permanent human inhabitants at all. Groups of scientists live and work on the frozen continent for brief periods to carry out their research.

Exploring the Region

1. **Which two continents lie in this region?**
2. **Why is Antarctica considered a desert?**
3. **Why is so little of Australia's land good for farming?**
4. **Where do most of the region's people live?**

◀ **Girl selling fruit, French Polynesia**

UNIT 9

The city of Melbourne, along the southeastern coast of Australia

Regional Atlas

Australia, Oceania, and Antarctica

Physical

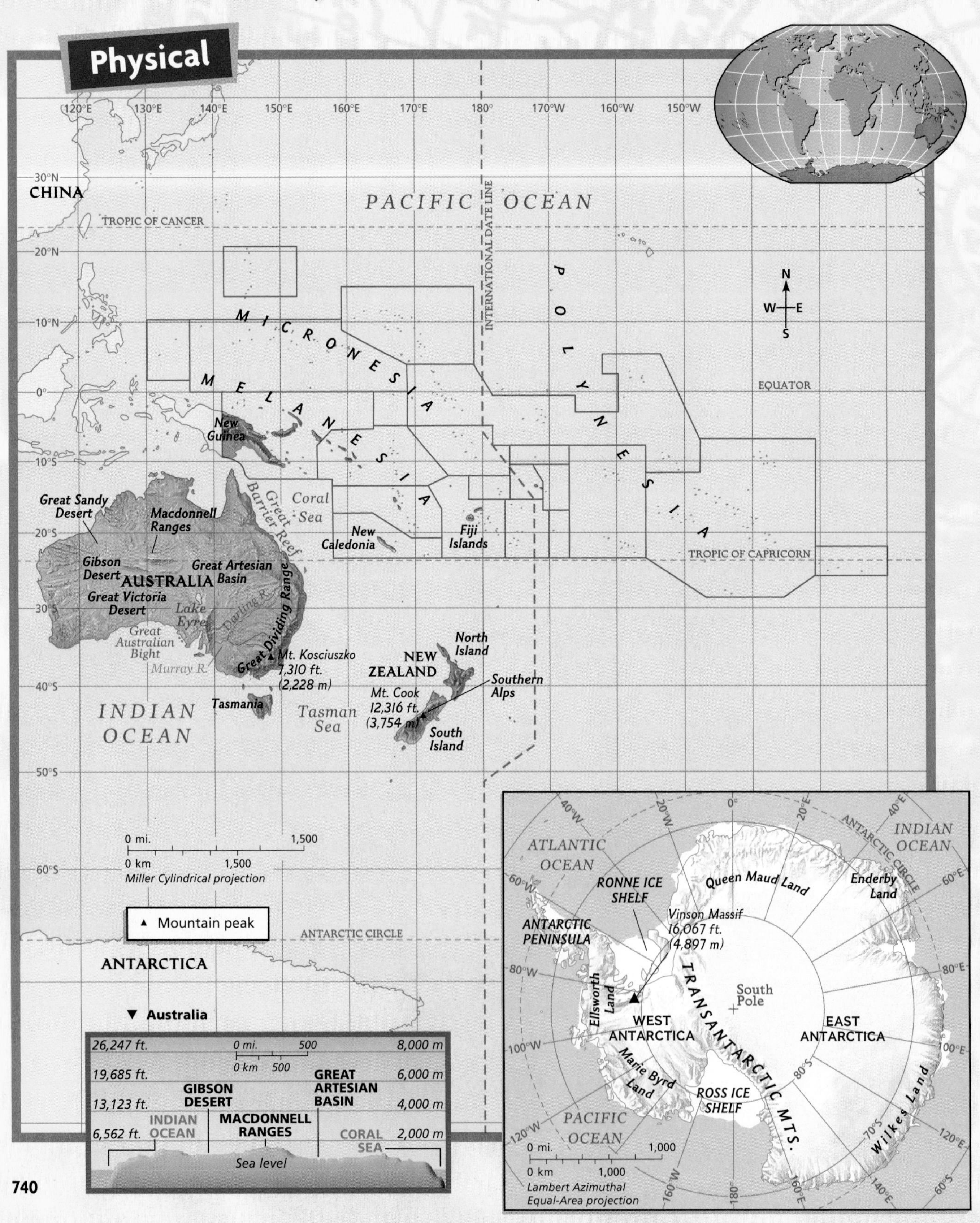

120°E
130°E
140°E
150°E
160°E
170°E
180°
170°W
160°W
150°W
30°N
20°N
10°N
0°
10°S
20°S
30°S
40°S
50°S
60°S
CHINA
TROPIC OF CANCER
PACIFIC OCEAN
INTERNATIONAL DATE LINE
MICRONESIA
MELANESIA
POLYNESIA
EQUATOR
New Guinea
Great Sandy Desert
Macdonnell Ranges
Great Barrier Reef
Coral Sea
New Caledonia
Fiji Islands
TROPIC OF CAPRICORN
Gibson Desert
AUSTRALIA
Great Artesian Basin
Great Victoria Desert
Lake Eyre
Darling R.
Great Dividing Range
Great Australian Bight
Murray R.
Mt. Kosciuszko 7,310 ft. (2,228 m)
North Island
NEW ZEALAND
Southern Alps
Mt. Cook 12,316 ft. (3,754 m)
South Island
Tasmania
Tasman Sea
INDIAN OCEAN
0 mi. 1,500
0 km 1,500
Miller Cylindrical projection
▲ Mountain peak
ANTARCTIC CIRCLE
ANTARCTICA
▼ Australia
26,247 ft.
19,685 ft.
13,123 ft.
6,562 ft.
8,000 m
6,000 m
4,000 m
2,000 m
0 mi. 500
0 km 500
GIBSON DESERT
GREAT ARTESIAN BASIN
MACDONNELL RANGES
INDIAN OCEAN
CORAL SEA
Sea level
ATLANTIC OCEAN
INDIAN OCEAN
ANTARCTIC CIRCLE
RONNE ICE SHELF
Queen Maud Land
Enderby Land
ANTARCTIC PENINSULA
Vinson Massif 16,067 ft. (4,897 m)
Ellsworth Land
TRANSANTARCTIC MTS.
South Pole
WEST ANTARCTICA
EAST ANTARCTICA
Marie Byrd Land
ROSS ICE SHELF
PACIFIC OCEAN
Wilkes Land
0 mi. 1,000
0 km 1,000
Lambert Azimuthal Equal-Area projection

UNIT 9

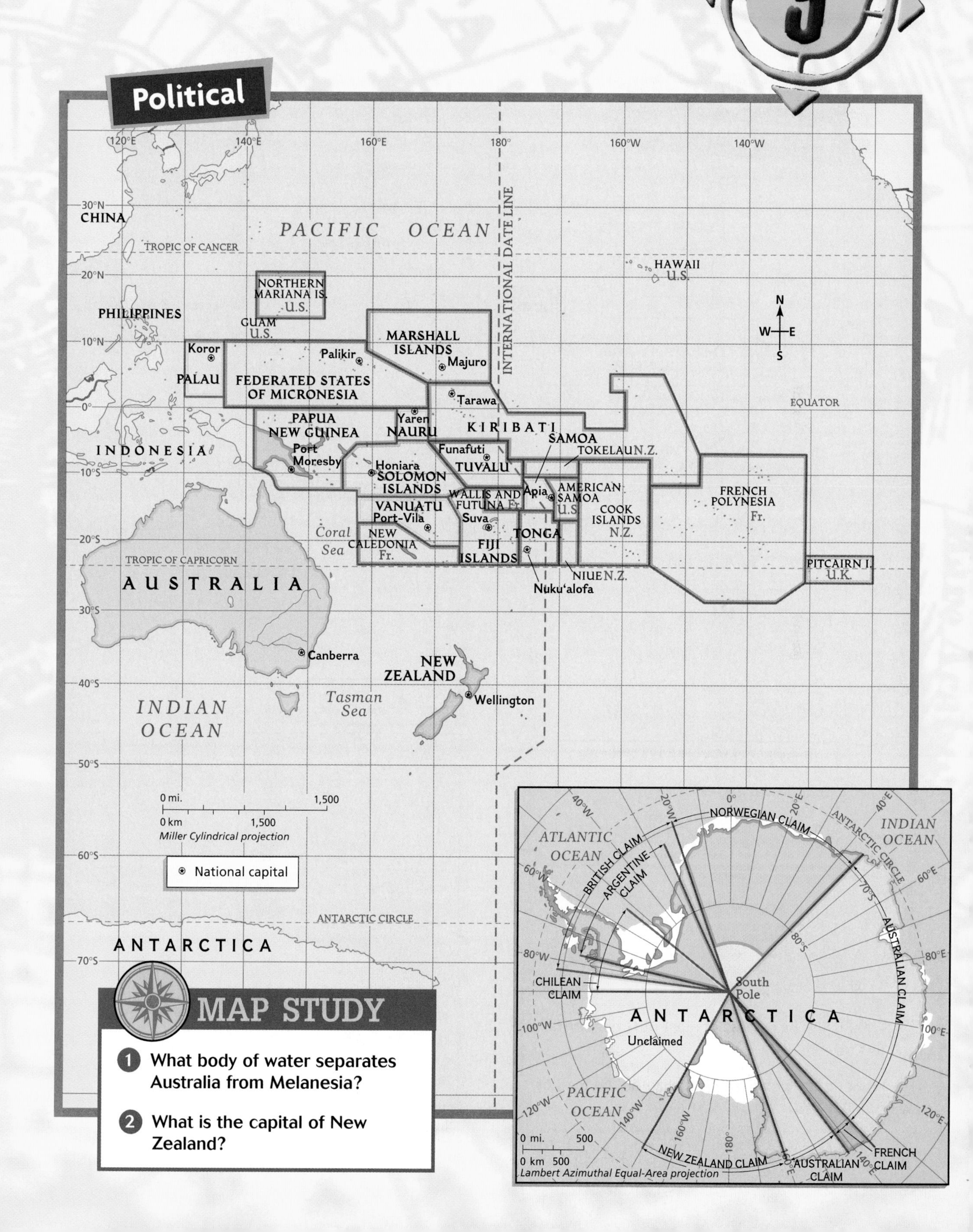

Political
PACIFIC OCEAN
INTERNATIONAL DATE LINE
TROPIC OF CANCER
TROPIC OF CAPRICORN
EQUATOR
ANTARCTIC CIRCLE
120°E
140°E
160°E
180°
160°W
140°W
30°N
20°N
10°N
0°
10°S
20°S
30°S
40°S
50°S
60°S
70°S
CHINA
PHILIPPINES
INDONESIA
NORTHERN MARIANA IS. U.S.
GUAM U.S.
Koror
PALAU
Palikir
FEDERATED STATES OF MICRONESIA
MARSHALL ISLANDS
Majuro
Tarawa
KIRIBATI
Yaren
NAURU
PAPUA NEW GUINEA
Port Moresby
Honiara
SOLOMON ISLANDS
Funafuti
TUVALU
SAMOA
Apia
TOKELAU N.Z.
AMERICAN SAMOA U.S.
WALLIS AND FUTUNA Fr.
VANUATU
Port-Vila
Suva
FIJI ISLANDS
TONGA
Nuku'alofa
NIUE N.Z.
COOK ISLANDS N.Z.
FRENCH POLYNESIA Fr.
PITCAIRN I. U.K.
HAWAII U.S.
NEW CALEDONIA Fr.
Coral Sea
AUSTRALIA
Canberra
Tasman Sea
NEW ZEALAND
Wellington
INDIAN OCEAN
ANTARCTICA
N
W
E
S
0 mi. 1,500
0 km 1,500
Miller Cylindrical projection
National capital
ATLANTIC OCEAN
INDIAN OCEAN
PACIFIC OCEAN
NORWEGIAN CLAIM
BRITISH CLAIM
ARGENTINE CLAIM
CHILEAN CLAIM
AUSTRALIAN CLAIM
FRENCH CLAIM
NEW ZEALAND CLAIM
Unclaimed
South Pole
ANTARCTICA
0°
20°W
40°W
60°W
80°W
100°W
120°W
140°W
160°W
180°
160°E
140°E
120°E
100°E
80°E
60°E
40°E
20°E
70°S
80°S
0 mi. 500
0 km 500
Lambert Azimuthal Equal-Area projection

MAP STUDY

1. What body of water separates Australia from Melanesia?
2. What is the capital of New Zealand?

Regional Atlas

Australia, Oceania, and Antarctica

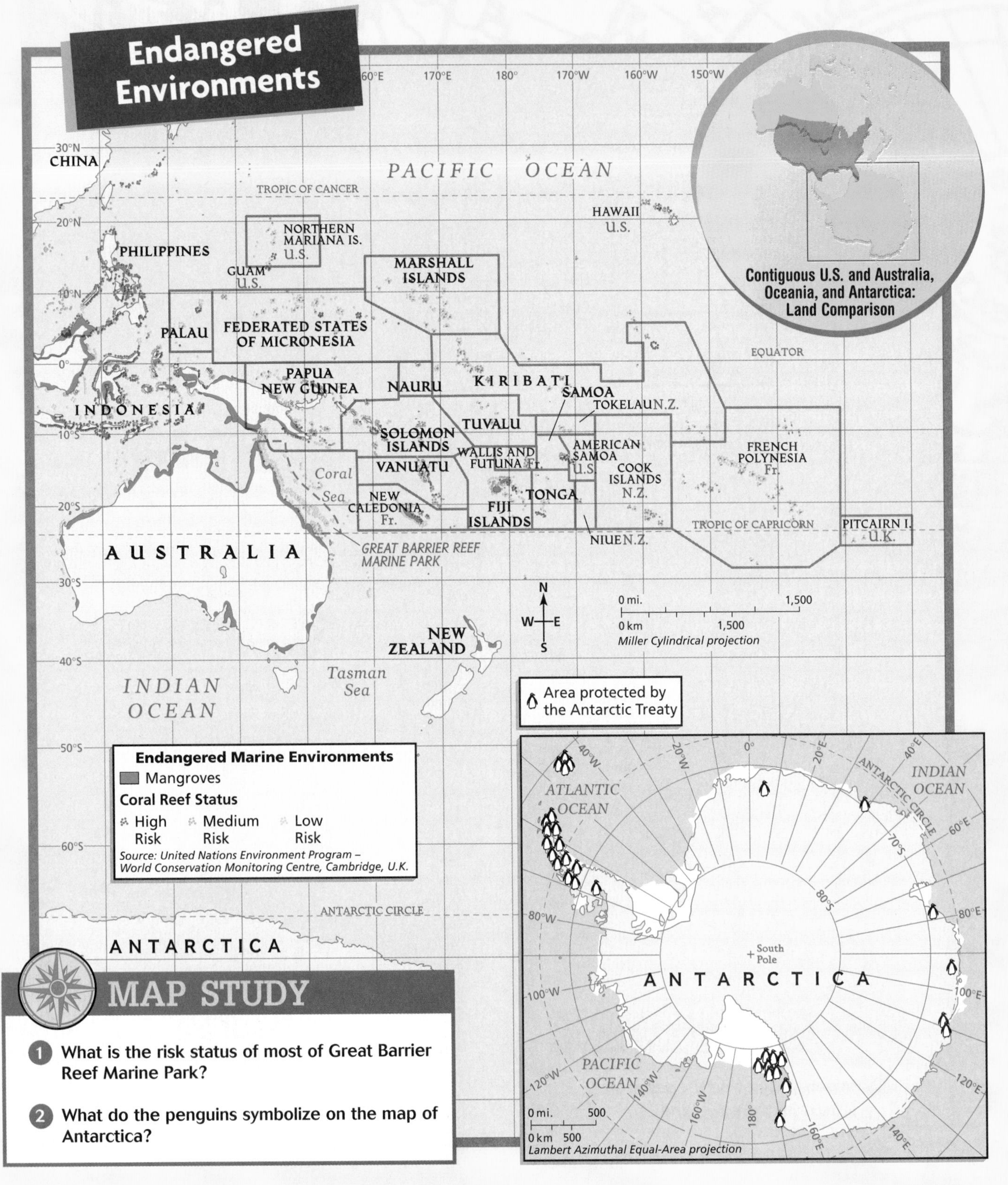

MAP STUDY

1. What is the risk status of most of Great Barrier Reef Marine Park?
2. What do the penguins symbolize on the map of Antarctica?

Geo Extremes

① HIGHEST POINT
Vinson Massif (Antarctica)
16,067 ft. (4,897 m) high

② LOWEST POINT
Bently Subglacial Trench (Antarctica)
8,366 ft. (2,550 m) below sea level

③ LONGEST RIVER
Murray-Darling (Australia)
2,310 mi. (3,718 km) long

④ LARGEST LAKE
Lake Eyre (Australia)
3,600 sq. mi. (9,324 sq. km)

⑤ LARGEST HOT DESERT
Great Victoria (Australia)
134,650 sq. mi. (348,742 sq. km)

⑥ LARGEST COLD DESERT
Antarctica
5,100,000 sq. mi. (13,209,000 sq. km)

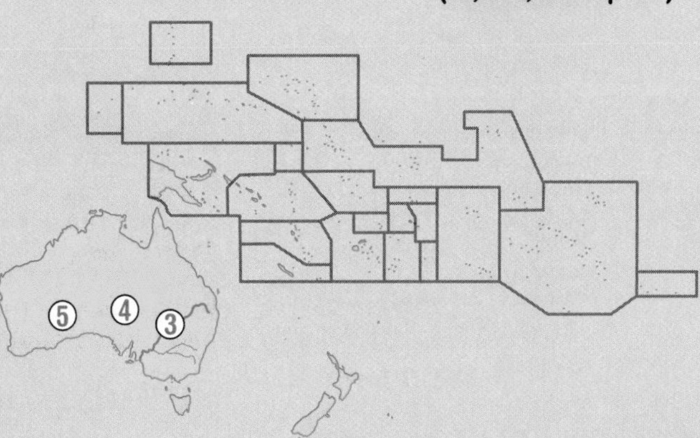

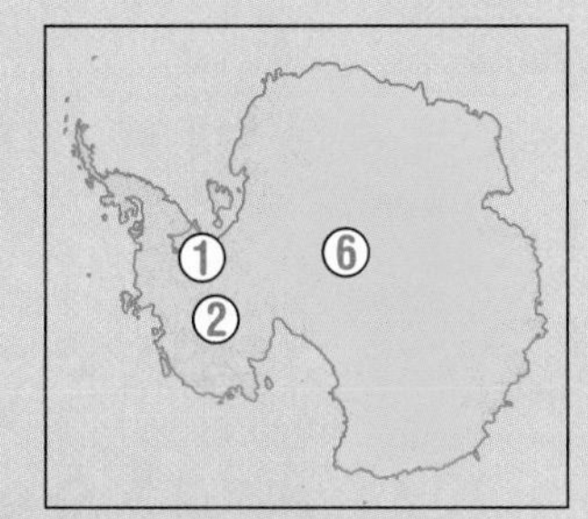

COMPARING POPULATION: United States and Selected Countries of Australia, Oceania, and Antarctica

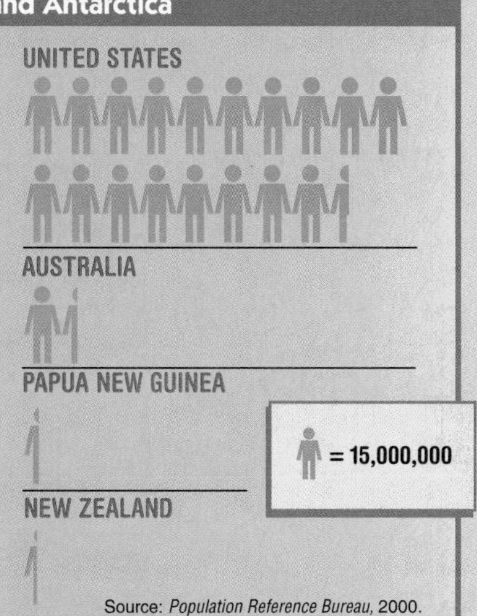

Source: *Population Reference Bureau*, 2000.

POPULATION GROWTH: Australia, 1958–2008

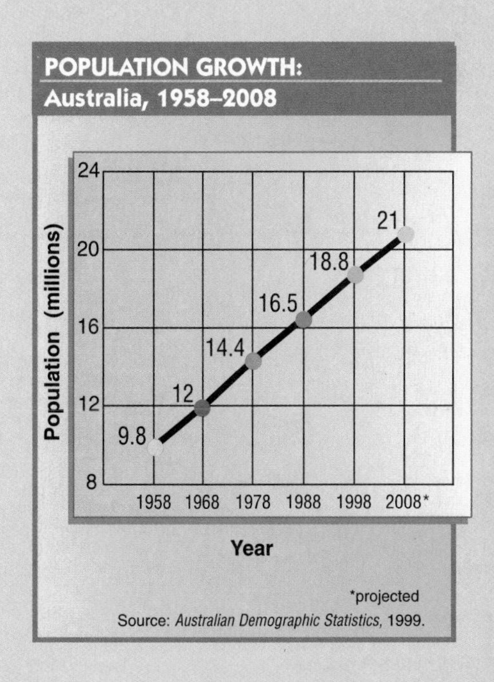

*projected
Source: *Australian Demographic Statistics*, 1999.

GRAPHIC STUDY

1. The largest cold desert in this region is also the largest desert in the *world*. What is it?
2. By how much is Australia's population expected to have grown between 1958 and 2008?

Country Profiles

AUSTRALIA

POPULATION:
18,981,000
6 per sq. mi.
2 per sq. km

LANGUAGE:
English

MAJOR EXPORT:
Coal

MAJOR IMPORT:
Machinery

CAPITAL:
Canberra

LANDMASS:
2,966,153 sq. mi.
7,682,300 sq. km

FEDERATED STATES of MICRONESIA

POPULATION:
117,000
432 per sq. mi.
167 per sq. km

LANGUAGES:
English, Local Languages

MAJOR EXPORT:
Fish

MAJOR IMPORT:
Foods

CAPITAL:
Palikir

LANDMASS:
271 sq. mi.
702 sq. km

FIJI ISLANDS

POPULATION:
794,000
113 per sq. mi.
43 per sq. km

LANGUAGES:
English, Fijian, Hindi

MAJOR EXPORT:
Sugar

MAJOR IMPORT:
Machinery

CAPITAL:
Suva

LANDMASS:
7,056 sq. mi.
18,274 sq. km

KIRIBATI

POPULATION:
82,000
296 per sq. mi.
114 per sq. km

LANGUAGES:
English, Gilbertese

MAJOR EXPORT:
Coconut Products

MAJOR IMPORT:
Foods

CAPITAL:
Tarawa

LANDMASS:
277 sq. mi.
717 sq. km

MARSHALL ISLANDS

POPULATION:
62,000
886 per sq. mi.
343 per sq. km

LANGUAGES:
English, Local Languages

MAJOR EXPORT:
Coconut Products

MAJOR IMPORT:
Foods

CAPITAL:
Majuro

LANDMASS:
70 sq. mi.
181 sq. km

NAURU

POPULATION:
11,000
1,357 per sq. mi.
524 per sq. km

LANGUAGES:
Nauruan, English

MAJOR EXPORT:
Phosphates

MAJOR IMPORT:
Foods

CAPITAL:
Yaren

LANDMASS:
8 sq. mi.
21 sq. km

NEW ZEALAND

POPULATION:
3,817,000
37 per sq. mi.
14 per sq. km

LANGUAGE:
English

MAJOR EXPORT:
Wool

MAJOR IMPORT:
Machinery

CAPITAL:
Wellington

LANDMASS:
103,883 sq. mi.
269,057 sq. km

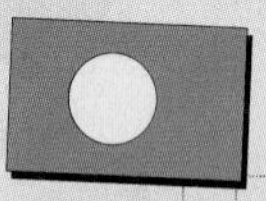

PALAU

POPULATION:
19,000
101 per sq. mi.
35 per sq. km

LANGUAGES:
English, Palauan

MAJOR EXPORT:
Fish

MAJOR IMPORT:
N/A

CAPITAL:
Koror

LANDMASS:
188 sq. mi.
487 sq. km

PAPUA NEW GUINEA

POPULATION:
4,669,000
26 per sq. mi.
10 per sq. km

LANGUAGES:
English, Local Languages

MAJOR EXPORT:
Gold

MAJOR IMPORT:
Machinery

CAPITAL:
Port Moresby

LANDMASS:
178,260 sq. mi.
461,691 sq. km

SAMOA

POPULATION:
195,000
178 per sq. mi.
69 per sq. km

LANGUAGES:
Samoan, English

MAJOR EXPORT:
Coconut Products

MAJOR IMPORT:
Foods

CAPITAL:
Apia

LANDMASS:
1,093 sq. mi.
2,831 sq. km

SOLOMON ISLANDS

POPULATION:
430,000
39 per sq. mi.
15 per sq. km

LANGUAGES:
English, Local Languages

MAJOR EXPORT:
Cocoa

MAJOR IMPORT:
Machinery

CAPITAL:
Honiara

LANDMASS:
10,985 sq. mi.
28,450 sq. km

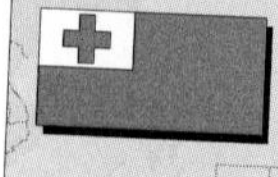

TONGA

POPULATION:
109,000
404 per sq. mi.
156 per sq. km

LANGUAGES:
Tongan, English

MAJOR EXPORT:
Squash

MAJOR IMPORT:
Foods

CAPITAL:
Nuku'alofa

LANDMASS:
270 sq. mi.
699 sq. km

Countries and flags not drawn to scale

UNIT 9

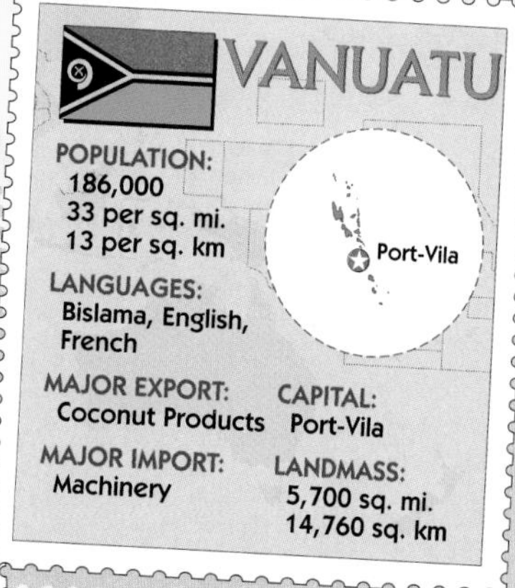

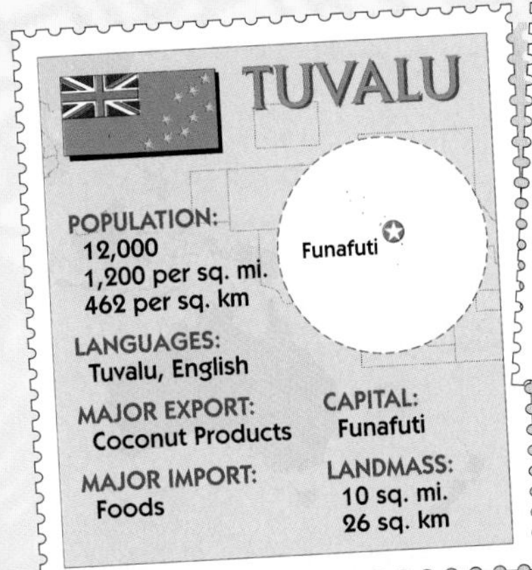

TUVALU
POPULATION:
12,000
1,200 per sq. mi.
462 per sq. km
LANGUAGES:
Tuvalu, English
MAJOR EXPORT:
Coconut Products
MAJOR IMPORT:
Foods
CAPITAL:
Funafuti
LANDMASS:
10 sq. mi.
26 sq. km
Funafuti

GEO BEE

Questions From Buzz Bee!

The following questions are taken from National Geographic GeoBees. Use your textbook, the Internet, and other library resources to find the answers.

1. **Dried coconut meat is the main export of many South Pacific islands. What is it called?**

2. **Ross, Amundsen, and Bellingshausen are names of seas that border which country?**

3. **The world's southernmost city with a metropolitan population greater than 1 million is located north of the Bass Strait. Name this city.**

4. **What U.S.-administered islands were named for their location about halfway between the continents of North America and Asia?**

5. **A 1991 international agreement prohibits mining on which continent for at least the next 50 years?**

6. **If you were swimming off the coast of Australia near Perth, you would be in what body of water?**

7. **Which continent receives the least amount of rainfall?**

Australia and New Zealand

The World and Its People NATIONAL GEOGRAPHIC

To learn more about the people and places of Australia and New Zealand, view ***The World and Its People*** **Chapter 27** video.

Geography Online

Chapter Overview Visit the ***Geography: The World and Its People*** Web site at gwip.glencoe.com and click on **Chapter 27–Chapter Overviews** to preview information about Australia and New Zealand.

Australia

Guide to Reading

Main Idea

Both a continent and a country, Australia has many natural resources but few people.

Terms to Know

- coral reef
- outback
- station
- marsupial
- immigrant
- boomerang
- bush

Places to Locate

- Australia
- Tasmania
- Great Barrier Reef
- Murray River
- Darling River
- Sydney
- Melbourne
- Canberra

Reading Strategy

Make a chart like this one. Then fill in two facts about Australia for each category.

Land	History
Climate	Government
Economy	People

◀ Ayers Rock in central Australia

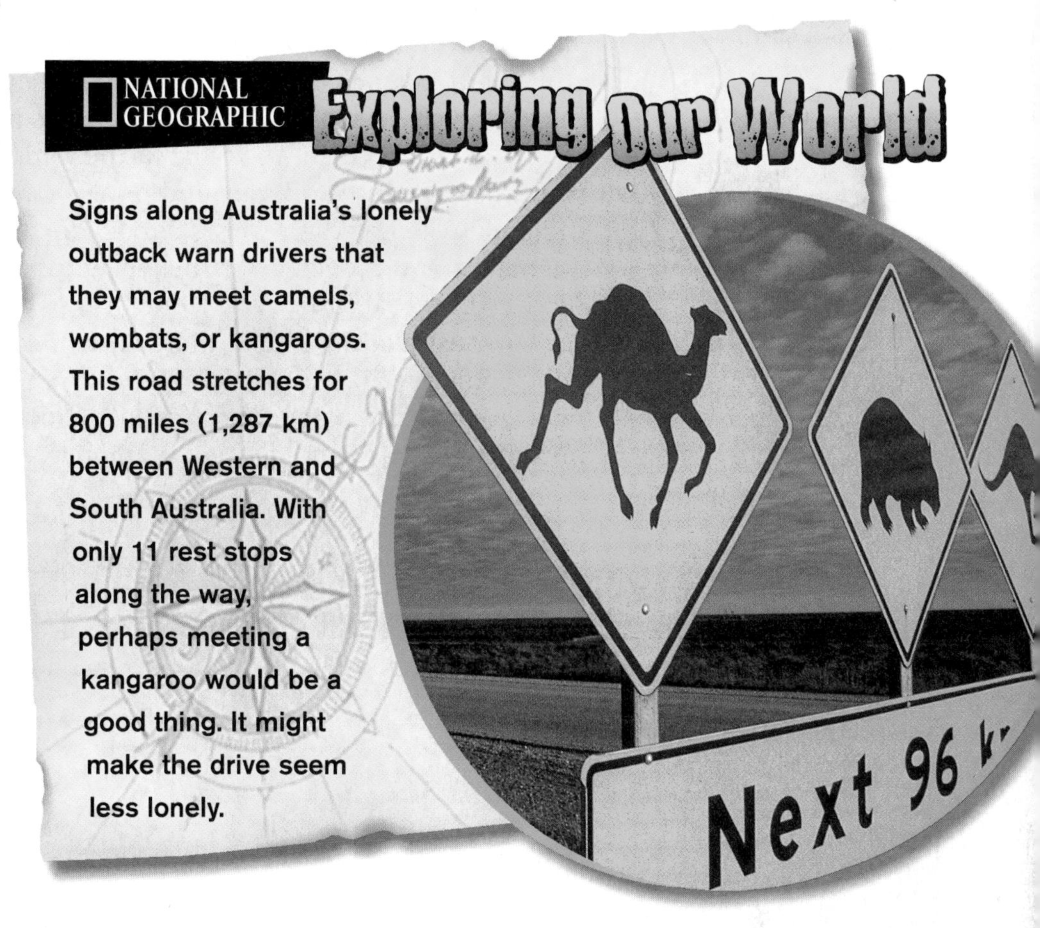

Signs along Australia's lonely outback warn drivers that they may meet camels, wombats, or kangaroos. This road stretches for 800 miles (1,287 km) between Western and South Australia. With only 11 rest stops along the way, perhaps meeting a kangaroo would be a good thing. It might make the drive seem less lonely.

Is **Australia** a country or a continent? It is both. Australia is the sixth-largest country in the world. Surrounded by water, Australia is too large to be called an island. So geographers call it a continent.

Australia's Land and Climate

Australia is sometimes referred to as the Land Down Under because it is located in the Southern Hemisphere. The Indian Ocean washes its western and southern shores. The Coral Sea, Pacific Ocean, and Tasman Sea border the eastern coast. The island of **Tasmania,** to the south, is part of Australia.

The **Great Barrier Reef** lies off Australia's northeastern coast. Coral formations have piled up for millions of years to create a colorful chain that stretches 1,250 miles (2,012 km). As you recall, a coral reef is a structure formed by the skeletons of small sea animals.

Plateaus and lowland plains spread across most of Australia. The longest and highest mountain range, the Great Dividing Range, runs along the eastern coast. Mt. Kosciuszko (KAH•zee•UHS•koh), the tallest mountain in Australia, measures 7,310 feet (2,228 km).

Narrow plains run along the south and southeast. These fertile flatlands hold Australia's best farmland and most of the country's people. Two major rivers, the **Murray** and the **Darling,** drain this region.

Australia's vast interior is pastureland. The people of Australia use the name **outback** for the inland regions of their country. Dry grasslands and mineral deposits are found here. Mining camps and cattle and sheep ranches called **stations** dot this region. Some stations are huge. One cattle station is almost twice as large as Delaware.

Water is scarce in Australia. In the Great Artesian Basin, however, water lies in deep, underground pools. Ranchers drill wells and bring the underground water to the surface for their cattle. Far to the west, Australia's land is even drier. Imagine a carpet of sand twice as large as Alaska, Texas, California, and New Mexico combined. That is about the size of Australia's western plateau, which includes the Macdonnell Ranges and the Hamersley Range. Most people who cross this vast, dry plateau do so by plane. The region is rich in resources such as gold, nickel, iron ore, diamonds, and uranium.

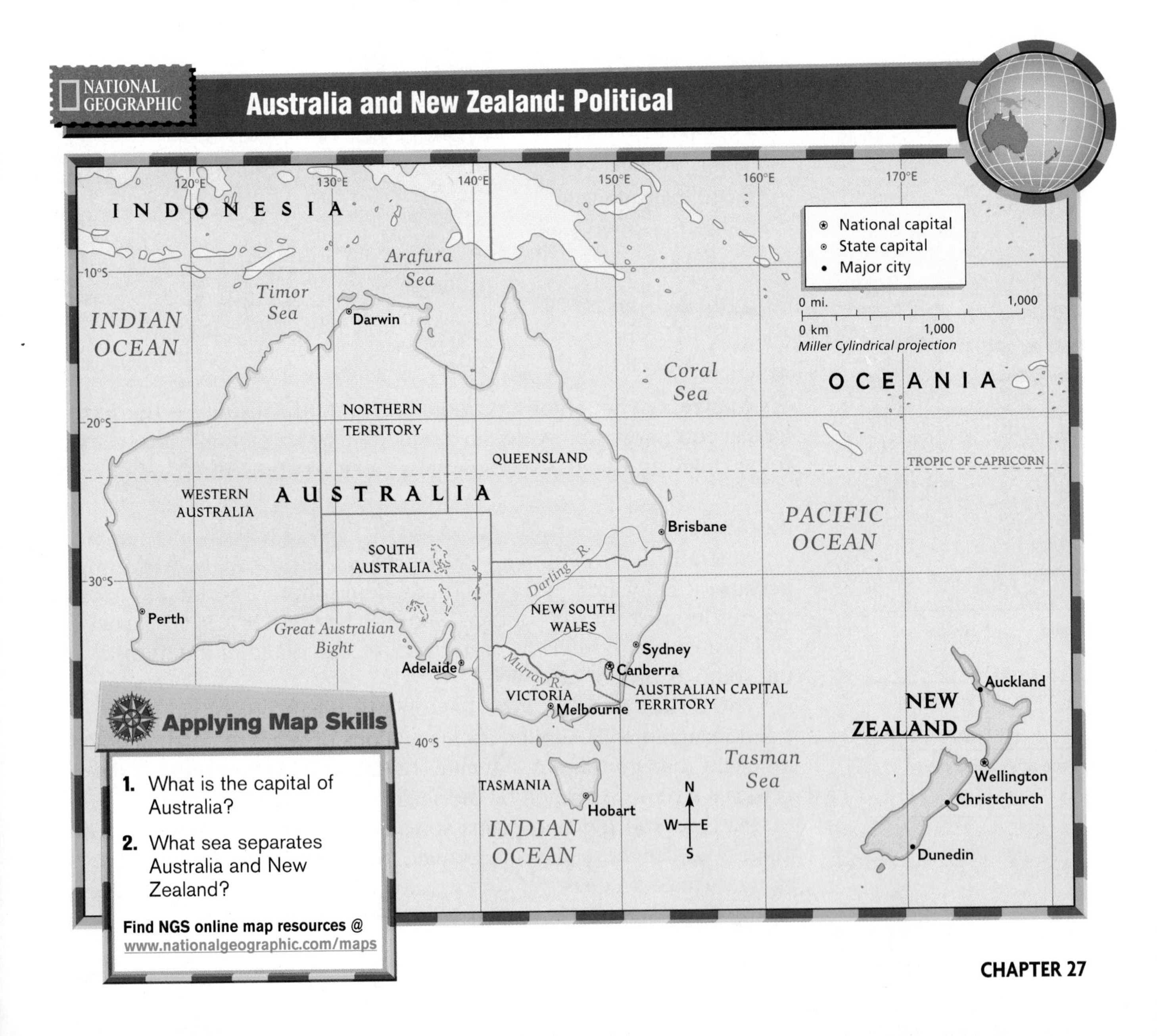

Desert and steppe climates are found in most of the country. Only the Great Dividing Range and Tasmania have winter temperatures that fall below freezing. Because of the country's location in the Southern Hemisphere, summer starts in December and winter starts in June.

Unusual Animals About 200 million years ago, the tectonic plate upon which Australia sits separated from the other continents. As a result, Australia's native plants and animals are not found elsewhere in the world. Two famous Australian animals are kangaroos and koalas. Both are **marsupials,** or mammals that carry their young in a pouch. Turn to page 753 to read about these and other amazing animals.

Reading Check Where do most of Australia's people live?

Australia's Economy

Australia has a strong, prosperous economy. Australia is a treasure chest overflowing with mineral resources. These riches include iron ore, zinc, bauxite, gold, silver, opals, diamonds, and pearls. Australia also has

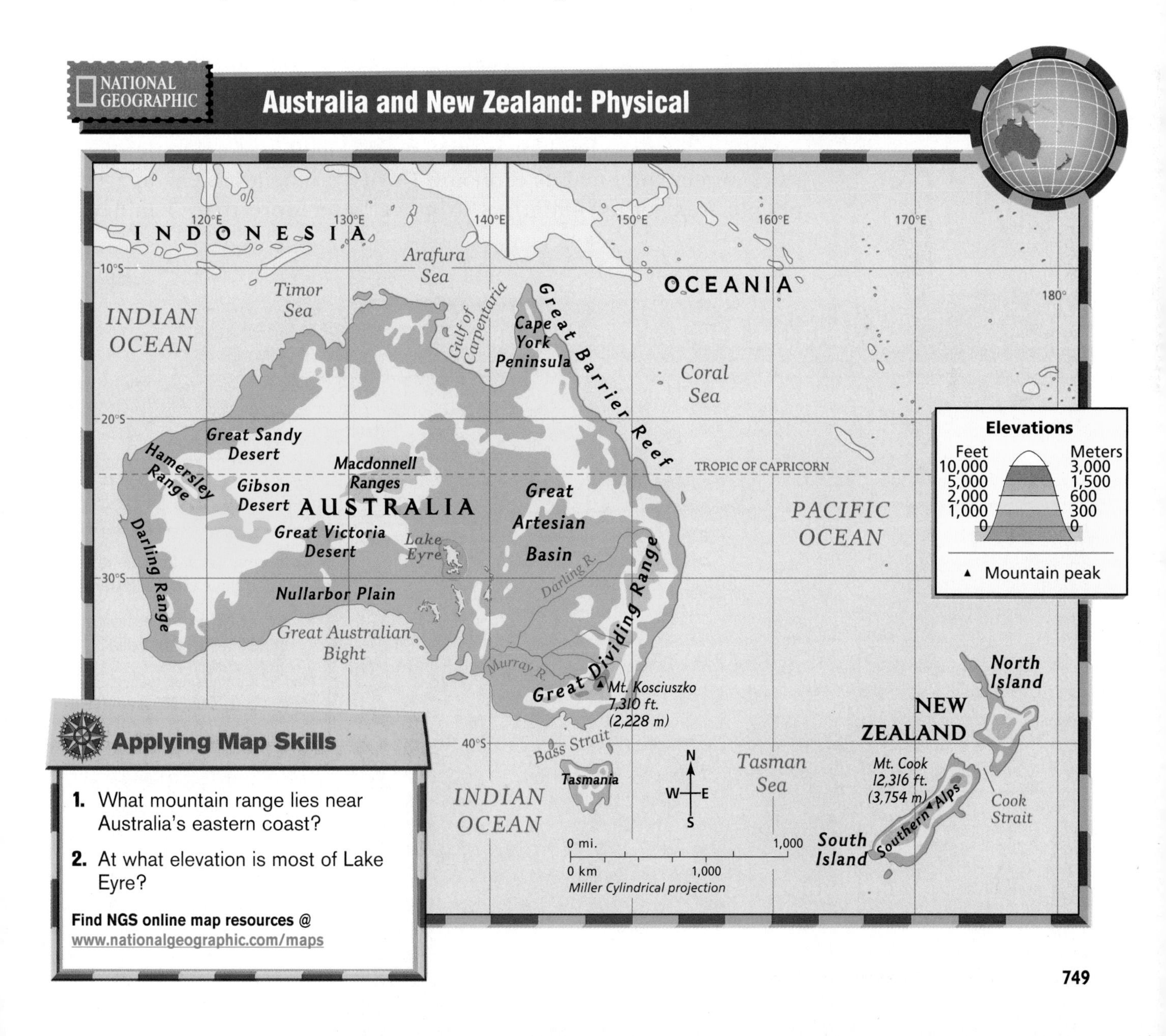

Applying Map Skills

1. What mountain range lies near Australia's eastern coast?
2. At what elevation is most of Lake Eyre?

Find NGS online map resources @ www.nationalgeographic.com/maps

energy resources, including coal, oil, and natural gas. Less than 1 percent of Australia's people work in the mining industry. Still, mineral and energy resources make up more than one-third of Australia's exports.

Australia's dry climate limits farming. With irrigation, however, farmers grow grains, sugarcane, cotton, fruits, and vegetables. Grains and sugar are exported, but most other crops are grown to feed Australia's people. The main agricultural activity is raising livestock, especially cattle and sheep. Australia is the world's top producer and exporter of wool. Ranchers also ship beef and cattle hides.

Australia's factories employ about 10 percent of the country's workers. Manufacturing, which is growing in importance, includes processed foods, transportation equipment, metals, cloth, and chemicals. High-tech industries, service industries, and tourism also play a large role in the economy. Modern ocean shipping enables Australia to export goods to very distant markets. More than half goes to Asia. The United States is also an important destination for exports.

Reading Check What is Australia's main agricultural activity?

Australia's History and People

Despite its huge area, Australia has few people—only 19 million. Most live in scattered areas along the coast, especially in the east and southeast. Australia has long needed more skilled workers to develop its resources and build its economy. The government has encouraged people from other countries to move here. More than 5 million

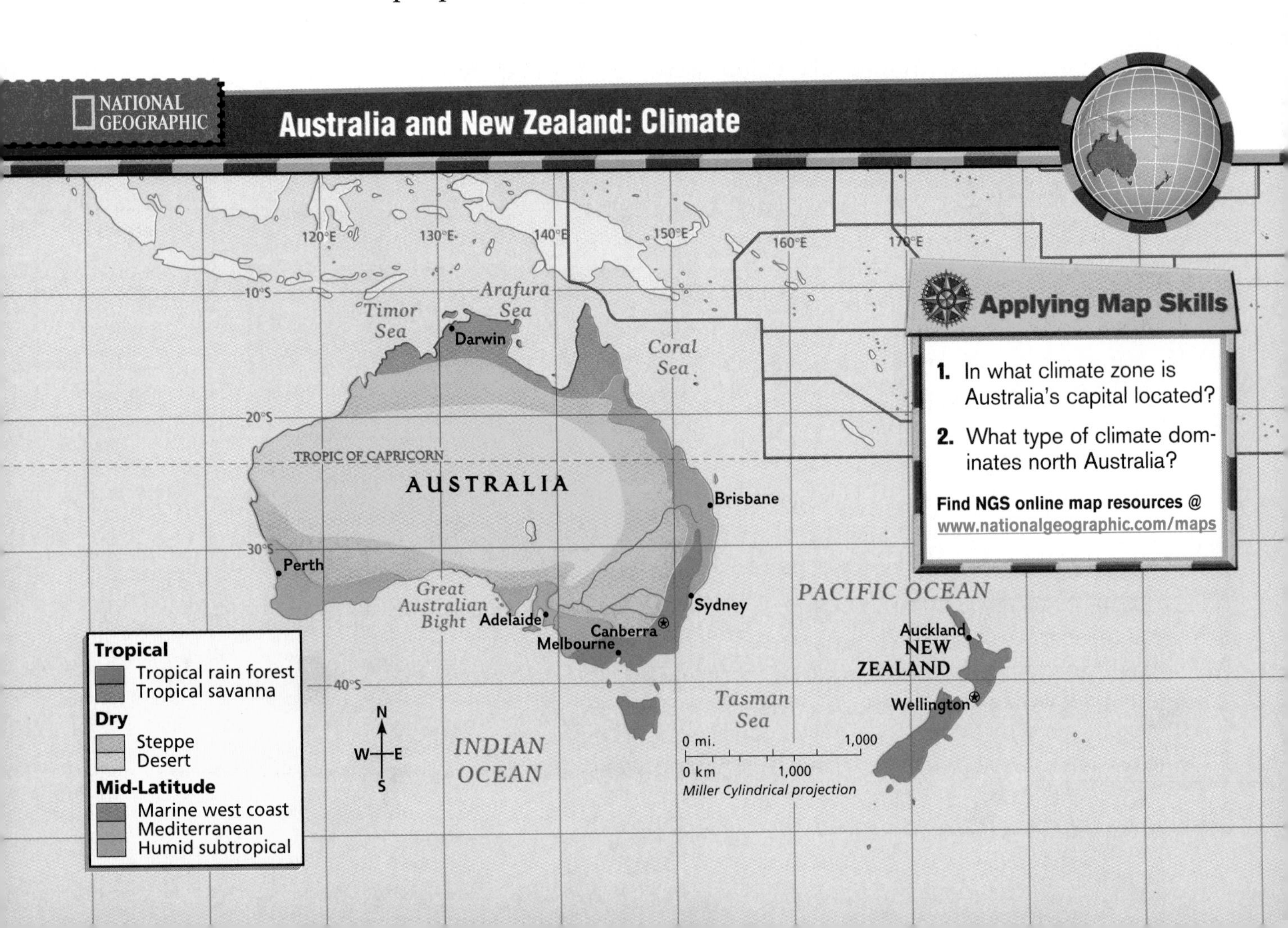

Music

The *didgeridoo* is the most famous musical instrument from Australia. In its original form, it was made when a eucalyptus branch fell to the ground and was hollowed out by termites. Someone playing a *didgeridoo* creates a variety of sounds by making a combination of lip, tongue, and mouth movements. Once you hear the eerie sounds, it is easy to understand why the Aborigines considered this instrument sacred and made it part of their ceremonies.

Looking Closer **What other instruments are similar to the *didgeridoo*?**

GO TO

World Music: A Cultural Legacy

Hear music of this region on Disc 2, Track 28.

immigrants, or people who move from one country to live in another, have arrived in recent decades.

A small part of Australia's population are Aborigines (A•buh•RIHJ•neez). They are the descendants of the first immigrants who came from Asia about 30,000 to 40,000 years ago. For centuries the Aborigines moved throughout Australia gathering plants, hunting, and searching for water. They developed a unique culture. You may have heard of one of their weapons—the **boomerang.** This wooden tool is shaped like a bent bird's wing. The hunter throws it to stun his prey. If the boomerang misses, it curves and sails back to the hunter.

In 1770 Captain James Cook reached Australia and claimed it for Great Britain. At first the British government used Australia as a place to send prisoners. Then other British people set up colonies, especially after gold was discovered in the outback in 1851. Land was taken from the Aborigines, and many of them died of European diseases.

Today about 300,000 Aborigines live in Australia. Growing numbers of them are moving to cities to find jobs. After years of harsh treatment and isolation in the outback, the Aborigines now are demanding more opportunities. In 1967 the Australian government finally recognized the Aborigines as citizens.

The Government In 1901 the Australian British colonies united to form the independent Commonwealth of Australia. Today Australia has a British-style parliamentary democracy. A prime minister is the head of government. The political party with the most seats in the Australian parliament, or legislature, chooses the prime minister.

Like the United States, Australia has a federal system of government. This means that political power is divided between a national government

Section 2

New Zealand

Guide to Reading

Main Idea

New Zealand is a small country with a growing economy based on trade.

Terms to Know

- geyser
- *manuka*
- fjord
- geothermal energy
- hydroelectric power

Places to Locate

- New Zealand
- North Island
- South Island
- Cook Strait
- Southern Alps
- Auckland
- Wellington

Reading Strategy

Make a time line like this one with at least four dates in New Zealand's history. Write the dates on one side of the line and the corresponding event on the opposite side.

Have you ever tasted a ripe green kiwifruit (KEE•wee•FROOT)? If so, it might have been grown on a New Zealand farm like the one shown here. After all, New Zealand is one of the world's leading producers of this tasty fruit. The kiwifruit, once known as the Chinese gooseberry, is now named for the kiwi bird—New Zealand's national symbol.

New Zealand lies in the Pacific Ocean about 1,200 miles (1,931 km) southeast of its nearest neighbor, Australia. In contrast to Australia's flat, dry land, New Zealand is mountainous and very green. Its marine west coast climate is mild and wet.

New Zealand's Land

New Zealand is about the size of Colorado. It includes two main islands—**North Island** and **South Island**—as well as many smaller islands. The **Cook Strait** separates North Island and South Island.

North Island A large plateau forms the center of North Island. Three active volcanoes and the inactive Mount Egmont are located here. You also find **geysers**, or hot springs that spout hot steam and water through a crack in the earth.

Small shrubs called ***manuka*** grow well in the plateau's fertile volcanic soil. Fertile lowlands, forested hills, and sandy beaches surround North Island's central plateau. On the plateau's slopes, sheep and cattle graze. Fruits and vegetables are grown on the coastal lowlands.

South Island The **Southern Alps** run along South Island's western coast. Snowcapped Mount Cook, the highest peak in New Zealand, soars 12,316 feet (3,754 m) here. Glaciers lie on mountain slopes above green forests and sparkling blue lakes. These glaciers once cut deep **fjords** (fee•AWRDS), or steep-sided valleys, into the mountains. The sea has filled these fjords with crystal-blue waters.

To the east of the Southern Alps stretch the Canterbury Plains. They form New Zealand's largest area of flat or nearly flat land. Farmers grow grains and ranchers raise sheep here.

Plants and Animals New Zealanders take pride in their unique wildlife. Their national symbol is a flightless bird called the kiwi. Giant kauri (KOWR•ee) trees once dominated all of North Island. About 100 years ago, European settlers cut down many of these trees, using the wood to build homes and ships. Today the government protects kauri trees. One of them is more than 2,000 years old.

Reading Check **Which island of New Zealand has glaciers and fjords?**

New Zealand's Economy

New Zealand has a thriving agricultural economy. Sheep are an important agricultural resource. New Zealand is the second-leading wool producer in the world. Lamb meat is another important export. Apples, barley, wheat, and corn are the main crops.

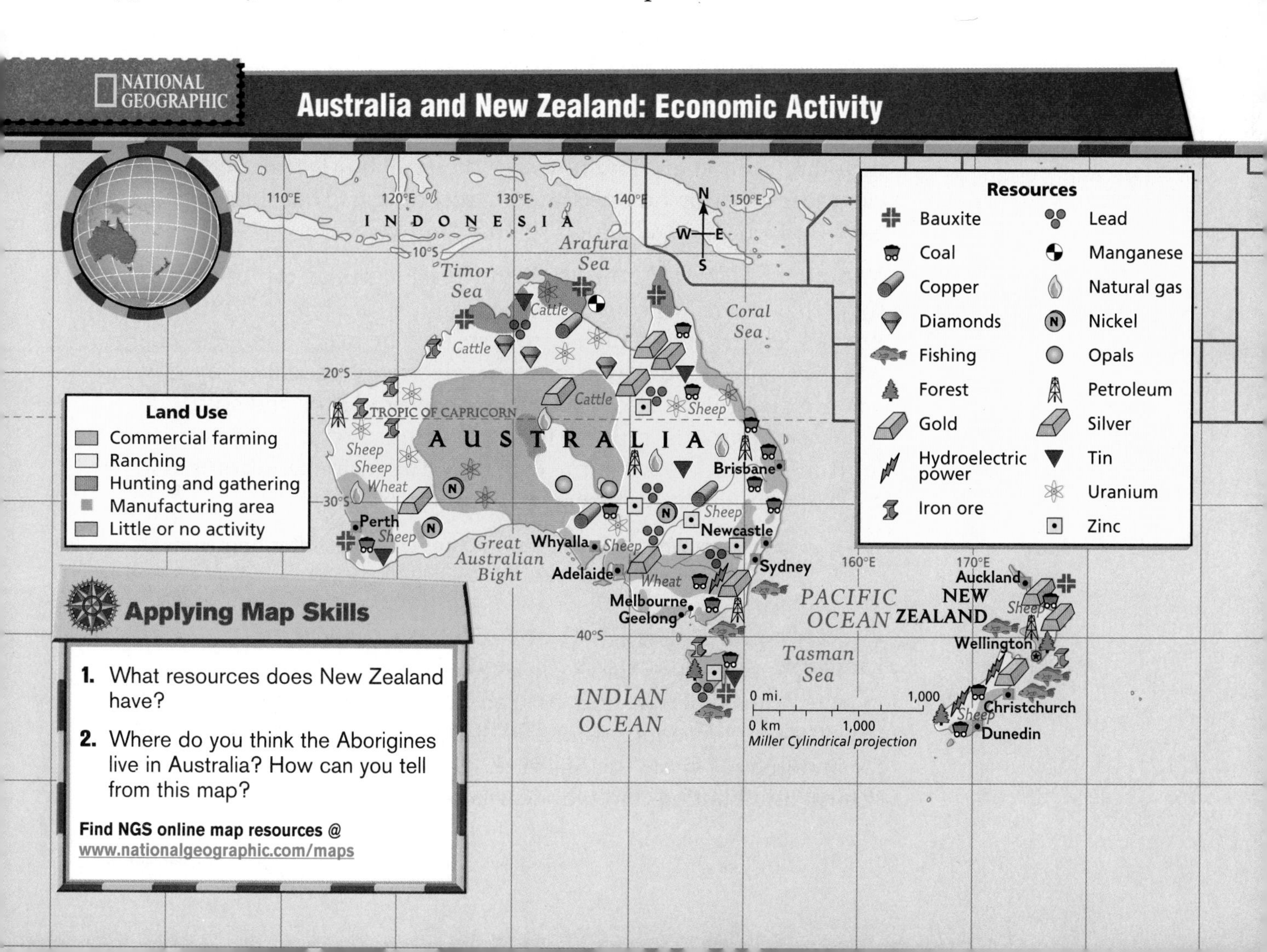

Applying Map Skills

1. What resources does New Zealand have?
2. Where do you think the Aborigines live in Australia? How can you tell from this map?

Find NGS online map resources @ www.nationalgeographic.com/maps

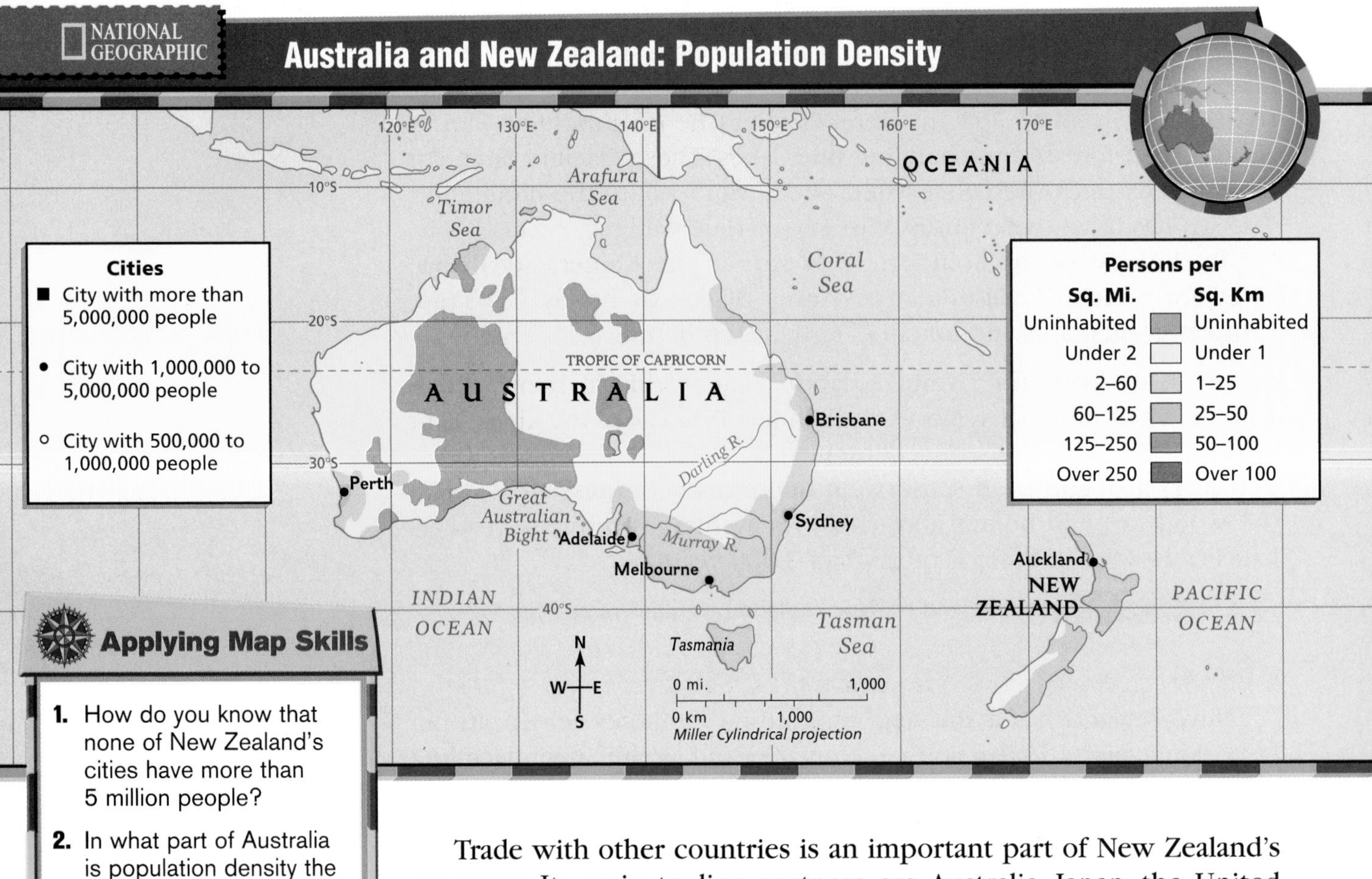

Applying Map Skills

1. How do you know that none of New Zealand's cities have more than 5 million people?
2. In what part of Australia is population density the highest?

Find NGS online map resources @ www.nationalgeographic.com/maps

Trade with other countries is an important part of New Zealand's economy. Its main trading partners are Australia, Japan, the United States, and the United Kingdom. Depending on trade brings both benefits and dangers to New Zealand. If the economies of other countries are growing quickly, demand for goods from New Zealand will rise. If the other economies slow, however, they will buy fewer products. This can cause hardship on the islands. In recent years, trade has grown, and New Zealanders enjoy a high standard of living.

Mining and Manufacturing New Zealand sits on top of the molten rock that forms volcanoes. As a result, it is rich in **geothermal energy,** electricity produced from steam. The major source of energy, however, is **hydroelectric power**—electricity generated by flowing water. New Zealand also has coal, oil, iron ore, silver, and gold.

The country is rapidly industrializing. Service industries and tourism play large roles in the economy. The main manufactured items are wood products, fertilizers, wool products, and shoes.

✓Reading Check **Why does trade with other countries offer both benefits and dangers to New Zealand?**

New Zealand's History and People

People called the Maori (MOWR•ee) are believed to have arrived in New Zealand between A.D. 950 and 1150. They probably crossed the Pacific Ocean in canoes from islands far to the northeast. Undisturbed for hundreds of years, the Maori developed skills in farming, weaving, fishing, bird hunting, and woodcarving.

The first European explorers came to the islands in the mid-1600s. Almost 200 years passed before settlers—most of them British—arrived. In 1840 British officials signed a treaty with Maori leaders. In this treaty, the Maori agreed to accept British rule in return for the right to keep their land. More British settlers eventually moved onto Maori land. War broke out in the 1860s—a war that the Maori lost.

In 1893 the colony became the first land to give women the right to vote. New Zealand was also among the first places in which the government gave help to people who were old, sick, or out of work.

New Zealand became independent in 1907. The country is a parliamentary democracy in which elected representatives choose a prime minister to head the government. Five seats in the parliament can be held only by Maoris. Today about 10 percent of New Zealand's 3.8 million people are Maori. Most of the rest are descendants of British settlers. Asians and Pacific islanders, attracted by the growing economy, have increased the diversity of New Zealand's society.

About 85 percent of the people live in urban areas. The largest cities are **Auckland,** an important port, and **Wellington,** the capital. Both are on North Island, where about 75 percent of the people live.

New Zealanders take advantage of the country's mild climate and beautiful landscapes. They enjoy camping, hiking, hunting, boating, and mountain climbing in any season. They also play cricket and rugby, sports that originated in Great Britain.

✓Reading Check **What group settled New Zealand 1,000 years ago?**

Assessment

Defining Terms

1. Define geyser, *manuka,* fjord, geothermal energy, hydroelectric power.

Recalling Facts

2. Region How do New Zealand's land and climate compare to Australia's?

3. Economics What two animal products are important exports for New Zealand?

4. History Most of New Zealand's people are descendants of settlers from what European country?

Critical Thinking

5. Analyzing Information Why do you think New Zealand's government guarantees the Maori a certain number of seats in the parliament?

6. Making Predictions With so many different peoples settling in New Zealand, how do you think the country's culture may change?

Graphic Organizer

7. Organizing Information Imagine that you are moving to New Zealand. Write a question you would ask for each topic in the chart below.

Physical features	Economy	Recreation
Climate	Government	Culture

Applying Geography Skills

8. Analyzing Maps Look at the economic activity map on page 755. Is more land in New Zealand used for farming or for ranching?

Study and Writing Skill

Outlining

Outlining may be used as a starting point for a writer. The writer begins with the rough shape of the material and gradually fills in the details in a logical manner. You may also use outlining as a method of note taking and organizing information as you read.

Learning the Skill

There are two types of outlines—formal and informal. An informal outline is similar to taking notes—you write words and phrases needed to remember main ideas. In contrast, a formal outline has a standard format. Follow these steps to formally outline information:

- Read the text to identify the main ideas. Label these with Roman numerals.
- Write subtopics under each main idea. Label these with capital letters.
- Write supporting details for each subtopic. Label these with Arabic numerals.
- Each level should have at least two entries and should be indented from the level above.
- All entries should use the same grammatical form, whether phrases or complete sentences.

▼ **A huge sheep herd pours down a ravine on North Island.**

Practicing the Skill

On a separate sheet of paper, copy the following outline for Section 2 of this chapter. Then use your textbook to fill in the missing subtopics and details.

I. New Zealand's Land
  A. North Island
    1. Central plateau surrounded by fertile lowlands
    2. Active volcanoes and geysers
  B. ____________
    1. Southern Alps on western coast
    2. ____________
  C. Plants and Animals
    1. ____________
    2. ____________
II. New Zealand's Economy
  A. Agriculture
    1. ____________
    2. ____________
  B. Trading Partners
    1. ____________
    2. ____________
    3. ____________
    4. ____________
  C. ____________
    1. ____________
    2. Wood products, fertilizers, wool products, and shoes
III. New Zealand's History and People
  A. ____________
  B. ____________

Applying the Skill

Following the guidelines above, prepare an outline for Section 1 of this chapter.

GO TO Practice key skills with **Glencoe Skillbuilder Interactive Workbook, Level 1.**

Chapter 27 Reading Review

Section 1 Australia

Terms to Know
- coral reef
- outback
- station
- marsupial
- immigrant
- boomerang
- bush

Main Idea

Both a continent and a country, Australia has many natural resources but few people.

✓ **Place** The land of Australia is mostly flat and dry, with little rainfall.

✓ **Location** Because Australia has been separated from other continents for millions of years, unique plants and animals developed here.

✓ **Economics** Most of Australia's wealth comes from minerals and the products of its ranches. It is the world's leading producer and exporter of wool.

✓ **Culture** Australia has relatively few people, most of whom live along the coasts.

Sydney Opera House in Sydney, Australia ▶

Section 2 New Zealand

Terms to Know
- geyser
- *manuka*
- fjord
- geothermal energy
- hydroelectric power

Main Idea

New Zealand is a small country with a growing economy based on trade.

✓ **Place** Most people live on New Zealand's two largest islands.

✓ **Place** New Zealand has volcanic mountains, high glaciers, deep-cut fjords, fertile hills, and coastal plains. The climate is mild and wet.

✓ **Economics** New Zealand's economy is built on trade. Sheepherding is an important activity, and wool and lamb meat are major exports.

✓ **History** The people called the Maori first came to New Zealand around 1,000 years ago.

✓ **Culture** Most people live on North Island, where the country's two main cities can be found.

Chapter 27 Assessment and Activities

Using Key Terms

Match the terms in Part A with their definitions in Part B.

A.

1. boomerang
2. bush
3. station
4. geothermal energy
5. outback
6. *manuka*
7. marsupial
8. hydroelectric power
9. coral reef
10. geyser

B.

a. electricity produced from steam
b. wooden weapon that returns to the thrower
c. mammal that carries its young in a pouch
d. hot spring that shoots hot water into the air
e. rural area in Australia
f. structure formed by the skeletons of small sea animals
g. name for entire inland region of Australia
h. cattle or sheep ranch in Australia
i. electricity generated by flowing water
j. small shrub found in New Zealand

Reviewing the Main Ideas

Section 1 Australia

11. **Location** Why is Australia called the Land Down Under?
12. **Place** For what is the outback used?
13. **Economics** What does Australia lead the world in producing and exporting?
14. **History** What country colonized Australia?
15. **Culture** What percentage of people live in Australia's cities?

Section 2 New Zealand

16. **Location** On which island do most New Zealanders live?
17. **Place** What kind of climate does New Zealand have?
18. **Economics** Why can New Zealand's economy suffer if other countries have economic problems?
19. **Culture** How many New Zealanders have Maori heritage?
20. **Human/Environment Interaction** What leisure activities do New Zealanders enjoy that are made possible by the country's climate?

Australia and New Zealand

Place Location Activity

On a separate sheet of paper, match the letters on the map with the numbered places listed below.

1. Auckland
2. Sydney
3. Tasmania
4. Great Barrier Reef
5. Great Dividing Range
6. Southern Alps
7. Darling River
8. Wellington
9. Canberra
10. Perth

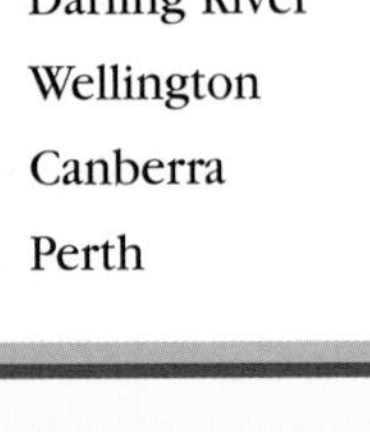

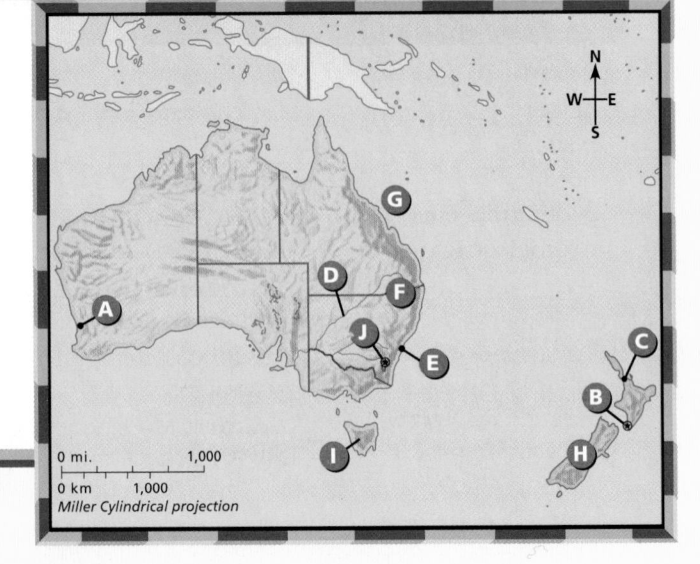

Self-Check Quiz Visit the ***Geography: The World and Its People*** Web site at gwip.glencoe.com and click on **Chapter 27–Self-Check Quizzes** to prepare for the Chapter Test.

Critical Thinking

21. **Understanding Cause and Effect** Why do most Australians and New Zealanders live in coastal areas?
22. **Organizing Information** Draw two ovals like these. In the outer ovals, write four facts about each country under their heading. Where the ovals overlap, write three facts that are true of both countries.

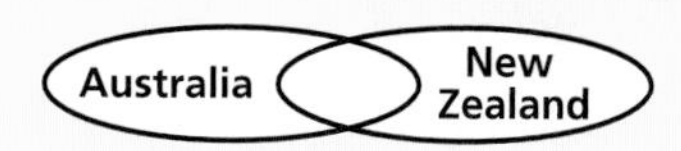

GeoJournal Activity

23. **Designing a Poster** Choose one of the unusual physical features found in Australia or New Zealand. You might choose the Great Barrier Reef, the Great Artesian Basin, or the geysers or glaciers of New Zealand. Research to learn more about this physical feature. Create an illustrated poster that includes a map, four photographs, and four facts about the feature.

Mental Mapping Activity

24. **Focusing on the Region** Draw a simple outline map of Australia and New Zealand, then label the following:

- North Island
- Sydney
- Melbourne
- Auckland
- Tasmania
- Tasman Sea
- Coral Sea
- Darling River
- Great Artesian Basin
- Hamersley Range

Technology Skills Activity

25. **Using the Internet** Use the Internet to find out more about one of Australia's or New Zealand's cities. Prepare a travel brochure aimed at a tourist who might visit the city. Describe the city's main attractions.

Standardized Test Practice

Directions: Study the graph below, then answer the following question.

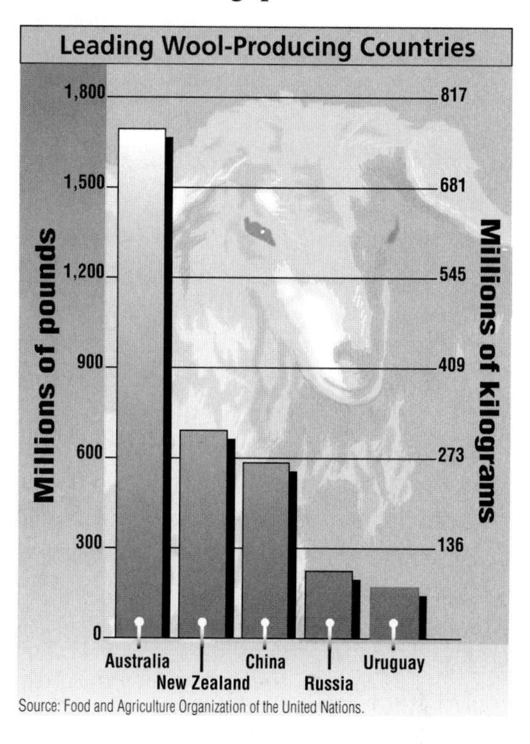

Source: Food and Agriculture Organization of the United Nations.

1. **How much wool does Australia produce a year?**
   A 1,800 pounds
   B 1,800,000 pounds
   C about 1,700 pounds
   D about 1,700,000,000 pounds

Test-Taking Tip: Remember to read the information along the sides of the graph to understand what the bars represent. In addition, eliminate answers that you know are wrong.

EYE on the Environment

OZONE

Earth's Natural Sunscreen

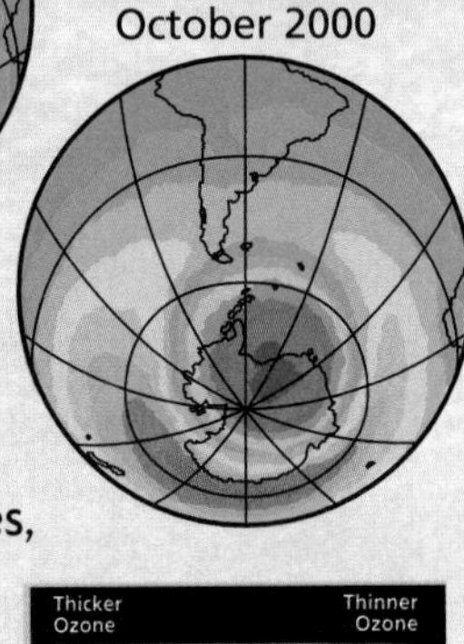

Source: Ozone Processing Team, NASA, GSFC.

The Ozone Hole If you spend lots of time outdoors, you probably know that "SPF 30" is a rating for sunscreen. The higher a sunscreen's Sun Protection Factor (SPF), the longer you can be exposed to sunlight before your skin begins to burn. Earth has a sunscreen, too. It is called ozone. Ozone is a kind of gas. A thin band of ozone high above the earth shields the planet from the sun's most harmful ultraviolet (UV) rays. The ozone layer, however, is being destroyed. The satellite images (above, right) show an expanding ozone hole above Antarctica. For several decades, the ozone layer has been in trouble.

Human-made chemicals, particularly chlorofluorocarbons (CFCs), destroy ozone and thin the ozone layer. CFCs were used for years in refrigerators, air conditioners, foam-insulated cups, aerosol sprays, and in some cleaning products.

Ozone losses of about 10 percent have occurred over Europe, Canada, and other parts of the Northern Hemisphere, too.

When ozone is destroyed, more UV rays strike the earth. Exposure to harmful rays can cause skin cancer in humans, destroy plants, and kill ocean plankton.

Reversing the Damage The good news is that ozone destruction can be reversed. Officials around the world are taking action.

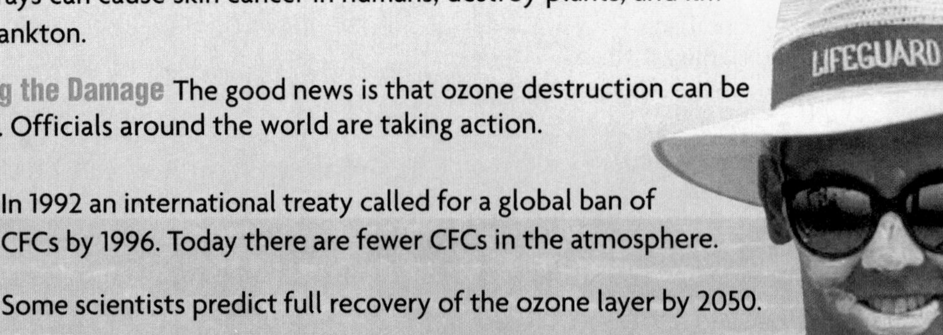

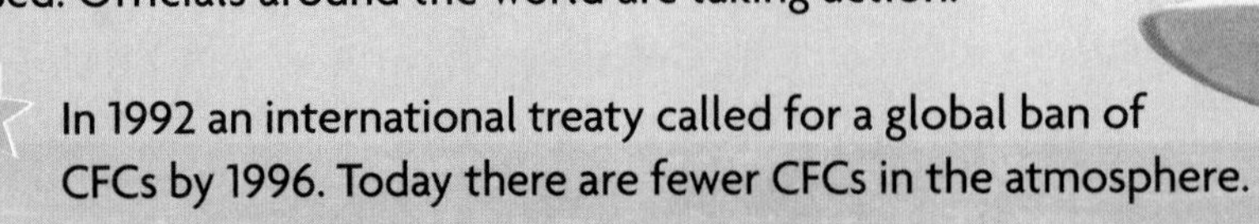

- In 1992 an international treaty called for a global ban of CFCs by 1996. Today there are fewer CFCs in the atmosphere.
- Some scientists predict full recovery of the ozone layer by 2050.

A lifeguard in Australia prepares for a day in the sun with hat, sunglasses, and zinc cream.

Making a Difference

Ozone Prizewinners Three scientists shared the 1995 Nobel Prize in chemistry for their research on ozone. Americans Mario Molina (photo, at right) and F. Sherwood Rowland and Dutch citizen Paul Crutzen shared the honor after describing the chemical processes by which ozone is formed and destroyed in the atmosphere. Before they explored the issue, little was known about how human-made chemicals affect ozone. The three scientists were able to show that the release of CFCs into the air damages the ozone layer. Their important research led governments around the world to ban the use of CFCs.

Mario Molina

Keeping Watch Antarctica has long been seen as a barometer of Earth's health. Scientists from all over the world live and work in research stations scattered throughout Antarctica. In 1985 scientists reported that the ozone layer over Antarctica had decreased dramatically. Since then, they have been closely watching the ozone layer, collecting data from special instruments that record ozone levels. Governments and environmental groups use this information to determine what should be done.

A scientist in Antarctica checks ozone levels.

What Can You Do?

Get Involved

Organize a "Sun Alert" campaign to warn students in younger grades about the dangers of overexposure to the sun.

Find Out More

On the Trail of the Missing Ozone, an online book, tells why we need the ozone layer and how to prevent ozone depletion. You can read it at www.epa.gov/ozone/science/missoz/index.html. Share what you learn with the class.

Use the Internet

Check current ozone levels worldwide by visiting NASA's Goddard Space Flight Center's Web site at http://jwocky.gsfc.nasa.gov

Chapter 28

Oceania and Antarctica

To learn more about the people and places of Oceania and Antarctica, view ***The World and Its People* Chapter 28** video.

Chapter Overview Visit the ***Geography: The World and Its People*** Web site at gwip.glencoe.com and click on **Chapter 28—Chapter Overviews** to preview information about Oceania and Antarctica.

Section 1 Oceania

Guide to Reading

Main Idea

Oceania is made up of thousands of Pacific Ocean islands organized into countries and territories.

Terms to Know

- continental island
- cacao
- copra
- pidgin language
- high island
- low island
- atoll
- typhoon
- phosphate
- trust territory

Places to Locate

- Oceania
- Melanesia
- Micronesia
- Polynesia
- Papua New Guinea

Reading Strategy

Make a chart like this one. In the right column, write two facts about each region.

Region	Facts
Melanesia	
Micronesia	
Polynesia	

◀ View of Tahiti, French Polynesia

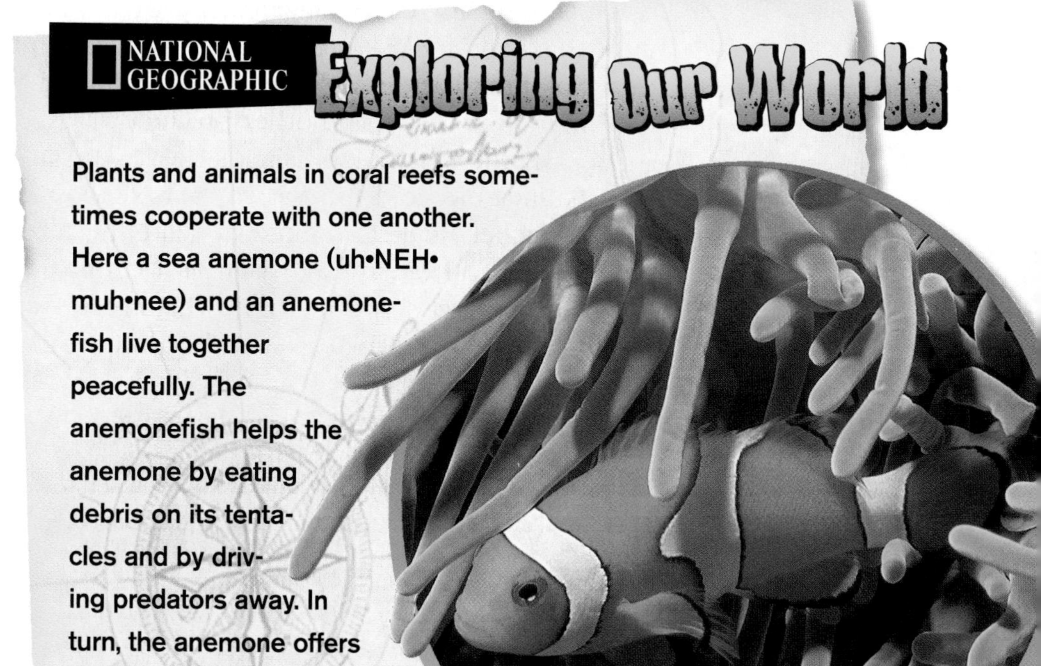

Plants and animals in coral reefs sometimes cooperate with one another. Here a sea anemone (uh•NEH•muh•nee) and an anemonefish live together peacefully. The anemonefish helps the anemone by eating debris on its tentacles and by driving predators away. In turn, the anemone offers the fish protection. Most anemonefish spend their lives inside the anemone.

Oceania is a culture region that includes about 25,000 islands in the Pacific Ocean. Geographers group Oceania into three main island regions—**Melanesia, Micronesia,** and **Polynesia.**

Melanesia

The islands of Melanesia lie just to the north and east of Australia. The largest country in size is **Papua New Guinea** (PA•pyu•wuh noo GIH•nee). It lies on the eastern half of the island of New Guinea. Slightly larger than California, the country's 4.7 million people also make it Oceania's most populous island. Southeast of Papua New Guinea are three other independent island countries: the Solomon Islands, the Fiji (FEE•jee) Islands, and Vanuatu (VAN•WAH•TOO). Near these countries is New Caledonia, a group of islands ruled by France.

Melanesia consists mostly of continental islands. A **continental island** is formed in two ways. Chunks of land may split off from a continent, or a land bridge erodes, breaking the island's connection to the continent. Rugged mountains and dense rain forests cover Melanesia's islands. Strips of fertile plains hug island coastlines. Most of Melanesia has a tropical climate. Temperatures seldom fall below 70°F (21°C) or rise above 80°F (27°C).

Most Melanesians work on subsistence farms. Papua New Guinea exports coffee, palm oil, and cacao. **Cacao** is a tropical tree whose seeds are used to make chocolate. The Solomon Islands also export palm oil and cacao. Sugarcane grown on the Fiji Islands is exported as sugar and molasses. **Copra,** or dried coconut meat, is produced on the Solomon Islands and Vanuatu. Coconut oil from copra is used to make margarine, soap, and other products.

Some Melanesian islands hold rich mineral resources. Papua New Guinea has gold, oil, and copper. New Caledonia mines large deposits of nickel. Rugged mountains make these resources difficult to reach.

Several Melanesian islands export timber and fish. Papua New Guinea has some factories that manufacture processed foods and machinery. Melanesia is also becoming a popular tourist destination.

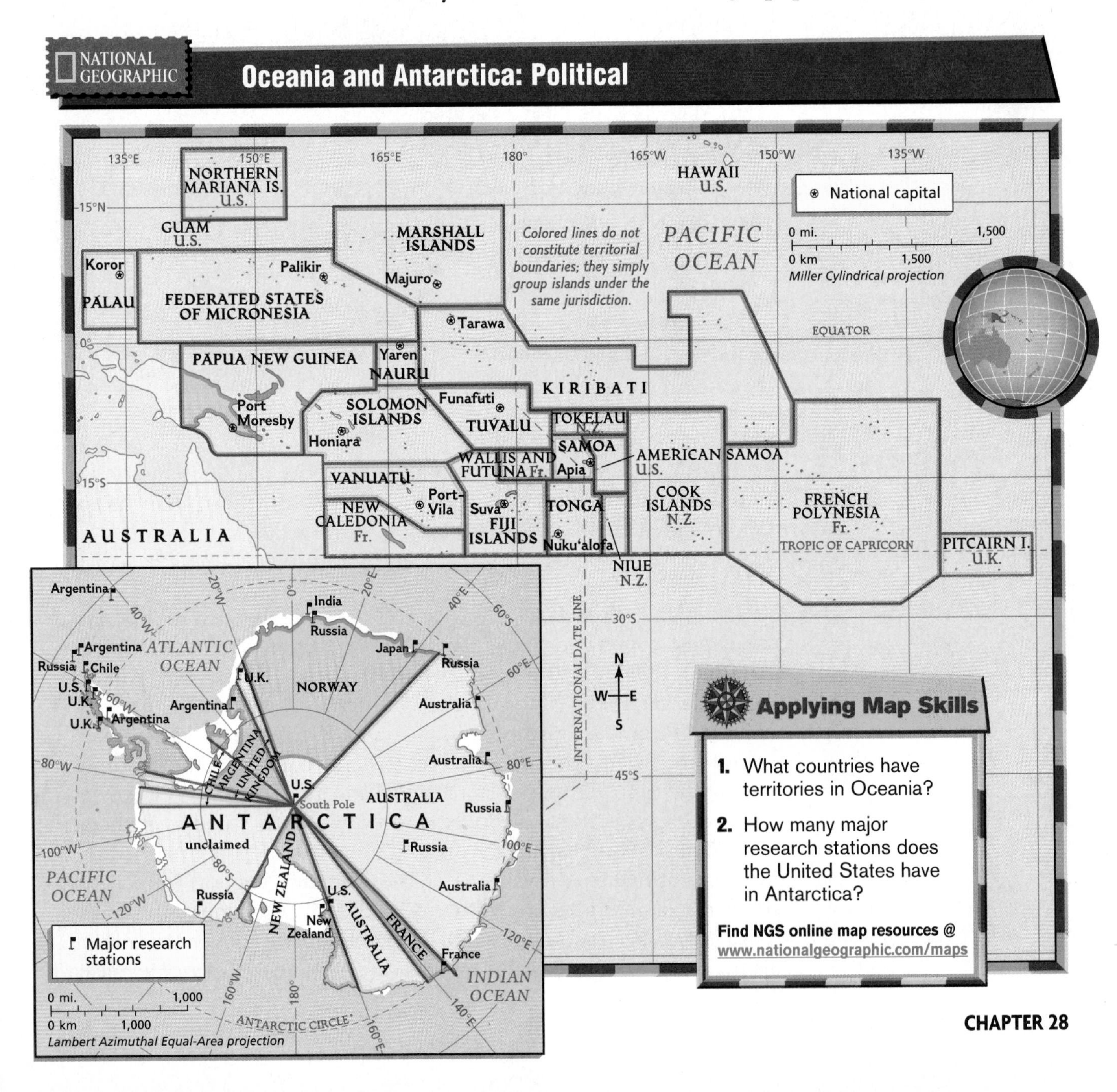

Melanesia's People Most Melanesians are originally from the Pacific islands. Exceptions include the people of New Caledonia—about one-third of whom are Europeans—and the Fiji Islands—half of whom are from South Asia. These South Asians are descendants of workers brought from British India in the late 1800s and early 1900s to work on sugarcane plantations. Today South Asians control much of the economy of the Fiji Islands. Fijians of Pacific descent own most of the land. Conflict often arises as the two different groups struggle for control of the government.

Languages and religions are diverse as well. More than 700 languages are spoken in Papua New Guinea alone. People here speak a **pidgin language** formed by combining parts of several different languages. People speak English in the Fiji Islands, while French is the main language of New Caledonia. Local traditional religions are practiced, but Christianity is widespread. The South Asian population is mostly Hindu.

Many Melanesians live in small villages in houses made of grass or other natural materials. In recent years, people have built concrete houses to protect themselves from tropical storms. Melanesians keep

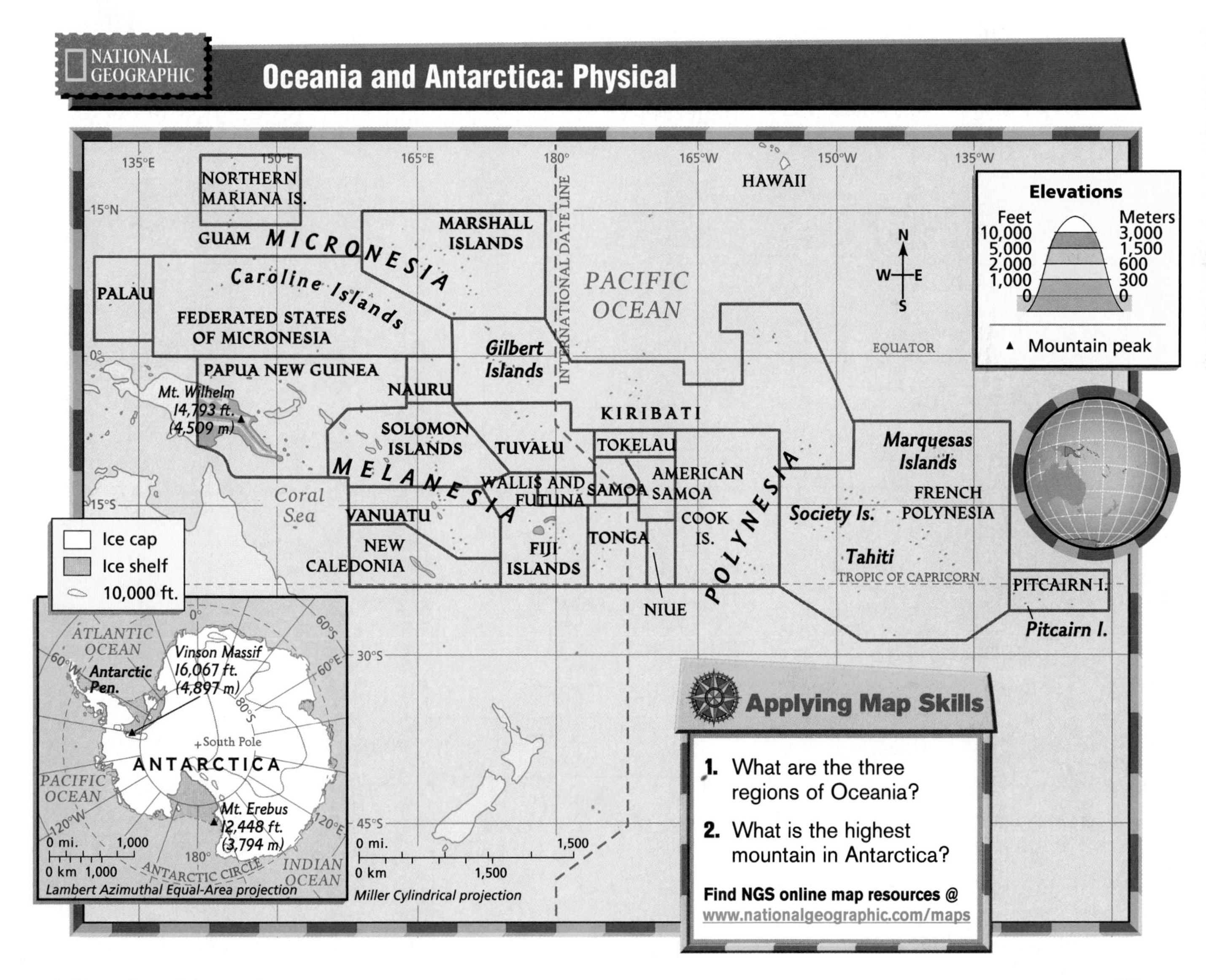

strong ties to their local group and often hold on to traditional ways. Only a small number live in cities. Port Moresby in Papua New Guinea and Suva in the Fiji Islands are the largest urban areas. Many people in these cities have jobs in business and government.

Reading Check **What is the largest country in Melanesia?**

Micronesia

The map on page 767 shows that the islands of Micronesia are scattered over a vast area of the ocean. Independent countries include the Federated States of Micronesia, the Marshall Islands, Palau (puh•LOW), Nauru (nah•OO•roo), and Kiribati (KIHR•uh•BAH•tee). The Northern Mariana Islands and Guam are territories of the United States.

Micronesia is made up of two types of islands—high islands and low islands. Volcanic activity formed the mountainous **high islands** many centuries ago. Coral, or skeletons of millions of tiny sea animals, formed the **low islands.** Most of the low islands are **atolls**—low-lying, ring-shaped islands that surround lagoons. The two types of islands have very different vegetation. The high islands have rich soil, towering waterfalls, and thick green vegetation. The low islands have maybe just a few palm trees.

Like Melanesia, Micronesia has a tropical climate. Daily temperatures average about 80°F (27°C) year-round, and rain is plentiful. From July to

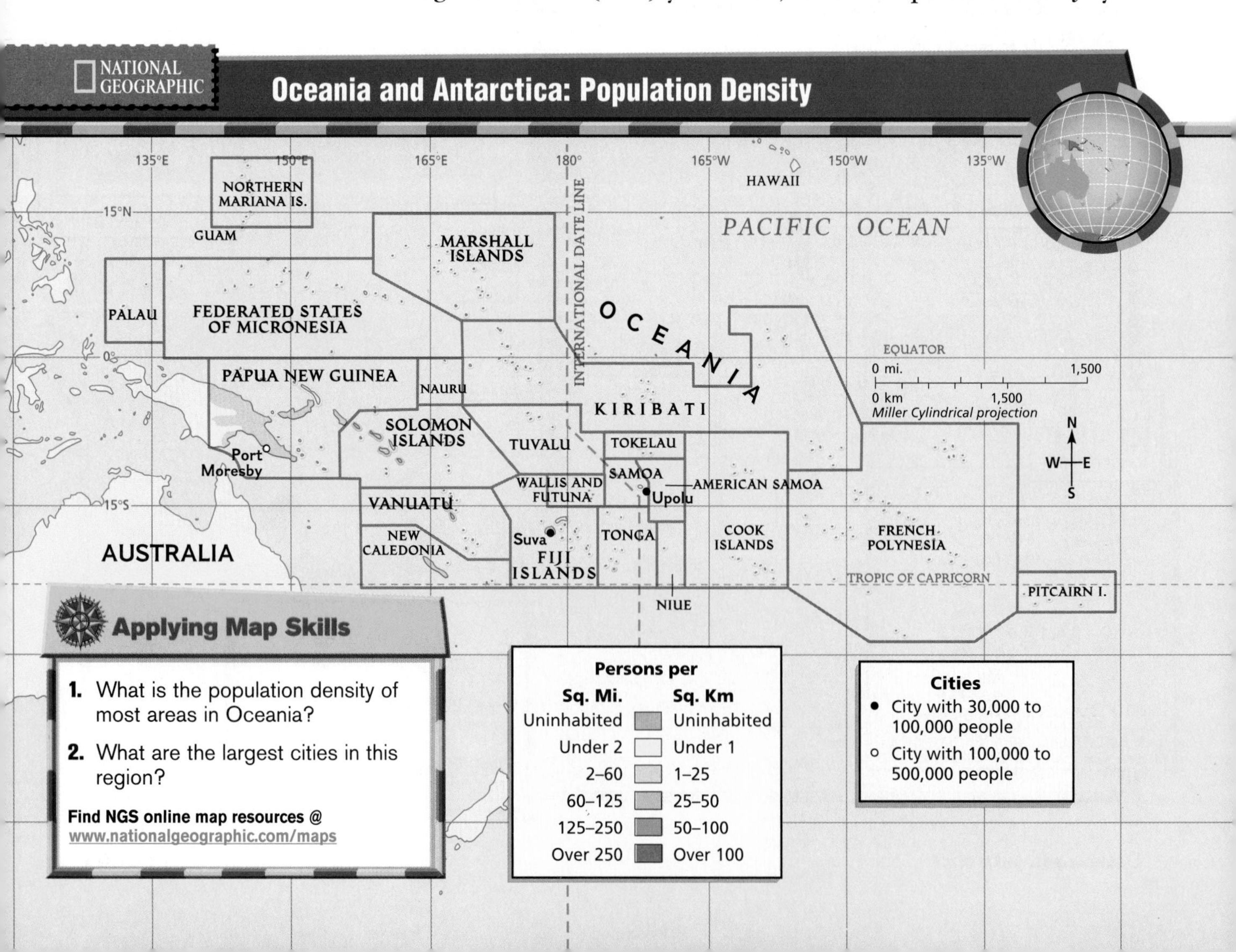

Applying Map Skills

1. What is the population density of most areas in Oceania?
2. What are the largest cities in this region?

Find NGS online map resources @ www.nationalgeographic.com/maps

October, typhoons sometimes strike the islands, causing loss of life and much destruction. A **typhoon** is another name for a hurricane, a fierce storm with winds of more than 74 miles (119 km) per hour.

On Micronesia's high islands, most people engage in subsistence farming. Farmers grow cassava, sweet potatoes, bananas, and coconuts. Some high island farmers also raise livestock. People in the low islands obtain food from the sea. On low islands, such as Palau, Guam, and the Northern Mariana Islands, recent population growth has resulted in the need to import food.

Micronesia receives financial aid from the United States, the European Union, and Australia. With this money, the Micronesians have built roads, ports, airfields, and small factories. Clothing is made on the Northern Mariana Islands. Beautiful beaches draw many tourists here, as well as to Palau, the Marshall Islands, and Guam. Several Micronesian islands have **phosphate,** a mineral salt used to make fertilizer. Phosphate supplies are now gone on Kiribati, and they have almost run out on Nauru. The Federated States of Micronesia and the Marshall Islands have phosphate but lack the money to mine this resource.

Micronesia's People Southeast Asians first settled Micronesia about 4,000 years ago. Explorers, traders, and missionaries from European countries came in the 1700s and early 1800s. By the early 1900s, many European countries, the United States, and Japan held colonies here.

During World War II, the United States and Japan fought a number of bloody battles on Micronesian islands. After World War II, most of Micronesia was turned over to the United States as **trust territories.** These territories were under temporary United States control. Since the 1970s, most have become independent.

Many of Micronesia's people are Pacific islanders. They speak local languages, although English is spoken on Nauru, the Marshall Islands, and throughout the rest of Micronesia. Christianity, brought by Western missionaries, is the most widely practiced religion. Micronesians generally live in villages headed by local chiefs. In recent years, many young people have left the villages to find jobs in towns.

Reading Check **In what two ways were Micronesia's islands formed?**

The oysters that produced these black pearls thrive in the warm waters of Oceania.

Movement **What is one of the fastest-growing industries in Polynesia?**

Polynesia

Polynesia includes three independent countries—Samoa, Tonga, and Tuvalu. Other island groups are under French rule and are known as French Polynesia. Tahiti, Polynesia's largest island, is part of this French-ruled area. American Samoa, a United States territory, is also part of this region.

Most Polynesian islands are high volcanic islands, some with tall, rugged mountains. Dense rain forests cover mountain valleys and coastal plains. Some of the islands are of the low atoll type. With little soil, the only vegetation is scattered coconut palms. Because Polynesia lies in the tropics, the climate is hot and humid.

Polynesians grow crops or fish for their food. Some farmers export coconuts and tropical fruits. The main manufacturing activity is food processing. The tuna you eat for lunch may have come from American Samoa. This island supplies about one-third of the tuna brought into the United States. Tonga exports squash and vanilla.

Tourism is one of the fastest-growing industries of Polynesia. Visitors come by air or sea to the emerald-green mountains and white, palm-lined beaches. New roads, hotels, shops, and restaurants serve these tourists.

Polynesia's People Settlers came to Polynesia later than they did to the other island regions. The first to arrive were probably Melanesians or Micronesians who crossed the vast Pacific Ocean in canoes.

During the late 1800s, several European nations divided Polynesia among themselves. They built military bases on the islands and later added airfields. The islands served as excellent refueling stops for long voyages across the Pacific. Beginning in the 1960s, several Polynesian territories chose independence, while others remained territories.

About 600,000 people live in Polynesia. Most Polynesians live in rural villages and practice traditional crafts. An increasing number live in towns and cities. Papeete (PAH•pay•AY•tay), located on Tahiti, is the capital of French Polynesia and the largest city in the region.

Reading Check What is the largest island in Polynesia?

Assessment

Defining Terms

1. **Define** continental island, cacao, copra, pidgin language, high island, low island, atoll, typhoon, phosphate, trust territory.

Recalling Facts

2. **Region** What three regions make up Oceania?
3. **Economics** What two kinds of economic activities are most important in these regions?
4. **History** What groups first settled the lands of Micronesia?

Critical Thinking

5. **Summarizing Information** What is copra, and what is it used for?
6. **Drawing Conclusions** Why do many people in Melanesia speak a pidgin language?

Graphic Organizer

7. **Organizing Information** Make a chart like this one. List all the islands of Oceania under their specific region, then note whether they are independent countries or territories.

Melanesia	Micronesia	Polynesia	Country/Territory of ?

Applying Geography Skills

8. **Analyzing Maps** Look at the economic activity map on page 774. Which territory has large deposits of nickel?

Study and Writing Skill

Writing a Report

Writing skills allow you to organize your ideas in a logical manner. The writing process involves using skills you have already learned, such as taking notes, outlining, and sequencing information.

Learning the Skill

Use the following guidelines to help you apply the writing process:

- Select an interesting topic. Do preliminary research to determine whether your topic is too broad or too narrow.
- Write a general statement that explains what you want to prove, discover, or illustrate in your writing. This will be the focus of your entire paper.
- Research your topic by coming up with a list of main ideas. Prepare note cards listing facts and source information for each main idea.
- Your report should have an introduction, a body, and a conclusion summarizing and restating your findings.
- Each paragraph should express one main idea in a topic sentence. Additional sentences should support or explain the main idea by using details and facts.

An atoll in the Pacific Ocean ▼

Practicing the Skill

Read the following paragraph, then answer the questions that follow.

> Most of Micronesia's low islands are atolls—low-lying, ring-shaped islands that surround lagoons. An atoll begins as a ring of coral that forms around the edge of a volcanic island. Over time, wind and water erode the volcano, wearing it down to sea level. Eventually only the atoll remains above the surface. The calm, shallow seawater inside the atoll is called a lagoon.

1. What is the main idea of this paragraph?
2. What are the supporting sentences?
3. What might be the topic of an additional paragraph that follows this one?

Applying the Skill

Suppose you are writing a report on Oceania. Answer the following questions about the writing process.

1. How could you narrow this topic?
2. What are three main ideas?
3. Name three possible sources of information.

Section 2

Antarctica

Guide to Reading

Main Idea

Antarctica is a harsh land of rock and ice, which the world's nations have agreed to leave open to scientific study.

Terms to Know

- crevasse
- ice shelf
- iceberg
- krill
- ozone

Places to Locate

- Antarctica
- Vinson Massif
- Antarctic Peninsula
- Mount Erebus

Reading Strategy

Make a chart like the one below. Under each heading, fill in at least one fact about Antarctica.

Antarctica	
Land	Climate
Resources	People

NATIONAL GEOGRAPHIC Exploring Our World

Whee! These Emperor penguins live on the harsh continent of Antarctica. Their shiny "tuxedos" and waddling walk fascinate people. Although they cannot fly, their feathers provide excellent insulation against the ice, snow, and freezing water. Emperor penguins often travel 30 miles (48 km) a day to bring food to their rookeries, or nests. Sometimes walking takes too long, so penguins simply slide on their bellies, which is called "tobogganing."

Antarctica sits on the southern end of the earth. Icy ocean water surrounds it. Freezing ice covers it. Cold winds blow over it. The least explored of all the continents, this frigid mysterious land is larger than either Europe or Australia.

A Unique Continent

Picture Antarctica—a rich, green land covered by forests and lush plants. Does this description match your mental image of the continent? Fossils discovered here reveal that millions of years ago, Antarctica's landscape was inhabited by dinosaurs and small mammals.

Today, however, a huge ice cap buries nearly 98 percent of Antarctica's land area. In some spots, this ice cap is 2 miles (3.2 km) thick—about the height of 10 tall skyscrapers stacked upon one another. This massive "sea" of ice holds about 70 percent of all the freshwater in the world.

The Antarctic ice cap is heavy and strong, but it also moves. In some areas, the ice cap forms **crevasses,** or cracks, that plunge more than 100 feet (30 m). At the Antarctic coast, the ice cap spreads past the land into the ocean. This layer of ice above the water is called an **ice shelf.** Huge chunks of ice sometimes break off, forming **icebergs** that float freely in the icy waters.

Highlands, Mountains, and Valleys Beneath the ice cap, Antarctica has highlands, valleys, and mountains—the same landforms you find on other continents. A long mountain range called the Transantarctic Mountains crosses the continent. The highest peak in Antarctica, the **Vinson Massif,** rises 16,067 feet (4,897 m). The Transantarctic Mountains sweep along the **Antarctic Peninsula,** which reaches within 600 miles (966 km) of South America's Cape Horn. East of the mountains is a high, flat plateau where you find the South Pole, the southernmost point of the earth. On an island off Antarctica's coast rises **Mount Erebus** (EHR•uh•buhs). It is Antarctica's most active volcano.

Climate Now that you have a mental picture of Antarctica's ice cap, think about this: Antarctica receives so little precipitation that it is the world's largest, coldest desert. Inland Antarctica receives no rain and hardly any new snow each year. The map below shows you that Antarctica has a polar ice cap climate. Imagine summer in a place where temperatures may fall as low as –30°F (–35°C) and climb to only

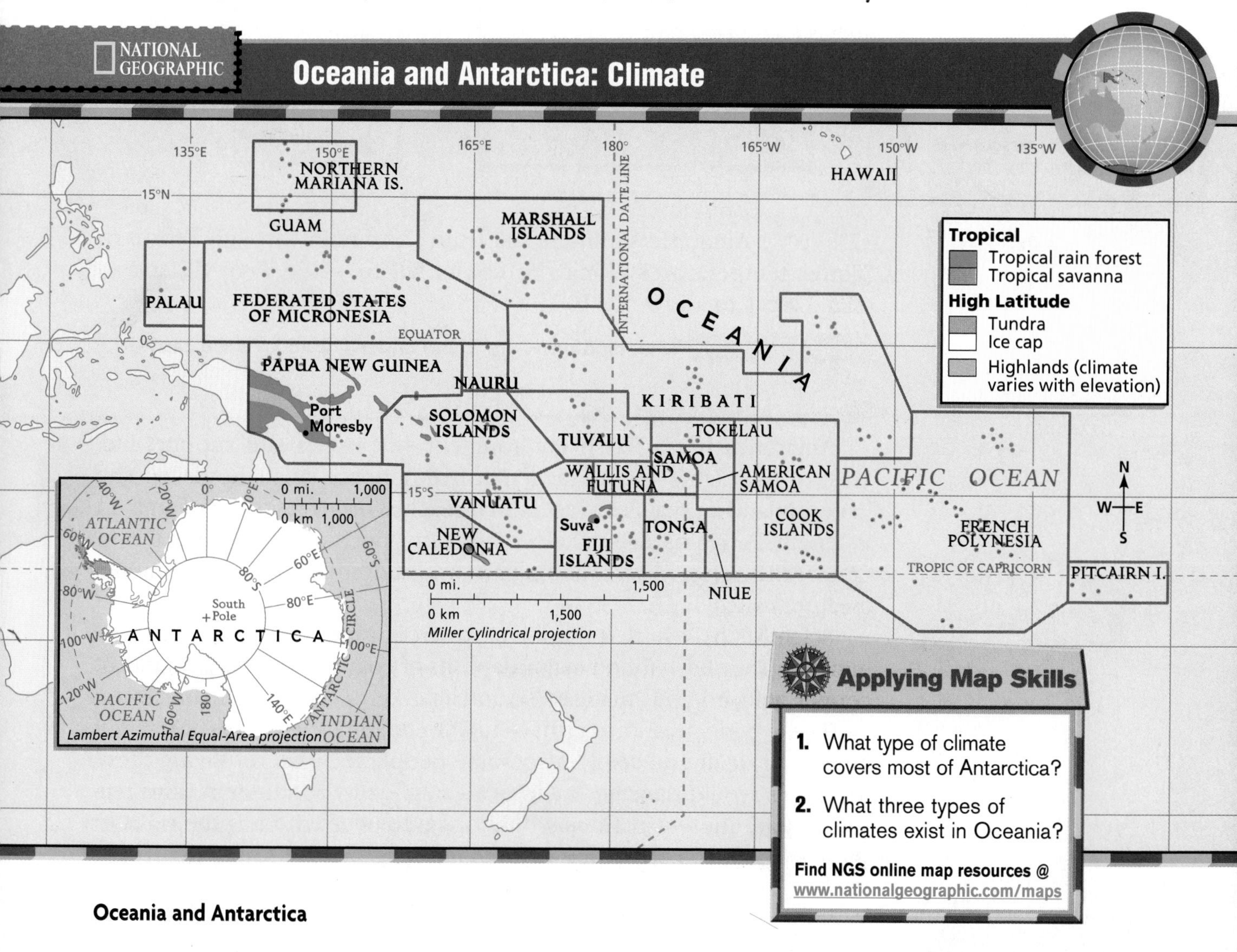

Applying Map Skills

1. What type of climate covers most of Antarctica?
2. What three types of climates exist in Oceania?

Find NGS online map resources @ www.nationalgeographic.com/maps

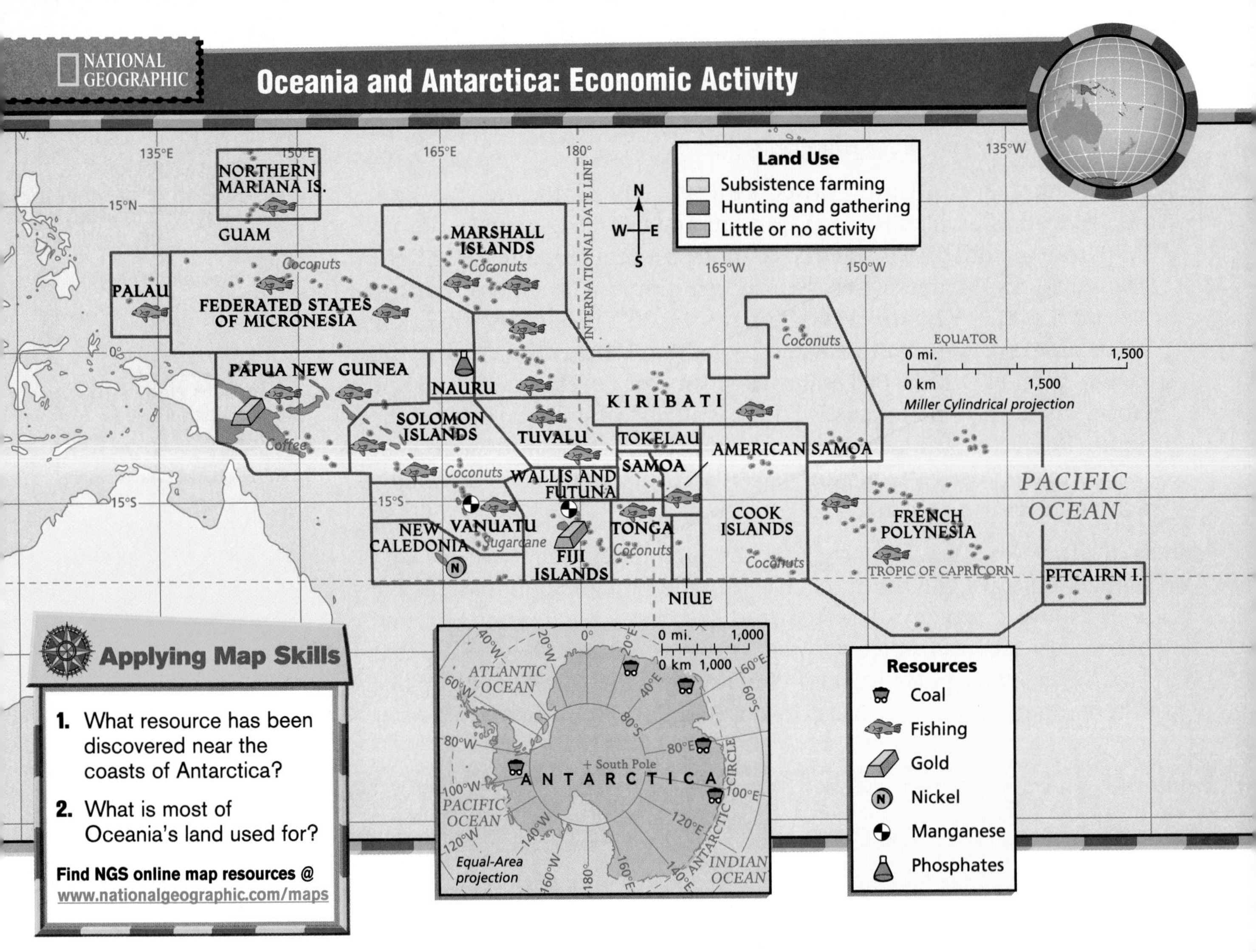

32°F (0°C). Antarctic summers last from December through February. Winter temperatures along the coasts fall to −40°F (−40°C), and in inland areas to a low of −100°F (−73°C).

Reading Check **What landforms are found under Antarctica's ice cap?**

Resources of Antarctica

Antarctica has a harsh environment, but it can still support life. Most of the plants and animals that live here are small, however. The largest inland animal is an insect that reaches only one-tenth of an inch in length. Penguins, fish, whales, and many kinds of flying birds live in or near the rich seas surrounding Antarctica. Many eat a tiny, shrimplike creature called **krill.**

Scientists believe that the ice of Antarctica hides a treasure chest of minerals. They have found major deposits of coal and smaller amounts of copper, gold, iron ore, manganese, and zinc. Petroleum may lie offshore.

These mineral resources have not yet been tapped. To do so would be very difficult and costly. Also, some people feel that removing these resources would damage Antarctica's fragile environment. A third reason is that different nations would disagree over who has the right to these resources. Forty-three nations have signed the Antarctic Treaty, which prohibits any nation from taking resources from the continent. It also bans weapons testing in Antarctica.

✓Reading Check **What is the Antarctic Treaty?**

A Vast Scientific Laboratory

The Antarctic Treaty does allow for scientific research in Antarctica. Many countries have scientific research stations here, but no single nation controls the vast continent. In January—summer in Antarctica—about 10,000 scientists come to study the land, plants, animals, and ice of this frozen land. Some 1,000 hardy scientists even stay during the harsh polar winter.

Much of the research focuses on ozone. **Ozone** is a type of oxygen that forms a layer in the atmosphere. The ozone layer protects all living things on the earth from certain harmful rays of the sun. In the 1980s, scientists discovered a weakening, or "hole," in this layer above Antarctica. If such weakening continues, some scientists say, climates around the world will get warmer. By studying this layer further, they hope to learn more about possible changes.

This frozen world attracts more than just scientists, though. Each year, a few thousand tourists come to Antarctica. With such a harsh environment, however, Antarctica is the only continent in the world that has no permanent population.

✓Reading Check **Why are scientists studying the ozone layer?**

The *Endurance*

In January 1915, Ernest Shackleton and his crew in the *Endurance* became trapped in Antarctica's freezing seawater. In late October, ice crushed the wooden ship, forcing the explorers to abandon it (below). They spent five more months drifting on the ice until they reached open water and used lifeboats to get away.

Assessment

Defining Terms

1. **Define** crevasse, ice shelf, iceberg, krill, ozone.

Recalling Facts

2. **Place** What landform covers central Antarctica?
3. **Location** Where in Antarctica would you find the most living things?
4. **Human/Environment Interaction** Why do scientists come to Antarctica?

Critical Thinking

5. **Summarizing Information** Why have countries agreed not to use the resources of Antarctica?
6. **Writing Questions** Imagine that you are planning a trip to Antarctica. What questions would you ask scientists working there?

Graphic Organizer

7. **Organizing Information** Draw a chart like the one below, and then look at the political map on page 766. In your chart, list the various national claims made in Antarctica by the world's countries. Then give the number of research stations for each country.

Countries With Claims in Antarctica	Number of Research Stations

Applying Geography Skills

8. **Analyzing Maps** Look at the physical map on page 767. What is the highest point in Antarctica?

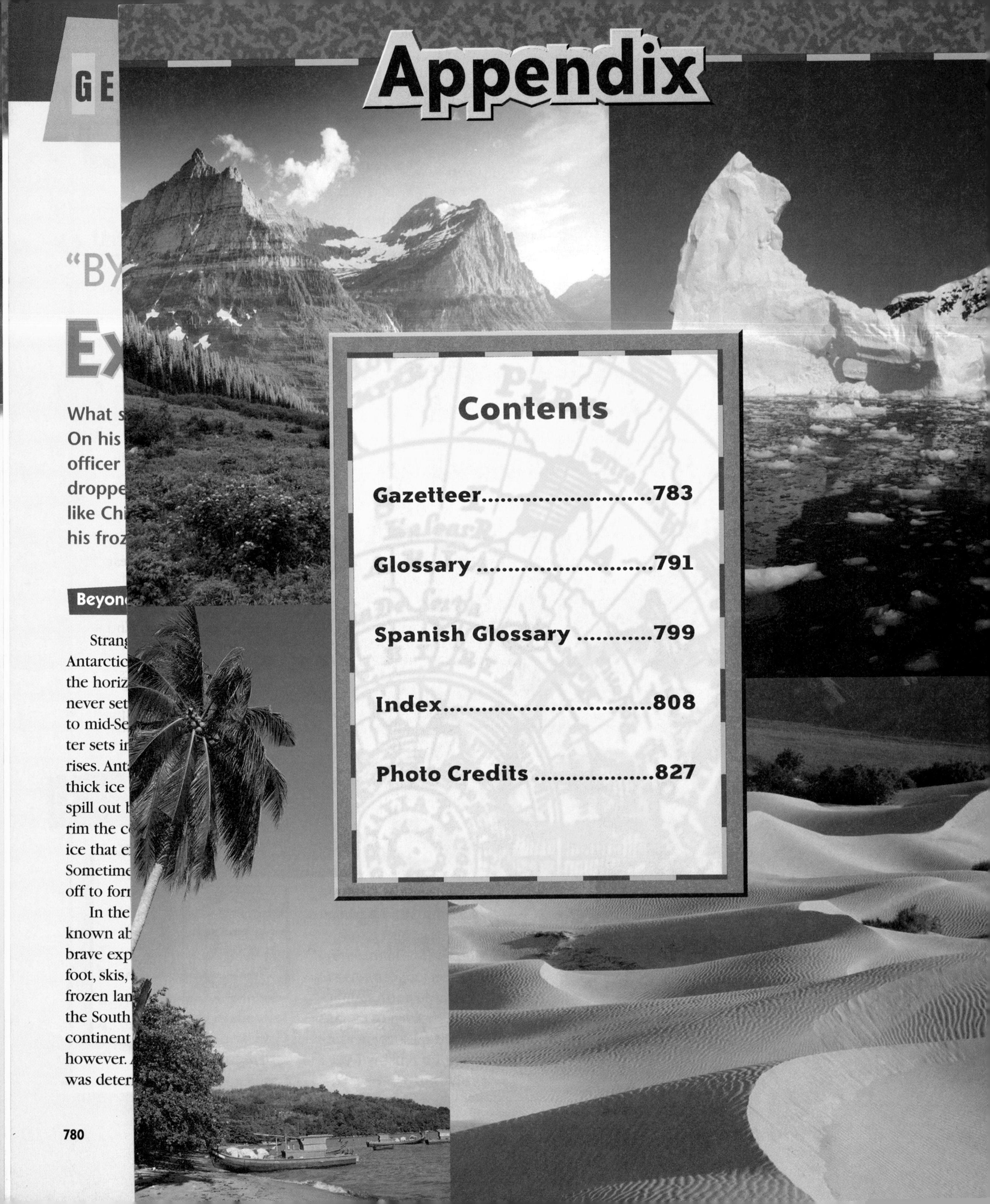

Appendix

Contents

Gazetteer............................783

Glossary 791

Spanish Glossary 799

Index.................................808

Photo Credits 827

GAZETTEER

Gazetteer

A Gazetteer (GA•zuh•TIHR) is a geographic index or dictionary. It shows latitude and longitude for cities and certain other places. Latitude and longitude are shown in this way: 48°N 2°E, or 48 degrees north latitude and two degrees east longitude. This Gazetteer lists most of the world's largest independent countries, their capitals, and several important geographic features. The page numbers tell where each entry can be found on a map in this book. As an aid to pronunciation, most entries are spelled phonetically.

Abidjan [AH•bee•JAHN] Capital of Côte d'Ivoire. 5°N 4°W (p. 531)

Abu Dhabi [AH•boo DAH•bee] Capital of the United Arab Emirates. 24°N 54°E (p. 449)

Abuja [ah•BOO•jah] Capital of Nigeria. 8°N 9°E (p. 531)

Accra [ah•KRUH] Capital of Ghana. 6°N 0° longitude (p. 531)

Addis Ababa [AHD•dihs AH•bah•BAH] Capital of Ethiopia. 9°N 39°E (p. 531)

Adriatic [AY•dree•A•tihk] **Sea** Arm of the Mediterranean Sea between the Balkan Peninsula and Italy. 44°N 14°E (p. 280)

Afghanistan [af•GA•nuh•STAN] Central Asian country west of Pakistan. 33°N 63°E (p. 449)

Albania [al•BAY•nee•uh] Country on the Adriatic Sea, south of Yugoslavia. 42°N 20°E (p. 281)

Algeria [al•JIHR•ee•uh] North African country east of Morocco. 29°N 1°E (p. 449)

Algiers [al•JIHRZ] Capital of Algeria. 37°N 3°E (p. 449)

Alps [ALPS] Mountain ranges extending through central Europe. 46°N 9°E (p. 280)

Amazon [A•muh•ZAHN] **River** Largest river in the world by volume and second-largest in length. 2°S 53°W (p. 172)

Amman [a•MAHN] Capital of Jordan. 32°N 36°E (p. 449)

Amsterdam [AHM•stuhr•DAHM] Capital of the Netherlands. 52°N 5°E (p. 281)

Andes [AN•DEEZ] Mountain system extending north and south along the western side of South America. 13°S 75°W (p. 172)

Andorra [an•DAWR•uh] Small country in southern Europe between France and Spain. 43°N 2°E (p. 281)

Angola [ang•GOH•luh] Southern African country north of Namibia. 14°S 16°E (p. 531)

Ankara [AHNG•kuh•ruh] Capital of Turkey. 40°N 33°E (p. 449)

Antananarivo [AHN•tah•NAH•nah•REE•voh] Capital of Madagascar. 19°S 48°E (p. 531)

Arabian [uh•RAY•bee•uhn] **Peninsula** Large peninsula extending into the Arabian Sea. 28°N 40°E (p. 448)

Argentina [AHR•juhn•TEE•nuh] South American country east of Chile on the Atlantic Ocean. 36°S 67°W (p. 173)

Armenia [ahr•MEE•nee•uh] Southeastern European country between the Black and Caspian Seas. 40°N 45°E (p. 449)

Ashgabat [AHSH•gah•BAHT] Capital of Turkmenistan. 38°N 58°E (p. 449)

Asmara [az•MAHR•uh] Capital of Eritrea. 16°N 39°E (p. 531)

Astana Capital of Kazakhstan. 51°N 72°E (p. 449)

Asunción [ah•SOON•see•OHN] Capital of Paraguay. 25°S 58°W (p. 173)

Athens [A•thuhnz] Capital of Greece. 38°N 24°E (p. 281)

Atlas [AT•luhs] **Mountains** Mountain range on the northern edge of the Sahara. 31°N 5°W (p. 448)

Australia [aw•STRAYL•yuh] Country and continent in Southern Hemisphere. 25°S 135°W (p. 741)

Austria [AWS•tree•uh] Western European country east of Switzerland and south of Germany and the Czech Republic. 47°N 12°E (p. 281)

Azerbaijan [A•zuhr•BY•JAHN] European-Asian country on the Caspian Sea. 40°N 47°E (p. 449)

Baghdad [BAG•DAD] Capital of Iraq. 33°N 44°E (p. 449)

Bahamas [buh•HAH•muhz] Country made up of many islands between Cuba and the United States. 23°N 74°W (p. 172)

Bahrain [bah•RAYN] Country located on the Persian Gulf. 26°N 51°E (p. 449)

Baku [bah•KOO] Capital of Azerbaijan. 40°N 50°E (p. 449)

Balkan [BAWL•kuhn] **Peninsula** Peninsula in southeastern Europe. 42°N 20°E (p. 280)

Baltic [BAWL•tihk] **Sea** Sea in northern Europe that is connected to the North Sea. 55°N 17°E (p. 280)

Bamako [BAH•mah•KOH] Capital of Mali. 13°N 8°W (p. 531)

Bangkok [BANG•KAHK] Capital of Thailand. 14°N 100°E (p. 637)

GAZETTEER

Bangladesh [BAHNG•gluh•DEHSH] South Asian country bordered by India and Myanmar. 24°N 90°E (p. 637)
Bangui [BAHNG•GEE] Capital of the Central African Republic. 4°N 19°E (p. 531)
Banjul [BAHN•JOOL] Capital of Gambia. 13°N 17°W (p. 531)
Barbados [bahr•BAY•duhs] Island country between the Atlantic Ocean and the Caribbean Sea. 14°N 59°W (p. 173)
Beijing [BAY•JIHNG] Capital of China. 40°N 116°E (p. 637)
Beirut [bay•ROOT] Capital of Lebanon. 34°N 36°E (p. 449)
Belarus [BEE•luh•ROOS] Eastern European country west of Russia. 54°N 28°E (p. 281)
Belgium [BEHL•juhm] Western European country south of the Netherlands. 51°N 3°E (p. 281)
Belgrade [BEHL•GRAYD] Capital of Yugoslavia. 45°N 21°E (p. 281)
Belize [buh•LEEZ] Central American country east of Guatemala. 18°N 89°W (p. 173)
Belmopan [BEHL•moh•PAHN] Capital of Belize. 17°N 89°W (p. 173)
Benin [buh•NEEN] West African country west of Nigeria. 8°N 2°E (p. 531)
Berlin [behr•LEEN] Capital of Germany. 53°N 13°E (p. 281)
Bern [BEHRN] Capital of Switzerland. 47°N 7°E (p. 281)
Bhutan [boo•TAHN] South Asian country northeast of India. 27°N 91°E (p. 637)
Bishkek [bihsh•KEHK] Capital of Kyrgyzstan. 43°N 75°E (p. 449)
Bissau [bihs•SOW] Capital of Guinea-Bissau. 12°N 16°W (p. 531)
Black Sea Large sea between Europe and Asia. 43°N 32°E (p. 281)
Bloemfontein [BLOOM•FAHN•TAYN] Judicial capital of South Africa. 26°E 29°S (p. 531)
Bogotá [BOH•goh•TAH] Capital of Colombia. 5°N 74°W (p. 173)
Bolivia [buh•LIHV•ee•uh] Country in the central part of South America, north of Argentina. 17°S 64°W (p. 173)
Bosnia and Herzegovina [BAHZ•nee•uh HEHRT•seh•GAW•vee•nuh] Southeastern European country between Yugoslavia and Croatia. 44°N 18°E (p. 281)
Botswana [bawt•SWAH•nah] Southern African country north of the Republic of South Africa. 22°S 23°E (p. 531)
Brasília [brah•ZEEL•yuh] Capital of Brazil. 16°S 48°W (p. 173)
Bratislava [BRAH•tih•SLAH•vuh] Capital of Slovakia. 48°N 17°E (p. 281)
Brazil [bruh•ZIHL] Largest country in South America. 9°S 53°W (p. 173)
Brazzaville [BRAH•zuh•VEEL] Capital of the Congo. 4°S 15°E (p. 531)
Brunei [bru•NY] Southeast Asian country on northern coast of the island of Borneo. 5°N 114°E (p. 636)
Brussels [BRUH•suhlz] Capital of Belgium. 51°N 4°E (p. 281)
Bucharest [BOO•kuh•REHST] Capital of Romania. 44°N 26°E (p. 281)
Budapest [BOO•duh•PEHST] Capital of Hungary. 48°N 19°E (p. 281)
Buenos Aires [BWAY•nuhs AR•eez] Capital of Argentina. 34°S 58°W (p. 173)
Bujumbura [BOO•juhm•BUR•uh] Capital of Burundi. 3°S 29°E (p. 531)
Bulgaria [BUHL•GAR•ee•uh] Southeastern European country south of Romania. 42°N 24°E (p. 281)
Burkina Faso [bur•KEE•nuh FAH•soh] West African country south of Mali. 12°N 3°E (p. 531)
Burundi [bu•ROON•dee] East African country at the northern end of Lake Tanganyika. 3°S 30°E (p. 531)

Cairo [KY•ROH] Capital of Egypt. 31°N 32°E (p. 449)
Cambodia [kam•BOH•dee•uh] Southeast Asian country south of Thailand and Laos. 12°N 104°E (p. 637)
Cameroon [KA•muh•ROON] Central African country on the northeast shore of the Gulf of Guinea. 6°N 11°E (p. 531)
Canada [KA•nuh•duh] Northernmost country in North America. 50°N 100°W (p. 109)
Canberra [KAN•BEHR•uh] Capital of Australia. 35°S 149°E (p. 741)
Cape Town Legislative capital of the Republic of South Africa. 34°S 18°E (p. 531)
Cape Verde [VUHRD] Island country off the coast of western Africa in the Atlantic Ocean. 15°N 24°W (p. 531)
Caracas [kah•RAH•kahs] Capital of Venezuela. 11°N 67°W (p. 173)
Caribbean [KAR•uh•BEE•uhn] **Sea** Part of the Atlantic Ocean bordered by the West Indies, South America, and Central America. 15°N 76°W (p. 172)
Caspian [KAS•pee•uhn] **Sea** Salt lake between Europe and Asia that is the world's largest inland body of water. 40°N 52°E (p. 448)
Caucasus [KAW•kuh•suhs] **Mountains** Mountain range between the Black and Caspian Seas. 43°N 42°E (p. 448)
Central African Republic Central African country south of Chad. 8°N 21°E (p. 531)
Chad [CHAD] Country west of Sudan in the African Sahel. 18°N 19°E (p. 530)
Chile [CHEE•lay] South American country west of Argentina. 35°S 72°W (p. 173)
China [CHY•nuh] Country in eastern and central Asia, known officially as the People's Republic of China. 37°N 93°E (p. 637)
Chişinău [KEE•shee•NOW] Capital of Moldova. 47°N 29°E (p. 281)
Colombia [kuh•LUHM•bee•uh] South American country west of Venezuela. 4°N 73°W (p. 173)
Colombo [kuh•LUHM•boh] Capital of Sri Lanka. 7°N 80°E (p. 637)

Comoros [KAH•muh•ROHZ] Small island country in Indian Ocean between the island of Madagascar and the southeast African mainland. 13°S 43°E (p. 531)
Conakry [KAH•nuh•kree] Capital of Guinea. 10°N 14°W (p. 531)
Congo [KAHNG•goh] Central African country east of the Democratic Republic of the Congo. 3°S 14°E (p. 531)
Congo, Democratic Republic of the Central African country north of Zambia and Angola. 1°S 22°E (p. 531)
Copenhagen [KOH•puhn•HAY•guhn] Capital of Denmark. 56°N 12°E (p. 281)
Costa Rica [KAWS•tah REE•kah] Central American country south of Nicaragua. 11°N 85°W (p. 173)
Côte d'Ivoire [KOHT dee•VWAHR] West African country south of Mali. 8°N 7°W (p. 531)
Croatia [kroh•AY•shuh] Southeastern European country on the Adriatic Sea. 46°N 16°E (p. 281)
Cuba [KYOO•buh] Island country in the West Indies. 22°N 79°W (p. 172)
Cyprus [SY•pruhs] Island country in the eastern Mediterranean Sea, south of Turkey. 35°N 31°E (p. 281)
Czech [CHEHK] **Republic** Eastern European country north of Austria. 50°N 15°E (p. 281)

Dakar [dah•KAHR] Capital of Senegal. 15°N 17°W (p. 531)
Damascus [duh•MAS•kuhs] Capital of Syria. 34°N 36°E (p. 449)
Dar es Salaam (DAHR EHS sah•LAHM] Capital of Tanzania. 7°S 39°E (p. 531)
Denmark [DEHN•MAHRK] Northern European country between the Baltic and North Seas. 56°N 9°E (p. 281)
Dhaka [DA•kuh] Capital of Bangladesh. 24°N 90°E (p. 637)
Djibouti [jih•BOO•tee] East African country on the Gulf of Aden. 12°N 43°E (p. 531)
Doha [DOH•huh] Capital of Qatar. 25°N 51°E (p. 449)
Dominican [duh•MIH•nih•kuhn] **Republic** Country in the West Indies on the eastern part of Hispaniola. 19°N 71°W (p. 206)
Dublin [DUH•blihn] Capital of Ireland. 53°N 6°W (p. 281)
Dushanbe [doo•SHAM•buh] Capital of Tajikistan. 39°N 69°E (p. 449)

East Timor [TEE•MOHR] Previous province of Indonesia, now under UN administration. 10°S 127°E (p. 637)
Ecuador [EH•kwuh•DAWR] South American country southwest of Colombia. 0° latitude 79°W (p. 173)
Egypt [EE•jihpt] North African country on the Mediterranean Sea. 27°N 27°E (p. 449)
El Salvador [ehl SAL•vuh•DAWR] Central American country southwest of Honduras. 14°N 89°W (p. 173)
Equatorial Guinea [EE•kwuh•TOHR•ee•uhl GIH•nee] Central African country south of Cameroon. 2°N 8°E (p. 531)
Eritrea [EHR•uh•TREE•uh] East African country north of Ethiopia. 17°N 39°E (p. 531)
Estonia [eh•STOH•nee•uh] Eastern European country on the Baltic Sea. 59°N 25°E (p. 281)
Ethiopia [EE•thee•OH•pee•uh] East African country north of Somalia and Kenya. 8°N 38°E (p. 531)
Euphrates [yu•FRAY•TEEZ] **River** River in southwestern Asia that flows through Syria and Iraq and joins the Tigris River. 36°N 40°E (p. 448)

Fiji [FEE•jee] **Islands** Country comprised of an island group in the southwest Pacific Ocean. 19°S 175°E (p. 741)
Finland [FIHN•luhnd] Northern European country east of Sweden. 63°N 26°E (p. 281)
France [FRANS] Western European country south of the United Kingdom. 47°N 1°E (p. 281)
Freetown [FREE•TOWN] Capital of Sierra Leone. 9°N 13°W (p. 531)
French Guiana [gee•A•nuh] French-owned territory in northern South America. 5°N 53°W (p. 173)

Gabon [ga•BOHN] Central African country on the Atlantic Ocean. 0° latitude 12°E (p. 531)
Gaborone [GAH•boh•ROH•NAY] Capital of Botswana. 24°S 26°E (p. 531)
Gambia [GAM•bee•uh] West African country along the Gambia River. 13°N 16°W (p. 531)
Georgetown [JAWRJ•TOWN] Capital of Guyana. 8°N 58°W (p. 173)
Georgia [JAWR•juh] Asian-European country bordering the Black Sea south of Russia. 42°N 43°E (p. 449)
Germany [JUHR•muh•nee] Western European country south of Denmark, officially called the Federal Republic of Germany. 52°N 10°E (p. 281)
Ghana [GAH•nuh] West African country on the Gulf of Guinea. 8°N 2°W (p. 531)
Great Plains The continental slope extending through the United States and Canada. 45°N 104°W (p. 108)
Greece [GREES] Southern European country on the Balkan Peninsula. 39°N 22°E (p. 280)
Greenland [GREEN•luhnd] Island in northwestern Atlantic Ocean and the largest island in the world. 74°N 40°W (p. 109)
Guatemala [GWAH•tay•MAH•lah] Central American country south of Mexico. 16°N 92°W (p. 173)
Guatemala City Capital of Guatemala. 15°N 91°W (p. 173)

Guinea [GIH•nee] West African country on the Atlantic coast. 11°N 12°W (p. 531)

Guinea-Bissau [GIH•nee bih•SOW] West African country on the Atlantic coast. 12°N [illegible]°W (p. 531)

Gulf of Mexico Gulf on part of the southern coast of North America. 25°N [illegible] (p. [illegible])

Guyana [gy•AH•nuh] South American country between Venezuela and Suriname. 8°N 59°W (p. 173)

Haiti [HAY•tee] Country in the West Indies on the western part of Hispaniola. 19°N 72°W (p. 206)

Hanoi [ha•NOY] Capital of Vietnam. 21°N 106°E (p. 637)

Harare [hah•RAH•RAY] Capital of Zimbabwe. 18°S 31°E (p. 531)

Havana [huh•VA•nuh] Capital of Cuba. 23°N 82°W (p. 173)

Helsinki [HEHL•SIHNG•kee] Capital of Finland. 60°N 24°E (p. 281)

Himalaya [HI•muh•LAY•uh] Mountain ranges in southern Asia, bordering the Indian subcontinent on the north. 30°N 85°E (p. 636)

Honduras [hahn•DUR•uhs] Central American country on the Caribbean Sea. 15°N 88°W (p. 173)

Hong Kong [HAWNG KAWNG] Port and industrial center in southern China. 22°N 115°E (p. 637)

Hungary [HUHNG•guh•ree] Eastern European country south of Slovakia. 47°N 18°E (p. 281)

Iberian [eye•BIHR•ee•uhn] **Peninsula** Peninsula in southwest Europe, occupied by Spain and Portugal. 41°N 1°W (p. 280)

Iceland [EYES•luhnd] Island country between the North Atlantic and the Arctic Oceans, 65°N 20°W (p. 281)

India [IHN•dee•uh] South Asian country south of China and Nepal. 23°N 78°E (p. 637)

Indonesia [IHN•duh•NEE•zhuh] Southeast Asian island country known as the Republic of Indonesia. 5°S 119°E (p. 637)

Indus [IHN•duhs] **River** River in Asia that begins in Tibet and flows through Pakistan to the Arabian Sea. 27°N 68°E (p. 636)

Iran [ih•RAN] Southwest Asian country that was formerly named Persia. 31°N 54°E (p. 449)

Iraq [ih•RAHK] Southwest Asian country west of Iran. 32°N 43°E (p. 449)

Ireland [EYER•luhnd] Island west of Great Britain occupied by the Republic of Ireland and Northern Ireland. 54°N 8°W (p. 281)

Islamabad [ihs•LAH•muh•BAHD] Capital of Pakistan. 34°N 73°E (p. 637)

Israel [IHZ•ree•uhl] Southwest Asian country south of Lebanon. 33°N 34°E (p. 449)

Italy [IHT•uhl•ee] Southern European country south of Switzerland and east of France. 44°N 11°E (p. 281)

Jakarta [juh•KAHR•tuh] Capital of Indonesia. 6°S 107°E (p. 637)

Jamaica [juh•MAY•kuh] Island country in the West Indies. 18°N 78°W (p. 173)

Japan [juh•PAN] East Asian country consisting of the four large islands of Hokkaido, Honshu, Shikoku, and Kyushu, plus thousands of small islands. 37°N 134°E (p. 636)

Jerusalem [juh•ROO•suh•luhm] Capital of Israel and a holy city for Christians, Jews, and Muslims. 32°N 35°E (p. 449)

Jordan [JAWRD•uhn] Southwest Asian country south of Syria. 30°N 38°E (p. 449)

Kabul [KAH•buhl] Capital of Afghanistan. 35°N 69°E (p. 449)

Kampala [kahm•PAH•lah] Capital of Uganda. 0° latitude 32°E (p. 531)

Kathmandu [KAT•MAN•DOO] Capital of Nepal. 28°N 85°E (p. 637)

Kazakhstan [kuh•ZAHK•STAHN] Large Asian country south of Russia and bordering the Caspian Sea. 48°N 59°E (p. 449)

Kenya [KEHN•yuh] East African country south of Ethiopia. 1°N 37°E (p. 531)

Khartoum [kahr•TOOM] Capital of Sudan. 16°N 33°E (p. 531)

Kiev [KEE•ihf] Capital of Ukraine. 50°N 31°E (p. 281)

Kigali [kee•GAH•lee] Capital of Rwanda. 2°S 30°E (p. 531)

Kingston [KIHNG•stuhn] Capital of Jamaica. 18°N 77°W (p. 173)

Kinshasa [kihn•SHAH•suh] Capital of the Democratic Republic of the Congo. 4°S 15°E (p. 531)

Kuala Lumpur [KWAH•luh LUM•PUR] Capital of Malaysia. 3°N 102°E (p. 637)

Kuwait [ku•WAYT] Country on the Persian Gulf between Saudi Arabia and Iraq. 29°N 48°E (p. 449)

Kyrgyzstan [KIHR•gih•STAN] Central Asian country on China's western border. 41°N 75°E (p. 449)

Laos [LOWS] Southeast Asian country south of China and west of Vietnam. 20°N 102°E (p. 637)

La Paz [lah PAHS] Administrative capital of Bolivia, and the highest capital in the world. 17°S 68°W (p. 173)

Latvia [LAT•vee•uh] Eastern European country west of Russia on the Baltic Sea. 57°N 25°E (p. 281)

GAZETTEER

Lebanon [LEH•buh•nuhn] Country south of Syria on the Mediterranean Sea. 34°N 34°E (p. 449)
Lesotho [luh•SOH•TOH] Southern African country within the borders of the Republic of South Africa. 30°S 28°E (p. 531)
Liberia [ly•BIHR•ee•uh] West African country south of Guinea. 7°N 10°W (p. 531)
Libreville [LEE•bruh•VIHL] Capital of Gabon. 1°N 9°E (p. 531)
Libya [LIH•bee•uh] North African country west of Egypt on the Mediterranean Sea. 28°N 15°E (p. 449)
Liechtenstein [LIHKT•uhn•SHTYN] Small country in central Europe between Switzerland and Austria. 47°N 10°E (p. 281)
Lilongwe [lih•LAWNG•GWAY] Capital of Malawi. 14°S 34°E (p. 531)
Lima [LEE•mah] Capital of Peru. 12°S 77°W (p. 173)
Lisbon [LIHZ•buhn] Capital of Portugal. 39°N 9°W (p. 281)
Lithuania [LIH•thuh•WAY•nee•uh] Eastern European country northwest of Belarus on the Baltic Sea. 56°N 24°E (p. 281)
Ljubljana [lee•OO•blee•AH•nuh] Capital of Slovenia. 46°N 14°E (p. 281)
Lomé [loh•MAY] Capital of Togo. 6°N 1°E (p. 531)
London [LUHN•duhn] Capital of the United Kingdom, on the Thames River. 52°N 0° longitude (p. 281)
Luanda [lu•AHN•duh] Capital of Angola. 9°S 13°E (p. 531)
Lusaka [loo•SAH•kah] Capital of Zambia. 15°S 28°E (p. 531)
Luxembourg [LUHK•suhm•BUHRG] Small European country between France, Belgium, and Germany. 50°N 7°E (p. 281)

Macau [muh•KOW] Port in southern China. 22°N 113°E (p. 637)
Macedonia [MA•suh•DOH•nee•uh] Southeastern European country north of Greece. 42°N 22°E (p. 281). Macedonia also refers to a geographic region covering northern Greece, the country Macedonia, and part of Bulgaria.
Madagascar [MA•duh•GAS•kuhr] Island in the Indian Ocean off the southeastern coast of Africa. 18°S 43°E (p. 531)
Madrid [muh•DRIHD] Capital of Spain. 41°N 4°W (p. 281)
Malabo [mah•LAH•boh] Capital of Equatorial Guinea. 4°N 9°E (p. 531)
Malawi [mah•LAH•wee] Southern African country south of Tanzania and east of Zambia. 11°S 34°E (p. 531)
Malaysia [muh•LAY•zhuh] Southeast Asian country with land on the Malay Peninsula and on the island of Borneo. 4°N 101°E (p. 636)
Maldives [MAWL•DEEVZ] Island country southwest of India in the Indian Ocean. 5°N 42°E (p. 637)
Mali [MAH•lee] West African country east of Mauritania. 16°N 0° longitude (p. 531)
Managua [mah•NAH•gwah] Capital of Nicaragua. 12°N 86°W (p. 173)
Manila [muh•NIH•luh] Capital of the Philippines. 15°N 121°E (p. 637)
Maputo [mah•POO•toh] Capital of Mozambique. 26°S 33°E (p. 531)
Maseru [MA•zuh•ROO] Capital of Lesotho. 29°S 27°E (p. 531)
Mauritania [MAWR•uh•TAY•nee•uh] West African country north of Senegal. 20°N 14°W (p. 531)
Mauritius [maw•RIH•shuhs] Island country in the Indian Ocean east of Madagascar. 21°S 58°E (p. 531)
Mbabane [uhm•bah•BAH•nay] Capital of Swaziland. 26°S 31°E (p. 531)
Mediterranean [MEH•duh•tuh•RAY•nee•uhn] **Sea** Large inland sea surrounded by Europe, Asia, and Africa. 36°N 13°E (p. 280)
Mekong [MAY•KAWNG] **River** River in southeastern Asia that begins in Tibet and empties into the South China Sea. 18°N 104°E (p. 636)
Mexico [MEHK•sih•KOH] North American country south of the United States. 24°N 104°W (p. 173)
Mexico City Capital of Mexico. 19°N 99°W (p. 173)
Minsk [MIHNSK] Capital of Belarus. 54°N 28°E (p. 281)
Mississippi [MIH•suh•SIH•pee] **River** Large river system in the central United States that flows southward into the Gulf of Mexico. 32°N 92°W (p. 108)
Mogadishu [MOH•guh•DEE•SHOO] Capital of Somalia. 2°N 45°E (p. 531)
Moldova [mawl•DAW•vuh] Small European country between Ukraine and Romania. 48°N 28°E (p. 281)
Monaco [MAH•nuh•KOH] Small country in southern Europe on the French Mediterranean coast. 44°N 8°E (p. 281)
Mongolia [mahn•GOHL•yuh] Country in Asia between Russia and China. 46°N 100°E (p. 637)
Monrovia [muhn•ROH•vee•uh] Capital of Liberia. 6°N 11°W (p. 531)
Montevideo [MAHN•tuh•vuh•DAY•OH] Capital of Uruguay. 35°S 56°W (p. 173)
Morocco [muh•RAH•KOH] North African country on the Mediterranean Sea and the Atlantic Ocean. 32°N 7°W (p. 449)
Moscow [MAHS•KOW] Capital of Russia. 56°N 38°E (p. 401)
Mount Everest [EHV•ruhst] Highest mountain in the world, in the Himalaya between Nepal and Tibet. 28°N 87°E (p. 636)
Mozambique [MOH•zahm•BEEK] Southern African country south of Tanzania. 20°S 34°E (p. 531)
Muscat [MUHS•KAHT] Capital of Oman. 23°N 59°E (p. 449)
Myanmar [MYAHN•MAHR] Southeast Asian country south of China and India, formerly called Burma. 21°N 95°E (p. 637)

Nairobi [ny•ROH•bee] Capital of Kenya. 1°S 37°E (p. 531)

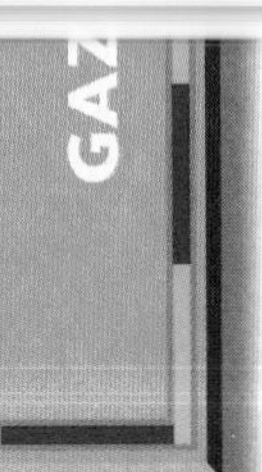

Namibia [nuh•MIH•bee•uh] Southern African country south of Angola on the Atlantic Ocean. 20°S 16°E (p. 531)

[illegible]

and China. 29°N 83°E (p. 637)

Netherlands [NEH•thuhr•lundz] Western European country north of Belgium. 53°N 4°E (p. 281)

New Delhi [NOO DEH•lee] Capital of India. 29°N 77°E (p. 637)

New Zealand [NOO ZEE•luhnd] Major island country southeast of Australia in the South Pacific. 42°S 175°E (p. 741)

Niamey [nee•AHM•ay] Capital of Niger. 14°N 2°E (p. 531)

Nicaragua [NIH•kuh•RAH•gwuh] Central American country south of Honduras. 13°N 86°W (p. 173)

Nicosia [NIH•kuh•SEE•uh] Capital of Cyprus. 35°N 33°E (p. 281)

Niger [NY•juhr] West African country north of Nigeria. 18°N 9°E (p. 531)

Nigeria [ny•JIHR•ee•uh] West African country along the Gulf of Guinea. 9°N 7°E (p. 530)

Nile [NYL] **River** Longest river in the world, flowing north through eastern Africa. 19°N 33°E (p. 448)

North Korea [kuh•REE•uh] East Asian country in the northernmost part of the Korean Peninsula. 40°N 127°E (p. 637)

Norway [NAWR•WAY] Northern European country on the Scandinavian peninsula. 64°N 11°E (p. 280)

Nouakchott [nu•AHK•SHAHT] Capital of Mauritania. 18°N 16°W (p. 531)

Oman [oh•MAHN] Country on the Arabian Sea and the Gulf of Oman. 20°N 58°E (p. 449)

Oslo [AHZ•loh] Capital of Norway. 60°N 11°E (p. 281)

Ottawa [AH•tuh•wuh] Capital of Canada. 45°N 76°W (p. 109)

Ouagadougou [WAH•gah•DOO•goo] Capital of Burkina Faso. 12°N 2°W (p. 531)

Pakistan [PA•kih•STAN] South Asian country northwest of India on the Arabian Sea. 28°N 68°E (p. 637)

Palau [puh•LOW) Island country in the Pacific Ocean. 7°N 135°E (p. 741)

Panama [PA•nuh•MAH] Central American country on the Isthmus of Panama. 9°N 81°W (p. 173)

Panama City Capital of Panama. 9°N 79°W (p. 173)

Papua New Guinea [PA•pyu•wuh NOO GIH•nee] Island country in the Pacific Ocean north of Australia. 7°S 142°E (p. 741)

Paraguay [PAR•uh•GWY] South American country northeast of Argentina. 24°S 57°W (p. 173)

Paramaribo [PAH•rah•MAH•ree•boh] Capital of [illegible]

Ecuador and Colombia. 10°S 75°W (p. 173)

Philippines [FIH•luh•PEENZ] Island country in the Pacific Ocean southeast of China. 14°N 125°E (p. 637)

Phnom Penh [puh•NAWM PEHN] Capital of Cambodia. 12°N 106°E (p. 637)

Poland [POH•luhnd] Eastern European country on the Baltic Sea. 52°N 18°E (p. 281)

Port-au-Prince [POHR•toh•PRIHNS] Capital of Haiti. 19°N 72°W (p. 173)

Port Moresby [MOHRZ•bee] Capital of Papua New Guinea. 10°S 147°E (p. 741)

Port-of-Spain [SPAYN] Capital of Trinidad and Tobago. 11°N 62°W (p. 173)

Porto-Novo [POHR•toh•NOH•voh] Capital of Benin. 7°N 3°E (p. 531)

Portugal [POHR•chih•guhl] Country west of Spain on the Iberian Peninsula. 39°N 8°W (p. 280)

Prague [PRAHG] Capital of the Czech Republic. 51°N 15°E (p. 281)

Pretoria [prih•TOHR•ee•uh] Executive capital of South Africa. 26°S 28°E (p. 531)

Puerto Rico [PWEHR•toh REE•koh] Island in the Caribbean Sea; U.S. Commonwealth. 19°N 67°W (p. 173)

Pyongyang [pee•AWNG•YAHNG] Capital of North Korea. 39°N 126°E (p. 637)

Qatar [KAH•tuhr] Country on the southwestern shore of the Persian Gulf. 25°N 53°E (p. 449)

Quito [KEE•toh] Capital of Ecuador. 0° latitude 79°W (p. 173)

Rabat [ruh•BAHT] Capital of Morocco. 34°N 7°W (p. 449)

Reykjavík [RAY•kyah•VEEK] Capital of Iceland. 64°N 22°W (p. 281)

Rhine [RYN] **River** River in western Europe that flows into the North Sea. 51°N 7°E (p. 280)

Riga [REE•guh] Capital of Latvia. 57°N 24°E (p. 281)

Rio Grande [REE•oh GRAND] River that forms part of the boundary between the United States and Mexico. 30°N 103°W (p. 108)

Riyadh [ree•YAHD] Capital of Saudi Arabia. 25°N 47°E (p. 449)

Rocky Mountains Mountain system in western North America. 50°N 114°W (p. 108)

Romania [ru•MAY•nee•uh] Eastern European country east of Hungary. 46°N 23°E (p. 281)
Rome [ROHM] Capital of Italy. 42°N 13°E (p. 281)
Russia [RUH•shuh] Largest country in the world, covering parts of Europe and Asia. 60°N 90°E (p. 401)
Rwanda [ruh•WAHN•duh] East African country south of Uganda. 2°S 30°E (p. 531)

Sahara [suh•HAR•uh] Desert region in northern Africa that is the largest hot desert in the world. 24°N 2°W (p. 448)
Saint Lawrence [LAWR•uhns] **River** River that flows from Lake Ontario to the Atlantic Ocean and forms part of the boundary between the United States and Canada. 48°N 70°W (p. 108)
Sanaa [sahn•AH] Capital of Yemen. 15°N 44°E (p. 449)
San José [SAHNG hoh•SAY] Capital of Costa Rica. 10°N 84°W (p. 173)
San Marino [SAN muh•REE•noh] Small European country located in the Italian peninsula. 44°N 13°E (p. 281)
San Salvador [san SAL•vuh•DAWR] Capital of El Salvador. 14°N 89°W (p. 173)
Santiago [SAN•tee•AH•goh] Capital of Chile. 33°S 71°W (p. 173)
Santo Domingo [SAN•toh duh•MIHNG•goh] Capital of the Dominican Republic. 19°N 70°W (p. 173)
Sao Tome and Principe [SOW•too•MAY PREEN•see•pee] Small island country in the Gulf of Guinea off the coast of central Africa. 1°N 7°E (p. 530)
Sarajevo [SAR•uh•YAY•voh] Capital of Bosnia and Herzegovina. 43°N 18°E (p. 281)
Saudi Arabia [SOW•dee uh•RAY•bee•uh] Country on the Arabian Peninsula. 23°N 46°E (p. 448)
Senegal [SEH•nih•GAWL] West African country on the Atlantic coast. 15°N 14°W (p. 531)
Seoul [SOHL] Capital of South Korea. 38°N 127°E (p. 637)
Seychelles [say•SHEHL] Small island country in the Indian Ocean off eastern Africa. 6°S 56°E (p. 531)
Sierra Leone [see•EHR•uh lee•OHN] West African country south of Guinea. 8°N 12°W (p. 531)
Singapore [SIHNG•uh•POHR] Southeast Asian island country near tip of Malay Peninsula. 2°N 104°E (p. 636)
Skopje [SKAW•PYAY] Capital of the country of Macedonia. 42°N 21°E (p. 281)
Slovakia [sloh•VAH•kee•uh] Eastern European country south of Poland. 49°N 19°E (p. 281)
Slovenia [sloh•VEE•nee•uh] Southeastern European country south of Austria on the Adriatic Sea. 46°N 15°E (p. 281)
Sofia [SOH•fee•uh] Capital of Bulgaria. 43°N 23°E (p. 281)
Solomon [SAH•luh•muhn] **Islands** Island country in the Pacific Ocean northeast of Australia. 7°S 160°E (p. 741)
Somalia [soh•MAH•lee•uh] East African country on the Gulf of Aden and the Indian Ocean. 3°N 45°E (p. 531)
South Africa [A•frih•kuh] Country at the southern tip of Africa, officially the Republic of South Africa. 28°S 25°E (p. 531)
South Korea [kuh•REE•uh] East Asian country on the Korean Peninsula between the Yellow Sea and the Sea of Japan. 36°N 128°E (p. 637)
Spain [SPAYN] Southern European country on the Iberian Peninsula. 40°N 4°W (p. 280)
Sri Lanka [SREE•LAHNG•kuh] Country in the Indian Ocean south of India, formerly called Ceylon. 9°N 83°E (p. 637)
Stockholm [STAHK•HOHLM] Capital of Sweden. 59°N 18°E (p. 281)
Sucre [SOO•kray] Constitutional capital of Bolivia. 19°S 65°W (p. 173)
Sudan [soo•DAN] East African country south of Egypt. 14°N 28°E (p. 531)
Suriname [SUR•uh•NAH•muh] South American country between Guyana and French Guiana. 4°N 56°W (p. 173)
Suva [SOO•vah] Capital of the Fiji Islands. 18°S 177°E (p. 741)
Swaziland [SWAH•zee•LAND] Southern African country west of Mozambique, almost entirely within the Republic of South Africa. 27°S 32°E (p. 531)
Sweden [SWEED•uhn] Northern European country on the eastern side of the Scandinavian peninsula. 60°N 14°E (p. 280)
Switzerland [SWIHT•suhr•luhnd] European country in the Alps south of Germany. 47°N 8°E (p. 280)
Syria [SIHR•ee•uh] Southwest Asian country on the east side of the Mediterranean Sea. 35°N 37°E (p. 449)

Taipei [TY•PAY] Capital of Taiwan. 25°N 122°E (p. 637)
Taiwan [TY•WAHN] Island country off the southeast coast of China, and the seat of the Chinese Nationalist government. 24°N 122°E (p. 637)
Tajikistan [tah•JIH•kih•STAN] Central Asian country east of Turkmenistan. 39°N 70°E (p. 449)
Tallinn [TA•luhn] Capital of Estonia. 59°N 25°E (p. 281)
Tanzania [TAN•zuh•NEE•uh] East African country south of Kenya. 7°S 34°E (p. 531)
Tashkent [tash•KEHNT] Capital of Uzbekistan. 41°N 69°E (p. 449)
T'bilisi [tuh•bih•LEE•see] Capital of the Republic of Georgia. 42°N 45°E (p. 449)
Tegucigalpa [tay•GOO•see•GAHL•pah] Capital of Honduras. 14°N 87°W (p. 173)
Tehran [TAY•uh•RAN] Capital of Iran. 36°N 52°E (p. 449)
Thailand [TY•LAND] Southeast Asian country east of Myanmar. 17°N 101°E (p. 637)

Thimphu [thihm•POO] Capital of Bhutan. 28°N 90°E (p. 637)

Tigris [TY•gruhs] **River** River in southeastern Turkey and Iraq that merges with the Euphrates [illegible]

[illegible]

and Ghana on the Gulf of Guinea. 8°N 1°E (p. 530)

Tokyo [TOH•kee•OH] Capital of Japan. 36°N 140°E (p. 637)

Trinidad and Tobago [TRIH•nuh•DAD tuh•BAY•goh] Island country near Venezuela between the Atlantic Ocean and the Caribbean Sea. 11°N 61°W (p. 173)

Tripoli [TRIH•puh•lee] Capital of Libya. 33°N 13°E (p. 449)

Tunis [TOO•nuhs] Capital of Tunisia. 37°N 10°E (p. 449)

Tunisia [too•NEE•zhuh] North African country on the Mediterranean Sea between Libya and Algeria. 35°N 10°E (p. 449)

Turkey [TUHR•kee] Country in southeastern Europe and western Asia. 39°N 32°E (p. 449)

Turkmenistan [tuhrk•MEH•nuh•STAN] Central Asian country on the Caspian Sea. 41°N 56°E (p. 449)

Uganda [yoo•GAHN•dah] East African country south of Sudan. 2°N 32°E (p. 531)

Ukraine [yoo•KRAYN] Eastern European country west of Russia on the Black Sea. 49°N 30°E (p. 281)

Ulaanbaatar [OO•LAHN•BAH•TAWR] Capital of Mongolia. 48°N 107°E (p. 637)

United Arab Emirates [EH•muh•ruhts] Country made up of seven states on the eastern side of the Arabian Peninsula. 24°N 54°E (p. 449)

United Kingdom Western European island country made up of England, Scotland, Wales, and Northern Ireland. 57°N 2°W (p. 281)

United States of America Country in North America made up of 50 states, mostly between Canada and Mexico. 38°N 110°W (p. 109)

Uruguay [YUR•uh•GWAY] South American country south of Brazil on the Atlantic Ocean. 33°S 56°W (p. 173)

Uzbekistan [UZ•BEH•kih•STAN] Central Asian country south of Kazakhstan. 42°N 60°E (p. 449)

Vanuatu [VAN•WAH•TOO] Country made up of islands in the Pacific Ocean east of Australia. 17°S 170°W (p. 741)

Vatican [VA•tih•kuhn] **City** Headquarters of the Roman Catholic Church, located in the city of Rome in Italy. 42°N 13°E (p. 281)

Venezuela [VEH•nuh•ZWAY•luh] South American country on the Caribbean Sea between Colombia and Guyana. 8°N 65°W (p. 173)

Vienna [vee•EH•nuh] Capital of Austria. 48°N 16°E (p. 281)

Vientiane [vyehn•TYAHN] Capital of Laos. 18°N [illegible]

[illegible]

Warsaw [WAWR•SAW] Capital of Poland. 52°N 21°E (p. 281)

Washington, D.C. Capital of the United States, in the District of Columbia. 39°N 77°W (p. 109)

Wellington [WEH•lihng•tuhn] Capital of New Zealand. 41°S 175°E (p. 741)

West Indies [IHN•deez] Islands in the Caribbean Sea between North America and South America. 19°N 79°W (p. 172)

Windhoek [VIHNT•HUK] Capital of Namibia. 22°S 17°E (p. 531)

Yamoussoukro [YAH•moo•SOO•kroh] Second capital of Côte d'Ivoire. 7°N 6°W (p. 531)

Yangon [YAHNG•GOHN] Capital of Myanmar. 17°N 96°E (p. 637)

Yangtze [YANG•SEE] **River** Principal river of China that begins in Tibet and flows into the East China Sea near Shanghai, also known as the Chang Jiang [CHAHNG jee•AHNG]. 31°N 117°E (p. 636)

Yaoundé [yown•DAY] Capital of Cameroon. 4°N 12°E (p. 531)

Yellow River River in northern and eastern China, also known as the Huang He [HWAHNG HUH]. 35°N 114°E (p. 636)

Yemen [YEH•muhn] Country south of Saudi Arabia on the Arabian Peninsula. 15°N 46°E (p. 449)

Yerevan [YEHR•uh•VAHN] Capital of Armenia. 40°N 44°E (p. 449)

Yugoslavia [YOO•goh•SLAH•vee•uh] Eastern European country south of Hungary; includes Serbia and Montenegro. 44°N 21°E (p. 281)

Zagreb [ZAH•GREHB] Capital of Croatia. 46°N 16°E (p. 281)

Zambia [ZAM•bee•uh] Southern African country north of Zimbabwe. 14°S 24°E (p. 531)

Zimbabwe [zihm•BAH•bway] Southern African country northeast of Botswana. 18°S 30°E (p. 531)

absolute location exact position of a place on the earth's surface (p. 6)
acid rain rain containing high amounts of chemical pollutants (pp. 97, 127, 154, 368)
adobe sun-dried clay bricks (p. 193)
alluvial plain area that is built up by rich fertile soil left by river floods (p. 497)
altiplano large highland plateau (p. 260)
altitude height above sea level (pp. 184, 242, 583)
anthracite type of hard coal (p. 705)
apartheid system of laws that separated racial and ethnic groups and limited the rights of blacks in South Africa (p. 611)
aquifer underground rock layer so rich in water that water actually flows through it (pp. 51, 465)
archipelago group of islands (pp. 213, 353, 695)
atoll low-lying, ring-shaped island that surrounds a lagoon (pp. 662, 768)
atmosphere layer of air surrounding the earth (p. 30)
autobahn superhighway (p. 307)
autonomy self-government (p. 593)
axis imaginary line that runs through the earth's center between the North and South poles (p. 31); *also* the horizontal (bottom) or vertical (side) line of measurement on a graph (p. 14)

bar graph graph in which vertical or horizontal bars represent quantities (p. 14)
basin low area surrounded by higher land (pp. 227, 564)
bauxite mineral used to make aluminum (pp. 215, 376, 414, 555)
bazaar marketplace (p. 462)
Bedouins nomadic desert peoples of Southwest Asia (p. 489)
bilingual referring to a country that has two official languages (pp. 158, 517)
birthrate number of children born each year for every 1,000 people (p. 85)
Boers name for the Dutch who were the first European settlers in South Africa (p. 610)
bog low swampy land (pp. 296, 365)
boomerang Australian weapon shaped like a bent wing that either strikes a target or curves and sails back to land on the ground near the person who threw it (p. 751)
bush rural areas of Australia (p. 752)

cacao tropical tree whose seeds are used to make chocolate and cocoa (pp. 543, 766)
calligraphy art of beautiful writing (p. 684)
campesino Colombian farmer (p. 259)
canopy umbrella-like covering formed by the tops of trees in a rain forest (pp. 64, 208, 564)
cardinal directions basic directions on the earth: north, south, east, west (p. 11)
casbah older section of Algerian cities (p. 468)
cash crop product grown to be sold for export (pp. 258, 513)
cassava plant with roots that can be ground into flour to make bread or eaten in other ways (pp. 552, 584)
caste social class based on a person's ancestry (p. 648)
caudillo military ruler (p. 242)
channel body of water wider than a strait between two pieces of land (p. 42)
chart graphic way of presenting information clearly (p. 16)
circle graph round or pie-shaped graph showing how a whole is divided (p. 15)
city-state city and its surrounding countryside (p. 333)
civilizations highly developed cultures (p. 81)
civil war fight among different groups within a country (pp. 467, 490, 545)
clan group of people related to one another (pp. 517, 600, 698)
climate usual, predictable pattern of weather in an area over a long period of time (p. 54)
climograph combination bar and line graph giving information about temperature and precipitation (p. 15)

See also the Geographic Dictionary on page 18.

GLOSSARY

coalition government government in which two or more political parties work together to run a country (pp. 334, 649)
cold war period between the late 1940s and late 1980s when the United States and the Soviet Union competed for world influence without actually fighting each other (p. 427)
collection process in the water cycle during which streams and rivers carry water back to the oceans (p. 50)
colony overseas territory or settlement tied to a parent country (pp. 131, 156, 190, 215, 327)
commonwealth partly self-governing territory (p. 219)
communist state country whose government has strong control over the economy and society as a whole (pp. 216, 308, 367, 426, 675)
compass rose device drawn on maps to show the directions (p. 11)
compound group of houses surrounded by walls (p. 544)
condensation process in which air rises and cools, which makes the water vapor it holds change back into a liquid (p. 50)
conservation careful use of resources so they are not wasted (p. 96)
constitutional monarchy government in which a king or queen is the official head of state, but elected officials run the government (pp. 293, 469, 491, 700)
consumer goods household products, clothing, and other goods people buy to use for themselves (pp. 382, 413, 676)
contiguous areas that are joined together inside a common boundary (p. 115)
continent massive land area (p. 35)
continental island island formed when chunks of land are split off from larger continents or when a piece of land that once linked an island to the mainland is eroded or covered by water (p. 765)
continental shelf plateau off each coast of a continent that lies under the ocean and stretches for several miles (p. 41)
contour line line connecting all points at the same elevation on a contour map (p. 12)
cooperative farm owned and operated by the government (p. 217)
copper belt large area of copper mines in northern Zambia (p. 618)
copra dried coconut meat, which is used to make margarine, soap, and other products (p. 776)
coral reef structure at or near the water's surface formed by the skeletons of small sea animals (pp. 119, 581, 747)
cordillera group of mountain ranges that run side by side (pp. 147, 255)
core center of the earth, formed of hot iron mixed with other metals (p. 35)
cottage industry home- or village-based industry in which family members supply their own equipment to make goods (p. 648)
crevasse deep crack in the Antarctic ice cap (p. 772)
crop rotation varying what is planted in a field to avoid using up all the minerals in the soil (p. 97)
crust uppermost layer of the earth (p. 35)
cultural diffusion the process of spreading new knowledge and skills to other cultures (p. 81)
culture way of life of a group of people who share similar beliefs and customs (p. 77)
culture region area of the world that includes many different countries that all have cultural traits in common (p. 82)
currency form of money (p. 292)
current moving streams of water in the world's oceans, which affect the climate of land areas (p. 57)
cyclone intense storm system with heavy rain and high winds (pp. 623, 654)
czar name for emperor in Russia's past (p. 424)

death rate number of people out of every 1,000 who die in a year (p. 84)
deforestation widespread cutting of forests (pp. 97, 572, 624, 716)
delta area formed from soil deposited by a river at its mouth (pp. 42, 457, 516, 654, 719)
democracy form of government in which citizens choose the nation's leaders by voting for them (pp. 79, 131, 431)
desalinization process used to make seawater drinkable (pp. 95, 493)
desertification process by which grasslands change to desert (p. 548)
developed country country in which a great deal of manufacturing is carried out (pp. 94, 609)
developing country country that is working toward industrialization (p. 94)

devolution transfer of certain powers from the central government to regional governments (p. 293)
diagram drawing that shows steps in a process or parts of an object (p. 16)
dialect local form of a language that differs from the main language in pronunciation or the meaning of words (pp. 78, 327)
Diaspora collective name for scattered Jewish settlements around the world (p. 485)
dictator individual who takes control of a government and rules the country as he or she wishes (pp. 79, 568)
dictatorship government under the control of one all-powerful leader (pp. 465, 729)
dike high banks of soil built along rivers to control floods (p. 672)
dominion self-governing nation that accepts the British monarch as head of state (p. 157)
drought long period of extreme dryness and water shortages (pp. 56, 547, 597)
dry farming method in which the land is left unplanted every few years so that it can store moisture (p. 324)
dynasty line of rulers from the same family (pp. 680, 703)
dzong Buddhist center of prayer and study in Bhutan (p. 659)

earthquake violent and sudden movement of the earth's crust (p. 36)
economic system system that sets rules for how people decide what goods and services to produce and how they are exchanged (p. 80)
ecosystem place where the plants and animals are dependent upon one another and their surroundings for survival (p. 96)
eco-tourist person who travels to another country to view its natural wonders (pp. 209, 589)
elevation height above sea level (pp. 12, 40, 336, 517)
elevation profile cutaway diagram showing changes in elevation of land (p. 17)
El Niño combination of temperature, wind, and water effects in the Pacific Ocean that causes heavy rains in some areas and drought in others (p. 57)
embargo order that restricts or prohibits trade with another country (pp. 218, 498)
emigrate to move to another country (pp. 88, 349)
empire group of lands under one ruler (pp. 261, 687)
enclave small territory entirely surrounded by a larger territory (pp. 511, 612)
endangered species plant or animal under the threat of completely dying out (p. 594)
environment natural surroundings (p. 25)
equinox day when day and night are of equal length in both hemispheres (p. 32)
erg huge area of shifting sand dunes in the Sahara (p. 466)
erosion process of moving weathered material on the earth's surface (pp. 37, 97)
escarpment steep cliff between higher and lower land (pp. 228, 582, 608)
estancia ranch (p. 236)
ethnic cleansing forcing people from a different ethnic group to leave their homes (p. 383)
ethnic group people who share a common culture, language, or history (pp. 78, 133, 427)
evaporation process in which the sun's heat turns liquid water into water vapor (p. 49)
exclave small part of a country that is separated from the main part (p. 614)
exile inability to live in one's own country because of political beliefs (p. 683)
export to trade goods to other countries (p. 92)

famine lack of food (pp. 85, 707)
fault crack in the earth's crust (pp. 37, 510, 382, 673)
favela slum area (p. 231)
federal republic government divided between national and state powers (pp. 131, 191, 309, 431)
fellahin farmers in Egypt who live in villages and work on small plots of land that they rent from landowners (p. 462)
fjord steep-sided valley cut into mountains by the action of glaciers (pp. 345, 755)
foothill low hill at the base of a mountain range (p. 261)
fossil fuel coal, oil, or natural gas (p. 127)
free enterprise system economic system in which people start and run businesses with limited government intervention (pp. 123, 430, 583)
free port place where goods can be loaded, stored, and shipped again without needing to pay any import taxes (p. 723)

free trade taking down trade barriers so that goods flow freely among countries (pp. 93, 128)

free trade zone area where people can buy goods from other countries without paying extra taxes (p. 218)

gaucho cowhand (p. 236)

geographic information systems (GIS) special software that helps geographers gather and use information about a place (pp. 10, 26)

geography the study of the earth in all its variety (p. 23)

geothermal energy electricity produced by natural underground sources of steam (pp. 356, 756)

geyser spring of water heated by molten rock inside the earth so that, from time to time, it shoots hot water into the air (pp. 356, 754)

glacier giant slow-moving sheets of ice (pp. 38, 50, 144)

Global Positioning System (GPS) group of satellites that travels around the earth, which can be used to locate exact places on the earth (p. 25)

globe spherical model of Earth (p. 4)

great circle route ship or airplane route following a great circle; the shortest distance between two points on the earth (p. 6)

greenhouse effect buildup of certain gases in the atmosphere that, like a greenhouse, hold more of the sun's warmth (p. 60)

grid system network of imaginary lines on the earth's surface, formed by the crisscrossing patterns of the lines of latitude and longitude (p. 6)

groundwater water that fills tiny cracks and holes in the rock layers below the earth's surface (p. 51)

habitat type of environment in which a particular animal species lives (p. 589)

hacienda large ranch (p. 190)

hajj religious journey to Makkah that Muslims are expected to make at least once during their lifetime if they are able to do so (p. 494)

harmattan dry, dusty wind that blows south from the Sahara (p. 541)

heavy industry manufactured goods such as machinery, mining equipment, and steel (pp. 350, 413)

hemisphere one-half of the globe; the Equator divides the earth into Northern and Southern Hemispheres; the Prime Meridian divides it into Eastern and Western Hemispheres (p. 5)

hieroglyphics form of writing that uses signs and symbols (pp. 189, 461)

high island Pacific island formed by volcanic activity (p. 768)

high-technology industry industry that produces computers and other kinds of electronic equipment (p. 685)

high veld flat, grass-covered plains on the interior plateau of South Africa (p. 608)

Holocaust systematic murder of more than 6 million European Jews by Adolf Hitler and his followers during World War II (pp. 308, 485)

human rights basic freedoms and rights that all people should enjoy (p. 683)

humid continental climate weather pattern characterized by long, cold, and snowy winters and short, hot summers (p. 66)

humid subtropical climate weather pattern characterized by hot, humid, rainy summers and short, mild winters (p. 67)

hurricane violent tropical storm with high winds and heavy rains (pp. 56, 185, 207)

hydroelectric power electricity generated by flowing water (pp. 242, 414, 565, 756)

iceberg chunk of a glacier that has broken away and floats free in the ocean (p. 773)

ice shelf layer of ice above water in Antarctica (p. 773)

immigrant person who moves to a new country to make a permanent home (pp. 132, 751)

import to buy goods from another country (p. 93)

industrialize to change an economy to rely more on manufacturing and less on farming (pp. 195, 425, 571)

inflation overall increase in the price of goods across the entire economy (p. 232)

infrastructure transportation and communication networks on which an economy depends (p. 309)

intensive cultivation growing crops on every available piece of land (p. 698)

invest to put money into a business (p. 676)
irrigation farming practice followed in dry areas to collect water and bring it to crops (p. 97)
Islamic republic government run by Muslim religious leaders (p. 499)
island body of land smaller than a continent and surrounded by water (p. 40)
isthmus narrow piece of land that connects two larger pieces of land (pp. 40, 205)

jute plant fiber used for making rope, burlap bags, and carpet backing (p. 647)

kibbutz settlement in Israel where the people share property and produce goods (p. 483)
krill tiny, shrimplike animal that lives in the waters off Antarctica and is the food for many other creatures (p. 774)

lagoon shallow pool of water surrounded by reefs, sandbars, or atolls (p. 662)
land bridge narrow strip of land that joins two larger landmasses (p. 181)
landfill area where trash companies dump the waste they collect (p. 127)
landform individual features of the land (p. 24)
landlocked country with no land bordering a sea or an ocean (pp. 247, 266, 375, 511)
La Niña pattern of unusual weather in the Pacific Ocean that has the opposite effects of El Niño (p. 57)
latitude location north or south of the Equator, measured by imaginary lines (parallels) numbered in degrees north or south (pp. 5, 184)
leap year year that has an extra day; occurs every fourth year (p. 31)
life expectancy the number of years that an average person is expected to live (p. 431)
light industry making of such goods as clothing, shoes, furniture, and household products (p. 413)

lines

read and write (p. 208)
llanos grassy plains (pp. 242, 256)
local wind pattern of wind caused by landforms in a particular area (p. 58)
loch narrow bay that reaches far inland (p. 290)
loess fertile, yellow-gray soil deposited by wind and water (p. 672)
longitude location east or west of the Prime Meridian, measured by imaginary lines (meridians) numbered in degrees east or west (p. 6)
low island Pacific island formed of coral and having little vegetation (p. 768)

magma hot melted rock that sometimes flows to the earth's surface in a volcanic eruption (p. 35)
mainland the major part of a country (p. 336)
mangrove tropical tree with roots that extend both above and beneath the water (p. 541)
mantle rock layer about 1,800 miles (2,897 km) thick between the core and the crust (p. 35)
manuka small shrub of New Zealand (p. 754)
map key code that explains the lines, symbols, and colors used on a map (p. 10)
maquiladora factory that assembles parts made in other countries (p. 185)
marine west coast climate weather pattern characterized by rainy and mild winters and cool summers (p. 65)
marsupial mammal that carries its young in a pouch (p. 749)
Mediterranean climate weather pattern characterized by mild, rainy winters and hot, dry summers (p. 66)
megalopolis pattern of heavy urban settlement over a large area (pp. 116, 700)
mestizo person with mixed Spanish and Native American or African background (pp. 190, 258)
migrant worker person who travels from place to place when extra help is needed to plant or harvest crops (p. 196)
migrate to move from one place to another (p. 480)

monarchy form of government in which a king or queen inherits the right to rule a country (p. 79)
monotheism belief in one God (p. 486)
monsoon seasonal wind that blows over a continent for months at a time (pp. 56, 646, 705, 715)
moor treeless, windy highland area with damp ground (pp. 290, 354)
moshav settlement in Israel where people share property but also own some private property (p. 484)
mosque place of worship for followers of Islam (pp. 384, 462, 479)
multilingual able to speak several languages (p. 316)
multinational firm that does business in several countries (p. 316)
mural wall painting (p. 189)

nationalism desire of a territory or colony to become an independent nation (p. 351)
national park area set aside to protect wilderness and wildlife and for recreation (p. 135)
nature preserve protected areas for plants and animals (pp. 378, 581)
natural resource product of the earth that people use to meet their needs (p. 90)
navigable describes a body of water wide and deep enough to allow the passage of ships (pp. 126, 260, 301, 324)
neutrality refusing to take sides in disagreements and wars between countries (p. 310)
newsprint type of paper used for printing newspapers (p. 153)
nomads people who move from place to place with herds of animals (pp. 377, 516, 687)
nonrenewable resource natural resource such as metals or minerals that cannot be replaced (p. 91)
nuclear energy power made by creating a controlled atomic reaction (p. 431)

oasis green area in a desert fed by underground water (pp. 459, 493, 518)
oil shale rock that contains oil (p. 372)
orbit path that a body in the solar system travels around the sun (p. 29)
outback inland regions of Australia (p. 748)
overgraze to allow animals to strip areas so bare that plants cannot grow back (p. 547)
ozone type of oxygen that forms a layer in the atmosphere and protects all living things on the earth from certain harmful rays of the sun (p. 775)

parliamentary democracy government in which voters elect representatives to a lawmaking body, which chooses a prime minister to head the government (pp. 157, 210, 292)
parliamentary republic see *parliamentary democracy* (p. 327)
peat wet ground with decaying plants that can be dried and used for fuel (pp. 296, 372)
peninsula piece of land with water on three sides (pp. 40, 146, 182, 408)
permafrost permanently frozen lower layers of soil in the tundra and subarctic regions (pp. 68, 410)
pesticides powerful chemicals that kill crop-destroying insects (pp. 96, 648)
phosphate mineral salt used in fertilizers (pp. 461, 555, 769)
pictograph graph in which small symbols represent quantities (p. 15)
pidgin language language formed by combining elements of several different languages (p. 767)
plain low-lying stretch of flat or gently rolling land (p. 40)
plantain kind of banana (p. 592)
plantation large farm that grows a single crop for sale (pp. 186, 207)
plateau flat land with higher elevation than a plain (pp. 40, 324)
plate huge slab of rock that makes up the earth's crust (pp. 599, 726)
plate tectonics theory that the earth's crust is not an unbroken shell but consists of plates, or huge slabs of rock, that move (p. 35)
plaza public square (p. 193)
poaching illegal hunting of protected animals (p. 581)
polder area of land reclaimed from the sea (p. 315)
pope head of the Roman Catholic Church (pp. 334, 369)

population density average number of people living in a square mile or square kilometer (p. 86)
potash type of mineral salt that is often used in fertilizers (p. 388)
prairie rolling, inland grassy area with very fertile soil (p. 146)
precipitation water that falls back to the earth as rain, snow, sleet, or hail (p. 50)
prime minister official who heads the government in a parliamentary democracy (p. 157)
privatize to transfer the ownership of factories from the government to individual citizens (p. 380)
projection in mapmaking, a way of drawing the round Earth on a flat surface (p. 7)
province regional political division similar to states (p. 143)

quota number limit on how many items of a particular product can be imported from a particular country (p. 93)

rain forest dense forest that receives high amounts of rain each year (p. 61)
rain shadow dry area on the inland side of coastal mountains (p. 59)
recycling reusing materials instead of throwing them out (p. 128)
refugee person who flees to another country to escape persecution or disaster (pp. 88, 383, 568, 595)
region area that shares common characteristics (p. 25)
relief differences in height in a landscape; how flat or rugged the surface is (p. 12)
Renaissance period of great achievement in art and learning that began in Italy in the 1300s and spread throughout Europe (p. 333)
renewable resource natural resource that cannot be used up or can fairly quickly be replaced naturally or grown again (p. 91)
republic strong national government headed by elected leaders (pp. 210, 233, 258, 303, 369, 461)
reunification bringing together the two parts of Germany under one government (p. 309)
revolution one complete orbit around the sun (p. 31)
rural area in the countryside (pp. 134, 433)

saga long story (p. 357)
samurai powerful land-owning warriors in Japan (p. 699)
sauna wooden room heated by water sizzling on hot stones (p. 351)
savanna broad grassland in the tropics with few trees (pp. 64, 541, 563)
scale relationship between distance on a map and actual distance on the earth (p. 11)
scale bar on a map, a divided line showing the map scale, usually in miles or kilometers (p. 11)
secede to withdraw from a national government (pp. 131, 154)
secular nonreligious (pp. 467, 480)
selva tropical rain forests in Brazil (p. 228)
serf farm laborer who could be bought and sold along with the land (p. 425)
service industry business that provides services to people instead of producing goods (pp. 124, 150, 195)
shah title given to kings who ruled Iran (p. 499)
shogun military leaders in Japan (p. 699)
silt small particles of rich soil (p. 458)
sirocco hot, dry winds that blow across Italy from North Africa (p. 331)
sisal plant fiber used to make rope and twine (p. 588)
skerry rocky island (p. 349)
slash-and-burn farming method of clearing land for planting by cutting and burning forest (p. 624)
smog thick haze of fog and chemicals (p. 198)
socialism economic system in which many businesses are owned and run by the government (p. 717)
solar system Earth, eight other planets, and thousands of smaller bodies that all revolve around the sun (p. 29)
sodium nitrate chemical used in fertilizer and explosives (p. 269)
sorghum tall grass with seeds that are used as grain and to make syrup (p. 619)
spa resort that has hot mineral springs that people bathe in to regain their health (p. 378)

GLOSSARY

strait narrow body of water between two pieces of land (pp. 42, 721)
subarctic climate weather pattern characterized by severely cold and bitter winters and short, cool summers (p. 67)
subcontinent large landmass that is part of another continent but distinct from it (p. 645)
subsistence farm small plot where a farmer grows only enough food to feed his own family (pp. 186, 207, 261, 542)
suburb smaller community that surrounds a city (pp. 134, 340, 432)
summer solstice day with the most hours of sunlight and the fewest hours of darkness (p. 32)

taiga huge forests of evergreen trees that grow in subarctic regions (p. 410)
tannin substance used in processing leather (p. 236)
tariff tax added to the value of goods that are imported (p. 93)
terraced field strips of land cut out of a hillside like stair steps so the land can hold water and be used for farming (pp. 678, 724)
timberline elevation along mountains above which no trees grow (p. 70)
tornado funnel-shaped windstorm that sometimes forms during a severe thunderstorm (p. 56)
townships crowded neighborhoods outside cities in South Africa where most nonwhites live (p. 611)
trench valley in the ocean floor (p. 41)
tributary small river that flows into a larger river (p. 653)
tropics low-latitude region between the Tropic of Cancer and the Tropic of Capricorn (p. 55)
trust territory area temporarily placed under control of another nation (p. 769)
tsetse fly insect whose bite can kill cattle or humans with a deadly disease called sleeping sickness (p. 571)
tungsten metal used in electrical equipment (p. 677)
typhoon name for hurricane in Asia (pp. 56, 673, 769)

urban area in the city (pp. 134, 432)
urbanization movement to cities (p. 87)

vaquero cowhand (p. 185)

wadi dry riverbed filled by rainwater from rare downpours (p. 493)
water cycle process in which water moves from the oceans to the air to the ground and finally back to the oceans (p. 49)
watershed region drained by a river (p. 594)
water vapor water in the form of gas (p. 49)
weather unpredictable changes in the air that take place over a short period of time (p. 53)
weathering process that breaks surface rocks into boulders, gravel, sand, and soil (p. 37)
welfare state country that uses tax money to support people who are sick, needy, jobless, or retired (pp. 246, 349)
winter solstice day with the fewest hours of sunlight and the most hours of darkness (p. 32)

yurt large circle-shaped tent made of animal skins that can be packed up and moved from place to place (p. 688)

absolute location/ubicación absoluta posición exacta de un lugar en la superficie de la Tierra (pág. 6)
acid rain/lluvia ácida lluvia que contiene grandes cantidades de contaminantes químicos (págs. 97, 127, 154, 368)
adobe/adobe ladrillos secados al Sol (pág. 193)
alluvial plain/llanura aluvial área creada por el suelo fértil que se acumula después de las inundaciones causadas por los ríos (pág. 497)
altiplano/altiplano meseta grande y muy elevada; también se llama altiplanicie (pág. 260)
altitude/altitud altura sobre el nivel del mar (págs. 184, 242, 583)
anthracite/antracita tipo de carbón mineral muy duro (pág. 705)
apartheid/apartheid sistema de leyes que separaba los grupos raciales y étnicos y limitaba los derechos de la población negra (pág. 611)
aquifer/manto acuífero capa de rocas subterránea en que el agua es tan abundante que corre entre las rocas (págs. 51, 465)
archipelago/archipiélago grupo de islas (págs. 213, 353, 695)
atoll/atolón isla de muy poca elevación que se forma alrededor de una laguna en la forma de un anillo (págs. 662, 768)
atmosphere/atmósfera capa de aire que rodea la Tierra (pág. 30)
autobahn/autobahn autopista muy rápida (pág. 307)
autonomy/autonomía gobernarse por sí mismo (pág. 593)
axis/eje terrestre línea imaginaria que atraviesa el centro de la Tierra entre el Polo Norte y el Polo Sur (pág. 31); también la línea vertical (del lado) u horizontal (de abajo) de una gráfica que se usa para medir (pág. 14)

bar graph/gráfica de barras gráfica en que franjas verticales u horizontales representan cantidades (pág. 14)
basin/cuenca área baja rodeada de tierras más elevadas (págs. 227, 564)

**Véase también* el Diccionario Geográfico en la página 18.

bauxite/bauxita mineral que se usa para hacer aluminio (págs. 215, 376, 414, 555)
bazaar/bazar mercado (pág. 462)
Bedouins/beduinos gente nomádica del desierto del sudoeste de Asia (pág. 489)
bilingual/bilingüe se refiere a un país que tiene dos idiomas oficiales (págs. 158, 517)
birthrate/índice de natalidad número de niños que nace cada año por cada mil personas (pág. 85)
Boers/bóers los holandeses que fueron los primeros colonos en Sudáfrica (pág. 610)
bog/ciénaga tierra baja y pantanosa (págs. 296, 365)
boomerang/bumerán arma australiana con forma de ala que se lanza para que golpee un objetivo o de la vuelta y caiga a los pies de la persona que la lanzó (pág. 751)
bush/campo áreas rurales de Australia (pág. 752)

cacao/cacao árbol tropical cuyas semillas se usan para hacer chocolate y cocoa (págs. 543, 766)
calligraphy/caligrafía el arte de escribir con letra muy bella (pág. 684)
campesino/campesino agricultor (pág. 259)
canopy/bóveda techo formado por las copas de los árboles en los bosques húmedos (págs. 64, 208, 564)
cardinal directions/puntos cardinales cuatro direcciones básicas en la Tierra: norte, sur, este, oeste (pág. 11)
casbah/casbah la sección antigua de las ciudades de Argelia; también se llama *alcazaba* (pág. 468)
cash crop/cultivo comercial producto que se cultiva para exportación (págs. 258, 513)
cassava/yuca planta con raíces que se pueden convertir en harina para hacer pan o que se pueden cocinar de otras formas (págs. 552, 584)
caste/casta clase social basada en la ascendencia de una persona (pág. 648)
caudillo/caudillo gobernante militar (pág. 242)
channel/canal una masa de agua entre dos tierras que tiene más anchura que un estrecho (pág. 42)
chart/cuadro manera gráfica de presentar información con claridad (pág. 16)

circle graph/gráfica de círculo gráfica redonda que muestra como un todo es dividido (pág. 15)
city-state/ciudad estado ciudad junto con las tierras que la rodean (pág. 333)
civilizations/civilizaciones culturas altamente desarrolladas (pág. 81)
civil war/guerra civil pelea entre distintos grupos dentro de un país (págs. 467, 490, 545)
clan/clan grupo de personas que están emparentadas (págs. 517, 600, 698)
climate/clima el patrón que sigue el estado del tiempo en un área durante muchos años (pág. 54)
climograph/gráfica de clima gráfica que combina barras y líneas para dar información sobre la temperatura y la precipitación (pág. 15)
coalition government/gobierno por coalición gobierno en que dos o más partidos trabajan juntos para dirigir un país (págs. 334, 649)
cold war/guerra fría período entre los fines de los 1940 y los fines de los 1980 en que los Estados Unidos y la Unión Soviética compitieron por tener influencia mundial sin pelear uno contra el otro (pág. 427)
collection/drenaje proceso durante el ciclo hidrológico en que los ríos llevan el agua de regreso a los océanos (pág. 50)
colony/colonia territorio o poblado con lazos a un país extranjero (págs. 131, 156, 190, 215, 327)
commonwealth/estado libre asociado territorio que en parte se gobierna por sí solo (pág. 219)
communist state/estado comunista país cuyo gobierno mantiene mucho control sobre la economía y la sociedad en su totalidad (págs. 216, 308, 367, 426, 675)
compass rose/rosa náutica dibujo en los mapas que muestra las direcciones; también se llama rosa de los vientos (pág. 11)
compound/complejo residencial grupo de viviendas rodeada por una muralla (pág. 544)
condensation/condensación proceso en que el aire sube y se enfría, lo cual hace que el vapor de agua que contiene se convierta de nuevo en líquido (pág. 50)
conservation/conservación uso juicioso de los recursos para no malgastarlos (pág. 96)
constitutional monarchy/monarquía constitucional gobierno en que un rey o reina es el jefe de estado oficial pero los gobernantes son elegidos (págs. 293, 469, 491, 700)
consumer goods/bienes de consumo productos para la casa, ropa y otras cosas que la gente compra para su uso personal (págs. 382, 413, 676)
contiguous/contiguas áreas adyacentes dentro de la misma frontera (pág. 115)
continent/continente masa de tierra inmensa (pág. 35)
continental island/isla continental isla formada cuando un pedazo de tierra se separa de un continente o cuando la erosión hace que el agua cubra la parte del continente que antes unía la isla a éste (pág. 765)
continental shelf/plataforma continental meseta formada por parte de un continente que se extiende por varias millas debajo del mar (pág. 41)
contour line/curva de nivel línea que conecta todos los puntos a la misma elevación en un mapa de relieve (pág. 12)
cooperative/cooperativa granja que es propiedad y es operada por el gobierno (pág. 217)
copper belt/cinturón de cobre área extensa de minas de cobre en el norte de Zambia (pág. 618)
copra/copra pulpa seca del coco que se usa para hacer margarina, jabón y otros productos (pág. 776)
coral reef/arrecife coralino estructura formada al nivel del mar o cerca de éste por los esqueletos de pequeños animales marinos (págs. 119, 581, 747)
cordillera/cordillera grupo de cadenas paralelas de montañas (págs. 147, 255)
core/núcleo centro de la Tierra, que está formado de hierro caliente y otros metales (pág. 35)
cottage industry/industria familiar industria basada en una casa o aldea en que los miembros de la familia usan sus propias herramientas para hacer productos (pág. 648)
crevasse/grieta rajadura profunda en el casquete de hielo de la Antártida (pág. 772)
crop rotation/rotación de cultivos variar lo que se siembra en un terreno para no agotar todos los minerales que tiene el suelo (pág. 97)
crust/corteza capa de afuera de la Tierra (pág. 35)
cultural diffusion/difusión cultural el proceso de esparcir nuevos conocimientos y habilidades a otras culturas (pág. 81)
culture/cultura modo de vida de un grupo de personas que comparten creencias y costumbres similares (pág. 77)
culture region/región cultural área del mundo que incluye muchos países que tienen los mismos rasgos culturales (pág. 82)
currency/moneda tipo de dinero (pág. 292)
current/corriente movimiento de las aguas del mar que afecta el clima de las masas de tierra (pág. 57)

muy fuertes y mucha lluvia (págs. 623, 654)

czar/zar título de los antiguos emperadores rusos (pág. 424)

death rate/índice de mortalidad número de personas de cada mil que mueren en un año (pág. 84)

deforestation/deforestación la extensa destrucción de los bosques (págs. 97, 572, 624, 716)

delta/delta área formada por el suelo que deposita un río en su desembocadura (págs. 42, 457, 516, 654, 719)

democracy/democracia tipo de gobierno en que los ciudadanos seleccionan los líderes de la nación por medio del voto (págs. 79, 131, 431)

desalinization/desalinización proceso de hacer el agua de mar potable (págs. 95, 493)

desertification/desertización proceso por el cual los pastos se convierten en desiertos (pág. 548)

developed country/país desarrollado país donde hay mucha manufactura de productos (págs. 94, 609)

developing country/país en vías de desarrollo país que está industrializándose (pág. 94)

devolution/devolución transferencia de ciertos poderes del gobierno central a los gobiernos regionales (pág. 293)

diagram/diagrama dibujo que muestra los pasos en un proceso o las partes de un objeto (pág. 16)

dialect/dialecto forma local de un idioma que se diferencia del idioma normal por su pronunciación o por el sentido de algunas palabras (págs. 78, 327)

Diaspora/diáspora nombre dado a los poblados judíos alrededor del mundo (pág. 485)

dictator/dictador individuo que toma control de un país y lo gobierna como quiere (págs. 79, 568)

dictatorship/dictadura gobierno bajo el control de un líder que tiene todo el poder (págs. 465, 729)

dike/dique muros de tierra muy altos construidos a lo largo de los ríos para controlar las inundaciones (pág. 672)

dominion/dominio naciones que se gobiernan por sí solas que aceptan al monarca británico como jefe de estado (pág. 157)

drought/sequía largos períodos de sequedad y de escasez de agua (págs. 56, 547, 597)

dry farming/agricultura en seco método de cultivar en que la tierra se deja sin sembrar cada varios años para que almacene humedad (pág. 324)

misma familia (págs. 680, 703)

dzong/dzong centro budista en Bután para rezar y estudiar (pág. 659)

earthquake/terremoto movimiento violento e inesperado de la corteza de la Tierra (pág. 36)

economic system/sistema económico sistema que establece reglas que determinan cómo las personas deciden cuáles bienes y servicios van a producir y cómo los van a intercambiar (pág. 80)

ecosystem/ecosistema lugar en el cual las plantas y animales dependen unos de otros y de sus alrededores para sobrevivir (pág. 96)

eco-tourist/ecoturista persona que viaja a otro país para ver sus bellezas naturales (págs. 209, 589)

elevation/elevación altura por encima del nivel del mar (págs. 12, 40, 336, 517)

elevation profile/perfil de elevaciones diagrama que muestra los cambios en la elevación de la tierra como si se hubiera hecho un corte vertical del área (pág. 17)

El Niño/El Niño combinación de la temperatura, los vientos y los efectos del agua en el océano Pacífico que causa lluvias fuertes en algunas áreas y sequía en otras (pág. 57)

embargo/embargo orden que limita o prohibe el comercio con otro país (págs. 218, 498)

emigrate/emigrar mudarse a otro país (págs. 88, 349)

empire/imperio grupo de países bajo un gobernante (págs. 261, 687)

enclave/enclave territorio pequeño totalmente rodeado por un territorio más grande (págs. 511, 612)

endangered species/especie en vías de extinción planta o animal que está en peligro de desaparecer completamente (pág. 594)

environment/medio ambiente alrededores naturales (pág. 25)

equinox/equinoccio día en que el día y la noche tienen la misa duración en los dos hemisferios (pág. 32)

erg/ergio inmensas áreas en el Sahara en que se mueven las dunas de arena (pág. 466)

erosion/erosión proceso de mover los materiales desgastados en la superficie de la Tierra (págs. 37, 97)

escarpment/escarpa acantilado empinado entre una área baja y una alta (págs. 228, 582, 608)

estancia/estancia rancho (pág. 236)
ethnic cleansing/limpieza étnica forzar a personas de un grupo étnico distinto a abandonar el lugar donde viven (pág. 383)
ethnic group/grupo étnico personas que comparten una cultura, idioma o historia en común (págs. 78, 133, 427)
evaporation/evaporación proceso mediante el cual el calor del sol convierte el agua líquida en vapor de agua (pág. 49)
exclave/territorio externo parte pequeña de un país que está separada de la parte principal (pág. 614)
exile/exilio tener que vivir fuera de su país nativo por causa de sus creencias políticas (pág. 683)
export/exportar comerciar y mandar bienes a otros países (pág. 92)

famine/hambruna falta de alimentos (págs. 85, 707)
fault/falla fractura en la corteza de la Tierra (págs. 37, 510, 582, 673)
favela/favela barrio pobre y deteriorado (pág. 231)
federal republic/república federal nación en que el poder está dividido entre el gobierno nacional y el de los estados (págs. 131, 191, 309, 431)
fellahin/felás granjeros en Egipto que viven en aldeas y cultivan pequeños terrenos que arriendan de un hacendado (pág. 462)
fjord/fiordo valle creado por el movimiento de glaciares en las montañas que deja laderas sumamente empinadas (págs. 345, 755)
foothill/estribaciones colinas bajas al pie de una cadena de montañas (pág. 261)
fossil fuel/combustibles fósiles carbón, petróleo o gas natural (pág. 127)
free enterprise system/sistema de libre empresa sistema económico en que la gente empieza y administra negocios con poca intervención del gobierno (págs. 123, 430, 583)
free port/puerto libre lugar donde las mercancías se pueden cargar, almacenar y embarcar de nuevo sin tener que pagar derechos de importación (pág. 723)
free trade/libre comercio eliminar las barreras al comercio para que se puedan mover productos libremente entre países (págs. 93, 128)
free trade zone/zona de cambio libre área donde la gente puede comprar bienes de otros países sin pagar impuestos adicionales (pág. 218)

gaucho/gaucho vaquero (pág. 236)
geographic information systems (GIS)/ sistemas de información geográfica (SIG) programas de computadoras especiales que ayudan a los geógrafos a obtener y usar la información geográfica sobre un lugar (págs. 10, 26)
geography/geografía el estudio de la Tierra y de toda su variedad (pág. 23)
geothermal energy/energía geotérmica electricidad producida por fuentes de vapor subterráneas naturales (págs. 356, 756)
geyser/géiser manantial de agua calentado por rocas fundidas dentro de la Tierra que, de vez en cuando, arroja agua caliente al aire (págs. 356, 754)
glacier/glaciar capa de hielo inmensa que se mueve muy lentamente (págs. 38, 50, 144)
Global Positioning System (GPS)/Sistema global de posición (GPS) grupo de satélites que le dan la vuelta a la Tierra y se usan para localizar lugares exactos en la Tierra (pág. 25)
globe/globo terráqueo modelo esférico de la Tierra (pág. 4)
great circle route/línea de rumbo ruta que sigue un círculo máximo; usada por aviones y barcos porque es la distancia más corta entre dos puntos en la Tierra (pág. 6)
greenhouse effect/efecto invernadero la acumulación de ciertos gases en la atmósfera que mantienen más del calor del Sol, como hace un invernadero (pág. 60)
grid system/sistema de coordenadas geográficas red de líneas imaginarias en la superficie de la Tierra, formada por las líneas de latitud y longitud que se cruzan (pág. 6)
groundwater/agua subterránea agua que llena las rajaduras y hoyos en las capas de roca debajo de la superficie de la Tierra (pág. 51)

habitat/hábitat tipo de ambiente en que vive una especie animal en particular (pág. 589)
hacienda/hacienda un rancho grande (pág. 190)
hajj/_hajj_ viaje religioso a La Meca que todo musulmán debe hacer por lo menos una vez en la vida si puede (pág. 494)
harmattan/harmattan viento seco y lleno de polvo que sopla hacia el sur desde el Sahara (pág. 541)

heavy industry/industria pesada manufactura de productos como maquinaria, equipo de minería y acero (págs. 350, 413)

hemisphere/hemisferio una mitad del globo terráqueo; el ecuador divide la Tierra en los hemisferios norte y sur; el primer meridiano la divide en hemisferios este y oeste (pág. 5)

hieroglyphics/jeroglíficos forma de escribir que usa signos y símbolos (págs. 189, 461)

high island/isla oceánica isla del Pacífico formada por actividad volcánica (págs. 768)

high-technology industry/industria de alta tecnología industria que produce computadoras y otras clases de equipo electrónico (pág. 685)

high veld/veld pastos llanos de la meseta interior de Sudáfrica (pág. 608)

Holocaust/Holocausto la matanza sistemática de más de 6 millones de judíos europeos por Adolfo Hitler y sus seguidores durante la Segunda Guerra Mundial (págs. 308, 485)

human rights/derechos humanos libertades y derechos básicos que todas las personas deben disfrutar (pág. 683)

humid continental climate/clima húmedo continental patrón del estado del tiempo con inviernos largos, fríos y con mucha nieve y veranos cortos y calurosos (pág. 66)

humid subtropical climate/clima húmedo subtropical patrón del estado del tiempo con veranos calurosos, húmedos y lluviosos e inviernos cortos y templados (pág. 67)

hurricane/huracán tormenta tropical violenta con vientos y lluvias fuertes (págs. 56, 185, 207)

hydroelectric power/energía hidroeléctrica electricidad generada por una corriente de agua (págs. 242, 414, 565, 756)

iceberg/iceberg pedazo de un glaciar que se ha desprendido y flota libremente en los océanos (pág. 773)

ice shelf/plataforma de hielo capa de hielo sobre el mar en la Antártida (pág. 773)

immigrant/inmigrante persona que se muda permanentemente a un país nuevo (págs. 132, 751)

import/importar comprar productos de otro país (pág. 93)

industrialize/industrializar cambiar una economía de manera que dependa más de la manufactura que de la agricultura (págs. 195, 425, 571)

inflation/inflación aumento en el precio de productos en toda la economía (pág. 232)

infrastructure/infraestructura redes de transporte y comunicación de las cuales depende una economía (pág. 309)

intensive cultivation/cultivo intensivo labrar toda la tierra posible (pág. 698)

intermediate direction/puntos intermedios cualquier dirección entre los puntos cardinales, como sudeste o noroeste (pág. 11)

invest/invertir poner dinero en un negocio (pág. 676)

irrigation/irrigación práctica agrícola en áreas secas de colectar agua y llevarla a los cultivos (pág. 97)

Islamic republic/república islámica gobierno dirigido por líderes musulmanes (pág. 499)

island/isla masa de tierra más pequeña que un continente, rodeada de agua (pág. 40)

isthmus/istmo lengua de tierra que conecta a dos masas de tierra más grandes (págs. 40, 205)

jute/yute fibras de una planta que se usan para hacer soga, sacos y el revés de alfombras (pág. 647)

kibbutz/kibutz poblado en Israel donde las personas comparten la propiedad y producen bienes (pág. 483)

krill/krill animales diminutos parecidos a los camarones que viven en las aguas alrededor de la Antártida y sirven de alimento para muchos otros animales (pág. 774)

lagoon/laguna masa de agua poco profunda rodeada por arrecifes, bancos de arena o un atolón (pág. 662)

land bridge/puente de tierra franja de tierra que une a dos masas de tierra mayores (pág. 181)

landfill/vertedero de basura lugar donde las compañías que recogen la basura botan los residuos que colectan (pág. 127)

landform/accidente geográfico característica particular de la tierra (pág. 24)

landlocked/rodeado de tierra país que no tiene tierras bordeadas por un mar u océano (págs. 247, 266, 375, 511)

La Niña/La Niña patrón infrecuente en el estado del tiempo del océano Pacífico que tiene los efectos contrarios a los de El Niño (pág. 57)

latitude/latitud posición al norte o al sur del ecuador, medida por medio de líneas imaginarias (paralelos) numeradas con grados norte o sur (págs. 5, 184)

leap year/año bisiesto año que tiene un día adicional; cada cuarto año (pág. 31)

life expectancy/expectativas de vida el número de años que se espera que viva la persona promedio (pág. 431)

light industry/industria ligera fabricación de productos como muebles, ropa, zapatos y artículos para el hogar (pág. 413)

line graph/gráfica lineal gráfica en que una o varias líneas representan cambios de cantidad a través del tiempo (pág. 14)

literacy rate/índice de alfabetización porcentaje de personas que saben leer y escribir (pág. 208)

llanos/llanos planicie cubierta de hierba (págs. 242, 256)

local wind/vientos locales patrones en los vientos causados por los accidentes geográficos de un área en particular (pág. 58)

loch/rías bahías estrechas que llegan hasta muy dentro de la tierra (pág. 290)

loess/loes suelo amarillento y fértil depositado por el viento y el agua (pág. 672)

longitude/longitud posición al este o el oeste del primer meridiano, medida por medio de líneas imaginarias (meridianos) numeradas con grados este u oeste (pág. 6)

low island/isla coralina isla del Pacífico formada por coral que tiene poca vegetación (pág. 768)

magma/magma roca caliente y fundida que a veces fluye hasta la superficie de la Tierra en erupciones volcánicas (pág. 35)

mainland/territorio continental la parte principal de un país (pág. 336)

mangrove/mangle árbol tropical con raíces que se extienden por encima y por debajo del agua (pág. 541)

mantle/manto capa de rocas de 1,800 millas (2,897 km.) de grueso entre el núcleo y la corteza (pág. 35)

manuka/manuka pequeño arbusto de Nueva Zelanda (pág. 754)

map key/leyenda explicación de las líneas, símbolos y colores usados en un mapa; también se llama clave del mapa (pág. 10)

***maquiladora*/maquiladora** fábrica donde se ensamblan piezas hechas en otros países (pág. 185)

marine west coast climate/clima húmedo marítimo patrón del estado del tiempo con inviernos lluviosos y templados y veranos frescos (pág. 65)

marsupial/marsupial mamífero que lleva a sus crías en una bolsa (pág. 749)

Mediterranean climate/clima húmedo mediterráneo patrón del estado del tiempo con inviernos lluviosos y templados y veranos calurosos y secos (pág. 66)

megalopolis/megalópolis área extensa de mucha urbanización (págs. 116, 700)

mestizo/mestizo persona cuya ascendencia incluye indios americanos o africanos y españoles (págs. 190, 258)

migrant worker/trabajador itinerante persona que viaja a distintos lugares donde hacen falta trabajadores para sembrar y cosechar cultivos (pág. 196)

migrate/migrar mudarse de un lugar a otro (pág. 480)

monarchy/monarquía tipo de gobierno en que un rey o reina hereda el derecho de gobernar un país (pág. 79)

monotheism/monoteísmo creencia en un solo Dios (pág. 486)

monsoon/monzón vientos que soplan en un continente por varios meses seguidos en ciertas estaciones del año (págs. 56, 646, 705, 715)

moor/páramo área elevada y sin árboles pero con mucho viento y tierra húmeda (págs. 290, 354)

moshav/*moshav* poblados en Israel en que la gente comparte alguna propiedad pero también tiene propiedad privada (pág. 484)

mosque/mezquita edificio de devoción islámico (págs. 384, 462, 479)

multilingual/multilingüe que puede hablar varios idiomas (pág. 316)

multinational/multinacional compañía que hace negocios en varios países (pág. 316)

mural/mural pintura hecha sobre una pared (pág. 189)

nationalism/nacionalismo deseo de un territorio o colonia de hacerse independiente (pág. 351)

national park/parque nacional área reservada para proteger la flora y la fauna y para la recreación (pág. 135)

nature preserve/santuario natural área protegida para las plantas y animales (págs. 378, 581)

la Tierra que la gente usa para satisfacer sus necesidades (pág. 90)

navigable/navegable describe una masa de agua ancha y profunda suficiente para que los barcos puedan viajar por ella (págs. 126, 260, 301, 324)

neutrality/neutralidad negarse a ponerse a favor de uno de los adversarios en un desacuerdo o una guerra entre países (pág. 310)

newsprint/papel de periódico tipo de papel en que se imprimen los periódicos (pág. 153)

nomads/nómadas gente que se muda de un lugar a otro con sus manadas o rebaños de animales (págs. 377, 516, 687)

nonrenewable resource/recurso no renovable recurso natural, como metales o minerales, que no puede reemplazarse (pág. 91)

nuclear energy/energía nuclear energía producida por medio de una reacción atómica controlada (pág. 431)

oasis/oasis área verde en medio de un desierto a donde llegan aguas subterráneas (págs. 459, 493, 518)

oil shale/esquistos grasos rocas que contienen aceite (pág. 372)

orbit/órbita trayectoria que los cuerpos en el sistema solar siguen alrededor del Sol (pág. 29)

outback/tierra adentro el interior de Australia (pág. 748)

overgraze/pastar excesivamente cuando el ganado despoja los pastos hasta tal punto que las plantas no pueden crecer de nuevo (pág. 547)

ozone/ozono tipo de oxígeno que forma una capa en la atmósfera que protege a todas las cosas vivas de ciertos rayos del Sol que son peligrosos (pág. 775)

parliamentary democracy/democracia parlamentaria gobierno en que los votantes eligen a representantes a un cuerpo que hace las leyes y que selecciona a un primer ministro para que sea el jefe del gobierno (págs. 157, 210, 292)

parliamentary republic/república parlamentaria *véase* parliamentary democracy/democracia parlamentaria (pág. 327)

descomposición que se puede secar y usar para combustible (págs. 296, 372)

peninsula/península masa de tierra con agua alrededor de tres lados (págs. 40, 146, 182, 408)

permafrost/permafrost capa de suelo congelada en la tundra y las regiones subárticas; también se llama permagel (págs. 68, 410)

pesticides/pesticidas sustancias químicas poderosas que matan a los insectos que destruyen los cultivos (págs. 96, 648)

phosphate/fosfato sal mineral que se usa en los abonos (págs. 461, 555, 769)

pictograph/pictograma gráfica en que pequeños símbolos representan cantidades (pág. 15)

pidgin language/idioma rudimentario lenguaje formado al combinar elementos de varios idiomas distintos (pág. 767)

plain/llanura extensión de tierra plana u ondulante a elevaciones bajas (pág. 40)

plantain/plátano de cocinar tipo de banano (pág. 592)

plantation/plantación granja grande en que se siembra un solo cultivo para venderse (págs. 186, 207)

plateau/meseta planicie a elevaciones más altas que las llanuras (págs. 40, 324)

plate/placa plancha de roca inmensa que forma parte de la corteza de la tierra (págs. 599, 726)

plate tectonics/tectónica de placas teoría que dice que la corteza de la Tierra no es una envoltura enteriza, sino que está formada por placas, o planchas de roca inmensas, que se mueven (pág. 35)

plaza/plaza sitio donde se reúne el público (pág. 193)

poaching/caza furtiva cacería ilegal de animales protegidos (pág. 581)

polder/pólder área de tierra ganada del mar (pág. 315)

pope/papa líder de la Iglesia Católica Apostólica Romana (págs. 334, 369)

population density/densidad de población promedio de personas que viven en una milla cuadrada o kilómetro cuadrado (pág. 86)

potash/potasa tipo de sal mineral que a menudo se usa en los abonos (pág. 388)

prairie/pradera área de pastos ondulantes en el interior con suelo muy fértil (pág. 146)

precipitation/precipitación agua que regresa a la Tierra en la forma de lluvia, nieve, aguanieve o granizo (pág. 50)

prime minister/primer ministro líder del gobierno en una democracia parlamentaria (pág. 157)

privatize/privatizar transferir la propiedad de fábricas de las manos del gobierno a las de individuos (pág. 380)

projection/proyección una de las maneras de dibujar la Tierra redonda en una superficie plana para hacer un mapa (pág. 7)

province/provincia división política regional, parecida a un estado (pág. 143)

quota/cuota límite en la cantidad de un producto que se puede importar de un país en particular (pág. 93)

rain forest/bosque húmedo bosque denso que recibe grandes cantidades de lluvia todos los años (pág. 61)

rain shadow/sombra pluviométrica área seca en el lado interior de montañas costeras (pág. 59)

recycling/reciclaje usar materiales de nuevo en vez de botarlos (pág. 128)

refugee/refugiado persona que huye de un país a otro para evitar la persecución o un desastre (págs. 88, 383, 568, 595)

region/región área destacada por ciertas características (pág. 25)

relief/relieve las diferencias en altitud de una zona; lo plana o accidentada que es una superficie (pág. 12)

Renaissance/Renacimiento período de grandes logros en el arte y el estudio de la antigüedad que comenzó en Italia en los años 1300 y continuó por toda Europa (pág. 333)

renewable resource/recurso renovable recurso natural que no se puede gastar, que la naturaleza puede reemplazar o que se puede cultivar de nuevo (pág. 91)

republic/república gobierno nacional fuerte encabezado por líderes elegidos (págs. 210, 233, 258, 303, 369, 461)

reunification/reunificación juntar de nuevo las dos partes de Alemania bajo un mismo gobierno (pág. 309)

revolution/revolución una órbita completa alrededor del Sol (pág. 31)

rural/rural área en el campo (págs. 134, 433)

saga/saga historia larga (pág. 357)

samurai/samurai propietarios y guerreros poderosos del Japón (pág. 699)

sauna/sauna cuarto de madera calentado por agua que hierve sobre piedras calientes (pág. 351)

savanna/sabana pastos extensos en los trópicos con pocos árboles (págs. 64, 541, 563)

scale/escala relación entre las distancias en un mapa y las distancias verdaderas en la Tierra (pág. 11)

scale bar/barra de medir la escala en un mapa, línea con divisiones que muestra la escala del mapa, generalmente en millas o kilómetros (pág. 11)

secede/secesión separarse de un gobierno nacional (págs. 131, 154)

secular/secular no religioso (págs. 467, 480)

***selva*/selva** bosque húmedo tropical, como el de Brasil (pág. 228)

serf/siervo labrador que podía ser comprado y vendido con la tierra (pág. 425)

service industry/industria de servicio negocio que proporciona servicios a la gente en vez de producir productos (págs. 124, 150, 195)

shah/sha título de los reyes que gobernaban Irán (pág. 499)

shogun/shogun líder militar en Japón (pág. 699)

silt/cieno pequeñas partículas de suelo fértil (pág. 458)

sirocco/siroco vientos calurosos y secos que soplan a través de Italia desde el norte de África (pág. 331)

sisal/sisal fibra de una planta que se usa para hacer soga y cordel (pág. 588)

skerry/*skerry guard* isla rocosa (pág. 349)

slash-and-burn farming/agricultura por tala y quema método de limpiar la tierra para el cultivo en que se cortan y se queman los bosques (pág. 624)

smog/smog neblina espesa compuesta de niebla y sustancias químicas (pág. 198)

socialism/socialismo sistema económico en que muchos negocios son propiedad y están dirigidos por el gobierno (pág. 717)

solar system/sistema solar la Tierra, ocho planetas adicionales y miles de astros más pequeños que giran alrededor del Sol (pág. 29)

sodium nitrate/nitrato de sodio sustancia química usada en abonos y explosivos (pág. 269)

sorghum/sorgo cereal de tallo alto cuyas semillas sirven de alimento y del cual se hace un jarabe para endulzar (pág. 619)

spa/termas balneario con manantiales de agua mineral caliente en que la gente se baña para recobrar su salud (pág. 378)

station/estación rancho donde se crían ganado vacuno u ovejas en Australia (pág. 748)

steppe/estepa pastos parcialmente secos que a menudo se encuentran en los bordes de un desierto (págs. 69, 386, 406, 515, 570, 687)

strait/estrecho masa de agua delgada entre dos masas de tierra (págs. 42, 721)

estado del tiempo con inviernos extremadamente fríos y veranos cortos y frescos (pág. 67)

subcontinent/subcontinente masa de tierra grande que forma parte de un continente pero se puede diferenciar de él (pág. 645)

subsistence farm/granja de subsistencia terreno pequeño en el cual un granjero cultiva sólo lo suficiente para alimentar a su propia familia (págs. 186, 207, 261, 542)

suburb/suburbio comunidad pequeña en los alrededores de una ciudad (págs. 134, 340, 432)

summer solstice/solsticio de verano día con más horas de sol y menos horas de oscuridad (pág. 32)

taiga/taiga bosques enormes de árboles de hoja perenne en regiones subárticas (pág. 410)

tannin/tanino sustancia usada en el procesamiento del cuero (pág. 236)

tariff/arancel impuesto sobre el valor de bienes importados (pág. 93)

terraced field/terrazas franjas, parecidas a escalones, que se cortan en la ladera de una colina para que el suelo aguante el agua y se pueda usar para la agricultura (págs. 678, 724)

timberline/límite de los árboles elevación por encima de la cual no crecen árboles en las montañas (pág. 70)

tornado/tornado tormenta en forma de un torbellino que a veces se forma durante una tormenta eléctrica fuerte (pág. 56)

townships/municipios barrios abarrotados de gente en las afueras de las ciudades de Sudáfrica donde viven la mayoría de las personas que no son blancas (pág. 611)

trench/fosa marina valle en el fondo del mar (pág. 41)

tributary/afluente río pequeño que desagua en un río más grande (pág. 653)

tropics/trópicos región entre el Trópico de Cáncer y el Trópico de Capricornio (pág. 55)

trust territory/territorio en fideicomiso área que está bajo el control temporario de otra nación (pág. 769)

tsetse fly/mosca tsetsé insecto cuya picada puede matar al ganado o a los seres humanos por medio de la enfermedad del sueño (pág. 571)

tsunami/tsunami ola inmensa causada por un terremoto en el fondo del mar (págs. 36, 695)

tundra/tundra inmensas planicies ondulantes y sin árboles en latitudes altas con climas en que sólo varias pulgadas del suelo de la superficie se deshielan (págs. 68, 146, 410)

eléctricos (pág. 677)

typhoon/tifón nombre para un huracán en Asia (págs. 56, 673, 769)

urban/urbano parte de una ciudad (págs. 134, 432)

urbanization/urbanización movimiento hacia las ciudades (pág. 87)

vaquero/vaquero pastor de ganado vacuno (pág. 185)

wadi/uadi lecho de un río seco que llenan los aguaceros poco frecuentes (pág. 493)

water cycle/ciclo hidrológico proceso mediante el cual el agua se mueve de los océanos al aire, del aire a la tierra y de la tierra a los océanos una vez más (pág. 49)

watershed/cuenca fluvial región drenada por un río (pág. 594)

water vapor/vapor de agua agua en forma de gas (pág. 49)

weather/estado del tiempo cambios en la atmósfera que son difíciles de pronosticar y tienen lugar durante un período de tiempo corto (pág. 53)

weathering/desgaste proceso que rompe la superficie rocosa en peñas, grava, arena y suelo (pág. 37)

welfare state/estado de bienestar social estado que usa el dinero recaudado por los impuestos para mantener a personas que están enfermas, pobres, sin trabajo o retiradas (págs. 246, 349)

winter solstice/solsticio de invierno día con menos horas de sol y más horas de oscuridad (pág. 32)

yurt/*yurt* tienda de campaña grande y circular hecha de pieles de animales que se puede desmantelar y llevar de un lugar a otro (pág. 688)

INDEX

c=chart *m*=map
d=diagram *p*=photo
g=graph *ptg*=painting

Abdullah II (King), 491
Abidjan, Côte d'Ivoire, 554, 556
Abomey, 553
Aborigines, 738, 751, *p751,* 752, *p752*
Abuja, Nigeria, *p541,* 544
Accra, Ghana, 554, 556
acid rain, 97, 127, 154, 346, *p362,* 362–63, *p363,* 368, 698
Acid Rain 2000 project, 363
Aconcagua, Mount, *g175,* 235, *g660*
acquired immune deficiency syndrome (AIDS) epidemic: in South Africa, 611; in Zimbabwe, 620
Addis Ababa, Ethiopa, 598
Aden, Gulf of, 599
Aden, Yemen, 496
adobe, 193
Adriatic Sea, 331, 383, 385
Aegean Islands, 338
Aegean Sea, 338, 477
Afghanistan, 444, *m449, m478,* 500, 669; climate, *m483;* economy, *m493;* landforms, *m448, m479;* population, 86–87, *g451, m498;* profile, *c452*
Africa, 35, *m531, c533, p539;* agriculture, 528, *p528;* climate, 526; desert in, *p24;* economy, 528, *p528;* endangered species in, 604–05, *m604, p604, p605;* gems and minerals, *m532;* independence dates for countries in, 616; landforms, 526, *m530;* population, *g533;* profiles, *c534–38;* salt trade in, 560, *p560, m561. See also* North Africa; *specific countries*
African National Congress (ANC), 610
Afrikaners, 610
Ahaggar, 466
AIDS. *See* acquired immune deficiency syndrome
air pollution, *p97,* 97–98; in Mexico, 198, *p198*
Alaska, 115, 119, 120, *p123,* 126, 127
Alaska Range, 119
Albania, *m281, m366,* 381, 384, *p384,* 385; climate, *m372;* economy, *m376;* landforms, *m280, m367;* language, *m282;* people, 384; population, *m387;* profile, *c284*
Albert, Lake, 563
Alberta, Canada, 144, 153
Albuquerque, New Mexico, 126
Alexander I (Russian czar), 440
Alexander II (Russian czar), 425
Alexandria, Egypt, 461, 462; population, *g451*
Algeria, 444, *m449, m458,* 466–67; civil war, 467; climate, *m465,* 466–67; economy, *m460;* history and people, 467–68; landforms, *m448, m459,* 466–67; population, *m467;* profile, *c452*
Alhambra, 327, *p327*
alluvial plain, 497
almanac, 358
Alpine countries, 309
Alps, 276, *p277,* 306, 307, 309, 311, *p311,* 330–31
Al-Qaddhafi, Muammar, 465
altiplano, 260, 266, 267, 268
altitude, 184–85, 242, 583
Amazon River, 168, *g175,* 227
Amazon Basin, 64, 230
American Samoa, 769
Amin, Idi, 593–94
Amman, Jordan, 491
Amsterdam, Netherlands, 315–16
Amu Darya River, 518
Anatolia, 477
Andean countries, 237, 254–69, *m256;* archeological sites, 264, *p264;* music, 263, *p263;* population, *m262. See also* Bolivia; Chile; Colombia; Ecuador; Peru
Andean highlands, 241–42
Andean plateau, 255
Andersen, Hans Christian, 355
Andes, 25, 41, 168, *p169,* 227, 235, 255, 263
Andorra, *m281,* 323, *m324,* 327; climate, *m331;* economy, *m338;* landforms, *m280,* language, *m282;* 323, *m325;* population, *m337;* profile, *c284*
anemonefish, 765, *p765*
Angel Falls (Venezuela), *g175,* 242, *p242*
Angkor, 719
Angkor Wat, 719, *p719*
Anglos, 293
Angola, *m531, m608,* 614–16; agriculture, 615; climate, 614–15, *m615;* economy, 615, *m620;* history, 615–16; industry, 615; landforms, *m530, m609,* 614; languages, 615; natural resources, 615; population, *m625;* profile, *c534;* religion, 615
Ankara, Turkey, 479; population, *g451*
Antananarivo, 624
Antarctica, 25, *m741, c743, m766,* 772–76; climate, 736, *m773,* 773–74; coral reefs, *m742;* economy, 738, *m774;* environmental stations, 776, *p776;* exploring, 780, *p780, p781;* landforms, 736, *m740, m767,* 772–73; natural resources, 738; ozone layer over, *p762,* 762–63, *p763;* people, 738, *p738;* population, *g743, m768;* profile, *c744–45;* resources, 774–75; scientific research, 775, *p775;* wildlife, *p734,* 772, *p773*
Antarctic Peninsula, 773
Antarctic Treaty, 774–75
anthracite, 705
Antigua and Barbuda, *m173, m206;* climate, *m214;* economy, *m209;* landforms, *m173, m207;* population, *m216;* profile, *c176*
apartheid, 611, 612, *p612*
Apennines, 331
Appalachian Mountains, 104, *p105,* 115, 117–18, 145
aquifers, 51, 465, 506
Arabian Desert, 458–59
Arabian Peninsula, 491, 492–96
Arabian Sea, 496, 645, 654
Aral Sea, 516, 518, 519, p519
archipelagos, 213, 695
architecture: Angkor Wat, 719, *p719;* earthquake-prone parts of the world, 37, *p37;* gothic, 303, *p303;* Leaning Tower of Pisa, 333, *p333;* Rotterdam, 315, *p315*
Arctic Circle, 346, 348, 352, 396, 405
Arctic Islands, 145, 146
Arctic Ocean, 51, 104, 143, 276, 405
Argentina, *m173, m228,* 235–39; climate, 236, *m236;* culture, 239; economy, 236–37, 238, *m238;* ethnic people, 237–38; farming, 235–36; history, 237–39; industry, 237; landforms, *m172, m229,* 235–36; languages, 239; literature, 240; minerals and mining, 237; people, 237–39; population, *m174, g175,* 239, *m246;* profile, *c176;* provinces, 239; religion, 239; size, 235
Aristotle, 340
Armenia, 446, *m449,* 509, *m510,* 510–11; climate, *m512;* earthquakes, 37; economy, *m516;* landforms, *m448, m511;* population,

Azerbaijan, 511
Armenian Orthodox Church, 510
arts, 79; in Canada, 158; in Japan, 700–01; in Europe, 335, *ptg335;* in Mexico, 193, *p193,* 194, *ptg194;* in Paraguay, 248; in Russia, 426, *p426,* 434; in United States, 134–35
Ashanti, 553, 556
Ashgabat, Turkmenistan, 519
Asia, 36, *c639, m637;* agriculture, 634; climate, 632; country profiles, *c640–42;* economy, 634, *p634;* industry, 634, *p634;* landforms, 632, *m636;* monsoons, *m638;* people, 634; population, *g639,* country profiles, *c452–54. See also specific countries*
Asia Minor, 477
Assai, Lake, 533
Astrakhan, Russia, 414
Asunción, Paraquay, 248
Aswan, Egypt, *p445*
Aswan High Dam, 460
Atacama Desert, 168, *g175,* 268
Atatürk, Kemal, 480
Athens, Greece, 336, 337, 339
Athos, Mount, *p341*
Atlanta, Georgia, 125
Atlantic Coastal Plain, 116
Atlantic Ocean, 51, 57, 104, 115, 227, 276, 303, 607
atlas, 358
Atlas Mountains, *p443,* 444, 466, *p468*
atmosphere, 30
atolls, 662, 768
Auckland, New Zealand, 757
Aung San Suu Kyi, 717
Austen, Jane, 294
Australia, *m741, c743, c744, m748;* agriculture, 750; city and rural life, 752; climate, 736, 749, 750, *m750;* coral reefs, *m742;* Dreamtime, 752, *p752;* economy, 738, 749–50, *m755;* government, 751–52; Great Barrier Reef, *p39,* 747; history, 750–51; industry, 750; landforms, 736, *m740,* 747–48, *m749;* music, 751, *p751;* natural resources, 738; outback, 747, *p747,* 748; people, 738, *p738,* 750–51; population, *g743, m756;* profile, *c744–45;* vegetation, *p735;* wildlife, 749, 753, *p753*
Austria, *m281, m290,* 309, 311–12; climate, *m297,* 312; culture, 312; economy, *m302,* 312; Hungary and, 377; landforms, *m280, m291,* 311–12; language, *m282;* people, 312; population, *m314;* profile, *c284;* religion, 312; in World War I, 377
autobahns, 307
autonomous underwater vehicle (AUV), 52, *p52*
Autosub, 52
autumnal equinox, 32
Azerbaijan, 444, *m449,* 509, *m510,* 513; Armenian war with, 511; climate, *m512;* economy, *m516;* landforms, *m448, m511;* people, 513; population, *m518;* profile, *c452*
Azores Islands, p34
Aztec civilization, 189, *m189,* 192, *p192*

Bach, Johann Sebastian, 308
Baghdad, Iraq, *p497,* 498
bagpipes, 379, *m379*
Bahamas, *m173, m207,* 213; climate, *m214;* economy, *m209;* government, 216; landforms, *m172, m207;* population, *m216;* profile, *c176*
Bahrain, *m449, m478,* 495; climate, *m483;* economy, *m493;* landforms, *m448, m479;* population, *m498;* profile, *c452*
Baikal, Lake, 396, *c403,* 409, 563
Baja California, 182, 185
Baku, Azerbaijan, 513
Balaton, Lake, 376
Bali, Indonesia, *p726,* 728; agriculture, *p633*
Balkan Mountains, 382
Balkan Peninsula, 336, 381, 477
Balkhash, Lake, 515
Baltic Republics, 371–74
Baltic Sea, 276, 349, 353, 365, 371, 396, 405, 414
Baltimore, Maryland, 116, 125
Bandar Seri Begawan, 724
Banff National Park, 143, p143, 147
Bangkok, Thailand, 718
Bangladesh, 41, 632, *m637,* 645, *m646,* 652, *m659;* agriculture, 655; climate, *m653,* 654–55; economy, 655, *m659;* education, 654, *p654;* industry, 655; landforms, *m636, m647,* 654; languages, 654; people, 655; population, *m661;* profile, *c640*
Bangui, Central African Republic, 571
Bantu people, 567
Barbados, *m173, m206;* climate, *m214;* economy, *m209;* government, 216; landforms, *m172, m207;* population, *m216;* profile, *c176*
Barbuda. *See* Antigua and Barbuda
Barcelona, Spain, 326, *p326,* 328
barriers to trade, 93
basins, 115, 119, 227, 570
Basque people, 328
bauxite, 215, 376, 414, 555
p308
Bay of Bengal, 645
Bay of Fundy, 111
bazaar, 462
Beatles, 294
Bedouins, *p488,* 489
bee hummingbird, 215, *p215*
Beethoven, Ludwig van, 308
Beijing, China, 677, *p679,* 680, 682, *p683*
Beirut, Lebanon, 490
Belarus, *m281, m366,* 388–89; climate, *m372;* communism in, 388–89; economy, *m376;* landforms, *m280, m367;* language, *m282;* population, *m387;* profile, *c284*
Belgian Congo, 567–68
Belgium, *m281, m290,* 313–15; climate, *m297;* economy, *m302,* 314; landforms, *m280, m291,* 313–14; language, *m282,* 314–15; people, 314–15; population, *m314;* profile, *c284;* religion, 314
Belize, *m173,* 205, *m206,* 210; climate, *m214;* daily life, 210–11; economy, *m209;* landforms, *m172, m207;* population, *m216;* profile, *c176*
Benelux countries, 313–16
Benghazi, Libya, 465
Ben-Gurion, David, 485
Benin, *m531, m542,* 552; climate, *m548;* economy, *m553;* landforms, *m530, m543;* population, *m554;* profile, *c534*
Bently Subglacial Trench (Antarctica), *c743*
Benue River, 544–45
Berbers, 465, 466, *p468*
Bergen, Norway, 347, 348
Bering Sea, 396
Bering Strait, 410
Berlin, Germany, 309
Berlin Wall, *p84, p306,* 309; fall of, 320, *m321, p321*
Bern, Switzerland, 310
Bhutan, 634, *m637,* 645, *m646,* 657, 658–60, *m659, m661;* agriculture, 659; climate, *m653,* 658–59; economy, 659, *m659;* industry, 659; landforms, *m636, m647,* 658–59; people, 659–60; population, *m661;* profile, *c640*
Bhutto, Benazir, 654
bilingual country: Canada as, 158
biographical dictionary, 358
Birmingham, England, 289
birthrate, 85
Bismarck, Otto von, 308
Black Forest, 307
Black Sea, 276, 382, 408, 410, 444, 477, 509
Blanc, Mont, *c283*
Blue Nile River, 596
Boers, 610

INDEX

Bogotá, Colombia, 255, 256, 259
Bohemia, 378
Bolívar, Simón, 242, 258
Bolivia, *m173,* 235, *m256,* 266–67; agriculture, 267; climate, *m261,* 266–67; culture, *m261,* 267; economy, 267, *m267;* landforms, *m172, m257,* 266–68; natural resources, 267; people, 266, 267; population, *m174, g175, m262;* profile, *c176*
Bonaparte, Napoleon, 303, 425, 440
Borneo, 721, 726; contour map, *m725*
Bosnia and Herzegovina, *m281, m366,* 384; climate, *m372;* economy, *m376;* landforms, *m280, m367;* language, *m282;* population, *m387;* profile, *c284*
Bosporus, 477, *p477*
Boston, Massachusetts, 116, 125
Bothnia, Gulf of, 349
Botswana, *m531, m608,* 618, 621; agriculture, 621; climate, *m615,* 621; economy, *m620;* history, 621; landforms, *m530, m609,* 621; language, 621; natural resources, 621; people, 621; population, *m625;* profile, *c534;* religion, 621
Bourdeaux, 304
Brahmaputra River in India, 632, 654
Brasília, Brazil, 229, 232, *p234*
Bratislava, Slovia, 380, 427, *p427*
Brazil, 92, 168, *m173,* 170, 227–33, *m228,* 235; climate, 227–28, 229, *m236;* culture, 233; economy, *g230,* 230–31, *m238;* ethnic groups, 233–34; farming, 230; government, 232–33; history, 231–33; landforms, *m172,* 227–28, 228, *m229;* languages, 231; leisure time, 233; natural resources, 230; population, *m174, g175, m246;* profile, *c176;* rain forests, *p226;* recreation and sports, 233; religion, 231; size, 227
Brazilian Highlands, 228
British Columbia, 144, 148; landforms, *m149*
British Isles, 276
Brontë, Charlotte, 294
Brontë, Emily, 294
Broz, Joseph, 383
Brunei, *m637, m716,* 723–24; climate, *m722;* economy, *m727;* landforms, *m636, m717;* population, *m728;* profile, *c640*
Brussels, Belgium, 314–15
Budapest, Hungary, 378, *p378*
Buddhism, 79; in Asia, 634; in Bhutan, 659; in Cambodia, 719; in Canada, 158; in China, 681, 683; in India, 649; in Indonesia, 728–29; in Japan, 699, 700; in Korea, *p703;* in Laos, 718; in Malaysia, 722; in Mongolia, 688; in Myanmar, 716; in Nepal, 658; in Singapore, 723; in South Korea, 706; spread of, 692; in Sri Lanka, 661; in Taiwan, 686; in Thailand, 718, *p718;* in United States, 134
Buenos Aires, Argentina, 88, 170, 236, 237, *p237,* 238, 239
Bukhara, 515, *p515,* 518
Bulgaria, *m281, c284, m366,* 381, 382–83; agriculture, 382; climate, *m372,* 382; communism in, 383; economy, *m376,* 382; history, 382–83; industry, 382; landforms, *m280, m367,* 382; language, *m282;* people, 382–83; population, *m387*
Burkina Faso, *m531, m542,* 547, 550; climate, *m548;* economy, *m553;* landforms, *m530, m543;* population, *m554;* profile, *c534*
Burma. *See* Myanmar
Burundi, *m531,* 568, 582, *m582,* 594–95; climate, *m588;* economy, *m593;* landforms, *m530, m583;* population, *m597;* profile, *c534*
bush (Australia), 752
Byrd, Richard E., 780, *p780*

Cabinda, 614
Cabora Bassa Dam, 623
Cabot, John, 156
cacao, 766
caimans, 253
Cairo, Egypt, 446, 459, 461, 462; population, *g451*
Calcutta, India, 650
Calgary, Alberta, 153
Cali, 259
California, 26, 126, 127, 131
calligraphy, 684
Cambodia, *m637, m716,* 718–19; architecture, 719, *p719;* climate, *m722;* communism in, 719; economy, *m727;* landforms, *m636, m717;* population, *m728;* profile, *c640*
camels, 492, *p492*
Cameroon, *m531, m564, p570,* 570–71; climate, *m572;* economy, *m566,* 571; landforms, *m530, m565;* people, 571; population, *m571;* profile, *c534*
campesinos, 259
Canada, 104–13, *m109, c111,* 115, 142–65, 143, *m144;* agriculture, 106, *m110,* 153; arts, 158; British Columbia (economic region), 153; climate, 104, *m147,* 148; economy, 106, 150–54, *m151;* ethnic groups, *g111;* farming, 151, *p153;* fishing, *p150,* 151, 153; foods, *m110,* 159; government, 157; Great Lakes region, 148; history, 155–57; industry, 106, *p106,* 152; Inside Passage, 147; landforms, 104, *m108, p143,* 144–47, *m145,* 145–46, 148; language, 155, 158; literature, 158; Native Americans, 144, 156, *p156,* 157; natural resources, 106, 146, 152, 153, 154; Newfoundland and the Maritime Provinces (economic region), 151–52; Northern regions, 154; oil, 153; Ontario (economic region), 152–53; people and culture, 106, *p106,* 158–59; population, *g111, m158;* prairie, 146; Prairie Provinces (economic region), 153; provinces, *c113,* 143–44, 151–53; Quebec (economic region), 152; size, 143; sports and recreation, 159, 160, *p160;* tourism, 152; tundra, 146
Canadian Arctic, *p105*
Canadian Shield, 145, 146, 153
Canadian Rockies, 147
Canberra, Australia, 752
Cancún, Mexico, 186
canopy, 64, 564
canyons, 40
Cape Canaveral, 125
Cape Horn, 269, 773
Cape Town, South Africa, *p524, p607,* 608, 611
Cape Verde, *m531, m542,* 552, 555; climate, *m548;* economy, *m553;* landforms, *m530, m543;* population, *m554;* profile, *c534*
Captiva Island, *p125*
capybaras, 243, *p243*
Caracas, Venezuela, 242, 243
card catalog, 358
cardinal directions, 33
Caribbean Islands, 168
Caribbean Lowlands, 206
Caribbean Sea, 56, 120, 213
Carpathian Mountains, 276, 365, 376, 379, 380, 381, 386
carpet weaving, 514, *p514*
Carr, Emily, 158
Cartagena, 255
Carthage, 466
Cartier, Jacques, 156, 164
Casablanca, Morocco, 468
Cascade Range, 119
cash crops, 258, 513
Caspian Sea, 396, *c403,* 408–09, 414, 432, 444, 446, 451, 499, 509, 510, 513
cassava, 552–53, 584
caste, 648
Castile, 327
Castro, Fidel, 218
Catalan, 327
Catalonia, *p326,* 327
Cather, Willa, 135
Catherine I, 425
Catholics: in Albania, 385; in Angola, 615; in Argentina, 239; in Austria, 312; in Belgium, 314; in

INDEX

210, 211; in Chile, 269; in Colombia, 258; in Croatia, 384; in Democratic Republic of the Congo, 567; in France, 303; in French Guiana, 244; in Germany, 308; in Hungary, 378; in Indonesia, 729; in Italy, 334; in Lithuania, 374; in Luxembourg, 316; in Mexico, 190; in the Netherlands, 316; in Northern Ireland, 298, 299; in Paraguay, 248; in Peru, 262; in Philippines, 724; in Poland, 369; in Portugal, 327; in Republic of Ireland, 296; in Slovakia, 380; in Slovenia, 383; in Spa, 327; in Sweden, 349; in Switzerland, 311; in Uruguay, 246; in Venezuela, 243; in West Indies, 216

Caucasus. *See* Armenia; Azerbaijan; Georgia

Caucasus Mountains, 396, 407–08, 444, 509

caudillos, 242–43

cause and effect, 436

Celebes, 726

Celts, 293, 296, 298

census, 89

Central African Republic, *m531, m564,* 570–71; climate, *m572;* economy, *m566,* 571; landforms, *m530, m565;* people, 571; population, *m571;* profile, *c534*

Central America, *m206;* Caribbean Lowlands, 206–07; Central Highlands, 205–06; climate, 206–07; countries, 205; daily life, 210–11; economy, 207–09, *m209;* farming, 207; history and people, 209–11; industry, 208; landforms, 205–07, *m207;* Pacific Lowlands, 206; Panama Canal, 209; rain forests, *p205,* 208; Roman Catholics, 210, 211; tourism, 208. *See also specific countries*

Central Asian Republics. *See* Kazakhstan; Kyrgyzstan; Tajikistan; Turkmenistan; Uzbekistan

Central Highlands, 205–06

Central Lowland (United States), 118

Central Siberian Plateau, 408, 409

Central Valley, California, 119

Ceylon. *See* Sri Lanka

Chad, *m531, m542,* 547, 550; climate, *m548;* economy, *m553;* landforms, *m530, m543;* people, *p524;* population, *m554;* profile, *c534*

Champlain, Samuel de, 164

channel, 42

Chao Phraya River, 717

Charlemagne, 303

Charleston, South Carolina, 116

Chavez, Hugo, 243

Chechnya, 432

Cheju Island, 707

Chennai, 650

Chernobyl, 386, *p386,* 431

Chiang Kai-shek, 681, 686

Chicago, Illinois, 106, *p118,* 164

Chihuahua Desert, *p49*

Chile, 168, *p169, m173, m256,* 268–69; agriculture, 269; climate, *m261,* 268–69; culture, 269; economy, *m267,* 269; landforms, *m172, m257, p268,* 268–69; natural resources, 269; population, *m174, m262;* profile, *c176;* religion, 269

China, Republic of, 81, 92, 634, *m637, m670;* climate, *m672,* 673, 677; communism in, 675, 681; culture, 683–84; earthquakes, 673; economy, *p675,* 675–78, *m676;* environment, 677; farming, 672, 675, 676, *g677,* 677–78, 683, 712, *p712;* foreign trade, 676; government, 682–83; history, 680–81; industry, *p675,* 676, 677, 678; landforms, 23, 24, *m636,* 669–73, *m671;* monsoons, 712; natural resources, 677; people, *p635,* 680, *p683;* population, *g639,* 680, *m682;* profile, *c640;* religion, 681, 683; Silk Road and, 692, *p693;* society, 683–84; wildlife, *p669;* workers, *p93*

Chipaya, 266

Chişinău, Moldova, 389

chocolate, 310, *p310*

Christianity, 79, 504, *p505;* in Asia, 634; in Botswana, 621; in Central America, 210, 211; in Democratic Republic of the Congo, 567; in Ethiopia, 598; in Georgia, 512; in Guyana, 243; in India, 649; in Indonesia, 728–29; in Israel, 484, 486; in Lebanon, 490; in Malawi, 619; in Malaysia, 722; in Melanesia, 767; in Micronesia, 769; in Mozambique, 624; in Nigeria, 544; in Norway, 348; in Philippines, 724; in Russia, 434; in Singapore, 723; in South Africa, 611; in South Korea, 706; in Sudan, 597; in Suriname, 244; in Uganda, 593; in United Kingdom, 293; in United States, 134; in West Africa, 554; in West Indies, 216; in Zimbabwe, 620. *See also specific religions*

Chunnel, 303

Churchill, Winston, 293–94

Cincinnati, Ohio, 126

circle graph, *g15, g656*

Cisneros, Sandra, 135

city-state, 333

Ciudad Juárez, 185

civilizations, 81–82

civil war: in Algeria, 467; in El Salvador, 210; in Guatemala, 210; in Lebanon, 490; in Nicaragua, 210; in Nigeria, 545; in Spain, 327

clans, 600, 698

clay warriors, 681, *p681*

Cleopatra, 461

Cleveland, Ohio, 126

climate, 41; defined, 53–54; dry, 69, 121; effect of wind on, 56; highland, 70; high latitude, 67–68, 122; humid subtropical, 120; impact of people on, 60–61; landforms and, 58–59; latitude and, 55; mid-latitude, 65–67, 120–21; sun and, 54–55; tropical, 64, 122; world regions, *m65. See also specific countries;* weather

climograph, 16, *g16*

coalition governments, 334

coal mining, 91, *p368,* 378

coastal lowlands, 184

Coastal Plains, 115, 116–17

Coast Mountains (Canada), 147, 148

Cod Wars, 357

Cold War, 427–28

Cologne, Germany, 307

Colombia, *m173,* 210, 255–59, *m256;* agriculture, 258; climate, 255–57, *m261;* coffee, 258; culture, 259; economy, 257–58, *m267;* government, 258–59; history and people, 258–59; landforms, *m172,* 255–56, *m257;* language, 258; manufacturing, 257; Native American groups, 256; natural resources, 255; population, *m174, m262;* ports, 255; profile, *c176;* religion, 258; size, 255

Colombo, Sri Lanka, 661

colonies, 131, 156, 327

Colorado Plateau, 119

Colorado River, 40, 119

Columbian Exchange, 215, 224, *p224, m225, p225*

Columbia Plateau, 119

Columbia River, 119

Columbus, Christopher, 215, 224, 255, 327

command economy, 80

commercial farming, 92

commonwealth: Puerto Rico as, 219

communism, 80; in Belarus, 388–89; in Bulgaria, 383; in Cambodia, 719; in China, 675, 681; in Czech Republic, 378; in East Germany, 308; in Hungary, 377; in Poland, 367, 369; in Romania, 382; in Soviet Union, 425–28, 430–32, *p431;* in Vietnam, 720; in Yugoslavia, 383

Comoros, *m531, m608,* 623, 625, 626; climate, *m615;* economy, *m620;* landforms, *m530, m609;* population, *m625;* profile, *c534, c537*

comparisons, making, 708
compass, 33
compound, 544
computer database, 358
conclusions, drawing, 546
condensation, 50
Confucianism, 681, 706
Confucius, 681
Congo, *c535, m531, m564;* climate, *m572;* economy, *m566,* 572; landforms, *m530, m565;* people, 572; population, *m571;* profile, *c534*
Congo, Democratic Republic of the, *m531,* 563–68, *m564,* 595; climate, 563–66, *m572;* economy, *m566,* 566–67; government, 567–68; history, 567–68; landforms, *m530,* 563–66, *m565;* people, 567–68; population, *m571;* profile, *c535*
Congo River, 526, 563, *p563,* 564–65, 614; watershed, 578
conservation, 96
constitutional monarchy: in Japan, 700; in Jordan, 491; in Morocco, 469; in Norway, 348; in Spain, 327; in United Kingdom, 293
consumer goods, 382, 413, 676
Continental Divide, 119
continental island, 765
continental shelf, 41
continents, 35, *m41*
contour map, reading, *m12, m725*
Cook, James, 751
Cook Strait, 754
cooperatives, 217
Copán, Honduras, 210
Copenhagen, Denmark, 349, 354, 355, *p355*
copper, 618–19
copra, 766
coral reefs, *m742;* in Australia, 747; in Hawaii, 119
cordillera, 147, 255
core, 34
Cortés, Hernán, 190
Costa Rica, *m173,* 205, *m206, p208;* climate, *m214;* daily life, *p208;* economy, *m209;* farming, 207; government, 210; landforms, *m172, m207;* population, *m216;* profile, *c176;* rain forests, 208; resources, 208; tourism, 208
Côte d'Ivoire, *m531, c535,* 550, 554, 555–56; landforms, *m530;* profile, *c535*
cottage industry, 648
Court of the Lions, 327, *p327*
Cranberries, 299
crevasses, 772–73
Crimean Peninsula, 386
critical thinking skills: drawing inferences and conclusions, 546; making comparisons, 708; making predictions, 586; sequencing and categorizing information, 234; understanding cause and effect, 436
Croatia, *m281, m366,* 383; climate, *m372;* economy, *m376;* language, *m282;* landforms, *m280, m367;* people, 384; population, *m387;* profile, *c284*
crocus, *p91*
crop rotation, 97
crust, Earth's, 34, 35–37
Crutzen, Paul, 763
Cuba, *m173, m206,* 217–18; climate, *m214;* economy, *m209;* government, 216; landforms, *m172, m207;* population, *m216;* profile, *c176;* size, 214
culture: changes in, 81–82; defined, 77–80; diffusion in, 81; regions, *p81,* 82. *See also specific countries*
Curitiba, 231
currency, 292
cyclones, 623, 654–55
Cyprus, *m281, m326,* 338; climate, *m331;* economy, *m338;* landforms, *m280, m325;* language, *c282;* population, *m337;* profile, *c284*
Cyrillic alphabet, 382–83
czars, 424
Czechoslovakia, 378–79; Soviet Union and, 427, *p427*
Czech Republic, *m281,* 365, *m366,* 378–79; agriculture, 378; climate, *m372,* 378; communism in, 378; culture, 379; economy, *m376,* 378; history, 378–79; industry, 378; landforms, *m280, m367,* 378; language, *m282;* music in, 379, *p379;* natural resources, 378; people, 378–79; population, *m387;* profile, *c284*

Dakar, Senegal, 555
Dalai Lama, 683
Dallas, Texas, 125
Dalol, Denakil Depression, 533
Damascus, Syria, 489
Danube River, 276, *c283,* 307, 311–12, 375–76, 377, *p378,* 379, 381, 383
Daoism, 681
Dardanelles, 477
Dar es Salaam, Tanzania, 590, 595
Darling River, 748
Darwin, Charles, 265
data, collecting, for mapping Earth, 25–26
database, using, 270
da Vinci, Leonardo, 335, *ptg335*
daylight, number of hours of, 352, *p352*
Dayton Peace Treaty (1995), 384
Dead Sea, 451, 482, *p484,* 491
death rate, 84
Death Valley (California), 111, 121
Deccan Plateau, 646, 647
deciduous trees, 66
deforestation, 97, 572, 624, 666, 716
Delhi, India, 650
Delphi, Greece, *p274,* 339
delta, 42, 457, 516, 719
Delta Plan Project, 315
democracy, 79, 131; in Russia, 431, 432
democratic republics: Argentina as, 238; Brazil as, 233; Chile as, 269; Poland as, 369; Romania as, 382
Deng Xiaoping, 682
Denmark, *m281, m346,* 348, 353–55; agriculture, 354; climate, 354, *m354;* economy, *m350,* 354; government, 355; history, 355; industry, 354; landforms, *m280, m347,* 353–54; language, *m282,* 355; Midnight Sun in, 352, *p352;* people, 355; population, *m356;* profile, *c284;* religion, 355
Denver, Colorado, 126
desalinization, 95, 493, 506, *p507*
desertification, 548
deserts, 69, 692
Detroit, Michigan, 126
developing countries, 94
Dhaka, Bangladesh, *p53,* 655
dialect, 78, 327
diamonds: in Angola, 615; in Botswana, 621; mining and cutting, 613, *p613*
Diaspora, 485
Dickens, Charles, 294
dictators, 79, 465, 729
didgeridoo, 751, *p751*
dikes, 672
Djenné, Great Mosque, 551, *p551*
Djibouti, *m531, c535, m582,* 599–600; climate, *m588;* economy, *m593;* landforms, *m530, m583;* population, *m597;* profile, *c535*
Dnieper River, 386
***Doctor Zhivago* (Pasternak),** 411
Dodoma, Tanzania, 590
Dominica, *m173, m206;* climate, *m214;* economy, *m209;* landforms, *m172, m207;* population, *m216;* profile, *c177*
Dominican Republic, *m173, m206,* 218–19; climate, *m214;* economy, *m209;* government, 216; industry, 215; landforms, *m172, m207;* population, *m216;* profile, *c177*
Dortmund, 307
Dostoyevsky, Fyodor, 434
Douala, 571
***Dracula* (Stoker),** 382, *p382*
Drakensberg Range, 608

Dresden, Germany, 307
drought, 56, 547, 597–98
drug traffic: Colombia and, 258
Druids, 295
dry climates, 69, *p69,* 121
dry farming, 324
Dubai, United Arab Emirates, *p442*
Dublin, Ireland, 297, *p298*
Duluth, Minnesota, 164
Durban, South Africa, 611
Dutch Guiana, 243
dynasties, 680, 703

Eakins, Thomas, 135
Earle, Sylvia, 75
Earth, 29; collecting data for mapping, 25; forces beneath crust, 35–37; layers of, 34–35, *d35;* movement of, 31; as viewed from *Endeavour, p23;* water on, 52
earthquakes, 36–37; architecture and, 37, *p37;* in Armenia, 510; in China, 673; incredible power of, 202–03; in Indonesia, 726; in Kobe, Japan, 695, *p695;* in Mexico, 183, 202, *p202*
East Berlin, 309, 320
East China Sea, 671, 672
Eastern Desert, 458–59
Eastern Ghats, 645
Eastern Orthodox Christianity: in Albania, 385; in Belarus, 389; in Bosnia, 384; in Bulgaria, 383; in Croatia, 384; in Romania, 382; in Russia, 423, 434; in Ukraine, 388
East Germany, 308–09, 320; communism in, 308
East Pakistan, 652
eclipse, 29, *p29*
economic development, differences in, 94
economic system, 80
economy, 80; command, 80; free market, 150–51; market, 80; traditional, 80; world, *m92. See also specific countries*
ecosystems, 96
eco-tourists, 208
Ecuador, *m173, m256,* 258, 263–64; agriculture, 263; climate, *m261,* 263; economy, 263, *m267;* landforms, *m172, m257,* 263; natural resources, 263; people, 264; population, *m174, m262;* profile, *c177*
Edinburgh, Scotland, 290
Edmonton, Alberta, 153
Edward, Lake, 563
Egypt, 81, *m449, m458;* agriculture, *p21,* 460–61; climate, 457–59, *m465;* economy, *m460,* 460–61; history, 461–62; industry, *m459;* Nile River, 578, *p578;* people, 461–62; population, *g451, m467;* profile, *c452;* pyramids in, 463, *p463;* religion, 461, 462; tourism, 461
Eiffel Tower, *p288,* 304
Elbrus, Mount, *c403,* 407–08, *g660*
Elburz Mountains, 499
elevation, 40, 336, 517
elevation profiles, *d17,* 220, *d220*
"El Gaucho Martin Fierro" (Hernández), 240
El Niño, 56, 57, d57
El Salvador, *m173,* 205, *m206;* civil war, 210; climate, *m214;* economy, *m209;* landforms, *m172, m207;* population, *m216;* profile, *c177*
embargo, 498
emeralds, 255, *p255*
emigrants, 88, 349
empire, 261, 687
Empty Quarter, 444
emu, 753, *p753*
enclaves, 511, 612
encyclopedia, 358
endangered species, 594, 604
Endeavour: Earth as viewed from, *p23*
Endurance, 775, *p775*
English Channel, 289, 303
environmental issues, 25; acid rain, 97, 127, 154, 346, *p362,* 362–63, *p363,* 368, 698; air pollution, *p97,* 97–98; in China, 677; deforestation, 97, 572, 624, 666, 716; greenhouse effect, 60–61; in Japan, 698; ocean problems, *p74,* 74–75, *p75;* ozone hole, *p762,* 762–63, *p763;* rain forests and, *m252, p252,* 252–53, *p253;* recycling and, 141, *p141;* in Russia, *p420, m420,* 420–21, *p421,* 431; soil erosion, 37–38, 97, 117, 666, 712, *p712;* testing of nuclear weapons and, 517; Three Gorges Dam and, 672, 674; trash disposal, *p140,* 140–41, *p141*
Equator, 55, 62
Equatorial Guinea, *m531, c535, m564,* 573; climate, *m572;* economy, *m566;* landforms, *m530, m565;* population, *m571;* profile, *c535*
equinoxes, 32
Erebus, Mount, 773
Erie, Lake, 118, 146, *p146,* 164
Eriksson, Leif, 347
Eritrea, *m531, c535, m582,* 598–99; climate, *m588;* economy, *m593;* landforms, *m530, m583;* population, *m597;* profile, *c535*
Ertis River, 516
escarpments, 228, 582, 608
estancias, 236, 258
371–72, 372; agriculture, 371–72; climate, 371, *m372;* economy, *m376;* industry, 372; landforms, *m280, m367,* 371; language, *m282;* natural resources, 372; people, 372; population, *m387;* profile, *c284;* religion, 372
Ethiopia, 41, *m531, c535, m582;* climate, *m588,* 598; economy, *m593;* history, 598; landforms, *m530, m583,* 598; people, 598; population, *m597;* profile, *c535*
Ethiopian Highlands, 41
ethnic cleansing, 383
ethnic groups, 78, 133
Etna, Mount, 331
Euphrates River, 444, 497, 506
Eurasian country, 405
Europe, 276–87, *m281, c283, c284, c285, c286, c287;* agriculture, 278; climate, 276; economy, 278; industry, 278, *p278;* landforms, 276, *m280;* languages, *m282;* natural resources, 278; people, 278; population, *g283;* profile, *c284–87;* religions, *g283;* vegetation, 276. *See also specific countries*
European Union (EU), 93, 292, 313, 332, 348
evaporation, 49
Everest, Mount, *p26,* 39, 41, 632, *c639,* 657, *g660,* 666, 667
exclave, 614
export, 92
Eyre, Lake (Australia), *c743*

Fabergé, Peter Carl, 426
Fabergé Easter eggs, 426, *p426*
facts, distinguishing opinion from, 679
Falklands, 238
famines, 85, 707
Fante, 556
farming: in Argentina, 235–36; in Brazil, 230; in Canada, 151, 153, *p153;* in Central America, 207; in Colombia, 258; commercial, 92; in Kenya, *p93;* livestock, 92; in Mexico, *p181,* 186; in Paraguay, 248; subsistence, 94; in United States, 125, *p126,* 127; in West Indies, 215
Faroe archipelago, 353
Faulkner, William, 135
fault line, 356
faults, 37, 582, 673; in Armenia, 510. *See also* earthquakes
favelas, 231
Federal Council, 311
federal republics: Germany as, 309; Mexico as, 191; Russia as, 431; Switzerland as, 311; United States as, 131

INDEX

Federated States of Micronesia, 768; profile, *c744*
fellahin, 462
Festival of San Fermin, *p323*
fiestas, 194
Fifth Republic, 303
Fiji Islands, 765, 766, 767; profile, *c744*
Finland, *m281, m346,* 350–51; agriculture, 350; climate, 350, *m354;* economy, 350, *m350;* history, 351; industry, 350; landforms, *m280, m347,* 350; language, *m282,* 351; Midnight Sun in, 352, *p352;* people, 351; population, *m356;* profile, *c285;* religion, 351
Finland, Gulf of, 414
fishing: in Canada, *p150,* 151, 153; in United States, 127
fjords, 345
Flemings, 314
Flemish, 314
Florence, Italy, 332
food and food production: Columbian Exchange in, 215, 224, *p224, m225, p225;* in Mexico, 194; in United States and Canada, *m110,* 159
foothills, 261
Former Yugoslav republics, 383–84
Fortaleza, 230
Fossey, Dian, 605
fossil fuels, 91, 127; burning, *p97*
Fox, Vicente, 191
France, *m281, m290,* 300–04; agriculture, 302–03; climate, *m297,* 301–02; economy, *m302,* 302–03; government, 303–04; history, 303–04; industry, 303; landforms, *p24, m285, m291,* 300–01; language, *m282;* natural resources, 303; people, 303–04; population, 304, *m314;* profile, *c285;* religion, 303; restaurants, 300, *p300;* vegetation, 305, *m305*
Franco, Francisco, 327
Frank, Anne, 316
free enterprise system, 80, 123, 430, 583
free market economy, 150–51, 367
free port, 723
free trade, 93, 154, 196, 218
French Guiana, *m173, m228,* 243, 244; climate, *m236;* economy, *m238;* landforms, *m172, m229;* population, *m246;* profile, *c177*
French Polynesia, 769, *p769*
French Revolution, 303
freshwater, 50–51
Fuentes, Carlos, 194
Fuji, Mount, 695
Fujimori, Alberto, 262
Fulani in Nigeria, 543

Gabon, *m531, c535, m564;* climate, *m572;* economy, *m566,* 572; landforms, *m530, m565;* people, 572; population, *m571;* profile, *c535*
Gaborone, Botswana, 621
Galápagos Islands, 263, 265, *p265*
Galicia, 327
Gambia, *m531, c535, m542,* 555; climate, *m548;* economy, *m553;* landforms, *m530, m543;* population, *m554;* profile, *c535*
Gambia River, 552
Gandhi, Mohandas, 649
Ganges Plain, 645, 647
Ganges River, 632, 645, *p645,* 648, 654
Garagum, 519
Garbage Project, 141
Garonne River, 301
gauchos, 236
Gautama, Siddartha, 658
Gaza Strip, 486
Gdańsk, Poland, 368
Geneva, Switzerland, 310
Genghis Khan, 687
Genoa, Italy, 332
geographers, use of databases by, 270
geographic information systems (GIS), 26, 28, *p28*
geography: defined, 23; tools, 25–26; uses, 26–27; view of places, 24–25
Georgetown, Guyana, 243
Georgia, 444, 446, *m449,* 509, *m510,* 512; climate, *m512;* economy, *m516;* landforms, *m448, m511;* population, *m518;* profile, *c452*
Georgian Orthodox Church, 512
Georgia (United States), 131
geothermal power, 356
Germany, *m281, m290,* 306–09; agriculture, 307; climate, *m297,* 307; culture, 308; economy, *m302,* 307, 309; government, 309; history, 308–09; landforms, *m280, m291,* 306–07; language, *m282,* 308; people, 308; pollution, 307; population, *m314;* profile, *c285;* religion, 308; unification of East and West, 320, *m321, p321*
geysers, 202, 356
Ghana, *m531, c535, m542,* 550, 552, 554, 556; climate, *m548;* economy, *m553;* landforms, *m530, m543;* population, *m554;* profile, *c535*
giant sequoia, *p63*
glaciers, 49, 50, 144; as cause of erosion, 38; melting, 98
Glasgow, Scotland, 290
glasnost, 428
Global Positioning System (GPS), 25
global warming, 60
globes, 4
Gobi Desert, 632, c639, 670, 687
Golan Heights, 482
Good Hope, Cape of, 607
Goode's interrupted projection, *m8*
Gorbachev, Mikhail, 428
gorillas, 604; protecting, 605, *p605*
government, as cultural trait, 79
Gran Chaco, 236, 247
Grand Banks, 74, 151
Grand Canyon, 40, *p40,* 111, 119, 126
Great Artesian Basin, 748
Great Barrier Reef, *p39,* 747
Great Basin, 119
Great Britain, *c283,* 289
great circle route, *m6*
Great Dividing Range, 736, 747
Greater Antilles, 213
Great Escarpment, 228, 608
Great Hungarian Plain, 375–76
Great Indian Desert, 645, 653
Great Karroo, 607
Great Lakes, 118, 120, 145–46, 154
Great Plains, 104, 118, 120, 121
Great Rift Valley, *p40,* 526, 582, 587–88, 592, 598, 619
Great Salt Lake, 119
Great Victoria, *c743*
Great Wall of China, *p668,* 681, 692
Great Zimbabwe, 618, *p618,* 620
Greece, *m281, m324,* 336–40; agriculture, 339; climate, *m331,* 336–38; culture, 339–40; economy, *m338,* 339; government, 340; history, 339–40; industry, 339; landforms, *m280, m325,* 336–38; language, *m282;* pollution, 339; population, *m337;* profile, *c285;* religion, 340; ruins, *p274;* tourism, 339
Greek Orthodox Christians, 340
greenhouse effect, 60, 61, 98
Greenland, 353
Grenada, *m173, m206;* climate, *m214;* economy, *m209;* government, 216; landforms, *m172, m207;* population, *m216;* profile, *c177*
Grenadines. *See* St. Vincent and the Grenadines
groundwater, 51
Group of Seven, 158
Guadalajara, Mexico, 186
Guadalquivir River, 324
Guadeloupe, *p204,* 216
Guam, 768, 769
Guangzhou, China, 678
Guaraní, 248

architecture, *p166;* civil war, 210; climate, *m214;* economy, *m209;* farming, 207; industry, 208; landforms, *m172, m207;* population, *m174, g175,* 211, *m216;* profile, *c177;* resources, 208; tourism, 208
Guatemala City, 211
Guatemalan lowlands, 170
Guayaquil, Ecuador, 263
Guiana Highlands, 242
Guinea, *m531, c535, m542,* 555; climate, *m548;* economy, *m553;* landforms, *m530, m543;* population, *m554;* profile, *c535*
Guinea, Gulf of, 541, 552
Guinea-Bissau, *m531, c535, m542,* 555; climate, *m548;* economy, *m553;* landforms, *m530, m543;* population, *m554;* profile, *c535*
Gulf Coastal Plain, 116
Gulf of Mexico, 74, 104, 115, 118, 120, 181, 195
Gulf of Rīga, 373
Gulf Stream, 57
GUM state department store, 434, *p434*
Guyana, *m173, m228,* 243; climate, *m236;* economy, *m238;* landforms, *m172, m229;* population, *m246;* profile, *c177*

habitat, 589
haciendas, 190
Haifa, Israel, 486
haiku, 700, 702
Haiti, *m173, m206,* 218; climate, *m214;* economy, *m209;* government, 215–16; industry, 215; landforms, *m172, m207;* population, *m216;* profile, *c177*
Halifax, Nova Scotia, 152
Hamersley Range, 748
Hanoi, Vietnam, 720
Harare, Zimbabwe, 620
harmattan, 541
Hashemite family, 491
Hausa people, 543
Havel, Vaclav, 379
Hawaii, 115, *p115,* 119, 120, 122, *p122,* 126
Hawthorne, Nathaniel, 135
Haydn, Joseph, 312
heat island, 60–61
heavy industry, 350, 413
Helsinki, Finland, 351
hemispheres, *d4*
Henry "the Navigator" (Prince of Portugal), 327
Hermitage Museum, 434, *p437*
Hernández, José, 240
Herzegovina. *See* Bosnia and Herzegovina.
hieroglyphics: Egyptian, 461; Mayan, 189
high islands, 768
highland climate, 70
highland vegetation, 70
high latitude climates, 67–68, 122
high latitude vegetation, *p68*
high-technology industries, 685–86
high veld, 608
Hijuelos, Oscar, 135
Hikmet, Nazim, 481
Himalaya, 36, 39, 632, 645, 657, 658–59, 669; protecting, *p666,* 666–67, *p667*
Hinduism, 79; in Asia, 634; in Bangladesh, 655; in Canada, 158; in Guyana, 243; in India, 645, *p645,* 648, *p649;* in Indonesia, 728–29; in Malaysia, 722; in Melanesia, 767; in Nepal, 658; in Singapore, 723; in Suriname, 244; in United States, 134
Hindu Kush, 444, 500, 597, 653
Hispaniola, 215, 218
history, as cultural trait, 79
Hitler, Adolf, 308, 333
Ho Chi Minh City, 720
Hokkaido, Japan, 695, 697
Holocaust, 308, 485
Homer, 339
Homer, Winslow, 135
Honduras, *m173,* 205, *m206, p208;* climate, *m214;* economy, *m209;* industry, 208; landforms, *m172, m207;* population, *m216;* profile, *c177;* tourism, 208
Hong Kong, *p20, p643,* 678
Honolulu, Hawaii, 122
Honshu, Japan, 695, 697
Horn of Africa, 596–600
Horseshoe Falls, *p146*
House of Commons, 292–93
House of Lords, 292–93
Houston, Texas, 125
Hughes, Langston, 135, 136
Hugo, Victor, 304
human rights, in China, 683
humid continental climate, 66, 67, 148
humidity, 49
humid subtropical climate, 120
Hungary, *p274, m281, m366,* 375–78; agriculture, 376; Austria and, 377; climate, *m372,* 375–76; communism in, 377; economy, *m376,* 376–77; history, 377–78; industry, *m376,* 376–77; landforms, *m280, m367,* 375–76; language, *m282;* natural resources, 376–77; people, 377–78; population, *m387;* profile, *c285;* religion, 378; Soviet Union and, 377; World War I in, 380
Huron, Lake, 118, 146
America, 207, *p208;* in West Indies, 214
Hussein (King), 491
Hussein, Saddam, 498
Hutu, 595
hydroelectric power, 91, 565; in Russia, 414–15

"I, Too" (Hughes), 136
Ibadan, Nigeria, 544
Iberian Peninsula, 323
Ibo people, 543, 545
icebergs, 773
ice cap climate, 68
ice caps, 772–73; melting, 98
Iceland, *m281, m346,* 353, *p353,* 356–57; climate, *m354,* 356; economy, *m350,* 357; fishing, 357; government, 357; history, 357; landforms, *m280, m347,* 356; language, *m282;* Midnight Sun in, 352, *p352;* people, 357; population, *m356;* profile, *c285*
ice sheets, 49
ice shelves, 773; exploring, 52
Ignashov, Andrey, 421
Iguazú Falls, Brazil, *p226*
immigrants, 88; to United States, 133–34
import, 93
Inca civilization, 261, 262
India, 81, 634, *m637,* 645–51, *m646, m659;* agriculture, 647; climate, 646, *m653;* daily life, 650; economy, 647–48, *m659;* history, 648–49; industry, 647–48; landforms, *m636, p645,* 645–46, *m647;* languages, 648; music, 649, *p649;* people, *p648,* 648–50; population, *g639, m661;* profile, *c640;* religion, 645, *p645,* 649–50; Taj Mahal, 651, *p651*
Indian Ocean, 51, 581, 607, 632
Indonesia, *m637, c639, m716,* 726–29; agriculture, 727; climate, *m722,* 727; economy, *m727,* 727–28; history, 728–29; industry, 728; landforms, *m636, m717,* 726–27; natural resources, 727; people, 728–29; population, *g639, m728;* profile, *c640;* religion, 728–29
Indus River, 632, 653
industry, 425, 571
inferences, drawing, 546
inflation, 232
information: analyzing, 520; sequencing and categorizing, 234
infrastructure, 309
Inland Sea, 696
Inside Passage, 147
Institutional Revolutionary Party, 191

intensive cultivation, 698
Interior Plains, 115, 118, *p118,* 145, 146–47, 148
intermediate directions, 33
Internet, using, 187
Inuit people, *p107,* 144, 154, 156, *p156,* 157
invest, 676
Ionian Islands, 338
Ionian Sea, 338
Iran, 444, *m449, m478,* 499–500, 692; climate, *m483;* economy, *m493;* landforms, *m448, m479;* people, 499–500; population, *m498;* profile, *c452*
Iraq, 81, 444, *m449, m478,* 497–98, 504, *p505,* 692; climate, *m483,* 497; economy, *m493,* 497; landforms, *m448, m479,* 497; people, 498; population, *m498;* profile, *c452*
Ireland, Republic of, *m281,* 289, *m290,* 296; agriculture, 297; climate, 297, *m297;* conflict with Northern Ireland, 299; culture, 299; economy, *m302;* history, 298; industry, 297–98; landforms, *m280, m291, p296,* 296–97; language, *m282,* 298; people, *p279,* 296, 298; population, *m314;* profile, *c285;* religion, 296. *See also* Northern Ireland
Irish Gaelic, 298
Iroquois, 164
Irrawaddy River, 715
irrigation, 97
Islam, *m78,* 79, 499–500, 504, *p505;* in Albania, 385; in Algeria, 467; in Asia, 634; in Azerbaijan, 513; in Bangladesh, 655; in Bosnia, 384; in Brunei, 723–24; in Canada, 158; in Egypt, 461, 462; in Iberian Peninsula, 327; in India, 649, 651, *p651;* in Indonesia, 728–29, 730; in Israel, 484, 486; in Lebanon, 490; in Maldives, 662; in Mozambique, 624; in the Netherlands, 316; in Nigeria, 544, 545; in Pakistan, 654; in Russia, 433–34; in Sahel countries, 549; in Singapore, 723; in Somalia, 600; in Sudan, 597; in Suriname, 244; in Syria, 489; in Tunisia, 466; in Turkey, 479, 480; in United Kingdom, 293; in United States, 134; in West Africa, 551, *p551,* 554
Islamabad, 654
Islamic republic, 499–500
islands, 40; limestone, 213; volcanic, 213. *See also* continental island; high island; low island
Ismail Samani Peak, 451
Israel, 446, *m449, m478;* agriculture, 483, *p484;* climate, 482–83, *m483;* culture, *p482;* economy, 483–84, *m493;* history, 484–86; industry, 483–84; landforms, *m448, m479,* 482–83; natural resources, 484; neighbors, *m485;* people, 484–86; population, *g451, m498;* profile, *c452*
Istanbul, Turkey, 446, 479, 480; population, *g451*
isthmus, 40
Isthmus of Tehuantepec, 181
Itaipu Dam, 247, *p247,* 248
Italy, *m281, m324,* 330–34; agriculture, 332; climate, 330–31, *m331;* culture, 334; economy, 332, *m338;* government, 333–34; history, 333; industry, 332; landforms, *m280, m325,* 330–31; language, *m282,* 334; natural resources, 332; people, 334; population, *m337;* profile, *c285;* religion, 334
Ivan III (Russian czar), 424
Ivan IV (Russian czar), 424

Jackson, Mount, 202
Jahan, Shah, 651
Jainism, in India, 650
Jakarta, Indonesia, 728
Jamaica, *m173, m206,* 218; climate, *m214;* economy, *m209;* elevation profile, *m220;* government, 216; industry, 215; landforms, *m172, m207;* population, *m216;* profile, *c177*
Japan, 634, *m637, m696, m697, m705;* agriculture, 698; climate, 696–97, *m706;* culture, 700–01, *p701,* 702, *p702;* earthquake, 695, *p695;* economy, 697–98, 700; education, 700, *p700,* 701; environment, 698; government, 700; history, 698–99; industry, 697–98; landforms, *m636,* 695–96, *m697;* natural resources, 697; people, 700; population, *g639, m699;* profile, *c640;* religion, 699, 700; sports in, 701, *p701*
Jasper National Park, 147
Java, 726; shadow puppets, 730, *p730*
Java Sea, 721, 726
Jerusalem, Israel, 484, 486, *p505*
Jesus, 504, *p504*
Jews. *See* Judaism
Johannesburg, South Africa, 609, 611
John Paul II, 369
Jordan, *m449, m478,* 482, 490–91; climate, *m483,* 490–91; economy, *m493;* landforms, *m448, m479,* 490–91; people, 491; population, *m498;* profile, *c452*
Jordan River, 482, 490, 506
Joyce, James, 299
Judaism, *m78,* 79, 446, 504, *p505;* in Canada, 158; in Israel, 484–85; in Russia, 433–34; in United States, 134
Jupiter, 29
Jutes, 293
Jutland, 353–54, 354, 355

Kabale, Uganda, *p539*
Kabuki theater, 701
Kabul, Afghanistan, 500
Kahlo, Frida, 194
Kalahari Desert, 526, 617, 621
Kaliningrad, Russia, 414, *p414*
Kamchatka Peninsula, 408
Kampala, 593; public transportation, *p594*
kangaroos, 749, 753, p753
Kano, Nigeria, 543, 544
Kanto Plain, 696
Karachi, Pakistan, 654
Karakoram Range, 645, 653
Kara Sea, 74
Kariba Dam, 618
Kasai River, 565
Kashmir, 652–53
Kathmandu, 658
Kazakh nomads, 516, 517, *p517*
Kazakhstan, *m449, m510,* 515–17; climate, *m512;* economy, *m516;* landforms, *m448, m511;* population, *g451, m518;* profile, *c453*
Kazakh Uplands, 515
Kazbek, Mount, 509
Kenya, *p93, m531, c536,* 581–84, *m582;* changing, 591; climate, 583, *m588;* economy, 583–84, *m593;* government, 584–85; history, 584–85; landforms, *m530,* 581–83, *m583;* people, 584–85; population, *m597;* profile, *c536*
Kenya, Mount, 582
Kenyatta, Jomo, 585
Khartoum, Sudan, 596
Khmer people, 719
Khyber Pass, 500, 653
kibbutz, 483–84
Kiev, Russia, 387, 423
Kikuyu people, 585
Kilimanjaro, Mount, 526, *p527,* 533, 587, *g660*
Kim Il Sung, 707
Kim Jong Il, 707
King, Martin Luther, Jr., 132
Kingwana, 567
Kinshasa, 568, *p575*
Kiribati, 768; profile, *c744*
Kivu, Lake, 563
kiwifruit, 754, *p754*
Kjølen Mountains, 346
koalas, 749, 753, *p753*
Kobe, Japan, 695, *p695*

Kongfuzi, 681
Kongo kingdom, 615
Korean Peninsula, 632, 703; history, 704, *p704;* landforms, 703–04. *See also* North Korea; South Korea
Kosciuszko, Mount, *g660,* 747
Kosovo, *p87,* 384; civil war, *p87*
krill, 774
Kuala Lumpur, 722–23
Kunlun Shan, 669
Kurds, 479, 498
Kuwait, 446, *m449, m478,* 495; climate, *m483;* economy, *m493;* landforms, *m448, m479;* population, *m498;* profile, *c453*
Kuzomenskikh Desert, 421
Kyrgyzstan, 444, *m449, m510,* 517; climate, *m512;* economy, *m516;* landforms, *m448, m511;* population, *m518;* profile, *c453*
Kyushu, Japan, 695

Ladysmith Black Mambazo, 612
Lagos, Nigeria, *p541,* 544
lakes, 42
land bridge, 181
landfills, 127–28
landforms, *d18,* 24; climate and, 58–59; forces shaping, 37–38; local winds and, 58; and people, 41; types, 39–40
landlocked country, 266, 375, 511
LANDSAT photos, 25
languages, *g14,* 78; families of, in Europe, *c388*
La Niña, 56, 57
Laos, 634, *m637, m716,* 718; climate, *m722;* economy, *m727;* landforms, *m636, m717;* population, *m728;* profile, *c640*
Laozi, 681
La Paz, Bolivia, 267
Latin America, 168–79, *p169, m173;* agriculture, 170; architecture, *p166;* climate, 168; country profiles, *c176–79;* economy, 170; ethnic groups, *g175;* industry, 170; landforms, 168, *m172;* natural resources, 170; people, *p166,* 170, *p170;* population, *g175;* religions, *p170;* urban population growth, *m174. See also* Central America; *specific countries*
latitude, *d5,* 62, *m62,* 184; climate and, 55
Latvia, *m281, m366,* 371, 373, *p373;* climate, *m372,* 373; economy, 373, *p373, m376;* landforms, *m280, m367,* 373; language, *m282;* people, 373; population, *m387;* profile, *c285;* religion, 373
Law of Return, 486
Leaning Tower of Pisa, 333, *p333*
leap year, 31
Lebanon, *m449, m478,* 490; civil war, 490; climate, *m483,* 490; culture, 490; economy, *m493;* landforms, *m448, m479,* 490; population, *m498;* profile, *c453*
leeward side of mountain, 59, *d59*
Leipzig, 307
Lena River, 396, 409
Lenin, Vladimir, 426
Leopold II (King of Belgium), 567
Lesotho, *m531, c536, m608,* 611, *p611,* 612; climate, *m615;* economy, *m620;* landforms, *m530, m609;* population, *m625;* profile, *c536*
Lesser Antilles, 213, 216
Liberia, *m531, c536, m542,* 555; climate, *m548;* economy, *m553;* landforms, *m530, m543;* population, *m554;* profile, *c536*
library resources, 358
Libreville, Gabon, 572
Libya, *m449, m458,* 464–65; climate, 464–65, *m465;* economy, *m460,* 474, *p474;* history, 465; landforms, *m448, m459,* 464–65; oil, 474; people, 465; population, *m467;* profile, *c453*
Libyan Desert, 458–59
lichens, 68
Liechtenstein, *m281, m290,* 309–10; climate, *m297;* economy, *m302;* landforms, *m280, m291;* language, *m282;* population, *m314;* profile, *c285*
life expectancy, 431
light industry, 413
Lima, Peru, 262
limestone islands, 213
Limpopo River, 619
Lingala, 567
Lisbon, Portugal, 328
literacy rates: in Central America, 208, in Iceland, 357
literature: in Argentina, 240; in Canada, 158; in Colombia, 259; early Greek, 339; in Japan, 700, 702; in Russia, 411, 434; in United States, 135, 136
Lithuania, *m281, c285, m366,* 371, 373–74; climate, *m372,* 373–74; economy, *m376;* landforms, *m280, m367,* 373–74; language, *m282;* people, 374; population, *m387;* Soviet Union and, 374
Little America, 780
Little Karroo, 607
livestock farming, 92
Livingstone, David, 619
llanos, 168, 242, 256
loess, 672
Logan, Mount, 147
Loginova, Ulya, 421, *p421*
London, 289, *p289*
longitude, *d5,* 62, *m62*
Los Angeles, California, 106, 127
Louisville, Kentucky, 126
Louis XIV (King of France), 303
Louvre museum, *p275,* 304
Lusaka, Zambia, 619
Luther, Martin, 308
Lutherans: in Denmark, 355; in Estonia, 372; in Finland, 351; in Latvia, 373; in Norway, 348
Luxembourg, *m281, m290,* 313, 316; climate, *m297;* economy, *m302;* landforms, *m280, m291;* language, *m282;* population, *m314;* profile, *c285*
Luzon, Philippines, 724
Lyon, France, 304

Macau, 678
Macdonnell Ranges, 748
Macedonia, Former Yugoslav Republic of, *m281, m366,* 384; climate, *m372;* economy, *m376;* landforms, *m280, m367;* language, *m282;* population, *m387;* profile, *c286*
Macedonia-Thrace Plain, 337
MacGregor, Ewan, 294
Machu Picchu, *p260,* 261
Madagascar, *m531,* 533, *c536, m608,* 623, 624–25, 625; agriculture, 624; climate, *m615,* 624; economy, *m620;* landforms, *m530, m609,* 624; music, 624–25; people, 624; population, *m625;* profile, *c536;* wildlife, *p623*
Madrid, Spain, 326, 328
Magdalena River, 255
Maghreb, 464, 466
magma, 35
Magyars, 377, 378
Mahal, Mumtaz, 651
Mahé, 626
Makkah, Saudi Arabia, 494–95, *p495,* 504, *p505,* 548, 651
Malabo, Equatorial Guinea, 573
Malawi, *m531, c536, m608,* 618, 619; agriculture, 619; climate, *m615;* economy, *m620;* government, 619; history, 619; landforms, *m530, m609,* 619; natural resources, 619; people, 619; population, *m625;* profile, *c536*
Malay Peninsula, 721
Malaysia, *m637, m716,* 721–23; agriculture, 721–22; climate, 721, *m722;* economy, *m727;* government, 722; landforms, *m636,*

m717, 721; natural resources, 722; people, 722, *m723;* population, *m728;* profile, *c641;* religion, 722
Maldives, *m637,* 645, *m646,* 657, *m659, m661,* 662; climate, *m653;* economy, *m659;* landforms, *m636, m647;* population, *m661;* profile, *c641*
Male, Maldives, 662
Mali, *m531, c536, m542,* 547, 548, 550; climate, *m548;* economy, *m553;* landforms, *m530, m543;* population, *m554;* profile, *c536*
Malmö, Sweden, 349
Malta, *m281, m324,* 330; climate, *m331;* economy, *m338;* landforms, *m280, m325;* language, *m282;* people, *p277;* population, *m337;* profile, *c286*
Malvinas, 238
Manchuria, 671
Manchurian Plain, 671
Mandela, Nelson, 611, 612, *p612*
mangrove, 541
Manila, Philippines, 724
Manitoba, Canada, 144, 153
Mansa Musa, 548
mantle, 34, 35
manuka, 754
Maori, 756–57
Mao Zedong, 681, 682
map key, 33
maps: projections, 7–9; reading contour, *m725;* reading physical, 12, *m12,* 149, *m149;* reading population, 329, *m329;* reading special purpose, 12, 83, *m83;* reading time zone, 487, *m487;* reading transportation, 416, *m416;* reading vegetation, 305, *m305;* scale on, 11, 129
Maputo, Mozambique, 624, *p624*
maquiladoras, 185
Maracaibo, Lake (Venezuela), *g175,* 241
Mardalsfossen, *c283*
Margarethe (Queen of Denmark), 355
Mariana Trench, 41
Mariinsky Theater, 434
marine west coast climate, 65–66, *p67*
Maritime Provinces, 151
market economy, 80
Márquez, Gabriel García, 259
Marrakech, Morocco, 468
Mars, 29
Marseille, France, 301, 304
Marshall Islands, 768, 769; profile, *c744*
marsupials, 749
Martinez, Maria, 134
Martinique, 216
Masai, 581, *p581*
Massachusetts, 131
Massif Central, 301
Mauritania, *m531, c536, m542,* 547, 549–50; climate, *m548;* economy, *m553;* landforms, *m530, m543;* population, *m554;* profile, *c536*
Mauritius, *p79, m531, c536, m608,* 623, 626; climate, *m615;* economy, *m620;* landforms, *m530, m609;* population, *m625;* profile, *c536*
Mawsmai, *c639*
Mawsynram, *c639*
Maya, *p167,* 170, 188–89, *m189,* 210
McKinley, Mount, 111, 119, *g660*
Mecca. *See* Makkah
Medellín, 259
Mediterranean climate, 66, *p67,* 121
Mediterranean Sea, 276, 303, 408, 444, 457, 477
megalopolis, 116, 700
Mekong River, 632, 718, 719
Melanesia, 765–68, *m766;* agriculture, 766; climate, 765–66, *m773;* economy, 766, *m774;* industry, 766; landforms, 765–66, *m767;* people, 767–68; population, *m768*
Melbourne, Australia, 738, *p739,* 752
Memphis, Tennessee, 126
mental mapping, 569
Mercator projection, *m7*
Mercury, 29
Mesa Verde, 126
mestizos, 258
Mexican Plateau, 168
Mexico, 170, *m173,* 180–98, 182, *m182;* altitude, 184–85, *d185;* arts, 193, *p193,* 194, *ptg194;* celebrations, 194–95, *p195;* challenges for, 196–98; Chihuahua Desert, *p49;* cities and villages, 193; climate, *m184,* 184–85; coastal lowlands, 184; culture, 194–95; earthquakes, 183, 202, *p202;* economy, 195–96, *m196,* 197; economic regions, 185–86; farming, *p181,* 186; foods, 194; foreign investment and foreign debt, 197; free trade, 196; government, 191; history, 188–91; industries, 195, *p195;* landforms, *p49, m172,* 181–84, *m183;* Mayan ruins, *p167;* mestizos, 190; migrant workers, 196–97; minerals and mining, 182; Native American civilizations, 188–89, *m189;* natural resources, 182, 195; Olmec people in, p77; Plateau of Mexico, 183; pollution, 198, *p198;* population, *m174, g175,* 196–97, *m197;* profile, *c178;* Sierra Madre, 182–83; tourism, 186; volcanoes, *p181,* 183, *p183. See also* Central America
Mexico City, 88, 170, 183, 186, 189; earthquake damage, *p202;* pollution, 198, *p198*
Miami, Florida, 125
Michigan, Lake, 118, 164
Micronesia, 765, *m766,* 768–69; agriculture, 769; climate, 768–69, *m773;* economy, 769, *m777;* industry, 769; landforms, *m767,* 768; people, 769; population, *m768*
mid-latitude climates, 65–67, 120–21
Midnight Sun, 352
migrant workers, 196–97
migrations, 480
Milan, Italy, 332
Mindanao, 724
mineral resources, 124
Minsk, Belarus, 389
Mir **(space station),** 429, *p429*
Mississippi River, 104, 111, 118, 126
Missouri River, 111
Mitchell, Mount, 117
Mittelland, 310
Mobutu Sese Seko, 568
Mogul Empire, 649
Moldova, *m281, m366;* climate, *m372;* economy, *m376;* landforms, *m280, m367;* language, *m282;* population, *m387;* profile, *c286*
Molina, Mario, 763, *p763*
Mombasa, Kenya, 585
Monaco, *m281, m290;* climate, *m297;* economy, *m302;* landforms, *m280, m291;* language, *m282;* population, *m314;* profile, *c286*
Mona Lisa, 335, *ptg335*
monarchy, 79
Monet, Claude, 304
Mongolia, *m637, m670,* 687–88; climate, *m672;* economy, *m676;* government, 687; history, 687; landforms, *m636, m671,* 687; people, 688, *p688;* population, *m682;* profile, *c641;* religion, 688
Mongols, 423–24
monotheism, 486, 504
Monroe, James, 555
Monrovia, Liberia, 555
monsoons, 56, 632, *m638,* 712; in Cambodia, 718; in China, 673, 712; in Myanmar, 715; in North Korea, 707; in South Korea, 705; in Vietnam, 719
Mont Blanc, 301
Montenegro. *See* Yugoslavia
Monterrey, Mexico, *p180,* 185
Montevideo, Uruguay, 247
Montreal, Canada, 106, 152, 156
Montserrat, 214
moon, 30
moors, 290, 468–69
Moravia, 378
Morocco, 444, *m449, m458,* 468–69; agriculture, 468; climate, *m465,* 468; culture, 469;

INDEX

ple, *p79,* 443, *p443,* 468–69; landforms, *m448, m459,* 468; population, *m467;* profile, *c453*
Morrison, Toni, 135
Moscow, Russia, *p394,* 398, *p404,* 406, 412–13, 424, 432
moshav, 484
mosques, 384, 462, 479. *See also* Islam
Mother Teresa, 385
mountains, 39–40, 115, 119; temperature, rainfall and, 59
Mozambique, 526, *m531, c536, m608,* 612, 623–24; agriculture, 624; climate, *m615,* 623; economy, *m620;* history, 624; landforms, *m530, m609,* 623; language, 624; people, 624; population, *m625;* profile, *c536;* religion, 624
Mozambique Channel, 42
Mozart, Wolfgang Amadeus, 312
Muhammad, 504, *p505*
multilingual, 316
multimedia presentations, 622
multinational companies, 316
Mumbai (Bombay), 650
Munich, Germany, 308
Murmansk, Russia, 414
Murray-Darling River (Australia), *c743*
Murray River, 748
Muscat, 496
Music: in Australia, 751, *p751;* in Bolivia, 267; in Colombia, 259; in Czech Republic, 379, *p379;* in Ecuador, 263, *p263;* in India, 649, *p649;* in Madagascar, 624–25; in Paraguay, 245, *p245;* in Portugal, 328; in Russia, 434; in South Africa, 610, *p610,* 612; in Spain, 328; in Syria, 489; in Ukraine, 388; in United States, 134, *p134;* in West Indies, 217, *p217*
Muslims. *See* Islam
Mussolini, Benito, 333
Muzo, 255
Myanmar, 634, *m637,* 715–17, *m716;* agriculture, 716; climate, 715, *m722;* economy, *m727;* government, 716–17; history, 716–17; industry, 716; landforms, *m636,* 715, *m717;* natural resources, 716; people, 716; population, *m728;* profile, *c641*

Naadam Festival, 688
Nagoya, Japan, 700
Nairobi, Kenya, 584, *p584*
Namib Desert, 526, 607, 614, 616; animals in, *p614*
616–17; agriculture, 617; climate, *m615,* 617; economy, 617, *m620;* industry, 617; landforms, *m530, m609,* 616–17; language, 617; natural resources, 617; people, 617; population, *m625;* profile, *c536*
Nasser, Gamal Abdel, 461
Nasser, Lake, 461
National Aeronautics and Space Administration (NASA), 776
nationalism, 351
National Marine Sanctuary system, 75
National Oceanic and Atmospheric Administration (NOAA), 75
National Palace, *p194*
national parks, 135
Native Americans: in Argentina, 237; in Canada, 144, 156, *p156,* 157; in Central America, 209–10; in Colombia, 256; in Mexico, 188–89, *m189;* religion and, 190
natural gas, 91; in Canada, 153; production and distribution, *m450*
natural resources, 90–91; balancing people and, 98; defined, 90
natural vegetation, world regions, *m66*
nature preserves, 378, 581
Nauru, 768, 769; profile, *c744*
navigable, 260
Nazi Party, 308
Ndebele people, 620
Negev, 482, 484
Nepal, 634, *m637,* 645, *m646,* 657–58, *m659, m661;* agriculture, 658, *p658;* climate, *m653,* 658; economy, 658, *p658, m659;* industry, 658; landforms, *m636, m647,* 657; people, 658; population, 86–87, *m661;* profile, *c641;* tourism, 666
Neptune, 29
Netherlands, *m281, m290,* 313, 315–16; agriculture, 315; climate, *m297,* 315; economy, *m302,* 315; landforms, *m280, m291,* 315; language, *m282,* 316; people, 315–16; population, *m314;* profile, *c286;* religion, 316
neutrality, 310
Neva River, 414, 421
Nevis. *See* St. Kitts and Nevis
New Brunswick, Canada, 144, 156–57
New Caledonia, 765, 766
New Delhi, India, 649
Newfoundland, 144, 151–52, 156
New France, 156
New Granada, 258
New Guinea, *c639,* 726, 765
New Orleans, Louisiana, 125
newsprint, 153
New York City, *p102,* 106, 116, 125, *p130*
m749, 754–57; agriculture, 750; animals, *p737;* climate, 750, *m750;* economy, *m755,* 755–56; history, 756–57; industry, 750; landforms, 754–55; people, 756–57; plants, 755; population, *g743, m756;* profile, *c744*
Niagara Falls, *p146,* 164
Nicaragua, *m173,* 205; civil war, 210; industry, 208; landforms, *m172;* Pacific Lowlands, 206; population, *m174;* profile, *c178*
Nicaragua, Lake, 206, 210
Nice, France, 301, *p301*
Nicholas II (Russian czar), 426
Nieuwerkerk aan, *c283*
Niger, *m531, m542,* 547; climate, *m548;* economy, *m553;* landforms, *m530, m543;* population, *m554;* profile, *c536*
Nigeria, 528, *m531, c537,* 541–45, *m542,* 547; agriculture, 542–43; civil war, 545; climate, 541, *m548;* economy, 542–43, *p544, m553;* government, 544–45; history, 544–45; landforms, *m530,* 541, *m543;* natural resources, 542; people, 543–44; population, *m554;* profile, *c537*
Niger River, *c536,* 541, 544–45, 545, 550, 552, *p555*
Nile River, 444, *p445,* 446, 451, 457–58, 460, 506, 533, 578, *p578,* 592, 596–97
Nok people, 544–45
nomadic cultures, 81, 377, 516
nonreligious policies, 467
nonrenewable resources, 91
Normans, 293
North Africa, 444, *m449, c451,* 456–74, *m458;* climate, 444, *m465;* country profiles, *c452–54;* economy, 446, *p446;* landforms, 444, *m448, m459;* need for water, 506, *p506, m506;* oil and gas production and distribution, *m450;* people, 446; population, *g451, m467. See also specific countries*
North American Free Trade Agreement (NAFTA), 93, 128, 154, 196
North Atlantic Current, 57, 276, 291, 297, 349, 350, 354, 356
North Carolina, 117
North China Plain, 671, 672
Northern Hemisphere, 32, 55
Northern Ireland, 290, 293, 298; conflict with Republic of Ireland, 299. *See also* Ireland, Republic of
Northern Mariana Islands, 768, 769
North European Plain, 40, 276, 301, 365, 396, *p397,* 406
North Island, 736, 754
North Korea, *m637,* 671, *m696, m697, m699,* 703, *m705, m706,*

INDEX

706–07; agriculture, 707; border with South Korea, 704, *p704;* climate, *m706,* 706–07; economy, 707; government, 707; industry, 707; landforms, *m636,* 706; natural resources, 707; people, 707; profile, *c641;* religion, 707. *See also* Korean Peninsula
North Pole, 31, 55
North Sea, 276, 292, 315
Northwest Territories, 144, 154
Norway, *m281,* 345–48, *m346;* agriculture, 346; climate, 346, *m354;* culture, 348; economy, 347, 348, *p348, m350;* government, 348; history, 347–48; landforms, *m280,* 345–46, *m347;* language, *m282;* Midnight Sun in, 352, *p352;* natural resources, 347; people, 347–48; population, *m356;* profile, *c286;* religion, 348
notes, taking, 370
Notre Dame, 304
Nouakchott, Mauritania, 550
Nova Scotia, 144, 156–57
nuclear energy, 91, 386, *p386,* 431
Nunavut, 144, 154, 157

oasis, 459, 518
Ob-Irtysh River, *c403*
Ob River, 396, 409
ocean currents, 57–58, *m58*
Oceania, *m741, c743,* 765–70, *m766;* climate, 736, *m773;* coral reefs, *m742;* country profiles, *c744–45;* economy, 738, *m774;* landforms, 736, *m740, m767;* natural resources, 738; people, 738, *p738;* population, *g743, m768*
oceans, 40–41, 42; exploring, 52, *p52,* 75, *p75;* problems, 74, *p74;* world, *m41*
Oder River, 366
oil, 91; in Canada, 153; on the ocean, 474; production and distribution, *m450;* shipment of, 474, *p474;* in Venezuela, 242, 243; world, *g494*
oil shale, 372
okapi, 568, *p568*
Okavango River, 621
O'Keeffe, Georgia, 135
Olmec people, *p77*
Oman, *m449, m478,* 496; climate, *m483;* economy, *m493;* landforms, *m448, m479;* population, *m498;* profile, *c453*
Ontario, Canada, 144, 152–53, 156–57, *p157*
Ontario, Lake, 118, 146, *p146,* 159
opinion, distinguishing fact from, 679
orbit, 29
Oregon, 126, 127
Organization of Petroleum Exporting Countries (OPEC), 493
Orinoco River, 242
Orlando, Florida, 125
Orozco, José Clemente, 194
Ortiz, Simon J., 136
Osaka, Japan, 700
Oslo, Norway, 346
Ottawa, Canada, 153
oud, 489, *p489*
outback, 748
outlining, 758
overgrazing, 547
Oyama Volcano, Mount, 700
ozone, 775
ozone Hole, *p762,* 762–63, *p763*

Pacific coast, 115, 119, 145, 147
Pacific Lowlands, 206
Pacific Ocean, 41, 51, 104, 115, 119, 181, 405, 632, 736
Padaung ethnic group, *p715*
Pakistan, 632, *m637,* 645, *m646,* 648, 652–54, *m653, m659, m661;* agriculture, 652, *p652,* 653–54; climate, *m653;* economy, 653–54, *m659;* industry, 654; landforms, *m636, m647,* 652–53; people, 654; population, *m661;* profile, *c641*
Palau, 768, 769; profile, *c744*
Palio, 330, *p330*
Pamirs, 692
pampas, 168, 236
Pamplona, *p323*
Panama, *p169, m173,* 205, *m206,* 210, 258; climate, *m214;* economy, *m209;* landforms, *m172, m207;* population, *m216;* profile, *c178*
Panama Canal, 209, 212, *p212,* 215
pandas, *p669*
Pangaea, 35
panpipe, 263, *p263*
Papeete, 770
Papua New Guinea, *c639,* 726, 765, 766; population, *g743;* profile, *c744*
papyrus, 461
Paraguay, *m173,* 235, 247–48; climate, 247; crafts, 248; economy, 247–48; farming, 248; government, 248; history, 248; landforms, *m172,* 247; music, 245, *p245;* people, 248; profile, *c178*
Paraguay River, 247
Paraná River, 228, 248
Paricutin, 181, *p181*
Paris, France, 301, 304
parliamentary democracy: in Belgium, 313; in Belize, 210; in Canada, 157; in Denmark, 355; in Finland, 351; in Luxembourg, 313; in Netherlands, 313; in Norway, 348; in United Kingdom, 292
parliamentary republic: in Portugal, 327
Parthenon (Athens), 336, *p336,* 339
Parthian Empire, 692
Pashtuns, 500
Pasternak, Boris, 411
Patagonia, *p235,* 236, 237
Paz, Octavio, 194
Pearl Harbor, Hawaii, 699
peat, 296–97, 372
Peloponnesus, 337
Pemba, 588
penguins, 772, *p772*
peninsula, 40, 146, 182, 408
people: balancing resources and, 98; impact of, on climate, 60–61
People's Republic of China. *See* China
periodical guide, 358
permafrost, 68, 396, 410
Perón, Eva, 238
Perón, Juan, 238
Perry, Matthew, 699
Persian Gulf, 446, 492, 497
Persian Gulf states, 495
Persian Gulf War, 498
Peru, *m173, m256,* 260–62; agriculture, 261; climate, 261, *m261;* culture, 262, 263, *p263;* economy, 261, *m267;* geography, 261; history, 261–62; Incas, *p260,* 261; landforms, *m172, m257,* 260–61; language, 262; Native Americans, 262; people, *p166;* population, *m174, m262;* profile, *c178;* religion, 262
Peru Current, 260, 261, 263, 268
pesticides, 96
Pestova, Maria, 421, *p421*
Peter I (Russian czar), 425
Peter the Great, 425, *p425*
Petra, 490, *p490*
Petronas Towers, *p714,* 722–23
Philadelphia, Pennsylvania, 116, 125
Philip II (king of Spain), 724
Philippines, 634, *m637, m716,* 724; climate, *m722;* economy, *m727;* landforms, *m636, m717;* population, *m728;* profile, *c641;* volcanic mountains in, *p56*
Phnom Penh, Cambodia, 719
Phoenicians, 466
Phoenix, Arizona, 126
phosphate, 461, 555, 769
physical characteristics, 24
pictographs, 16, *g16*
pidgin language, 767
Pinatubo, Mount, 56, *p56*
Pindus Mountains, 337
pirarucu, *p227*
Plain of Attica, 337

plains, 40
planets, 29–30
plantains, 592
plantations, 186, 207
plateaus, 40, 41, 324; of Mexico, 183; of Tibet, 40, 632, 669–70, 672, 673, 678
plate tectonics, 35. *See also* tectonic plates
Plato, 340
Platte, Arkansas, 119
platypus, 753
Pluto, 29
poaching, 581
Pogues, 299
Poland, *m281, m366;* agriculture, 368; climate, 365–66, *m366, m372;* communism in, 367, 369; economy, 367–68, *p368, m376;* environment, 368; history, 368–69; industry, 368, *p368;* landforms, *m280,* 365–66, *m366, m367;* language, *m282;* natural resources, 368; people, 368–69; population, *m387;* profile, *c286;* religion, 369; Soviet Union and, 369
polar bear, *p105*
polders, 315
pollution: in Germany, 307. *See also* Air pollution
Polynesia, 765, *m766,* 769–70; agriculture, 770; climate, 770, *m773;* economy, 770, *m777;* industry, 770; landforms, *m767,* 769–70; people, 770; population, *m768*
Pompeii, 332
Pontic Mountains, 477
Pope, 334, 369
Popocatépetl, 183
population: census in counting, 89, *p89;* distribution of, 86; movement of, 87–88; world, *g85, c85*
population density, 86–87; world, *m86*
population growth, 84–85; challenge of, 85; reasons for, 84–85
population map: reading, 329, *m329*
Po River, 331, 332, 334
Port-au-Prince, Haiti, 218
porteños, 239
Port Moresby, Papua New Guinea, 768
Porto, 328
Porto Alegre, 231
Port Said, 462
Portugal, *m281,* 323, *m324;* agriculture, 324, 325–26; climate, 324–25, *m331;* culture, 328; economy, 325–26, *m338;* government, 327; history, 326–27; industry, 326; landforms, *m280,* 323–25, *m325;* language, *m282,* 326–27; people, 327–28; population, gion, 327
Postel, Sandra, 507, *p507*
potash, 388
Potato Famine, 297
Prague, Czech Republic, *p364,* 378
prairie, 146–47
Prairie Provinces, Canada, 153
precipitation, 50, 54
predictions, making, 586
Preseli Mountains, 295
Pretoria, South Africa, 611
prevailing wind patterns, *m55*
Prime Meridian, 62
prime minister: in Canada, 157
Prince Edward Island, 144, 152, 157
Principe. *See* Sao Tome and Principe
privatize, 380
Protestant Reformed Church in Switzerland, 311
Protestants: in Canada, 158; in Germany, 308; in Hungary, 378; in Northern Ireland, 298, 299; in Sweden, 349; in Switzerland, 311; in United Kingdom, 293; in West Indies, 216. *See also* Christianity
Puerto Rico, *m173, m206,* 213, 216, 219; climate, *m214;* economy, *m209;* industry, 215; landforms, *m172, m207;* population, *m216;* profile, *c178*
Puerto Vallarta, Mexico, 186
Pushkin, Alexander, 434
Putin, Vladimir, 432
Pyongyang, North Korea, 707
Pyrenees, 276, 301, 323, 328

Qatar, *m449, m478,* 495; climate, *m483;* economy, *m493;* landforms, *m448, m479;* population, *m498;* profile, *c453*
Qin Shi Huang, soldiers from tomb of, 681, *p681*
Qomolangma National Nature Preserve, 667, *m666*
Quebec, Canada, 144, 152, 154, *p155,* 156, 157, 160, 164
quechua, 262
Quito, Ecuador, 264
quota, 93
Quran, 494–95, 500, 504, *p505. See also* Islam

Rabat, Morocco, 469
rain forests, 168, *p169;* animals, *p205;* in Bolivia, 266–67; in Brazil, *p220;* in Central America, *p205,* 208; clearing, 61; in Costa Rica, 208; vanishing, *p96, m252, p252,* 252–53, *p253;* in Venezuela, *p241*
rain shadow, 59, *d59,* 268
Ramblas, 326
Ramses II, 461, *p461*
Rangoon. *See* Yangon
Rathje, William, 141, *p141*
recession, 131, 154
Recife, 230
recycling, 128, 141, *p141*
Red River, 119, 719
Red Sea, 461, 496, 598
refugees, 88, 568; ethnic, 383, 384, 385
regions, 25
Reichstag, 309
Reims Cathedral, 303, *p303*
religion, 79; influence of, on American life, 134; world, *m78. See also* Buddhism; Christianity; Hinduism; Islam; Judaism; traditional religions; *specific religions*
Renaissance, 333
renewable resources, 91
Renoir, Pierre-Auguste, 304
report, writing, 771
representative democracy, 131
republics, 210, 369, 461–62; Columbia as, 258
reunification, 309
revolution, 31, 293
Reykjavík, Iceland, 356, 357
Rhine River, 276, 301, 310
Rhodes, Cecil, 620
Rhodesia, 620
Rhodope Mountains, 382
Rhône River, 301, 310
Rīga, Latvia, 373
Rimsky-Korsakov, Nikolay, 434
Ring of Fire, 408, 632, 673, 695
Rio de Janeiro, Brazil, 170, *p171,* 229, 230, 232, *p232*
Río de la Plata, 236, 245
Rio Grande, 119, 168, 198
Rivera, Diego, 194, *ptg194*
Riyadh, Saudi Arabia, 494
Robinson projection, *m8*
Rocky Mountains, *p102,* 104, 119, 121, 145, 147, 148, 153
Roman Catholic. *See* Catholics
Romania, *m281, m366,* 381–82; climate, *m372,* 381; communism in, 382; economy, *m376,* 381–89; history, 382; landforms, *m280, m367,* 381; language, *m282;* natural resources, 381–82; people, 382, *p382;* population, *m387;* profile, *c286*
Rome, Italy, 332, *p332,* 333
Roosmalen, Marc van, 253, *p253*
Rotterdam, *p278,* 315, *p315*
Rowland, Sherwood, 763
Royal Danish Ballet, *p357*
Rub' al Khali, 444
Ruhr, 307

running of the bulls, 323, *p323*
Russia, 396–403, *m401, m406;* agriculture, 396, 398, 413; animals, *p394–95;* arts, 426, *p426,* 434; celebrations, 435; climate, 396, *m402, m409,* 410, *p411;* culture, *p394,* 433–35; economy, 398, *p412,* 412–15, *m413, p414,* 425–28, 430–31, *p431;* environmental issues, 398, *p420, m420,* 420–21, *p421,* 431; foods, 435; German invasion of, 440, *p441, m441;* government, 431–32; history, 351, 423–28, *m424, p427;* industry, 398, *p398,* 413; landforms, 396, *m400,* 405–09, literature, 411, 434; music, 434; natural resources, 398, 412, 414–15; people, *p394, p397,* 398, *p399, p432,* 432–33; *m407;* population, *g403, c403, m433;* profile, *c401, c403;* religion, *p397,* 433–34; Revolution in, 426; space exploration and, 429, *p429;* sports and recreation, 435; time zones, 396; transportation, *m416;* vegetation, *p397. See also* Soviet Union
Russian Academy of Fine Arts, 423
Rwanda, *m531, c537,* 568, 582, *m582, p592,* 594–95; climate, *m588;* economy, *m593;* landforms, *m530, m583;* population, *m597;* profile, *c537*

Sabah, 721
saffron, *p91*
sagas, 357
Sahara, 444, 451, 459, 464, *p464,* 466; people, *p445*
Sahel countries, *m542, p547,* 547–50; climate, 547–48, *m548;* economy, *m553;* history, 548–49; people, 549–50; landforms, *m543,* 547–48; population, *m554*
Saigon, Vietnam, 720
St. Kitts and Nevis, *m173, m206;* climate, *m214;* economy, *m209;* landforms, *m172, m207;* population, *m216;* profile, *c178*
St. Lawrence River, 118, 145–46, 146, 152, 153, 156, 160, 164, *p165, m165*
St. Lawrence Seaway, *d152,* 153, 164
St. Louis, Missouri, 126
St. Lucia, *m173, m206;* climate, *m214;* economy, *m209;* landforms, *m172, m207;* population, *m216;* profile, *c178*
St. Peter's Basilica, 334
St. Petersburg, Russia, 398, 406, 414, *p422, p423,* 425, 434, 440
St. Vincent and the Grenadines, *m173, m206;* climate, *m214;* economy, *m209;* landforms, *m172, m207;* population, *m216;* profile, *c178*
Sakhalin, *c403*
Salt Lake City, Utah, 126
salt trade in Africa, 560, *p560, m561*
salt water, 51
Salvador, 230
Salween River, 715
Samarqand, Uzbekistan, 518
Samburu, *p585*
Sami, 348, 349, *p349*
Samoa, 769; people, *p734;* profile, *c744*
samurai, 699
Sanaa, 496
sand, use of, 96–97
San Francisco, California, *p90,* 127
San José, Costa Rica, *p208*
San Marino, *m281,* 330, *m324;* climate, *m331;* economy, *m338;* landforms, *m280, m325;* language, *m282;* population, *m337;* profile, *c286*
San Martín, José de, 238
San Salvador, 215
Santiago, Chile, 268, *p268*
San Xavier del Bac, 131, *p131*
São Francisco River, 228
São Paulo, Brazil, 231
Sao Tome and Principe, *m531, c537, m564,* 573; climate, *m572;* economy, *m566;* landforms, *m530, m565;* population, *m571;* profile, *c537*
Sarajevo, 384
Sarawak, 721
Sardinia, *m324,* 330; climate, *m331;* economy, *m338;* landforms, *m325;* population, *m337*
Saro-Wiwa, Ken, 545
Saskatchewan, Canada, 144, 153, *p153*
Satpura Range, 645, 646
Saturn, 29
Saudi Arabia, 95, 446, *m449, m478,* 492–95; climate, *m483,* 492–93; economy, 493, *m493;* landforms, *m448, m479,* 492–93; population, *m498;* profile, *c453*
savannas, 64, 541, 563
Saxons, 293
scale, 129
scale bar, 129
Schubert, Franz, 312
Scotland, 289–90, 293
sea anemone, 765, *p765*
sea ice, 52
Sea of Japan, 695
Sea of Marmara, 477
seasonal affective disorder (SAD), 352
seasons, *d31,* 31–32
Seattle, Washington, 127, 147
secular policies, 467
seed bank, *p95*
Seine River, 301, 304
Senegal, *m531, c537, m542,* 555; climate, *m548;* economy, *m553;* landforms, *m530, m543;* population, *m554;* profile, *c537*
Sénégal River, 552
Seoul, Korea, 704, 705
Serbia. *See* Yugoslavia
Serbs, 383, 384
Serengeti National Park, 589
Serengeti Plain, 587
serfs, 425
service industries, 124, 125, 195
Seychelles, *m531, c537, m608,* 623, 626; climate, *m615;* economy, *m620;* landforms, *m530, m609;* population, *m625;* profile, *c537*
shadow puppets, 730, *p730*
shahs, 499
Shakespeare, William, 294
Shanghai, China, *p60,* 678
Shannon River, 297
Shaw, George Bernard, 299
Shenyang, China, *p675*
Shikoku, Japan, 695
Shinto religion in Japan, *p698,* 699, 700
shoguns, 699
Shona, 620
Siam. *See* Thailand
Siberia, 396, *p397,* 406–07, 415, *p430*
Siberian tigers, *p394,* 405, *p405*
Sicily, *m324,* 330, 331; climate, *m331;* economy, *m338;* landforms, *m325;* population, *m337*
Siena, Italy, 330, *p330*
Sierra Leone, *m531, m542,* 555; climate, *m548;* economy, *m553;* landforms, *m530, m543;* population, *m554;* profile, *c537*
Sierra Madre, 168, 182–83
Sierra Madre del Sur, 182
Sierra Madre Occidental, 182
Sierra Madre Oriental, 182
Sierra Nevada, 119
Sikhism in India, 649
silk, *p692*
Silk Road, 515, 518, 692, *p693*
Silla, 703
silt, 458
Sinai Peninsula, 458
Singapore, *m637, m716, p721,* 723; climate, *m722;* economy, *m727;* landforms, *m636, m717;* population, *m728;* profile, *c641*
Sinhalese, 661–62
siroccos, 331
Skopje, Macedonia, 384
Skorohodova, Natalia, 421, *p421*
slash-and-burn farming, 624
Slavic people, 398
sleeping sickness, 571, 574

INDEX

Slovakia, *m281,* 365, *m366, p375,* 379–80, 380; agriculture, 379; climate, *m372,* 379; economy, *m376,* 379–80; industry, 379–80; landforms, *m280, m367,* 379; language, *m282;* people, 380; population, *m387;* profile, *c286*
Slovaks, 380
Slovenia, *m281,* 383; language, *m282;* landforms, *m280;* profile, *c286*
smog, 198, *p198*
social groups, 77–78
socialism, 80; in Myanmar, 717
sodium nitrate, 269
Sofia, Bulgaria, 383
Sognefjord, Norway, *p359*
soil, 96–97
soil erosion, 37–38, 97, 117, 666, 712, *p712*
Sokkuram, *p703,* 706
solar eclipse, 30, *d30*
solar energy, 91
solar system, 29–31, *d30*
Solidarity, 369
Solomon Islands, 765, 766; profile, *c744*
solstices, 32
Solzhenitsyn, Alexander, 434
Somalia, *m531, c537, m582,* 600; climate, *m588;* economy, *m593;* landforms, *m530, m583;* population, *m597;* profile, *c537*
Songhai, 548–49
Sophocles, 340
sorghum, 619
Sotho people, 610
source, 42
South Africa, Republic of, *c537,* 607–12, *m608;* agriculture, 608, 609, 610; AIDS in, 611; apartheid, 611; climate, 608, *m615;* culture, 612; diamond production, 609; economy, 609–10, *m620;* history, 610–11; industry, 609; landforms, 607–08, *m609;* languages, 611; music, 610, *p610,* 612; natural resources, 609; people, 610–12; population, *m625;* profile, *c537;* religion, 611; size, 607; townships, 611; wildlife, 607
South America, 35, 36, 226–69. *See also* Latin America; *specific countries*
South China Sea, 671, 719
Southern Hemisphere, 32
South Island, 736, 754
South Korea, *m637, m696, m697, m699,* 703, *m705, m706;* agriculture, 705; border with North Korea, 704, *p704;* climate, 705, *m706;* economy, 705, *m705;* industry, 705; landforms, *m636,* 704–05; martial arts, 706; natural resources, 705; people, 705–06; profile, *c642;* religion, 706. *See also* Korean Peninsula
South Pole, 31, 35, 736, 760
Southwest Asia, *m449, c451, m478;* climate, 444, *m483;* country profiles, *c452–54;* economy, 446, *p446, m493;* landforms, 444, *m448, m479;* need for water, 506, *m506;* oil and gas production and distribution, *m450;* people, 446; population, *g451, m498. See also specific countries*
Soviet Union: Caucasus republics and, 509; collapse of, 428; communism in, 425–28, 430–32, *p431;* Czechoslovakia and, 427, *p427;* Hungary and, 377; Lithuania and, 374; Poland and, 369. *See also* Russia
Soyinka, Wole, 544
space exploration, 429, *p429*
Spain, *m281,* 323, *m324;* agriculture, 324, 325–26; climate, 324–25, *m331;* culture, 328; economy, 325–26, *m338;* government, 327; history, 326–27; industry, 326; influence on Mexico, 190; landforms, *m280,* 323–25, *m325;* language, *m282,* 326–27; people, 327–28; population, *m329, m337;* profile, *c287;* religion, 327
Spanish-American War (1898), 219
spas, 378
special purpose maps, 83, *m83*
spreadsheet, using, 470
Sri Lanka, *m637,* 645, *m646,* 657, *m659,* 660–62, *m661;* agriculture, 660; climate, *m653,* 660; economy, *m659,* 660–61; fishing, 657, *p657;* industry, 660–61; landforms, *m636, m647,* 660; people, *p80,* 661; population, *m661;* profile, *c642*
Stalin, Joseph, 387–88, 426
steppe climate, 69
steppes, 69, 386, 406, *p408,* 515–16, 570
Stockholm, Sweden, *p344,* 349
Stoker, Bram, 382
Stonehenge, 295, *p295*
storms, 56
strait, 42, 721
Strait of Gibraltar, 323, 326, 327, 468
Strait of Magellan, 42, 268–69
Strait of Malacca, 721
Stravinsky, Igor, 434
study and writing skills: outlining, 758; taking notes as, 370; using library resources, 358
subarctic climate, 67–68
subsistence farms, 94, 186, 207, 261
suburbs, 134, 340; in Russia, 432
Sucre, Bolivia, 267
Sudan, *m531, c537, m582,* 596–98; climate, *m588,* 596–97; economy, *m593;* history, 597; landforms, *m530, m583,* 596–97; people, 597; population, *m597;* profile, *c537*
Sudeten Mountains, 365
Suez Canal, 458
Suharto (General), 729
Sumatra, 726
summer solstice, 32
sun, 29, 30; climate and, 54–55; seasons and, 31–32
Superior, Lake, 111, 118, 164
Suriname, *m173, m228,* 243, 244; climate, *m173, m236;* economy, *m238;* landforms, *m172, m229;* population, *m246;* profile, *c178*
"Survival This Way" (Ortiz), 136
Swahili, 584
Swaziland, *m531, c537, m608,* 612; climate, *m615;* economy, *m620;* landforms, *m530, m609;* population, *m625;* profile, *c537*
Sweden, *m281,* 345, *m346, p345,* 349; agriculture, 349; climate, 349, *m354;* economy, 349, *m350;* history, 349; industry, 349; landforms, *m280, m347,* 349; language, *m282;* Midnight Sun in, 352, *p352;* people, 349; population, *m356;* profile, *c287*
Switzerland, *m281, m290,* 309–11; climate, *m297,* 310; culture, 310–11; economy, *m302,* 310; government, 310; industry, 310; landforms, *m280, m291,* 309–10; languages, *m282,* 310–11; population, *m314;* profile, *c287;* religion, 311
Sydney, Australia, *p20,* 738, 752, *p759*
Syr Darya River, 516, 518
Syria, 444, *m449, m478, p488,* 488–89; agriculture, 488–89; climate, *m483,* 488; economy, 488–89, *m493;* landforms, *m448, m479,* 488; music, 489; people, *p488,* 489; population, *m498;* profile, *c453*
Syrian Desert, 488

Table Mountain, *p607*
Taebaek Mountains, 704
Tagus River, 324
Tahiti, *p764,* 769, 770
taiga, 68, 396, 410
Táino, 214, 215
Taipei, 686
Taiwan, *m637, m670,* 681, 685–86; agriculture, 686; climate, *m672,* 685; economy, *m676,* 685–86; history, 686; industry, 685–86, *p686;* landforms, *m636, m671,* 685; people, 686; population, *m682;* profile, *c642;* religion, 686

INDEX

Tajikistan, *m449, m510,* 517; climate, *m512;* economy, *m516;* landforms, *m448, m511;* people, 500; population, *m518;* profile, *c453*
Taj Mahal, 650, 651, *p651*
Taklimakan Desert, 632, 670, 692
Tallinn, Estonia, 372
Tamils, 661–62
Tan, Amy, 135
Tanganyika, Lake, 563, 588, 595
tannin, 236
Tanzania, *m531, c537,* 582, *m582,* 587, *p587,* 595; climate, *m588;* economy, 588–89, *m593;* government, 589–90; history, 589–90; landforms, *m530, m583,* 587–88, *p589;* people, 590; population, *m597;* profile, *c537*
Taoudenni, 560
tariff, 93
Tashkent, Uzbekistan, 518
Tasmania, 747
Tasman Sea, 736, 747
Taurus Mountains, 477
tax, 93
T'bilisi, Georgia, 512
Tchaikovsky, Peter, 425, 434
technology skills: using a database, 270; using a spreadsheet, 470; using the Internet, 187
tectonic activity, 35, *m36*
tectonic plates, 39, 119, 202, 599, 726; boundaries of, *d36;* meeting of, 36–37; movements of, 35
Tehran, Iran, 446, 499; population, *g451*
Tel Aviv-Yafo, 484, 486
telenovelas, 233
temperate climates, 65
temperature, 54
Temple of Apollo (Delphi), 339
Tenochtitlán, 189, 192
Terhazza, 560
terraced farming, 678, 712, *p712*
Texas, *p118,* 131
Thailand, *m637, m716,* 717–18; agriculture, 718; climate, 717, *m722;* economy, *m727;* landforms, *m636, m717,* 717–18; natural resources, 717; people, 717; population, *m728;* profile, *c642;* religion, *p631*
Thailand, Gulf of, 719
Thames River, 294
Thimphu, Bhutan, 659
Three Gorges Dam, 672, 674, *p674*
Tiananmen Square, China, 682
Tian Shan, 444, 517, 669
Tibet, 683
Tibetan New Year, *p680*
tierra caliente, 184, *d185*
Tierra del Fuego, 42, 168, *p268* 269
tierra fría, 185, *d185*
tierra templada, 185, *d185*
Tigris River, 444, 497, 506
Tijuana, Mexico, 185
Tikal, Guatemala, 210
timberline, 70
time zones map, *m487*
Tirana, 385
Titicaca, Lake, *g175,* 260
Tivoli Gardens, 355
Tobago. *See* Trinidad and Tobago
Togo, *m531, c538, m542,* 552; climate, *m548;* economy, *m553;* landforms, *m530, m543;* population, *m554;* profile, *c538*
Tolstoy, Leo, 434
Tombouctou, 548
Tonga, 769; profile, *c744*
Tonkin, Gulf of, 719
"Tornado Alley," 46
tornadoes, *p46,* 46–47, 56
Toronto, Canada, 106, 153, *p157,* 159, *p159*
Toulouse, 304
townships, 611
trade: barriers to, 93; free, 93, 154, 218; salt, 560, *p560, m561;* Silk Road, 692, *p693;* St. Lawrence River and, 164–65; in United States, 128; in West Indies, 215; world, 91–93, 128
traditional economy, 80
traditional religions: in Botswana, 621; in Malaysia, 722; in Melanesia, 767; in Mozambique, 624; in Singapore, 723; in South Africa, 611; in Sudan, 597; in Uganda, 593; in Zimbabwe, 620
traits, 77
Transantarctic Mountains, 773
Transdanubia, 376
transportation map, 416
Trans-Siberian Railroad, 425
Transylvania, 382, *p382*
trash, problem of, *p140,* 140–41, *p141*
trenches, 41
tributary, 653
Trinidad and Tobago, *m173, m206;* climate, *m214;* economy, *m209;* industry, 215; landforms, *m172, m207;* population, *m216;* profile, *c178*
Tripoli, Libya, 465, 474, *p474*
tropical climates, 64, 122
tropical rain forest climate, 64, *p64,* 206
tropical savanna climate, 64, 206–7
Tropic of Cancer, 32, 55, 184
Tropic of Capricorn, 32, 55
tropics, 55
tsetse flies, 571, 574
tsunamis, 36, 695
Tswana, 621
Tuareg people, *p549*
tundra, 68, 146, 410
tungsten, 677
Tunisia, *m449, m458,* 466; climate, *m465,* 466; economy, *m460;* history and people, 466, *p466;* landforms, *m448, m459,* 466; population, *m467;* profile, *c454*
Turin, Italy, 332
Turkey, 444, *m449,* 477–80, *m478;* agriculture, 478; climate, 478, *m483;* culture, 480, *p480;* economy, 478, *m493;* history, 480; industry, 478; landforms, *m448,* 477, *m479;* natural resources, 478; people, 479; population, *m498;* profile, *c454*
Turkish Straits, 477
Turkmenistan, 444, *m449, m510,* 519; climate, *m512;* economy, *m516;* landforms, *m448, m511;* population, *m518;* profile, *c454;* women, *p442*
Turpan Depression, *c639,* 670
Tutankhamun, 457, *p457*
Tutsi, 595
Tuvalu, 769; profile, *c745*
Twain, Mark, 135
typhoons, 56, 673, 769

U2, 299
Ubangi River, 572
Uganda, *m531, c538,* 582, *m582,* 592–94; climate, *m588,* 592; economy, *m593;* ethnic conflict, 595; government, 593–94; history, 593–94; landforms, *m530, m583,* 592–93; people, 593; population, *m597;* profile, *c538*
Ukraine, *m281, m366,* 386–88; agriculture, 387; climate, *m372,* 386; economy, *m376;* history, 387–88; industry, 387; landforms, *m280, m367,* 386; language, *m282;* people, 387–88; population, *m387;* profile, *c287*
Ukrainian Easter eggs, 390, *p390*
Ulaanbaatar, *p687,* 688
Union of Soviet Socialist Republics (U.S.S.R.), 398, 426. *See also* Russia; Soviet Union
United Arab Emirates, *m449, m478,* 495; climate, *m483;* economy, *m493;* landforms, *m448, m479;* population, *m498;* profile, *c454*
United Kingdom, *m281,* 289–95, *m290;* agriculture, 292; climate, 289–91, *m297;* culture, 294; economy, 292, *m302;* government, 292–93; history, 293–94; landforms, *m280,* 289–91; language, *m282;* natural resources, 292; *m291;* population, *m314;* profile, *c287;* Stonehenge, 295,

INDEX

Scotland; Wales
United States, 25, 104–13, *m109, c111, m116;* agriculture, 106, *m110;* arts, 134–35; boundaries, 115; climate, 104, *m120,* 120–22; culture, *p134,* 134–35; earthquakes, 202, *p202;* economy, 106, 123–28, *m124;* ethnic groups, *g111, p130;* food production, *m110;* government branches, *d132;* history, 130–35; immigrants to, 133–34; industry, 106, *p106;* Interior Plains, 118, *p118;* landforms, 104, *m108,* 115–19, *m117;* language, 134; literature, 135, 136; music, 134, *p134;* natural resources, 106, 124; people, *p102,* 106, *p106;* population, *g111, m133, g175;* religion, 134; size, 115; sports and recreation, 135; state profiles, *c112–13;* states, 115; technology, 128; trade, 128; weather, 46, *p46*
Ural Mountains, 396, 405, 407, 414, 432
Ural River, 516
uranium, 617
Uranus, 29
urbanization, 87
Uruguay, *m173, m228,* 235, 245–47; climate, *m236,* 245–46; economy, *m238,* 246; government, 246; history, 246; landforms, *m172, m229,* 245; people, 246–47; population, *m174, m246;* profile, *c179*
Uruguay River, 245
Uzbekistan, 444, *m449, m510,* 518; climate, *m512;* economy, *m516;* landforms, *m448, m511;* people, 517; population, *m518;* profile, *c454*

Valdés Peninsula (Argentina), *g175*
valley, 40
Vancouver, British Columbia, 106, 153
Vänern, Lake, *c283*
vanilla beans, 624
Vanuatu, 765, 766; profile, *c745*
vaqueros, 185–86
Vasa, Gustav, 349
Vatican City, *m281, m324,* 330, 334; profile, *c287*
vegetation, 305, *m305;* highland, 70; high latitude, *p68*
Venezuela, 170, *m173, m228,* 241–43, 258; climate, *m236,* 242; economy, *m238,* 242; government, 242–43; history, 242–43; landforms, *m172, m229,* 241–42; people, 243; population, forests, *p241*
Venice, Italy, 332, 334, *p334*
Venus, 29
Vermont, *p121*
vernal equinox, 32
Victoria, British Columbia, 148
Victoria, Lake, 533, 588, 592, 626
Victoria Falls, 526, *p527,* 618
Vienna, Austria, 311, *p311,* 312
Vientiane, Laos, 718
Vietnam, 634, *m637, m716,* 719–20; agriculture, 719; climate, *m722;* communism in, 720; economy, 720, *m727;* government, 720; history, 720; landforms, *m636, m717,* 719; national resources, 720; people, 720; population, *m728;* profile, *c642*
Vietnam War, 720
Vikings, 347, 351, 357
Vilnius, Lithuania, 374
Vinson Massif, (Antarctica), *g660, c743,* 773
Virgin Islands, *m173, m206,* 216; climate, *m214;* economy, *m209;* landforms, *m172, m207;* population, *m216;* profile, *c179*
Vistula River, 366
Vladivostok, Russia, 414, 425
volcanic island, 213
volcanoes, 36, 39; in Azores Islands, *p34;* in creation of Hawaii, 119; in Indonesia, 726; in Mexico, *p181,* 183, *p183;* in Philippines, 56, *p56*
Volga River, 396, 409, 414
Volgograd, Russia, 406
Volta, Lake, 556
Volta River, 552

Wales, 289–90, 293
Walesa, Lech, 369
Walloons, 314
Warsaw, Poland, 369
Warta River, 366
Washington, D.C., *p54,* 116, 125
Washington, George, 131
Washington State, 126, 127
water: bodies, 42; management, 96; salt, 51; use of, 95–96
water cycle, 49–50, *d50*
Water Demand Management Research Network (WDMRN), 507
water resources, 50–51
watershed, 578, *p578,* 594
water vapor, 49
weather: cyclones, 623; defined, 53; El Niño, 56, 57, *d57;* hurricanes, 56, 185, 207, 214; La Niña, 56, 57; monsoons, 56, 632, 673, 705, 707, 712, 715, 718, tsunamis, 36, 695; typhoons, 56, 673, 769. *See also* climate
weathering, 37
Webber, Andrew Lloyd, 294
weddings, *p79*
welfare state: Sweden as, 349; Uruguay as, 246
Welland Canal, 164
Wellington, New Zealand, 757
West Berlin, 309, 320
Western Ghats, 645
Western Sahara, 469
West Germany, 308–09, 320
West Indies, *m206;* climate, 214, *m214;* daily life, 216–17; economy, 214–15, *m216;* farming, 215; history, 215–16; hurricanes, 214; island profiles, 217–19; landforms, *m207,* 213–14; manufacturing, 215; music, 217, *p217;* natural resources, 215; population, *m216;* religion, 216; sports and recreation, 217; tourism, *p213;* trade, 215
Westminster Abbey, 294
West Pakistan, 652
West Siberian Plain, 406–07
West Virginia, *p105*
White Nile River, 596
Wieliczka, Poland, 365, *p365*
wildlife: in Africa, *p580,* 604, 605, *p605;* in Antarctica, *p734,* 772, *p772;* in Australia, 749, 753, *p753;* in China, *p669;* in Latin America, *p169;* in New Zealand, 755; in Russia, *p394–95*
Willamette Valley, 119
winds: as cause of erosion, 38; effect on climate, 56; landforms and, 58; prevailing patterns, *m55*
windward side of mountains, 59, *d59*
Winkel Tripel projection, *m9*
winter solstice, 32
Witwatersrand, 609
Wojtyla, Karol, 369
women: in Bolivia, 266, *p266;* diving skills of, on Cheju Island, 707; in government, 654; in Hungary, *p274;* as leaders in Southeast Asia, 717; in New Zealand, 757; as Olympic athletes, *p20;* as scientists, 507, *p507,* 605; in Southeast Asia, *p715*
World War I: in Austria, 377; in Germany, 308; in Hungary, 380; in Lithuania, 374; in Russia, 425; in Turkey, 480
World War II: in Belgium, 314; in Denmark, 355; in Germany, 308; in Greece, 340; in Italy, 333; in Netherlands, 316; in Poland, 368–69; in Russia, 426; United States in, 132
World Wide Web, 358
Wudi, 692
Wuhan, China, 678

Xhosa people, 610
Xi River, 671, 672

Yalu River, 706
Yangon, Myanmar, 716
Yangtze River, 23, 24, 632, *c639,* 671, 672, 674
Yaoundé, Cameroon, 571
Yarapa River, 253
Yeats, William Butler, 299
Yellow River, 632, 671, 672, 712
Yellow Sea, 74
Yellowstone National Park, 126, 202
Yeltsin, Boris, 432
Yemen, *m449, m478,* 496; climate, *m483;* economy, *p446;* landforms, *m448, m479;* population, *m497;* profile, *c454*
Yenisey River, 396, 409
Yerevan, Armenia, 510
Yokohama, Japan, 700
Yoruba people, 543
Yucatán Peninsula, 170, 182, 186, *p188,* 188–89
Yugoslavia, *m281, m366,* 383–84; climate, *m372;* communism in, 383; economy, *m376;* landforms, *m280, m367;* language, *m282;* population, *m387;* profile, *c287*
Yugoslav republics, 381; climate, *m372;* economy, *m376;* landforms, *m367;* population, *m387. See also* Former Yugoslav Republics
Yukon River, 119
Yukon Territory, 144, 154
yurt, 517, *p517*

Zagreb, Croatia, 383–84
Zagros Mountains, 444, 499
Zaire, 568
Zambezi River, 526, *p527,* 618, 619, 623
Zambia, *m531, c538, m608,* 618–19; agriculture, 619; climate, *m615;* economy, 618–19, *m620;* history, 619; landforms, *m530, m609,* 618; language, 619; natural resources, 618–19; people, 619; population, *m625, m625;* profile, *c538*
Zanzibar, 588, 589–90
Zapata, Emiliano, 190–91
Zealand, 354
Zeppelin, Led, 294
Zhang Qian, 692
Zimbabwe, *m531, c538, m608,* 618, 619–20; AIDS in, 620; agriculture, 619–20; climate, *m615;* economy, 619–20, *m620;* history, 620; landforms, *m530, m609,* 619; natural resources, 619–20; people, 620; population, *m625;* profile, *c538*
Zionists, 485
Zulu people, 610
Zurich, Switzerland, 310

Cover (tl)PhotoDisc, (tr)Carol Beckwith & Angela Fisher/Robert Estall Photo Agency, (c)PhotoDisc, (bl)PhotoDisc, (br)CORBIS; **iv–v** Ken Stimpson/Panoramic Images, Chicago, All Rights Reserved; **ix** Owen Franken/CORBIS; **vi–vii** PhotoDisc; **xvi; 1** PhotoDisc; **2** Craig Aurness/CORBIS, (bl)Todd Gipstein/CORBIS, (br)Craig Lovell/CORBIS; **3** (tr)Robert Caputo/AURORA, (bl)Peter Turnley/CORBIS, (br)Roger Ressmeyer/CORBIS; **20** (l)Robert Landau/CORBIS, (r)Arne Dedert/AFP; **20–21**Stone/S. Purdy Matthews; **22** Norman Kent/oi2.com; **23** NASA/National Geographic Image Collection; **24** (l)Richard T. Nowitz/National Geographic Image Collection, (r)Yann Arthus-Bertrand/CORBIS; **26** Galen Rowell/CORBIS; **29** Maria Stenzel/National Geographic Image Collection; **30** Timothy G. Laman/National Geographic Image Collection; **34** Natalie Fobes/National Geographic Image Collection; **37** Michael K. Nichols/National Geographic Image Collection; **39** David Doubilet/National Geographic Image Collection; **40** (l)Kenneth Garrett/National Geographic Image Collection, (t)Michael K. Nichols/National Geographic Image Collection; **46** AP/Wide World Photos; **47** Matt Meadows; **48** Annie Griffiths Belt/National Geographic Image Collection; **49** George Grall/National Geographic Image Collection; **52** Southampton Oceanography Centre; **53** J. Blair/National Geographic Image Collection; **54** Medford Taylor/National Geographic Image Collection; **60** (l)Bruce Dale/National Geographic Image Collection, (r)Jodi Cobb/National Geographic Image Collection; **63** Phil Schermeister/National Geographic Image Collection; **64** (l)Michael K. Nichols/National Geographic Image Collection, (r)Beverly Joubert/National Geographic Image Collection; **67** (t)Annie Griffiths Belt/National Geographic Image Collection, (l)James P. Blair/National Geographic Image Collection, (bc)Jodi Cobb/National Geographic Image Collection, (br)Raymond K. Gehman/National Geographic Image Collection; **68** (t)Natalie Fobes/National Geographic Image Collection, (c)George F. Mobley/National Geographic Image Collection, (l)Maria Stenzel/National Geographic Image Collection; **69** (l)James L. Stanfield/National Geographic Image Collection, (r)Phil Schemeister/National Geographic Image Collection; **70** Pat Jerrold, Papilio/CORBIS; **71** Bill Curtsinger/National Geographic Image Collection; **74** (l)Calvin Larsen/Photo Researchers, (r)Stone/Joseph Sohm; **74–75** Planet Earth Pictures/FPG International/PictureQuest; **75** (tl)Francois Gohier/Photo Researchers, (tr, b)©1995 Jan Sonnenmair/AURORA; **76** Steve McCurry/National Geographic Image Collection; **77** Kenneth Garrett/National Geographic Image Collection; **79** (l)Black Star/National Geographic Image Collection, (r)Steven L. Raymer/National Geographic Image Collection; **80** (l)Steve McCurry/National Geographic Image Collection, (r)Steve McCurry/National Geographic Image Collection; **84** Gerd Ludwig/National Geographic Image Collection; **87** AFP/CORBIS; **89** Bettmann/CORBIS; **90** Jim Sugar Photography/CORBIS; **91** (t)Ric Ergenbright/CORBIS, (b)Steve McCurry/National Geographic Image Collection; **93** (l)H.Edward Kim/National Geographic Image Collection, (r)George F. Mobley/National Geographic Image Collection; **95** Bryan & Cherry Alexander; **96** James P. Blair/National Geographic Image Collection; **97** Jodi Cobb/National Geographic Image Collection; **102** (l)Steve McCurry, (r)Michael Lewis; **102–103** David R. Stoecklein; **105** Susie Post, (l)Norbert Rosing; **106** Richard Nowitz/PhotoTake.NYC/Picture Quest; **107** Eugene Fisher & Barbara Brundege; **114** Owen Franken/CORBIS; **115** David Hiser/National Geographic Image Collection; **118** (l)Steven L. Raymer/National Geographic Image Collection, (r)Vincent Musl/National Geographic Image Collection; **121** Michael S. Yamashita/National Geographic Image Collection; **122** Chris Johns/National Geographic Image Collection; **123** Karen Kasmauski/Matrix; **125** Gregory Scott Doramus/Omnigraphix; **126** Annie Griffiths Belt/CORBIS; **127** Joel Satore/National Geographic Image Collection; **130** The Stock Market; **131** Ira Block/National Geographic Image Collection; **134** Mike Habermann, courtesy Music of the World; **136** CORBIS; **140** Ray Pfortner, Peter Arnold Inc.; **140–141** Michael Mathers, Peter Arnold Inc.; **141** (l)Stone/David Young-Wolff, (r)David Schmidt; **142** David A. Harvey/National Geographic Image Collection; **143** Raymond K. Gehman/National Geographic Image Collection; **146** Richard T. Nowitz/National Geographic Image Collection; **150** Anna Susan/National Geographic Image Collection; **153** David A. Harvey/National Geographic Image Collection; **155** Marie-Louise Brimberg/National Geographic Image Collection; **156** Michael Evan Sewell/Visual Pursuit; **157** Marie-Louise Brimberg/National Geographic Image Collection; **159** Paul A. Souders/CORBIS; **160** (l)Daniel J. Wiener/National Geographic Image Collection, (r)Bruce Dale/National Geographic Image Collection; **164** Giraudon/Art Resource, New York; **165** Michael S. Yamashita; **166** (l)Stone/David Levy, (r)Stone/Oliver Benn; **166–167** Kenneth Garrett/NGS; **169** Stone/Norbert Wu, Stone/William J. Hebert; **170** Sisse Brimberg; **171** Stone/Chad Ehlers; **180** Randy Faris/CORBIS; **181** Bettmann/CORBIS; **188** Tomasz Tomaszewski/National Geographic Image Collection; **190** Cotton Coulson/National Geographic Image Collection; **192** James L. Amos/National Geographic Image Collection; **193** David A. Harvey/National Geographic Image Collection; **194** Nik Wheeler/CORBIS; **195** (l)Joel Satore/National Geographic Image Collection, (r)Tomas Tomaszewski/National Geographic Image Collection; **198** Albert Moldvay/National Geographic Image Collection; **202** Nik Wheeler/CORBIS; **203** Matt Meadows; **204** SuperStock; **205** Art Wolfe; **208** (t)Jan Butchofsky-Houser/CORBIS, (b)Vincent Musl/National Geographic Image Collection; **210** Michael S. Yamashita/CORBIS; **213** Jonathan Blair/National Geographic Image Collection; **215** ©Robert A. Tyrrell; **217** (l)Tony Arruza/CORBIS, (r)Michael K. Nichols/National Geographic Image Collection; **221** George Mobley/National Geographic Image Collection; **224** (l)Artville, (r)Giraudon/Art Resource, New York; **225** Loren McIntyre; **226** PhotoDisc; **227** Alex Webb/Magnum; **232** (l)Jim Zuckerman/CORBIS, (r)Yann Arthus-Bertrand/CORBIS; **234** Jeremy Horner/CORBIS; **235** Robert van der Hilst/CORBIS; **237** Winfield I. Parks, Jr./National Geographic Image Collection; **239** Pablo Corral V/CORBIS; **240** Kit Houghton Photography/CORBIS; **241** Michael K. Nichols/National Geographic Image Collection; **242** Pablo Corral V/CORBIS; **243** Robert Caputo/Aurora & Quanta Productions; **245** Louis O. Mazzatenta/National Geographic Image Collection; **247** (l)Jack Fields/CORBIS, (r)Louis O. Mazzatenta/National Geographic Image Collection; **252** (t)Michael & Patricia Fogden, (b)William Albert Allard; **252–253** Stuart Franklin; **253** (l)Michael Doolittle, (r)Marc Van Roosmalen Bat Conservation International; **254** Owen Franken/CORBIS; **258** Richard S. Durrance/National Geographic Image Collection; **260** Frank & Helen Schreider/National Geographic Image Collection; **263** Tiziana and Gianni Baldizzone/CORBIS; **264** Johan Reinhard/National Geographic Image Collection; **265** Art Wolfe; **266** Maria Stenzel/National Geographic Image Collection; **268** (l)Richard T. Nowitz/National Geographic Image Collection, (r)James L. Stanfield/National Geographic Image Collection; **271** William A. Allard/National Geographic Image Collection; **274** (l)IFA-Bilderteam-Travel/Bruce Coleman Inc., (r)Stone/Robert Everts; **274–275** SuperStock; **277** Stone/D.C. Lowe, (l)Stone/Chris Haigh; **278** Stone/Bert Blokhuis; **279** Stone/Ron Sanford; **288** James L. Stanfield/National Geographic Image Collection; **289** London Aerial Photo Library/CORBIS; **293** Emory Kristof/National Geographic Image Collection; **295** Adam Woolfitt/CORBIS; **296** Thad Samuels Abell II/National Geographic Image Collection; **298** (l)Thad Samuels Abell II/National Geographic Image Collection, (r)Tim Thompson/CORBIS; **299** Michael St. Maur Sheil/CORBIS; **300** James L. Stanfield/National Geographic Image Collection; **301** (l)James P. Blair/National Geographic Image Collection, (r)Felix Zaska/CORBIS; **303** Ric Ergenbright/CORBIS; **306** Owen Franken/CORBIS; **308** Ric Ergenbright/CORBIS; **310** Sisse Brimberg/National Geographic Image Collection; **311** Bob Krist/CORBIS; **313** Michael John Kielty/CORBIS; **315** SuperStock; **320** AP/Wide World Photos; **321** Reuters/CORBIS; **322** Mark L. Stephenson/CORBIS; **323** ©David Cumming, Eye Ubiquitous/CORBIS; **326** (l)Medford Taylor/National Geographic Image Collection, (r)Charles O'Rear/CORBIS; **327** Adam Woolfitt/CORBIS; **330** National Geographic Image Collection; **332** (t)Richard T. Nowitz/National Geographic Image Collection, (b)Vittoriano Rastelli/CORBIS; **333** Louis O. Mazzatenta/National Geographic Image Collection; **334** Thad Samuels Abell II/National Geographic Image Collection; **335** (l)Bettmann/CORBIS, (r)Gianni Dagli Orti/CORBIS; **336** Ira Block/National Geographic Image Collection; **339** Panoramic Images; **341** James L. Stanfield/National Geographic Image Collection; **344 345** Tomasz Tomaszewski/National Geographic Image Collection; **348** (l)©Buddy May/CORBIS, (r)Richard S. Durrance/National Geographic Image Collection; **349** Bryan & Cherry Alexander; **352** Jonathan Blair/CORBIS; **353** ©Ian Yates; Eye Ubiquitous/CORBIS; **355** SuperStock; **357** Sisse Brimberg/National Geographic Image Collection; **358** SuperStock; **359** Richard T. Nowitz/CORBIS; **362** (l)Robert Winslow, (r)Stone/Johan Elzenga; **362–363** Stone/Oliver Strewe; **363** (l)Martin Bond/Science Photo Library/Photo Researchers, (r)Mike Lewis, Northamptonshire Grammar School, United Kingdom; **364** Bob Krist/CORBIS; **365** Tomasz Tomaszewski/National Geographic Image Collection; **368 369 370** James L. Stanfield/National Geographic Image Collection; **371** Priit J. Vesilind/National Geographic Image Collection; **373** Steven L. Raymer/National Geographic Image Collection; **375** James Stanfield/National Geographic Image Collection;

377 Dean Conger/CORBIS; **379** Owen Franken/CORBIS; **381** Steve Raymer/CORBIS; **382** Peter Wilson/CORBIS; **384** Catherine Karnow/CORBIS; **386** Gerd Ludwig/National Geographic Image Collection; **390** (l)Kelly-Mooney Photography/CORBIS, (r)Craig Aurness/CORBIS; **394** (l)Bruce Dale, (r)Stone/Alain Le Garsmeur; **394–395** Marc Moritsch/National Geographic Image Collection; **397** Stone/Simeone Huber, (bl)Stone/Paul Harris; **398** B. Klipinitsen, M. Moshkov, Sovfoto/Eastfoto/PictureQuest; **399** Jay Dickman; **404** Gerd Ludwig/National Geographic Image Collection; **405** Tom Brakefield/CORBIS; **408** Gerd Ludwig/National Geographic Image Collection; **410** Dmitry Shparo/National Geographic Image Collection; **411** Michael Boys/CORBIS; **412** Gerd Ludwig/National Geographic Image Collection; **414** Dennis Chamberlin/National Geographic Image Collection; **417** Wolfgang Kaehler/CORBIS; **420** (l)Bios/M. Gunther/Peter Arnold Inc., (r)Gerd Ludwig; **420–421** Peter Turnley/CORBIS; **421** Stiftung Jugend forscht e.V./Deutsche Bank AG; **422** Michael Nichols/National Geographic Image Collection; **423** Sisse Brimberg/National Geographic Image Collection; **425** Roger Tidman/CORBIS; **426** Kremlin Museums, Moscow, Russia/The Bridgeman Art Library; **427** Marc Garanger/CORBIS; **430** Dean Conger/CORBIS; **431 432** Peter Turnley/CORBIS; **434** (l)Marc Garanger/CORBIS, (r)David Turnley/CORBIS; **436** Farrell Grehan/CORBIS; **437** Steve Raymer/CORBIS; **440** Giraudon/Art Resource, New York; **441** Sovfoto/Eastfoto/PictureQuest; **442** (l)Stone/Hugh Sitton, (r)Zefa/The Stock Market; **442 443** X. Richer/Hoaqui, Photo Researchers; **445** Stone/James Strachan, (bl)David Coulson; **446** Stone; **447** Stone/Wayne Eastep; **455** Stone/Jeff Rotman; **456** Reza/National Geographic Image Collection; **457** Kenneth Garrett/National Geographic Image Collection; **461** Otto Lang/CORBIS; **463** SuperStock; **464** George Steinmetz/National Geographic Image Collection; **466 468** K.M. Westermann/CORBIS; **471** Kenneth Garrett/National Geographic Image Collection; **474** P. Boulat/Cosmos/Woodfin Camp and Associates; **475** (l)Matt Meadows, (r)Science Pictures Limited/CORBIS; **476** James L. Stanfield/National Geographic Image Collection; **477** Patrick Ward/CORBIS; **480** James L. Stanfield/National Geographic Image Collection; **481** Dean Conger/CORBIS; **482** I. Talby/Index Stock; **484** (l)Steve Kaufman/CORBIS, (r)Duby Tal/Albatross/Index Stock; **488** Dean Conger/National Geographic Image Collection; **489** Dave Bartruff/CORBIS; **490** Annie Griffiths Belt/National Geographic Image Collection; **492** James L. Stanfield/National Geographic Image Collection; **495** AP/Wide World Photos; **497** Charles & Josette Lenars/CORBIS; **499** (l)Alexandra Avakian/National Geographic Image Collection, (r)Jon Spaull/CORBIS; **504** Art & Immagini srl/CORBIS; **505** Stone/Sylvain Grandadam; **506–507** James L. Stanfield, (br)©1993 Ed Kashi; **507** (l)©1993 Ed Kashi, (r)courtesy Sandra Postel; **508** Dugald Bremner/National Geographic Image Collection; **509** Reza/National Geographic Image Collection; **514** James L. Stanfield/National Geographic Image Collection; **515** Wolfgang Kaehler; **517 519 520** Gerd Ludwig/National Geographic Image Collection; **521** Michael S. Yamashita/CORBIS; **524** (l)Stone/Hugh Sitton, (r)Stone/Jacques Jangoux; **524–525** Stone/Manoj Shah; **527** Stone/Ian Murphy, (bl)Stone/Renee Lynn; **528** Stone/Ian Murphy; **529** Stone/Will Curtis; **539** Stone/Nicholas Parfitt; **540** Wolfgang Kaehler; **541** AP/Wide World Photos; **544** (l)Robert W. Moore/National Geographic Image Collection, (r)AP/Wide World Photos; **545** ©1993 Greenpeace/Lambon; **546** (l)Davis Factor/CORBIS, (r)Bowers Museum of Cultural Art/CORBIS; **547** Steve McCurry/Magnum Photos, Inc.; **549** Carol Beckwith & Angela Fisher/Robert Estall Photo Agency; **551** Nik Wheeler/CORBIS; **552** Carol Beckwith & Angela Fisher/Robert Estall Photo Agency; **555** Robert W. Moore/National Geographic Image Collection; **557** AP/Wide World Photos; **560** Michael A. Hampshire; **561** James L. Stanfield; **562** SuperStock; **563** Robert Caputo/Aurora & Quanta Productions; **568** Michael K. Nichols/National Geographic Image Collection; **570** AP/Wide World Photos; **574** David Turnley/CORBIS; **575** AP/Wide World Photos; **578** Robert Caputo/Stock Boston; **579** Matt Meadows; **580** Sharna Balfour, Gallo Images/CORBIS; **581** Carol Beckwith & Angela Fisher/Robert Estall Photo Agency; **584** The Purcell Team/CORBIS; **585** Frank Lane Picture Agency/CORBIS; **586** Darrell Gulin/CORBIS; **587** J. Jafferji/OSF/Animals Animals; **589** Wolfgang Kaehler; **591** Robert Caputo/Aurora & Quanta Productions; **592** Art Wolfe; **594** Chinch Gryniewicz, Ecoscene/CORBIS; **596** Dave Bartruff/CORBIS; **598** AP/Wide World Photos; **604** (l)Gerry Ellis/ENP Images, (r)Jose Azel/Aurora/PictureQuest; **604–605** Michael Nichols/NGS; **605** (tl)Lisa Hoffner/Wildeye Photography, (tr)Lisa Hoffner/WildEye Photography, (b)Stone/Art Wolfe; **606–607** Wolfgang Kaehler; **610** Jack Vartoogian; **611** Nik Wheeler/CORBIS; **612** Turnley Collection/CORBIS; **614** Des & Jen Bartlett/National Geographic Image Collection; **618** Walter Edwards/National Geographic Image Collection; **619** Chris Johns/National Geographic Image Collection; **622** Chris Johns/National Geographic Image Collection; **623** Michael K. Nichols/National Geographic Image Collection; **624** AP/Wide World Photos; **630** (l)Stone/Tim Davis, (r)Stone/David Sutherland; **630–631** Waranun Chutchawan-Tipakorn; **633** Stone/Hilarie Kavanagh, (bl)Stone/Martin Puddy; **634** Stone/Paul Chesley; **635** Stone/Keren Su; **643** Stone/John Lamb; **644** Steve McCurry/National Geographic Image Collection; **645** George F. Mobley/National Geographic Image Collection; **648** Steve McCurry/National Geographic Image Collection; **649** Rudi Von Briel/PhotoEdit; **651** Brian Vikander/CORBIS; **652** Jonathan Blair/National Geographic Image Collection; **654** Brian Vikander/CORBIS; **655** Ed Kashi/National Geographic Image Collection; **657** Steve McCurry/National Geographic Image Collection; **658** Robert Caputo/National Geographic Image Collection; **663** Robert Holmes/CORBIS; **666** (l)Stone/Chris Noble, (r)Robert Weight, Ecoscene/CORBIS; **666–667** Takehide Kazami, Peter Arnold Inc.; **667** (l)Future Generations, (r)Galen Rowell/Mountain Light; **668** Stone; **669** Keren Su/CORBIS; **674–680** AFP/CORBIS; **681** The Telegraph Colour Library/FPG; **683** (t)Owen Franken/CORBIS, (b)Joseph Sohm, ChromoSohm Inc./CORBIS; **685** How-Man Wong/CORBIS; **686** Marc Garanger/CORBIS; **687** Nik Wheeler/CORBIS; **688** James L. Stanfield/National Geographic Image Collection; **692** Cary Wolinsky/Stock Boston; **693** Keren Su/CORBIS; **694** Dallas & John Heaton/CORBIS; **695** Reuters NewMedia Inc./CORBIS; **698** Karen Kasmauski/Matrix; **700** Roger Ressmeyer/CORBIS; **701** Michael S. Yamashita/CORBIS; **702** Asian Art & Archaeology, Inc./CORBIS; **703** Carmen Redondo/CORBIS; **704** (l)Nathan Benn/CORBIS, (r)Wolfgang Kaehler/CORBIS; **707** Catherine Karnow/CORBIS; **709** Neil Beer/CORBIS; **712** Yann Layma/Stone; **713** Matt Meadows; **714** AFP/CORBIS; **715 718** Paul Chesley/National Geographic Image Collection; **719** Kevin R. Morris/CORBIS; **721** AFP/CORBIS; **723** Earl & Nazima Kowall/CORBIS; **726** Roger Ressmeyer/CORBIS; **729** Wolfgang Kaehler/CORBIS; **730** David Hanson/Stone; **731** AP/Wide World Photos; **734** (l)Stone/David Madison, (r)Stone/David Hiser; **734–735** Stone/Oliver Strewe; **737** Stone/Oliver Strewe, (bl)Stone/Johnny Johnson; **738** Stone/Nicholas DeVore; **739** Stone/Glen Allison; **746** Charles & Josette Lenars/CORBIS; **747** R. Ian Productions P Lloyd/National Geographic Image Collection; **751** Penny Tweedie/CORBIS; **752** Paul A. Souders/CORBIS; **753** (tr)Australian Picture Library/CORBIS, (br)Earl & Nazima Kowall/CORBIS, (l)Charles Philip Cangialosi/CORBIS; **754 758** Kevin Fleming/CORBIS; **759** Australian Picture Library/CORBIS; **762** (r)Christine Osborne/CORBIS; **762–763** Penny Tweedie/CORBIS; **763** (tl)PhotoDisc, (tr)Bettmann/CORBIS, (b)George F. Mobley; **764** Douglas Peebles/CORBIS; **765** Hal Beral/CORBIS; **769 771** David Doubllet/National Geographic Image Collection; **772** Wolfgang Kaehler/CORBIS; **775** Underwood & Underwood/CORBIS; **776** Galen Rowell/CORBIS; **777** AP/Wide World Photos; **780** Byrd Antarctic Expedition; **781** Andrew H. Brown; **782** PhotoDisc.

TIME REPORTS

FOCUS ON WORLD ISSUES

A New Kind of War

Battling Terrorism in the Land of the Free

THOMAS E. FRANKLIN/RECORD/CORBIS SABA

TIME REPORTS

FOCUS ON WORLD ISSUES

1995
Terror strikes Oklahoma City.

1996
Terrorists in Saudi Arabia blow up housing for U.S. troops.

A Day for Heroes

September 11, 2001, was a day John Jonas will never forget. A New York City firefighter, Jonas was the captain of Ladder Company 6. That morning two **hijacked** commercial airplanes were deliberately flown into the World Trade Center's twin towers.

Jonas and five other men in his company rushed to the scene. An hour later they were walking down the fire stairs of the south tower. An older woman they were rescuing told them she couldn't go on. They told her they wouldn't leave her. And just then the building collapsed around them in clouds of dust. "I'm thinking," Jonas said later, "'I can't believe this is how it ends for me.'"

But life didn't end either for Jonas or for the five firefighters with him. Nor did it end for Josephine Harris, the woman they saved. Above and below them and in the nearby north tower, more than 2,600 people died. But somehow the part of the stairway they were on didn't collapse. "It was a freak of timing," said Jonas. Another minute, either way, and the crumpling building would have crushed them like the others.

JAMES NACHTWEY—VII FOR TIME

▲ New York City firefighters battle the World Trade Center disaster on September 11, 2001.

More than 300 rescue workers lost their lives trying to save others that day. All of them were heroes.

No Ordinary Crime

On September 11, hijackers also flew an airplane into the Pentagon—the headquarters of the U.S. military. Another 125 people died. Other hijackers seized a fourth airplane, but the passengers heroically fought back. Instead of hitting a building, the plane crashed in a field in Pennsylvania leaving no survivors.

Clearly, the horrible crashes were not ordinary crimes. They were acts of **terrorism**. Terrorism is the illegal use of violence against people or property to make a point. The point may involve a particular belief, such as religion or politics.

EXPLORING THE ISSUES

1. **Interpreting Points of View** Do you agree with the article's view of heroes? Why or why not?
2. **Making Inferences** How is the violence of terrorism different from the violence of war?

1998
Terrorists destroy the U.S. Embassy in Nairobi, Kenya.

2000
Terrorists in Yemen batter a U.S. warship, the USS Cole.

2001
Suicide pilots level the World Trade Center's twin towers.

Al-Qaeda

The terrorists who hijacked the airplanes belonged to a group called al-Qaeda (al•KAY•dah). The group was founded by Osama bin Laden, a wealthy Saudi Arabian.

Al-Qaeda was created to fight the Russian invasion of Afghanistan. After the Russians left Afghanistan, al-Qaeda members changed their goals. They wanted to force all non-Muslims out of the Middle East. They hated the U.S. troops based in Saudi Arabia and the Jewish people living in Israel.

Al-Qaeda's members also believed Muslims were being changed too much by modern ideas. They hated freedom of religion and wanted strict religious leaders to control Muslim countries.

Al-Qaeda's beliefs were not shared by all Muslims. The attacks on the United States horrified people around the world, including millions of Muslims who live in the Middle East, the United States, and elsewhere.

Closing in on America

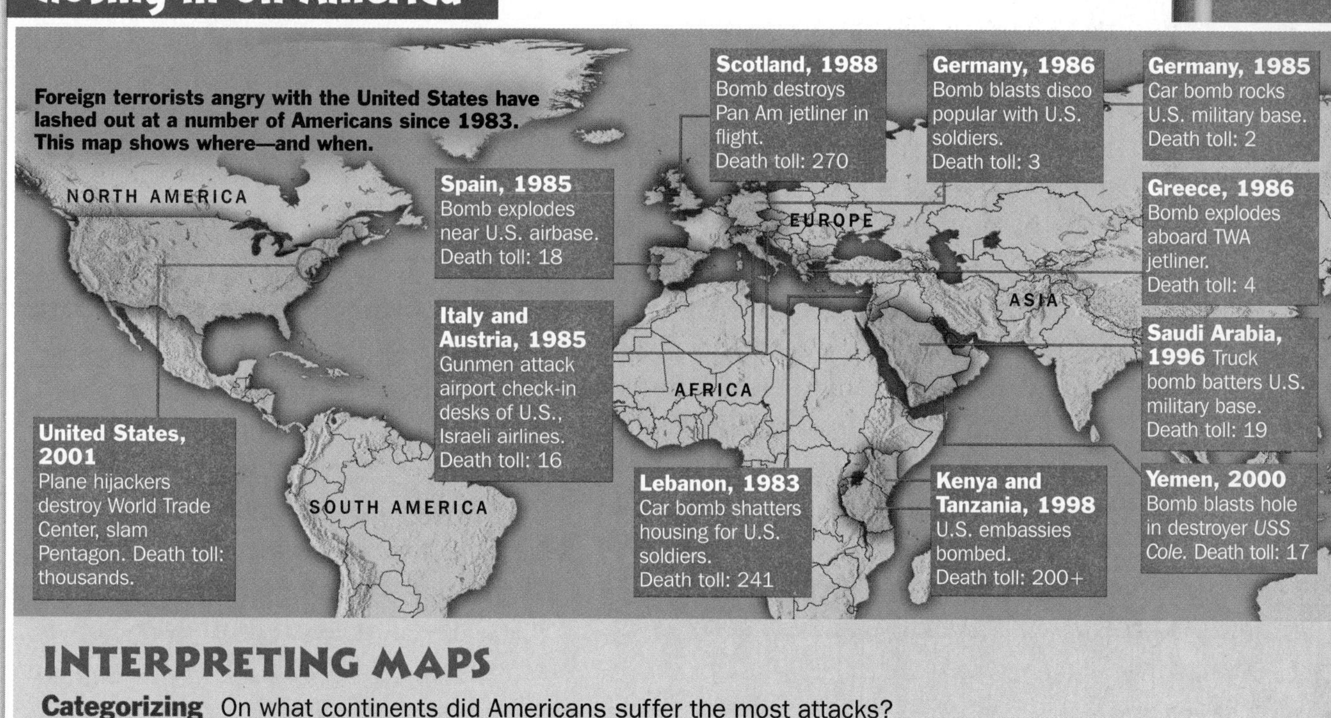

Foreign terrorists angry with the United States have lashed out at a number of Americans since 1983. This map shows where—and when.
NORTH AMERICA
SOUTH AMERICA
EUROPE
AFRICA
ASIA
Scotland, 1988 Bomb destroys Pan Am jetliner in flight. Death toll: 270
Germany, 1986 Bomb blasts disco popular with U.S. soldiers. Death toll: 3
Germany, 1985 Car bomb rocks U.S. military base. Death toll: 2
Spain, 1985 Bomb explodes near U.S. airbase. Death toll: 18
Greece, 1986 Bomb explodes aboard TWA jetliner. Death toll: 4
Italy and Austria, 1985 Gunmen attack airport check-in desks of U.S., Israeli airlines. Death toll: 16
Saudi Arabia, 1996 Truck bomb batters U.S. military base. Death toll: 19
United States, 2001 Plane hijackers destroy World Trade Center, slam Pentagon. Death toll: thousands.
Lebanon, 1983 Car bomb shatters housing for U.S. soldiers. Death toll: 241
Kenya and Tanzania, 1998 U.S. embassies bombed. Death toll: 200+
Yemen, 2000 Bomb blasts hole in destroyer USS Cole. Death toll: 17

INTERPRETING MAPS

Categorizing On what continents did Americans suffer the most attacks? On which continents did they suffer the least?

TIME REPORTS

Behind the Hatred

What makes the United States the target of so much deadly anger? One answer is its support for Israel. Israel was founded in 1948. Soon afterwards, an Arab-Israeli war forced about 750,000 Palestinian Arabs from their homes.

Today many of those Palestinians live in refugee camps. So do their children and their grandchildren. Those 4 million Palestinians want a nation of their own. Israel has offered to exchange land for a promise of peace. But so far the Palestinians have rejected that offer.

U.S. Troops in Saudi Arabia

Another source of anger is the presence of U.S. troops in Saudi Arabia. The Saudi government asked the United States to station troops there. But the holy cities of Makkah (Mecca) and Madinah (Medina) are in Saudi Arabia. To many Muslims, U.S. troops on Saudi soil are an insult to Islam.

To terrorist Osama bin Laden and his followers, the solution to these problems was violence. In 1996, he urged Muslims to kill U.S. troops in Saudi Arabia. In 1998, he called for attacks on American **civilians**. Civilians are people not in the armed forces or diplomatic services. By the end of 2001, several thousand people had been killed.

The United States responded to September 11 with a determination and resolve bin Laden surely didn't expect. "Our war on terror begins with al-Qaeda," President George W. Bush said. "It will not end until every terrorist group of global reach has been found, stopped, and defeated."

ALEXANDRA BOULAT—VII

▲ Fanatics hail terrorist Osama bin Laden in 2001.

EXPLORING THE ISSUES

1. **Drawing Conclusions** What are some reasons many Americans support Israel's presence in the Middle East?
2. **Making Inferences** Why are acts of terror against civilians often effective?

War on All Fronts

After the terrorist attacks on September 11, 2001, President George W. Bush spoke to the nation and announced a war on terrorism. He warned the world, "We will make no distinction between the terrorists who committed these acts and those who harbor them."

Al-Qaeda had followers all over the world, but it was based in Afghanistan. The Taliban, a strict religious party that controlled Afghanistan's government, protected Osama bin Laden and al-Qaeda. They refused to help the U.S. So in October, the President ordered the U.S. military to attack Afghanistan.

Aid for Children

The U.S. was not at war with the Afghan people, but with the Taliban and al-Qaeda. During the attack, U.S. planes dropped food and medicine to the men, women, and children in the civilian population.

Nations around the world backed the United States and began arresting al-Qaeda members hiding in their countries. Some sent troops to help the Americans fight in Afghanistan.

A few weeks after the attack began, the Taliban government collapsed. With the aid of the U.S. and its allies, the Afghan people created a new government. Meanwhile, American and allied troops began hunting for al-Qaeda forces in Afghanistan. The U.S. also sent troops to the Philippines, Yemen, and the nation of Georgia to train local troops to fight terrorists.

MARK RICHARDS

▲ **Safety checks help prevent terrorism. But the cost—less freedom—worries many Americans.**

Liberty and Security

At home, federal agencies stepped up their efforts to find terrorists. President Bush ordered banks to hold money belonging to groups linked to terrorists. Congress passed a new antiterrorist law making it easier to tap phones, intercept e-mail, and search homes.

Some people worried that the new antiterrorist law would chip away at our **liberties**—such as freedom of speech and the right to privacy. For this reason, Congress set a five-year time limit on parts of the new law.

EXPLORING THE ISSUE

1. **Analyzing Information** Shortly after September 11, 2001, President Bush said, "No one should be singled out for unfair treatment or unkind words because of their ethnic background or religious faith." What do you think he meant by that statement?
2. **Problem Solving** What liberties, if any, might you be willing to give up in order to ensure national security?

Stopping Terrorism: What Can One Person Do?

The rescue workers who responded to the attacks on the World Trade Center and the Pentagon were true heroes. In the months that followed, Americans honored them for their courage and sacrifice.

The response of Americans to tragedy showed the world the nation's hidden strengths—its people. Wherever they lived, Americans reacted. They gave blood. They held candlelight **vigils** to honor the victims. They flew flags to show their unity. They cut deeply into their budgets, contributing more than $200 million in the first week to help victims' families.

They all made it clear, as a girl from Ohio told TIME For Kids, that no terrorist can weaken the nation's spirit. "They bent steel," said Danielle, 12, of the World Trade Center murderers, "but they can't break the U.S."

Be a Local Hero

Wherever you live, you can help keep that spirit alive. And you can do it even years after the disasters of September 2001 took place.

Learn all you can about terrorism. Learn what it is, why it exists, and how people at all levels of government are fighting it.

Then join that fight any way you can. With posters and letters, report successful efforts to combat this evil. Raise money for groups that help out the victims of terrorism everywhere.

CHRIS USHER

▲ **Terrorist attacks in September 2001 trigger a burst of patriotism everywhere in America.**

Finally, refuse to give in to fear. Terrorists use fear as a weapon. If you can keep fear from changing your life, you will have taken a big bite out of terrorism.

The novelist Stephen King—who often writes about human fears—agreed. "If everybody continues working," he said, "they [the terrorists] don't win."

EXPLORING THE ISSUE

1. **Problem Solving** What might people do to stop the fear of terrorism from keeping them from doing what they want to do?
2. **Summarizing the Main Idea** Write a new title for this piece. Share it with your classmates. Explain why you think your title fits the story.

REVIEW AND ASSESS

UNDERSTANDING THE ISSUE

1. **Defining Key Terms** In your own words, define the following terms: *terrorism, hijacker, the Pentagon, al-Qaeda, Taliban, liberty, security, principles, ideals,* and *vigil.*

2. **Writing to Inform** In a 300-word article, describe a terrorist act you heard or read about. Describe how you reacted when you heard about it.

3. **Writing to Persuade** What do you think Americans should know about terrorism? Put your answer in a 250-word letter to the editor of your local newspaper. Support your answer with facts. Use at least five of the terms listed above.

INTERNET RESEARCH ACTIVITY

4. Use Internet resources to find information on what individuals and organizations are doing today to help victims of terrorism. Use what you learn to write a report on current efforts, and share it with the class.

5. With your teacher's help, use Internet resources to learn more about how the tragedies of September 11 resulted in an increase of visible patriotism in the United States. Focus your research on finding personal stories of how the attacks increased individual Americans' beliefs in and loyalty to the United States. Prepare a brief report on your findings.

BEYOND THE CLASSROOM

6. **Study the map** on page 513. Research one of the terrorist attacks noted there. What does the attack tell you about the goals, thoughts, and methods terrorists have? Describe the attack and answer those questions in a brief oral report.

TERRY BARNER/SILVER IMAGE

▲ **Muslims mourn victims of the World Trade Center attack.**

7. **Visit your school or local library.** Research a country, such as Israel, Northern Ireland, or Bosnia where the people have suffered from terrorist attacks. Find out what programs have been started by groups or individuals to bring an end to the violence. Present your findings to the whole class.

TIME/CNN Poll
Fighting Terrorism: How Far Would You Go?

What are Americans willing to do to fight terrorism? These pie graphs show what a TIME/CNN Poll found out.

To prevent terrorist attacks, would you favor or oppose the government doing each of the following?

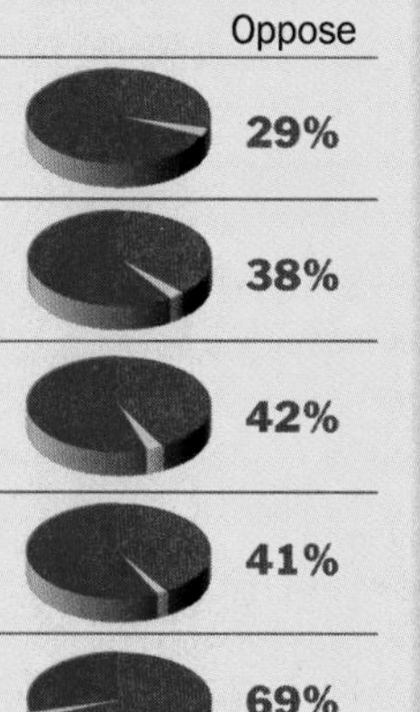

	Favor	Oppose
1. Allow police to wire-tap phone conversations of suspected terrorists without a court's okay:	68%	29%
2. Let courts jail, for as long as they want, people suspected of links to terrorist groups:	59%	38%
3. Let police intercept e-mail messages sent by anyone in the United States and scan them for suspicious words or phrases:	55%	42%
4. Require everyone in the United States to carry an identification card issued by the Federal Government:	57%	41%
5. Let police stop people on the street and search them:	29%	69%

Source: TIME magazine, October 8, 2001; Gray slices indicate respondents who were not sure.

BUILDING GRAPH READING SKILLS

1. **Analyzing the Data** The U.S. Constitution bars the government from making "unreasonable searches and seizures" of citizens. Which of the graphs show how Americans think about this right? What's your thinking on this issue?

2. **Making Inferences** The U.S. Constitution bars the government from taking away a person's "life, liberty, or property" without a fair trial. What do most people who took part in this poll seem to think about this right? Why do you think they hold that view?

FOR UPDATES ON WORLD ISSUES GO TO www.timeclassroom.com/glencoe

P/N G15404.50